283/0600/580/10/5/07 *

20 062 800 74

AF791597

WITHDRAWN

University

Handbook of Power System Engineering

Handbook of Power System Engineering

Yoshihide Hase

BICENTENNIAL 1807 WILEY 2007 BICENTENNIAL

John Wiley & Sons, Ltd

NORTHUMBRIA UNIVERSITY
LIBRARY
ITEM No. 2006280074
CLASS No. 621.319 HAS

Copyright © 2007 John Wiley & Sons Ltd, The Atrium, Southern Gate, Chichester,
West Sussex PO19 8SQ, England

Telephone (+44) 1243 779777

Email (for orders and customer service enquiries): cs-books@wiley.co.uk
Visit our Home Page on www.wiley.com

All Rights Reserved. No part of this publication may be reproduced, stored in a retrieval system or transmitted in any form or by any means, electronic, mechanical, photocopying, recording, scanning or otherwise, except under the terms of the Copyright, Designs and Patents Act 1988 or under the terms of a licence issued by the Copyright Licensing Agency Ltd, 90 Tottenham Court Road, London W1T 4LP, UK, without the permission in writing of the Publisher. Requests to the Publisher should be addressed to the Permissions Department, John Wiley & Sons Ltd, The Atrium, Southern Gate, Chichester, West Sussex PO19 8SQ, England, or emailed to permreq@wiley.co.uk, or faxed to (+44) 1243 770620.

Designations used by companies to distinguish their products are often claimed as trademarks. All brand names and product names used in this book are trade names, service marks, trademarks or registered trademarks of their respective owners. The Publisher is not associated with any product or vendor mentioned in this book.

This publication is designed to provide accurate and authoritative information in regard to the subject matter covered. It is sold on the understanding that the Publisher is not engaged in rendering professional services. If professional advice or other expert assistance is required, the services of a competent professional should be sought.

Other Wiley Editorial Offices

John Wiley & Sons Inc., 111 River Street, Hoboken, NJ 07030, USA

Jossey-Bass, 989 Market Street, San Francisco, CA 94103-1741, USA

Wiley-VCH Verlag GmbH, Boschstr. 12, D-69469 Weinheim, Germany

John Wiley & Sons Australia Ltd, 42 McDougall Street, Milton, Queensland 4064, Australia

John Wiley & Sons (Asia) Pte Ltd, 2 Clementi Loop #02-01, Jin Xing Distripark, Singapore 129809

John Wiley & Sons Canada Ltd, 6045 Freemont Blvd, Mississauga, ONT, L5R 4J3, Canada

Wiley also publishes its books in a variety of electronic formats. Some content that appears in print may not be available in electronic books.

Anniversary Logo Design: Richard J. Pacifico

Library of Congress Cataloging-in-Publication Data

Hase, Yoshihide, 1937-
Handbook of power system engineering / Yoshihide Hase.
p. cm.
Includes bibliographical references and index.
ISBN-13: 978-0-470-02742-4 (cloth : alk. paper)
1. Electric power systems. I. Title.
TK3001.H63 2007
621.319–dc22 2006038136

British Library Cataloguing in Publication Data

A catalogue record for this book is available from the British Library

ISBN-13 978-0-470-02742-4 (HB)

Typeset in 9/11 pt Times by Thomson Digital
Printed and bound in Great Britain by Antony Rowe Ltd, Chippenham, Wiltshire
This book is printed on acid-free paper responsibly manufactured from sustainable forestry
in which at least two trees are planted for each one used for paper production.

To Keiko for her endurance and stimulation

Contents

Preface

This book deals with the art and science of power system engineering for those engineers who work in electricity-related industries such as power utilities, manufacturing enterprises, engineering companies, or for students of electrical engineering in universities and colleges. Each engineer's relationship with power system engineering is extremely varied, depending on the types of companies they work for and their positions. We expect readers to study the characteristics of power systems theoretically as a multi-dimensional concept by means of this book, regardless of readers' business roles or specialties.

We have endeavoured to deal with the following three points as major features of the book:

First, as listed in the Contents, the book covers the theories of several subsystems, such as generating plants, transmission lines and substations, total network control, equipment-based local control, protection, and so on, as well as phenomena ranging from power (fundamental) frequency to lightning and switching surges, as the integrally unified art and science of power systems. Any equipment in a power system network plays its role by closely linking with all other equipment, and any theory, technology or phenomenon of one network is only a viewpoint of the profound dynamic behaviour of the network. This is the reason why we have covered different categories of theories combined in a single hierarchy in this book.

Secondly, readers can learn about the essential dynamics of power systems mostly through mathematical approaches. We explain our approach by starting from physically understandable equations and then move on to the final solutions that illustrate actual phenomena, and never skip explanations or adopt half-measures in the derivations.

Another point here is the difference in meaning between 'pure mathematically solvable' and 'engineering analytically solvable'. For example, a person (even if expert in transient analysis) cannot derive transient voltage and current solutions of a simple circuit with only a few *LCR* constants connected in series or parallel because the equational process is too complicated, except in special cases. Therefore only solutions of special cases are demonstrated in books on transient analysis. However, engineers often have to find solutions of such circuits by manual calculation. As they usually know the actual values of *LCR* constants in such cases, they can derive 'exact solutions' by theoretically justified approximation. Also, an appropriate approximation is an important technique to find the correct solution. Readers will also find such approximation techniques in this book.

Thirdly, the book deals with scientific theories of power system networks that will essentially never change. We intentionally excluded descriptions of advanced technologies, expecting such technologies to continue to advance year by year.

In recent years, analytical computation or simulation of the behaviour of large power system or complicated circuits has been executed by the application of powerful computers with outstanding software. However, it is quite easy to mishandle the analysis or the results because of the number of so many influential parameters. In this book, most of the theoretical explanation is based on typical simple circuits with one or two generators and one or two transmission lines. Precise understanding of the phenomena in such simple systems must always be the basis of understanding actual large systems and the incidents that may occur on them. This is the reason why power system behaviour is studied using small models.

Acknowledgements

This book contains the various experiences and knowledge of many people. I am deeply indebted to these people, although I can only humbly acknowledge them in a general way.

Also, I wish to acknowledge all my former colleagues and friends who gave me various opportunities to work and study together over many years throughout my engineering career.

I would also like to deeply thank Simone Taylor, Lucy Bryan, Emily Bone and all the other associated staff of John Wiley & Sons, Ltd, and in particular Kelly Board and Wendy Hunter for their encouragement and patient support. Finally, I wish to sincerely acknowledge Neville Hankins who accomplished the editing and amendments and Neetu Kalra who accomplished the typesetting, both hard tasks to my work.

About the author

Yohshihide Hase was born in Gifu Prefecture, Japan, in 1937. After graduating in electrical engineering from Kyoto University, he joined the Toshiba Corporation in 1960 and took charge of various power system projects, both at home and abroad, including the engineering of generating station equipment, substation equipment, as well as power system control and protection, until 1996. During that time, he held the positions of general manager, senior executive of technology for the energy systems sector, and chief fellow. In 1996, he joined Showa Electric Wire & Cable Company as the senior managing director and representative director and served on the board for eight years. He has been a lecturer at Kokushikan University since 2004. He was the vice president of the IEEJ (1995–96) and was also the representative officer of the Japanese National Committee of CIGRE (1987–1996) and has been bestowed as a distinguished member of CIGRE.

Introduction

'Utilization of fire', 'agricultural cultivation' and 'written communication': these three items are sometimes quoted as the greatest accomplishments of humankind. As a fourth item, **'social structures based on an electrical infrastructure**', which was created by humans mostly within the twentieth century, may be added.

Within the last hundred years, we have passed through the era of 'electricity as a convenient tool' to the point where electricity has become an inevitable part of our infrastructure as a means of energy acquisition, transport and utilization as well as in communication media. Today, without electricity we cannot carry out any of our living activities such as 'making fire', 'getting food and water' 'manufacturing tools', 'moving', 'communicating with others', and so on. Humans in most part of the world have thus become very dependent on electricity. Of course, such an important electrical infrastructure means our modern power system network.

A power system network can be likened to the human body. A trial comparison between the two may be useful for a better understanding of the essential characteristics of the power system.

First, the human body is composed of a great many subsystems (individual organs, bones, muscles, etc.), and all are composed in turn of an enormous number of minute cells. A power system network of a large arbitrary region is composed of a single unified system. Within this region, electricity is made available in any town, public utility, house and room by means of metal wires as a totally integrated huge network.

Generating plants, substations and transmission lines; generators, transformers, switchgear and other high-voltage equipment; several types of control equipment, protection equipment and auxiliary equipment; control and communication facilities in a dispatching or control centre; and the various kinds of load facilities – all these are also composed of a very large number of small parts or members. Individual parts play their important roles by linking with the rest of the network system. Human operators at any part of the network can be added as important members of the power system. We might say that a power system network is the **largest and greatest artificial system** ever produced by people in the modern era.

Secondly, the human body maintains life by getting energy from the external environment, and by processing and utilizing this energy. New cellular tissue is consequently created and old tissue is discarded. In such a procedure, the human body continues to grow and change.

A power system can be compared in the same way. A prerequisite condition of a power system network is that it is operated continuously as a single unified system, always adding new parts and discarding old ones. Since long-distance power transmission was first established about a hundred years ago, power systems have been operating and continuing to grow and change in this way, and, apart from the failure of localized parts, have never stopped. Further, no new power system isolated from the existing system in the same region has ever been constructed. A power system is the **ultimate inheritance succeeded by every generation** of humankind.

Thirdly, humans experience hunger in just a few hours after their last meal; their energy storage capacity is negligible in comparison with their lifetimes. In a power system such as a pumped-storage hydro-station, for example, the capacity of any kind of battery storage system is a very small part of the total capacity. The power generation balance has to be maintained every second to correspond to fluctuations or sudden changes in total load consumption. In other words, '**Simultaneity and Equality of energy generation and energy consumption**' is a vital characteristic of power system as well as of human body.

Fourthly, humans can continue to live even if parts of the body or organs are removed. At the other extreme, a minute disorder in cellular tissue may be life-threatening. Such opposites can be seen in power systems.

A power system will have been planned and constructed, and be operated, to maintain reasonable redundancy as an essential characteristic. Thus the system may continue to operate successfully in most cases even if a large part of it is suddenly cut off. On the contrary, the rare failure of one tiny part, for example a protective relay (or just one of its components), may trigger a kind of domino effect leading to a **black-out**.

Disruption of large part of power system network by '**domino-effect**' means big power failure leaded by abrupt segmentation of power system network, which may be probably caused by cascade trips of generators caused by total imbalance of power generation and consumption which leads to 'abnormal power frequency exceeding over or under frequency capability limits (OF/UF) of individual generators', 'cascade trips of generators caused by power stability limits, Q-V stability limits or by any other operational capability limits', 'cascade trips of trunk-lines/stations equipment caused by abnormal current flow exceeding individual current capacity limits (OC), or by over or under voltage limits (OV/UV)', ' succeeding cascade trips after fault tripping failure due to a breaker set back or caused by mal-operation of a protective relay' and so on, and may be perhaps caused as of 'these composite phenomena'. These nature of power systems is the outcome that **all the equipment and parts of the power system, regardless of their size, are closely linked and coordinated. The opposites of toughnees with well redundancy and delicacy** are the essential nature of power systems.

Fifthly, as with the human body, a power system cannot tolerate maltreatment, serious system disability or damage, which may cause **chronic power cuts**, and moreover would probably causes extremely fatal social damages. Recovery of a damaged power system is not easy. It takes a very long time and is expensive, or may actually be impossible. Power systems can be kept sound only by the endeavours of dedicated engineers and other professional people.

Sixthly, and finally, almost as elaborate as the human body, all the parts of power system networks today (including all kinds of loads) are **masterpieces of the latest technology**, based on a century of accumulated knowledge, something which all electrical engineers can share proudly together with mechanical engineers. Also all these things **have to be succeeded to our next generations** as the indispensable social structures.

1

Overhead Transmission Lines and Their Circuit Constants

In order to understand fully the nature of power systems, we need to study the nature of transmission lines as the first step. In this chapter we examine the characteristics and basic equations of three-phase overhead transmission lines. However, the actual quantities of the constants are described in Chapter 2.

1.1 Overhead Transmission Lines with *LR* Constants

1.1.1 Three-phase single circuit line without overhead grounding wire

1.1.1.1 Voltage and current equations, and equivalent circuits

A three-phase single circuit line between a point m and a point n with only L and R and without an overhead grounding wire (OGW) can be written as shown in Figure 1.1a. In the figure, r_g and L_g are the equivalent resistance and inductance of the earth, respectively. The outer circuits I and II connected at points m and n can theoretically be three-phase circuits of any kind.

All the voltages V_a, V_b, V_c and currents I_a, I_b, I_c are vector quantities and the symbolic arrows show the measuring directions of the three-phase voltages and currents which have to be written in the same direction for the three phases as a basic rule to describe the electrical quantities of three-phase circuits.

In Figure 1.1, the currents I_a, I_b, I_c in each phase conductor flow from left to right (from point m to point n). Accordingly, the composite current $I_a + I_b + I_c$ has to return from right to left (from point n to m) through the earth–ground pass. In other words, the three-phase circuit has to be treated as the set of 'three phase conductors + one earth circuit' pass.

In Figure 1.1a, the equations of the transmission line between m and n can be easily described as follows. Here, voltages V and currents I are complex-number vector values:

$$\left.\begin{aligned}
{}_mV_a - {}_nV_a &= (r_a + j\omega L_{aag})I_a + j\omega L_{abg}I_b + j\omega L_{acg}I_c - {}_{mn}V_g \quad ①\\
{}_mV_b - {}_nV_b &= j\omega L_{bag}I_a + (r_b + j\omega L_{bbg})I_b + j\omega L_{bcg}I_c - {}_{mn}V_g \quad ②\\
{}_mV_c - {}_nV_c &= j\omega L_{cag}I_a + j\omega L_{cbg}I_b + (r_c + j\omega L_{ccg})I_c - {}_{mn}V_g \quad ③\\
\text{where } {}_{mn}V_g &= (r_g + j\omega L_g)I_g = -(r_g + j\omega L_g)(I_a + I_b + I_c) \quad ④
\end{aligned}\right\} \qquad (1.1)$$

Handbook of Power System Engineering Yoshihide Hase
© 2007 John Wiley & Sons, Ltd

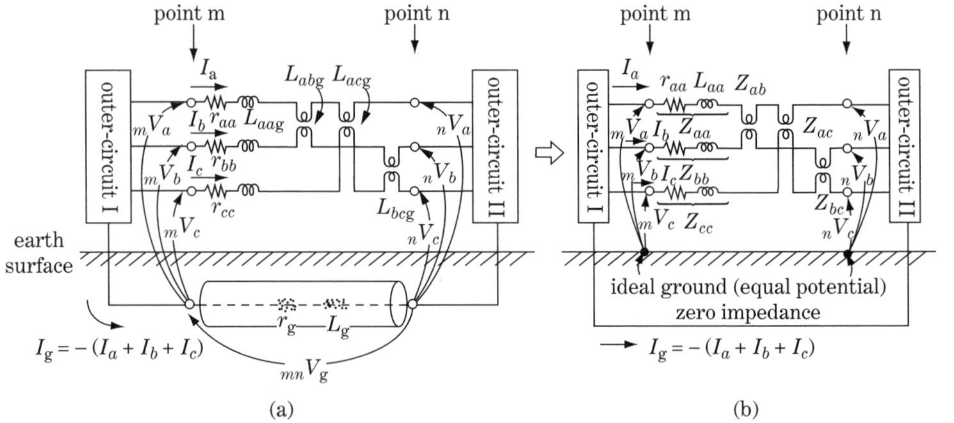

Figure 1.1 Single circuit line with *LR* constants

Substituting ④ into ①, and then eliminating $_{mn}V_g$, I_g,

$$_mV_a - {}_nV_a = (r_a + r_g + j\omega\overline{L_{aag} + L_g})I_a + (r_g + j\omega\overline{L_{abg} + L_g})I_b + (r_g + jw\overline{L_{acg} + L_g})I_c \quad ⑤$$

Substituting ④ into ② and ③ in the same way,

$$_mV_b - {}_nV_b = (r_g + j\omega\overline{L_{bag} + L_g})I_a + (r_b + r_g + j\omega\overline{L_{bbg} + L_g})I_b + (r_g + j\omega\overline{L_{bcg} + L_g})I_c \quad ⑥$$

$$_mV_c - {}_nV_c = (r_g + j\omega\overline{L_{cag} + L_g})I_a + (r_g + j\omega\overline{L_{cbg} + L_g})I_b + (r_c + r_g + j\omega\overline{L_{ccg} + L_g})I_c \quad ⑦$$

(1.2)

Now, the original Equation 1.1 and the derived Equation 1.2 are the equivalent of each other, so Figure 1.1b, showing Equation 1.2, is also the equivalent of Figure 1.1a.

Equation 1.2 can be expressed in the form of a matrix equation and the following equations are derived accordingly (refer to Appendix B for the matrix equation notation):

$$\begin{bmatrix} _mV_a \\ _mV_b \\ _mV_c \end{bmatrix} - \begin{bmatrix} _nV_a \\ _nV_b \\ _nV_c \end{bmatrix}$$

$$= \begin{bmatrix} r_a + r_g + j\omega\overline{L_{aag} + L_g} & r_g + j\omega\overline{L_{abg} + L_g} & r_g + j\omega\overline{L_{acg} + L_g} \\ r_g + j\omega\overline{L_{bag} + L_g} & r_b + r_g + j\omega\overline{L_{bbg} + L_g} & r_g + j\omega\overline{L_{bcg} + L_g} \\ r_g + j\omega\overline{L_{cag} + L_g} & r_g + j\omega\overline{L_{cbg} + L_g} & r_c + r_g + j\omega\overline{L_{ccg} + L_g} \end{bmatrix} \cdot \begin{bmatrix} I_a \\ I_b \\ I_c \end{bmatrix} \quad (1.3)$$

$$\equiv \begin{bmatrix} r_{aa} + j\omega L_{aa} & r_{ab} + j\omega L_{ab} & r_{ac} + j\omega L_{ac} \\ r_{ba} + j\omega L_{ba} & r_{bb} + j\omega L_{bb} & r_{bc} + j\omega L_{bc} \\ r_{ca} + j\omega L_{ca} & r_{cb} + j\omega L_{cb} & r_{cc} + j\omega L_{cc} \end{bmatrix} \cdot \begin{bmatrix} I_a \\ I_b \\ I_c \end{bmatrix}$$

$$\equiv \begin{bmatrix} Z_{aa} & Z_{ab} & Z_{ac} \\ Z_{ba} & Z_{bb} & Z_{bc} \\ Z_{ca} & Z_{cb} & Z_{cc} \end{bmatrix} \cdot \begin{bmatrix} I_a \\ I_b \\ I_c \end{bmatrix}$$

$$\left.\begin{array}{l}\text{where} \quad Z_{aa} = r_{aa} + j\omega L_{aa} = (r_a + r_g) + j\omega(L_{aag} + L_g) \\ Z_{bb}, Z_{cc} \text{ are written in similar equation forms} \\ \text{and } Z_{ac}, Z_{bc} \text{ are also written in similar forms}\end{array}\right\} \quad (1.4)$$

Now, we can apply symbolic expressions for the above matrix equation as follows:

$$_m\boldsymbol{V}_{abc} - {}_n\boldsymbol{V}_{abc} = \boldsymbol{Z}_{abc} \cdot \boldsymbol{I}_{abc} \quad (1.5)$$

where

$$_m\boldsymbol{V}_{abc} = \begin{bmatrix} _mV_a \\ _mV_b \\ _mV_c \end{bmatrix}, \quad _n\boldsymbol{V}_{abc} = \begin{bmatrix} _nV_a \\ _nV_b \\ _nV_c \end{bmatrix}, \quad \boldsymbol{Z}_{abc} = \begin{bmatrix} Z_{aa} & Z_{ab} & Z_{ac} \\ Z_{ba} & Z_{bb} & Z_{bc} \\ Z_{ca} & Z_{cb} & Z_{cc} \end{bmatrix}, \quad \boldsymbol{I}_{abc} = \begin{bmatrix} I_a \\ I_b \\ I_c \end{bmatrix} \quad (1.6)$$

Summarizing the above equations, Figure 1.1a can be described as Equations 1.3 and 1.6 or Equations 1.5 and 1.6, in which the resistance r_g and inductance L_g of the earth return pass are already reflected in all these four equations, although I_g and $_{mn}V_g$ are eliminated in Equations 1.5 and 1.6. We can consider Figure 1.1b as the equivalent circuit of Equations 1.3 and 1.4 or Equations 1.5 and 1.6. In Figure 1.1b, earth resistance r_g and earth inductance L_g are already included in the line constants Z_{aa}, Z_{ab}, etc., so the earth in the equivalent circuit of Figure 1.1b is 'the ideal earth' with zero impedance. Therefore the earth can be expressed in the figure as the equal-potential (zero-potential) earth plane at any point. It is clear that the mutual relation between the constants of Figure 1.1a and Figure 1.1b is defined by Equation 1.4. It should be noted that the self-impedance Z_{aa} and mutual impedance Z_{ab} of phase a, for example, involve the earth resistance r_g and earth inductance L_g.

Generally, in actual engineering tasks, Figure 1.1b and Equations 1.3 and 1.4 or Equations 1.5 and 1.6 are applied instead of Figure 1.1a and Equations 1.1 and 1.2; in other words, the line impedances are given as Z_{aa}, Z_{ab}, etc., instead of Z_{aag}, Z_{abg}. The line impedances Z_{aa}, Z_{bb}, Z_{cc} are named **'the self-impedances of the line including the earth–ground effect'**, and Z_{ab}, Z_{ac}, Z_{bc}, etc., are named **'the mutual impedances of the line including the earth–ground effect'**.

1.1.1.2 Measurement of line impedances Z_{aa}, Z_{ab}, Z_{ac}

Let us consider how to measure the line impedances taking the earth effect into account.

As we know from Figure 1.1b and Equations 1.3 and 1.4, the impedances Z_{aa}, Z_{ab}, Z_{ac}, etc., can be measured by the circuit connection shown in Figure 1.2a.

The conductors of the three phases are grounded to earth at point n, and the phase b and c conductors are opened at point m. Accordingly, the boundary conditions $_nV_a = {}_nV_b = {}_nV_c = 0, I_b = I_c = 0$ can be adopted for Equation 1.3:

$$\left.\begin{array}{l}\begin{bmatrix} _mV_a \\ _mV_b \\ _mV_c \end{bmatrix} - \begin{bmatrix} 0 \\ 0 \\ 0 \end{bmatrix} = \begin{bmatrix} Z_{aa} & Z_{ab} & Z_{ac} \\ Z_{ba} & Z_{bb} & Z_{bc} \\ Z_{ca} & Z_{cb} & Z_{cc} \end{bmatrix} \cdot \begin{bmatrix} I_a \\ 0 \\ 0 \end{bmatrix} \quad ① \\ \therefore \quad _mV_a/I_a = Z_{aa}, \quad _mV_b/I_a = Z_{ba}, \quad _mV_c/I_a = Z_{ca} \quad ② \end{array}\right\} \quad (1.7)$$

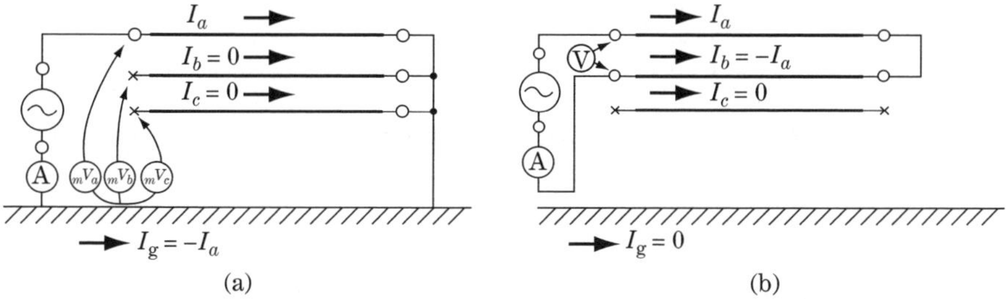

Figure 1.2 Measuring circuit of line impedance

Therefore the impedances Z_{aa}, Z_{ab}, Z_{ac} can be calculated from the measurement results of ${}_mV_a$, ${}_mV_b$, ${}_mV_c$ and I_a.

All the impedance elements in the impedance matrix $\boldsymbol{Z_{abc}}$ of Equation 1.7 can be measured in the same way.

1.1.1.3 Working inductance ($L_{aa} - L_{ab}$)

Figure 1.2b shows the case where the current I flows along the phase a conductor from point m to n and comes back from n to m only through the phase b conductor as the return pass. The equation is with boundary conditions $I_a = -I_b = I, \quad I_c = 0, \quad {}_nV_a = {}_nV_b$:

$$\begin{bmatrix} {}_mV_a \\ {}_mV_b \\ {}_mV_c \end{bmatrix} - \begin{bmatrix} {}_nV_a \\ {}_nV_a \\ {}_nV_c \end{bmatrix} = \begin{bmatrix} Z_{aa} & Z_{ab} & Z_{ac} \\ Z_{ba} & Z_{bb} & Z_{bc} \\ Z_{ca} & Z_{cb} & Z_{cc} \end{bmatrix} \cdot \begin{bmatrix} I \\ -I \\ 0 \end{bmatrix} \qquad (1.8a)$$

Therefore

$$\left.\begin{array}{l} \left.\begin{array}{l} {}_mV_a - {}_nV_a = (Z_{aa} - Z_{ab})I : \text{ voltage drop of the phase a conductor between points m and n} \\ {}_mV_b - {}_nV_b = -(Z_{bb} - Z_{ba})I : \text{ voltage drop of the phase b conductor between points m and n} \end{array}\right\} ① \\ \left.\begin{array}{l} V = {}_mV_a - {}_mV_b = \{(Z_{aa} - Z_{ab}) + (Z_{bb} - Z_{ba})\}I \\ V/I = ({}_mV_a - {}_mV_b)/I = (Z_{aa} - Z_{ab}) + (Z_{bb} - Z_{ba}) = \{\text{twice values of working impedance}\} \end{array}\right\} ② \end{array}\right\} \qquad (1.8b)$$

Equation 1.8① indicates the voltage drop of the parallel circuit wires a, b under the condition of the 'go-and-return-current' connection. The current I flows out at point m on the phase a conductor and returns to m only through the phase b conductor, so any other current flowing does not exist on the phase c conductor or earth–ground pass. In other words, Equation 1.8b① is satisfied regardless of the existence of the third wire or earth–ground pass. Therefore the impedance $(Z_{aa} - Z_{ab})$ as well as $(Z_{bb} - Z_{ba})$ should be specific values which are determined only by the relative condition of the phase a and b conductors, and they are not affected by the existence or absence of the third wire or earth–ground pass. $(Z_{aa} - Z_{ab})$ is called the **working impedance** and the corresponding $(L_{aa} - L_{ab})$ is called the **working inductance** of the phase a conductor with the phase b conductor.

Furthermore, as the conductors a and b are generally of the same specification (the same dimension, same resistivity, etc.), the ipedance drop between m and n of the phase a and b conductors should be the same. Accordingly, the working inductances of both conductors are clearly the same, namely $(L_{aa} - L_{ab}) = (L_{bb} - L_{ba})$.

The value of the working inductance can be calculated from the well-known equation below, which is derived by an electromagnetic analytical approach as a function only of the conductor radius r and the parallel distance s_{ab} between the two conductors:

$$L_{aa} - L_{ab} = L_{bb} - L_{ba} = 0.4605 \log_{10}\frac{s_{ab}}{r} + 0.05 \qquad [\text{mH/km}] \qquad (1.9)$$

This is the equation for the working inductance of the parallel conductors a and b, which can be quoted from analytical books on electromagnetism. The equation shows that the working inductance $L_{aa} - L_{ab}$ for the two parallel conductors is determined only by the relative distance between the two conductors s_{ab} and the radius r, so it is not affected by any other conditions such as other conductors or the distance from the earth surface.

The working inductance can also be measured as the value $(1/2)V/I$ by using Equation 1.8b②.

1.1.1.4 Self- and mutual impedances including the earth–ground effect L_{aa}, L_{ab}

Now we evaluate the actual numerical values for the line inductances contained in the impedance matrix of Equation 1.3.

The currents I_a, I_b, I_c flow through each conductor from point m to n and $I_a + I_b + I_c$ returns from n to m through the ideal earth return pass. All the impedances of this circuit can be measured by the method of Figure 1.2a. However, these measured impedances are experimentally a little larger than those obtained by pure analytical calculation based on the electromagnetic equations with the assumption of an ideal, conductive, earth plane surface.

In order to compensate for these differences between the analytical result and the measured values, we can use an imaginary ideal conductive earth plane at some deep level from the ground surface as shown in Figure 1.3.

In this figure, the imaginary perfect conductive earth plane is shown at the depth H_g, and the three imaginary conductors α, β, γ are located at symmetrical positions to conductors a, b, c, respectively, based on this datum plane.

The inductances can be calculated by adopting the equations of the electromagnetic analytical approach to Figure 1.3.

1.1.1.4.1 Self-inductances L_{aa}, L_{bb}, L_{cc} In Figure 1.3, the conductor a (radius r) and the imaginary returning conductor α are symmetrically located on the datum plane, and the distance between a and α is $h_a + H_a$. Thus the inductance of conductor a can be calculated by the following equation which is a special case of Equation 1.9 under the condition $s_{ab} \to h_a + H_a$:

$$L_{aag} = 0.4605 \log_{10} \frac{h_a + H_a}{r} + 0.05 \qquad [\text{mH/km}] \tag{1.10a}$$

Conversely, the inductance of the imaginary conductor α (the radius is H_a, because the actual grounding current reaches up to the ground surface), namely the inductance of earth, is

$$L_g = 0.4605 \log_{10} \frac{h_a + H_a}{H_a} + 0.05\,[\text{mH/km}] \fallingdotseq 0.05 \quad [\text{mH/km}] \tag{1.10b}$$

Therefore,

$$L_{aa} = L_{aag} + L_g = 0.4605 \log_{10} \frac{h_a + H_a}{r} + 0.1 \quad [\text{mH/km}] \tag{1.11}$$

L_{bb}, L_{cc} can be derived in the same way.

Incidentally, the depth of the imaginary datum plane can be checked experimentally and is mostly within the range of $H_g = 300 - 1000$ m. On the whole H_g is rather shallow, say $300 - 600$ m in the

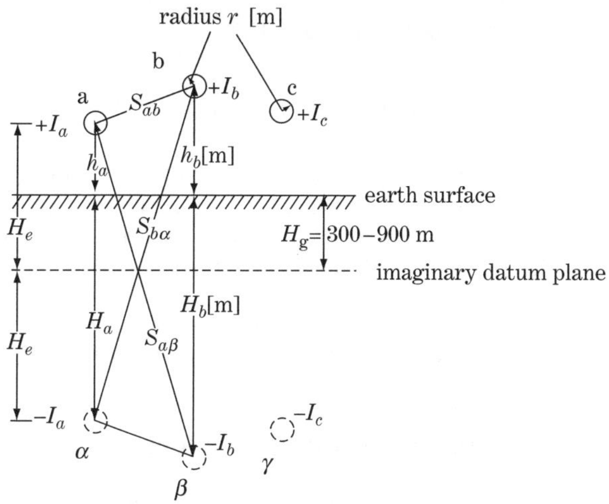

Figure 1.3 Earth–ground as conductor pass

geological younger strata after the Quaternary period, but is generally deep, say 800 – 1000 m, in the older strata of the Tertiary period or earlier.

1.1.1.4.2 Mutual inductances L_{ab}, L_{bc}, L_{ca}

The mutual inductance L_{ab} can be derived by subtracting L_{aa} from Equation 1.11 and the working inductance $(L_{aa} - L_{ab})$ from Equation 1.9:

$$\begin{aligned} L_{ab} &= L_{aa} - (L_{aa} - L_{ab}) = 0.4605 \log_{10} \frac{h_a + H_a}{s_{ab}} + 0.05 \quad [\mathrm{mH/km}] \\ &\fallingdotseq 0.4605 \log_{10} \frac{s_{a\beta}}{s_{ab}} + 0.05 \quad [\mathrm{mH/km}] \end{aligned} \tag{1.12a}$$

Similarly

$$\begin{aligned} L_{ba} &= 0.4605 \log_{10} \frac{h_b + H_b}{s_{ab}} + 0.05 \quad [\mathrm{mH/km}] \\ &\fallingdotseq 0.4605 \log_{10} \frac{s_{b\alpha}}{s_{ab}} + 0.05 \quad [\mathrm{mH/km}] \end{aligned} \tag{1.12b}$$

where $h_a + H_a = 2H_e \fallingdotseq 2H_g$, and so on.

Incidentally, the depth of the imaginary datum plane $H_g \fallingdotseq H_e = (h_a + H_a)/2$ would be between 300 and 1000 m, while the height of the transmission tower h_a is within the range of 10–100 m (UHV towers of 800–1000 kV would be approximately 100 m). Furthermore, the phase-to-phase distance S_{ab} is of order 10 m, while the radius of conductor r is a few centimetres (the equivalent radius r_{eff} of EHV/UHV multi-bundled conductor lines may be of the order of 10–50 cm).

Accordingly,

$$\left. \begin{aligned} & H_a \fallingdotseq H_b \fallingdotseq H_c \fallingdotseq 2H_e \gg h_a \fallingdotseq h_b \fallingdotseq h_c \gg s_{ab} \fallingdotseq s_{bc} \fallingdotseq s_{ca} \gg r, r_{\mathrm{eff}} \\ & s_{a\beta} \fallingdotseq s_{b\alpha} \fallingdotseq h_a + H_a = 2H_e \fallingdotseq h_b + H_b \end{aligned} \right\} \tag{1.13a}$$

Then, from Equations 1.9, 1.11 and 1.12,

$$L_{aa} \fallingdotseq L_{bb} \fallingdotseq L_{cc}, \quad L_{ab} \fallingdotseq L_{bc} \fallingdotseq L_{ca} \tag{1.13b}$$

1.1.1.4.3 Numerical check

Let us assume conditions $s_{ab} = 10\,\mathrm{m}$, $r = 0.05\,\mathrm{m}$, $H_e = (h_a + H_a)/2 \fallingdotseq H_g = 900\,\mathrm{m}$.

Then calculating the result by Equation 1.11 and 1.12,

$$L_{aa} = 2.20\,\mathrm{mH/km}, \quad L_{ab} = 1.09\,\mathrm{mH/km}$$

If $H_e = (h_a + H_a)/2 = 300\,\mathrm{m}$, then $L_{aa} = 1.98\,\mathrm{mH/km}$, $L_{ab} = 0.87\,\mathrm{mH/km}$. As $h_a + H_a$ is contained in the logarithmic term of the equations, constant values L_{aa}, L_{ab} and so on are not largely affected by $h_a + H_a$, neither is radius r nor r_{eff} as well as the phase-to-phase distance s_{ab}. Besides, 0.1 and 0.05 in the second term on the right of Equations 1.9–1.12 do not make a lot of sense.

Further, if transmission lines are reasonably transpositioned, $Z_{aa} \fallingdotseq Z_{bb} \fallingdotseq Z_{cc}$, $Z_{ab} \fallingdotseq Z_{bc} \fallingdotseq Z_{ca}$ can be justified so that Equation 1.3 is simplified into Equation 2.13 of Chapter 2.

1.1.1.5 Reactance of multi-bundled conductors

For most of the recent large-capacity transmission lines, multi-bundled conductor lines ($n = 2 - 8$ per phase) are utilized as shown in Figure 1.4. In the case of n conductors (the radius of

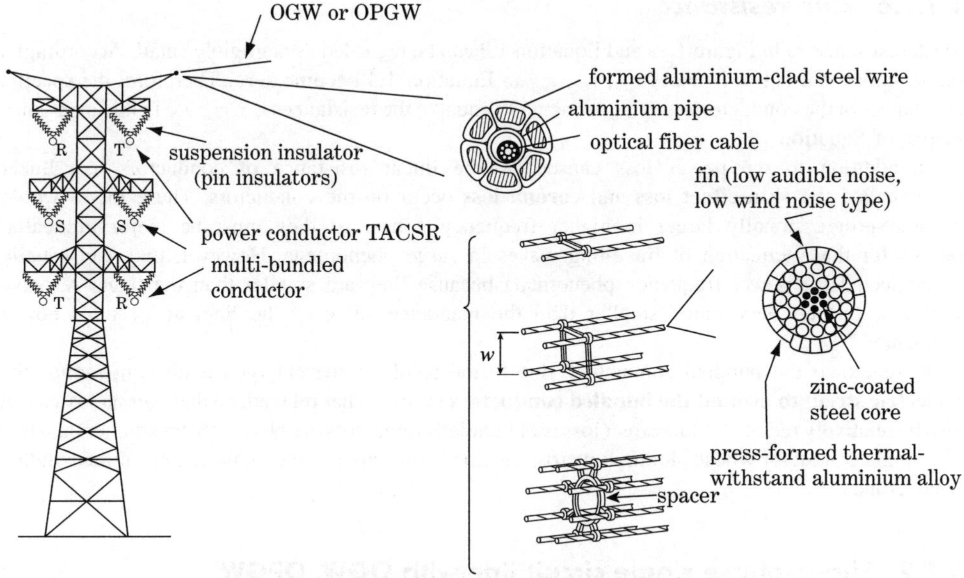

Figure 1.4 Overhead double circuit transmission line

each conductor is r), L_{aag} of Equation 1.10a can be calculated from the following modified equation:

$$\left.\begin{aligned} L_{aag} &= 0.4605 \log_{10} \frac{h_a + H_a}{r^{1/n} \times w^{(n-1)/n}} + \frac{0.05}{n} \quad [\text{mH/km}] \\ &\equiv 0.4605 \log_{10} \frac{h_a + H_a}{r_{\text{eff}}} + \frac{0.05}{n} \quad [\text{mH/km}] \\ &\text{where} \quad r_{\text{eff}} = r^{1/n} \times w^{(n-1)/n} \text{ is the equivalent radius and} \\ &w\,[\text{m}] \text{ is the geometrical averaged distance of bundled conductors} \end{aligned}\right\} \quad (1.14a)$$

Since the self-inductance L_g of the virtual conductor α given by Equation 1.10b is not affected by the adoption of multi-bundled phase a conductors, accordingly

$$L_{aa} = L_{aag} + L_g = 0.4605 \log_{10} \frac{h_a + H_a}{r_{\text{eff}}} + 0.05\left(1 + \frac{1}{n}\right) \quad [\text{mH/km}] \Big\} \quad (1.14b)$$

1.1.1.5.1 Numerical check

Using TACSR = 810 mm^2 (see Chapter 2), $2r = 40$ mm and four bundled conductors ($n = 4$), with the square allocation $w = 50$ cm averaged distance

$$\left.\begin{aligned} w &= (w_{12} \cdot w_{13} \cdot w_{14} \cdot w_{23} \cdot w_{24} \cdot w_{34})^{1/6} \\ &= (50 \cdot 50\sqrt{2} \cdot 50 \cdot 50 \cdot 50\sqrt{2} \cdot 50)^{1/6} = 57.24\,\text{cm} \\ r_{\text{eff}} &= r^{1/n} \cdot w^{(n-1)/n} = 20^{1/4} \cdot 57.25^{3/4} = 44.0\,\text{mm} \end{aligned}\right\} \quad (1.14c)$$

The equivalent radius $r_{\text{eff}} = 44$ mm is 2.2 times $r = 20$ mm, so that the line self-inductance L_{aa} can also be reduced by the application of bundled conductors. The mutual inductance L_{ab} of Equation 1.12a is not affected by the adoption of multi-bundled conductor lines.

1.1.1.6 Line resistance

Earth resistance r_g in Figure 1.1a and Equation 1.2 can be regarded as negligibly small. Accordingly, the so-called mutual resistances r_{ab}, r_{bc}, r_{ca} in Equation 1.4 become zero. Therefore, the specific resistances of the conductors r_a, r_b, r_c are actually equal to the resistances r_{aa}, r_{bb}, r_{cc} in the impedance matrix of Equation 1.3.

In addition to the power loss caused by the linear resistance of conductors, non-linear losses called the **skin-effect loss** and **corona loss** occur on the conductors. These losses would become progressionally larger in higher frequency zones, so they must be major influential factors for the attenuation of travelling waves in surge phenomena. However, they can usually be neglected for power frequency phenomena because they are smaller than the linear resistive loss and, further, very much smaller than the reactance value of the line, at least for power frequency.

In regard to the bundled conductors, due to the result of the enlarged equivalent radius, the **dielectric strength around the bundled conductors** is somewhat relaxed, so that corona losses can also be relatively reduced. Skin-effect losses of bundled conductors are obviously far smaller than that of a single conductor whose aluminium cross-section is the same as the total sections of the bundled conductors.

1.1.2 Three-phase single circuit line with OGW, OPGW

Most high-voltage transmission lines are equipped with **OGW (overhead grounding wires)** and/or **OPGW (OGW with optical fibres** for communication use).

In the case of a single circuit line with single OGW, the circuit includes four conductors and the fourth conductor (x in Figure 1.5) is earth grounded at all the transmission towers. Therefore, using the figure for the circuit, Equation 1.3 has to be replaced by the following equation:

$$\begin{bmatrix} {}_mV_a \\ {}_mV_b \\ {}_mV_c \\ {}_mV_x = 0 \end{bmatrix} - \begin{bmatrix} {}_nV_a \\ {}_nV_b \\ {}_nV_c \\ {}_nV_x = 0 \end{bmatrix} = \begin{bmatrix} Z_{aa} & Z_{ab} & Z_{ac} & Z_{ax} \\ Z_{ba} & Z_{bb} & Z_{bc} & Z_{bx} \\ Z_{ca} & Z_{cb} & Z_{cc} & Z_{cx} \\ Z_{xa} & Z_{xb} & Z_{xc} & Z_{xx} \end{bmatrix} \cdot \begin{bmatrix} I_a \\ I_b \\ I_c \\ I_x \end{bmatrix} \quad (1.15a)$$

Extracting the fourth row,

$$I_x = -\frac{1}{Z_{xx}}(Z_{xa}I_a + Z_{xb}I_b + Z_{xc}I_c) \quad (1.15b)$$

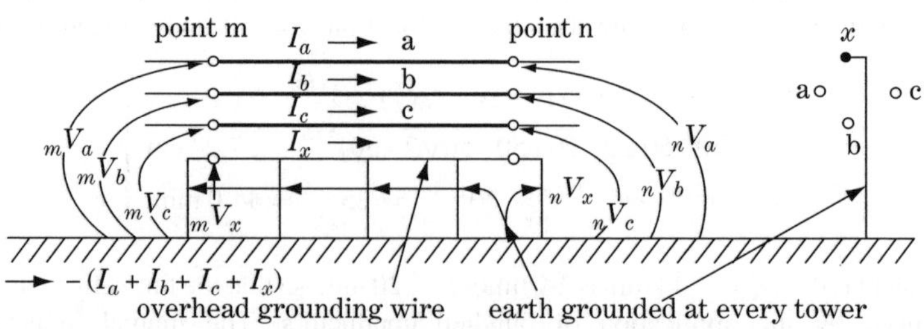

Figure 1.5 Single circuit line with OGW

Substituting I_x into the first, second and third rows of Equation 1.15a,

$$
\left.
\begin{aligned}
&\begin{bmatrix} {}_mV_a \\ {}_mV_b \\ {}_mV_c \end{bmatrix} - \begin{bmatrix} {}_nV_a \\ {}_nV_b \\ {}_nV_c \end{bmatrix} = \begin{bmatrix} Z_{aa} & Z_{ab} & Z_{ac} \\ Z_{ba} & Z_{bb} & Z_{bc} \\ Z_{ca} & Z_{cb} & Z_{cc} \end{bmatrix} \cdot \begin{bmatrix} I_a \\ I_b \\ I_c \end{bmatrix} + \begin{bmatrix} Z_{ax}I_x \\ Z_{bc}I_x \\ Z_{cx}I_x \end{bmatrix} \\
&= \begin{bmatrix} Z_{aa} - \dfrac{Z_{ax}Z_{xa}}{Z_{xx}} & Z_{ab} - \dfrac{Z_{ax}Z_{xb}}{Z_{xx}} & Z_{ac} - \dfrac{Z_{ax}Z_{xc}}{Z_{xx}} \\ Z_{ba} - \dfrac{Z_{bx}Z_{xa}}{Z_{xx}} & Z_{bb} - \dfrac{Z_{bx}Z_{xb}}{Z_{xx}} & Z_{bc} - \dfrac{Z_{bx}Z_{xc}}{Z_{xx}} \\ Z_{ca} - \dfrac{Z_{cx}Z_{xa}}{Z_{xx}} & Z_{cb} - \dfrac{Z_{cx}Z_{xb}}{Z_{xx}} & Z_{cc} - \dfrac{Z_{cx}Z_{xc}}{Z_{xx}} \end{bmatrix} \cdot \begin{bmatrix} I_a \\ I_b \\ I_c \end{bmatrix} \\
&\equiv \begin{bmatrix} Z'_{aa} & Z'_{ab} & Z'_{ac} \\ Z'_{ba} & Z'_{bb} & Z'_{bc} \\ Z'_{ca} & Z'_{cb} & Z'_{cc} \end{bmatrix} \cdot \begin{bmatrix} I_a \\ I_b \\ I_c \end{bmatrix} \\
&\text{where } Z_{ax} = Z_{xa}, \quad Z_{bx} = Z_{xb}, \quad Z_{cx} = Z_{xc} \\
&Z'_{aa} = Z_{aa} - \delta_{aa}, \quad Z'_{ab} = Z_{ab} - \delta_{ab} \\
&\delta_{aa} = -\frac{Z_{ax}Z_{xa}}{Z_{xx}}, \quad \delta_{ab} = -\frac{Z_{ax}Z_{xb}}{Z_{xx}}
\end{aligned}
\right\} \quad (1.16)
$$

This is the fundamental equation of the three-phase single circuit line with OGW in which I_x has already been eliminated and the impedance elements of the grounding wire are slotted into the three-phase impedance matrix. Equation 1.16 is obviously of the same form as Equation 1.3, while all the elements of the rows and columns in the impedance matrix have been revised to smaller values with corrective terms $\delta_{ax} = -Z_{ax}Z_{xa}/Z_{xx}$ etc.

The above equations indicate that the three-phase single circuit line with OGW can be expressed as a 3×3 impedance matrix equation in the form of Equation 1.16 regardless of the existence of OGW, as was the case with Equation 1.3. Also, we can comprehend that OGW has roles not only to shield lines against lightning but also to reduce the self- and mutual reactances of transmission lines.

1.1.3 Three-phase double circuit line with *LR* constants

The three-phase double circuit line can be written as in Figure 1.6 and Equation 1.17 regardless of the existence or absence of OGW:

$$
\begin{bmatrix} {}_mV_a \\ {}_mV_b \\ {}_mV_c \\ {}_mV_A \\ {}_mV_B \\ {}_mV_C \end{bmatrix} - \begin{bmatrix} {}_nV_a \\ {}_nV_b \\ {}_nV_c \\ {}_nV_A \\ {}_nV_B \\ {}_nV_C \end{bmatrix} = \begin{bmatrix} Z_{aa} & Z_{ab} & Z_{ac} & Z_{aA} & Z_{aB} & Z_{aC} \\ Z_{ba} & Z_{bb} & Z_{bc} & Z_{bA} & Z_{bB} & Z_{bC} \\ Z_{ca} & Z_{cb} & Z_{cc} & Z_{cA} & Z_{cB} & Z_{cC} \\ Z_{Aa} & Z_{Ab} & Z_{Ac} & Z_{AA} & Z_{AB} & Z_{AC} \\ Z_{Ba} & Z_{Bb} & Z_{Bc} & Z_{BA} & Z_{BB} & Z_{BC} \\ Z_{Ca} & Z_{Cb} & Z_{Cc} & Z_{CA} & Z_{CB} & Z_{CC} \end{bmatrix} \cdot \begin{bmatrix} I_a \\ I_b \\ I_c \\ I_A \\ I_B \\ I_C \end{bmatrix} \quad (1.17)
$$

In addition, if the line is appropriately phase balanced, the equation can be expressed by Equation 2.17 of Chapter 2.

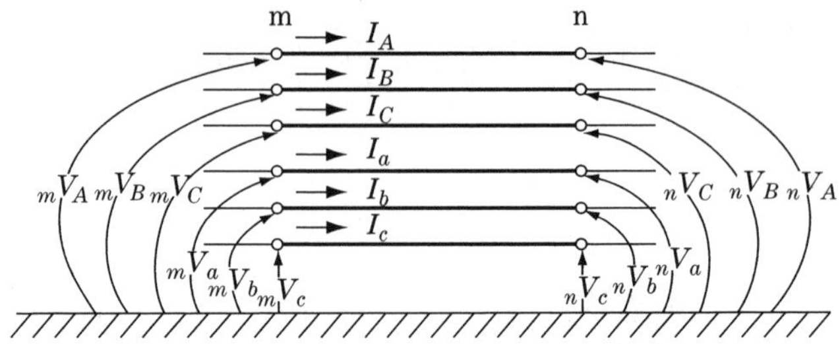

Figure 1.6 Three-phase double circuit line with *LR* constants

1.2 Stray Capacitance of Overhead Transmission Lines

1.2.1 Stray capacitance of three-phase single circuit line

1.2.1.1 Equation for electric charges and voltages on conductors

Figure 1.7a shows a single circuit line, where electric charges q_a, q_b, q_c [C/m] are applied to phase a, b, c conductors and cause voltages v_a, v_b, v_c [V], respectively. The equation of this circuit is given by

$$\left.\begin{array}{l} \underbrace{\begin{bmatrix} v_a \\ v_b \\ v_c \end{bmatrix}}_{\boldsymbol{v}_{abc}} = \underbrace{\begin{bmatrix} p_{aa} & p_{ab} & p_{ac} \\ p_{ba} & p_{bb} & p_{bc} \\ p_{ca} & p_{cb} & p_{cc} \end{bmatrix}}_{\boldsymbol{p}_{abc}} \cdot \underbrace{\begin{bmatrix} q_a \\ q_b \\ q_c \end{bmatrix}}_{\boldsymbol{q}_{abc}} \quad \therefore \boldsymbol{v}_{abc} = \boldsymbol{p}_{abc} \cdot \boldsymbol{q}_{abc} \\ \text{where } q\,[\text{C/m}],\ v\,[\text{V}] \text{ are instantaneous real numbers} \end{array}\right\} \tag{1.18}$$

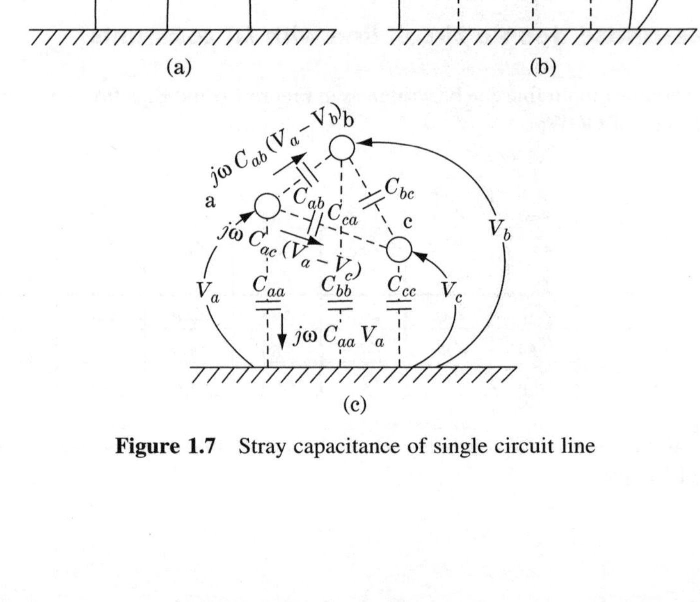

Figure 1.7 Stray capacitance of single circuit line

The inverse matrix equation can be derived from the above equation as

$$\underbrace{\begin{bmatrix} q_a \\ q_b \\ q_c \end{bmatrix}}_{\boldsymbol{q}_{abc}} = \underbrace{\begin{bmatrix} k_{aa} & k_{ab} & k_{ac} \\ k_{ba} & k_{bb} & k_{bc} \\ k_{ca} & k_{cb} & k_{cc} \end{bmatrix}}_{\boldsymbol{k}_{abc}} \cdot \underbrace{\begin{bmatrix} v_a \\ v_b \\ v_c \end{bmatrix}}_{\boldsymbol{v}_{abc}} \quad \therefore \boldsymbol{q}_{abc} = \boldsymbol{k}_{abc} \cdot \boldsymbol{v}_{abc} \tag{1.19}$$

Here, $\boldsymbol{p}_{abc}$ and $\boldsymbol{k}_{abc}$ are inverse 3×3 matrices of each other, so that $\boldsymbol{p}_{abc} \cdot \boldsymbol{k}_{abc} = \mathbf{1}$ ($\mathbf{1}$ is the 3×3 unit matrix; refer to Appendix B).

Accordingly,

$$\left.\begin{aligned} k_{aa} &= (p_{bb}p_{cc} - p_{bc}^2)/\Delta && [\text{F/m}] \\ k_{bb} &= (p_{cc}p_{aa} - p_{ca}^2)/\Delta && [\text{F/m}] \\ k_{cc} &= (p_{aa}p_{bb} - p_{ab}^2)/\Delta && [\text{F/m}] \\ k_{ab} &= k_{ba} = -(p_{ab}p_{cc} - p_{ac}p_{bc})/\Delta && [\text{F/m}] \\ k_{bc} &= k_{cb} = -(p_{bc}p_{aa} - p_{ba}p_{ca})/\Delta && [\text{F/m}] \\ k_{ca} &= k_{ac} = -(p_{ca}p_{bb} - p_{cb}p_{ab})/\Delta && [\text{F/m}] \\ \Delta &= p_{aa}p_{bb}p_{cc} + 2p_{ab}p_{bc}p_{ac} - (p_{aa}p_{bc}^2 + p_{bb}p_{ca}^2 + p_{cc}p_{ab}^2) && [\text{m/F}]^3 \end{aligned}\right\} \tag{1.20}$$

where p [m/F] are the **coefficients of the potential** and k [F/m] are the electrostatic **coefficients of static capacity**.

Modifying Equation 1.19 a little,

$$\left.\begin{aligned} q_a &= k_{aa}v_a + k_{ab}v_b + k_{ac}v_c \\ &= (k_{aa} + k_{ab} + k_{ac})v_a + (-k_{ab})(v_a - v_b) + (-k_{ac})(v_a - v_c) && [\text{C/m}] \\ q_b &= (k_{ba} + k_{bb} + k_{bc})v_b + (-k_{bc})(v_b - v_c) + (-k_{ba})(v_b - v_a) && [\text{C/m}] \\ q_c &= (k_{ca} + k_{cb} + k_{cc})v_c + (-k_{ca})(v_c - v_a) + (-k_{ca}) + (v_c - v_b) && [\text{C/m}] \end{aligned}\right\} \tag{1.21}$$

then

$$\left.\begin{aligned} q_a &= C_{aa}v_a + C_{ab}(v_a - v_b) + C_{ac}(v_a - v_c) && [\text{C/m}] \\ q_b &= C_{bb}v_b + C_{bc}(v_b - v_c) + C_{ba}(v_b - v_a) && [\text{C/m}] \\ q_c &= C_{cc}v_c + C_{ca}(v_c - v_a) + C_{cb}(v_c - v_b) && [\text{C/m}] \end{aligned}\right\} \tag{1.22}$$

with q_a, q_b, q_c [C/m], v_b, v_b, v_c [V] and

$$\left.\begin{aligned} C_{aa} &= k_{aa} + k_{ab} + k_{ac} \quad [\text{F/m}] \quad & C_{ab} &= -k_{ab} \quad [\text{F/m}] \\ C_{bb} &= k_{ba} + k_{bb} + k_{bc} \quad [\text{F/m}] \quad & C_{bc} &= -k_{bc} \quad [\text{F/m}] \\ C_{cc} &= k_{ca} + k_{cb} + k_{cc} \quad [\text{F/m}] \quad & C_{ca} &= -k_{ca} \quad [\text{F/m}] \\ C_{ac} &= -k_{ac} \quad [\text{F/m}] \\ C_{ba} &= -k_{ba} \quad [\text{F/m}] \\ C_{cb} &= -k_{cb} \quad [\text{F/m}], \end{aligned}\right\} \tag{1.23}$$

Equations 1.22 and 1.23 are the fundamental equations of stray capacitances of a three-phase single circuit overhead line. Noting the form of Equation 1.22, Figure 1.7b can be used for another expression of Figure 1.7a: C_{aa}, C_{bb}, C_{cc} are the phase-to-ground capacitances and $C_{ab} = C_{ba}$, $C_{bc} = C_{cb}$, $C_{ca} = C_{ac}$ are the phase-to-phase capacitances between two conductors.

1.2.1.2 Fundamental voltage and current equations

It is usually convenient in actual engineering to adopt current $i(= dq/dt)$ [A] instead of charging value q [C], and furthermore to adopt effective (rms: root mean square) voltage and current of complex-number V, I instead of instantaneous value $v(t)$, $i(t)$.

As electric charge q is the integration over time of current i, the following relations can be derived:

$$\left.\begin{array}{ll}
q = \int i dt, \quad i = \dfrac{dq}{dt} & ① \\
i(t) = \mathrm{Re}(\sqrt{2}I) = \mathrm{Re}(\sqrt{2}|I| \cdot \mathrm{e}^{j(\omega t+\theta_1)}) = \sqrt{2}|I|\cos(\omega t + \theta_1) & ② \\
\text{Re() shows the real part of the complex number}(\mathrm{Re}(a+jb)=a). & \\
v(t) = \mathrm{Re}(\sqrt{2}V) = \mathrm{Re}(\sqrt{2}|V| \cdot \mathrm{e}^{j(\omega t+\theta_2)}) & \\
\quad = \sqrt{2}|V|\cos(\omega t + \theta_2) & ③ \\
\therefore \quad q(t) = \int i dt = \int \mathrm{Re}(\sqrt{2}|I| \cdot \mathrm{e}^{j(\omega t+\theta_1)}) dt & \\
\quad = \mathrm{Re}(\sqrt{2}|I| \cdot \int \mathrm{e}^{j(\omega t+\theta_1)} dt) & \text{(note that, in this book,} \\
\quad = \mathrm{Re}\left(\sqrt{2}|I| \cdot \dfrac{\mathrm{e}^{j(\omega t+\theta_1)}}{j\omega}\right) = \mathrm{Re}\left(\dfrac{\sqrt{2}I}{j\omega}\right) & \text{the exponential function} \\
 & ④ \text{ will be denoted by 'e').}
\end{array}\right\} \quad (1.24)$$

Equation 1.22 can be modified to the following form by adopting Equation 1.24④ and by replacement of $v_a \rightarrow \sqrt{2}V_a$ etc.:

$$\left.\begin{array}{l}
\mathrm{Re}\left(\dfrac{\sqrt{2}I_a}{j\omega}\right) = \mathrm{Re}\{C_{aa} \cdot \sqrt{2}V_a + C_{ab} \cdot \sqrt{2}(V_a - V_b) + C_{ac} \cdot \sqrt{2}(V_a - V_c)\} \\
\mathrm{Re}\left(\dfrac{\sqrt{2}I_b}{j\omega}\right) = \mathrm{Re}\{C_{bb} \cdot \sqrt{2}V_b + C_{bc} \cdot \sqrt{2}(V_b - V_c) + C_{ba} \cdot \sqrt{2}(V_b - V_a)\} \\
\mathrm{Re}\left(\dfrac{\sqrt{2}I_c}{j\omega}\right) = \mathrm{Re}\{C_{cc} \cdot \sqrt{2}V_c + C_{ca} \cdot \sqrt{2}(V_c - V_a) + C_{cb} \cdot \sqrt{2}(V_c - V_b)\}
\end{array}\right\} \quad (1.25)$$

Therefore

$$\left.\begin{array}{l}
I_a = j\omega C_{aa}V_a + j\omega C_{ab}(V_a - V_b) + j\omega C_{ac}(V_a - V_c) \\
I_b = j\omega C_{bb}V_b + j\omega C_{bc}(V_b - V_c) + j\omega C_{ba}(V_b - V_a) \\
I_c = j\omega C_{cc}V_c + j\omega C_{ca}(V_c - V_a) + j\omega C_{cb}(V_c - V_b)
\end{array}\right\} \quad (1.26a)$$

or, with a small modification,

$$\begin{bmatrix} I_a \\ I_b \\ I_c \end{bmatrix} = j\omega \begin{bmatrix} C_{aa}+C_{ab}+C_{ac} & -C_{ab} & -C_{ac} \\ -C_{ba} & C_{ba}+C_{bb}+C_{bc} & -C_{bc} \\ -C_{ca} & -C_{cb} & C_{ca}+C_{cb}+C_{cc} \end{bmatrix} \cdot \begin{bmatrix} V_a \\ V_b \\ V_c \end{bmatrix} \quad (1.26b)$$

This is the fundamental equation for stray capacitances of a three-phase single circuit transmission line. Also Figure 1.7c is derived from one-to-one correspondence with Equation 1.26.

1.2.1.3 Coefficients of potential (p_{aa}, p_{ab}), coefficients of static capacity (k_{aa}, k_{ab}) and capacitances (C_{aa}, C_{ab})

The earth surface can be taken as a perfect equal-potential plane, so that we can use Figure 1.8, in which the three imaginary conductors α, β, γ are located at symmetrical positions of conductors a, b, c, respectively, based on the earth surface plane. By assuming electric charges $+q_a$, $+q_b$, $+q_c$ and $-q_a$, $-q_b$, $-q_c$ per unit length on conductors a, b, c, and α, β, γ respectively, the following voltage equation can be derived:

$$\begin{aligned}
v_a = & \left(\text{voltage of conductor a due to} \pm q_a \text{ of conductor a, } \alpha : 2q_a \log_\mathrm{e} \frac{2h_a}{r} \times 9 \times 10^9 \text{ [V]}\right) \\
& + \left(\text{voltage of conductor a due to} \pm q_b \text{ of conductor b, } \beta : 2q_b \log_\mathrm{e} \frac{s_{a\beta}}{s_{ab}} \times 9 \times 10^9 \text{ [V]}\right) \\
& + \left(\text{voltage of conductor a due to} \pm q_c \text{ of conductor c, } \gamma : 2q_c \log_\mathrm{e} \frac{s_{a\gamma}}{s_{ac}} \times 9 \times 10^9 \text{ [V]}\right) \quad ①
\end{aligned}$$

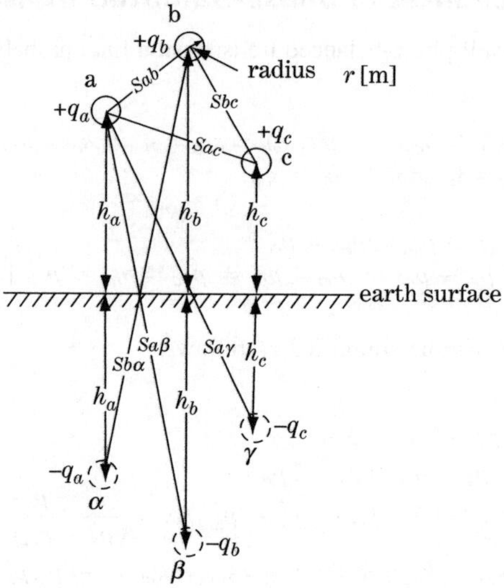

Figure 1.8 Three parallel overhead conductors

Equations for v_b, v_c can be derived in the same way. Then

$$\begin{bmatrix} v_a \\ v_b \\ v_c \end{bmatrix} = \begin{bmatrix} p_{aa} & p_{ab} & p_{ac} \\ p_{ba} & p_{bb} & p_{bc} \\ p_{ca} & p_{cb} & p_{cc} \end{bmatrix} \cdot \begin{bmatrix} q_a \\ q_b \\ q_c \end{bmatrix}$$

$$= 2 \times 9 \times 10^9 \times \begin{bmatrix} \log_e \dfrac{2h_a}{r} & \log \dfrac{s_{a\beta}}{s_{ab}} & \log_e \dfrac{s_{a\gamma}}{s_{ac}} \\ \log_e \dfrac{s_{b\alpha}}{s_{ba}} & \log_e \dfrac{2h_b}{r} & \log_e \dfrac{s_{b\gamma}}{s_{bc}} \\ \log_e \dfrac{s_{c\alpha}}{s_{ca}} & \log_e \dfrac{s_{c\beta}}{s_{cb}} & \log_e \dfrac{2h_c}{r} \end{bmatrix} \cdot \begin{bmatrix} q_a \\ q_b \\ q_c \end{bmatrix} \quad ② \qquad (1.27)$$

where $s_{a\beta} = s_{b\alpha} = \sqrt{\{s_{ab}^2 - (h_a - h_b)^2\} + (h_a + h_b)^2} = \sqrt{s_{ab}^2 + 4h_a h_b}$.
Refer to the supplement at the end of this chapter for the derivation of Equation 1.27①.

The equation indicates that the coefficients of potential (p_{aa}, p_{ab}, etc.) are calculated as a function of the conductor's radius r, height (h_a, h_b, h_c) from the earth surface, and phase-to-phase distances (s_{ab}, s_{ac}, etc.) of the conductors. p_{aa}, p_{ab}, etc., are determined only by physical allocations of each phase conductor (in other words, by the structure of towers), and relations like $p_{ab} = p_{ba}$ are obvious.

In conclusion, the coefficients of potential (p_{aa}, p_{ab}, etc.), the coefficients of static capacity (k_{aa}, k_{ab}, etc.) and the capacitance (C_{aa}, C_{ab}, etc.) are calculated from Equations 1.27, 1.20 and 1.23, respectively. Again, all these values are determined only by the physical allocation of conductors and are not affected by the applied voltage.

1.2.1.4 Stray capacitances of phase-balanced transmission lines

Referring to Figure 1.8, a well-phase-balanced transmission line, probably by transposition, can be assumed. Then

$$\left.\begin{array}{l} h \equiv h_a \fallingdotseq h_b \fallingdotseq h_c, \qquad s_{ll} \equiv s_{ab} = s_{ba} \fallingdotseq s_{bc} = s_{cb} \fallingdotseq s_{ca} = s_{ac} \\ s_{a\beta} \fallingdotseq s_{b\alpha} \fallingdotseq s_{a\gamma} \fallingdotseq s_{c\alpha} \fallingdotseq s_{b\gamma} \fallingdotseq s_{c\beta} \end{array}\right\} \tag{1.28}$$

$$\left.\begin{array}{l} p_s \equiv p_{aa} \fallingdotseq p_{bb} \fallingdotseq p_{cc} \\ p_m \equiv p_{ab} = p_{ba} \fallingdotseq p_{ac} = p_{ca} \fallingdotseq p_{bc} = p_{cb} \end{array}\right\} \tag{1.29}$$

Accordingly, Equation 1.20 can be simplified as follows:

$$\left.\begin{array}{ll} \Delta & = p_s^3 + 2p_m^3 - 3p_s p_m^2 \\ & = (p_s - p_m)^2 (p_s + 2p_m) \\ k_s & \equiv k_{aa} \fallingdotseq k_{bb} \fallingdotseq k_{cc} \fallingdotseq (p_s^2 - p_m^2)/\Delta = \dfrac{p_s + p_m}{(p_s - p_m)(p_s + 2p_m)} \\ k_m & \equiv k_{ab} = k_{ba} \fallingdotseq k_{ac} = k_{ca} \fallingdotseq k_{bc} = k_{cb} = -(p_m p_s - p_m^2)/\Delta \\ & = \dfrac{-p_m}{(p_s - p_m)(p_s + 2p_m)} \\ k_s + 2k_m & = \dfrac{1}{p_s + 2p_m} \end{array}\right\} \tag{1.30}$$

and from Equation 1.23

$$\left.\begin{array}{l} C_s \equiv C_{aa} \fallingdotseq C_{bb} \fallingdotseq C_{cc} = k_s + 2k_m = \dfrac{1}{p_s + 2p_m} \\ C_m \equiv C_{ab} = C_{ba} \fallingdotseq C_{ac} = C_{ca} \fallingdotseq C_{bc} = C_{cb} = -k_m \\ = \dfrac{p_m}{(p_s - p_m)(p_s + 2p_m)} = \dfrac{p_m}{p_s - p_m} \cdot C_s \end{array}\right\} \tag{1.31}$$

and from Equation 1.27

$$p_s \equiv p_{aa} \fallingdotseq p_{bb} \fallingdotseq p_{cc} = 2 \times 9 \times 10^9 \log_e \frac{2h}{r} \quad [\text{m/K}] \quad ①$$

$$p_m \equiv p_{ab} \fallingdotseq p_{bc} \fallingdotseq p_{ca} = 2 \times 9 \times 10^9 \log_e \frac{s_{b\alpha}}{s_{ll}} \quad [\text{m/F}]$$

$$\fallingdotseq 2 \times 9 \times 10^9 \log_e \frac{\sqrt{s_{ll}^2 + (2h)^2}}{s_{ll}} \tag{1.32}$$

$$= 2 \times 9 \times 10^9 \log_e \left\{ 1 + \left(\frac{2h}{s_{ll}} \right)^2 \right\}^{1/2} \quad [\text{m/F}] \quad ②$$

where generally

$$h > s_{ll}, \quad \left(\frac{2h}{s_{ll}} \right)^2 \gg 1$$

and

$$\therefore p_m \fallingdotseq 2 \times 9 \times 10^9 \log_e \frac{2h}{s_{ll}} \quad [\text{m/F}] \quad ②'$$

Substituting p_s, p_m from Equation 1.32 into Equation 1.31,

$$C_s = \frac{1}{p_s + 2p_m} \doteqdot \frac{1}{2\times 9\times 10^9\left(\log_e \frac{2h}{r} + 2\log_e \frac{2h}{s_{ll}}\right)} = \frac{1}{2\times 9\times 10^9 \log_e \frac{8h^3}{rs_{ll}^2}}$$

$$= \frac{0.02413}{\log_{10}\frac{8h^3}{rs_{ll}^2}} \times 10^{-9}[\text{F/m}] = \frac{0.02413}{\log_{10}\frac{8h^3}{rs_{ll}^2}}[\mu\text{F/km}] \quad ①$$

(zero-sequence capacitance)

while

$$\frac{p_m}{p_s - p_m} \doteqdot \frac{\log_e \frac{2h}{s_{ll}}}{\log_e \frac{2h}{r} - \log_e \frac{2h}{s_{ll}}} = \frac{\log_{10}\frac{2h}{s_{ll}}}{\log_{10}\frac{s_{ll}}{r}}$$

$$\therefore C_m = C_s \cdot \frac{p_m}{p_s - p_m} \doteqdot C_s \cdot \frac{\log_{10}\frac{2h}{s_{ll}}}{\log_{10}\frac{s_{ll}}{r}} = \frac{0.02413}{\log_{10}\frac{8h^3}{rs_{ll}^2}} \cdot \frac{\log_{10}\frac{2h}{s_{ll}}}{\log_{10}\frac{s_{ll}}{r}}[\mu\text{F/km}] \quad ② \qquad (1.33)$$

In conclusion, a well-phase-balanced transmission line can be expressed by Figure 1.9a1 and Equation 1.26b is simplified into Equation 1.34, where the stray capacitances C_s, C_m can be calculated from Equation 1.33:

$$\underbrace{\begin{bmatrix} I_a \\ I_b \\ I_c \end{bmatrix}}_{\boldsymbol{I}_{abc}} = j\omega \underbrace{\begin{bmatrix} C_s + 2C_m & -C_m & -C_m \\ -C_m & C_s + 2C_m & -C_m \\ -C_m & -C_m & C_s + 2C_m \end{bmatrix}}_{\boldsymbol{C}_{abc}} \cdot \underbrace{\begin{bmatrix} V_a \\ V_b \\ V_c \end{bmatrix}}_{\boldsymbol{V}_{abc}} \qquad (1.34)$$

$$\therefore \boldsymbol{I}_{abc} = j\omega \boldsymbol{C}_{abc} \cdot \boldsymbol{V}_{abc}$$

Incidentally, Figure 1.9a1 can be modified to Figure 1.9a2, where the total capacitance of one phase $C \equiv C_s + 3C_m$ is called the **working capacitance** of single circuit transmission lines, and can be calculated by the following equation:

$$C \equiv C_s + 3C_m = (k_s + 2k_m) + 3(-k_m) = k_s - k_m = \frac{1}{p_s - p_m}$$

$$= \frac{1}{2\times 9\times 10^9\left(\log_e \frac{2h}{r} - \log_e \frac{2h}{s_{ll}}\right)} = \frac{1}{2\times 9\times 10^9 \log_e \frac{s_{ll}}{r}} \quad [\text{F/m}]$$

$$= \frac{0.02413}{\log_{10}\frac{s_{ll}}{r}} \quad [\mu\text{F/km}] \quad \text{(positive sequence capacitance)} \quad ①$$

In case of multi-bundled (n) conductor lines, the radius r is replaced by the equivalent radius r_{eff},

$$r_{\text{eff}} = r^{1/n} \times w^{(n-1)/n} \quad [\text{m}] \quad ② \qquad (1.35)$$

where w is the geometrical averaged distance between bundled conductors.

1.2.1.4.1 *Numerical check* Taking the conditions conductor radius $r = 0.05$ m, averaged phase-to-phase distance $s_{ll} = 10$ m and average height $h = 60$ m, then by Equations 1.33 and 1.35, we have

$$C_s = 0.00436\,\mu\text{F/km},\ C_m = 0.00204\,\mu\text{F/km and } C = C_s + 3C_m = 0.01048\,\mu\text{F/km}$$

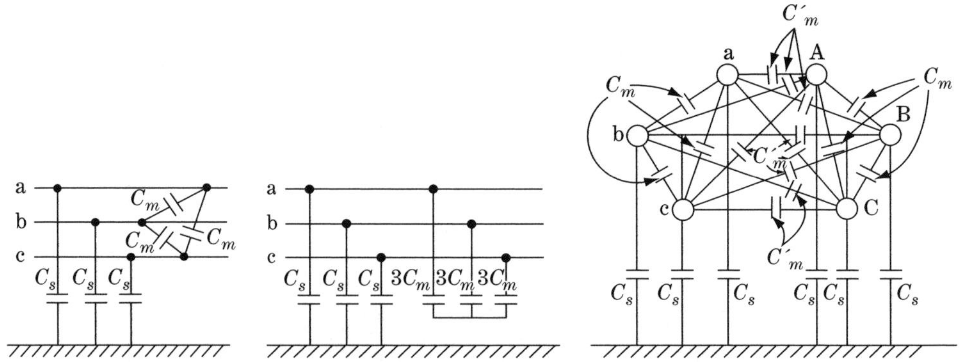

(a1) single circuit line (a2) single circuit line (b) double circuit line

Figure 1.9 Stray capacitances of overhead line (well balanced)

1.2.2 Three-phase single circuit line with OGW

Four conductors of phase names a, b, c, x exist in this case, so the following equation can be derived as an extended form of Equation 1.26a:

$$I_a = j\omega C_{aa}V_a + j\omega C_{ab}(V_a - V_b) + j\omega C_{ac}(V_a - V_c) + j\omega C_{ax}(V_a - V_x) \tag{1.36a}$$

where $V_x = 0$, because OGW is earth grounded at every tower. Accordingly,

$$\begin{bmatrix} I_a \\ I_b \\ I_c \end{bmatrix} = j\omega \begin{bmatrix} C_{aa}+C_{ab}+C_{ac}+C_{ax} & -C_{ab} & -C_{ac} \\ -C_{ba} & C_{ba}+C_{bb}+C_{bc}+C_{bx} & -C_{bc} \\ -C_{ca} & -C_{cb} & C_{ca}+C_{cb}+C_{cc}+C_{cx} \end{bmatrix} \cdot \begin{bmatrix} V_a \\ V_b \\ V_c \end{bmatrix} \tag{1.36b}$$

This matrix equation is again in the same form as Equation 1.26b. However, the phase-to-ground capacitance values (diagonal elements of the matrix $\boldsymbol{C}$) are increased (the value of C_{ax} is increased for the phase a conductor, from $C_{aa} + C_{ab} + C_{ac}$ *to* $C_{aa} + C_{ab} + C_{ac} + C_{ax}$).

1.2.3 Three-phase double circuit line

Six conductors of phase names a, b, c, A, B, C exist in this case, so the following equation can be derived as an extended form of Equation 1.26a:

$$\begin{aligned} I_a = j\omega[&C_{aa}V_a + C_{ab}(V_a - V_b) + C_{ac}(V_a - V_c) + C_{aA}(V_a - V_A) + C_{aB}(V_a - V_B) \\ &+ C_{aC}(V_a - V_C)] \end{aligned} \tag{1.37a}$$

Then

$$\begin{bmatrix} I_a \\ I_b \\ I_c \\ I_A \\ I_B \\ I_C \end{bmatrix} = j\omega \begin{bmatrix} \begin{matrix} C_{aa}+C_{ab}+ \\ C_{ac}+C_{aA}+ \\ C_{aB}+C_{aC} \end{matrix} & -C_{ab} & -C_{ac} & -C_{aA} & -C_{aB} & -C_{aC} \\ -C_{ba} & \begin{matrix} C_{ba}+C_{bb}+ \\ C_{bc}+C_{bA}+ \\ C_{bB}+C_{bC} \end{matrix} & -C_{bc} & -C_{bA} & -C_{bB} & -C_{bC} \\ -C_{ca} & -C_{cb} & \begin{matrix} C_{ca}+C_{cb}+ \\ C_{cc}+C_{cA}+ \\ C_{cB}+C_{cC} \end{matrix} & -C_{cA} & -C_{cB} & -C_{cC} \\ -C_{Aa} & -C_{Ab} & -C_{Ac} & \begin{matrix} C_{AA}+C_{AB}+ \\ C_{AC}+C_{Aa}+ \\ C_{Ab}+C_{Ac} \end{matrix} & -C_{AB} & -C_{AC} \\ -C_{Ba} & -C_{Bb} & -C_{Bc} & -C_{BA} & \begin{matrix} C_{BA}+C_{BB}+ \\ C_{BC}+C_{Ba}+ \\ C_{Bb}+C_{Bc} \end{matrix} & -C_{BC} \\ -C_{Ca} & -C_{Cb} & -C_{Cc} & -C_{CA} & -C_{CB} & \begin{matrix} C_{CA}+C_{CB}+ \\ C_{CC}+C_{Ca}+ \\ C_{Cb}+C_{Cc} \end{matrix} \end{bmatrix} \cdot \begin{bmatrix} V_a \\ V_b \\ V_c \\ V_A \\ V_B \\ V_C \end{bmatrix} \tag{1.37b}$$

It is obvious that the double circuit line with OGW can be expressed in the same form.

The case of a well-transposed double circuit line is as shown in Figure 1.9b:

$$\begin{bmatrix} I_a \\ I_b \\ I_c \\ I_A \\ I_B \\ I_C \end{bmatrix} = j\omega \begin{bmatrix} \begin{matrix} C_s+2C_m \\ +3C'_m \end{matrix} & -C_m & -C_m & -C'_m & -C'_m & -C'_m \\ -C_m & \begin{matrix} C_s+2C_m \\ +3C'_m \end{matrix} & -C_m & -C'_m & -C'_m & -C'_m \\ -C_m & -C_m & \begin{matrix} C_s+2C_m \\ +3C'_m \end{matrix} & -C'_m & -C'_m & -C'_m \\ -C'_m & -C'_m & -C'_m & \begin{matrix} C_s+2C_m \\ +3C'_m \end{matrix} & -C_m & -C_m \\ -C'_m & -C'_m & -C'_m & -C_m & \begin{matrix} C_s+2C_m \\ +3C'_m \end{matrix} & -C_m \\ -C'_m & -C'_m & -C'_m & -C_m & -C_m & \begin{matrix} C_s+2C_m \\ +3C'_m \end{matrix} \end{bmatrix} \cdot \begin{bmatrix} V_a \\ V_b \\ V_c \\ V_A \\ V_B \\ V_C \end{bmatrix}$$

$$\begin{aligned} &C_s \equiv C_{aa} \fallingdotseq C_{bb} \fallingdotseq C_{cc} \fallingdotseq C_{AA} \fallingdotseq C_{BB} \fallingdotseq C_{CC} && : \text{one phase-to-ground capacitance} \\ &C_m \equiv C_{ab} \fallingdotseq C_{bc} \fallingdotseq \cdots \fallingdotseq C_{AB} \fallingdotseq C_{BC} \fallingdotseq \cdots && : \text{capacitance between two conductors of the same circuit} \\ &C'_m \equiv C_{aA} \fallingdotseq C_{bC} \fallingdotseq \cdots \fallingdotseq C_{Aa} \fallingdotseq C_{Bb} \fallingdotseq \cdots && : \text{capacitance between two conductors of a different circuit} \end{aligned} \tag{1.38}$$

Above, we have studied the fundamental equations and circuit models of transmission lines and the actual calculation method for the L, C, R constants. Concrete values of the constants are investigated in Chapter 2.

1.3 Supplement: Additional Explanation for Equation 1.27

Equation 1.27① can be explained by the following steps.

Step-1: The induced voltage v at arbitrary point y in Figure a

Figure a shows two parallel conductors x, x' (radius r) whose mutual distance is s [m] and the conductor length is l [m], where $l \gg s$.

When charges $+q, -q$ are applied to per unit length of the conductors x, x' respectively, the voltage potential v at the arbitrary single point y is given by the following equation (expressed in MKS units):

$$v = \frac{2q}{4\pi\varepsilon_0} \cdot \log_e \frac{s_2}{s_1} = 2q \cdot 9 \times 10^9 \log_e \frac{s_2}{s_1} \tag{1}$$

where

$$4\pi\varepsilon_0 = \frac{1}{9 \times 10^9} \quad \text{(refer the Equation (4) in the next page.)}$$

The voltage of the centre line g is obviously zero.

Step-2: The induced voltage v at arbitrary point y when $+q_a$ is applied to the overhead conductor a in Figure b

This is a special case shown in Figure b in which the names of the conductors have been changed ($x \to$ a, $x' \to \alpha$). The upper half zones of Figures a and b (the open space above the earth surface) are completely the same. Accordingly, under the state of a single overhead conductor a with an existing charge $+q$, the voltage v at the arbitrary point y in the open space can be calculated from Equation 1.

Step-3: The induced voltage v_a on the conductor a when $+q_a$ is applied to conductor a

This case corresponds to choosing the arbitrary point y on the conductor surface in Figure b. Therefore the voltage v_a can be derived by replacing $s_1 \to r$, $s_2 \to 2h$ in Equation 1:

$$\therefore v_a = 2q_a \cdot 9 \times 10^9 \log_e \frac{2h}{r} \tag{2}$$

Step-4: The induced voltage v_a on the conductor a when $+q_b$ is applied to conductor b

This case corresponds to replacement of $x \to$ b, $x' \to \beta$, y $\to$ a in Figure a. Accordingly, $s_1 \to s_{ab}$, $s_2 \to_{a\beta}$ in Equation 1,

$$\therefore v_a = 2q_b \cdot 9 \times 10^9 \log_e \frac{s_{a\beta}}{s_{ab}} \tag{3}$$

Equations 2 and 3 are the first and second terms on the right-hand side of Equation 1.27①. The equation is the expanded case for three conductors by applying the theorem of superposition. Clearly, the equation can be expanded to the cases of parallel multi-conductors of arbitrary number n. (Note that $4\pi\varepsilon_0 = 1/(9 \times 10^9)$.)

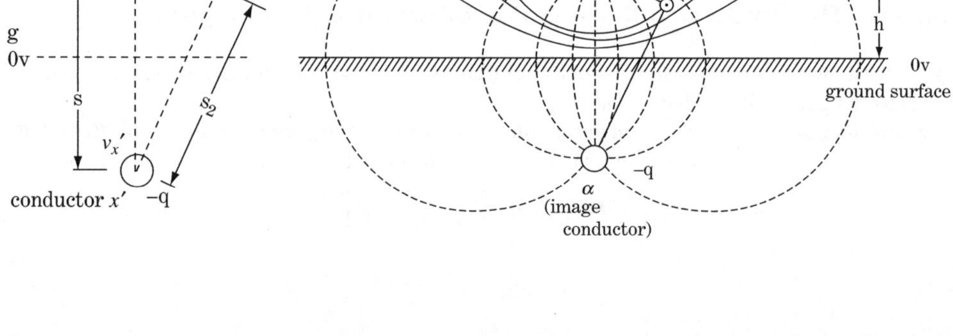

In the rationalized MKS system of units, $4\pi\varepsilon_0$ is given by the equation

$$4\pi\varepsilon_0 = \frac{1}{c_0^2} \cdot 10^7 = \frac{1}{(3 \times 10^8)^2} \cdot 10^7 = \frac{1}{9 \times 10^9} \tag{4}$$

where
c_0 is the velocity of electromagnetic waves (light), $c_0 = 3 \times 10^8$ [m/s]
10^7 is the coefficient of translation from CGS to MKS units, namely

$$\text{energy} = (\text{force}) \cdot (\text{distance}) = ((\text{kg} \cdot \text{m/s}^2) \cdot \text{m}) = ((\text{g} \cdot \text{cm/s}^2) \cdot \text{cm}) \times 10^7$$

Coffee break 1: Electricity, its substance and methodology

The new steam engine of **James Watt** (1736–1819) ushered in the great dawn of the Industrial Revolution in the 1770s. Applications of the steam engine began to appear quickly in factories, mines, railways, and so on, and the curtain of modern mechanical engineering was raised. The first steam locomotive, designed by **George Stephenson** (1781–1848), appeared in 1830.

Conversely, electrical engineering had to wait until Volta began to provide 'stable electricity' from his voltaic pile to other electrical scientists in the 1800s. Since then, scientific investigations of the unseen electricity on one hand and practical applications for **telegraphic communication** on the other hand have been conducted by scientists or electricians simultaneously, often the same people. In the first half of the nineteenth century, the worth of electricity was recognized for telegraphic applications, but its commercial application was actually realized in the 1840s. Commercial telegraphic communication through wires between New York and Boston took place in 1846, followed at Dover through a submarine cable in 1851. However, it took another 40 years for the realization of commercial applications of electricity as the replacement energy for steam power or in lighting.

2

Symmetrical Coordinate Method (Symmetrical Components)

The three-phase circuit generally has four electric conducting passes (phase a, b, c passes and an earth pass) and these four electric passes are closely coupled by mutual inductances L and mutual capacitances C. Therefore phenomena on any pass of a three-phase circuit cannot be independent of phenomena on the other passes. For this reason, the three-phase circuit is always very complicated, even for smaller system models. Furthermore, rotating machines including generators cannot be treated as adequate circuit elements to be combined with transmission line or transformers. Accordingly, the analysis of three-phase circuits by straightforward methods is not easy, even for only small models. Symmetrical components is the vital method to describe transmission lines, solid-state machines, rotating machines and combined total power systems as 'precise and simple circuits' instead of 'connection diagrams' by which circuit analysis can be conducted. Surge phenomena as well as power frequency phenomena of total networks or partial three-phase circuits cannot actually be solved without symmetrical components regardless of the purposes of analysis or the sizes of the networks.

In this chapter, the essential concept of the symmetrical coordinate method is examined first, followed by a circuit description of three-phase transmission lines and other equipment by symmetrical components.

2.1 Fundamental Concept of Symmetrical Components

It should be noted that the direct three-phase analytical circuits of power systems cannot be obtained even for a small, local part of a network, although their connection diagrams can be obtained. First, mutual inductances/mutual capacitances existing between different phases (typically of generators) cannot be adequately drawn as analytical circuits of phases a, b, c. Furthermore, the analytical solution of such circuits, including some mutual inductances or capacitances, is quite hard and even impossible for smaller circuits. In other words, straightforward analysis of three-phase circuit quantities is actually impossible regardless of steady-state phenomena or transient phenomena of small circuits. The symmetrical coordinate method can give us a good way to draw the analytical circuit of a three-phase system and to solve the transient phenomena (including surge phenomena) as well as steady-state phenomena.

The symmetrical coordinate method (symmetrical components) is a kind of variable transformation technique from a mathematical viewpoint. That is, three electrical quantities on a, b, c phases are always handled as one set in the **a–b–c domain**, and these three variables are then transformed into another set of three variables named **positive (1), negative (2) and zero (0) sequence quantities** in the newly defined **0–1–2 domain**. An arbitrary set of three variables in the a–b–c domain and the

Handbook of Power System Engineering Yoshihide Hase
© 2007 John Wiley & Sons, Ltd

transformed set of three variables in the 0–1–2 domain are mathematically in one-to-one correspondence with for each other. Therefore, the phenomena of a–b–c phase quantities in any frequency zone can be transformed into the 0–1–2 domain and can be observed, examined and solved from the standpoint in the defined 0–1–2 domain. Then the obtained behaviour or the solution in the 0–1–2 domain can be retransformed into the original a–b–c domain.

It can be safely said that the symmetrical coordinate method is an essential analytical tool for any kind of three-phase circuit phenomenon, and inevitably utilized in every kind of engineering work of power systems. Only symmetrical components can provide ways to obtain the large and precise analytical circuits of integrated power systems including generators, transmission lines, station equipment as well as loads.

Figure 2.1 shows the concept of such a transformation between the two domains in one-to-one correspondence. One set of a, b, c phase currents $I_a,\ I_b,\ I_c$ (or phase voltages $V_a,\ V_b,\ V_c$) at an arbitrary point in the three-phase network based on the a–b–c domain is transformed to another set of three variables named I_0, I_1, I_2 (or V_0, V_1, V_2) in the 0–1–2 domain, by the particularly defined transformation rule. The equations of the original a–b–c domain will be changed into new equations of the 0–1–2 domain, by which three-phase power systems can be described as precise and quite simple circuits. Therefore, rather complex subjects in the a–b–c domain can be treated and resolved easily in the 0–1–2 domain, and the solution in this domain is easily inverse transformed as the correct solution in the original a–b–c domain.

There are two other important transformation methods:

a) **$\alpha - \beta - 0$ transformation method**, $(I_a, I_b, I_c) \Leftrightarrow (I_\alpha, I_\beta, I_0)$: This is also useful as a complementary analytical tool of symmetrical components. In some special circuits, $\alpha - \beta - 0$ components provide easier solutions for the problems for which symmetrical components may not give good solutions.

b) ***d–q–0* transformation method**, $(I_a, I_b, I_c) \Leftrightarrow (I_d, I_q, I_0)$: This is a very powerful transformation specialized for the treatment of generators and other rotating machinery. Rotating machines can be described as precise and simple circuits only by the d–q–0 method. Due to the precise description of generator characteristics by the d–q–0 method, dynamic system behaviour can be analysed.

We will learn more about these methods in later chapters.

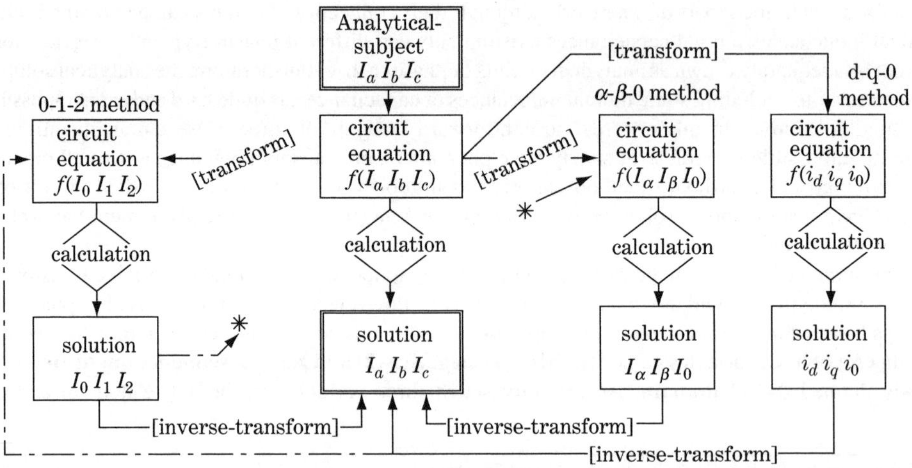

Figure 2.1 Concept of transformation

2.2 Definition of Symmetrical Components

2.2.1 Definition

Let us imagine a.c. voltages and currents at an arbitrary point of a three-phase circuit and name these quantities by complex-number variables V_a, V_b, V_c and I_a, I_b, I_c. In association with this set of voltages and currents, we introduce a new set of complex-number voltages and currents V_0, V_1, V_2 and I_0, I_1, I_2, defining them in the following equations:

$$\left.\begin{aligned}
V_0 &= \frac{1}{3}(V_a + V_b + V_c)\\
V_1 &= \frac{1}{3}(V_a + aV_b + a^2V_c)\\
V_2 &= \frac{1}{3}(V_a + a^2V_b + aV_c)
\end{aligned}\right\} \tag{2.1}$$

or

$$\underbrace{\begin{bmatrix} V_0 \\ V_1 \\ V_2 \end{bmatrix}}_{\boldsymbol{V}_{012}=} = \frac{1}{3}\underbrace{\begin{bmatrix} 1 & 1 & 1 \\ 1 & a & a^2 \\ 1 & a^2 & a \end{bmatrix}}_{\boldsymbol{a}} \cdot \underbrace{\begin{bmatrix} V_a \\ V_b \\ V_c \end{bmatrix}}_{\cdot \boldsymbol{V}_{abc}} \quad \therefore \boldsymbol{V}_{012} = \boldsymbol{a} \cdot \boldsymbol{V}_{abc}$$

$$\left.\begin{aligned}
I_0 &= \frac{1}{3}(I_a + I_b + I_c)\\
I_1 &= \frac{1}{3}(I_a + aI_b + a^2I_c)\\
I_2 &= \frac{1}{3}(I_a + a^2I_b + aI_c)
\end{aligned}\right\} \tag{2.2}$$

or

$$\underbrace{\begin{bmatrix} I_0 \\ I_1 \\ I_2 \end{bmatrix}}_{\boldsymbol{I}_{012}=} = \frac{1}{3}\underbrace{\begin{bmatrix} 1 & 1 & 1 \\ 1 & a & a^2 \\ 1 & a^2 & a \end{bmatrix}}_{\boldsymbol{a}} \cdot \underbrace{\begin{bmatrix} I_a \\ I_b \\ I_c \end{bmatrix}}_{\cdot \boldsymbol{I}_{abc}} \quad \therefore \boldsymbol{I}_{012} = \boldsymbol{a} \cdot \boldsymbol{I}_{abc}$$

a and a^2 are called **vector operators** and are defined as follows:

$$\left.\begin{aligned}
a &= -\frac{1}{2}+j\frac{\sqrt{3}}{2} = e^{j120^\circ} = \underline{\lvert 120^\circ} = \cos 120^\circ + j\sin 120^\circ\\
a^2 &= -\frac{1}{2}-j\frac{\sqrt{3}}{2} = e^{-j120^\circ} = \underline{\lvert -120^\circ} = \cos 120^\circ - j\sin 120^\circ\\
&= \overline{\lvert 120^\circ} = \left(-\frac{1}{2}+j\frac{\sqrt{3}}{2}\right)^2
\end{aligned}\right\} \tag{2.3a}$$

where $120^\circ = 2\pi/3$ [rad].

The vector operators a, a^2 can be modified into the following equations and written as vectors as shown in Figure 2.2:

$$\left.\begin{array}{ll}
a = e^{j120^\circ} = \underline{|120^\circ} & a^2 = e^{-j120^\circ} = \underline{|-120^\circ} = \overline{|+120^\circ} \\
a^2 + a + 1 = 0 & a^3 - 1 = (a-1)(a^2+a+1) = 0 \\
a^3 = 1 & a^2 + a = -1 \\
a^2 + 1 = -a & a + 1 = -a^2 \\
a^4 = a^3 \cdot a = a & a^5 = a^3 \cdot a^2 = a^2 \\
a^{-1} = a^{-1} \cdot a^3 = a^2 & a^{-2} = a^{-2} \cdot a^3 = a \\
|a| = |a^2| = 1 & a - a^2 = j\sqrt{3} \\
\multicolumn{2}{l}{1 - a = a^3(1-a) = a^2(a - a^2) = a^2 \cdot j\sqrt{3}} \\
\multicolumn{2}{l}{a^2 - 1 = (a+1)(a-1) = -a^2(a-1) = -a^2(-a^2 \cdot j\sqrt{3}) = a \cdot j\sqrt{3}}
\end{array}\right\} \quad (2.3b)$$

where $j = e^{j90^\circ} = \underline{|90^\circ}$, $-j = e^{-j90^\circ} = \overline{|90^\circ}$.

The defined set of voltages V_0, V_1, V_2 are named the zero (0), positive (1) , negative (2) sequence voltages, respectively, and the set of currents I_0, I_1, I_2 are also named zero (0), positive (1), negative (2) sequence currents in the newly defined 0–1–2 domain.

As V_a, V_b, V_c, I_a, I_b, I_c are expressed as complex-number quantities (effective valued or peak valued) in the a–b–c domain, then V_0, V_1, V_2, I_0, I_1, I_2 are consequently complex-number quantities (effective valued or peak valued) in the 0–1–2 domain.

The inverse matrix equations of Equations 2.1 and 2.2 can be easily introduced as follows:

$$\left.\begin{array}{l}
V_a = V_0 + V_1 + V_2 \\
V_b = V_0 + a^2 V_1 + a V_2 \\
V_c = V_0 + a V_1 + a^2 V_2
\end{array}\right\} \text{ or } \underbrace{\begin{bmatrix} V_a \\ V_b \\ V_c \end{bmatrix}}_{\boldsymbol{V}_{abc}=} = \underbrace{\begin{bmatrix} 1 & 1 & 1 \\ 1 & a^2 & a \\ 1 & a & a^2 \end{bmatrix}}_{\boldsymbol{a}^{-1}} \cdot \underbrace{\begin{bmatrix} V_0 \\ V_1 \\ V_2 \end{bmatrix}}_{\cdot\, \boldsymbol{V}_{012}} \quad (2.4)$$

$$\left.\begin{array}{l}
I_a = I_0 + I_1 + I_2 \\
I_b = I_0 + a^2 I_1 + a I_2 \\
I_c = I_0 + a I_1 + a^2 I_2
\end{array}\right\} \text{ or } \underbrace{\begin{bmatrix} I_a \\ I_b \\ I_c \end{bmatrix}}_{\boldsymbol{I}_{abc}=} = \underbrace{\begin{bmatrix} 1 & 1 & 1 \\ 1 & a^2 & a \\ 1 & a & a^2 \end{bmatrix}}_{\boldsymbol{a}^{-1}} \cdot \underbrace{\begin{bmatrix} I_0 \\ I_1 \\ I_2 \end{bmatrix}}_{\cdot\, \boldsymbol{I}_{012}} \quad (2.5)$$

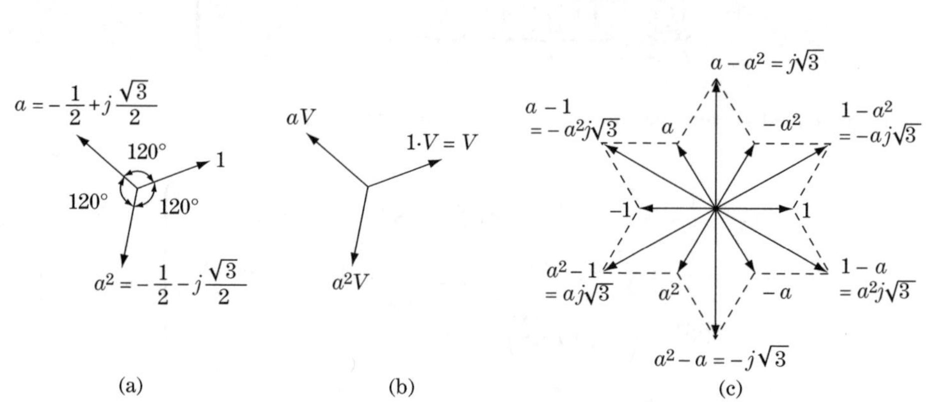

Figure 2.2 Vector operators a, a^2

The equations for transformation to the 0–1–2 domain, Equations 2.1 and 2.2, and the equations for inverse transformation to the a–b–c domain, Equations 2.3 and 2.4, are the basic definitions of the symmetrical components transformation. Incidentally, the vector operator matrices $\boldsymbol{a}$ and $\boldsymbol{a}^{-1}$ are inverse matrices of each other, namely

$$\left.\begin{aligned}
\boldsymbol{a}\cdot\boldsymbol{a}^{-1} &= \frac{1}{3}\underbrace{\begin{bmatrix}1 & 1 & 1\\ 1 & a & a^2\\ 1 & a^2 & a\end{bmatrix}}_{\boldsymbol{a}}\cdot\underbrace{\begin{bmatrix}1 & 1 & 1\\ 1 & a^2 & a\\ 1 & a & a^2\end{bmatrix}}_{\boldsymbol{a}^{-1}}\\
&= \frac{1}{3}\begin{bmatrix}1+1+1 & 1+a^2+a & 1+a+a^2\\ 1+a+a^2 & 1+a^3+a^3 & 1+a^2+a^4\\ 1+a^2+a & 1+a^4+a^2 & 1+a^3+a^3\end{bmatrix} = \begin{bmatrix}1 & 0 & 0\\ 0 & 1 & 0\\ 0 & 0 & 1\end{bmatrix} = \mathbf{1}\\
\boldsymbol{a}^{-1}\times\boldsymbol{a} &= \boldsymbol{a}\cdot\boldsymbol{a}^{-1} = \mathbf{1}
\end{aligned}\right\} \tag{2.3c}$$

All the quantities in the above defined equations are assigned as complex-number quantities; however, any assignment does not exist in the definition with regard to frequency or waveforms. In other words, the quantities may contain d.c. and/or higher harmonics. It should be noted that the symmetrical components transformation can be applied not only for power frequency steady-state phenomena but also for transient phenomena of any kind or even for travelling surges.

The voltage and current quantities are assigned as complex numbers in the above definitions, so that the corresponding real-number equations (or imaginary-number equations) can be extracted from them, which indicates the real behaviour of the actual voltage and current quantities in the a–b–c domain as well as the 0–1–2 domain.

Lastly, needless to say, all the electrical quantities in the a–b-c domain such as electric charge q, electric lines E, flux ϕ, etc., can be transformed into the 0–1–2 domain using the same definitions with the above vector operators.

2.2.2 Implication of symmetrical components

We need to examine more aspects of the symmetrical components defined by the above equations. The explanation below is followed by the current $\boldsymbol{I}$, and obviously the same analogy can be applied to all other quantities.

2.2.2.1 *Transformation from a–b–c quantities to 0–1–2 quantities*

Equation 2.2 can be transformed into Equation 2.2′ by multiplying by 3 both sides of the equations:

$$\left.\begin{aligned}
3I_0 &= \boxed{I_a} + \boxed{I_b} + \boxed{I_c}\\
3I_1 &= \boxed{I_a} + \boxed{aI_b} + \boxed{a^2I_c}\\
3I_2 &= \boxed{I_a} + \boxed{a^2I_b} + \boxed{aI_b}
\end{aligned}\right\} \tag{2.2′}$$

- For the first term: the same current components
- For the second term: counterclockwise balanced current components
- For the third term: clockwise balanced current components.

Whenever the current quantities are composed of only power frequency components (sinusoidal waveform), they can be visualized by drawing them as vectors in complex-number domain coordinates. Figure 2.3a shows the composition process of I_0, I_1, I_2 from I_a, I_b, I_c.

2.2.2.2 Inverse transformation from 0–1–2 quantities to a–b–c quantities

Equation 2.5 can be examined as follows:

$$\left.\begin{array}{l} I_a = \boxed{I_0} + \boxed{I_1} + \boxed{I_2} \\ I_b = \boxed{I_0} + \boxed{a^2 I_1} + \boxed{a I_2} \\ I_c = \boxed{I_0} + \boxed{a I_1} + \boxed{a^2 I_2} \end{array}\right\} \tag{2.6}$$

- Clockwise balanced complex-number currents I_1, a^2I_1, aI_1 are the components of the phase a, phase b, phase c currents, respectively (**positive-sequence components**).
- Counterclockwise balanced complex-number currents I_2, aI_2, a^2I_2 are the components of the phase a, phase b, phase c currents, respectively. (**negative-sequence components**).
- The three same-value quantities I_0, I_0, I_0 are the components of the phase a, phase b, phase c currents, respectively (**zero-sequence components**).

Figure 2.3b shows the composition process of I_a, I_b, I_c from I_0, I_1, I_2. (Figure 2.3a and b are drawn as a mutually paired case; however, the vectors in Figure 2.3a are drawn in half-dimensional length.)

Again, the quantities of the a–b–c and 0–1–2 domains are bilaterally transformable by the above definition.

2.2.2.3 Three-phase-balanced condition

Figure 2.4 shows the special case where three phase currents are balanced with a sinusoidal waveform. As I_a, I_b, I_c are clockwise phase balanced, then

$$I_a = I_a, \qquad I_b = a^2 I_a, \qquad I_c = aI_a \tag{2.7a}$$

and

$$\left.\begin{aligned} I_0 &= \frac{1}{3}(I_a + I_b + I_c) = \frac{1}{3}I_a(1 + a^2 + a) = 0 \\ I_1 &= \frac{1}{3}(I_a + aI_b + a^2I_c) = \frac{1}{3}I_a(1 + a \cdot a^2 + a^2 \cdot a) = I_a \\ I_2 &= \frac{1}{3}(I_a + a^2I_b + aI_c) = \frac{1}{3}I_a(1 + a^2 \cdot a^2 + a \cdot a) \\ &= \frac{1}{3}I_a(1 + a + a^2) = 0 \end{aligned}\right\} \tag{2.7b}$$

Under the three-phase-balanced condition, the zero-phase current I_0 and negative-phase currents I_2 are zero (or 'do not exist') and only positive current I_1 exists with the same vector value as I_a.

The three phase quantities I_a, I_b, I_c or the transformed I_0, I_1, I_2 under steady-state conditions (i.e. including only power frequency terms) can be visualized as vectors in Gauss coordinates whether balanced or unbalanced. Although transient quantities or multi-frequency quantities may not be simply visualized, the equational relations between the a–b–c and 0–1–2 domains are always justified.

Note that the currents I_a, I_b, I_c at an arbitrary point in a three-phase circuit and the corresponding currents I_1, I_2, I_0 in positive, negative and zero circuits should be marked by arrows ($\rightarrow$) in the same direction as the symbolic rule. The arrows of voltage polarities have to be selected analogously.

2.3 Conversion of Three-phase Circuit into Symmetrical Coordinated Circuit

As the first step in studying the symmetrical components transformation, we need to study how the equations and the related drawn circuits in the a–b–c domain are transformed into those in the symmetrical 0–1–2 domain.

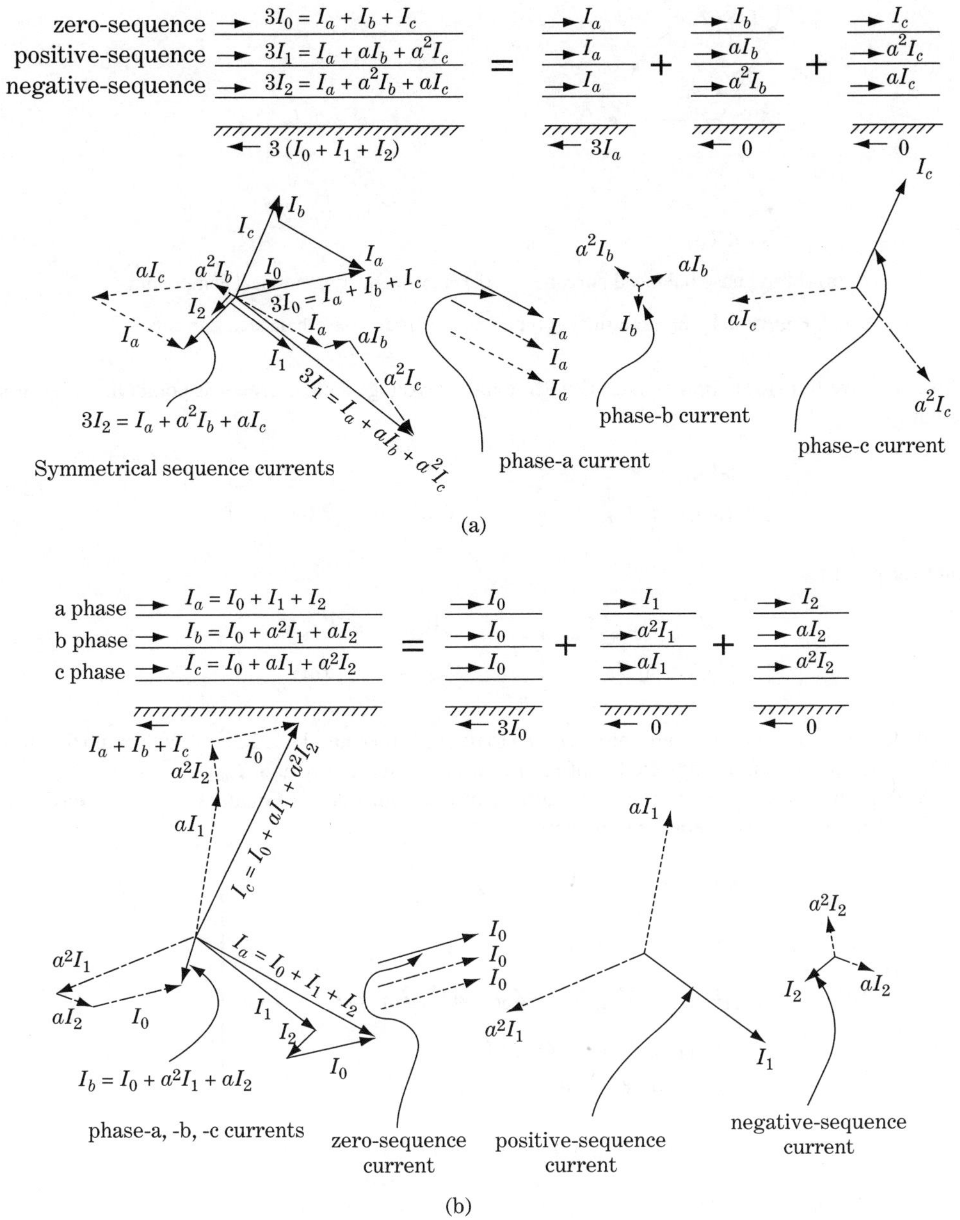

Figure 2.3 (a) Composition of I_0, I_1, I_2 from I_a, I_b, I_c. (b) Composition of I_a, I_b, I_c from I_0, I_1, I_2

Let us try to transform Equations 1.3, 1.5 and 1.6 of the transmission line and Figure 1.1b in Chapter 1 into symmetrical components. The equations of the transmission line (between points m and n), Equations 1.5 and 1.6, are written again here:

$$_m\boldsymbol{V}_{abc} - {}_n\boldsymbol{V}_{abc} = \boldsymbol{Z}_{abc} \cdot \boldsymbol{I}_{abc} \tag{2.8}$$

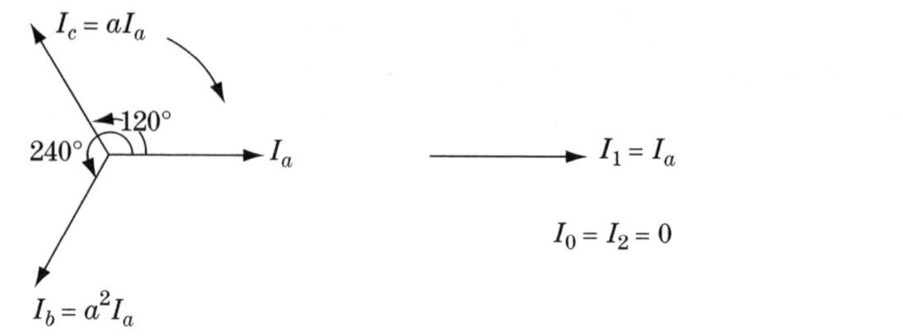

Figure 2.4 Symmetrical components of three-phase-balanced currents

Also we have transformation equations with regard to voltages and currents at points m and n. For point m

$$\left.\begin{array}{l} {}_m\boldsymbol{V}_{012} = \boldsymbol{a} \cdot {}_m\boldsymbol{V}_{abc} \\ \boldsymbol{I}_{012} = \boldsymbol{a} \cdot \boldsymbol{I}_{abc} \end{array}\right\} \qquad \left.\begin{array}{l} {}_m\boldsymbol{V}_{abc} = \boldsymbol{a}^{-1} \cdot {}_m\boldsymbol{V}_{012} \\ \boldsymbol{I}_{abc} = \boldsymbol{a}^{-1} \cdot \boldsymbol{I}_{012} \end{array}\right\} \tag{2.9}$$

and for point n

$$\left.\begin{array}{l} {}_n\boldsymbol{V}_{012} = \boldsymbol{a} \cdot {}_n\boldsymbol{V}_{abc} \\ \boldsymbol{I}_{012} = \boldsymbol{a} \cdot \boldsymbol{I}_{abc} \end{array}\right\} \qquad \left.\begin{array}{l} {}_n\boldsymbol{V}_{abc} = \boldsymbol{a}^{-1} \cdot {}_n\boldsymbol{V}_{012} \\ \boldsymbol{I}_{abc} = \boldsymbol{a}^{-1} \cdot \boldsymbol{I}_{012} \end{array}\right\} \tag{2.10}$$

As the currents at point m and n are assumed to be equal (because leakage current through the stray capacitance of the line is neglected), suffix m or n is omitted for symbol $\boldsymbol{I}_{abc}$.

Multiplying by $\boldsymbol{a}$ at the top (i.e. left-multiplying) of both sides in Equation 2.8, it can easily be changed into a symmetric equation, namely

$$\left.\begin{array}{l} \underbrace{\boldsymbol{a} \cdot {}_m\boldsymbol{V}_{abc}}_{\uparrow \; {}_m\boldsymbol{V}_{012}} - \underbrace{\boldsymbol{a} \cdot {}_n\boldsymbol{V}_{abc}}_{\uparrow \; {}_n\boldsymbol{V}_{012}} = \boldsymbol{a}\boldsymbol{Z}_{abc} \cdot \underbrace{\boldsymbol{I}_{abc}}_{\uparrow \; \boldsymbol{a}^{-1} \cdot \boldsymbol{I}_{012}} \\ \therefore \; {}_m\boldsymbol{V}_{012} - {}_n\boldsymbol{V}_{012} = \boldsymbol{a} \cdot \boldsymbol{Z}_{abc} \cdot \boldsymbol{a}^{-1} \cdot \boldsymbol{I}_{012} \equiv \boldsymbol{Z}_{012} \cdot \boldsymbol{I}_{012} \\ \text{i.e.} \quad \left.\begin{array}{l} {}_m\boldsymbol{V}_{012} - {}_n\boldsymbol{V}_{012} = \boldsymbol{Z}_{012} \cdot \boldsymbol{I}_{012} \\ \boldsymbol{Z}_{012} = \boldsymbol{a} \cdot \boldsymbol{Z}_{abc} \cdot \boldsymbol{a}^{-1} \end{array}\right\} \end{array}\right\} \tag{2.11}$$

Equation 2.8 was transformed into Equation 2.11 by symmetrical coordinates. The a–b–c impedance matrix $\boldsymbol{Z}_{abc}$ was transformed into the 0–1–2 sequence impedance matrix $\boldsymbol{Z}_{012}$ which is defined by Equation 2.11.

2.4 Transmission Lines by Symmetrical Components

2.4.1 Single circuit line with *LR* constants

Assuming that the transmission line of Figure 1.1b is well phase transposed,

$$\left.\begin{array}{l} Z_{aa} \fallingdotseq Z_{bb} \fallingdotseq Z_{cc} \equiv Z_s \\ Z_{ab} \fallingdotseq Z_{ba} \fallingdotseq Z_{bc} \fallingdotseq Z_{cb} \fallingdotseq Z_{ca} \fallingdotseq Z_{ac} \equiv Z_m \end{array}\right\} \tag{2.12}$$

then Equations 1.5 and 1.6 can be simplified as follows:

$$\begin{bmatrix} {}_mV_a \\ {}_mV_b \\ {}_mV_c \end{bmatrix} - \begin{bmatrix} {}_nV_a \\ {}_nV_b \\ {}_nV_c \end{bmatrix} = \begin{bmatrix} Z_s & Z_m & Z_m \\ Z_m & Z_s & Z_m \\ Z_m & Z_m & Z_s \end{bmatrix} \cdot \begin{bmatrix} I_a \\ I_b \\ I_c \end{bmatrix}$$
$$_m\boldsymbol{V}_{abc} \quad - \quad {}_n\boldsymbol{V}_{abc} = \quad \boldsymbol{Z}_{abc} \quad \cdot \boldsymbol{I}_{abc} \tag{2.13}$$

Accordingly $\boldsymbol{Z}_{012}$ of Equation 2.11 is

$$\boldsymbol{Z}_{012} = \boldsymbol{a} \cdot \boldsymbol{Z}_{abc} \cdot \boldsymbol{a}^{-1} = \boldsymbol{a} \underbrace{\begin{bmatrix} Z_s & Z_m & Z_m \\ Z_m & Z_s & Z_m \\ Z_m & Z_m & Z_s \end{bmatrix}}_{\boldsymbol{Z}_{abc}} \cdot \underbrace{\begin{bmatrix} 1 & 1 & 1 \\ 1 & a^2 & a \\ 1 & a & a^2 \end{bmatrix}}_{\boldsymbol{a}^{-1}}$$

$$= \boldsymbol{a} \begin{bmatrix} Z_s + 2Z_m & Z_s + (a^2 + a)Z_m & Z_s + (a + a^2)Z_m \\ Z_s + 2Z_m & a^2Z_s + (1 + a)Z_m & aZ_s + (1 + a^2)Z_m \\ Z_s + 2Z_m & aZ_s + (1 + a^2)Z_m & a^2Z_s + (1 + a)Z_m \end{bmatrix}$$

$$= \frac{1}{3} \underbrace{\begin{bmatrix} 1 & 1 & 1 \\ 1 & a & a^2 \\ 1 & a^2 & a \end{bmatrix}}_{\boldsymbol{a}} \cdot \underbrace{\begin{bmatrix} Z_s + 2Z_m & Z_s - Z_m & Z_s - Z_m \\ Z_s + 2Z_m & a^2(Z_s - Z_m) & a(Z_s - Z_m) \\ Z_s + 2Z_m & a(Z_s - Z_m) & a^2(Z_s - Z_m) \end{bmatrix}}_{\boldsymbol{Z}_{abc} \cdot \boldsymbol{a}^{-1}} \tag{2.14}$$

$$= \begin{bmatrix} Z_s + 2Z_m & 0 & 0 \\ 0 & Z_s - Z_m & 0 \\ 0 & 0 & Z_s - Z_m \end{bmatrix}$$

Namely,

$$\underbrace{\begin{bmatrix} {}_mV_0 \\ {}_mV_1 \\ {}_mV_2 \end{bmatrix}}_{{}_m\boldsymbol{V}_{012}} - \underbrace{\begin{bmatrix} {}_nV_0 \\ {}_nV_1 \\ {}_nV_2 \end{bmatrix}}_{{}_n\boldsymbol{V}_{012}} = \underbrace{\begin{bmatrix} Z_s + 2Z_m & 0 & 0 \\ 0 & Z_s - Z_m & 0 \\ 0 & 0 & Z_s - Z_m \end{bmatrix}}_{\boldsymbol{Z}_{012}} \cdot \underbrace{\begin{bmatrix} I_0 \\ I_1 \\ I_2 \end{bmatrix}}_{\boldsymbol{I}_{012}} \equiv \underbrace{\begin{bmatrix} Z_0 & 0 & 0 \\ 0 & Z_1 & 0 \\ 0 & 0 & Z_1 \end{bmatrix}}_{\boldsymbol{Z}_{012}} \cdot \underbrace{\begin{bmatrix} I_0 \\ I_1 \\ I_2 \end{bmatrix}}_{\boldsymbol{I}_{012}}$$

or

$$\left.\begin{aligned} {}_mV_0 - {}_nV_0 &= (Z_s + 2Z_m)I_0 = Z_0I_0 \\ {}_mV_1 - {}_nV_1 &= (Z_s - Z_m)I_1 = Z_1I_1 \\ {}_mV_2 - {}_nV_2 &= (Z_s - Z_m)I_2 = Z_1I_2 \end{aligned}\right\} \\ \text{where} \quad Z_0 = Z_s + 2Z_m, \; Z_1 = Z_s - Z_m \tag{2.15}$$

This is the equation of a single circuit transmission line in the symmetrical components domain. $\boldsymbol{Z}_{012}$ is a simple diagonal matrix in which all the off-diagonal elements vanish (become zero). This means that the positive-, negative- and zero-sequence equations are mutually independent of each other because mutual impedances do not exist among them. Now we can conclude that, if the original three-phase circuit is phase balanced (this assumption is acceptable for most cases only with small errors), positive (1) sequence, negative (2) sequence and zero (0) sequence circuits can be independently handled. Figure 2.5 shows the equivalent circuit of a three-phase (single circuit) transmission line by symmetrical components, which is drawn from Equation 2.15.

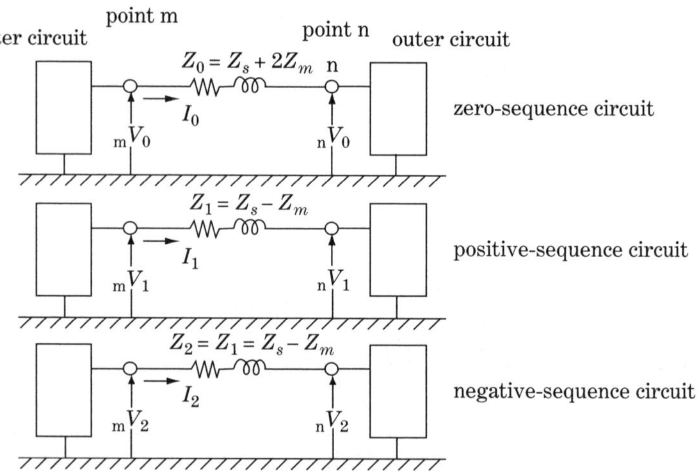

Figure 2.5 The equivalent circuit of three-phase single circuit transmission line (impedances)

Symmetrical impedances Z_0, Z_1, Z_2 are defined by Equation 2.15 in relation to the original impedances Z_s, Z_m, while from the relation

$$Z_0 > Z_1 = Z_2 \tag{2.16}$$

That is, for the transmission line, the positive-sequence impedance Z_1 and the negative-sequence impedance Z_2 are equal and smaller than the zero-sequence impedance Z_0.

Note that, as the transmission line is not perfectly phase balanced, very small off-diagonal elements may exist in the impedance matrix $\boldsymbol{Z}_{012}$, so that positive-, negative- and zero- sequence circuits are mutually linked by small mutual inductances. If necessary, we can examine the strict impedance matrix by calculating equation $\boldsymbol{Z}_{012} = \boldsymbol{a} \cdot \boldsymbol{Z}_{abc} \cdot \boldsymbol{a}^{-1}$ without any assumption of Equation 2.12.

2.4.2 Double circuit line with *LR* constants

Let us examine a double circuit transmission line as shown in Figure 1.5, assuming that the first and second circuits are well phase transposed. The symbols ${}^1V$, ${}^1I$ and ${}^2V$, ${}^2I$ below refer to the voltages and currents of the first and second circuits:

$$\begin{bmatrix} {}^1_mV_a \\ {}^1_mV_b \\ {}^1_mV_c \\ {}^2_mV_a \\ {}^2_mV_b \\ {}^2_mV_c \end{bmatrix} - \begin{bmatrix} {}^1_nV_a \\ {}^1_nV_b \\ {}^1_nV_c \\ {}^2_nV_a \\ {}^2_nV_b \\ {}^2_nV_c \end{bmatrix} = \begin{bmatrix} Z_s & Z_m & Z_m & Z'_m & Z'_m & Z'_m \\ Z_m & Z_s & Z_m & Z'_m & Z'_m & Z'_m \\ Z_m & Z_m & Z_s & Z'_m & Z'_m & Z'_m \\ Z'_m & Z'_m & Z'_m & Z_s & Z_m & Z_m \\ Z'_m & Z'_m & Z'_m & Z_m & Z_s & Z_m \\ Z'_m & Z'_m & Z'_m & Z_m & Z_m & Z_s \end{bmatrix} \cdot \begin{bmatrix} {}^1I_a \\ {}^1I_b \\ {}^1I_c \\ {}^2I_a \\ {}^2I_b \\ {}^2I_c \end{bmatrix}$$

or

$$\begin{bmatrix} {}^1_m\boldsymbol{V}_{abc} \\ {}^2_m\boldsymbol{V}_{abc} \end{bmatrix} - \begin{bmatrix} {}^1_n\boldsymbol{V}_{abc} \\ {}^2_n\boldsymbol{V}_{abc} \end{bmatrix} = \begin{bmatrix} \boldsymbol{Z}_{sm} & \boldsymbol{Z}'_m \\ \boldsymbol{Z}'_m & \boldsymbol{Z}_{sm} \end{bmatrix} \cdot \begin{bmatrix} {}^1\boldsymbol{I}_{abc} \\ {}^2\boldsymbol{I}_{abc} \end{bmatrix} = \begin{bmatrix} \boldsymbol{Z}_{sm} \cdot {}^1\boldsymbol{I}_{abc} + \boldsymbol{Z}'_m \cdot {}^2\boldsymbol{I}_{abc} \\ \boldsymbol{Z}'_m \cdot {}^1\boldsymbol{I}_{abc} + \boldsymbol{Z}_{sm} \cdot {}^2\boldsymbol{I}_{abc} \end{bmatrix} \tag{2.17}$$

where we assume

$Z_s \equiv Z_{aa} \doteqdot Z_{bb} \doteqdot Z_{cc} \doteqdot Z_{AA} \doteqdot$: self-impedance
$Z_m \equiv Z_{ab} \doteqdot Z_{bc} \doteqdot Z_{ca} \doteqdot Z_{AB} \doteqdot Z_{BC} \doteqdot$: mutual impedance between the conductors of the same circuit
$Z'_m \equiv Z_{aA} \doteqdot Z_{aB} \doteqdot Z_{aC} \doteqdot Z_{bA}$: mutual impedance between the conductors of another circuit

Symmetrical quantities of double circuit line ${}^1_m\boldsymbol{V}_{012}$, ${}^1_n\boldsymbol{V}_{012}$, ${}^2_m\boldsymbol{V}_{012}$, ${}^2_n\boldsymbol{V}_{012}$, ${}^1\boldsymbol{I}_{012}$, ${}^2\boldsymbol{I}_{012}$ are introduced in conjunction with a–b–c domain quantities ${}^1_m\boldsymbol{V}_{abc}$, ${}^1_n\boldsymbol{V}_{abc}$, ${}^2_m\boldsymbol{V}_{abc}$, ${}^2_n\boldsymbol{V}_{abc}$, ${}^1\boldsymbol{I}_{abc}$, ${}^2\boldsymbol{I}_{abc}$:

$$\left.\begin{array}{l}\left.\begin{array}{lll}{}^1_m\boldsymbol{V}_{012} = \boldsymbol{a}\cdot{}^1_m\boldsymbol{V}_{abc}, & {}^1_n\boldsymbol{V}_{012} = \boldsymbol{a}\cdot{}^1_n\boldsymbol{V}_{abc}, & {}^1\boldsymbol{I}_{012} = \boldsymbol{a}\cdot{}^1\boldsymbol{I}_{abc} \\ {}^2_m\boldsymbol{V}_{012} = \boldsymbol{a}\cdot{}^2_m\boldsymbol{V}_{abc}, & {}^2_n\boldsymbol{V}_{012} = \boldsymbol{a}\cdot{}^2_n\boldsymbol{V}_{abc}, & {}^2\boldsymbol{I}_{012} = \boldsymbol{a}\cdot{}^2\boldsymbol{I}_{abc}\end{array}\right\} \quad ① \\ \left.\begin{array}{lll}{}^1_m\boldsymbol{V}_{abc} = \boldsymbol{a}^{-1}\cdot{}^1_m\boldsymbol{V}_{012}, & {}^1_n\boldsymbol{V}_{abc} = \boldsymbol{a}^{-1}\cdot{}^1_n\boldsymbol{V}_{012}, & {}^1\boldsymbol{I}_{abc} = \boldsymbol{a}^{-1}\cdot{}^1\boldsymbol{I}_{012} \\ {}^2_m\boldsymbol{V}_{abc} = \boldsymbol{a}^{-1}\cdot{}^2_m\boldsymbol{V}_{012}, & {}^2_n\boldsymbol{V}_{abc} = \boldsymbol{a}^{-1}\cdot{}^2_n\boldsymbol{V}_{012}, & {}^2\boldsymbol{I}_{abc} = \boldsymbol{a}^{-1}\cdot{}^2\boldsymbol{I}_{012}\end{array}\right\} \quad ②\end{array}\right\} \tag{2.18}$$

The equation of circuit 1 in Equation 2.17 can be transformed to the 0–1–2 domain by utilizing Equation 2.18:

$$\begin{array}{l}{}^1_m\boldsymbol{V}_{abc} - {}^1_n\boldsymbol{V}_{abc} = \boldsymbol{Z}_{sm}\cdot{}^1\boldsymbol{I}_{abc} + \boldsymbol{Z}'_m\cdot{}^2\boldsymbol{I}_{abc} \\ \therefore\quad \boldsymbol{a}^{-1}\cdot{}^1_m\boldsymbol{V}_{012} - \boldsymbol{a}^{-1}\cdot{}^1_n\boldsymbol{V}_{012} = \boldsymbol{Z}_{sm}\cdot\boldsymbol{a}^{-1}\cdot{}^1\boldsymbol{I}_{012} + \boldsymbol{Z}'_m\cdot\boldsymbol{a}^{-1}\cdot{}^2\boldsymbol{I}_{012}\end{array}$$

Left-multiplying by $\boldsymbol{a}$ and recalling that $\boldsymbol{a}\cdot\boldsymbol{a}^{-1} = \boldsymbol{1}$,

$$\left.\begin{array}{l}\text{for circuit 1}\quad {}^1_m\boldsymbol{V}_{012} - {}^1_n\boldsymbol{V}_{012} = (\boldsymbol{a}\cdot\boldsymbol{Z}_{sm}\cdot\boldsymbol{a}^{-1})\cdot{}^1\boldsymbol{I}_{012} + (\boldsymbol{a}\cdot\boldsymbol{Z}'_m\cdot\boldsymbol{a}^{-1})\cdot{}^2\boldsymbol{I}_{012} \\ \text{and for circuit 2 analogously} \\ \quad {}^2_m\boldsymbol{V}_{012} - {}^1_n\boldsymbol{V}_{012} = (\boldsymbol{a}\cdot\boldsymbol{Z}'_m\cdot\boldsymbol{a}^{-1})\cdot{}^1\boldsymbol{I}_{012} + (\boldsymbol{a}\cdot\boldsymbol{Z}_{sm}\cdot\boldsymbol{a}^{-1})\cdot{}^2\boldsymbol{I}_{012}\end{array}\right\} \tag{2.19}$$

$\boldsymbol{a}\cdot\boldsymbol{Z}_{sm}\cdot\boldsymbol{a}^{-1}$ in the above equation is equal to $\boldsymbol{Z}_{012}$ of the single circuit line in Equation 2.14, so

$$\boldsymbol{a}\cdot\boldsymbol{Z}_{sm}\cdot\boldsymbol{a}^{-1} = \begin{bmatrix} Z_s+2Z_m & 0 & 0 \\ 0 & Z_s-Z_m & 0 \\ 0 & 0 & Z_s-Z_m \end{bmatrix}$$

and

$$\boldsymbol{a}\cdot\boldsymbol{Z}'_m\cdot\boldsymbol{a}^{-1} = \boldsymbol{a}\cdot\begin{bmatrix} Z'_m & Z'_m & Z'_m \\ Z'_m & Z'_m & Z'_m \\ Z'_m & Z'_m & Z'_m \end{bmatrix}\cdot\begin{bmatrix} 1 & 1 & 1 \\ 1 & a^2 & a \\ 1 & a & a^2 \end{bmatrix}$$

$$= \underbrace{\frac{1}{3}\begin{bmatrix} 1 & 1 & 1 \\ 1 & a & a^2 \\ 1 & a^2 & a \end{bmatrix}}_{\boldsymbol{a}}\cdot\underbrace{\begin{bmatrix} 3Z'_m & 0 & 0 \\ 3Z'_m & 0 & 0 \\ 3Z'_m & 0 & 0 \end{bmatrix}}_{\boldsymbol{Z}'_m\cdot\,\boldsymbol{a}^{-1}} = \begin{bmatrix} 3Z'_m & 0 & 0 \\ 0 & 0 & 0 \\ 0 & 0 & 0 \end{bmatrix}$$

Accordingly,

$$\begin{bmatrix} {}^1_m\boldsymbol{V}_{012} \\ {}^2_m\boldsymbol{V}_{012} \end{bmatrix} - \begin{bmatrix} {}^1_n\boldsymbol{V}_{012} \\ {}^2_n\boldsymbol{V}_{012} \end{bmatrix} = \begin{bmatrix} \boldsymbol{a}\cdot\boldsymbol{Z}_{sm}\cdot\boldsymbol{a}^{-1} & \boldsymbol{a}\cdot\boldsymbol{Z}'_{sm}\cdot\boldsymbol{a}^{-1} \\ \boldsymbol{a}\cdot\boldsymbol{Z}'_{sm}\cdot\boldsymbol{a}^{-1} & \boldsymbol{a}\cdot\boldsymbol{Z}_{sm}\cdot\boldsymbol{a}^{-1} \end{bmatrix} \cdot \begin{bmatrix} {}^1\boldsymbol{I}_{012} \\ {}^2\boldsymbol{I}_{012} \end{bmatrix} = \begin{bmatrix} \boldsymbol{Z}_{012} & \boldsymbol{Z}_{0M} \\ \boldsymbol{Z}_{0M} & \boldsymbol{Z}_{012} \end{bmatrix} \cdot \begin{bmatrix} {}^1\boldsymbol{I}_{012} \\ {}^2\boldsymbol{I}_{012} \end{bmatrix}$$

or

$$\begin{bmatrix} {}^1_m\boldsymbol{V}_0 \\ {}^1_m\boldsymbol{V}_1 \\ {}^1_m\boldsymbol{V}_2 \\ {}^2_m\boldsymbol{V}_0 \\ {}^2_m\boldsymbol{V}_1 \\ {}^2_m\boldsymbol{V}_2 \end{bmatrix} - \begin{bmatrix} {}^1_n\boldsymbol{V}_0 \\ {}^1_n\boldsymbol{V}_1 \\ {}^1_n\boldsymbol{V}_2 \\ {}^2_n\boldsymbol{V}_0 \\ {}^2_n\boldsymbol{V}_1 \\ {}^2_n\boldsymbol{V}_2 \end{bmatrix} = \begin{bmatrix} Z_s+2Z_m & 0 & 0 & 3Z'_m & 0 & 0 \\ 0 & Z_s-Z_m & 0 & 0 & 0 & 0 \\ 0 & 0 & Z_s-Z_m & 0 & 0 & 0 \\ 3Z'_m & 0 & 0 & Z_s+2Z_m & 0 & 0 \\ 0 & 0 & 0 & 0 & Z_s-Z_m & 0 \\ 0 & 0 & 0 & 0 & 0 & Z_s-Z_m \end{bmatrix} \cdot \begin{bmatrix} {}^1I_0 \\ {}^1I_1 \\ {}^1I_2 \\ {}^2I_0 \\ {}^2I_1 \\ {}^2I_2 \end{bmatrix}$$

$$\equiv \begin{bmatrix} Z_0 & 0 & 0 & Z_{0M} & 0 & 0 \\ 0 & Z_1 & 0 & 0 & 0 & 0 \\ 0 & 0 & Z_1 & 0 & 0 & 0 \\ Z_{0M} & 0 & 0 & Z_0 & 0 & 0 \\ 0 & 0 & 0 & 0 & Z_1 & 0 \\ 0 & 0 & 0 & 0 & 0 & Z_1 \end{bmatrix} \cdot \begin{bmatrix} {}^1I_0 \\ {}^1I_1 \\ {}^1I_2 \\ {}^2I_0 \\ {}^2I_1 \\ {}^2I_2 \end{bmatrix} \qquad (2.20a)$$

where

$$Z_1 = Z_s - Z_m, \qquad Z_0 = Z_s + 2Z_m, \qquad Z_{0M} = 3Z'_m$$

Equation 2.20a can be recast into the following equation:

$$\left.\begin{aligned} \begin{bmatrix} {}^1_mV_0 \\ {}^2_mV_0 \end{bmatrix} - \begin{bmatrix} {}^1_nV_0 \\ {}^2_nV_0 \end{bmatrix} &= \begin{bmatrix} Z_0 & Z_{0M} \\ Z_{0M} & Z_0 \end{bmatrix} \cdot \begin{bmatrix} {}^1I_0 \\ {}^2I_0 \end{bmatrix} \\ \begin{bmatrix} {}^1_mV_1 \\ {}^2_mV_1 \end{bmatrix} - \begin{bmatrix} {}^1_nV_1 \\ {}^2_nV_1 \end{bmatrix} &= \begin{bmatrix} Z_1 & 0 \\ 0 & Z_1 \end{bmatrix} \cdot \begin{bmatrix} {}^1I_1 \\ {}^2I_1 \end{bmatrix} \\ \begin{bmatrix} {}^1_mV_2 \\ {}^2_mV_2 \end{bmatrix} - \begin{bmatrix} {}^1_nV_2 \\ {}^2_nV_2 \end{bmatrix} &= \begin{bmatrix} Z_1 & 0 \\ 0 & Z_1 \end{bmatrix} \cdot \begin{bmatrix} {}^1I_2 \\ {}^2I_2 \end{bmatrix} \\ \text{where } Z_0 = Z_s + 2Z_m, \quad Z_{0M} &= 3Z'_m, \quad Z_1 = Z_s - Z_m \end{aligned}\right\} \qquad (2.20b)$$

Figure 2.6 shows the equivalent circuit of the three-phase double circuit transmission line by symmetrical components, which is drawn from Equation 2.20a or 2.20b.

The positive-, negative- and zero-sequence circuits are independent (mutual inductances do not exist) of each other. In the positive- and negative-sequence circuits, the mutual inductances do not exist between lines 1 and 2. However in the zero-sequence circuit, lines 1 and 2 are mutually coupled together by $Z_{0M} = 3Z'_m$.

Z_s and Z_m are actually the averaged values of Z_{aa}, Z_{bb}, Z_{cc} and Z_{ab}, Z_{bc}, Z_{ca} respectively, so that Z_s and Z_m can be calculated by using Equations 1.10 and 1.12. The positive-sequence impedance $Z_1 = Z_s - Z_m$ is derived from the working inductance $L_{aa} - L_{ab}$ given by Equation 1.9.

Also we learned in Chapter 1 that the values of L_{aa}, L_{ab} are not so largely affected by rated voltage classes because of the logarithmic term of these equations. Consequently $Z_1 = Z_2, Z_0$ are not also largely affected by rated voltage classes. Typical examples are shown in Table 2.1.

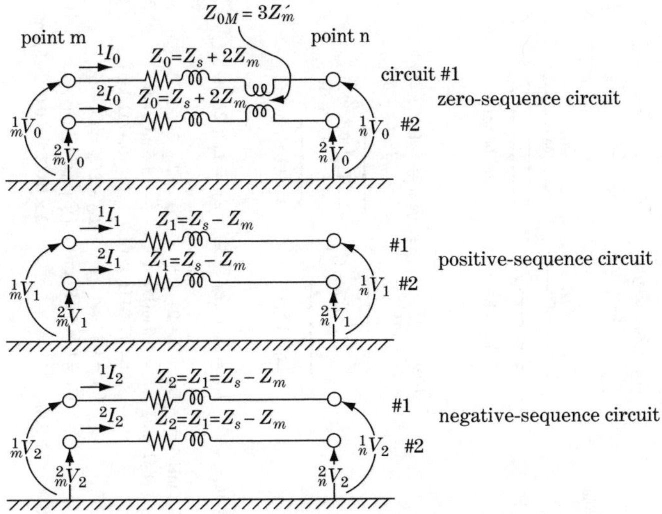

Figure 2.6 The equivalent circuit of three-phase double circuit transmission line (impedance)

2.4.3 Single circuit line with stray capacitance *C*

The stray capacitances of a well-phase-transposed single circuit line are shown by Figures 1.9a and b and Equation 1.34. Equation 1.34 is repeated here:

$$\left.\begin{aligned}\begin{bmatrix} I_a \\ I_b \\ I_c \end{bmatrix} &= j\omega \cdot \begin{bmatrix} C_s + 2C_m & -C_m & -C_m \\ -C_m & C_s + 2C_m & -C_m \\ -C_m & -C_m & C_s + 2C_m \end{bmatrix} \cdot \begin{bmatrix} V_a \\ V_b \\ V_c \end{bmatrix} \\ \boldsymbol{I}_{abc} &= j\omega \times \qquad \boldsymbol{C}_{abc} \qquad \times \boldsymbol{V}_{abc}\end{aligned}\right\} \tag{2.21}$$

The transformation of this equation into symmetrical components is as follows:

$$\left.\begin{aligned} &I_{012} = \boldsymbol{a} \cdot I_{abc} = \boldsymbol{a} \cdot j\omega \cdot C_{abc} \cdot V_{abc} = j\omega(\boldsymbol{a} \cdot C_{abc} \cdot \boldsymbol{a}^{-1})V_{012} \equiv j\omega C_{012} V_{012} \\ &\text{where } \boldsymbol{C}_{012} = \boldsymbol{a} \cdot C_{abc} \cdot \boldsymbol{a}^{-1} \\ &C_{012} = \underbrace{\frac{1}{3}\begin{bmatrix} 1 & 1 & 1 \\ 1 & a & a^2 \\ 1 & a^2 & a \end{bmatrix}}_{\boldsymbol{a}} \cdot \underbrace{\begin{bmatrix} C_s + 2C_m & -C_m & -C_m \\ -C_m & C_s + 2C_m & -C_m \\ -C_m & -C_m & C_s + 2C_m \end{bmatrix}}_{C_{abc}} \cdot \underbrace{\begin{bmatrix} 1 & 1 & 1 \\ 1 & a^2 & a \\ 1 & a & a^2 \end{bmatrix}}_{\boldsymbol{a}^{-1}} \\ &\quad = \begin{bmatrix} C_s & 0 & 0 \\ 0 & C_s + 3C_m & 0 \\ 0 & 0 & C_s + 3C_m \end{bmatrix} \end{aligned}\right\} \tag{2.22a}$$

then

$$\left.\begin{aligned} &\begin{bmatrix} I_0 \\ I_1 \\ I_2 \end{bmatrix} = j\omega \underbrace{\begin{bmatrix} C_s & 0 & 0 \\ 0 & C_s + 3C_m & 0 \\ 0 & 0 & C_s + 3C_m \end{bmatrix}}_{} \cdot \begin{bmatrix} V_0 \\ V_1 \\ V_2 \end{bmatrix} = j\omega \underbrace{\begin{bmatrix} C_0 & 0 & 0 \\ 0 & C_1 & 0 \\ 0 & 0 & C_1 \end{bmatrix}}_{} \cdot \underbrace{\begin{bmatrix} V_0 \\ V_1 \\ V_2 \end{bmatrix}}_{} \\ &\boldsymbol{I}_{012} = j\omega \times \qquad \boldsymbol{C}_{012} \qquad \times \qquad \boldsymbol{V}_{012} \qquad\qquad \boldsymbol{C}_{012} \qquad \boldsymbol{V}_{012} \\ &\text{where } C_0 = C_s, \quad C_1 = C_2 = C_s + 3C_m \quad C_1 : \text{working capacitance of single circuit line} \end{aligned}\right\} \tag{2.22b}$$

Table 2.1 Typical L, R, C constants of overhead transmission lines

		Single circuit line		Double circuit line					
		Positive-sequence	**Zero-sequence**	**Positive-sequence**	**Zero-sequence**		**Zero-sequence (first lane)**	**Zero-sequence (second lane)**	**Surge impedance**
Circuit equations		$Z_1 = Z_s - Z_m$	$Z_0 = Z_s + 2Z_m$	$Z_1 = Z_s - Z_m$	$Z_0 = Z_s + 2Z_m$	$Z_{0M} = 3Z'_m$	$Z_{00} = Z_0 + Z_{0M} = Z_s + 2Z_m + 3Z'_m$	$Z_{01} = Z_0 - Z_{0M} = Z_s + 2Z_m - 3Z'_m$	$\sqrt{\frac{L}{C}}$
		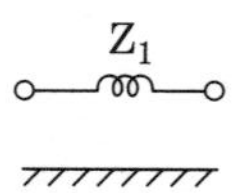 Eq. (2.15)	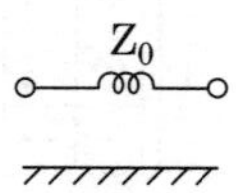 Eq. (2.15)	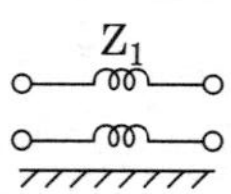 Eq. (2.20)	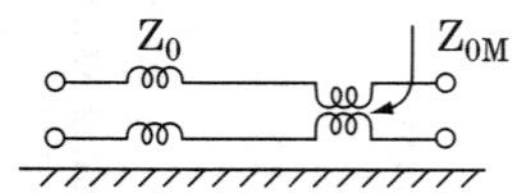 Eq. (2.20)		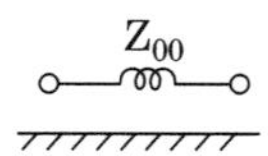 Eq. (4.5c)	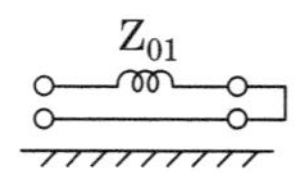 Eq. (4.5c)	
		$jY_1 = j\omega C_1$	$jY_0 = j\omega C_0$	$C_1 = C_s + 3C_m + 3C'_m$ $jY_1 = j\omega C_1$	$jY_0 = j\omega C_0 = j\omega C_s$ $jY'_0 = j\omega C'_0 = j\omega 3C'_m$		$jY_{00} = jwC_0 = jwC_s$	$jY_{01} = jw(C_0 + 2C'_0) = jw(C_s + 6C'_m)$	
		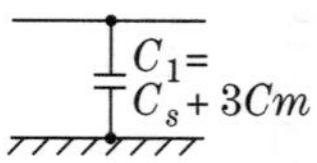 Eq. (2.22)	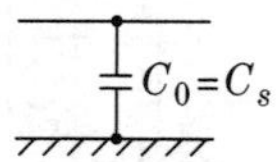 Eq. (2.22)	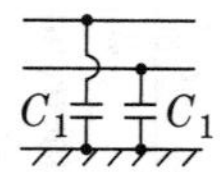 Eq. (2.24)	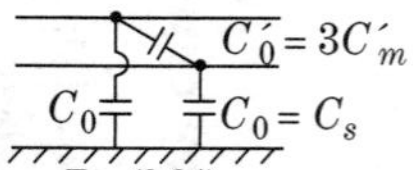 Eq. (2.24)		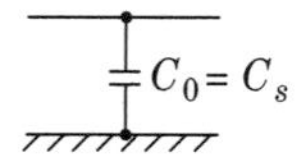 Eq. (4.5c)	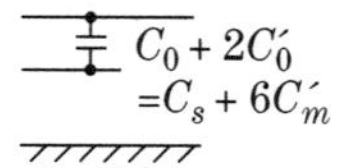 Eq. (4.5c)	
Typical value	Inductance [mH/km]	$L_1 = L_s - L_m$	$L_0 = L_s + 2L_m$	$L_1 = L_s - L_m$	$L_0 = L_s + 2L_m$	$L_{0M} = 3L'_s$	$L_{00} = L_s + 2L_m + 3L'_m$	$Z_{01} = L_s + 2L_m - 3L'_m$	
	$L_s = 2\,\text{mH/km}$, $L_m = 1\,\text{mH/km}$, $3L'_m = 2.5\,\text{mH/km}$	$L_1 = 2 - 1 = 1\,\text{mH/km}$ $(= j0.314\,\Omega/\text{km})$	$L_0 = 2 + 2 \times 1 = 4\,\text{mH/km}$ $(= j1.26\,\Omega/\text{km})$	$L_1 = 2 - 1 = 1\,\text{mH/km}$ $(= j0.314\,\Omega/\text{km})$	$L_0 = 2 + 2 \times 1 = 4\,\text{mH/km}$ $(= j1.26\,\Omega/\text{km})$	$L_0 = 2.5\,\text{mH/km}$ $(= j0.78\,\Omega/\text{km})$	$L_{00} = 2 + 2 + 2.5 = 6.5\,\text{mH/km}$ $(= j2.04\,\Omega/\text{km})$	$L_{01} = 2 + 2 - 2.5 = 1.5\,\text{mH/km}$ $(= j0.47\,\Omega/\text{km})$	Single line 258 Ω/km
	Capacitance [μF/km]	$C_1 = C_s + 3C_m$	$C_0 = C_s$	$C_1 = C_s + 3C_m + 3C'_m$	$C_0 = C_s$	$C'_0 = 3C'_m$	$C_0 = C_s$	$C_0 + 2C'_0 = C_s + 6C'_m$	
	$C_s = 0.005\,\mu\text{F/km}$, $3C_m = 0.01\,\mu\text{F/km}$, $3C'_m = 0.0025\,\mu\text{F/km}$	$C_1 = 0.005 + 0.010 = 0.015\,\mu\text{F/km}$	$C_0 = 0.005\,\mu\text{F/km}$	$C_1 = 0.005 + 0.010 + 0.0025 = 0.0175\,\mu\text{F/km}$	$C_0 = 0.005\,\mu\text{F/km}$	$C'_0 = 3C'_m = 0.0025\,\mu\text{F/km}$	$C_0 = 0.005\,\mu\text{F/km}$	$C_s + 6C'_m = 0.005 + 2 \times 0.0025 = 0.010\,\mu\text{F/km}$	Double line 239 Ω/km

Example 1	500 kV, double circuit line STACSR ($810\,mm^2$) × 4	(0.86 mH/km) $0.019 + j0.27\,\Omega/km$	(3.34 mH/km) $0.30 + j1.05\,\Omega/km$	(0.86 mH/km) $0.019 + j0.27\,\Omega/km$	(3.34 mH/km) $0.30 + j1.05\,\Omega/km$	(2.20 mH/km) $0.28 + j0.69\,\Omega/km$	(5.54 mH/km) $0.58 + j1.74\,\Omega/km$	(1.15 mH/km) $0.02 + j0.36\,\Omega/km$	S 257 Ω/km
		0.013 μF/km 1.18 A/km	0.0085 μF/km	0.016 μF/km 1.41 A/km		0.0029 μF/km	0.0062 μF/km	0.012 μF/km	D 232 Ω/km
Example 2	275 kV, double circuit line STACSR ($810\,mm^2$) ×2	(0.73 mH/km) $0.0085 + j0.23$	$0.26 + j1.10\,\Omega/km$	(0.73 mH/km) $0.0085 + j0.23\,\Omega/km$	$0.26 + j1.10\,\Omega/km$	(2.42 mH/km) $0.25 + j0.76\,\Omega/km$	(5.92 mH/km) $0.51 + j1.86\,\Omega/km$	(1.08 mH/km) $0.10 + j0.34\,\Omega/km$	S 216 Ω/km
		0.0156 μF/km 0.778 A/km	0.0054 μF/km	0.0185 μF/km 0.92 A/km		0.0029 μF/km	0.0058 μF/km	0.0116 μF/km	D 199 Ω/km
Example 3	154 kV, double circuit line ACSR ($610\,mm^2$) ×1	(1.21 mH/km) $0.005 + j0.38\,\Omega/km$	$0.30 + j1.43\,\Omega/km$	(1.21 mH/km) $0.005 + j0.38\,\Omega/km$	$0.30 + j1.43\,\Omega/km$	(2.90 mH/km) $0.25 + j0.91\,\Omega/km$	(7.45 mH/km) $0.55 + j2.34\,\Omega/km$	(1.66 mH/km) $0.05 + j0.52\,\Omega/km$	S 358 Ω/km
		0.0094 μF/km 0.262 A/km	0.0055 μF/km	0.0109 μF/km 0.148 A/km		0.0015 μF/km	0.0044 μF/km	0.0074 μF/km	D 333 Ω/km
Example 4	66 kV, double circuit line ACSR ($330\,mm^2$) ×1	(1.15 mH/km) $0.05 + j0.36\Omega/km$	$0.31 + j1.47\,\Omega/km$	(1.15 mH/km) $0.05 + j0.36\Omega/km$	$0.31 + j1.47\,\Omega/km$	(3.22 mH/km) $0.25 + j1.01\,\Omega/km$	(7.90 mH/km) $0.56 + j2.48\,\Omega/km$	(1.46 mH/km) $0.06 + j0.46\,\Omega/km$	S 337 Ω/km
		0.0101 μF/km 0.120 A/km	0.0055 μF/km	0.0122 μF/km 0.065 A/km		0.0021 μF/km	0.0042 μF/km	0.0082 μF/km	D 307 Ω/km

Notes: The reactances are the derived values from $j2\pi fL = 2\pi \cdot 50 \cdot L$ on a 50 Hz base, so that 1 mH/km corresponds to $jX = 2\pi(50) \cdot 1 \times 10^{-3} = 0.314\,\Omega/km$. Therefore jX for 60 Hz systems should be modified to 1.2 times larger values.

The charging current is derived from $I = j2\pi fC \cdot (1/\sqrt{3})\,V$ based on 50 Hz as rms values. Accordingly, the charging current I for Example 1 is calculated from $I = 2\pi \cdot 50 \cdot (0.015 \cdot 10^{-6})(1/\sqrt{3}) \cdot 500 \cdot 10^3 = 1.36$ A/km per phase. The charging current for 60 Hz systems should be modified to 1.2 times larger values.

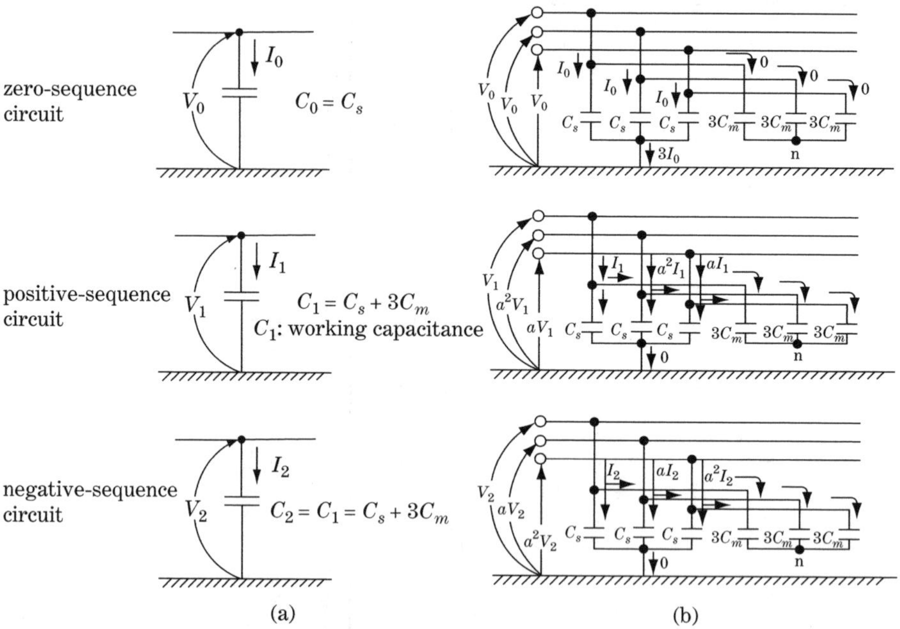

Figure 2.7 The equivalent circuit of three-phase single circuit transmission line (capacitance)

This is the equation of stray capacitances of a single circuit transmission line in the symmetrical components domain. $\boldsymbol{C}_{012}$ is a simple diagonal matrix in which all the off-diagonal elements vanish (become zero).

Figure 2.7a shows the equivalent circuit of a three-phase (single circuit) transmission line by symmetrical components, which is drawn from Equation (2.22b). The positive (1), negative (2) and zero (0) sequence circuits of the phase-balanced three-phase transmission line are obviously independent of each other.

Symmetrical capacitances C_0, C_1, C_2 are defined by Equation (2.22b) in relation to the original impedances C_s, C_m shown by Equation 1.33.

The physical meaning of the relations $C_0 = C_s$, $C_1 = C_s + 3C_m$ can be understood by Figure 2.7b, where zero-sequence current cannot flow in the circuit branch of $3C_m$ because point n is not earth grounded.

2.4.4 Double circuit line with *C* constants

The stray capacitances of a well-phase-transposed double circuit line are shown by Figure 1.9. The symbols ${}^1V$, ${}^1I$ and ${}^2V$, ${}^2I$ are adopted as quantities of circuits 1 and 2, respectively, below.

Concerning the phase a current of circuit 1,

$$\begin{aligned}{}^1I_a &= j\omega C_s{}^1V_a + \{j\omega C_m({}^1V_a - {}^1V_b) + j\omega C_m({}^1V_a - {}^1V_c)\} + \{j\omega C'_m({}^1V_a - {}^2V_a)\\ &\quad + j\omega C'_m({}^1V_a - {}^2V_a) + j\omega C'_m({}^1V_a - {}^2V_c)\} \quad ①\\ \therefore&\\ {}^1I_a &= j\omega(C_s + 3C_m + 3C'_m){}^1V_a - C_m({}^1V_a + {}^1V_b + {}^1V_c) - C'_m({}^2V_a + {}^2V_b + {}^2V_c)\\ &= j\omega(C_s + 3C_m + 3C'_m){}^1V_a - 3C_m{}^1V_0 - 3C'_m{}^2V_0 \quad ②\end{aligned} \tag{2.23a}$$

Similar equations are derived for the phase b and c currents. Accordingly,

$$\begin{bmatrix} {}^1I_a \\ {}^1I_b \\ {}^1I_c \end{bmatrix} = j\omega(C_s+3C_m+3C'_m)\begin{bmatrix} {}^1V_a \\ {}^1V_b \\ {}^1V_c \end{bmatrix} - 3C_m\begin{bmatrix} {}^1V_0 \\ {}^1V_0 \\ {}^1V_0 \end{bmatrix} - 3C'_m\begin{bmatrix} {}^2V_0 \\ {}^2V_0 \\ {}^2V_0 \end{bmatrix} \tag{2.23b}$$

This equation is easily transformed into symmetrical components:

$$\left.\begin{aligned} {}^1I_0 &= j\omega(C_s+3C_m+3C'_m)\,{}^1V_0 - j3\omega C_m\,{}^1V_0 - j3\omega C'_m\,{}^2V_0 \\ &= j\omega(C_s+3C'_m)\,{}^1V_0 - j3\omega C'_m\,{}^2V_0 = j\omega C_s\,{}^1V_0 + j3\omega C'_m({}^1V_0 - {}^2V_0) \\ {}^1I_1 &= j\omega(C_s+3C_m+3C'_m)\,{}^1V_1 \\ {}^1I_2 &= j\omega(C_s+3C_m+3C'_m)\,{}^1V_2 \end{aligned}\right\} \tag{2.24a}$$

Accordingly,

$$\begin{bmatrix} {}^1I_0 \\ {}^1I_1 \\ {}^1I_2 \\ {}^2I_0 \\ {}^2I_1 \\ {}^2I_2 \end{bmatrix} = j\omega \left[\begin{array}{ccc|ccc} C_s+3C'_m & 0 & 0 & -3C'_m & 0 & 0 \\ 0 & C_s+3C_m+3C'_m & 0 & 0 & 0 & 0 \\ 0 & 0 & C_s+3C_m+3C'_m & 0 & 0 & 0 \\ \hline -3C'_m & 0 & 0 & C_s+3C'_m & 0 & 0 \\ 0 & 0 & 0 & 0 & C_s+3C_m+3C'_m & 0 \\ 0 & 0 & 0 & 0 & 0 & C_s+3C_m+3C'_m \end{array}\right] \begin{bmatrix} {}^1V_0 \\ {}^1V_1 \\ {}^1V_2 \\ {}^2V_0 \\ {}^2V_1 \\ {}^2V_2 \end{bmatrix} \tag{2.24b}$$

Namely,

$$\begin{aligned} \begin{bmatrix} {}^1I_0 \\ {}^2I_0 \end{bmatrix} &= j\omega \begin{bmatrix} C_s+3C'_m & -3C'_m \\ -3C'_m & C_s+3C'_m \end{bmatrix} \cdot \begin{bmatrix} {}^1V_0 \\ {}^2V_0 \end{bmatrix} \\ &= j\omega \begin{bmatrix} C_s\cdot{}^1V_0 + 3C'_m({}^1V_0 - {}^2V_0) \\ C_s\cdot{}^2V_0 + 3C'_m({}^2V_0 - {}^1V_0) \end{bmatrix} \\ &\equiv j\omega \begin{bmatrix} C_0\cdot{}^1V_0 + C'_0({}^1V_0 - {}^2V_0) \\ C_0\cdot{}^2V_0 + C'_0({}^2V_0 - {}^1V_0) \end{bmatrix} = j\omega \begin{bmatrix} C_0+C'_0 & -C'_0 \\ -C'_0 & C_0+C'_0 \end{bmatrix} \begin{bmatrix} {}^1V_0 \\ {}^2V_0 \end{bmatrix} \qquad ① \\ \begin{bmatrix} {}^1I_1 \\ {}^2I_1 \end{bmatrix} &= j\omega \begin{bmatrix} C_s+3C_m+3C'_m & 0 \\ 0 & C_s+3C_m+3C'_m \end{bmatrix} \cdot \begin{bmatrix} {}^1V_1 \\ {}^2V_1 \end{bmatrix} \\ &\equiv j\omega \begin{bmatrix} C_1 & 0 \\ 0 & C_1 \end{bmatrix} \cdot \begin{bmatrix} {}^1V_1 \\ {}^2V_1 \end{bmatrix} \qquad ② \\ \begin{bmatrix} {}^1I_2 \\ {}^2I_2 \end{bmatrix} &= j\omega \begin{bmatrix} C_s+3C_m+3C'_m & 0 \\ 0 & C_s+3C_m+3C'_m \end{bmatrix} \cdot \begin{bmatrix} {}^1V_2 \\ {}^2V_2 \end{bmatrix} \\ &\equiv j\omega \begin{bmatrix} C_1 & 0 \\ 0 & C_1 \end{bmatrix} \cdot \begin{bmatrix} {}^1V_2 \\ {}^2V_2 \end{bmatrix} \qquad ③ \end{aligned}$$

where

$$\left.\begin{aligned} &\text{positive-sequence capacitance: } C_1 = C_2 = C_s + 3C_m + 3C'_m \\ &\text{zero-sequence capacitance: } C_0 = C_s,\ C'_0 = 3C'_m \end{aligned}\right\} \qquad ④ \qquad (2.24c)$$

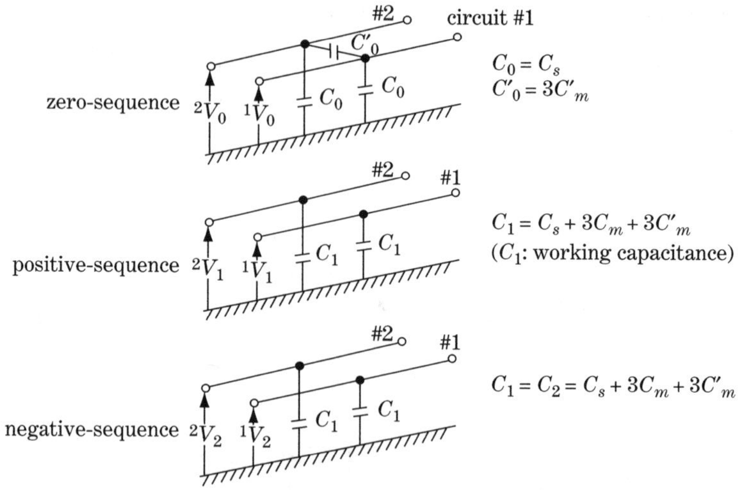

Figure 2.8 The equivalent circuit of three-phase double circuit transmission line (capacitance)

This is the equation of the stray capacitances of double circuit lines, from which the equivalent circuit of Figure 2.8 is derived.

2.5 Typical Transmission Line Constants

2.5.1 Typical line constants

L, *C* constant values of individual overhead transmission lines are different because the allocations of conductors (in other words, the physical length *r*, *h*, S_{ll}, etc., of the tower design) and the geological characteristics of the earth–ground are individually different. However, the line constants are not so different for lines of similar voltage classes, because the physical dimensions of the conductors are not so different, at least for the same voltage class of transmission lines. In addition, the constants would not be so different even for lines of different voltages, because the variables of the physical length *r*, *h*, S_{ll}, etc., of the tower design would be included in the logarithmic terms of the equations for *L*, *C*.

Table 2.1 shows typical *L*, *C* values of single circuit lines and double circuit lines. In Table 2.1, in addition to the quoted four cases of real measured examples, we have indicated typically 'easy-to-remember *L*, *C* values' which would be applicable as approximate values for most high-voltage transmission lines. At least, readers can consider that the orders of the *L*, *C* constants of individual lines could be appropriate as common values, regardless of the differences of area, utility companies or countries. (The zero-sequence constants for circuit lane 1 and 2 shown in Table 2.1 will be explained in Chapter 4.)

Typical constants of power cables are shown in Table 2.2 for convenience of comparison with overhead lines; the details of power cables will be examined in Chapter 23.

Further, besides *R*, *L*, *C* constants, the leakage resistance *G* exists as the fourth line constant. This is typically the creepage resistance of insulators of transmission lines or station equipment that are parallel resistances with stray admittance $j\omega C$, and usually has extraordinarily large ohmic values. G is an important constant which would be largely affected by the insulation characteristics of individual high-voltage insulators, attenuation ratio of surge phenomena, and so on. However, *G* can be neglected for most ordinary circuit analysis (except for surge analysis), because it has a quite large resistance of, say, megaohm order.

Table 2.2 Typical line constants of power cables (see also Chapter 23)
CV cables

Line voltages (kV)	**Sectional area** (mm^2)	**Conductor thickness** $2r$ (mm)	**Insulation-thickness** (mm)	**Sheath diameter** D (mm)	**Cable diameter** S (mm)	**Resistance** R (Ω/km)	**Working inductance** $L_s - L_m$ (mH/km)	**Working capacitance** C (μF/km)	**Reactance** jX (Ω/km)	**Leakage-current** Ic (A/km/ϕ)	**Surge-impedance** $\sqrt{L/C}$ (Ω)
500	2500	61.2	27	142	163	0.00746	0.383	0.25	0.112	22.7	39.1
	2000	53.8	27	134	155	0.00933	0.400	0.23	0.116	20.9	41.7
275	2500	61.2	23	133	160	0.00746	0.381	0.28	0.108	14.0	36.9
	2000	53.8	23	125	149	0.00933	0.392	0.25	0.112	12.5	39.6
	1200	41.7	23	112	134	0.01560	0.422	0.21	0.122	10.5	44.8
154	2000	53.8	17	108	122	0.00933	0.352	0.26	0.103	7.3	36.8
	1200	41.7	17	96	110	0.01560	0.382	0.22	0.112	8.7	41.7
	800	34.0	17	88	100	0.02310	0.404	0.19	0.119	5.3	46.1
66	2000	53.8	10	95	95	0.00933	0.302	0.53	0.086	6.3	23.9
	1200	41.7	10	82	82	0.0156	0.324	0.43	0.092	5.1	29.6
	800	340	10	73	73	0.0231	0.340	0.37	0.097	4.4	30.3
33	1200	41.7	8	73	73	0.0156	0.301	0.46	0.086	2.8	25.6
	600	29.5	8	58	58	0.0308	0.324	0.38	0.092	2.3	29.2
	200	17.0	8	45	45	0.0915	0.383	0.26	0.108	1.6	38.4
6.6	600	29.5	5	47	47	0.0308	0.282	0.71	0.089	0.8	19.9
	200	17.0	4	32	32	0.0915	0.315	0.51	0.102	0.6	24.9

OF cables

(kV)	(mm^2)	(mm)	(mm)	(mm)	(mm)	(Ω/km)	(mH/km)	(μF/km)	(Ω/km)	(A/km/φ)	(Ω)
500	2500	68.0	25.0	132	153	0.00732	0.305	0.37	0.101	33.5	28.7
	2000	59.1	33.0	139	160	0.00915	0.388	0.27	0.113	24.5	37.9
275	2000	57.5	19.5	107	137	0.00915	0.363	0.41	0.098	20.4	29.8
	1200	45.7	19.5	94	124	0.001510	0.389	0.34	0.105	17.0	33.8
154	2000	57.5	13.5	94	119	0.00915	0.333	0.57	0.09	15.9	24.2
	1200	45.7	13.5	81	106	0.01510	0.367	0.45	0.095	12.6	28.6
	800	40.6	12.5	74	96	0.02260	0.361	0.44	0.097	12.3	28.6
66	2000	57.0	8.0	82	106	0.00910	0.312	0.96	0.082	11.5	18.0
	1200	45.2	8.0	69	92	0.01510	0.331	0.80	0.086	9.6	20.3
	800	39.6	7.0	61	82	0.02230	0.334	0.79	0.087	9.5	20.6

Notes: The working inductance is calculated under the three-phase allocation of touched triangles. Accordingly, $(L_s - L_m) = 0.4605 \log(D/r) + 0.05\,\text{mH/km}$
where $(S_{ab} \cdot S_{bc} \cdot S_{ca})^{1/3} = (D \cdot D \cdot D)^{1/3} = D$.
If the averaged phase-to-phase distance S is larger, the inductance would become slightly larger.
The reactance is calculated from $jX = j2\pi \cdot 50(L_s - L_m)$ based on 50 Hz. Then the values should be multiplied 1.2 times for the 60 Hz system.
The leakage current is calculated from $I_c = 2\pi \cdot 50 \cdot C(1/\sqrt{3})V$.
The surge impedance is calculated from $\sqrt{(L_s - L_m)/C}$.

The transmission lines are described as distributed-constant circuits in a strict expression. However, provided that the evaluation of accuracy or percentage error is adequately investigated, approximation by concentrated-constant circuits can be justified for most analytical work. Approximation techniques including accuracy (or error percentage) estimation are essential in actual engineering activities. This theme is investigated in Chapter 18 in more detail.

2.5.2 *L*, *C* constant values derived from typical travelling-wave velocity and surge impedance

The velocity of travelling-wave propagation on transmission lines and the surge impedance are defined by the following equations whose reasons are investigated in detail in Chapter 18:

$$\left.\begin{array}{ll}\text{velocity of travelling-wave propagation} & : \ u = 1/\sqrt{LC}\,[\mathrm{m/s}] \\ \text{surge impedance} & : \ Z_{\mathrm{surge}} = \sqrt{\dfrac{L}{C}}\,[\Omega]\end{array}\right\} \tag{2.25a}$$

The inverse forms are

$$\left.\begin{array}{l} L = \dfrac{Z_{\mathrm{surge}}}{u} \\ C = \dfrac{1}{Z_{\mathrm{surge}} \cdot u}\end{array}\right\} \tag{2.25b}$$

There are typical values for velocity u and surge impedance Z_{surge} of overhead transmission lines and power cables that are very easy to remember. Therefore we can find typical L and C values from these typical u and Z_{surge} values by an inverse process.

For **overhead transmission lines,**

$$u = 300000\,[\mathrm{km/s}] \ = \ 3 \times 10^8\,[\mathrm{m/s}]\ (\text{velocity of light in air, } 300\,\mathrm{m}/\mu\mathrm{s})$$
$$\cdot Z_{\mathrm{surge}} = 300\,[\Omega]\ (\text{typically } 200 - 500\ \Omega)$$

Accordingly,

$$L = \frac{300}{3 \times 10^8}\,[\mathrm{H/m}] = 10^{-6}\,[\mathrm{H/m}] = 1\,[\mathrm{mH/km}]$$
$$C = \frac{1}{300 \times 3 \times 10^8} = 0.011 \times 10^{-9}\,[\mathrm{F/m}] = 0.011\,[\mu\mathrm{F/km}]$$

This is almost the same as the typical *L* and *C* values in Table 2.1.

For **power cables,**

$$u = 150000\,[\mathrm{km/s}] \ = \ 1.5 \times 10^8\,[\mathrm{m/s}]\ (1/2 \text{ the velocity of light in air typically, } 135000 - 150000\ \mathrm{km/s},\ 135 - 150\ \mathrm{m}/\mu\mathrm{s})$$

$$Z_{\mathrm{surge}} = 30\,[\Omega]\,(\text{typically } 20 - 30\ \Omega)$$

Accordingly,

$$L = \frac{30}{1.5 \times 10^8}\,[\mathrm{H/m}] = 0.2 \times 10^{-6}\,[\mathrm{H/m}]$$
$$= 0.2\,[\mathrm{mH/km}]\,(\text{about } 1/5 \text{ of overhead transmission line})$$

Table 2.3 Large-current-capacity types of conductors for overhead transmission lines (typical example)

	Continuous		Temporary	
	Maximum temperature [°C]	**Maximum current [A]**	**Maximum temperature [°C]**	**Maximum current [A]**
ACSR[a]	90	829	120	1125
TACSR	150	1323	180	1508
ZTACSR	210	1675	240	1831
XTACIR	230	1715	290	2004

[a]Aluminium alloy metal conductors.

$$C = \frac{1}{30 \times 1.5 \times 10^8} = 0.22 \times 10^{-9}[\text{F/m}]$$
$$= 0.22[\mu\text{F/km}] \text{ (about 20 times overhead transmission line)}$$

This is also very close to the typical L and C values in Table 2.2.

In total, the inductance L of the cable is smaller by about 1/2 or 1/5 while capacitance C is larger by about 20 times in comparison with that of the overhead line.

Table 2.3 and Figure 2.9 show typical advanced **ACSR (Aluminium Conductor Steel Reinforced)** conductors for overhead transmission lines. Due to recent advanced metal–alloy production and wire-drawing technology, large-current-capacity conductors with high-temperature-withstanding characteristics even at 230°C and of light weight have been realized as is shown in the table. Furthermore, ACFR conductors (where the tension member in steel twisted wires is replaced in carbon fibre string twisted wires) have been experimentally adopted in order to realize lightweight conductors.

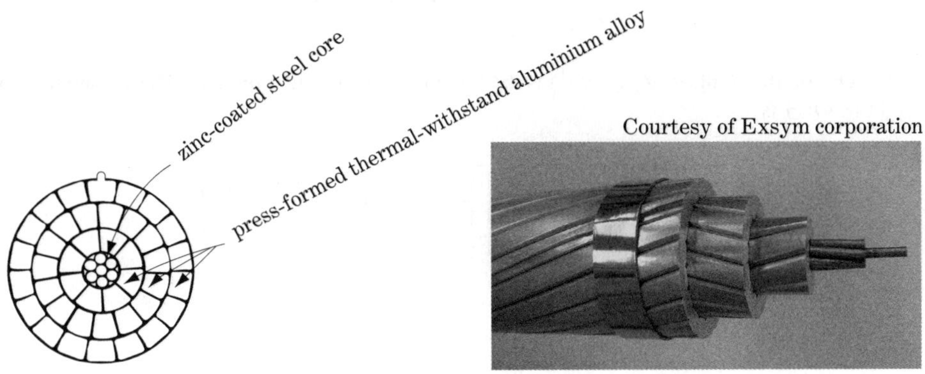

Figure 2.9 High-temperature-withstandable aluminium-clad steel wire (TACSR)

2.6 Generator by Symmetrical Components (Easy Description)

2.6.1 Simplified symmetrical equations

A synchronous generator (or synchronous motor) may be considered as a machine containing three-phase-balanced ideal power sources and three-phase-balanced leakage impedances, so that the generator may be simply expressed by Figure 2.10 as an approximate circuit. (Detailed approaches are discussed in Chapter 10.)

Now, we have from Figure 2.10

$$\left.\begin{aligned}
&\begin{bmatrix} E_a \\ E_b = a^2E_a \\ E_c = aE_a \end{bmatrix} - \begin{bmatrix} V_a \\ V_b \\ V_c \end{bmatrix} = \begin{bmatrix} Z_s & Z_m & Z_m \\ Z_m & Z_s & Z_m \\ Z_m & Z_m & Z_s \end{bmatrix} \cdot \begin{bmatrix} I_a \\ I_b \\ I_c \end{bmatrix} - \begin{bmatrix} V_n \\ V_n \\ V_n \end{bmatrix} \quad ① \\
&\quad \boldsymbol{E}_{abc} \qquad -\boldsymbol{V}_{abc} \quad = \qquad \boldsymbol{Z}_{abc} \qquad \cdot \boldsymbol{I}_{abc} \quad - \boldsymbol{V}_n \\
&V_n = -Z_n(I_a + I_b + I_c) = -Z_n(3I_0) = -3Z_n \cdot I_0 \qquad ② \\
&E_a, E_b, E_c : \text{the generated source voltages of three-phase-balanced design}
\end{aligned}\right\} \qquad (2.26)$$

Equation (2.26) can be transformed into symmetrical components by left-hand multiplication of the symmetric operator $\boldsymbol{a}$:

$$\begin{aligned}
&\boldsymbol{a} \cdot \boldsymbol{E}_{abc} - \boldsymbol{a} \cdot \boldsymbol{V}_{abc} = \boldsymbol{a}\boldsymbol{Z}_{abc} \cdot \boldsymbol{I}_{abc} - \boldsymbol{a} \cdot \boldsymbol{V}_n \\
\therefore\ &\boldsymbol{E}_{012} - \boldsymbol{V}_{012} = \boldsymbol{a}\boldsymbol{Z}_{abc} \cdot \boldsymbol{a}^{-1} \cdot \boldsymbol{I}_{012} - \boldsymbol{a} \cdot \boldsymbol{V}_n
\end{aligned} \qquad (2.27a)$$

where

$$\boldsymbol{E}_{012} = \begin{bmatrix} E_0 \\ E_1 \\ E_2 \end{bmatrix} = \boldsymbol{a} \cdot \boldsymbol{E}_{abc} = \frac{1}{3}\begin{bmatrix} 1 & 1 & 1 \\ 1 & a & a^2 \\ 1 & a^2 & a \end{bmatrix} \cdot \begin{bmatrix} E_a \\ a^2E_a \\ aE_a \end{bmatrix} = \begin{bmatrix} 0 \\ E_a \\ 0 \end{bmatrix}$$

The first term on the right ($\boldsymbol{a} \cdot \boldsymbol{Z}_{abc} \cdot \boldsymbol{a}^{-1}$) is the same form as in Equation 2.14. The second term on the right ($\boldsymbol{a} \cdot \boldsymbol{V}_n$) is

$$\boldsymbol{a} \cdot \boldsymbol{V}_n = \frac{1}{3}\begin{bmatrix} 1 & 1 & 1 \\ 1 & a & a^2 \\ 1 & a^2 & a \end{bmatrix} \cdot \begin{bmatrix} V_n \\ V_n \\ V_n \end{bmatrix} = \begin{bmatrix} V_n \\ 0 \\ 0 \end{bmatrix} = \begin{bmatrix} -3Z_n \cdot I_0 \\ 0 \\ 0 \end{bmatrix}$$

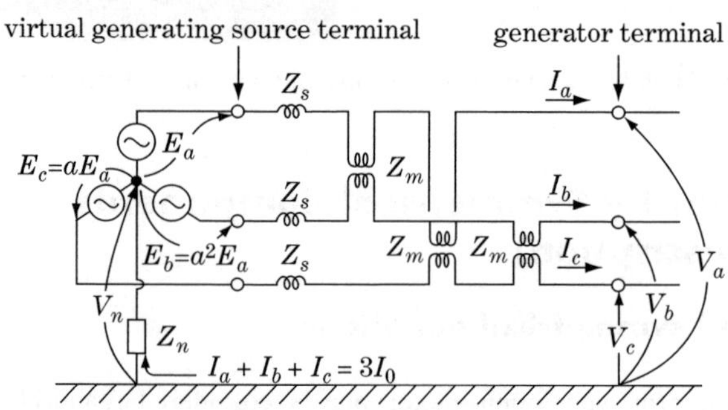

Figure 2.10 Generator (easy concept)

Accordingly,

$$\left.\begin{array}{l}\begin{bmatrix}0\\E_a\\0\end{bmatrix}-\begin{bmatrix}V_0\\V_1\\V_2\end{bmatrix}=\begin{bmatrix}Z_0&0&0\\0&Z_1&0\\0&0&Z_2\end{bmatrix}\cdot\begin{bmatrix}I_0\\I_1\\I_2\end{bmatrix}+\begin{bmatrix}3Z_n\cdot I_0\\0\\0\end{bmatrix}\\ \text{or}\\ -V_0=(Z_0+3Z_n)I_0\\ E_a-V_1=Z_1I_1\\ -V_2=Z_2I_2\end{array}\right\}\qquad(2.27\text{b})$$

This is the transformed symmetrical equation of the generator and Table 2.1 shows the symmetrical equivalent circuits of Equation (2.27b). The figure shows that a power source exists only in the positive-sequence circuit, and the negative- and zero-sequence circuits are only made of passive impedances. A generator may be theoretically named a 'positive-sequence power generator'.

2.6.2 Reactance of generator

Equation 2.27b derived from Figure 2.10 shows that the generator has time-independent constant symmetrical reactances and the positive- and negative-sequence reactances are the same quantities. However, this is not correct.

The generator reactances will change from time to time under transient conditions, and, moreover, the positive- and negative-sequence reactances as well as the zero-sequence reactances are different. The generator can strictly be treated only by the d–q–0 **transformation method** in which the new concept of **direct-axis reactances** (x''_d, x'_d, x_d) and **quadrature-axis reactances** (x''_q, x'_q, x_q) are introduced.

Now, by applying the reactances (x''_d, x'_d, x_d) as positive-sequence reactances, Figure 2.11 can be treated as the mostly correct equivalent circuit of the generator while the positive-sequence reactance will change from time to time as shown in Table 2.1 under transient conditions.

For most analyses of mainly power frequency phenomena (fault analysis, for example), Equation 2.27b and Figure 2.11 can be applied as the satisfactory equivalent circuit of the synchronous

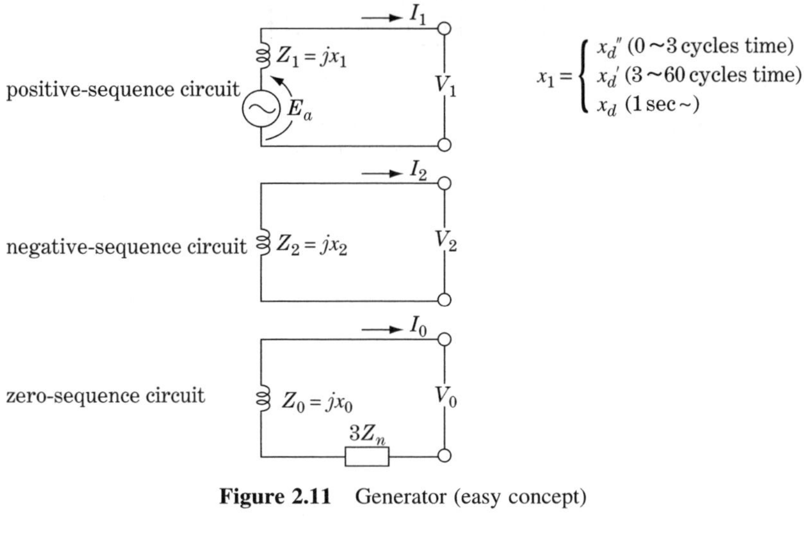

Figure 2.11 Generator (easy concept)

generator while the following reactances are used in equivalent circuits (details are examined in Chapter 10).

For the positive-sequence reactance,

$$x_1 = \begin{cases} x''_d: & \text{direct-axis subtransient reactance } (0 - 3\,\text{cycle time}, 0 \sim 45 \text{ or } 60\,\text{ms}) \\ x'_d: & \text{direct-axis transient reactance } (3 - 50 \text{ or } 60 \text{ cycle time}, \sim 1\,\text{sec}) \\ x_d: & \text{direct-axis steady-state reactance } (1\,\text{sec} \sim) \end{cases}$$

The time in parentheses means duration just after a sudden change of circuit condition.

For the negative-sequence reactance, x_2 can be treated as constant for most cases, although it may change slightly just after a sudden change of circuit condition.

And for the zero–sequence reactance, x_0 can always be treated as constant.

The values of x''_d, x'_d, x_d, x_2, x_0 for a synchronous generator are given on the name-plate in terms of ratings. Figure 10.1 in Chapter 10 shows typical values of generator reactances.

2.7 Description of Three-phase Load Circuit by Symmetrical Components

In power-receiving substations, feeder lines are connected to one of the HV, MV or LV buses, some of them are connected to other generating stations and substations through the lines, and others to load stations.

The equation for the totalized load is approximately written as follows.

$$\begin{bmatrix} V_a \\ V_b \\ V_c \end{bmatrix} = \begin{bmatrix} Z_s & Z_m & Z_m \\ Z_m & Z_s & Z_m \\ Z_m & Z_m & Z_s \end{bmatrix} \cdot \begin{bmatrix} I_a \\ I_b \\ I_c \end{bmatrix} \qquad (2.28)$$

$$\boldsymbol{V}_{abc} = \boldsymbol{Z}_{abc} \cdot \boldsymbol{I}_{abc}$$

or by symmetrical components

$$\left.\begin{bmatrix} V_0 \\ V_1 \\ V_2 \end{bmatrix} = \begin{bmatrix} Z_0 & 0 & 0 \\ 0 & Z_1 & 0 \\ 0 & 0 & Z_2 \end{bmatrix} \cdot \begin{bmatrix} I_0 \\ I_1 \\ I_2 \end{bmatrix}\right\} \qquad (2.29)$$

where $Z_1 = Z_2 = Z_s - Z_m$, $Z_0 = Z_s + 2Z_m > Z_1 = Z_2$.

It is obvious that the 1-, 2-, 0-sequence networks are mutually independent and the load can be approximately expressed simply by Z_1, Z_2, Z_0, respectively.

3

Fault Analysis by Symmetrical Components

We learned in the previous chapters that three-phase power systems can be described as simple equations and simple equivalent circuits by symmetrical components transformation. In this chapter we will study fault analysis using symmetrical components.

The analytical method explained in this chapter is called traditionally **fault analysis**. However, this is a very important analytical method invariably applied for the analysis of 'all' (instead of 'most') kinds of phenomena such as normal states/irregular states (including faults, switching, etc.), steady states/transient states, d.c./power frequency/higher harmonic frequency/surge (switching and lightning surges). In addition, this method is also applied for analysis by manual calculation by simple model as well as by computer-based detail analysis for large systems.

3.1 Fundamental Concept of Symmetrical Coordinate Method

Electric quantities in three phases are phase balanced in normal states because every part of the power system is more or less three-phase balanced. The balanced states are broken whenever line-to-line faults or line-to-ground faults occur. Straightforward calculation of such an imbalanced condition and, further, the transient condition in the a–b–c domain is impossible not only by manual calculation but also by using computers. One serious reason is the existence of many mutual inductances on lines and equipment; however, theoretically the reason is the fact that generators cannot be actually described as accurate circuits in the a–b–c domain. Synchronous generators can be described as accurate circuits only by application of the transformation technique of symmetrical components together with the d–q–0 method (refer to Chapter 10). As a matter of fact, a power system can be written as one circuit including various lines and machines only in the symmetrical domain. In other words, the symmetrical coordinate method is an essential analytical technique not only for drastic simplification to handle circuits but also for precise analysis.

Figure 3.1 shows the process flow of fault analysis using symmetrical components. The first step is to transform the power system connection and fault condition into the 0–1–2 domain circuit. The second step is to find the circuit solution in the 0–1–2 domain. The last step is to inverse-transform the solution into the a–b–c domain.

Handbook of Power System Engineering Yoshihide Hase
© 2007 John Wiley & Sons, Ltd

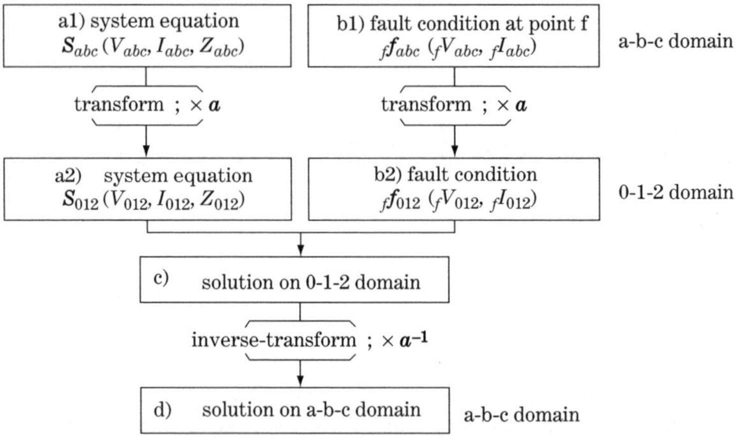

Figure 3.1 Procedure of fault analysis

3.2 Line-to-ground Fault (Phase a to Ground Fault: 1ϕG)

It should be recalled that a three-phase power system can be drawn only as a connection diagram in the a–b–c domain and not as a circuit in the a–b–c domain.

Let us examine a phase a to ground short-circuit fault (say, phase a 1ϕG) at an arbitrary point f on a transmission line. Figure 3.2a shows the partial situation of the connection diagram of the power system including point f, where virtual a–b–c terminals branch out at fault point f.

The power system before the fault at point f can be drawn as a symmetrical circuit in the 1–2–0 domain as shown within the dashed lines of Figure 3.2b, where the corresponding virtual terminals branch out at point f.

The related equations are

$$\left.\begin{array}{l}\left.\begin{array}{l}{}_fV_1 = E_1' - {}_fZ_1'I_1' = E_1'' - {}_fZ_1''I_1'' \\ {}_fV_2 = -{}_fZ_2'I_2' = -{}_fZ_2''I_2'' \\ {}_fV_0 = -{}_fZ_0'I_0' = -{}_fZ_0''I_0''\end{array}\right\} ① \\ \left.\begin{array}{l}{}_fI_1 = I_1' + I_1'' \\ {}_fI_2 = I_2' + I_2'' \\ {}_fI_0 = I_0' + I_0''\end{array}\right\} ② \\ \left.\begin{array}{l}{}_fZ_1 = ({}_fZ_1' // {}_fZ_1'') = \dfrac{{}_fZ_1'\,{}_fZ_1''}{{}_fZ_1' + {}_fZ_1''} \\ {}_fZ_2 = ({}_fZ_2' // {}_fZ_2'') = \dfrac{{}_fZ_2'\,{}_fZ_2''}{{}_fZ_2' + {}_fZ_2''} \\ {}_fZ_0 = ({}_fZ_0' // {}_fZ_0'') = \dfrac{{}_fZ_0'\,{}_fZ_0''}{{}_fZ_0' + {}_fZ_0''}\end{array}\right\} ③\end{array}\right\} \tag{3.1}$$

where

Z_1 : positive-sequence impedance looking into the circuit at point f

${}_fZ_1'$: positive-sequence impedance looking into the left-hand side at point f

${}_fZ''$: positive-sequence impedance looking into the right-hand side at point f

(the // symbol means parallel impedance values)

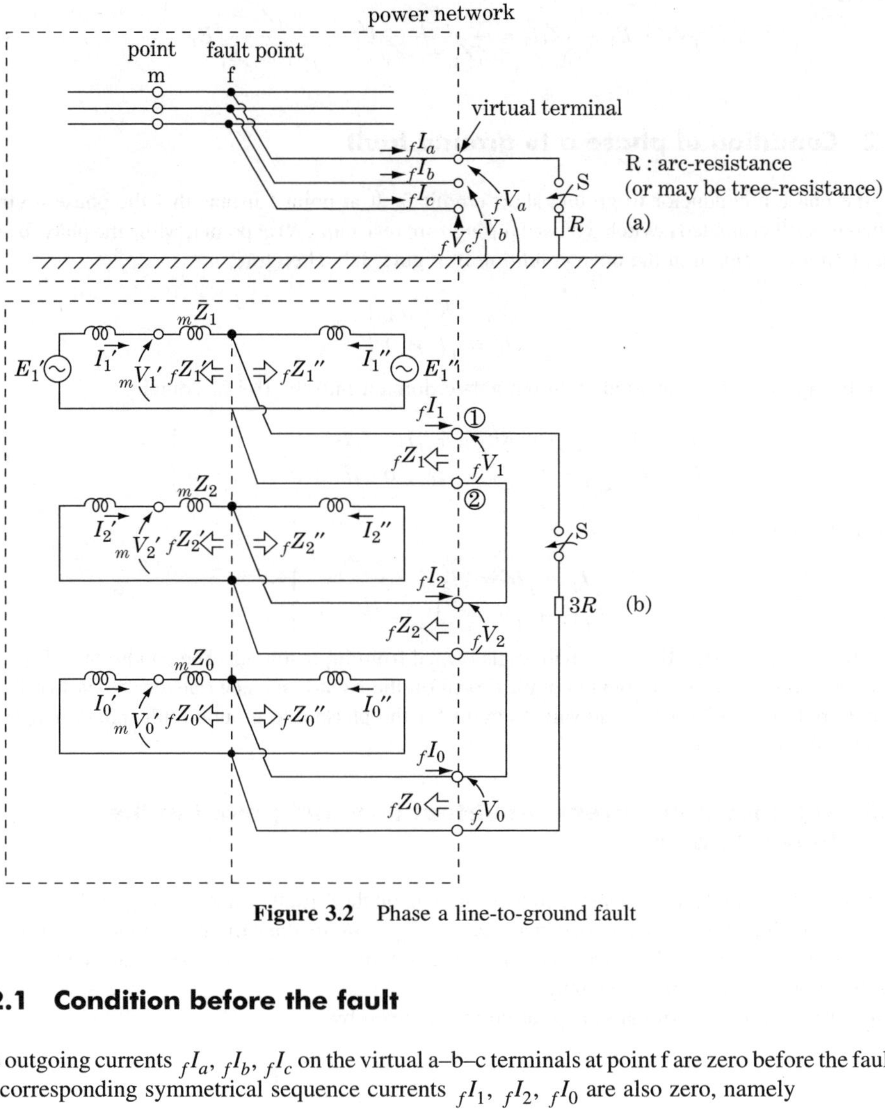

Figure 3.2 Phase a line-to-ground fault

3.2.1 Condition before the fault

The outgoing currents ${}_fI_a$, ${}_fI_b$, ${}_fI_c$ on the virtual a–b–c terminals at point f are zero before the fault, so the corresponding symmetrical sequence currents ${}_fI_1$, ${}_fI_2$, ${}_fI_0$ are also zero, namely

$$\left.\begin{aligned} {}_fI_a &= {}_fI_b = {}_fI_c = 0 \\ {}_fI_0 &= {}_fI_1 = {}_fI_2 = 0 \\ {}_fV_0 &= {}_fV_2 = 0 \end{aligned}\right\} \tag{3.2}$$

In the negative- and zero-sequence circuits, because any power source does not exist and the virtual terminals are open, all the quantities including ${}_fI_2$, ${}_fV_2$, ${}_fI_0$, ${}_fV_0$ at point f are therefore zero before the fault.

In the positive-sequence circuit, we have $I_f = I_1' + I_1'' = 0$. Accordingly, for the three-phase-balanced load current flowing through point f before the fault,

$$\frac{E_1' - E_1''}{{}_fZ_1' + {}_fZ_1''} = I_1' = -I_1'' \tag{3.3}$$

Therefore

$$_fV_1 = E_1' - {}_fZ_1'I_1' = \frac{_fZ_1''}{_fZ_1' + {}_fZ_1''}E_1' + \frac{_fZ_1'}{_fZ_1' + {}_fZ_1''}E_1'' \tag{3.4}$$

3.2.2 Condition of phase a to ground fault

Now, the phase a conductor to ground short-circuit fault at point f means that the phase a virtual terminal is earth grounded (switch S closed through arc resistance R) at point f, while the phase b and c virtual terminals remain in the open condition in Figure 3.2. Therefore,

$$\left.\begin{aligned} {}_fV_a &= R \cdot {}_fI_a \\ {}_fI_b &= {}_fI_c = 0 \end{aligned}\right\} \tag{3.5}$$

Transforming the above equation from the a–b–c domain into the 0–1–2 domain,

$$\left.\begin{aligned} {}_fV_0 + {}_fV_1 + {}_fV_2 = R({}_fI_0 + {}_fI_1 + {}_fI_2) \\ {}_fI_0 + a^2 \cdot {}_fI_1 + a \cdot {}_fI_2 = {}_fI_0 + a \cdot {}_fI_1 + a^2 \cdot {}_fI_2 = 0 \end{aligned}\right\} \tag{3.6}$$

Utilizing the relation $a^2 + a = -1$,

$$\left.\begin{aligned} {}_fI_0 = {}_fI_1 = {}_fI_2 \\ {}_fV_0 + {}_fV_1 + {}_fV_2 = 3R \cdot {}_fI_0 \end{aligned}\right\} \tag{3.7}$$

This is the equation in the 0–1–2 domain transformed from Equation 3.5. The condition of Equation 3.7 can be expressed as the drawing circuit shown on the right-hand side outside the dashed line in Figure 3.2b. Figure 3.2b is the equivalent circuit for the phase a to ground fault (phase a 1ϕG) by symmetrical components.

3.2.3 Voltages and currents at virtual terminal point f in the 0–1–2 domain

Now, phase a to ground faults are realized by switching on the virtual switch S in Figure 3.2b; in other words, connecting the outside impedance $_fZ_2 + {}_fZ_0 + 3R$ to the virtual terminals ①② of the positive-sequence circuit. The current flowing through terminals ①② and the voltage can be easily found by applying Thévenin's theorem.

The current through terminals ①② at point f is given by

$$\left.\begin{aligned} {}_fI_1 &= \frac{_fV_1}{_fZ_1 + ({}_fZ_2 + {}_fZ_0 + 3R)} \\ {}_fI_1 &= {}_fI_2 = {}_fI_0 = \frac{1}{_fZ_{\text{total}}}\,{}_fV_1 \\ {}_fZ_{\text{total}} &= {}_fZ_1 + {}_fZ_2 + {}_fZ_0 + 3R \end{aligned}\right\} \tag{3.8}$$

and the voltage at point f by

$$\left.\begin{aligned} {}_fV_0 &= -{}_fZ_0 \cdot {}_fI_0 = -{}_fZ_0 \cdot {}_fI_1 \\ {}_fV_2 &= -{}_fZ_2 \cdot {}_fI_2 = -{}_fZ_2 \cdot {}_fI_1 \\ {}_fV_1 &= -({}_fV_0 + {}_fV_2) + 3R \cdot {}_fI_1 = ({}_fZ_0 + {}_fZ_2 + 3R)\,{}_fI_1 \end{aligned}\right\} \tag{3.9}$$

The voltage $_fV_1$ in Equation 3.4 and 3.9 is the voltage between terminals ①② and is given by Equation 3.4 as an already known initial quantity before the switch S closes, where E_1' and E_1'' are known quantities.

Finally, the above solution of symmetrical voltages and currents on virtual terminals at point f are inverse transformed into a–b–c phase quantities at point f:

$$\begin{bmatrix} {}_fI_a \\ {}_fI_b \\ {}_fI_c \end{bmatrix} = \begin{bmatrix} 1 & 1 & 1 \\ 1 & a^2 & a \\ 1 & a & a^2 \end{bmatrix} \cdot \begin{bmatrix} {}_fI_0(= {}_fI_1) \\ {}_fI_1 \\ {}_fI_2(= {}_fI_1) \end{bmatrix} = \begin{bmatrix} 3\,{}_fI_1 \\ 0 \\ 0 \end{bmatrix} = \begin{bmatrix} \dfrac{3}{{}_fZ_{\text{total}}}\,{}_fV_1 \\ 0 \\ 0 \end{bmatrix} \quad ①$$

$$\begin{bmatrix} {}_fV_a \\ {}_fV_b \\ {}_fV_c \end{bmatrix} = \begin{bmatrix} 1 & 1 & 1 \\ 1 & a^2 & a \\ 1 & a & a^2 \end{bmatrix} \cdot \begin{bmatrix} -\,{}_fZ_0 \cdot {}_fI_1 \\ ({}_fZ_0 + {}_fZ_2 + 3R)\,{}_fI_1 \\ -\,{}_fZ_2 \cdot {}_fI_1 \end{bmatrix} = \begin{bmatrix} 3R \\ (a^2-1)\,{}_fZ_0 + (a^2-a)\,{}_fZ_2 + a^2 \cdot 3R \\ (a-1)\,{}_fZ_0 + (a-a^2)\,{}_fZ_2 + a \cdot 3R \end{bmatrix} \cdot {}_fI_1$$

$$= \begin{bmatrix} \dfrac{3R}{{}_fZ_{\text{total}}}\,{}_fV_1 \\ \dfrac{(a^2-1)\,{}_fZ_0 + (a^2-a)\,{}_fZ_2 + a^2 \cdot 3R}{{}_fZ_{\text{total}}}\,{}_fV_1 \\ \dfrac{(a-1)\,{}_fZ_0 + (a-a^2)\,{}_fZ_2 + a \cdot 3R}{{}_fZ_{\text{total}}}\,{}_fV_1 \end{bmatrix} \quad ②$$

$$\text{where } {}_fZ_{\text{total}} = {}_fZ_1 + {}_fZ_2 + {}_fZ_0 + 3R \quad ③$$

$${}_fV_1 = \frac{{}_fZ_1''}{{}_fZ_1' + {}_fZ_1''}E_1' + \frac{{}_fZ_1'}{{}_fZ_1' + {}_fZ_1''}E_1'' \quad ④$$

(3.10)

All the solutions ${}_fI_a$, ${}_fI_b$, ${}_fI_c$, ${}_fV_a$, ${}_fV_b$, ${}_fV_c$ in the a–b–c domain were found.

Incidentally, ${}_fV_1$ is the positive-sequence voltage (i.e. the phase a voltage) at point f before the fault. If the load flow current on the line at point f before the fault is zero, the voltage at point f is of the same value as that of the generator source voltage, namely $E' = E'' = {}_fV_1$.

3.2.4 Voltages and currents at an arbitrary point under fault conditions

Let us examine the voltages and currents at point m under the phase a 1ϕG fault condition at point f shown in Figure 3.2.

Figure 3.2b is the mathematical representation at any point of the system connection diagram Figure 3.2a. Therefore, voltages and currents at points $m({}_mV_1, {}_mV_2, {}_mV_0, {}_mI_1, {}_mI_2, {}_mI_0)$ and $({}_mV_a, {}_mV_b, {}_mV_c, {}_mI_a, {}_mI_b, {}_mI_c)$ are in correspondence to each other by the symmetrical transformation:

$$\left.\begin{aligned} {}_fV_1 &= E_1' - {}_fZ_1'I_1' = E_1'' - {}_fZ_1''I_1'' \\ {}_fI_1 &= I_1' + I_1'' \end{aligned}\right\} \quad (3.11a)$$

$$\therefore\ I_1' = \underbrace{\frac{{}_fZ_1''}{{}_fZ_1' + {}_fZ_1''}\,{}_fI_1}_{\substack{\text{the fault current supplied from the} \\ \text{left-hand side through point m to point f}}} + \underbrace{\frac{E_1' - E_1''}{{}_fZ_1' + {}_fZ_1''}}_{\text{the load current before fault}} \equiv C_1 \cdot {}_fI_1 + I_{\text{load}} \quad ①$$

The load current is not included in the negative- and zero-sequence circuits, so

$$\left.\begin{array}{ll} I'_2 = \dfrac{{}_fZ''_2}{{}_fZ'_2 + {}_fZ''_2}\,{}_fI_2 \equiv C_2 \times {}_fI_2 & ② \\ I'_0 = \dfrac{{}_fZ''_0}{{}_fZ'_0 + {}_fZ''_0}\,{}_fI_0 \equiv C_0 \cdot {}_fI_0 & ③ \\ \text{where } C_1 = \dfrac{{}_fZ''_1}{{}_fZ'_1 + {}_fZ''_1} \ (C_2, C_0 \text{ are defined by the same equation forms}) & \end{array}\right\} \quad (3.11b)$$

C_1 is the coefficient of the branched current ${}_mI_1/{}_fI_1$ from the left hand-side through point m from the total current ${}_fI_1$, and is the vector value of $0-1.0\angle\delta$. C_2, C_0 are also defined in the same way.

As we know already the value of ${}_fI_0 = {}_fI_1 = {}_fI_2$, the currents I'_1, I'_2, I'_0 at point m are calculated by Equation 3.11. Finally the currents in the a–b–c domain at point m are

$$\begin{bmatrix} I'_a \\ I'_b \\ I'_c \end{bmatrix} = \underbrace{\begin{bmatrix} 1 & 1 & 1 \\ 1 & a^2 & a \\ 1 & a & a^2 \end{bmatrix} \cdot \begin{bmatrix} C_0 \\ C_1 \\ C_2 \end{bmatrix} {}_fI_1}_{\text{fa\textbackslash ult current term}} + \underbrace{\begin{bmatrix} 1 & 1 & 1 \\ 1 & a^2 & a \\ 1 & a & a^2 \end{bmatrix} \cdot \begin{bmatrix} 0 \\ I_{load} \\ 0 \end{bmatrix}}_{\text{load current term}} \quad (3.12)$$

The second term on the right-hand side is the load current components that existed before the fault, and the first term is the fault current components caused by the fault at point f.

This equation explains the fact that the fault current component at any point of the system is not affected by the load current component just before the fault. In other words, we can calculate any fault under the condition of zero load current, and then use vectors to superpose the load current if necessary.

The voltages at point m can be calculated from the voltages and current quantities already found at point f by utilizing the following equations:

$$\begin{bmatrix} {}_mV'_0 \\ {}_mV'_1 \\ {}_mV'_2 \end{bmatrix} = \begin{bmatrix} {}_fV_0 \\ {}_fV_1 \\ {}_fV_2 \end{bmatrix} + \begin{bmatrix} {}_mZ_0 & 0 & 0 \\ 0 & {}_mZ_1 & 0 \\ 0 & 0 & {}_mZ_2 \end{bmatrix} \cdot \begin{bmatrix} I'_0 \\ I'_1 \\ I'_2 \end{bmatrix} \quad (3.13)$$

Finally, the voltages can be inverse transformed into the a–b–c domain.

3.2.5 Fault under no-load conditions

A fault under no-load conditions is a special case of $E'_1 = E''_1 = {}_fV_1$ in Figure 3.2b and Equation 3.10 ④. The power system looking from point f under this condition can be regarded as a black box with an internal power source, whose voltage across the terminals ①② is ${}_fV_1(= E'_1 = E''_1)$ and internal impedance is ${}_fZ_1 = ({}_fZ'_1 // {}_fZ''_1)$. On the other hand, as shown in Figure 3.3, the equivalent circuit of the phase a to ground fault at point f connects the outer impedance ${}_fZ_2 + {}_fZ_0 + 3R$ to the terminals ①② of the black box. Accordingly, the flow current at the terminals ①② is easily found by Thévenin's theorem. That is, the current is ${}_fI_1 = {}_fV_1/\{{}_fZ_1 + ({}_fZ_1 + {}_fZ_2 + 3R)\}$. This is of course in accordance with Equation 3.8.

In conclusion of the above explanation, we can apply Figure 3.3 as the equivalent circuit, instead of Figure 3.2b, whenever we need fault current components only (without load currents). Then we can superpose load currents if necessary.

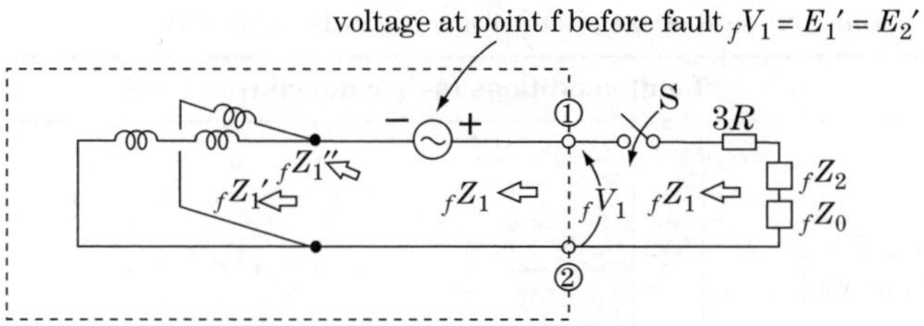

Figure 3.3 No-load fault calculation

3.3 Fault Analysis at Various Fault Modes

Voltage and current equations for cases of different mode faults are summarized in Tables 3.1a and b in which equations in the 1–2–0-sequence domain and a–b–c phase domain, as well as the equivalent circuits, are indicated. Case 7 is that of the phase a to ground fault which we have already examined in detail. The voltage and current equations and equivalent circuits for different fault modes can be derived by the same procedure.

Note, incidentally, that the generator impedances are $jx_{g1} \neq jx_{g2}$ in a strict sense, while transmission line impedances are exactly $Z_1 = Z_2$. Therefore, in case of a line fault at a great distance from the generator, the condition $jx_{g1}, jx_{g2} \ll Z_1, Z_2$ and the approximation $jx_{g1} + Z_1 \fallingdotseq jx_{g2} + Z_2$ would be justified, so the accuracy of calculation would be improved by the dominant line impedances.

Currents at point m for the case of a different mode fault at point f can be found by the following procedure:

$$ {}_mI_1 = C_1 \cdot {}_fI_1, \quad {}_mI_2 = C_2 \cdot {}_fI_2, \quad {}_mI_0 = C_0 \cdot {}_fI_0 \qquad (3.14)$$

where $C_1 = {}_fZ''/({}_fZ_1' + {}_fZ_1'')$ etc. The inverse transformed currents are

$$\begin{bmatrix} {}_mI_a \\ {}_mI_b \\ {}_mI_c \end{bmatrix} = \begin{bmatrix} C_0 \cdot {}_fI_0 + C_1 \cdot {}_fI_1 + C_2 \cdot {}_fI_2 \\ C_0 \cdot {}_fI_0 + a^2C_1 \cdot {}_fI_1 + aC_2 \cdot {}_fI_2 \\ C_0 \cdot {}_fI_0 + aC_1 \cdot {}_fI_1 + a^2C_2 \cdot {}_fI_2 \end{bmatrix} \qquad (3.15)$$

Voltage equations are derived analogously.

3.4 Conductor Opening

The cases of one- and two-phase conductor openings are examined in this section.

Conductor openings (or cut-offs) of one or two phases seldom happen as accidents in actual power systems. However, the state of a single phase breaker tripping as a procedure of single phase reclosing is a kind of one-phase conductor opening. Moreover, in the case of three-phase tripping by a circuit-breaker, current tripping by breaker–pole opening of each phase occurs sequentially in time and the timing of each phase tripping is different. In other words, a three-phase circuit is opened by the breaker through the transient states of trip-start→1ϕ opening→2ϕ opening and 3ϕ opening. Furthermore, breaker tripping failure may occur and give rise to severe situations. Phase-imbalanced opening occurs often and at various places in practical engineering. (Breaker tripping phenomena will be explained in Chapter 19.)

3.4.1 Single phase (phase a) conductor opening

Referring to Table 3.2(1A), the phase a conductor is opened between the points p and q; v_a, v_b, v_c, are the voltages across the points p and q of each phase, and i_a, i_b, i_c are the phase currents at points p and q.

Table 3.1a Equations and equivalent circuit for various fault modes

			Fault conditions (a–b–c domain)	
Metallic fault	#1	3ϕS three-phase line-to-line fault	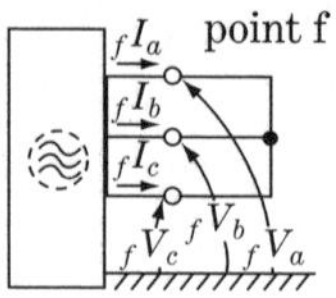	$\left.\begin{aligned} {}_fI_a + {}_fI_b + {}_fI_c &= 0 \\ {}_fV_a = {}_fV_b &= {}_fV_c \end{aligned}\right\}$ (1A)
	#2	3ϕG three-phase line-to-ground fault	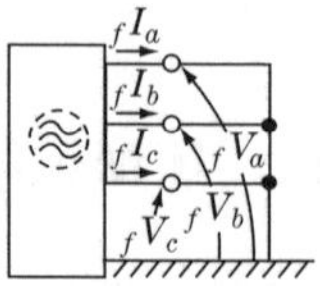	${}_fV_a = {}_fV_b = {}_fV_c = 0$ (2A)
	#3	1ϕG phase a line-to-ground fault	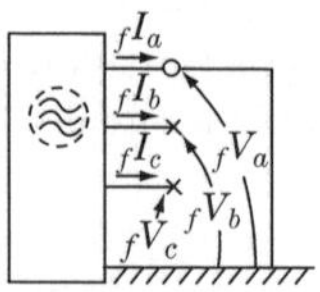	$\left.\begin{aligned} {}_fI_b = {}_fI_c &= 0 \\ {}_fV_a &= 0 \end{aligned}\right\}$ (3A)
	#4	2ϕS phase b to c line-to-line fault	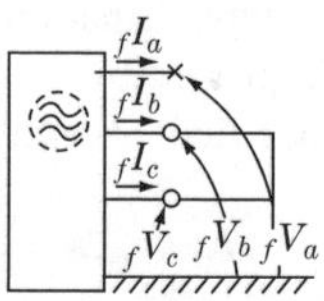	$\left.\begin{aligned} {}_fI_a &= 0 \\ {}_fI_b + {}_fI_c &= 0 \\ {}_fV_b &= {}_fV_c \end{aligned}\right\}$ (4A)
	#5	2ϕG phase b, c double line-to-ground fault	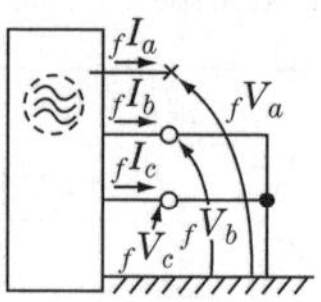	$\left.\begin{aligned} {}_fI_a &= 0 \\ {}_fV_b = {}_fV_c &= 0 \end{aligned}\right\}$ (5A)
Arc fault (arc resistance R)	#6	3ϕG three-phase line-to-ground fault	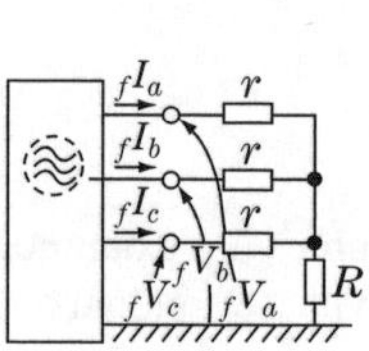	$\left.\begin{aligned} {}_fV_a - r\cdot{}_fI_a &= {}_fV_b - r\cdot{}_fI_b \\ &= {}_fV_c - r\cdot{}_fI_c \\ &= R({}_fI_a + {}_fI_b + {}_fI_c) \end{aligned}\right\}$ (6A)
	#7	1ϕG phase a line-to-ground fault	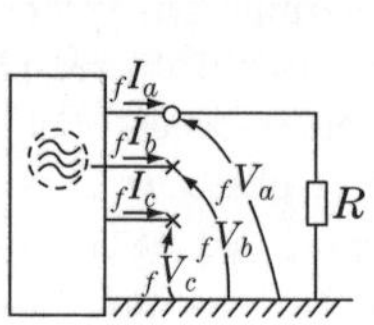	$\left.\begin{aligned} {}_fI_b = {}_fI_c &= 0 \\ {}_fV_a &= R\cdot{}_fI_a \end{aligned}\right\}$ (7A)
	#8	2ϕG phase b, c line-to-ground fault	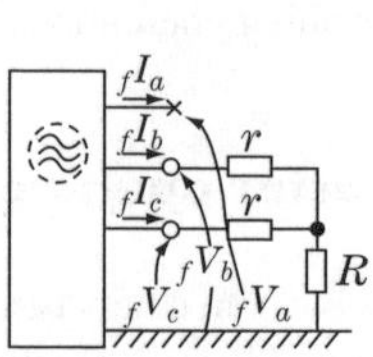	$\left.\begin{aligned} &{}_fI_a = 0 \\ &{}_fV_b - r\cdot{}_fI_b \\ &= {}_fV_c - r\cdot{}_fI_c \\ &= R({}_fI_b + {}_fI_c) \end{aligned}\right\}$ (8A)

Table 3.1b Equations and equivalent circuit for various fault modes

Fault condition at point f and the equivalent circuits: Metallic fault

#1 3ϕS
phase a, b, c

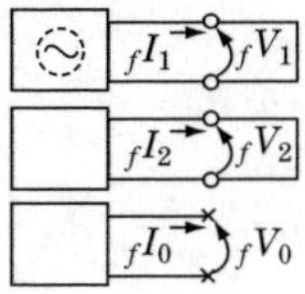

$$\left.\begin{array}{l} {}_fI_0 = 0 \\ {}_fV_1 = {}_fV_2 = 0 \\ {}_fI_0 = 0,\ {}_fI_2 = 0,\ {}_fV_0 = 0 \end{array}\right\}(1B) \Rightarrow \left.\begin{array}{l} {}_fI_0 = {}_fI_2 = 0,\ {}_fI_1 = \dfrac{{}_fE_a}{{}_fZ_1} \\ {}_fV_0 = {}_fV_1 = {}_fV_2 = 0 \end{array}\right\} \quad (1C)$$

#2 3ϕG
phase a, b, c

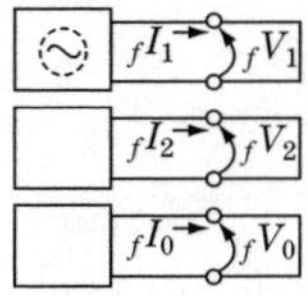

$$\left.\begin{array}{l} {}_fV_0 = {}_fV_1 = {}_fV_2 = 0 \\ {}_fI_2 = {}_fI_0 = 0 \end{array}\right\}(2B) \Rightarrow \left.\begin{array}{l} {}_fI_0 = {}_fI_2 = 0,\ {}_fI_1 = \dfrac{{}_fE_a}{{}_fZ_1} \\ {}_fV_0 = {}_fV_1 = {}_fV_2 = 0 \end{array}\right\} \quad (2C)$$

#3 1ϕG
phase a

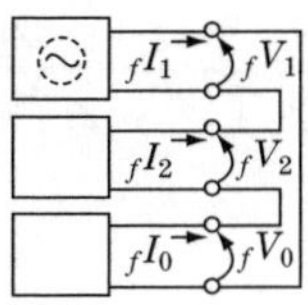

$$\left.\begin{array}{l} {}_fI_0 = {}_fI_1 = {}_fI_2 \\ {}_fV_0 + {}_fV_1 + {}_fV_2 = 0 \end{array}\right\}(3B) \Rightarrow \left.\begin{array}{l} {}_fI_0 = {}_fI_1 = {}_fI_2 = \dfrac{{}_fE_a}{\Delta} \\ {}_fV_0 = -{}_fZ_0 \cdot {}_fI_0 = \dfrac{-{}_fZ_0}{\Delta} \cdot {}_fE_a \\ {}_fV_1 = -({}_fV_0 + {}_fV_2) = \dfrac{{}_fZ_0 + {}_fZ_2}{\Delta} \cdot {}_fE_a \\ {}_fV_2 = -{}_fZ_2 \cdot {}_fI_2 = \dfrac{-{}_fZ_2}{\Delta} \cdot {}_fE_a \\ \text{where } \Delta = {}_fZ_0 + {}_fZ_1 + {}_fZ_2 \end{array}\right\} \quad (3C)$$

#4 2ϕS
phase b, c

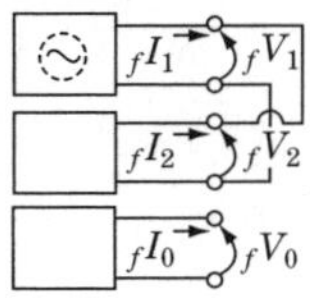

$$\left.\begin{array}{l} {}_fI_0 = 0 \\ {}_fI_1 = -{}_fI_2 \\ {}_fV_1 = {}_fV_2 \\ {}_fV_0 = 0 \end{array}\right\}(4B) \Rightarrow \left.\begin{array}{l} {}_fI_0 = 0,\ I_1 = -{}_fI_2 = \dfrac{{}_fE_a}{{}_fZ_1 + {}_fZ_2} \\ {}_fV_0 = 0 \\ {}_fV_1 = {}_fV_2 = -{}_fZ_2 \cdot {}_fI_2 = \dfrac{{}_fZ_2}{{}_fZ_1 + {}_fZ_2} \cdot {}_fE_a \end{array}\right\} \quad (4C)$$

#5 2ϕG
phase b, c

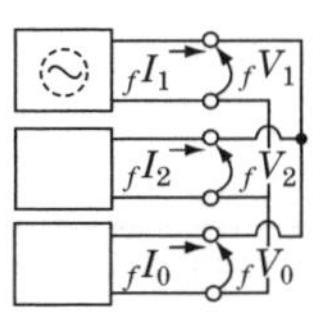

$$\left.\begin{array}{l} {}_fI_0 + {}_fI_1 + {}_fI_2 = 0 \\ {}_fV_0 = {}_fV_1 = {}_fV_2 \end{array}\right\}(5B) \Rightarrow \left.\begin{array}{l} {}_fI_1 = \dfrac{{}_fE_a}{{}_fZ_1 + ({}_fZ_2 // {}_fZ_0)} \\ \text{where } ({}_fZ_2 // {}_fZ_0) = \dfrac{{}_fZ_2 \cdot {}_fZ_0}{{}_fZ_2 + {}_fZ_0} \\ {}_fI_2 = \dfrac{-{}_fZ_0}{{}_fZ_2 + {}_fZ_0} \cdot {}_fI_1,\ {}_fI_0 = \dfrac{-{}_fZ_2}{{}_fZ_2 + {}_fZ_0} \cdot {}_fI_1 \\ {}_fV_0 = {}_fV_1 = {}_fV_2 = -{}_fZ_2 \cdot {}_fI_2 \\ = \dfrac{{}_fZ_2 \cdot {}_fZ_0}{{}_fZ_2 + {}_fZ_0} \cdot {}_fI_1 = ({}_fZ_2 // {}_fZ_0) \cdot {}_fI_1 \end{array}\right\} \quad (5C)$$

Continued

Table 3.1b *Continued*

Fault condition at point f and the equivalent circuits: Arc fault (arc resistance *R*)

#6 $3\phi G$
phase a, b, c

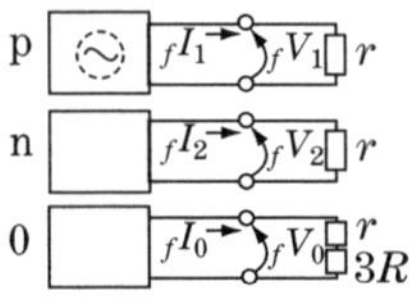

$$\left.\begin{array}{l} {}_fV_0 - r\cdot {}_fI_0 = 3R\cdot {}_fI_0 \\ {}_fV_1 = r\cdot {}_fI_1 \\ {}_fV_2 = r\cdot {}_fI_2 \\ {}_fI_2 = {}_fI_0 = 0 \\ {}_fV_2 = {}_fV_0 = 0 \end{array}\right\}(6B) \Rightarrow \left.\begin{array}{l} {}_fI_0 = {}_fI_2 = 0,\ {}_fI_1 = \dfrac{{}_fE_a}{{}_fZ_1 + r} \\ {}_fV_0 = {}_fV_2 = 0 \\ {}_fV_1 = r\cdot {}_fI_1 = \dfrac{r}{{}_fZ_1 + r}\cdot {}_fE_a \end{array}\right\} \quad (6C)$$

#7 $1\phi G$
phase a

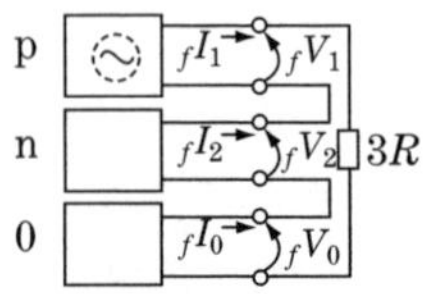

$$\left.\begin{array}{l} {}_fI_0 = {}_fI_1 = {}_fI_2 \\ {}_fV_0 + {}_fV_1 + {}_fV_2 = 3R\cdot {}_fI_0 \end{array}\right\}(7B) \Rightarrow \left.\begin{array}{l} {}_fI_0 = {}_fI_1 = {}_fI_2 = \dfrac{{}_fE_a}{\Delta} \\ {}_fV_0 = -{}_fZ_0\cdot {}_fI_0 = \dfrac{-{}_fZ_0}{\Delta}\cdot {}_fE_a \\ {}_fV_1 = -({}_fV_0 + {}_fV_2) + 3R\cdot {}_fI_1 \\ \quad = \dfrac{{}_fZ_0 + {}_fZ_2 + 3R}{\Delta}\cdot {}_fE_a \\ {}_fV_2 = -{}_fZ_2\cdot {}_fI_2 = \dfrac{-{}_fZ_2}{\Delta}\cdot {}_fE_a \\ \Delta = {}_fZ_0 + {}_fZ_1 + {}_fZ_2 + 3R \end{array}\right\} \quad (7C)$$

#8 $2\phi G$
phase b, c

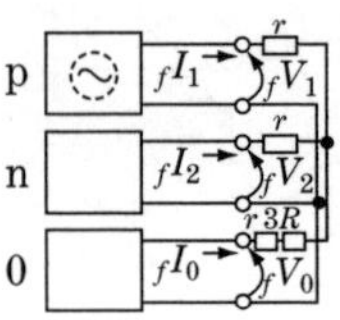

$$\left.\begin{array}{l} {}_fI_0 = {}_fI_1 = {}_fI_1 = 0 \\ {}_fV_0 - (r + 3R)\,{}_fI_0 \\ = {}_fV_1 - r\cdot {}_fI_1 \\ = {}_fV_2 - r\cdot {}_fI_2 \end{array}\right\}(8B) \Rightarrow \left.\begin{array}{l} {}_fI_1 = \dfrac{{}_fE_a}{\Delta_1 + \dfrac{\Delta_2\cdot\Delta_0}{\Delta_2 + \Delta_0}},\ {}_fI_0 = \dfrac{-\Delta_2}{\Delta_2 + \Delta_0}\cdot {}_fI_1 \\ {}_fI_2 = \dfrac{-\Delta_0}{\Delta_2 + \Delta_0}\cdot {}_fI_1 \\ {}_fV_0 = -{}_fZ_0\cdot {}_fI_0 = \dfrac{{}_fZ_0\Delta_2}{\Delta_2 + \Delta_0}\cdot {}_fI_1 \\ {}_fV_1 = (r + \dfrac{\Delta_2\cdot\Delta_0}{\Delta_2 + \Delta_0})\cdot {}_fI_1 \\ {}_fV_2 = -{}_fZ_2\cdot {}_fI_2 = -\dfrac{{}_fZ_0\Delta_2}{\Delta_2 + \Delta_0}\cdot {}_fI_1 \\ \text{where} \\ \Delta_1 = {}_fZ_1 + r \\ \Delta_2 = {}_fZ_2 + r \\ \Delta_0 = {}_fZ_0 + r + 3R \end{array}\right\}(8C)$$

Notes: All the quantities of the negative- and zero-sequence circuits becomes zero in cases $1(3\phi S)$, $2(3\phi G)$ and $6(3\phi G)$, because power sources do not exist in these circuits.
$_fE_a$ is the voltage at point f before the fault.

Table 3.1c Equations and equivalent circuit for various fault modes

	Case	Phase voltages and currents	
Metallic fault	#1 3ϕS phase a, b, c	$_fI_a = {_fI_1},\ {_fI_b} = a^2\,{_fI_1},\ {_fI_c} = a\,{_fI_1}$ ${_fI_1} = {_fE_a}/\,{_fZ_1}$ ${_fV_a} = {_fV_b} = {_fV_c} = 0$	(1D)
	#2 3ϕG phase a, b, c	Same as above	(2D)
	#3 1ϕG phase a	${_fI_a} = 3\,{_fI_0} = 3\,{_fE_a}/\Delta$ where $\Delta = {_fZ_0} + {_fZ_1} + {_fZ_2}$ ${_fI_b} = {_fI_c} = 0,\quad {_fI_a} = 0$ ${_fV_b} = \frac{(a^2-1)\,{_fZ_0} + (a^2-a)\,{_fZ_2}}{\Delta}\cdot{_fE_a},\ {_fV_c} = \frac{(a-1)\,{_fZ_0} + (a-a^2)\,{_fZ_2}}{\Delta}\cdot{_fE_a}$	(3D)
	#4 2ϕS phase b, c	${_fI_a} = 0,\ I_b = -{_fI_c} = (a^2-a)\,{_fI_1} = (a^2-a)\cdot\frac{_fE_a}{{_fZ_1} + {_fZ_2}}$ ${_fV_a} = 2\,{_fV_1},\ {_fV_b} = {_fV_c} = -{_fV_1}$, where ${_fV_1} = \frac{_fZ_2}{{_fZ_1} + {_fZ_2}}\cdot{_fE_a}$	(4D)
	#5 2ϕG phase b, c	${_fI_a} = 0,\ {_fI_b} = \frac{(a^2-a)\,{_fZ_0} + (a^2-1)\,{_fZ_2}}{{_fZ_0} + {_fZ_2}}\cdot{_fI_1}$ ${_fI_c} = \frac{(a-a^2)\,{_fZ_0} + (a-1)\,{_fZ_2}}{{_fZ_0} + {_fZ_2}}\cdot{_fI_1}$ ${_fV_a} = \frac{3\,{_fZ_2}\cdot{_fZ_0}}{{_fZ_2} + {_fZ_0}}\cdot{_fI_1},\ {_fV_b} = {_fV_c} = 0$, where ${_fI_1} = \frac{_fE_a}{{_fZ_1} + ({_fZ_2}//{_fZ_0})}$	(5D)
	#6 3ϕG phase a, b, c	${_fI_a} = {_fI_1},\ {_fI_b} = a^2\,{_fI_1},\ {_fI_c} = a\,{_fI_2}$, where ${_fI_1} = \frac{_fE_a}{{_fZ_1} + r}$ ${_fV_a} = {_fV_1},\ {_fV_b} = a^2\,{_fV_1},\ {_fV_c} = a\,{_fV_2}$, where ${_fV_1} = \frac{r}{{_fZ_1} + r}\cdot{_fE_a}$	(6D)
Arc fault (arc resistance R)	#7 1ϕG phase a	${_fI_a} = 3\,{_fI_1} = 3\,{_fE_a}/\Delta,\quad {_fI_b} = {_fI_c} = 0$ ${_fV_a} = 3R\cdot{_fI_1} = \frac{3R}{\Delta}\,{_fE_a},\ {_fV_b} = \frac{(a^2-1)\,{_fZ_0} + (a^2-a)\,{_fZ_2} + a^2\cdot 3R}{\Delta}\cdot{_fE_a}$ ${_fV_c} = \frac{(a-1)\,{_fZ_0} + (a-a^2)\,{_fZ_2} + a\cdot 3R}{\Delta}\cdot{_fE_a}$, where $\Delta = {_fZ_0} + {_fZ_1} + {_fZ_2} + 3R$	(7D)
	#8 2ϕG phase b, c	${_fI_a} = 0,\ {_fI_b} = \frac{(a^2-a)\Delta_0 + (a^2-1)\Delta_2}{\Delta_0 + \Delta_2}\cdot{_fI_1}$ ${_fI_c} = \frac{(a-a^2)\Delta_0 + (a-1)\Delta_2}{\Delta_0 + \Delta_2}\cdot{_fI_1}$ ${_fV_a} = \left\{\frac{{_fZ_0}\Delta_2 + \Delta_0\Delta_2 + {_fZ_2}\Delta_0}{\Delta_0 + \Delta_2} + r\right\}{_fI_1}$ where ${_fI_1} = {_fE_a}/\left(\Delta_1 + \frac{\Delta_2\cdot\Delta_0}{\Delta_2 + \Delta_0}\right)$ ${_fV_b} = \left\{\frac{{_fZ_0}\cdot\Delta_2 + a^2\Delta_0\Delta_2 + a\,{_fZ_2}\cdot\Delta_0}{\Delta_0 + \Delta_2} + a^2 r\right\}{_fI_1}$ $\Delta_1 = {_fZ_1} + r$ $\Delta_2 = fZ_2 + r$ ${_fV_c} = \left\{\frac{{_fZ_0}\cdot\Delta_2 + a\Delta_0\Delta_2 + a^2\,{_fZ_2}\cdot\Delta_0}{\Delta_0 + \Delta_2} + ar\right\}{_fI_1}$ $\Delta_0 = {_fZ_0} + r + 3R$	(8D)

where $a - a^2 = j\sqrt{3},\ a^2 - 1 = j\sqrt{3}a,\ 1 - a = j\sqrt{3}a^2$.

This condition can be described by the following equations:

$$\left.\begin{aligned} v_b &= v_c = 0 \\ i_a &= 0 \end{aligned}\right\} \tag{3.16}$$

Table 3.2 Phase opening modes (equations and equivalent circuits)

Phase a opening	Phase b, c opening
[1A] 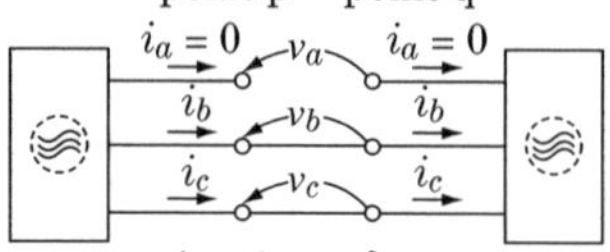 $\left.\begin{array}{l} i_a = 0 \\ v_b = v_c = 0 \end{array}\right\}$ (1A)	[2A] 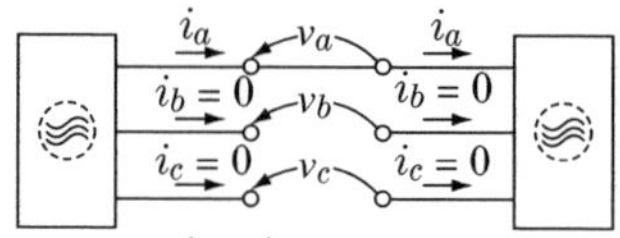 $\left.\begin{array}{l} i_b = i_c = 0 \\ v_a = 0 \end{array}\right\}$ (2A)
[1B] 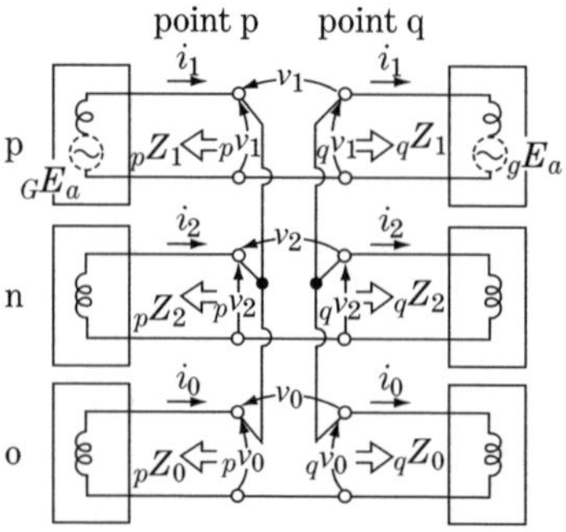 $\left.\begin{array}{l} i_0 + i_1 = i_2 = 0 \\ v_0 = v_1 = v_2 \end{array}\right\}$ (1B)	[2B] 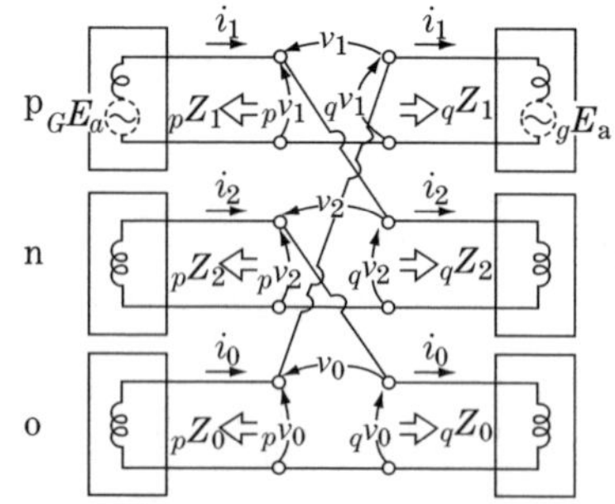 $\left.\begin{array}{l} i_0 = i_1 = i_2 \\ v_0 + v_1 + v_2 = 0 \end{array}\right\}$ (2B)
$\left.\begin{array}{l} i_1 = \dfrac{{}_GE_a - {}_gE_a}{Z_1 + \dfrac{Z_2 \cdot Z_0}{Z_2 + Z_0}} \\ i_2 = \dfrac{-Z_0}{Z_2 + Z_0} \cdot i_1 \\ i_0 = \dfrac{-Z_2}{Z_2 + Z_0} \cdot i_1 \\ v_0 = v_1 = v_2 = \dfrac{Z_2 \cdot Z_0}{Z_2 + Z_0} \cdot i_1 \\ Z_1 = {}_pZ_1 + {}_qZ_1 \\ Z_2 = {}_pZ_2 + {}_qZ_2 \\ Z_0 = {}_pZ_0 + {}_qZ_0 \end{array}\right\}$ (1C)	$\left.\begin{array}{l} i_0 = i_1 = i_2 = \dfrac{{}_GE_a - {}_gE_a}{Z_1 + Z_2 + Z_0} \\ v_1 = (Z_2 + Z_0) i_1 \\ v_2 = -Z_2 \cdot i_1 \\ v_0 = -Z_0 \cdot i_1 \\ \text{where} \\ Z_1 = {}_pZ_1 + {}_qZ_1 \\ Z_2 = {}_pZ_2 + {}_qZ_2 \\ Z_0 = {}_pZ_0 + {}_qZ_0 \end{array}\right\}$ (2C)
$\left.\begin{array}{l} i_a = 0 \\ i_b = \dfrac{(a^2 - a)Z_0 + (a^2 - 1)Z_2}{Z_2 + Z_0} \cdot i_1 \\ i_c = \dfrac{(a - a^2)Z_0 + (a - 1)Z_2}{Z_2 + Z_0} \cdot i_1 \\ v_a = \dfrac{3Z_2 \cdot Z_0}{Z_2 + Z_0} \cdot i_1 \\ i_1 = \dfrac{{}_GE_a - {}_gE_a}{Z_1 + \dfrac{Z_2 \cdot Z_0}{Z_2 + Z_0}} \end{array}\right\}$ (1D)	$\left.\begin{array}{l} i_a = 3i_1 = \dfrac{3({}_GE_a - {}_gE_a)}{Z_1 + Z_2 + Z_0} \\ i_b = i_c = 0,\ v_a = 0 \\ v_b = \{(a^2 - 1)Z_0 + (a^2 - a)Z_2\} i_1 \\ v_c = \{(a - 1)Z_0 + (a - a^2)Z_2\} i_1 \\ i_1 = \dfrac{{}_GE_a - {}_gE_a}{Z_1 + Z_2 + Z_0} \end{array}\right\}$ (2D)

The transformed equation in the 0–1–2 domain is

$$\left.\begin{array}{c} v_0 = v_1 = v_2 \\ i_0 + i_1 + i_2 = 0 \end{array}\right\} \tag{3.17}$$

This equation can be exactly described as the figure of equivalent circuits in Table 3.2 (1B). In the figure, the negative- and zero-sequence circuits are connected in parallel to the positive-sequence circuit. ${}_pZ_1$, ${}_pZ_2$, ${}_pZ_0$ are the impedances of the left-hand side circuit at point p, and ${}_qZ_1$, ${}_qZ_2$, ${}_qZ_0$ are the impedances of the right-hand side circuit at point q. Then, from the equivalent circuit,

$$\left.\begin{array}{l} i_1 = \dfrac{{}_GE_a - {}_gE_a}{Z_1 + \dfrac{Z_2 \cdot Z_0}{Z_2 + Z_0}}, \quad i_2 = \dfrac{-Z_0}{Z_2 + Z_0} i_1, \quad i_0 = \dfrac{-Z_2}{Z_2 + Z_0} \cdot i_1 \\ v_0 = v_1 = v_2 = \dfrac{Z_2 \cdot Z_0}{Z_2 + Z_0} \cdot i_1 \\ Z_1 = {}_pZ_1 + {}_qZ_1, \; Z_2 = {}_pZ_2 + {}_qZ_2, \; Z_0 = {}_pZ_0 + {}_qZ_0 \end{array}\right\} \tag{3.18}$$

This equation is written again in Table 3.2 (1C), and the inverse transformed equation for a–b–c phases is shown in Table 3.2 (1D).

3.4.2 Two-phases (phase b, c) conductor opening

Voltages and currents in this case are found in a similar way and the resulting equations as well as the equivalent circuits are shown in Table 3.2 (2A, 2B, 2C, 2D).

Coffee break 2: Dawn of the world of electricity, from Coulomb to Ampère and Ohm

It may be said that the modern history of electricity actually began with the great character Coulomb. Any review of the legacy left by the great scientists Coulomb, Ampère and Ohm cannot be omitted from the historical stories of electricity.

Charles Augustin Coulomb (1736–1806) wrote seven important treatises on electricity and magnetism between 1785 and 1791. He obtained some remarkable results by using the torsion balance method on 'electric point charges', 'magnetic poles', the distribution of electricity on the surface of charged bodies and others, and in particular the 'law of attraction and repulsion', which was the theory of attraction and repulsion between bodies of the same and opposite electrical charge. He demonstrated an **inverse square law for attractive and repulsive forces** ($F = q_1 \cdot q_2 / r^2$) using accurate measures of his own design. He also suggested that there was no perfect dielectric, proposing that every substance has a limit above which it will conduct electricity.

Charles Augustin Coulomb (1736–1806)

In 1800, **Alessandro Volta** (1745–1827) built the **voltaic pile**, which was the first battery to produce a reliable, steady current of electricity. He discovered, so to speak, the first practical **method of generating electricity**. Needless to say, Volta was a great benefactor to many electrical scientists as the person who provided stable electricity for their laboratory experiments at that time.

Alessandro Volta (1745–1827)

Hans Christian Oersted (1777–1851) discovered in 1820 that a compass needle deflects from magnetic north when an electric current is switched on or off in a nearby wire. This showed that electricity and magnetism were related phenomena. This eventually led him to the conclusion that 'an electric current creates a magnetic field' and thus 'electromagnetism' was born.

André Marie Ampère (1775–1836), a mathematician, immediately on hearing about Oersted's experimental results, formulated a **circuit force law** and treated magnetism by postulating small closed circuits inside a magnetized substance. He also discovered 'electro-dynamical forces' between linear wires through his experiment in 1820, the same year as Oersted's discovery. Ampère proved that **electric current also creates flux and furthermore mechanical force**. '**Ampère's corkscrew rule**' and '**Ampère's circuit law**' clearly indicate that current and flux are equivalent to each other because one can create the other.

Hans Christian Oersted (1777–1851)

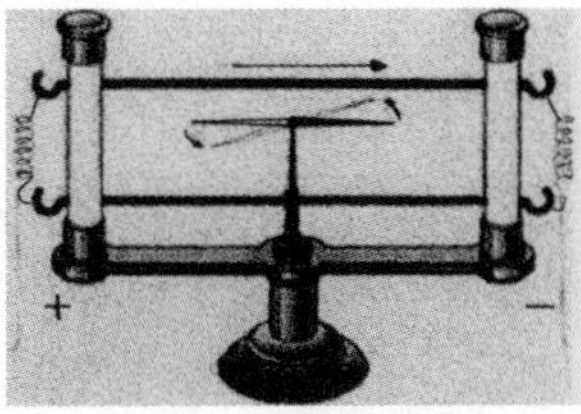

Oersted proved that electric current set up a magnetic field

André Marie Ampère (1775–1836)

Georg Simon Ohm (1787–1854) in 1825 was convinced of the truth of what we today call '**Ohm's law**' and gave its mathematical description in his 1826 paper. His law showed that the current through most materials is directly proportional to the potential difference applied across the material. He also published a fully mathematical approach to his complete theory of electricity in a book published in 1827, although the physics at that time rested mostly on a non-mathematical approach. His mathematical approach also had an impact in showing the true scientific method.

Now, the facts that 'electricity and magnetism are likely to be mutually related' and 'current produced mechanical force' were almost recognized in the works of Coulomb, Oersted, Ampèere and Ohm. However, no one knew then that 'magnetism can produce electricity', much less that 'mechanical power can make electrical power by moving magnetism'.

Georg Simon Ohm (1787–1854)

4

Fault Analysis of Parallel Circuit Lines (Including Simultaneous Double Circuit Fault)

Simultaneous line faults are often caused in power systems and can become serious, so a detailed examination and appropriate countermeasures are required to prevent serious power outages. Fault analysis for double circuit lines is rather complicated because mutual inductances as well as mutual capacitances exist between the double circuits. Moreover, fault analysis of simultaneous double faults is very hard work. Study of the principles of analogue methods for such complicated system behaviour is important regardless of whether we approach networks using computational or manual calculations.

So-called **two-phase circuit theory** is introduced as an effective approach in this chapter, and then the principles of faults analysis on double circuit lines, including double faults, is examined.

4.1 Two-phase Circuit and its Symmetrical Coordinate Method

4.1.1 Definition and meaning

Figures 4.1a and b show the two-phase circuit in comparison with the three-phase circuit. Although the two-phase circuit has not been utilized as a practical power system, positive-, negative- and zero-sequence circuits of double circuit transmission lines as are shown in Figures 2.6 and 2.8 for example are types of two-phase circuit lines, because they are the same as in Figure 4.1(b) if the double circuits are connected to the same single bus at the substation terminal.

It is assumed below that parallel circuits 1 and 2 of the same double circuit line are well balanced and, furthermore, each circuit is also well phase balanced by transposition. As is shown in Figures 2.6 and 2.8 of Chapter 2, mutual inductance and mutual capacitance between the first circuit 1 and the second circuit 2 exist on the zero-sequence circuit, but do not exist on the positive and negative-sequence circuits.

We have already learned that mutual inductances and capacitances of three-phase single circuit transmission lines are extinguished by symmetrical coordinate transformation. Analogously, mutual inductances and capacitances of the two-phase circuit must be extinguished by adopting a two-phase symmetrical coordinate transformation. This is the reason why we are going to apply two-phase symmetrical components as analytical tools for double circuit transmission lines. The so-called **double phase circuit theory** is indeed **the theory of symmetrical coordinates for double phase circuits** and is mathematically a kind of two-variable transformation.

Handbook of Power System Engineering Yoshihide Hase
© 2007 John Wiley & Sons, Ltd

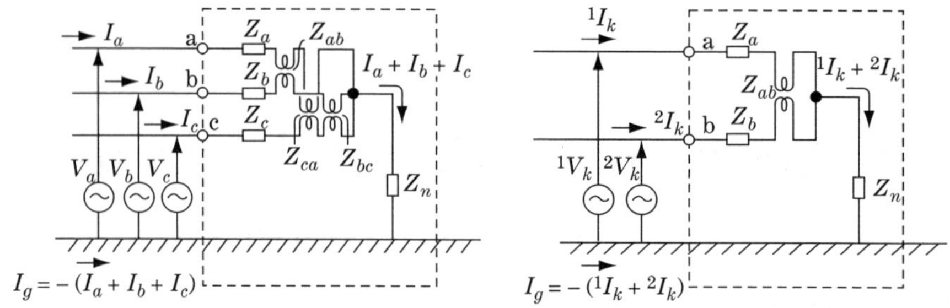

(a) three phase circuit (b) two phase circuit

Figure 4.1 Three-phase and two-phase circuits

The relevant equations are defined by the following equation.

For the transformation

$$\underbrace{\begin{bmatrix} V_{k0} \\ V_{k1} \end{bmatrix}}_{V_{k01}} = \underbrace{\frac{1}{2}\begin{bmatrix} 1 & 1 \\ 1 & a \end{bmatrix}}_{a_{2\phi}} \underbrace{\begin{bmatrix} {}^1V_k \\ {}^2V_k \end{bmatrix}}_{^{12}V_k} = \underbrace{\frac{1}{2}\begin{bmatrix} 1 & 1 \\ 1 & -1 \end{bmatrix}}_{a_{2\phi}} \cdot \underbrace{\begin{bmatrix} {}^1V_k \\ {}^2V_k \end{bmatrix}}_{^{12}V_k} \quad \therefore \boldsymbol{V}_{k01} = \boldsymbol{a}_{2\phi} \cdot {}^{12}\boldsymbol{V}_k \tag{4.1a}$$

where $a = e^{j360°/2} = \cos 180° + j\sin 180° = -1$

For the inverse transformation

$$\underbrace{\begin{bmatrix} {}^1V_k \\ {}^2V_k \end{bmatrix}}_{^{12}V_k} = \underbrace{\begin{bmatrix} 1 & 1 \\ 1 & -1 \end{bmatrix}}_{a_{2\phi}^{-1}} \underbrace{\begin{bmatrix} V_{k0} \\ V_{k1} \end{bmatrix}}_{V_{k01}} \quad \therefore {}^{12}\boldsymbol{V}_k = \boldsymbol{a}_{2\phi}^{-1} \cdot \boldsymbol{V}_{k01} \tag{4.1b}$$

This definition by Equations 4.1a , 4.1b is in the same form as in Chapter 2, with the size of matrix equations being changed and the size of the operational matrix $\boldsymbol{a}, \boldsymbol{a}^{-1}$ changed from 3 × 3 to 2 × 2; the vector operator is changed from $a = e^{j120°}$ to $a_{2\phi} = e^{j180°} = -1$.

The transformation equation above is applied to the quantities of the double circuits 1 and 2 on the positive-, negative- and zero-sequence double circuit lines. The suffix $k = 1, 2$ or 0 corresponds to 1-, 2- or 0-sequence quantities.

For the transformation

Zero-sequence components

$$\begin{bmatrix} V_{00} \\ V_{01} \end{bmatrix} = \frac{1}{2}\begin{bmatrix} 1 & 1 \\ 1 & -1 \end{bmatrix} \cdot \begin{bmatrix} {}^1V_0 \\ {}^2V_0 \end{bmatrix} \quad \text{or} \quad \left.\begin{aligned} V_{00} &= \frac{1}{2}({}^1V_0 + {}^2V_0) \\ V_{01} &= \frac{1}{2}({}^1V_0 - {}^2V_0) \end{aligned}\right\} ①$$

Positive-sequence components

$$\begin{bmatrix} V_{10} \\ V_{11} \end{bmatrix} = \frac{1}{2}\begin{bmatrix} 1 & 1 \\ 1 & -1 \end{bmatrix} \cdot \begin{bmatrix} {}^1V_1 \\ {}^2V_1 \end{bmatrix} \quad \text{or} \quad \left.\begin{aligned} V_{10} &= \frac{1}{2}({}^1V_1 + {}^2V_1) \\ V_{11} &= \frac{1}{2}({}^1V_1 - {}^2V_1) \end{aligned}\right\} ② \tag{4.2a}$$

Negative-sequence components

$$\begin{bmatrix} V_{20} \\ V_{21} \end{bmatrix} = \frac{1}{2}\begin{bmatrix} 1 & 1 \\ 1 & -1 \end{bmatrix} \cdot \begin{bmatrix} {}^1V_2 \\ {}^2V_2 \end{bmatrix} \quad \text{or} \quad \left.\begin{aligned} V_{20} &= \frac{1}{2}({}^1V_2 + {}^2V_2) \\ V_{21} &= \frac{1}{2}({}^1V_2 - {}^2V_2) \end{aligned}\right\} ③$$

For the inverse transformation

Zero-sequence components

$$\begin{bmatrix} {}^1V_0 \\ {}^2V_0 \end{bmatrix} = \begin{bmatrix} 1 & 1 \\ 1 & -1 \end{bmatrix} \cdot \begin{bmatrix} V_{00} \\ V_{01} \end{bmatrix} \quad \left.\begin{aligned} {}^1V_0 &= V_{00} + V_{01} \\ {}^2V_0 &= V_{00} - V_{01} \end{aligned}\right\} ①$$

Positive-sequence components

$$\begin{bmatrix} {}^1V_1 \\ {}^2V_1 \end{bmatrix} = \begin{bmatrix} 1 & 1 \\ 1 & -1 \end{bmatrix} \cdot \begin{bmatrix} V_{10} \\ V_{11} \end{bmatrix} \quad \left.\begin{aligned} {}^1V_1 &= V_{10} + V_{11} \\ {}^2V_1 &= V_{10} - V_{11} \end{aligned}\right\} ②$$

Negative-sequence components

$$\begin{bmatrix} {}^1V_2 \\ {}^2V_2 \end{bmatrix} = \begin{bmatrix} 1 & 1 \\ 1 & -1 \end{bmatrix} \cdot \begin{bmatrix} V_{20} \\ V_{21} \end{bmatrix} \quad \left.\begin{aligned} {}^1V_2 &= V_{20} + V_{21} \\ {}^2V_2 &= V_{20} - V_{21} \end{aligned}\right\} ③ \qquad (4.2b)$$

where

V_{00}, V_{01} : first- and second-lane voltages on the zero-sequence circuit
V_{10}, V_{11} : first- and second-lane voltages on the positive-sequence circuit
V_{20}, V_{21} : first- and second-lane voltages on the negative-sequence circuit
${}^1V_0, {}^2V_0$: first- and second-circuit voltages in the zero-sequence domain
${}^1V_1, {}^2V_1$: first- and second-circuit voltages in the positive-sequence domain
${}^1V_2, {}^2V_2$: first- and second-circuit voltages in the negative-sequence domain

The equations for the current or any other quantities are defined similarly.

Let us refer to the transformed new circuits as the '1st-lane circuit' and '2nd-lane circuit'.

4.1.2 Transformation process of double circuit line

The symmetrical equations of the double circuit line are quoted from Equations 2.20b and 2.24c. The positive-sequence circuit (the negative-sequence circuit is of the same form) is

$$\underbrace{\begin{bmatrix} {}^1_mV_1 \\ {}^2_mV_1 \end{bmatrix}}_{{}^{12}_m\boldsymbol{V}_1} - \underbrace{\begin{bmatrix} {}^1_nV_1 \\ {}^2_nV_1 \end{bmatrix}}_{{}^{12}_n\boldsymbol{V}_1} = \underbrace{\begin{bmatrix} Z_1 & 0 \\ 0 & Z_1 \end{bmatrix}}_{\boldsymbol{Z}_1} \cdot \underbrace{\begin{bmatrix} {}^1I_1 \\ {}^2I_1 \end{bmatrix}}_{{}^{12}\boldsymbol{I}_1} \quad \text{or} \quad {}^{12}_m\boldsymbol{V}_1 - {}^{12}_n\boldsymbol{V}_1 = \boldsymbol{Z}_1 \cdot {}^{12}\boldsymbol{I}_1 \quad ①$$

where $Z_1 = Z_s - Z_m$

$$\underbrace{\begin{bmatrix} {}^1I_1 \\ {}^2I_1 \end{bmatrix}}_{{}^{12}\boldsymbol{I}_1} = j\omega \underbrace{\begin{bmatrix} C_1 & 0 \\ 0 & C_1 \end{bmatrix}}_{\boldsymbol{C}_1} \cdot \underbrace{\begin{bmatrix} {}^1V_1 \\ {}^2V_1 \end{bmatrix}}_{{}^{12}\boldsymbol{V}_1} \quad \text{or} \quad {}^{12}\boldsymbol{I} = j\omega\boldsymbol{C}_1 \cdot {}^{12}\boldsymbol{V}_1 \quad ② \qquad (4.3a)$$

where $C_1 = C_s + 3C_m + 3C'_m$

and the zero-sequence circuit is

$$\underbrace{\begin{bmatrix} {}^1_mV_0 \\ {}^2_mV_0 \end{bmatrix}}_{{}^{12}_m\boldsymbol{V}_0} - \underbrace{\begin{bmatrix} {}^1_nV_0 \\ {}^2_nV_0 \end{bmatrix}}_{{}^{12}_n\boldsymbol{V}_0} = \underbrace{\begin{bmatrix} Z_0 & Z_{0M} \\ Z_{0M} & Z_0 \end{bmatrix}}_{\boldsymbol{Z}_0} \cdot \underbrace{\begin{bmatrix} {}^1I_0 \\ {}^2I_0 \end{bmatrix}}_{{}^{12}\boldsymbol{I}_0} \quad \text{or} \quad {}^{12}_m\boldsymbol{V}_0 - {}^{12}_n\boldsymbol{V}_0 = \boldsymbol{Z}_0 \cdot {}^{12}\boldsymbol{I}_0$$

where $Z_0 = Z_s + 2Z_m, \quad Z_{0M} = 3Z'_m$

$$\underbrace{\begin{bmatrix} {}^1I_0 \\ {}^2I_0 \end{bmatrix}}_{{}^{12}\boldsymbol{I}_0} = j\omega \underbrace{\begin{bmatrix} C_0 + C'_0 & -C'_0 \\ -C'_0 & C_0 + C'_0 \end{bmatrix}}_{\boldsymbol{C}_0} \cdot \underbrace{\begin{bmatrix} {}^1V_0 \\ {}^2V_0 \end{bmatrix}}_{{}^{12}\boldsymbol{V}_0} \quad \text{or} \quad {}^{12}\boldsymbol{I}_0 = j\omega\boldsymbol{C}_0 \cdot {}^{12}\boldsymbol{V}_0 \qquad (4.3b)$$

where $C_0 = C_s, \quad C'_0 = 3C'_m$

The transformation of the above symmetrical equations into 1st- and 2nd-lane circuit equations can be done using the following process and modifications:

$$\left.\begin{array}{l} {}_m\boldsymbol{V}_{k01} - {}_n\boldsymbol{V}_{k01} = (\boldsymbol{a}_{2\phi} \times \boldsymbol{Z}_k \cdot \boldsymbol{a}_{2\phi}^{-1}) \cdot \boldsymbol{I}_{k01} \\ {}^{12}\boldsymbol{I}_{k01} = j\omega(\boldsymbol{a}_{2\phi} \cdot \boldsymbol{C}_k \cdot \boldsymbol{a}_{2\phi}^{-1}) \cdot \boldsymbol{V}_{k01} \\ \text{where} \quad k = 1, 2, 0 \end{array}\right\} \tag{4.4}$$

$\boldsymbol{a}_{2\phi} \cdot \boldsymbol{Z}_k \cdot \boldsymbol{a}_{2\phi}^{-1}$ and $\boldsymbol{a}_{2\phi} \cdot \boldsymbol{C}_k \cdot \boldsymbol{a}_{2\phi}^{-1}$ can easily be calculated and the following transformed equations are derived:

For the positive-sequence circuit

$$\left.\begin{array}{l} \begin{bmatrix} {}_mV_{10} \\ {}_mV_{11} \end{bmatrix} - \begin{bmatrix} {}_nV_{10} \\ {}_nV_{11} \end{bmatrix} = \begin{bmatrix} Z_1 & 0 \\ 0 & Z_1 \end{bmatrix} \cdot \begin{bmatrix} I_{10} \\ I_{11} \end{bmatrix} = \begin{bmatrix} Z_s - Z_m & 0 \\ 0 & Z_s - Z_m \end{bmatrix} \cdot \begin{bmatrix} I_{10} \\ I_{11} \end{bmatrix} \\ \begin{bmatrix} I_{10} \\ I_{11} \end{bmatrix} = j\omega \begin{bmatrix} C_1 & 0 \\ 0 & C_1 \end{bmatrix} \begin{bmatrix} V_{10} \\ V_{11} \end{bmatrix} = j\omega \begin{bmatrix} C_s + 3C_m + 3C'_m & 0 \\ 0 & C_s + 3C_m + 3C'_m \end{bmatrix} \cdot \begin{bmatrix} V_{10} \\ V_{11} \end{bmatrix} \end{array}\right\} \tag{4.5a}$$

For the negative-sequence circuit

$$\left.\begin{array}{l} \begin{bmatrix} {}_mV_{20} \\ {}_mV_{21} \end{bmatrix} - \begin{bmatrix} {}_nV_{20} \\ {}_nV_{21} \end{bmatrix} = \begin{bmatrix} Z_1 & 0 \\ 0 & Z_1 \end{bmatrix} \cdot \begin{bmatrix} I_{20} \\ I_{21} \end{bmatrix} = \begin{bmatrix} Z_s - Z_m & 0 \\ 0 & Z_s - Z_m \end{bmatrix} \cdot \begin{bmatrix} I_{20} \\ I_{21} \end{bmatrix} \\ \begin{bmatrix} I_{20} \\ I_{21} \end{bmatrix} = j\omega \begin{bmatrix} C_1 & 0 \\ 0 & C_1 \end{bmatrix} \cdot \begin{bmatrix} V_{20} \\ V_{21} \end{bmatrix} = j\omega \begin{bmatrix} C_s + 3C_m + 3C'_m & 0 \\ 0 & C_s + 3C_m + 3C'_m \end{bmatrix} \cdot \begin{bmatrix} V_{20} \\ V_{21} \end{bmatrix} \end{array}\right\} \tag{4.5b}$$

For the zero-sequence circuit

$$\left.\begin{array}{l} \begin{bmatrix} {}_mV_{00} \\ {}_mV_{01} \end{bmatrix} - \begin{bmatrix} {}_nV_{00} \\ {}_nV_{01} \end{bmatrix} = \begin{bmatrix} Z_0 + Z_{0M} & 0 \\ 0 & Z_0 - Z_{0M} \end{bmatrix} \cdot \begin{bmatrix} I_{00} \\ I_{01} \end{bmatrix} = \begin{bmatrix} Z_s + 2Z_m + 3Z'_m & 0 \\ 0 & Z_s + 2Z_m - 3Z'_m \end{bmatrix} \cdot \begin{bmatrix} I_{00} \\ I_{01} \end{bmatrix} \\ \begin{bmatrix} I_{00} \\ I_{01} \end{bmatrix} = j\omega \begin{bmatrix} C_0 & 0 \\ 0 & C_0 + 2C'_0 \end{bmatrix} \begin{bmatrix} V_{00} \\ V_{01} \end{bmatrix} = j\omega \begin{bmatrix} C_s & 0 \\ 0 & C_s + 6C'_M \end{bmatrix} \cdot \begin{bmatrix} V_{00} \\ V_{01} \end{bmatrix} \end{array}\right\} \tag{4.5c}$$

These derived equations are in coincidence with the figures in Table 2.1. The transformed zero-sequence equation shows that the mutual inductance as well as the mutual capacitance vanished, so that the 1st- and 2nd-lane circuits can be treated as circuits with self-impedance and self-capacitance only.

Figure 4.2 shows the vector relations between $({}^1I_0, {}^2I_0)$ and (I_{00}, I_{01}).

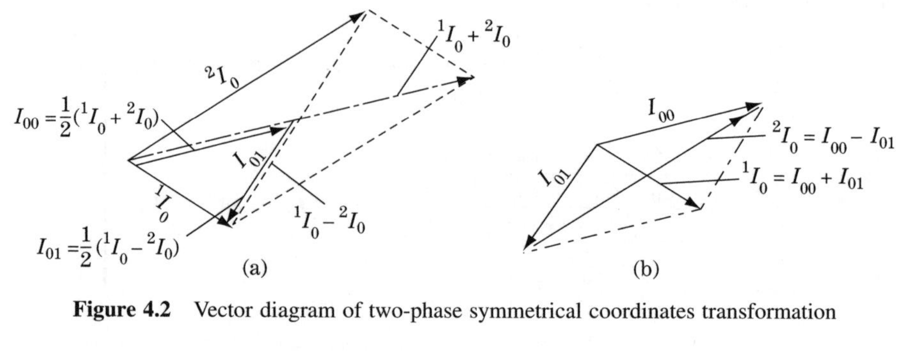

Figure 4.2 Vector diagram of two-phase symmetrical coordinates transformation

4.2 Double Circuit Line by Two-phase Symmetrical Transformation

4.2.1 Transformation of typical two-phase circuits

Figure 4.3(a) is a typical two-phase circuit. The circuit can be regarded as a zero-sequence circuit of the double circuit line (putting suffix $k = 0$), or as the positive-/negative-sequence circuits by further omitting the mutual inductance and mutual capacitance between the first and second circuits (putting $k = 1,2$). The circuit equations are

$$\left.\begin{array}{ll}
\begin{bmatrix} {}^{1}_{l}V_k \\ {}^{2}_{l}V_k \end{bmatrix} = \begin{bmatrix} {}_{\alpha}Z & {}_{\alpha}Z_M \\ {}_{\alpha}Z_M & {}_{\alpha}Z \end{bmatrix} \cdot \begin{bmatrix} {}^{1}_{l}I_k \\ {}^{2}_{l}I_k \end{bmatrix} + \begin{bmatrix} {}^{1}_{m}V_k \\ {}^{2}_{m}V_k \end{bmatrix} & ① \\
\begin{bmatrix} {}^{1}_{m}V_k \\ {}^{2}_{m}V_k \end{bmatrix} = \begin{bmatrix} Z_c & 0 \\ 0 & Z_c \end{bmatrix} \cdot \begin{bmatrix} {}^{1}_{m}I_k \\ {}^{2}_{m}I_k \end{bmatrix} = \begin{bmatrix} {}_{\beta}Z & {}_{\beta}Z_M \\ {}_{\beta}Z_M & {}_{\beta}Z \end{bmatrix} \cdot \begin{bmatrix} {}^{1}_{n}I_k \\ {}^{2}_{n}I_k \end{bmatrix} + \begin{bmatrix} {}_{n}V \\ {}_{n}V \end{bmatrix} & ② \\
\begin{bmatrix} {}_{n}V \\ {}_{n}V \end{bmatrix} = \begin{bmatrix} {}_{n}Z & {}_{n}Z \\ {}_{n}Z & {}_{n}Z \end{bmatrix} \cdot \begin{bmatrix} {}^{1}_{n}I_k \\ {}^{2}_{n}I_k \end{bmatrix} + \begin{bmatrix} E \\ E \end{bmatrix} & ③ \\
\begin{bmatrix} {}^{1}_{l}I_k \\ {}^{2}_{l}I_k \end{bmatrix} = \begin{bmatrix} {}^{1}_{n}I_k \\ {}^{2}_{n}I_k \end{bmatrix} + \begin{bmatrix} {}^{1}_{m}I_k \\ {}^{2}_{m}I_k \end{bmatrix} + \begin{bmatrix} {}^{1}_{m}I'_k \\ {}^{2}_{m}I'_k \end{bmatrix} & ④ \\
{}^{1}_{m}I'_k + {}^{2}_{m}I'_k = 0 & ⑤ \\
{}^{1}_{m}V_k - {}^{2}_{m}V_k = Z'_c \cdot {}^{1}_{m}I'_k & ⑥
\end{array}\right\} \tag{4.6}$$

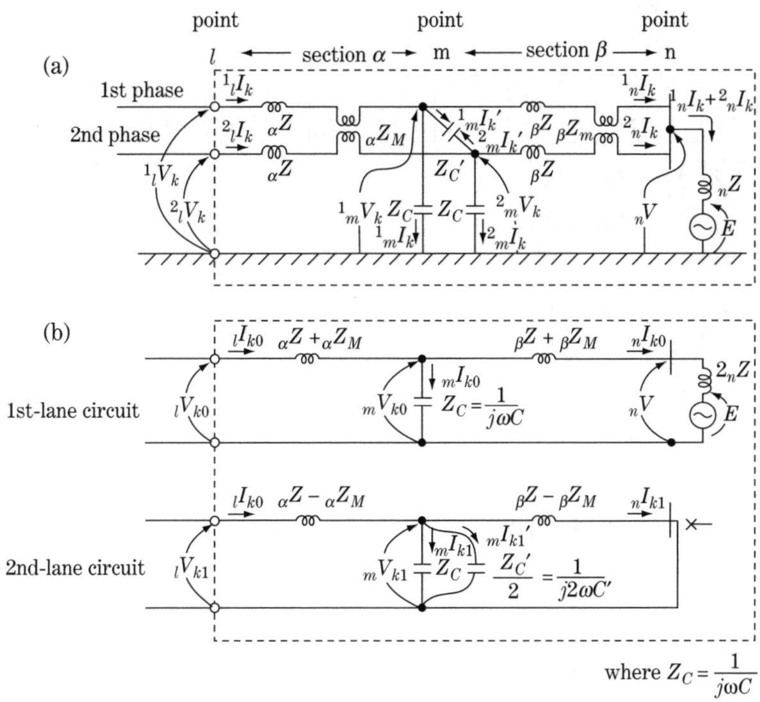

Figure 4.3 Circuit transformation by two-phase symmetrical components

The equations can be transformed into that of 1st- and 2nd-lane variables by the same transformation process as Equation 4.3:

from Equation (4.6) ①

$$\begin{bmatrix} {}_lV_{k0} \\ {}_lV_{k1} \end{bmatrix} = \frac{1}{2}\begin{bmatrix}1&1\\1&-1\end{bmatrix}\cdot\begin{bmatrix}{}_\alpha Z & {}_\alpha Z_M\\ {}_\alpha Z_M & {}_\alpha Z\end{bmatrix}\cdot\begin{bmatrix}1&1\\1&-1\end{bmatrix}\cdot\begin{bmatrix}{}_lI_{k0}\\ {}_lI_{k1}\end{bmatrix}+\begin{bmatrix}{}_mV_{k0}\\ {}_mV_{k1}\end{bmatrix}$$

$$= \begin{bmatrix}{}_\alpha Z + {}_\alpha Z_M & 0\\ 0 & {}_\alpha Z - {}_\alpha Z_M\end{bmatrix}\cdot\begin{bmatrix}{}_lI_{k0}\\ {}_lI_{k1}\end{bmatrix}+\begin{bmatrix}{}_mV_{k0}\\ {}_mV_{k1}\end{bmatrix} \qquad ①$$

from Equation (4.6) ②

$$\begin{bmatrix} {}_mV_{k0} \\ {}_mV_{k1} \end{bmatrix} = \frac{1}{2}\begin{bmatrix}1&1\\1&-1\end{bmatrix}\cdot\begin{bmatrix}Z_c & 0\\ 0 & Z_c\end{bmatrix}\cdot\begin{bmatrix}1&1\\1&-1\end{bmatrix}\cdot\begin{bmatrix}{}_mI_{k0}\\ {}_mI_{k1}\end{bmatrix}$$

$$= \frac{1}{2}\begin{bmatrix}1&1\\1&-1\end{bmatrix}\cdot\begin{bmatrix}{}_\beta Z & {}_\beta Z_M\\ {}_\beta Z_M & {}_\beta Z\end{bmatrix}\cdot\begin{bmatrix}1&1\\1&-1\end{bmatrix}\cdot\begin{bmatrix}{}_nI_{k0}\\ {}_nI_{k1}\end{bmatrix}+\frac{1}{2}\begin{bmatrix}1&1\\1&-1\end{bmatrix}\cdot\begin{bmatrix}{}_nV\\ {}_nV\end{bmatrix}$$

$$\therefore \begin{bmatrix} {}_mV_{k0} \\ {}_mV_{k1} \end{bmatrix} = \begin{bmatrix}Z_c & 0\\ 0 & Z_c\end{bmatrix}\cdot\begin{bmatrix}{}_mI_{k0}\\ {}_mI_{k1}\end{bmatrix}$$

$$= \begin{bmatrix}{}_\beta Z + {}_\beta Z_M & 0\\ 0 & {}_\beta Z - {}_\beta Z_M\end{bmatrix}\cdot\begin{bmatrix}{}_nI_{k0}\\ {}_nI_{k1}\end{bmatrix}+\begin{bmatrix}{}_nV\\ 0\end{bmatrix} \qquad ②$$

from Equation (4.6) ③

$$\frac{1}{2}\begin{bmatrix}1&1\\1&-1\end{bmatrix}\cdot\begin{bmatrix}{}_nV\\ {}_nV\end{bmatrix}$$

$$= \frac{1}{2}\begin{bmatrix}1&1\\1&-1\end{bmatrix}\cdot\begin{bmatrix}{}_nZ & {}_nZ\\ {}_nZ & {}_nZ\end{bmatrix}\cdot\begin{bmatrix}1&1\\1&-1\end{bmatrix}\cdot\begin{bmatrix}{}_nI_{k0}\\ {}_nI_{k1}\end{bmatrix}+\frac{1}{2}\begin{bmatrix}1&1\\1&-1\end{bmatrix}\cdot\begin{bmatrix}E\\ E\end{bmatrix}$$

$$\therefore \begin{bmatrix}{}_nV\\ 0\end{bmatrix} = \begin{bmatrix}2{}_nZ & 0\\ 0 & 0\end{bmatrix}\cdot\begin{bmatrix}{}_nI_{k0}\\ {}_nI_{k1}\end{bmatrix}+\begin{bmatrix}E\\ 0\end{bmatrix} \qquad ③$$

from Equation (4.6) ④

$$\begin{bmatrix}{}_lI_{k0}\\ {}_lI_{k1}\end{bmatrix} = \begin{bmatrix}{}_nI_{k0}\\ {}_nI_{k1}\end{bmatrix}+\begin{bmatrix}{}_mI_{k0}\\ {}_mI_{k1}\end{bmatrix}+\begin{bmatrix}{}_mI'_{k0}\\ {}_mI'_{k1}\end{bmatrix} \qquad ④$$

from Equation (4.6) ⑤

$${}_mI'_{k0} = \frac{1}{2}({}^1_mI'_k + {}^2_mI'_k) = 0 \qquad \therefore {}_mI'_{k0} = 0 \qquad ⑤$$

from Equation (4.6) ⑥

$$({}_mV_{k0} + {}_mV_{k1}) - ({}_mV_{k0} - {}_mV_{k1}) = Z'_c \cdot ({}_mI'_{k0} + {}_mI'_{k1})$$

$$\therefore {}_mV_{k1} = \frac{Z'_c}{2}\cdot {}_mI'_{k1} \qquad ⑥$$

(4.7)

The equivalent circuit in Figure 4.3(b) can be written from the above transformed equations. It consists of two independent circuits which are named **the 1st-lane circuit** and **the 2nd-lane circuit** of the positive-, negative- or zero-sequence domain for $k = 1,2,0$ respectively. In Figure 4.3(b), mutual inductances as well as mutual capacitances between the 1st lane and 2nd lane have already vanished. It should be noted that in the 1st-lane circuit $2_n Z$ is inserted, while in the 2nd-lane circuit, the first and second circuits are short-circuited at the bus terminal n and so $C_0 + 2C_0'$ is inserted.

4.2.2 Transformation of double circuit line

Figure 4.4a is the symmetrical equivalent circuit of the double circuit transmission line, in that the mutual inductance and capacitance between the first and second circuits exist only for the zero sequence. The figure can be easily transformed into the 1st- and 2nd-lane circuits of Figure 4.4b in the same way as described by Figure 4.3.

The original three-phase double circuit line (of a–b–c and A–B–C phases, Figure 1.5 in Chapter 1) has been transformed into Figure 4.4b, in which there are six mutually independent circuits. Mutual constants between the 1st and 2nd lanes for the zero sequence have already vanished. Each 2nd-lane circuit of the 1–2–0 sequence domain is a closed-circuit composed only of parallel line part constants.

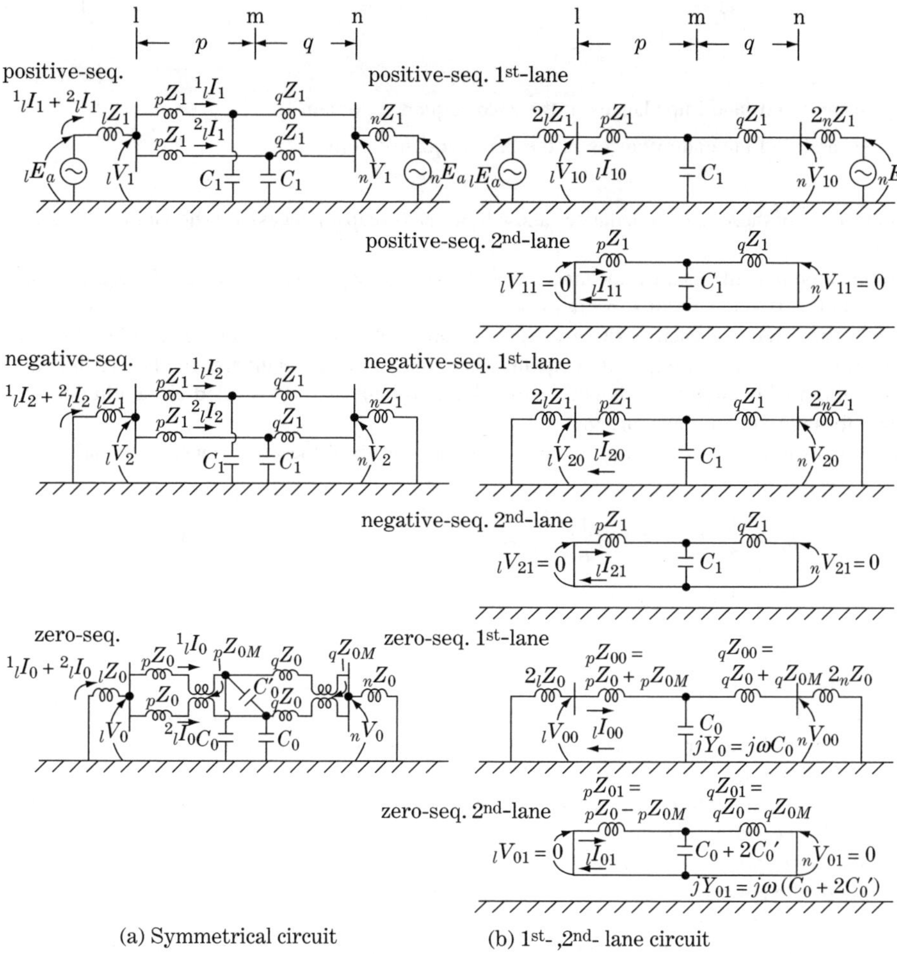

Figure 4.4 Equivalent circuit of double circuit line in 1st- and 2nd-lane circuit domain

The circuit constants in Figure 4.4(b) are given by the following equations:

$$\left.\begin{array}{lll} Z_1 = Z_s - Z_m & Z_{10} = Z_1 = Z_s - Z_m & Z_{11} = Z_1 = Z_s - Z_m \\ jY_1 = j\omega C_1 & jY_{10} = j\omega C_1 & jY_{11} = j\omega C_1 \\ \quad = j\omega(C_s + 3C_m & \quad = j\omega(C_s + 3C_m & \quad = j\omega(C_s + 3C_m \\ \quad\quad + 3C'_m) & \quad\quad + 3C'_m) & \quad\quad + 3C'_m) \\ & & \\ Z_0 = Z_s + 2Z_m & Z_{00} = Z_0 + Z_{0M} & Z_{01} = Z_0 - Z_{0M} \\ Z_{0M} = 3Z'_m & \quad = Z_s + 2Z_m + 3Z'_m & \quad = Z_0 + 2Z_m - 3Z'_m \\ jY_0 = j\omega C_0 & jY_{00} = j\omega C_0 & jY_{01} = j\omega(C_0 + 2C'_0) \\ \quad = j\omega C_s & \quad = j\omega C_s & \quad = j\omega(C_s + 6C'_m) \\ jY'_0 = j\omega C'_0 & & \\ \quad = j\omega \cdot 3C'_m & & \end{array}\right\} \quad (4.8)$$

where

Z_{00}, Z_{01}: 1st- and 2nd-lane impedances in the zero-sequence domain

Y_{00}, Y_{01}: 1st- and 2nd-lane admittances in the zero-sequence domain

Typical values of these lane impedances and admittances (capacitances) are shown in Table 2.1 in Chapter 2.

It must be remembered that the impedances of $2\,{}_lZ_0, 2\,{}_lZ_1, 2\,{}_nZ_1, 2\,{}_nZ_0$ (instead of ${}_lZ_1, {}_lZ_0, {}_nZ_1, {}_nZ_0$) are inserted in the 1st-lane circuits in Figure 4.4(b).

The 2nd-lane circuit of each positive-, negative- and zero-sequence circuit is of closed circuit without power source. Accordingly all the quantities in the 2nd-lane circuit are zero before the fault. However, if the double circuit line is not necessarily well balanced, a so-called circulating current would flow through the 2nd-lane circuits.

Let us now examine I_{00}, I_{01} in the 1st- and 2nd-lane circuit of the zero-sequence domain:

$$\left.\begin{array}{l} I_{00} = \frac{1}{2}({}^1I_0 + {}^2I_0) = \frac{1}{6}\{({}^1I_a + {}^1I_b + {}^1I_c) + ({}^2I_a + {}^2I_b + {}^2I_c)\} \\ I_{01} = \frac{1}{2}({}^1I_0 - {}^2I_0) = \frac{1}{6}\{({}^1I_a + {}^1I_b + {}^1I_c) - ({}^2I_a + {}^2I_b + {}^2I_c)\} \end{array}\right\} \quad (4.9)$$

Accordingly, if zero-sequence current exists under normal load conditions, then I_{01} of the 2nd lane also exists and flows through the 2nd-lane closed loop circuit. I_{01} is the so-called **circulating current** of the double or multiple circuit line and is the zero-sequence current component which actually circulates through the first and second circuits.

Further, I_{00}, I_{01} are quantities which can be measured as the addition or subtraction of the **current transformer (CT) residual currents** ${}^1I_0 = (1/3)({}^1I_a + {}^1I_b + {}^1I_c), {}^2I_0 = (1/3)({}^2I_a + {}^2I_b + {}^2I_c)$ at the CT secondary terminals of the first and second circuits.

In the practical engineering field of protective relaying, the zero-sequence circulating current I_{01} sometimes causes severe problems for certain types of protective relays, in particular for double circuit lines of a highly resistive neutral grounding system, for which special countermeasures may be required to prevent malfunction of the relays.

4.3 Fault Analysis of Double Circuit Line (General Process)

Figure 4.5 is the process flow diagram of fault analysis for the double circuit line. The steps with the marks *1, *2 in the diagram correspond to the Figure 4.4(a), (b) respectively.

Tables 4.1a and b summarize the related equations and the corresponding equivalent circuits for the cases of a single circuit fault and a double circuit fault at the same point f on a double circuit transmission line. The double circuit transmission line before the fault is shown as Figure 1 in Table 4.1, where a set of virtual terminals is prepared at point f for the connection of fault conditions (Figure 1 corresponds to the process step *1).

In relation to Figure 1 in Table 4.1, Figure 2 shows the circuits in the symmetrical coordinate domain (corresponding to *2) and Figure 3 the 1st- and 2nd-lane circuits in the symmetrical coordinate domain (corresponding to *3). Capacitances can be of course added to these circuits if necessary.

The 1st- and 2nd-lane circuits in the symmetrical coordinate domain of Figure 3 can be described by the following equation:

$$\left.\begin{aligned} {}_fV_{10} &= {}_fE_a - {}_fZ_{10}\cdot {}_fI_{10} \\ {}_fV_{11} &= -{}_fZ_{11}\cdot {}_fI_{11} \end{aligned}\right\} ① \qquad \left.\begin{aligned} {}_fV_{20} &= -{}_fZ_{20}\cdot {}_fI_{20} \\ {}_fV_{21} &= -{}_fZ_{21}\cdot {}_fI_{21} \end{aligned}\right\} ② \qquad \left.\begin{aligned} {}_fV_{00} &= -{}_fZ_{00}\cdot {}_fI_{00} \\ {}_fV_{01} &= -{}_fZ_{01}\cdot {}_fI_{01} \end{aligned}\right\} ③ \tag{4.10}$$

where ${}_fZ_{10}$, ${}_fZ_{11}$, ${}_fZ_{20}$, ${}_fZ_{21}$, ${}_fZ_{00}$, ${}_fZ_{01}$ are the 1st- and 2nd-lane impedances looking into the circuit from point f in Figure 3, all of which can be found from Figure 3 as known quantities.

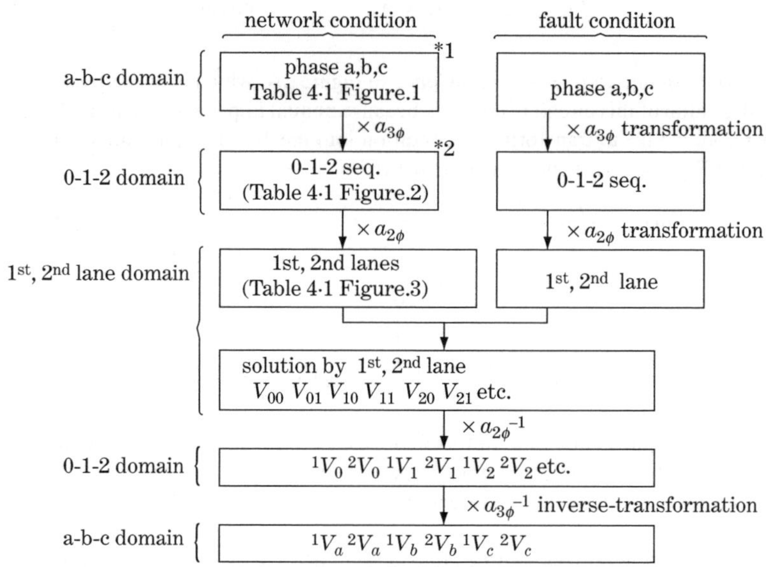

Figure 4.5 General procedure for fault analysis of double circuit line

For example,

$$
\begin{aligned}
{}_fZ_{10} &= \{\text{the parallel impedance of } ({}_pZ_1 + 2_mZ_1),\ ({}_qZ_1 + 2_nZ_1)\} \\
&= ({}_pZ_1 + 2_mZ_1)//({}_qZ_1 + 2_nZ_1) \\
{}_fZ_{11} &= \{\text{the parallel impedance of } {}_pZ_1, {}_qZ_1\} = {}_pZ_1//{}_qZ_1
\end{aligned}
$$

and ${}_fE_a$ is the known voltage at point f before the fault in Figure 4.3.

Now we complete our arrangement of circuit conditions before the fault.

4.4 Single Circuit Fault on the Double Circuit Line

Let us examine the phase b to phase c line-to-line fault at point f on the first circuit of the double circuit transmission line.

4.4.1 Line-to-ground fault (1ϕG) on one side circuit

The fault condition of this case is connection of the virtual terminals t_1, t_7 through the arc resistance R (see Figure 1A):

$$
\left.\begin{aligned}
{}^1_fI_b = {}^1_fI_c = 0 \\
{}^1_fV_a = R\cdot{}^1_fI_a
\end{aligned}\right\} \quad : \text{circuit 1}
$$
$$
{}^2_fI_a = {}^2_fI_b = {}^2_fI_c = 0 \quad : \text{circuit 2} \tag{4.11a}
$$

and in the 1–2–0 domain

$$
\left.\begin{aligned}
{}^1_fI_0 = {}^1_fI_1 = {}^1_fI_2 \\
{}^1_fV_0 + {}^1_fV_1 + {}^1_fV_2 = 3R\cdot{}^1_fI_0
\end{aligned}\right\} \quad : \text{circuit 1}
$$
$$
{}^2_fI_0 = {}^2_fI_1 = {}^2_fI_2 = 0 \quad : \text{circuit 2} \tag{4.11b}
$$

The equation is in one-to-one correspondence to Figure 2A, which is the equivalent circuit of this case. The calculation of this circuit is not easy, because mutual impedance exists in the zero-sequence circuit. Therefore we try to transform the condition into the 1st- and 2nd-lane circuits.

Substituting Equation 4.2b into Equation 4.11b,

$$
\left.\begin{aligned}
&({}_fI_{00} + {}_fI_{01}) = ({}_fI_{10} + {}_fI_{11}) = ({}_fI_{20} + {}_fI_{21}) \\
&({}_fV_{00} + {}_fV_{01}) + ({}_fV_{10} + {}_fV_{11}) + ({}_fV_{20} + {}_fV_{21}) = 3R\cdot({}_fI_{00} + {}_fI_{01}) \\
&({}_fI_{00} - {}_fI_{01}) = ({}_fI_{10} - {}_fI_{11}) = ({}_fI_{20} - {}_fI_{21}) = 0 \\
&\therefore {}_fI_{00} = {}_fI_{01} = {}_fI_{10} = {}_fI_{11} = {}_fI_{20} = {}_fI_{21} \\
&({}_fV_{00} + {}_fV_{01}) + ({}_fV_{10} + {}_fV_{11}) + ({}_fV_{20} + {}_fV_{21}) = 6R\cdot{}_fI_{00}
\end{aligned}\right\} \tag{4.12}
$$

Figure 3A is the equivalent circuit of this fault case because it is strictly in one-to-one correspondence to the above equation. All the mutual impedance has already disappeared in Figure 3A, so the voltage and current quantities at point f under the terminal condition of Figure 3A can easily be found by calculation.

The resolved quantities are transformed into the symmetrical quantities by applying Equation 4.2b for the inverse transformation, and are finally transformed into the three-phase quantities.

Table 4.1a Single & double circuits faults at one point

	Single circuit #1 fault				Double circuit #1,#2 fault at the same point
	#1: a–to–g fault	#1: b,c –to–g fault	#1: b –to–c fault	#1: three–phase fault	#1: a–to–g, #2: b–to–c fault
Figure-1 three-phase circuit 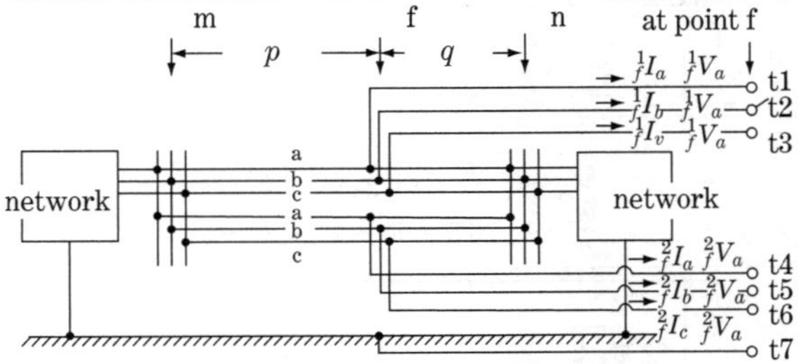	Figure- 1A 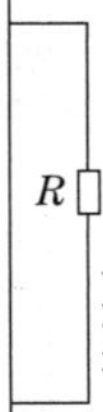 ${}^1_fI_b = {}^1_fI_c = 0,\quad {}^1_fV_a = R \cdot {}^1_fI_a$ ${}^2_fI_a = {}^2_fI_b = {}^2_fI_c = 0$ (4.11a)	Figure- 1B 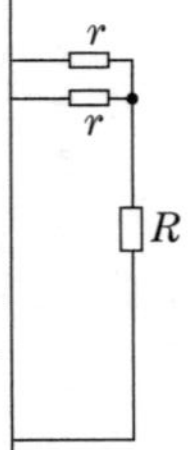	Figure- 1C 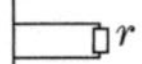	Figure- 1D 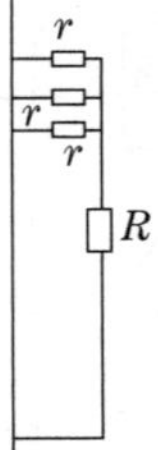	Figure- 1E 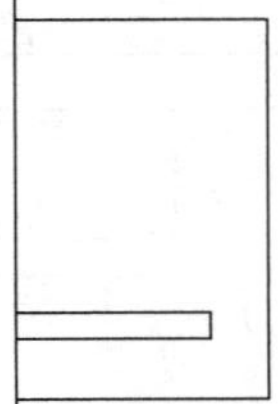 (4.13a)
Figure-2 0-1-2 three-phase circuit 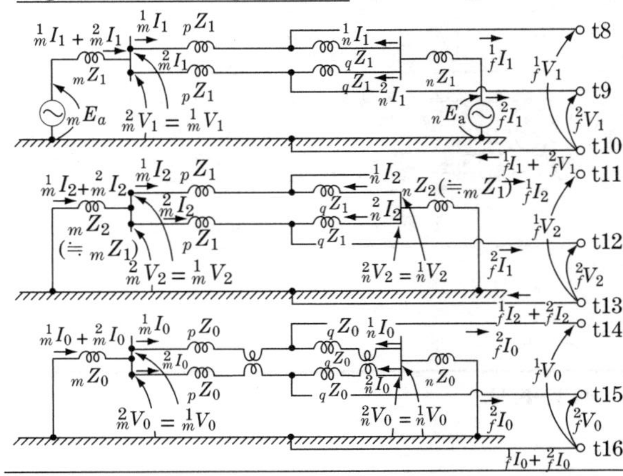	Figure- 2A 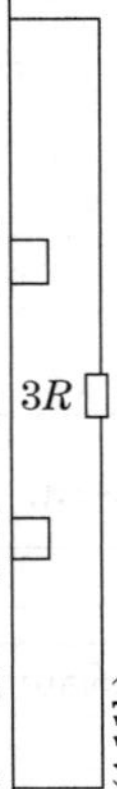 ${}^1_fI_0 = {}^1_fI_1 = {}^1_fI_2,\quad {}^1_fV_0 = {}^1_fV_1 = {}^1_fV_2 = 3R \cdot {}^1_fI_0$ ${}^2_fI_0 = {}^2_fI_1 = {}^2_fI_2 = 0$ (4.11b)	Figure- 2B 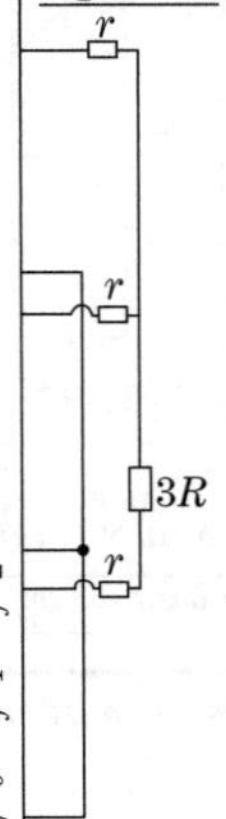	Figure- 2C 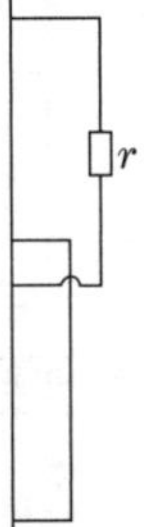	Figure- 2D 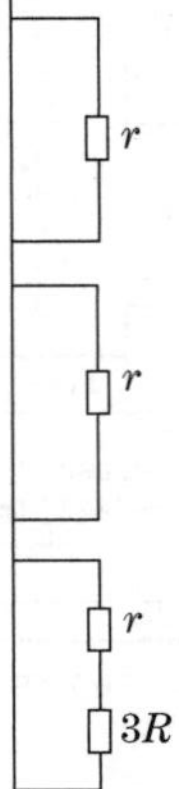	Figure- 2E 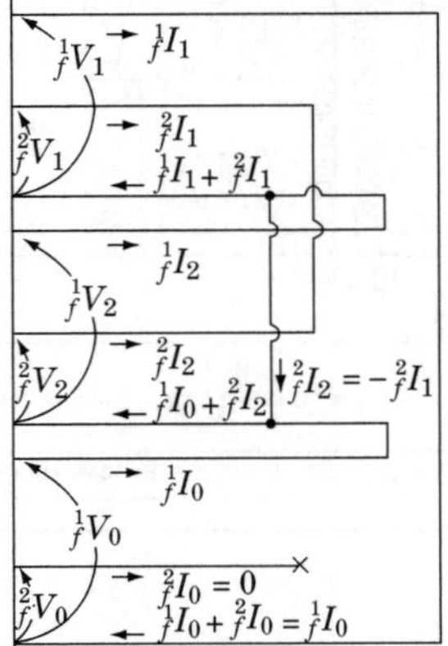 (4.13b)

(*Continued*)

Table 4.1b Single & double circuits faults at one point

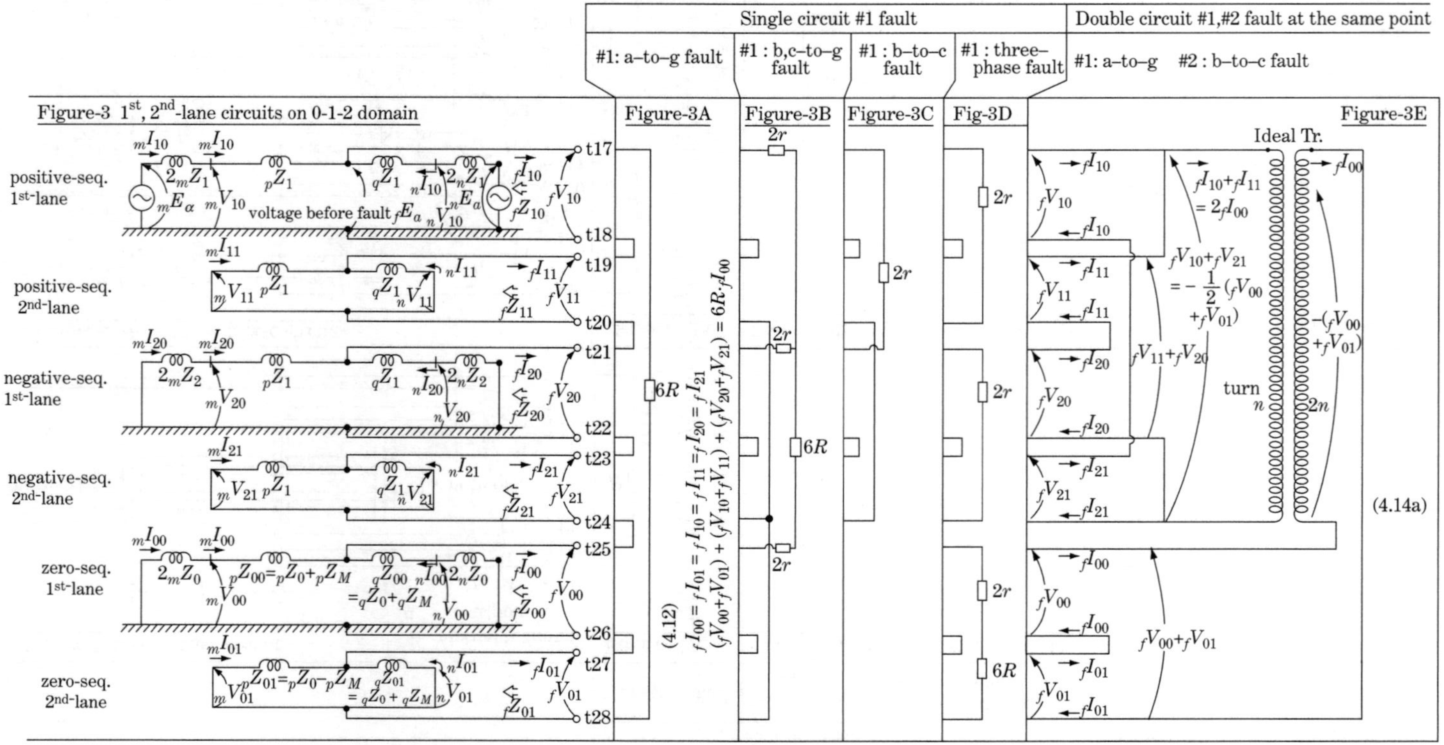

4.4.2 Various one-side circuit faults

We can solve one-side circuit faults of various other modes by the same method. The related equations and the equivalent circuits for these are also described in Table 4.1.

In the case of a phase b to c line-to-ground fault (Figure 1B), for example, the equivalent circuit in the symmetrical domain and the further transformed circuit are given by Figures 2B and 3B, respectively. The calculation of Figure 3B is easy because the circuit is a single loop circuit without mutual inductances. On the other hand, the manual calculation of Figure 2B is quite hard.

In the case of a phase b to c line-to-line fault (Figure 1C), as another example, the calculation by Figure 3C is easier than by Figure 2C in spite of the fact that the zero-sequence circuit with mutual inductances is not even included.

4.5 Double Circuit Fault at Single Point f

4.5.1 Circuit 1 phase a line-to-ground fault and circuit 2 phases b and c line-to-line faults at point f

The fault condition in this case is shown in Figure 1E, where arc resistance is neglected.

The fault condition in the three-phase domain (see Figure 1E) is

$$\left.\begin{array}{llll} {}^1_fI_b = {}^1_fI_c = 0, & {}^1_fV_a = 0 & & : \text{circuit 1} \\ {}^2_fI_a = 0, & {}^2_fI_b + {}^2_fI_c = 0, & {}^2_fV_b = {}^2_fV_c & : \text{circuit 2} \end{array}\right\} \tag{4.13a}$$

The fault condition in the 1–2–0 domain (see Figure 2E) is

$$\left.\begin{array}{lll} {}^1_fI_0 = {}^1_fI_1 = {}^1_fI_2, & {}^1_fV_0 + {}^1_fV_1 + {}^1_fV_2 = 0 & : \text{ circuit 1} \\ {}^2_fI_0 = 0, \quad {}^2_fI_1 + {}^2_fI_2 = 0, & {}^2_fV_1 = {}^2_fV_2 & : \text{ circuit 2} \end{array}\right\} \tag{4.13b}$$

and the fault condition in the 1st- and 2nd-lane domain (see Figure 3E) is, on substituting Equation 4.2b into Equation 4.13b,

$$\begin{array}{l} ({}_fI_{00} + {}_fI_{01}) = ({}_fI_{10} + {}_fI_{11}) = ({}_fI_{20} + {}_fI_{21}) \\ ({}_fV_{00} + {}_fV_{01}) + ({}_fV_{10} + {}_fV_{11}) + ({}_fV_{20} + {}_fV_{21}) = 0, \qquad ({}_fI_{00} - {}_fI_{01}) = 0 \\ ({}_fI_{10} - {}_fI_{11}) + ({}_fI_{20} - {}_fI_{21}) = 0, \qquad ({}_fV_{10} - {}_fV_{11}) = ({}_fV_{20} - {}_fV_{21}) \end{array}$$

Then

$$\left.\begin{array}{l} \left.\begin{array}{ll} {}_fI_{00} = {}_fI_{01} & {}_fI_{10} = {}_fI_{21} \\ {}_fI_{11} = {}_fI_{20} & 2\,{}_fI_{00} = ({}_fI_{10} + {}_fI_{11}) \end{array}\right\} ① \\ \left.\begin{array}{l} {}_fV_{10} + {}_fV_{21} = {}_fV_{11} + {}_fV_{20} \\ ({}_fV_{00} + {}_fV_{01}) = -2({}_fV_{10} + {}_fV_{21}) \end{array}\right\} ② \end{array}\right\} \tag{4.14a}$$

The equivalent circuit for Equation 4.14a is shown in Figure 3E. In this figure, one ideal transformer (of turn ratio 1:2) is inserted in the equivalent circuit in order to satisfy Equation 4.14a completely. Figure 3E can be redrawn as a quite simple circuit so that the calculation is easy, but on the other hand, manual calculation of Figure 2E is almost impossible.

For any other simultaneous double circuit fault modes, we can introduce solutions like Equation 4.14a in the same manner. However, the corresponding equivalent circuits would not be able to be drawn for most of the cases. We have to bear in mind that those cases where we can draw the equivalent circuits are actually exceptional.

In other words, we have to find solutions only from the related equations for most of the cases without having equivalent circuits. Therefore let us examine the above fault further as a typical example and find the solution only from the related equations.

We have the six equations of Equation (4.10) as system side conditions and the other six equations of Equation 4.14a as fault terminal conditions. Twelve variables (six for currents and six for voltages) and twelve equations exist in total, so the equations can be solved as a set of simultaneous equations of twelve dimensions and single order. Next, the simultaneous equations can be easily modified into a set of six dimensions and single order only, with six current variables in this case.

Eliminating variables V by substituting Equation 4.10 into Equation 4.14a ②,

$$\left.\begin{aligned} &{}_fZ_{10}\cdot{}_fI_{10}-{}_fZ_{11}\cdot{}_fI_{11}-{}_fZ_{20}\cdot{}_fI_{20}+{}_fZ_{21}\cdot{}_fI_{21}={}_fE_a \\ &{}_fZ_{00}\cdot{}_fI_{00}+{}_fZ_{01}\cdot{}_fI_{01}+2{}_fZ_{10}\cdot{}_fI_{10}+2{}_fZ_{21}\cdot{}_fI_{21}=2\cdot{}_fE_a \end{aligned}\right\} \tag{4.14b}$$

All the variables for voltages have vanished, so Equations 4.14a and 4.14b can be rearranged as simultaneous equations of six current variables:

$$\begin{bmatrix} 1 & -1 & & & & \\ 2 & & -1 & -1 & & \\ & & 1 & & & -1 \\ & & & -1 & -1 & \\ & & {}_fZ_{10} & -{}_fZ_{11} & -{}_fZ_{20} & {}_fZ_{21} \\ {}_fZ_{00} & {}_fZ_{01} & 2\cdot{}_fZ_{10} & & & 2\cdot{}_fZ_{21} \end{bmatrix}\cdot\begin{bmatrix} I_{00} \\ {}_fI_{01} \\ {}_fI_{10} \\ {}_fI_{11} \\ {}_fI_{20} \\ {}_fI_{21} \end{bmatrix}=\begin{bmatrix} 0 \\ 0 \\ 0 \\ 0 \\ {}_fE_a \\ 2\cdot{}_fE_a \end{bmatrix} \tag{4.15}$$

This set of simultaneous equations is of six dimensions and single order only. These particular equations can be solved manually, although this is not generally easy.

4.5.2 Circuit 1 phase a line-to-ground fault and circuit 2 phase b line-to-ground fault at point f (method 1)

A plain equivalent circuit would not exist in most cases of double circuit faults, so we have to execute a complicated step-by-step calculation. As a typical case let us try to solve the circuit 1 phase a $1\phi G$ and circuit 2 phase b $1\phi G$ fault. The fault condition in this case is shown as follows where we neglect arc resistance.

The fault condition in the three-phase domain is

$$\left.\begin{aligned} &\text{circuit 1 : Phase a to ground fault} \quad {}^1_fI_b={}^1_fI_c=0, \quad {}^1_fV_a=0 \\ &\text{circuit 2 : Phase b to ground fault} \quad {}^2_fI_a={}^2_fI_c=0, \quad {}^2_fV_b=0 \end{aligned}\right\} \tag{4.16}$$

The fault condition in the 1–2–0 domain is

$$\left.\begin{aligned} &\text{circuit 1} \quad {}^1_fI_0={}^1_fI_1={}^1_fI_2, \quad {}^1_fV_0+{}^1_fV_1+{}^1_fV_2=0 \\ &\text{circuit 2} \quad {}^2_fI_0=a^2\cdot{}^2_fI_1=a\cdot{}^2_fI_2, \quad {}^2_fV_0+a^2\cdot{}^2_fV_1+a\cdot{}^2_fV_2=0 \end{aligned}\right\} \tag{4.17}$$

The fault condition in the 1st- and 2nd-lane domain is, substituting Equation (4.2b) into Equation (4.17),

$$\left.\begin{array}{l}({}_fI_{00}+{}_fI_{01})=({}_fI_{10}+{}_fI_{11})=({}_fI_{20}+{}_fI_{21})\\({}_fV_{00}+{}_fV_{01})+({}_fV_{10}+{}_fV_{11})+({}_fV_{20}+{}_fV_{21})=0\\({}_fI_{00}-{}_fI_{01})=a^2({}_fI_{10}-{}_fI_{11})=a({}_fI_{20}-{}_fI_{21})\\({}_fV_{00}-{}_fV_{01})+a^2({}_fV_{10}-{}_fV_{11})+a({}_fV_{20}-{}_fV_{21})=0\end{array}\right\} \tag{4.18}$$

Equation (4.10) as the system condition and Equation 4.18 as the fault condition include 12 equations in total so that a set of simultaneous equations of 12 dimensions and single order can be prepared.

The six voltage variables can be eliminated by substituting Equation (4.10) into Equation (4.18) to obtain the following equations:

$$\left.\begin{array}{l}{}_fI_{00}+{}_fI_{01}-{}_fI_{10}-{}_fI_{11}=0\\{}_fI_{00}+{}_fI_{01}-{}_fI_{20}-{}_fI_{21}=0\\{}_fI_{00}-{}_fI_{01}-a^2{}_fI_{10}+a^2{}_fI_{11}=0\\{}_fI_{00}-{}_fI_{01}-a{}_fI_{20}+a{}_fI_{21}=0\\{}_fZ_{00}\cdot{}_fI_{00}+{}_fZ_{01}\cdot{}_fI_{01}+{}_fZ_{10}\cdot{}_fI_{10}+{}_fZ_{11}\cdot{}_fI_{11}+{}_fZ_{20}\cdot{}_fI_{20}+{}_fZ_{21}\cdot{}_fI_{21}={}_fE_a\\{}_fZ_{00}\cdot{}_fI_{00}-{}_fZ_{01}\cdot{}_fI_{01}+a^2({}_fZ_{10}\cdot{}_fI_{10}-{}_fZ_{11}\cdot{}_fI_{11})+a({}_fZ_{20}\cdot I_{20}-{}_fZ_{21}\cdot I_{21})=a^2\cdot{}_fE_a\end{array}\right\} \tag{4.19a}$$

namely

$$\begin{bmatrix}1&1&-1&-1&&\\1&1&&&-1&-1\\1&-1&-a^2&+a^2&&\\1&-1&&&-a&+a\\{}_fZ_{00}&{}_fZ_{01}&{}_fZ_{10}&{}_fZ_{11}&{}_fZ_{20}&{}_fZ_{21}\\{}_fZ_{00}&-{}_fZ_{01}&a^2\cdot{}_fZ_{10}&-a^2\cdot{}_fZ_{11}&a\cdot{}_fZ_{20}&-a\cdot{}_fZ_{21}\end{bmatrix}\cdot\begin{bmatrix}{}_fI_{00}\\{}_fI_{01}\\{}_fI_{10}\\{}_fI_{11}\\{}_fI_{20}\\{}_fI_{21}\end{bmatrix}=\begin{bmatrix}0\\0\\0\\0\\{}_fE_a\\a^2\cdot{}_fE_a\end{bmatrix} \tag{4.19b}$$

This set of simultaneous equations is of six dimensions and single order for the current-variables where all the impedances are known. We can solve the equations perhaps by using a PC, because solution by hand may be too hard.

Incidentally, whenever voltage and current quantities (${}_fV_{00}$, ${}_fI_{00}$, etc.) at point f are found, quantities at different arbitrary points can be found in the 1st- and 2nd-lane domain by straightforward additional calculation.

4.5.3 Circuit 1 phase a line-to-ground fault and circuit 2 phase b line-to-ground fault at point f (method 2)

The current values of Equation 4.19b can be easily calculated by a computational approach as the problem to obtain a 6 × 6 inverse matrix equation. However, considering the purpose of this book, a method to find a solution manually is demonstrated here.

The fault condition in the three-phase domain is

$$\left.\begin{array}{lll}\text{circuit 1 : Phase a to ground fault} & {}^1_fI_b = {}^1_fI_c = 0, & {}^1_fV_a = 0 \\ \text{circuit 2 : Phase b to ground fault} & {}^2_fI_a = {}^2_fI_c = 0, & {}^2_fV_b = 0\end{array}\right\} \quad (4.20)$$

The transformed symmetrical current equations are

$$\left.\begin{array}{ll}3\cdot{}^1_fI_0 = {}^1_fI_a + 0 + 0 = {}^1_fI_a, & 3\cdot{}^2_fI_0 = 0 + {}^2_fI_b + 0 = {}^2_fI_b \\ 3\cdot{}^1_fI_1 = {}^1_fI_a + a\cdot 0 + a^2\cdot 0 = {}^1_fI_a, & 3\cdot{}^2_fI_1 = 0 + a\cdot{}^2_fI_b + a^2\cdot 0 = a\cdot{}^2_fI_b \\ 3\cdot{}^1_fI_2 = {}^1_fI_a + a^2\cdot 0 + a\cdot 0 = {}^1_fI_a, & 3\cdot{}^2_fI_2 = 0 + a^2\cdot{}^2_fI_b + a\cdot 0 = a^2\cdot{}^2_fI_b\end{array}\right\} \quad (4.21)$$

Accordingly,

$$\left.\begin{array}{l}{}_fI_{00} = \frac{1}{2}({}^1_fI_0 + {}^2_fI_0) = \frac{1}{6}({}^1_fI_a + {}^2_fI_b) \\ {}_fI_{01} = \frac{1}{2}({}^1_fI_0 - {}^2_fI_0) = \frac{1}{6}({}^1_fI_a - {}^2_fI_b) \\ {}_fI_{10} = \frac{1}{2}({}^1_fI_1 + {}^2_fI_1) = \frac{1}{6}({}^1_fI_a + a\cdot{}^2_fI_b) \\ {}_fI_{11} = \frac{1}{2}({}^1_fI_1 - {}^2_fI_1) = \frac{1}{6}({}^1_fI_a - a\cdot{}^2_fI_b) \\ {}_fI_{20} = \frac{1}{2}({}^1_fI_2 + {}^2_fI_2) = \frac{1}{6}({}^1_fI_a + a^2\cdot{}^2_fI_b) \\ {}_fI_{21} = \frac{1}{2}({}^1_fI_2 - {}^2_fI_2) = \frac{1}{6}({}^1_fI_a - a^2\cdot{}^2_fI_b)\end{array}\right\} \quad (4.22)$$

The transformed symmetrical voltage equations of Equation 4.20 are

$$\left.\begin{array}{l}0 = {}^1_fV_a = {}^1_fV_0 + {}^1_fV_1 + {}^1_fV_2 \\ \quad = ({}_fV_{00} + {}_fV_{01}) + ({}_fV_{10} + {}_fV_{11}) + ({}_fV_{20} + {}_fV_{21}) \\ 0 = {}^2_fV_b = {}^2_fV_0 + a^2\cdot{}^2_fV_1 + a\cdot{}^2_fV_2 \\ \quad = ({}_fV_{00} - {}_fV_{01}) + a^2({}_fV_{10} - {}_fV_{11}) + a({}_fV_{20} - {}_fV_{21})\end{array}\right\} \quad (4.23)$$

Substituting Equation 4.10 into Equation 4.23,

$$\left.\begin{array}{l}({}_fZ_{00}\cdot{}_fI_{00} + {}_fZ_{01}\cdot{}_fI_{01}) + ({}_fZ_{10}\cdot{}_fI_{10} + {}_fZ_{11}\cdot{}_fI_{11}) + ({}_fZ_{20}\cdot{}_fI_{20} + {}_fZ_{21}\cdot{}_fI_{21}) = {}_fE_a \\ ({}_fZ_{00}\cdot{}_fI_{00} - {}_fZ_{01}\cdot{}_fI_{01}) + a^2({}_fZ_{10}\cdot{}_fI_{10} - {}_fZ_{11}\cdot{}_fI_{11}) + a({}_fZ_{20}\cdot{}_fI_{20} - {}_fZ_{21}\cdot{}_fI_{21}) = a^2\cdot{}_fE_a\end{array}\right\} \quad (4.24)$$

In Equation 4.22, all the currents in 1st- and 2nd-lane circuits in the 0–1–2 domain are shown as functions only of ${}^1_fI_a$ and ${}^2_fI_b$. Then, substituting Equation 4.22 into Equation 4.24,

$$\left.\begin{array}{l}\{({}_fZ_{00} + {}_fZ_{01}) + ({}_fZ_{10} + {}_fZ_{11}) + ({}_fZ_{20} + {}_fZ_{21})\}\cdot{}^1_fI_a \\ \quad + \{({}_fZ_{00} - {}_fZ_{01}) + a({}_fZ_{10} - {}_fZ_{11}) + a^2({}_fZ_{20} - {}_fZ_{21})\}\cdot{}^2_fI_b = 6\cdot{}_fE_a \\ \{({}_fZ_{00} - {}_fZ_{01}) + a^2({}_fZ_{10} - {}_fZ_{11}) + a({}_fZ_{20} - {}_fZ_{21})\}\cdot{}^1_fI_a \\ \quad + \{({}_fZ_{00} + {}_fZ_{01}) + ({}_fZ_{10} + {}_fZ_{11}) + ({}_fZ_{20} + {}_fZ_{21})\}\cdot{}^2_fI_b = 6a^2\cdot{}_fE_a\end{array}\right\} \quad (4.25)$$

This set of simultaneous equations is of two dimensions and single order in only the two variables ${}^1_fI_a$, ${}^2_fI_b$, and all other variables have vanished. Equation 4.25 can be solved easily by hand:

$$
\left.\begin{aligned}
&{}^1_fI_a = \frac{A_1 - A_2a^2}{A_1^2 - A_2B_1}\cdot 6_fE_a, \quad {}^2_fI_b = \frac{-B_1 + A_1\cdot a^2}{A_1^2 - A_2B_1}\\
&\text{where}\\
&A_1 = ({}_fZ_{00} + {}_fZ_{01}) + ({}_fZ_{10} + {}_fZ_{11}) + ({}_fZ_{20} + {}_fZ_{21})\\
&A_2 = ({}_fZ_{00} - {}_fZ_{01}) + a({}_fZ_{10} - {}_fZ_{11}) + a^2({}_fZ_{20} - {}_fZ_{21})\\
&B_1 = ({}_fZ_{00} - {}_fZ_{01}) + a^2({}_fZ_{10} - {}_fZ_{11}) + a({}_fZ_{20} - {}_fZ_{21})
\end{aligned}\right\} \quad (4.26a)
$$

Furthermore, ${}_fZ_{10} \fallingdotseq {}_fZ_{20}$ and ${}_fZ_{11} = {}_fZ_{21}$ in Figure 3 of Table 4.1b so that the terms including vector operators a and a^{-1} disappear as follows:

$$
\left.\begin{aligned}
&A_1 \fallingdotseq ({}_fZ_{00} + {}_fZ_{01}) + 2({}_fZ_{10} + {}_fZ_{11})\\
&A_2 \fallingdotseq B_1 \fallingdotseq ({}_fZ_{00} - {}_fZ_{01}) - ({}_fZ_{10} - {}_fZ_{11})
\end{aligned}\right\} \quad (4.26b)
$$

We have found directly the fault phase currents ${}^1_fI_a$ and ${}^2_fI_b$ on the virtual terminals at point f.

On the other hand, the sound phase currents other than ${}^1_fI_a$ and ${}^2_fI_b$ on the virtual terminals at point f are zero. Therefore all the phase currents at the virtual terminals of point f have been found. Quantities at other arbitrary points can be found consequently by additional calculation.

The characteristic of this method is to express all the lane circuit quantities of Figure 3 of Table 4.1b as parameters of the fault phase currents only. The method is generally a very valuable calculation technique to study various complicated fault conditions in double circuit lines. Furthermore, the analogy may be applied to other types of calculation in practical engineering.

4.5.4 Various double circuit faults at single point f

Double circuit faults of other modes at point f can be analysed by method 1 or method 2, regardless of the existence of visual equivalent circuits.

4.6 Simultaneous Double Circuit Faults at Different Points f, F on the Same Line

4.6.1 Circuit condition before fault

If lightning strikes a phase a conductor at point f, for example, it may cause flashover of other phases or other circuits at different point F. These cascade flashover phenomena at different two points are actually simultaneous faults at different points on the basis of a millisecond-order timescale. In other words, simultaneous faults of various modes would occur very often in power system networks. These phenomena have to be investigated from various engineering viewpoints. In particular, the behaviour of directional distance relays has to be carefully examined in order to prevent malfunction.

The analogy of fault analysis in case of a simultaneous fault at two different points is the same as that shown in Figures 1 of Table 4.1. However, we have to imagine virtual terminals at the two different points f, F as shown in Figures 4.6a–c.

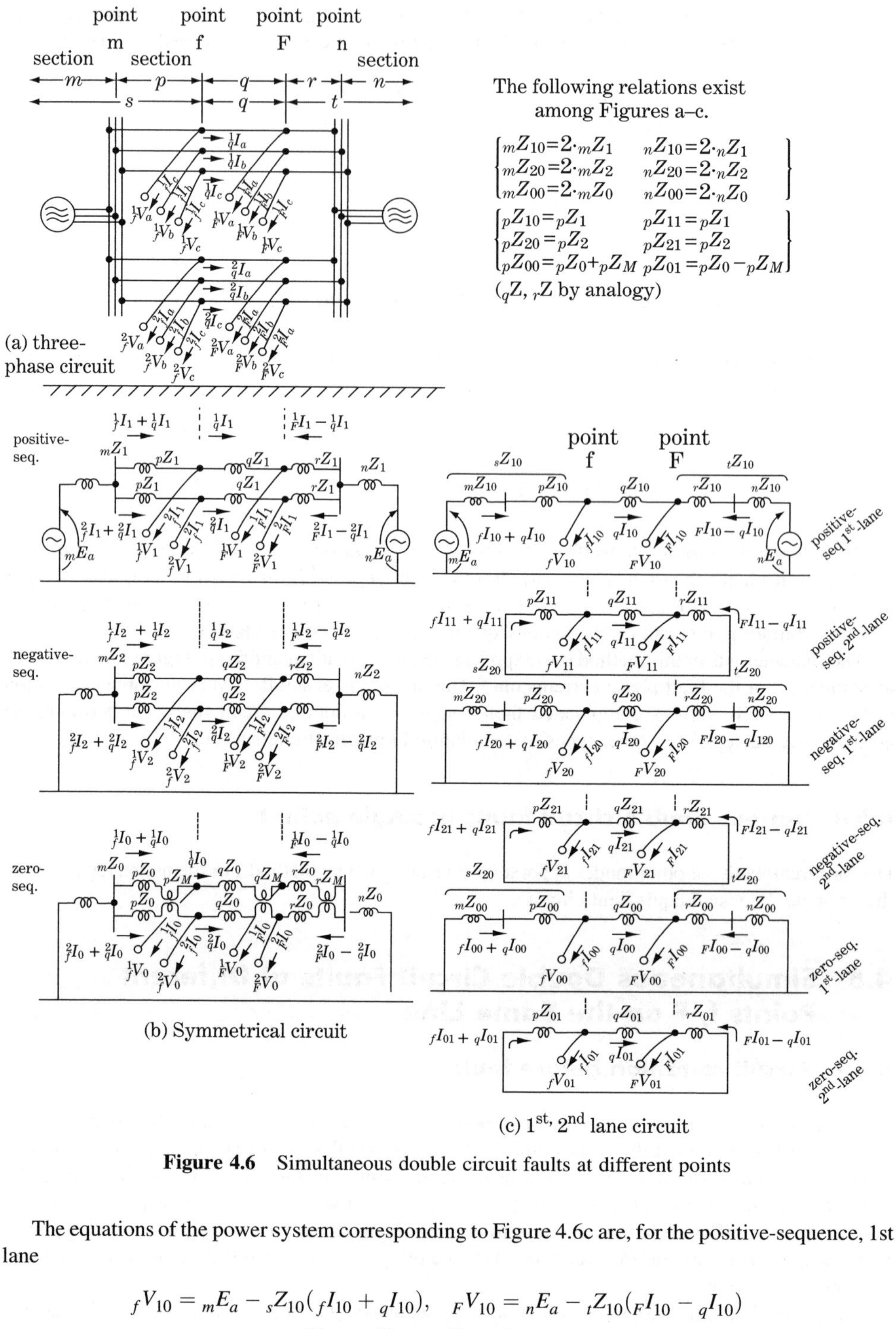

Figure 4.6 Simultaneous double circuit faults at different points

The equations of the power system corresponding to Figure 4.6c are, for the positive-sequence, 1st lane

$$_fV_{10} = {}_mE_a - {}_sZ_{10}({}_fI_{10} + {}_qI_{10}), \quad {}_FV_{10} = {}_nE_a - {}_tZ_{10}({}_FI_{10} - {}_qI_{10})$$

$$_fV_{10} - {}_FV_{10} = {}_qZ_{10} \cdot {}_qI_{10}$$

accordingly,

$$\left.\begin{aligned}
&{}_fV_{10} = {}_mE_a - {}_sZ_{10}({}_fI_{10} + {}_qI_{10})\\
&{}_FV_{10} = {}_nE_a - {}_tZ_{10}({}_FI_{10} - {}_qI_{10})\\
&({}_qZ_{10} + {}_sZ_{10} + {}_tZ_{10}){}_qI_{10} = {}_mE_a - {}_nE_a - {}_sZ_{10}\cdot {}_fI_{10} + {}_tZ_{10}\cdot {}_FI_{10}
\end{aligned}\right\} \textcircled{1}$$

for the positive-sequence, 2nd lane

$$\left.\begin{aligned}
&{}_fV_{11} = -{}_pZ_{11}({}_fI_{11} + {}_qI_{11})\\
&{}_FV_{11} = -{}_rZ_{11}({}_FI_{11} - {}_qI_{11})\\
&({}_qZ_{11} + {}_qZ_{11} + {}_rZ_{11}){}_qI_{11} = -{}_pZ_{11}\cdot {}_fI_{11} + {}_rZ_{11}\cdot {}_FI_{11}
\end{aligned}\right\} \textcircled{2}$$

for the negative-sequence, 1st lane

$$\left.\begin{aligned}
&{}_fV_{20} = -{}_sZ_{20}({}_fI_{20} + {}_qI_{20})\\
&{}_FV_{20} = -{}_tZ_{20}({}_FI_{20} - {}_qI_{20})\\
&({}_qZ_{20} + {}_sZ_{20} + {}_tZ_{20}){}_qI_{20} = -{}_sZ_{20}\cdot {}_fI_{20} + {}_tZ_{20}\cdot {}_FI_{20}
\end{aligned}\right\} \textcircled{3} \qquad (4.27)$$

for the negative-sequence, 2nd lane

$$\left.\begin{aligned}
&{}_fV_{21} = -{}_pZ_{21}({}_fI_{21} + {}_qI_{21})\\
&{}_FV_{21} = -{}_rZ_{21}({}_FI_{21} - {}_qI_{21})\\
&({}_qZ_{21} + {}_pZ_{21} + {}_rZ_{21}){}_qI_{21} = -{}_pZ_{21}\cdot {}_fI_{21} + {}_rZ_{21}\cdot {}_FI_{21}
\end{aligned}\right\} \textcircled{4}$$

for the zero-sequence, 1st lane

$$\left.\begin{aligned}
&{}_fV_{00} = -{}_sZ_{00}({}_fI_{00} + {}_qI_{00})\\
&{}_FV_{00} = -{}_tZ_{00}({}_FI_{00} - {}_qI_{00})\\
&({}_qZ_{00} + {}_sZ_{00} + {}_tZ_{00}){}_qI_{00} = -{}_sZ_{00}\cdot {}_fI_{00} + {}_tZ_{00}\cdot {}_FI_{00}
\end{aligned}\right\} \textcircled{5}$$

and for the zero-sequence, 2nd lane

$$\left.\begin{aligned}
&{}_fV_{01} = -{}_pZ_{01}({}_fI_{01} + {}_qI_{01})\\
&{}_FV_{01} = -{}_rZ_{01}({}_FI_{01} - {}_qI_{01})\\
&({}_qZ_{01} + {}_pZ_{01} + {}_rZ_{01}){}_qI_{01} = -{}_pZ_{01}\cdot {}_fI_{01} + {}_rZ_{01}\cdot {}_FI_{01}
\end{aligned}\right\} \textcircled{6}$$

Equation 4.27 and Figure 4.6(c) are in one-to-one correspondence to each other. Equation (4.27) ①, ${}_qI_{10}$, can be deleted so that voltages ${}_fV_{10}$ and ${}_FV_{10}$ can be written as functions of current variables ${}_fI_{10}$ and ${}_FI_{10}$ and the given source voltages ${}_mE_a$ and ${}_nE_a$.

By a similar treatment, Equation 4.27 can be reformed as follows by the general forms for the equations of a power system.

for the positive-sequence, 1st lane

$$\left.\begin{array}{l} {}_fV_{10} = {}_ff_{10}({}_mE_a, {}_nE_a, {}_fI_{10}, {}_FI_{10}) \\ {}_FV_{10} = {}_Ff_{10}({}_mE_a, {}_nE_a, {}_fI_{10}, {}_FI_{10}) \end{array}\right\}$$

for the positive-sequence, 2nd lane

$${}_fV_{11} = {}_ff_{11}({}_fI_{11}, {}_FI_{11}), \quad {}_FV_{11} = {}_Ff_{11}({}_fI_{11}, {}_FI_{11})$$

for the nagative-sequence, 1st lane

$${}_fV_{20} = {}_ff_{20}({}_fI_{20}, {}_FI_{20}), \quad {}_FV_{20} = {}_Ff_{20}({}_fI_{20}, {}_FI_{20}) \tag{4.28}$$

for the nagative-sequence, 2nd lane

$${}_fV_{21} = {}_ff_{21}({}_fI_{21}, {}_FI_{21}), \quad {}_FV_{21} = {}_Ff_{21}({}_fI_{21}, {}_FI_{21})$$

for the 0-sequence, 1st lane

$${}_fV_{00} = {}_ff_{00}({}_fI_{00}, {}_FI_{00}), \quad {}_FV_{00} = {}_Ff_{00}({}_fI_{00}, {}_FI_{00})$$

for the 0-sequence, 2nd lane

$${}_fV_{01} = {}_ff_{01}({}_fI_{01}, {}_FI_{01}), \quad {}_FV_{01} = {}_Ff_{01}({}_fI_{01}, {}_FI_{01})$$

These are the equations of the power system before the fault.

4.6.2 Circuit 1 phase a line-to-ground fault and circuit 2 phase b line-to-ground fault at different points f, F

Fault analysis of this sort of double fault is very challenging. To understand the logical analogy for the solution of such cases it is essential that readers can find solutions by either computer analysis or manual calculation.

Now, the process of the double fault will be demonstrated using method 2, which was discussed above.

The fault condition in the three-phase domain is,

at point f,

$$\left.\begin{array}{l} {}^1_fI_b = {}^1_fI_c = 0 \\ {}^2_fI_a = {}^2_fI_b = {}^2_fI_c = 0 \\ {}^1_fV_a = 0 \end{array}\right.$$

at point F

$$\left.\begin{array}{l} {}^1_FI_a = {}^1_FI_b = {}^1_FI_c = 0 \\ {}^2_FI_a = {}^2_FI_c = 0 \\ {}^2_FV_b = 0 \end{array}\right\} \tag{4.29}$$

The fault condition in the 1–2–0 domain is, by transforming Equation 4.29,

$$\left.\begin{aligned}
&\text{for point f} && {}^{1}_{f}I_0 = {}^{1}_{f}I_1 = {}^{1}_{f}I_2 = \tfrac{1}{3}\,{}^{1}_{f}I_a \\
& && {}^{2}_{f}I_0 = {}^{2}_{f}I_1 = {}^{2}_{f}I_2 = 0 \\
& && {}^{1}_{f}V_0 + {}^{1}_{f}V_1 + {}^{1}_{f}V_2 = 0 \\
&\text{for point F} && \\
& && {}^{1}_{F}I_0 = {}^{1}_{F}I_1 = {}^{1}_{F}I_2 = 0 \\
& && {}^{2}_{F}I_0 = \tfrac{1}{3}\,{}^{2}_{F}I_b, \quad {}^{2}_{F}I_1 = \frac{1}{3}a \cdot {}^{2}_{F}I_b, \quad {}^{2}_{F}I_2 = \frac{1}{3}a^2 \cdot {}^{2}_{F}I_b \\
& && {}^{2}_{F}V_0 + a^2 \cdot {}^{2}_{F}V_1 + a \cdot {}^{2}_{F}V_2 = 0
\end{aligned}\right\} \tag{4.30}$$

Then

$$ {}_{f}I_{00} = \frac{1}{2}\left({}^{1}_{f}I_0 + {}^{2}_{f}I_0\right) = \frac{1}{6}\,{}^{1}_{f}I_a, \quad {}_{f}I_{01} = \frac{1}{2}\left({}^{1}_{f}I_0 - {}^{2}_{f}I_0\right) = \frac{1}{6}\,{}^{1}_{f}I_a \tag{4.31}$$

In the same way, all the equations of Equation 4.30 can be transformed into the equations of the 1st- and 2nd-lane circuits in the 0–1–2 domain, where the 1st- and 2nd-lane currents are expressed only by the parameters of fault phase currents ${}^{1}_{f}I_a$ and ${}^{2}_{F}I_b$. That is,

$$\left.\begin{aligned}
& {}_{f}I_{00} = {}_{f}I_{01} = {}_{f}I_{10} = {}_{f}I_{11} = {}_{f}I_{20} = {}_{f}I_{21} = \frac{1}{6}\,{}^{1}_{f}I_a && ① \\
& \left.\begin{aligned}
& {}_{F}I_{00} = \frac{1}{6}\,{}^{2}_{F}I_b, \quad {}_{F}I_{01} = \frac{-1}{6}\,{}^{2}_{F}I_b \\
& {}_{F}I_{10} = \frac{1}{6}a \cdot {}^{2}_{F}I_b, \quad {}_{F}I_{11} = \frac{-1}{6}a \cdot {}^{2}_{F}I_b \\
& {}_{F}I_{20} = \frac{1}{6}a^2 \cdot {}^{2}_{F}I_b, \quad {}_{F}I_{21} = \frac{-1}{6}a^2 \cdot {}^{2}_{F}I_b
\end{aligned}\right\} && ② \\
& ({}_{f}V_{00} + {}_{f}V_{01}) + ({}_{f}V_{10} + {}_{f}V_{11}) + ({}_{f}V_{20} + {}_{f}V_{21}) = 0 && ③ \\
& ({}_{F}V_{00} - {}_{F}V_{01}) + a^2({}_{F}V_{10} - {}_{F}V_{11}) + a({}_{F}V_{20} - {}_{F}V_{21}) = 0 && ④
\end{aligned}\right\} \tag{4.32}$$

Now we have an equation showing the fault condition, namely Equation 4.32, and equations showing the system, namely Equation 4.27 or its modified Equation 4.28. Therefore we can solve the problem by combining all these equations. By substituting Equation 4.32 ① ② into Equation 4.27 or its modified Equation 4.28, all the 1st- and 2nd-lane voltages can be expressed as parameters of ${}^{1}_{f}I_a$ and ${}^{2}_{F}I_b$ only. Next, by substituting the six 1st- and 2nd-lane voltages into Equation 4.32 ③ ④, we obtain simultaneous equations of two dimensions in only two variables, ${}^{1}_{f}I_a$ and ${}^{2}_{F}I_b$. Then we can obtain the final solution.

4.6.3 Various double circuit faults at different points

In conclusion of the chapter, double circuits fault at different points of various modes can be solved by utilizing the three-phase and two-phase symmetrical components together. It must be remembered that actual power system analyses, even by large computers, are conducted mostly by utilizing these transformations in order to eliminate mutual inductances of the lines.

5

Per Unit Method and Introduction of Transformer Circuit

The per unit (PU) method (or % method) is a technique for handling any kind of quantity with its particular dimensions as quantities of dimensionless ratio value based on 1.0 pu or 100%. This practice is a very useful approach applied widely in many engineering fields, eliminating the troublesome handling of several different kinds of quantities.

However, in power system engineering, the PU method has various meanings such as a 'technique for describing electrical circuits, and far exceeding the simple meaning of the only convenient method to remove troublesome dimensions'. Many individual structuring members of power systems can be combined together as one circuit (instead of a connection diagram) only by using the PU method. Furthermore, transformers can be handled by PU expressions as equipment in which Kirchhoff's law are applied.

In this chapter we study the fundamental concept of the PU method first, and then study the circuit description of transformers. Finally we try to describe the circuit for a typical power system model containing several lines and various equipment.

5.1 Fundamental Concept of the PU Method

The PU method is quite important in power system engineering, the reasons for which are summarized as follows:

a) Kirchhoff's law is satisfied among currents of transformer primary, secondary and tertiary windings so that transformers can be described as very simple circuits.

b) Generators can also be described as accurate and simple circuits (see Chapter 10).

c) Transmission lines, generators, transformers, loads and other equipment of different types and ratings can all be combined together as one circuit. practically only by applying PU method.

d) Relief from troublesome handling of practical dimensions (V, A, MVA, Ω, Wb, etc.).

For power system engineers, the first three items are the essential reasons and the last item is just a supplementary reason.

Handbook of Power System Engineering Yoshihide Hase
© 2007 John Wiley & Sons, Ltd

5.1.1 PU method of single phase circuit

Let us consider the PU method for a single phase circuit first. The basic equations of voltage, current and apparent power are

$$\left.\begin{array}{l} V\,[\text{volt}] = Z\,[\text{ohm}]\cdot I\,[\text{ampere}] \\ VA\,[\text{volt}\cdot\text{ampere}] = P + jQ\,[\text{volt}\cdot\text{ampere}] = V\,[\text{volt}]\cdot I^*[\text{ampere}] \\ \text{where } V,\ I,\ Z,\ VA\,(\text{or } S) : \text{complex-number quantities},\ I^* : \text{the conjugate of } I \end{array}\right\} \quad (5.1)$$

Now, in order to unitize the V, Z, I, VA quantities, the base quantities by sign of V_{base}, I_{base}, Z_{base}, VA_{base} are introduced. All the base quantities are scalars (real numbers, or a vector of $\angle 0°$) and have to satisfy the equations below:

$$\left.\begin{array}{l} V_{base}\,[\text{volt}] = Z_{base}\,[\text{ohm}]\cdot I_{base}\,[\text{ampere}] \\ VA_{base}\,[\text{volt}\cdot\text{ampere}] = V_{base}\,[\text{volt}]\cdot I_{base}\,[\text{ampere}] \end{array}\right\} \quad (5.2a)$$

or

$$\left.\begin{array}{l} I_{base}\,[\text{ampere}] = \dfrac{VA_{base}\,[\text{volt}\cdot\text{ampere}]}{V_{base}\,[\text{volt}]} \\ Z_{base}\,[\text{ohm}] = \dfrac{V_{base}\,[\text{volt}]}{I_{base}\,[\text{ampere}]} = \dfrac{V_{base}^2}{VA_{base}}[\text{ohm}] \end{array}\right\} \quad (5.2b)$$

We can select any arbitrary value for voltage base V_{base} and capacity base VA_{base}, but the current base I_{base} and impedance base Z_{base} have to be decided as depending on V_{base} and VA_{base} to satisfy Equation 5.2a 5.2b.

Equation 5.1 can be unitized by the base quantities of Equation 5.2a as follows:

$$\left.\begin{array}{l} \dfrac{V}{V_{base}} = \dfrac{Z}{Z_{base}}\cdot\dfrac{I}{I_{base}} \\ \dfrac{VA}{VA_{base}} = \dfrac{P+jQ}{VA_{base}} = \dfrac{P}{VA_{base}} + j\dfrac{Q}{VA_{base}} = \dfrac{V}{V_{base}}\cdot\dfrac{I^*}{I_{base}} \end{array}\right\} \quad (5.3)$$

By using an overbar as the symbol for unitized quantities,

$$\left.\begin{array}{ll} \overline{V} = \overline{Z}\cdot\overline{I} & ① \\ \overline{VA} = \overline{P} + \overline{jQ} = \overline{V}\cdot\overline{I}^* & ② \\ \text{where } \overline{V} = \dfrac{V}{V_{base}},\quad \overline{Z} = \dfrac{Z}{Z_{base}},\quad \overline{I}^* = \dfrac{I^*}{I_{base}},\quad \overline{V} = \dfrac{V}{V_{base}} & ③ \\ \overline{VA} = \dfrac{VA}{VA_{base}},\quad \overline{P} = \dfrac{P}{VA_{base}},\quad \overline{Q} = \dfrac{Q}{VA_{base}} & ④ \end{array}\right\} \quad (5.4)$$

The unitized quantities $\overline{V}, \overline{Z}, \overline{I}^*$, $\overline{VA}$, $\overline{P} + \overline{jQ}$ are non-dimensional complex numbers.

Equation 5.4 is the same as the original Equation 5.1, and the vector phase relations in Equation 5.1 are preserved in Equation 5.4 because all the base quantities are selected as scalars (namely, a vector of $\angle 0°$).

Unitized quantities can obviously be changed into actual values with individual dimensions using the equations below:

$$\left.\begin{array}{lll} V(\text{volt}) = \overline{V}\cdot V_{base}, & Z = \overline{Z}\cdot Z_{base}, & I = \overline{I}\cdot I_{base} \\ VA = \overline{VA}\cdot VA_{base}, & P = \overline{P}\cdot VA_{base}, & Q = \overline{Q}\cdot VA_{base} \end{array}\right\} \quad (5.5)$$

Figure 5.1 summarizes the PU method.

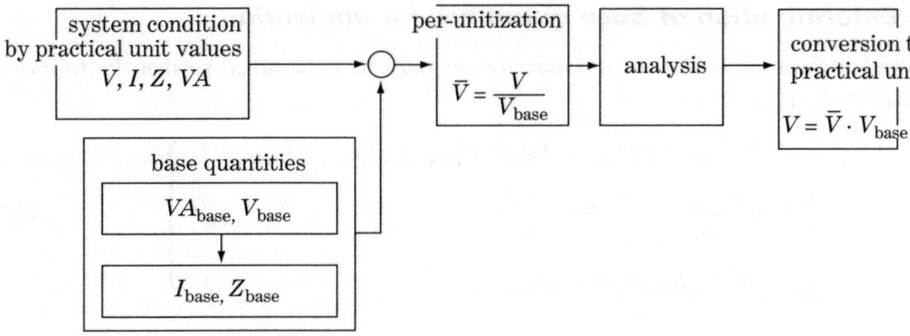

Figure 5.1 The concept of the PU method

5.1.2 Unitization of a single phase three-winding transformer and its equivalent circuit

5.1.2.1 The fundamental equations before unitization

A single phase three-winding transformer can be written as the circuit of Figure 5.2a, at least for power frequency phenomena, where ${}_PN \cdot {}_SN \cdot {}_TN$ are the numbers of turns of the primary (P), secondary (S), tertiary (T) windings, respectively. The transformer excitation current under a no-load condition can usually be neglected (the **excitation impedance** is large enough) except under the situation of core saturation caused by abnormally higher charging voltages. Therefore the relation of voltages and currents in this transformer may be described by the following equation in which **leakage impedances** of only three windings are taken into consideration:

$$\left.\begin{array}{ll} \begin{bmatrix} {}_PV \\ {}_SV \\ {}_TV \end{bmatrix} = \begin{bmatrix} Z_{PP} & Z_{PS} & Z_{PT} \\ Z_{SP} & Z_{SS} & Z_{ST} \\ Z_{TP} & Z_{TS} & Z_{PN} \end{bmatrix} \cdot \begin{bmatrix} {}_PI \\ {}_SI \\ {}_TI \end{bmatrix} & ① \\ {}_PI \cdot {}_PN + {}_SI \cdot {}_SN + {}_TI \cdot {}_TN = 0 & ② \\ Z_{PS} = Z_{SP}, \quad Z_{PT} = Z_{TP}, \quad Z_{ST} = Z_{TS} & ③ \end{array}\right\} \quad (5.6)$$

where Z_{PP}, Z_{SS}, Z_{TT} are the self-impedances of the primary (P), secondary (S) and tertiary (T) windings and Z_{PS}, Z_{PT}, Z_{ST} are the mutual impedances between the three windings.

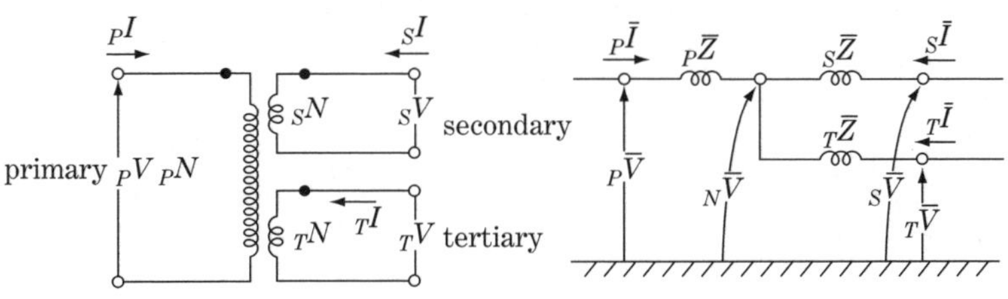

(a)

(b) The equivalent circuit by Pu basis

$_P\bar{Z} = ({}_{P\text{-}S}\bar{Z} + {}_{P\text{-}T}\bar{Z} - {}_{S\text{-}T}\bar{Z})/2$

$_S\bar{Z} = ({}_{P\text{-}S}\bar{Z} + {}_{S\text{-}T}\bar{Z} - {}_{P\text{-}T}\bar{Z})/2$

$_T\bar{Z} = ({}_{P\text{-}T}\bar{Z} + {}_{S\text{-}T}\bar{Z} - {}_{P\text{-}S}\bar{Z})/2$

${}_{P\text{-}S}\bar{Z}$, ${}_{P\text{-}T}\bar{Z}$, ${}_{S\text{-}T}\bar{Z}$, are given on the name-plate

Figure 5.2 Single phase three-winding transformer

5.1.2.2 Determination of base quantities for unitization

Each base quantity for unitization of this transformer must be determined so that the following equations are satisfied:

$$\left.\begin{aligned}
&VA_{\text{base}} = {}_PV_{\text{base}} \cdot {}_PI_{\text{base}} = {}_SV_{\text{base}} \cdot {}_SI_{\text{base}} = {}_TV_{\text{base}} \cdot {}_TI_{\text{base}} && ① \\
&\frac{{}_PV_{\text{base}}}{{}_PN} = \frac{{}_SV_{\text{base}}}{{}_SN} = \frac{{}_TV_{\text{base}}}{{}_TN} && ② \\
&{}_PI_{\text{base}} \cdot {}_PN = {}_SI_{\text{base}} \cdot {}_SN = {}_TI_{\text{base}} \cdot {}_TN && ③
\end{aligned}\right\} \tag{5.7}$$

In other words:

*1 Capacity bases (VA_{base}) of the primary (P), secondary (S) and tertiary (T) windings are selected to have equal value (Equation ①).

*2 Voltage bases of the primary (P), secondary (S) and tertiary (T) windings are proportional to the turns ratio (transformation ratio) of three windings (Equation ②).

*3 Ampere bases of the primary (P), secondary (S) and tertiary (T) windings are dependently determined as values of 'capacity base (*1) divided by each voltage base (*2)'. That is, the ampere-turn bases of the primary (P), secondary (S) and tertiary (T) windings have the same value (Equation ③).

5.1.2.3 Unitization of the original equation

Let us unitize the original Equation 5.6 using the base quantities of Equation 5.7 (recall that unitized quantities are indicated by an overbar, e.g. ${}_PV \rightarrow {}_P\overline{V}$):

$$\left.\begin{aligned}
&{}_PV = {}_P\overline{V} \cdot {}_PV_{\text{base}}, \quad {}_SV = {}_S\overline{V} \cdot {}_SV_{\text{base}}, \quad {}_TV = {}_T\overline{V} \cdot {}_TV_{\text{base}} \\
&{}_PI = {}_P\overline{I} \cdot {}_PI_{\text{base}} = {}_P\overline{I} \cdot \frac{VA_{\text{base}}}{{}_PV_{\text{base}}}, \quad {}_SI = {}_S\overline{I} \cdot {}_SI_{\text{base}} = {}_S\overline{I} \cdot \frac{VA_{\text{base}}}{{}_SV_{\text{base}}} \\
&{}_TI = {}_T\overline{I} \cdot {}_TI_{\text{base}} = {}_T\overline{I} \cdot \frac{VA_{\text{base}}}{{}_TV_{\text{base}}}
\end{aligned}\right\} \tag{5.8}$$

The equation for ${}_PV$ from Equation 5.6 ① can be unitized as shown below:

$$\left.\begin{aligned}
{}_PV &= {}_P\overline{V} \cdot {}_PV_{\text{base}} = Z_{PP} \cdot {}_P\overline{I} \cdot {}_PI_{\text{base}} + Z_{PS} \cdot {}_S\overline{I} \cdot {}_SI_{\text{base}} + Z_{PT} \cdot {}_T\overline{I} \cdot {}_TI_{\text{base}} \\
&= Z_{PP} \cdot \frac{VA_{\text{base}}}{{}_PV_{\text{base}}} \cdot {}_P\overline{I} + Z_{PS} \cdot \frac{VA_{\text{base}}}{{}_SV_{\text{base}}} \cdot {}_S\overline{I} + Z_{PT} \cdot \frac{VA_{\text{base}}}{{}_TV_{\text{base}}} \cdot {}_T\overline{I} \\
\therefore \ {}_P\overline{V} &= \left(Z_{PP} \cdot \frac{VA_{\text{base}}}{{}_PV^2_{\text{base}}}\right) \cdot {}_P\overline{I} + \left(Z_{PS} \cdot \frac{VA_{\text{base}}}{{}_PV_{\text{base}} \cdot {}_SV_{\text{base}}}\right) \cdot {}_S\overline{I} + \left(Z_{PT} \cdot \frac{VA_{\text{base}}}{{}_PV_{\text{base}} \cdot {}_TV_{\text{base}}}\right) \cdot {}_T\overline{I} \\
&\equiv \overline{Z}_{PP} \cdot {}_P\overline{I} + \overline{Z}_{PS} \cdot {}_S\overline{I} + \overline{Z}_{PT} \cdot {}_T\overline{I}
\end{aligned}\right\} \tag{5.9a}$$

${}_S\overline{V}, {}_T\overline{V}$ can be unitized analogously.

Next, Equation 5.6 ② can be unitized by Equation 5.7 ③ as follows:

$$\left.\begin{aligned}
&\frac{{}_PI \cdot {}_PN}{{}_PI_{\text{base}} \cdot {}_PN} + \frac{{}_SI \cdot {}_SN}{{}_SI_{\text{base}} \cdot {}_SN} + \frac{{}_TI \cdot {}_TN}{{}_TI_{\text{base}} \cdot {}_TN} = 0 \\
&\therefore \quad {}_P\overline{I} + {}_S\overline{I} + {}_T\overline{I} = 0
\end{aligned}\right\} \tag{5.9b}$$

Accordingly,

$$
\left.
\begin{aligned}
&\begin{bmatrix} {}_P\overline{V} \\ {}_S\overline{V} \\ {}_T\overline{V} \end{bmatrix}
=
\begin{bmatrix}
Z_{PP}\cdot\dfrac{VA_{\text{base}}}{{}_PV^2_{\text{base}}} & Z_{PS}\cdot\dfrac{VA_{\text{base}}}{{}_PV_{\text{base}}\cdot {}_SV_{\text{base}}} & Z_{PT}\cdot\dfrac{VA_{\text{base}}}{{}_PV_{\text{base}}\cdot {}_TV_{\text{base}}} \\
Z_{SP}\cdot\dfrac{VA_{\text{base}}}{{}_PV_{\text{base}}\cdot {}_SV_{\text{base}}} & Z_{SS}\cdot\dfrac{VA_{\text{base}}}{{}_SV^2_{\text{base}}} & Z_{ST}\cdot\dfrac{VA_{\text{base}}}{{}_SV_{\text{base}}\cdot {}_TV_{\text{base}}} \\
Z_{TP}\cdot\dfrac{VA_{\text{base}}}{{}_PZ_{\text{base}}\cdot {}_TZ_{\text{base}}} & Z_{TS}\cdot\dfrac{VA_{\text{base}}}{{}_SV_{\text{base}}\cdot {}_TV_{\text{base}}} & Z_{TT}\cdot\dfrac{VA_{\text{base}}}{{}_TV^2_{\text{base}}}
\end{bmatrix}
\cdot
\begin{bmatrix} {}_P\overline{I} \\ {}_S\overline{I} \\ {}_T\overline{I} \end{bmatrix} \\
&\equiv
\begin{bmatrix}
\overline{Z}_{PP} & \overline{Z}_{PS} & \overline{Z}_{PT} \\
\overline{Z}_{SP} & \overline{Z}_{SS} & \overline{Z}_{ST} \\
\overline{Z}_{TP} & \overline{Z}_{TS} & \overline{Z}_{PP}
\end{bmatrix}
\cdot
\begin{bmatrix} {}_P\overline{I} \\ {}_S\overline{I} \\ {}_T\overline{I} \end{bmatrix} \qquad ① \\
&{}_P\overline{I} + {}_S\overline{I} + {}_T\overline{I} = 0 \qquad ② \\
&\text{where } \overline{Z}_{PP} = \frac{VA_{\text{base}}}{{}_PV^2_{\text{base}}} \\
&\overline{Z}_{PS} \equiv Z_{PS}\cdot\frac{VA_{\text{base}}}{{}_PV_{\text{base}}\cdot {}_SV_{\text{base}}} = Z_{SP}\cdot\frac{VA_{\text{base}}}{{}_PV_{\text{base}}\cdot {}_SV_{\text{base}}} \equiv \overline{Z}_{SP} \text{ etc.}
\end{aligned}
\right\}
\tag{5.10}
$$

In conclusion, Equation 5.10 is the unitized equation of Equation 5.6 by the base quantities of Equation 5.8. In Equation 5.10, the summation of the unitized vector currents of the primary (P), secondary (S) and tertiary (T) windings is zero. In other words, the unitized transformer circuit equations are as if able to satisfy Kirchhoff's law.

5.1.2.4 Introduction of unitized equivalent circuit

We have introduced Equation 5.10 as the unitized fundamental equations of a transformer in which the vector sum of the currents is zero. Therefore, it would be useful if the equation could be written as the one-to-one corresponding equivalent circuit of Figure 5.2b. We can indeed do that. It is clear that Figure 5.2b satisfies Equation 5.10 ②. Then, if we define the impedances ${}_P\overline{Z}, {}_S\overline{Z}, {}_T\overline{Z}$ in the figure so that the circuits satisfies Equation 5.10 ①, the figure is the perfect equivalent circuit of the transformer which satisfies Equation 5.10. Now let us find such a condition below:

(i) Under the condition ${}_T\overline{I} = 0$ (with the tertiary terminal opened), Figure 5.2b and Equation 5.10 have to coincide (with the tertiary terminal opened). Putting ${}_T\overline{I} = 0$ in Equation 5.10, we have

$$
\begin{aligned}
{}_P\overline{V} - {}_S\overline{V} &= (\overline{Z}_{PP}\cdot {}_P\overline{I} + \overline{Z}_{PS}\cdot {}_S\overline{I}) - (\overline{Z}_{SP}\cdot {}_P\overline{I} + \overline{Z}_{SS}\cdot {}_S\overline{I}) \\
&= (\overline{Z}_{PP} + \overline{Z}_{SS} - 2\overline{Z}_{PS})\cdot {}_P\overline{I}
\end{aligned}
$$

On the other hand, putting ${}_T\overline{I} = 0$ in Figure 5.2b, we have

$$
\begin{aligned}
&{}_P\overline{V} - {}_S\overline{V} = ({}_P\overline{Z} + {}_S\overline{Z})\cdot {}_P\overline{I} = {}_{P-S}\overline{Z}\cdot {}_P\overline{I}, \qquad {}_P\overline{Z} + {}_S\overline{Z} = {}_{P-S}\overline{Z} \\
&{}_P\overline{I} + {}_S\overline{I} = 0
\end{aligned}
$$

The following equation has to be satisfied in order for the two equations above to coincide with each other under the tertiary terminal open condition:

$$
{}_{P-S}\overline{Z} = {}_P\overline{Z} + {}_S\overline{Z} = \overline{Z}_{PP} + \overline{Z}_{SS} - 2\overline{Z}_{PS}
$$

In the same way, the following conditions have to be satisfied.

(ii) The required condition in order that Figure 5.2b and Equation 5.10 coincide under the secondary terminal open condition ($_S\overline{I} = 0$) is

$$_{P-T}\overline{Z} = {_P\overline{Z}} + {_T\overline{Z}} = \overline{Z}_{PP} + \overline{Z}_{TT} - 2Z_{PT}$$

(iii) The required condition in order that Figure 5.2b and Equation 5.10 coincide with each other under the primary terminal open condition ($_PI = 0$) is

$$_{S-T}\overline{Z} = {_S\overline{Z}} + {_T\overline{Z}} = \overline{Z}_{SS} + Z_{TT} - 2\overline{Z}_{ST}$$

Summarizing, Figure 5.2b can be the precise equivalent circuit of the transformer by satisfying the above three equations for the impedances.

Accordingly, for the transformer equations,

$$\left.\begin{array}{ll}\text{tertiary terminal open: } {_P\overline{V}} - {_S\overline{V}} = ({_P\overline{Z}} + {_S\overline{Z}}) \cdot {_P\overline{I}} = {_{P-S}\overline{Z}} \cdot {_P\overline{I}}, & {_P\overline{I}} + {_S\overline{I}} = 0 \\ \text{secondary terminal open: } {_P\overline{V}} - {_T\overline{V}} = ({_P\overline{Z}} + {_T\overline{Z}}) \cdot {_P\overline{I}} = {_{P-T}\overline{Z}} \cdot {_P\overline{I}}, & {_P\overline{I}} + {_T\overline{I}} = 0 \\ \text{terminal open primary: } {_S\overline{V}} - {_T\overline{V}} = ({_S\overline{Z}} + {_T\overline{Z}}) \cdot {_S\overline{I}} = {_{S-T}\overline{Z}} \cdot {_S\overline{I}}, & {_S\overline{I}} + {_T\overline{I}} = 0\end{array}\right. \quad (5.11)$$

where the definitions of impedances are

$$\left.\begin{array}{l}\text{leakage impedance between P and S under the condition } {_T\overline{I}} = 0: \\ {_{P-S}\overline{Z}} = {_P\overline{Z}} + {_S\overline{Z}} = \overline{Z}_{PP} + \overline{Z}_{SS} - 2\overline{Z}_{PS} \\ \text{leakage impedance between P and T under the condition } {_S\overline{I}} = 0: \\ {_{P-S}\overline{Z}} = {_P\overline{Z}} + {_T\overline{Z}} = \overline{Z}_{PP} + \overline{Z}_{TT} - 2\overline{Z}_{PT} \\ \text{leakage impedance between S and T under the condition } {_P\overline{I}} = 0: \\ {_{S-T}\overline{Z}} = {_S\overline{Z}} + {_T\overline{Z}} = \overline{Z}_{SS} + \overline{Z}_{TT} - 2\overline{Z}_{ST}\end{array}\right. \quad (5.12a)$$

or, using the definition of $_P\overline{Z}, {_S\overline{Z}}, {_T\overline{Z}}$ in the equivalent circuit in Figure 5.2b,

$$\left.\begin{array}{l}{_P\overline{Z}} = \dfrac{{_{P-S}\overline{Z}} + {_{P-T}\overline{Z}} - {_{S-T}\overline{Z}}}{2} = \overline{Z}_{PP} + \overline{Z}_{ST} - \overline{Z}_{PS} - \overline{Z}_{PT} \\ {_S\overline{Z}} = \dfrac{{_{P-S}\overline{Z}} + {_{S-T}\overline{Z}} - {_{P-T}\overline{Z}}}{2} = \overline{Z}_{SS} + \overline{Z}_{PT} - \overline{Z}_{PS} - \overline{Z}_{ST} \\ {_T\overline{Z}} = \dfrac{{_{P-T}\overline{Z}} + {_{S-T}\overline{Z}} - {_{S-T}\overline{Z}}}{2} = \overline{Z}_{TT} + \overline{Z}_{PS} - \overline{Z}_{PT} - \overline{Z}_{ST}\end{array}\right\} \quad (5.12b)$$

Figure 5.2b with the impedances $_P\overline{Z}, {_S\overline{Z}}, {_T\overline{Z}}$ becomes the unitized equivalent circuit of the transformer by defining the impedances as in Equation 5.12b. The equivalent circuit of course satisfies Kirchhoff's law by unitization.

The impedances $\overline{Z}_{PP}$, $\overline{Z}_{PS}$, etc., are the self- and mutual impedances (actually reactances) so that the physical concept can be imagined from the winding structures, and the values can be estimated by engineers in their transformer designs.

The impedance $_{P-S}\overline{Z}$ can be measured as the leakage reactance between the primary and secondary terminal under the tertiary winding open condition, and $_{P-T}\overline{Z}, {_{S-T}\overline{Z}}$ can also be measured similarly.

On the other hand, $_P\overline{Z}, {_S\overline{Z}}, {_T\overline{Z}}$ are the impedances defined only by Equations 5.12a and 5.12b in order to obtain the equivalent circuit of Figure 5.2b, and we cannot find any other physical meaning for that. However, transformers can be treated as kinds of black boxes by utilizing the above defined equivalent circuits at least for power frequency phenomena of the power system networks.

Incidentally, the resistances of the transformer windings are negligibly small so that the above-described Z can be replaced by jX or $j\omega L$. $Z_{PP} = jX_{PP}$, $Z_{PS} = jX_{PS}$ as well as $j_{P-S}\overline{X}$, $j_{P-T}\overline{X}$, $j_{S-T}\overline{X}$, etc., and have positive values (namely reactances). However, one of $j_P\overline{X}$, $j_S\overline{X}$, $j_T\overline{X}$ could even have negative values, just like a series capacitive element in the equivalent circuit.

In regard to practical engineering, the **percentage impedance drop voltages (%IZ)** of individual transformers are indicated on their name-plates, and are actually the percentage expression of leakage reactances ${}_{P-S}\overline{X}, {}_{P-T}\overline{X}, {}_{S-T}\overline{X}$. Accordingly, utilizing these values, ${}_{P}\overline{Z}, {}_{S}\overline{Z}, {}_{T}\overline{Z}$ can be derived from Equation 5.12b. In practical engineering, the percentage value ${}_{P-S}\overline{X}$ is usually given by the MVA base of the primary winding side, while ${}_{P-S}\overline{X}, {}_{S-T}\overline{X}$ may be given by the MVA base of the tertiary winding side on a name-plate, so that the base value conversion is required to derive the equivalent circuit. This matter will be discussed in Sections 5.4 and 5.5.

The treatment for a two-winding transformer without tertiary winding can be done only by omitting ${}_{T}\overline{Z}$ in the equivalent circuit.

5.2 PU Method for Three-phase Circuits

Now the PU method for three-phase circuits needs to be introduced, followed by the unitized equations and equivalent circuit of three-phase transformers and other power system members.

5.2.1 Base quantities by PU method for three-phase circuits

In regard to the PU method for three-phase circuits, **the line-to-line (l–l) base quantities** and **line-to-ground (l–g) base quantities** are defined and both of them have to be strictly distinguished as the premise of three-phase circuit analysis for any investigation purpose. These base quantities are defined as follows:

$$\left.\begin{aligned} VA_{3\phi\text{base}} &= 3\cdot VA_{1\phi\text{base}} = 3\cdot V_{l-g\text{base}}\cdot I_{l-g\text{base}} \\ &= 3\cdot V_{l-l\text{base}}\cdot I_{l-l\text{base}} = \sqrt{3}\cdot V_{l-l\text{base}}\cdot I_{l-g\text{base}} & \text{①} \\ V_{l-l\text{base}} &= \sqrt{3}\cdot V_{l-g\text{base}} & \text{②} \\ \sqrt{3}\cdot I_{l-l\text{base}} &= I_{l-g\text{base}} & \text{③} \end{aligned}\right\} \tag{5.13a}$$

Bases of capacity (VA or MVA) and voltage (V or kV) are defined first, and then bases for currents [A], impedances Z [ohm], admittances [mho], etc., are dependently defined as follows:

$$\left.\begin{aligned} I_{l-g\text{base}} &= \frac{VA_{1\phi\text{base}}}{V_{l-g\text{base}}} = \frac{VA_{3\phi\text{base}}}{\sqrt{3}\cdot V_{l-l\text{base}}} \\ &= \frac{kVA_{3\phi\text{base}}}{\sqrt{3}\cdot KV_{l-l\text{base}}} = \frac{MVA_{3\phi\text{base}}}{\sqrt{3}\cdot KV_{l-l\text{base}}}\times 10^3 & \text{④} \\ Z_{l-g\text{base}} &= \frac{V_{l-g\text{base}}}{I_{l-g\text{base}}} = \frac{(V_{l-l\text{base}})^2}{VA_{3\phi\text{base}}} = \frac{(kV_{l-l\text{base}})^2}{kVA_{3\phi\text{base}}}\times 10^3 \\ &= \frac{(kV_{l-l\text{base}})^2}{MVA_{3\phi\text{base}}} & \text{⑤} \\ Y_{l-g\text{base}} &= \frac{1}{Z_{l-g\text{base}}} = \frac{MVA_{3\phi\text{base}}}{(kV_{l-l\text{base}})^2} & \text{⑥} \\ Z_{l-l\text{base}} &= \frac{V_{l-l\text{base}}}{I_{l-l\text{base}}} & \text{⑦} \\ Y_{l-l\text{base}} &= \frac{1}{Z_{l-l\text{base}}} = \frac{I_{l-l\text{base}}}{V_{l-l\text{base}}} & \text{⑧} \end{aligned}\right\} \tag{5.13b}$$

The values of all the unitized quantities based on the l–l bases are written as variables with the suffix $l-l$, and those based on the l–g bases with the suffix symbol $l-g$ as the description rule.

5.2.2 Unitization of three-phase circuit equations

Let us try to unitize Equation 2.26 and Figure 2.10 for the generator in Chapter 2 as a typical example. The generator's voltage equation from Equation 2.26 is

$$\left.\begin{array}{l}
\underbrace{\begin{bmatrix} E_a \\ E_b = a^2E_a \\ E_c = aE_a \end{bmatrix}}_{\boldsymbol{E}_{abc}} - \underbrace{\begin{bmatrix} V_a \\ V_b \\ V_c \end{bmatrix}}_{-\boldsymbol{V}_{abc}} = \underbrace{\begin{bmatrix} Z_s & Z_m & Z_m \\ Z_m & Z_s & Z_m \\ Z_m & Z_m & Z_s \end{bmatrix}}_{\boldsymbol{Z}_{abc}} \cdot \underbrace{\begin{bmatrix} I_a \\ I_b \\ I_c \end{bmatrix}}_{\cdot \boldsymbol{I}_{abc}} - \underbrace{\begin{bmatrix} V_n \\ V_n \\ V_n \end{bmatrix}}_{\boldsymbol{V}n} \\
V_n = -Z_n(I_a + I_b + I_c) = -Z_n(3I_0) = -3Z_n \cdot I_0
\end{array}\right\} \quad ①$$

the voltage base quantity equation

$$V_{l\text{-gbase}} = Z_{l\text{-gbase}} \cdot I_{l\text{-gbase}} \quad ②$$

the unitized generator equation

$$\left.\begin{array}{l}
\begin{bmatrix} \overline{E}_a \\ \overline{E}_b = a^2 \cdot \overline{E}_a \\ \overline{E}_c = a \cdot \overline{E}_a \end{bmatrix} = \begin{bmatrix} \overline{Z}_s & \overline{Z}_m & \overline{Z}_m \\ \overline{Z}_m & \overline{Z}_s & \overline{Z}_m \\ \overline{Z}_m & \overline{Z}_m & \overline{Z}_s \end{bmatrix} \cdot \begin{bmatrix} \overline{I}_a \\ \overline{I}_b \\ \overline{I}_c \end{bmatrix} - \begin{bmatrix} \overline{V}_n \\ \overline{V}_n \\ \overline{V}_n \end{bmatrix} \\
\overline{V}_n = -\overline{Z}_n(\overline{I}_a + \overline{I}_b + \overline{I}_c) = -\overline{Z}_n \cdot (3\overline{I}_0) = -3\overline{Z}_n \cdot \overline{I}_0 \\
\overline{E}_a = \dfrac{E_a}{V_{l\text{-gbase}}},\ \overline{E}_b = \dfrac{E_b}{V_{l\text{-gbase}}},\ \overline{E}_c = \dfrac{E_c}{V_{l\text{-gbase}}},\ \overline{V}_n = \dfrac{V_n}{V_{l\text{-gbase}}} \\
\overline{Z}_s = \dfrac{Z_s}{Z_{l\text{-gbase}}},\ \overline{Z}_m = \dfrac{Z_m}{Z_{l\text{-gbase}}},\ \overline{Z}_n = \dfrac{Z_n}{V_{l\text{-gbase}}} \\
\overline{I}_a = \dfrac{I_a}{I_{l\text{-gbase}}},\ \overline{I}_b = \dfrac{I_b}{I_{l\text{-gbase}}},\ \overline{I}_c = \dfrac{I_c}{I_{l\text{-gbase}}}
\end{array}\right\} \quad ③ \qquad (5.14)$$

The unitized generator Equation ③ is derived by dividing all the terms of Equation ① by the above base quantities ②. The unitized equation has the same form as that before unitization.

Equation 5.14 can obviously be transformed into the following equation as the one in the symmetrical coordinate domain:

$$\left.\begin{array}{l}
\begin{bmatrix} 0 \\ \overline{E}_a \\ 0 \end{bmatrix} - \begin{bmatrix} \overline{V}_0 \\ \overline{V}_1 \\ \overline{V}_2 \end{bmatrix} = \begin{bmatrix} \overline{Z}_0 & 0 & 0 \\ 0 & \overline{Z}_1 & 0 \\ 0 & 0 & \overline{Z}_2 \end{bmatrix} \cdot \begin{bmatrix} \overline{I}_0 \\ \overline{I}_1 \\ \overline{I}_2 \end{bmatrix} + \begin{bmatrix} 3\overline{Z}_n \cdot \overline{I}_0 \\ 0 \\ 0 \end{bmatrix} \\
\text{or} \quad \left.\begin{array}{r} -\overline{V}_0 = \overline{Z}_0 \cdot \overline{I}_0 + 3\overline{Z}_n \cdot \overline{I}_0 \\ \overline{E}_a - \overline{V}_1 = \overline{Z}_1 \cdot \overline{I}_1 \\ -\overline{V}_2 = \overline{Z}_2 \cdot \overline{I}_2 \end{array}\right\}
\end{array}\right\} \qquad (5.15)$$

Equation 5.15 is also of the same form as Equation 2.27b.

As demonstrated in the above example, the base quantities of the PU method for three-phase circuits are defined by Equations 5.13a and 5.13b and the unitized equations can be written in the same form as that before unitization. In other words, the forms of equations and the equivalent circuits of the usual three-phase circuits (generators, transmission line loads, etc.) are preserved unchanged by unitization.

5.3 Three-phase Three-winding Transformer, its Symmetrical Components Equations and the Equivalent Circuit

5.3.1 人 – 人 – Δ-connected three-phase transformer

Figure 5.3b shows a typical three-phase three-winding transformer with 人 – 人 – Δ-connected windings, whose connection diagram with the terminal code names is printed on the name-plate as shown in Figure 5.3a. This connection is called the 'tertiary 30° lagging connection', because the phase angle of the low-tension bushing terminal a is 30° lagging in comparison with the bushing U and u terminals.

The code names of all the bushing terminals have been changed in Figure 5.3b because the special names of the terminals are used only for analytical purposes, as is shown below:

primary (U, V, W → R, S, T), secondary (u, v, w → r, s, t), tertiary (a, b, c → b, c, a)

The tertiary terminal names a, b, c are intentionally changed by a 120° rotation, so the vector directions of newly named a, b, c terminals (the original c, a, b terminals, respectively) are rectangular to the phases R, S, T and r, s, t, respectively. Moreover, the quantities inside each tertiary winding (with suffix Δ) and the quantities outside each tertiary bushing (with suffix T) have to be strictly distinguished from each other.

5.3.1.1 *The fundamental equations before unitization*

There are three of the single phase three-winding transformers with the same ratings, whose winding connection is written in Figure 5.2. These three single phase transformers can be composed as one bank of three-phases transformer as is shown in Figure 5.3b by simply connecting the bushing terminals. Accordingly, Equation 5.16 is introduced as the fundamental equation of the three-phase transformer in Figure 5.3b:

$$
\begin{bmatrix} {}_PV_a \\ {}_SV_a \\ {}_\Delta V_a \\ {}_PV_b \\ {}_SV_b \\ {}_\Delta V_b \\ {}_PV_c \\ {}_SV_c \\ {}_\Delta V_c \end{bmatrix}
-
\begin{bmatrix} {}_PV_n \\ {}_SV_n \\ 0 \\ {}_PV_n \\ {}_SV_n \\ 0 \\ {}_PV_n \\ {}_SV_n \\ 0 \end{bmatrix}
=
\begin{bmatrix}
Z_{PP} & Z_{PS} & Z_{P\Delta} & 0 & 0 & 0 & 0 & 0 & 0 \\
Z_{SP} & Z_{SS} & Z_{S\Delta} & 0 & 0 & 0 & 0 & 0 & 0 \\
Z_{\Delta P} & Z_{\Delta S} & Z_{\Delta\Delta} & 0 & 0 & 0 & 0 & 0 & 0 \\
0 & 0 & 0 & Z_{PP} & Z_{PS} & Z_{P\Delta} & 0 & 0 & 0 \\
0 & 0 & 0 & Z_{SP} & Z_{SS} & Z_{S\Delta} & 0 & 0 & 0 \\
0 & 0 & 0 & Z_{\Delta P} & Z_{\Delta S} & Z_{\Delta\Delta} & 0 & 0 & 0 \\
0 & 0 & 0 & 0 & 0 & 0 & Z_{PP} & Z_{PS} & Z_{P\Delta} \\
0 & 0 & 0 & 0 & 0 & 0 & Z_{SP} & Z_{SS} & Z_{S\Delta} \\
0 & 0 & 0 & 0 & 0 & 0 & Z_{\Delta P} & Z_{\Delta S} & Z_{\Delta\Delta}
\end{bmatrix}
\cdot
\begin{bmatrix} {}_PI_a \\ {}_SI_a \\ {}_\Delta I_a \\ {}_PI_b \\ {}_SI_b \\ {}_\Delta I_b \\ {}_PI_c \\ {}_SI_c \\ {}_\Delta I_c \end{bmatrix}
\quad ①
$$

where $Z_{PS} = V_{SP}$, $Z_{P\Delta} = Z_{\Delta P}$

$$ {}_PV_n = {}_PZ_n \cdot {}_PI_n = {}_PZ_n({}_PI_a + {}_PI_b + {}_PI_c) = {}_PZ_n \cdot 3{}_PI_0 \quad ② $$

$$ {}_SV_n = {}_SZ_n \cdot {}_SI_n = {}_SZ_n({}_SI_a + {}_SI_b + {}_SI_c) = {}_SZ_n \cdot 3{}_SI_0 \quad ③ $$

$$ \begin{bmatrix} {}_TI_a \\ {}_TI_b \\ {}_TI_c \end{bmatrix} = \begin{bmatrix} {}_\Delta I_c \\ {}_\Delta I_a \\ {}_\Delta I_b \end{bmatrix} - \begin{bmatrix} {}_\Delta I_b \\ {}_\Delta I_c \\ {}_\Delta I_a \end{bmatrix} \quad ④ $$

$$ \begin{bmatrix} {}_\Delta V_a \\ {}_\Delta V_b \\ {}_\Delta V_c \end{bmatrix} = \begin{bmatrix} {}_TV_b \\ {}_TV_c \\ {}_TV_a \end{bmatrix} - \begin{bmatrix} {}_TV_c \\ {}_TV_a \\ {}_TV_b \end{bmatrix} \quad ⑤ $$

$$ \left. \begin{array}{l} {}_PI_a \cdot {}_PN + {}_SI_a \cdot {}_SN + {}_\Delta I_a \cdot {}_\Delta N = 0 \\ {}_PI_b \cdot {}_PN + {}_SI_b \cdot {}_SN + {}_\Delta I_b \cdot {}_\Delta N = 0 \\ {}_PI_c \cdot {}_PN + {}_SI_c \cdot {}_SN + {}_\Delta I_c \cdot {}_\Delta N = 0 \end{array} \right\} \quad ⑥ $$

(5.16)

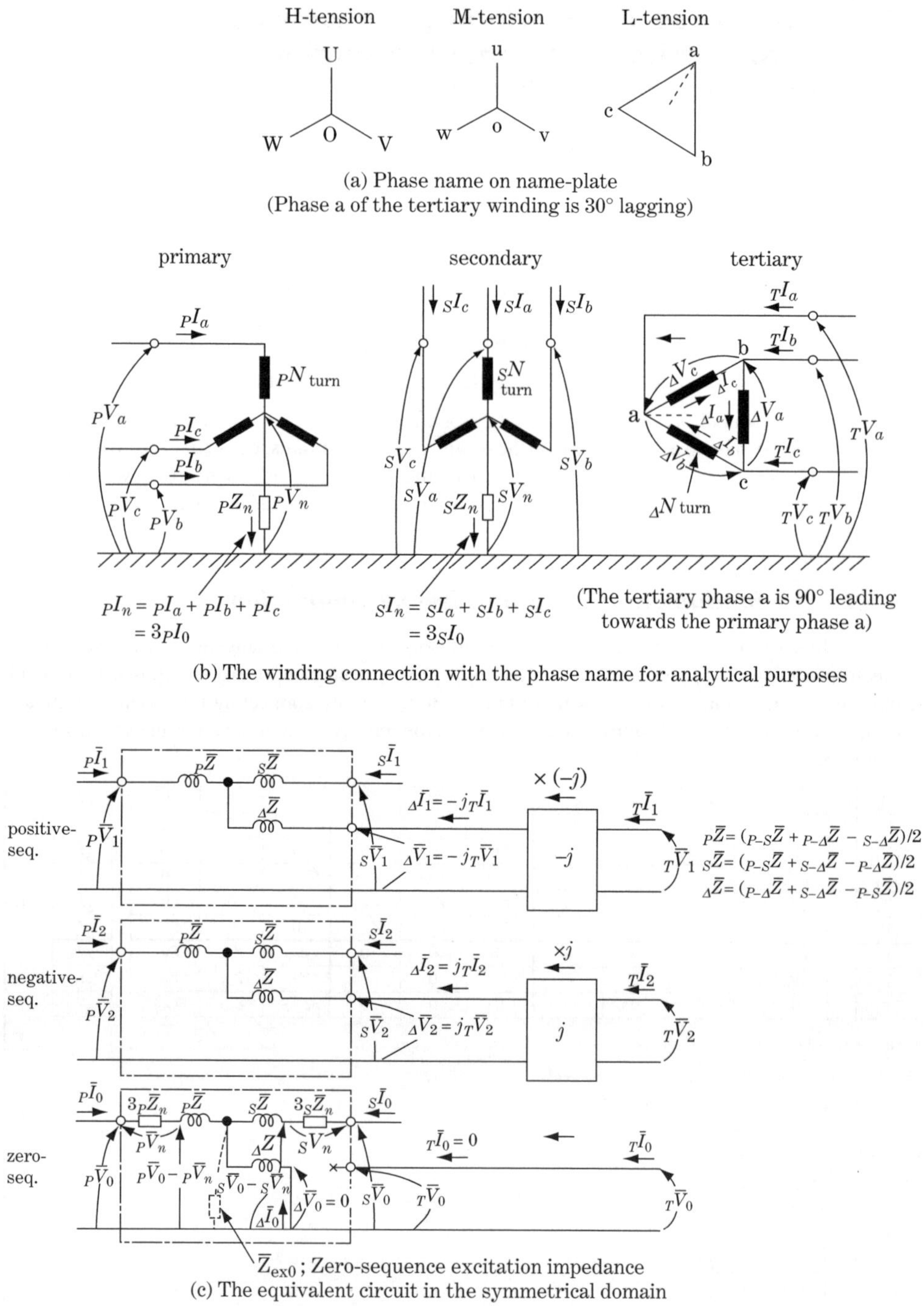

Figure 5.3 ⅄ – ⅄ – Δ transformer (low-tension winding 30° lagging connection)

The submatrix of Equation ① is equal to Equation 5.6. Equations ② and ③ correspond to the neutral connection of primary and secondary windings; equations ④ and ⑤ correspond to the delta connection of tertiary windings. Equation ⑥ is a physical premise for any transformer.

5.3.1.2 Determination of base quantities for unitization

The base quantities for the unitization of this transformer are determined so that the following equations are satisfied:

$$
\left.
\begin{array}{ll}
\frac{1}{3} VA_{3\phi\text{base}} = VA_{1\phi\text{base}} = {}_PV_{l-\text{gbase}} \cdot {}_PI_{l-\text{gbase}} = {}_SV_{l-\text{gbase}} \cdot {}_SI_{l-\text{gbase}} & \\
\quad = {}_\Delta V_{l-l\text{base}} \cdot {}_\Delta I_{l-l\text{base}} = (\sqrt{3} \cdot {}_TV_{l-\text{gbase}}) \left(\frac{1}{\sqrt{3}} \cdot {}_TI_{l-\text{gbase}} \right) & \\
\quad = {}_TV_{l-\text{gbase}} \cdot {}_TI_{l-\text{gbase}} (\equiv k_1 k_2) & ① \\
{}_\Delta V_{l-l\text{base}} = \sqrt{3} \cdot {}_TV_{l-\text{gbase}} & ② \\
{}_\Delta I_{l-l\text{base}} = \frac{1}{\sqrt{3}} \cdot {}_TI_{l-\text{gbase}} & ③ \\
\frac{{}_PV_{l-\text{gbase}}}{{}_PN} = \frac{{}_SV_{l-\text{gbase}}}{{}_SN} = \frac{{}_\Delta V_{l-l\text{base}}}{{}_\Delta N} = \frac{\sqrt{3} \cdot {}_TV_{l-\text{gbase}}}{{}_\Delta N} (\equiv k_1) & ④ \\
{}_PI_{l-\text{gbase}} \cdot {}_PN = {}_SI_{l-\text{gbase}} \cdot {}_SN = {}_\Delta I_{l-l\text{base}} \cdot {}_\Delta N & \\
\quad = \frac{1}{\sqrt{3}} \cdot {}_TI_{l-\text{gbase}} \cdot {}_\Delta N (\equiv k_2) & ⑤ \\
{}_PZ_{l-\text{gbase}} = \frac{({}_PV_{l-\text{gbase}})^2}{VA_{1\phi\text{base}}} & ⑥ \\
{}_SZ_{l-\text{gbase}} = \frac{({}_SV_{l-\text{gbase}})^2}{VA_{1\phi\text{base}}} & ⑦ \\
{}_\Delta Z_{l-l\text{base}} = \frac{({}_\Delta V_{l-\text{gbase}})^2}{VA_{1\phi\text{base}}} = \frac{(\sqrt{3} \cdot {}_TV_{l-\text{gbase}})^2}{VA_{1\phi\text{base}}} = 3 \cdot \frac{({}_TV_{l-\text{gbase}})^2}{VA_{1\phi\text{base}}} & \\
\quad = 3 \cdot {}_TZ_{l-\text{gbase}} & ⑧
\end{array}
\right\} \quad (5.17)
$$

In other words:

*1 Capacity bases (VA_{base}) of the primary (P), secondary (S) and tertiary (T) windings have the same value (Equation ⑦).

*2 Voltage bases of the primary (P), secondary (S) and tertiary (T) windings are proportional to the turns ratio (transformation ratio) of H(high)-/M(medium)-/L(low)-tension windings (Equation ④). This condition is satisfied simply by applying the rated voltages of each winding as base voltages.

*3 Ampere bases of the primary (P), secondary (S) and tertiary (T) windings are dependently determined as values of 'capacity base (*1) divided by each voltage base (*2)'. In other words, the ampere-turn bases of the primary (P), secondary (S) and tertiary (T) windings are to be of equal value (Equations ②③④⑤).

*4 Impedance bases are dependently determined (Equations ⑥⑦⑧).

5.3.1.3 Unitization of the original equation

Equation 5.16 can be divided by the appropriate base quantities which are defined by Equation 5.17, so that the unitized equations are derived as follows:

$$\begin{bmatrix} {}_P\overline{V}_a \\ {}_S\overline{V}_a \\ {}_\Delta\overline{V}_a \\ {}_P\overline{V}_b \\ {}_S\overline{V}_b \\ {}_\Delta\overline{V}_b \\ {}_P\overline{V}_c \\ {}_S\overline{V}_c \\ {}_\Delta\overline{V}_c \end{bmatrix} - \begin{bmatrix} {}_P\overline{V}_n \\ {}_S\overline{V}_n \\ 0 \\ {}_P\overline{V}_n \\ {}_S\overline{V}_n \\ 0 \\ {}_P\overline{V}_n \\ {}_S\overline{V}_n \\ 0 \end{bmatrix} = \begin{bmatrix} \overline{Z}_{PP} & \overline{Z}_{PS} & \overline{Z}_{P\Delta} & & & & & & \\ \overline{Z}_{SP} & \overline{Z}_{SS} & \overline{Z}_{S\Delta} & & & & & & \\ \overline{Z}_{\Delta P} & \overline{Z}_{\Delta S} & \overline{Z}_{\Delta\Delta} & & & & & & \\ & & & \overline{Z}_{PP} & \overline{Z}_{PS} & \overline{Z}_{P\Delta} & & & \\ & & & \overline{Z}_{SP} & \overline{Z}_{SS} & \overline{Z}_{S\Delta} & & & \\ & & & \overline{Z}_{\Delta P} & \overline{Z}_{\Delta S} & \overline{Z}_{\Delta\Delta} & & & \\ & & & & & & \overline{Z}_{PP} & \overline{Z}_{PS} & \overline{Z}_{P\Delta} \\ & & & & & & \overline{Z}_{SP} & \overline{Z}_{SS} & \overline{Z}_{S\Delta} \\ & & & & & & \overline{Z}_{\Delta P} & \overline{Z}_{\Delta S} & \overline{Z}_{\Delta\Delta} \end{bmatrix} \cdot \begin{bmatrix} {}_P\overline{I}_a \\ {}_S\overline{I}_a \\ {}_\Delta\overline{I}_a \\ {}_P\overline{I}_b \\ {}_S\overline{I}_b \\ {}_\Delta\overline{I}_b \\ {}_P\overline{I}_c \\ {}_S\overline{I}_c \\ {}_\Delta\overline{I}_c \end{bmatrix} \qquad ①$$

where

$$\begin{bmatrix} \overline{Z}_{PP} & \overline{Z}_{PS} & \overline{Z}_{P\Delta} \\ \overline{Z}_{SP} & \overline{Z}_{SS} & \overline{Z}_{S\Delta} \\ \overline{Z}_{\Delta P} & \overline{Z}_{\Delta S} & \overline{Z}_{\Delta\Delta} \end{bmatrix} = \begin{bmatrix} Z_{PP}\cdot\dfrac{VA_{1\phi\text{base}}}{({}_PV_{l\text{-gbase}})^2} & Z_{PS}\cdot\dfrac{VA_{1\phi\text{base}}}{{}_PV_{l\text{-gbase}}\cdot{}_SV_{l\text{-gbase}}} & Z_{P\Delta}\cdot\dfrac{VA_{1\phi\text{base}}}{{}_PV_{l\text{-gbase}}\cdot{}_\Delta V_{l\text{-gbase}}} \\ Z_{SP}\cdot\dfrac{VA_{1\phi\text{base}}}{{}_PV_{l\text{-gbase}}\cdot{}_SV_{l\text{-gbase}}} & Z_{SS}\cdot\dfrac{VA_{1\phi\text{base}}}{({}_SV_{l\text{-gbase}})^2} & Z_{S\Delta}\cdot\dfrac{VA_{1\phi\text{base}}}{{}_SV_{l\text{-gbase}}\cdot{}_\Delta V_{l\text{-gbase}}} \\ Z_{\Delta P}\cdot\dfrac{VA_{1\phi\text{base}}}{{}_PV_{l\text{-gbase}}\cdot{}_\Delta V_{l\text{-gbase}}} & Z_{\Delta S}\cdot\dfrac{VA_{1\phi\text{base}}}{{}_SV_{l\text{-gbase}}\cdot{}_\Delta V_{l\text{-gbase}}} & Z_{\Delta\Delta}\cdot\dfrac{VA_{1\phi\text{base}}}{({}_\Delta V_{l\text{-gbase}})^2} \end{bmatrix}$$

and ②

$\overline{Z}_{PS} = \overline{Z}_{SP}$, $\overline{Z}_{P\Delta} = \overline{Z}_{\Delta P}$, etc.,

$${}_P\overline{V}_n = {}_P\overline{Z}_n\cdot{}_P\overline{I}_n = {}_P\overline{Z}_n\cdot({}_P\overline{I}_a + {}_P\overline{I}_b + {}_P\overline{I}_c) = {}_P\overline{Z}_n\cdot 3{}_P\overline{I}_0 \qquad ③$$

$${}_S\overline{V}_n = {}_S\overline{Z}_n\cdot{}_S\overline{I}_n = {}_S\overline{Z}_n\cdot({}_S\overline{I}_a + {}_S\overline{I}_b + {}_S\overline{I}_c) = {}_S\overline{Z}_n\cdot 3{}_S\overline{I}_0 \qquad ④$$

$$\begin{bmatrix} \sqrt{3}\cdot{}_T\overline{I}_a \\ \sqrt{3}\cdot{}_T\overline{I}_b \\ \sqrt{3}\cdot{}_T\overline{I}_c \end{bmatrix} = \begin{bmatrix} {}_\Delta\overline{I}_c \\ {}_\Delta\overline{I}_a \\ {}_\Delta\overline{I}_b \end{bmatrix} - \begin{bmatrix} {}_\Delta\overline{I}_b \\ {}_\Delta\overline{I}_c \\ {}_\Delta\overline{I}_a \end{bmatrix} = \begin{bmatrix} 0 & -1 & 1 \\ 1 & 0 & -1 \\ -1 & 1 & 0 \end{bmatrix} \cdot \begin{bmatrix} {}_\Delta\overline{I}_a \\ {}_\Delta\overline{I}_b \\ {}_\Delta\overline{I}_c \end{bmatrix} \qquad ⑤$$

$$\begin{bmatrix} \sqrt{3}\cdot{}_\Delta\overline{V}_a \\ \sqrt{3}\cdot{}_\Delta\overline{V}_b \\ \sqrt{3}\cdot{}_\Delta\overline{V}_c \end{bmatrix} = \begin{bmatrix} {}_T\overline{V}_b \\ {}_T\overline{V}_c \\ {}_T\overline{V}_a \end{bmatrix} - \begin{bmatrix} {}_T\overline{V}_c \\ {}_T\overline{V}_a \\ {}_T\overline{V}_b \end{bmatrix} = \begin{bmatrix} 0 & 1 & -1 \\ -1 & 0 & 1 \\ 1 & -1 & 0 \end{bmatrix} \cdot \begin{bmatrix} {}_T{}^{v}\overline{V}_a \\ {}_T\overline{V}_b \\ {}_T\overline{V}_c \end{bmatrix} \qquad ⑥$$

$$\left.\begin{aligned} {}_P\overline{I}_a + {}_S\overline{I}_a + {}_\Delta\overline{I}_a = 0 \\ {}_P\overline{I}_b + {}_S\overline{I}_b + {}_\Delta\overline{I}_b = 0 \\ {}_P\overline{I}_c + {}_S\overline{I}_c + {}_\Delta\overline{I}_c = 0 \end{aligned}\right\} \qquad ⑦$$

(5.18)

5.3.1.4 *Symmetrical equations and the equivalent circuit*

The fundamental equations of Equation 5.18 for this transformer can be transformed into the symmetrical domain to derive Equation 5.19 below. (The process of transformation is shown in the supplement at the end of this chapter.)

$$\begin{bmatrix} {}_P\overline{V}_0 \\ {}_S\overline{V}_0 \\ {}_\Delta\overline{V}_0 \\ {}_P\overline{V}_1 \\ {}_S\overline{V}_1 \\ {}_\Delta\overline{V}_1 \\ {}_P\overline{V}_2 \\ {}_S\overline{V}_2 \\ {}_\Delta\overline{V}_2 \end{bmatrix} - \begin{bmatrix} {}_P\overline{V}_n \\ {}_S\overline{V}_n \\ 0 \\ 0 \\ 0 \\ 0 \\ 0 \\ 0 \\ 0 \end{bmatrix} = \begin{bmatrix} \overline{Z}_{PP} & \overline{Z}_{PS} & \overline{Z}_{P\Delta} & & & & & & \\ \overline{Z}_{SP} & \overline{Z}_{SS} & \overline{Z}_{S\Delta} & & & & & & \\ \overline{Z}_{\Delta P} & \overline{Z}_{\Delta S} & \overline{Z}_{\Delta\Delta} & & & & & & \\ & & & \overline{Z}_{PP} & \overline{Z}_{PS} & \overline{Z}_{P\Delta} & & & \\ & & & \overline{Z}_{SP} & \overline{Z}_{SS} & \overline{Z}_{S\Delta} & & & \\ & & & \overline{Z}_{\Delta P} & \overline{Z}_{\Delta S} & \overline{Z}_{\Delta\Delta} & & & \\ & & & & & & \overline{Z}_{PP} & \overline{Z}_{PS} & \overline{Z}_{P\Delta} \\ & & & & & & \overline{Z}_{SP} & \overline{Z}_{SS} & \overline{Z}_{S\Delta} \\ & & & & & & \overline{Z}_{\Delta P} & \overline{Z}_{\Delta S} & \overline{Z}_{\Delta\Delta} \end{bmatrix} \cdot \begin{bmatrix} {}_P\overline{I}_0 \\ {}_S\overline{I}_0 \\ {}_\Delta\overline{I}_0 \\ {}_P\overline{I}_1 \\ {}_S\overline{I}_1 \\ {}_\Delta\overline{I}_1 \\ {}_P\overline{I}_2 \\ {}_S\overline{I}_2 \\ {}_\Delta\overline{I}_2 \end{bmatrix} \qquad ①$$

$${}_P\overline{V}_n = {}_P\overline{Z}_n \cdot {}_P\overline{I}_n = 3{}_P\overline{Z}_n \cdot {}_P\overline{I}_0 \qquad ②$$

$${}_S\overline{V}_n = {}_S\overline{Z}_n \cdot {}_S\overline{I}_n = 3{}_S\overline{Z}_n \cdot {}_S\overline{I}_0 \qquad ③$$

$$\begin{bmatrix} {}_T\overline{I}_0 \\ {}_T\overline{I}_1 \\ {}_T\overline{I}_2 \end{bmatrix} = \begin{bmatrix} 0 \\ j_\Delta\overline{I}_1 \\ -j_\Delta\overline{I}_2 \end{bmatrix} \quad \text{or} \quad \begin{bmatrix} {}_T\overline{I}_0 \\ -j_T\overline{I}_1 \\ j_T\overline{I}_2 \end{bmatrix} = \begin{bmatrix} 0 \\ {}_\Delta\overline{I}_1 \\ {}_\Delta\overline{I}_2 \end{bmatrix} \qquad ④$$

$$\begin{bmatrix} 0 \\ {}_T\overline{V}_1 \\ {}_T\overline{V}_2 \end{bmatrix} = \begin{bmatrix} {}_\Delta\overline{V}_0 \\ j_\Delta\overline{V}_1 \\ -j_\Delta\overline{V}_2 \end{bmatrix} \quad \text{or} \quad \begin{bmatrix} 0 \\ -j_T\overline{V}_1 \\ j_T\overline{V}_2 \end{bmatrix} = \begin{bmatrix} {}_\Delta\overline{V}_0 \\ {}_\Delta\overline{V}_1 \\ {}_\Delta\overline{V}_2 \end{bmatrix} \qquad ⑤$$

$$\left.\begin{aligned} {}_P\overline{I}_0 + {}_S\overline{I}_0 + {}_\Delta\overline{I}_0 = 0 \\ {}_P\overline{I}_1 + {}_S\overline{I}_1 + {}_\Delta\overline{I}_1 = 0 \\ {}_P\overline{I}_2 + {}_S\overline{I}_2 + {}_\Delta\overline{I}_2 = 0 \end{aligned}\right\} \qquad ⑥$$

(5.19)

It is clear from Equation 5.19 that mutual inductances do not exist between positive-, negative- and zero-sequence quantities. The above equations can be recast as follows, and the positive-, negative- and zero-sequence quantities can be treated independently.

For the positive sequence

$$\begin{bmatrix} {}_P\overline{V}_1 \\ {}_S\overline{V}_1 \\ {}_\Delta\overline{V}_1 \end{bmatrix} = \begin{bmatrix} \overline{Z}_{PP} & \overline{Z}_{PS} & \overline{Z}_{P\Delta} \\ \overline{Z}_{SP} & \overline{Z}_{SS} & \overline{Z}_{S\Delta} \\ \overline{Z}_{\Delta P} & \overline{Z}_{\Delta S} & \overline{Z}_{\Delta\Delta} \end{bmatrix} \cdot \begin{bmatrix} {}_P\overline{I}_1 \\ {}_S\overline{I}_1 \\ {}_\Delta\overline{I}_1 \end{bmatrix} \quad ① \quad \left.\begin{aligned} {}_P\overline{I}_1 + {}_S\overline{I}_1 + {}_\Delta\overline{I}_1 = 0 \quad ② \\ \left.\begin{aligned} {}_\Delta\overline{I}_1 = -j_T\overline{I}_1 \\ {}_\Delta\overline{V}_1 = -j_T\overline{V}_1 \end{aligned}\right\} \quad ③ \end{aligned}\right\} \qquad (5.20a)$$

for the negative sequence

$$\begin{bmatrix} {}_P\overline{V}_2 \\ {}_S\overline{V}_2 \\ {}_\Delta\overline{V}_2 \end{bmatrix} = \begin{bmatrix} \overline{Z}_{PP} & \overline{Z}_{PS} & \overline{Z}_{P\Delta} \\ \overline{Z}_{SP} & \overline{Z}_{SS} & \overline{Z}_{S\Delta} \\ \overline{Z}_{\Delta P} & \overline{Z}_{\Delta S} & \overline{Z}_{\Delta\Delta} \end{bmatrix} \cdot \begin{bmatrix} {}_P\overline{I}_2 \\ {}_S\overline{I}_2 \\ {}_\Delta\overline{I}_2 \end{bmatrix} \quad ① \quad \left.\begin{aligned} {}_P\overline{I}_2 + {}_S\overline{I}_2 + {}_\Delta\overline{I}_2 = 0 \quad ② \\ \left.\begin{aligned} {}_\Delta\overline{I}_2 = j_T\overline{I}_2 \\ {}_\Delta\overline{V}_2 = j_T\overline{V}_2 \end{aligned}\right\} \quad ③ \end{aligned}\right\} \qquad (5.20b)$$

and for the zero sequence

$$\begin{bmatrix} {}_P\overline{V}_0 \\ {}_S\overline{V}_0 \\ {}_\Delta\overline{V}_0 \end{bmatrix} - \begin{bmatrix} {}_P\overline{V}_n \\ {}_S\overline{V}_n \\ {}_\Delta\overline{V}_n \end{bmatrix} = \begin{bmatrix} \overline{Z}_{PP} & \overline{Z}_{PS} & \overline{Z}_{P\Delta} \\ \overline{Z}_{SP} & \overline{Z}_{SS} & \overline{Z}_{S\Delta} \\ \overline{Z}_{\Delta P} & \overline{Z}_{\Delta S} & \overline{Z}_{\Delta\Delta} \end{bmatrix} \cdot \begin{bmatrix} {}_P\overline{I}_0 \\ {}_S\overline{I}_0 \\ {}_\Delta\overline{I}_0 \end{bmatrix} \quad ① \quad \left.\begin{aligned} \left.\begin{aligned} {}_P\overline{I}_0 + {}_S\overline{I}_0 + {}_\Delta\overline{I}_0 = 0 \\ {}_T\overline{I}_0 = 0,\ {}_\Delta\overline{V}_0 = 0 \end{aligned}\right\} \quad ② \\ \left.\begin{aligned} {}_P\overline{V}_n = 3{}_P\overline{Z}_n \cdot {}_P\overline{I}_0 \\ {}_S\overline{V}_n = 3{}_S\overline{Z}_n \cdot {}_S\overline{I}_0 \end{aligned}\right\} \quad ③ \end{aligned}\right\}$$

(5.20c)

These are the unitized equations in the symmetrical coordinate domain.

Positive-sequence Equation 5.20a is completely the same as Equation 5.10 for a single-phase transformer, so the equivalent circuit must be the same as in Figure 5.2b by the same analogy described in Section 5.1.2. The negative- and zero-sequence quantities can be treated in the same way.

The symmetrical equivalent circuits corresponding to Equations 5.20a–c can be written as in Figure 5.3c, where impedances ${}_P\overline{Z}, {}_S\overline{Z}, {}_\Delta\overline{Z}$ are defined by the equation

$$\left.\begin{aligned}
{}_P\overline{Z} &= \frac{{}_{P-S}\overline{Z} + {}_{P-\Delta}\overline{Z} - {}_{S-\Delta}\overline{Z}}{2} \\
{}_S\overline{Z} &= \frac{{}_{P-S}\overline{Z} + {}_{S-\Delta}\overline{Z} - {}_{P-\Delta}\overline{Z}}{2} \\
{}_\Delta\overline{Z} &= \frac{{}_{P-\Delta}\overline{Z} + {}_{S-\Delta}\overline{Z} - {}_{P-S}\overline{Z}}{2} \\
{}_{P-S}\overline{Z} &= {}_P\overline{Z} + {}_S\overline{Z} = \overline{Z}_{PP} + \overline{Z}_{SS} - 2\overline{Z}_{PS} \\
{}_{P-\Delta}\overline{Z} &= {}_P\overline{Z} + {}_\Delta\overline{Z} = \overline{Z}_{PP} + \overline{Z}_{\Delta\Delta} - 2\overline{Z}_{P\Delta} \\
{}_{S-\Delta}\overline{Z} &= {}_S\overline{Z} + {}_\Delta\overline{Z} = \overline{Z}_{SS} + \overline{Z}_{\Delta\Delta} - 2\overline{Z}_{S\Delta}
\end{aligned}\right\} \tag{5.21}$$

The expression on the right-hand side and the neutral grounding terminal in Figure 5.3c are strictly in one-to-one correspondence to Equations ②③ of Equation 5.20a–c.

Numerical check

As a typical example for a 1000 MVA, 500 kV transformer for substation use with:

- Rated capacity H: 1000 MVA, M:1000 MVA, L: 300 MVA
- Rated voltage 500 kV/275 kV/63 kV

percentage impedances ${}_{P-S}X = 14\%$ (1000 MVA_{base}), ${}_{P-\Delta}X = 44\%$ (1000 MVA_{base}) and ${}_{S-\Delta}X = 26\%$ (1000 MVA_{base}),
the equivalent circuit reactance of the transformer can be calculated as follows by Equation 5.21, where ${}_SX$ takes capacitive values

$${}_PZ = j_PX = 16\% = j0.16 \text{ pu}, \quad {}_SZ = j_SX = 2\% = -j0.02 \text{ pu}, \quad {}_SZ = j_SX = 28\% = j0.28 \text{ pu}$$

Now let us consider Equation 5.20c and the corresponding zero-sequence equivalent circuit. As we have the equations ${}_\Delta\overline{V}_0 = 0$ and ${}_T\overline{I}_0 = 0$, the Δ terminal is earth grounded and the tertiary (T) terminal is open. This means that the zero-sequence current from the tertiary (T) outside circuit cannot flow into the (delta windings of the) transformer, though the zero-sequence current from the primary (P) or secondary (S) outside circuit can flow into the (delta windings of the) transformer.

On the other hand, the equations for ${}_P\overline{V}_n$ and ${}_S\overline{V}_n$ in Equation 5.20c require us to insert $3{}_P\overline{Z}_n$ and $3{}_S\overline{Z}_n$ into the primary and secondary branches respectively.

Therefore, if the primary and secondary neutral terminals are solidly earth grounded, the zero-sequence current inflow from outside to the primary terminal flows partly into the delta winding (as the circulating current) and partly out through the secondary terminal. If the neutral terminal on the secondary side is opened or highly resistive grounded (${}_S\overline{Z}_n = \infty$), all the zero-sequence inflow current from the primary side circulates through the delta windings.

Equations 5.20a–5.20c and the equivalent circuit in Figure 5.3 as the expression for the three-phase three-winding transformer in the symmetrical sequence domain are important because:

*1 The 1–2–0 sequence circuits are mutually independent.

*2 The unitized simple circuits allow the use of Kirchhoff's law.

*3 They contain common reactances for 1–2–0 sequence circuits.

We know that an actual large power system can be expressed as a precise large single circuit, combining both lines and equipment with various rated capacities and voltages. It is satisfying to think that the largest key factor of such a technique owes much to the above-mentioned Equations 5.20a–c and the equivalent circuit in Figure 5.3 for three-phase transformers, realized by the symmetrical coordinate transformation and appropriate unitization.

Finally, the zero-sequence excitation impedance $\overline{Z}_{ex0}$ in Figure 5.3 will be discussed later.

5.3.2 Three-phase transformers with various winding connections

Three-phase transformers with various different winding connection and their unitized equations and equivalent circuits are shown in Table 5.1. Figure a in the table is just the case of Figure 5.3. The equations and the equivalent circuits for transformers of other winding connections can be described in the same way as Figure a. Autotransformers can be expressed by the same equivalent circuits.

5.3.3 Core structure and the zero-sequence excitation impedance

Table 5.2 shows a typical core structure of a transformer bank.

Say we want to impose three-phase-balanced voltages (i.e. positive- or negative-sequence voltages) from primary terminals. The induced fluxes by the balanced voltage charging are also three-phase balanced, so that any flux pass on the laminate steel core will not be saturated under normal voltage operation. This is the reason why the excatition impedance $\overline{Z}_{ex}$ can be neglected as very large impedance values under the condition three-phase-balanced voltages and currents.

Next, let us impose zero-sequence voltages from the primary terminals shown in the figures of Table 5.2. In the case of the transformers of Figures B and C, the caused zero-sequence flux ϕ_0 may be saturated because the return pass of ϕ_0 is absent or of high magnetic reluctance, so that flux saturation would be caused, and an abnormal temperature rise on the saturated flux pass would occur, if saturation by ϕ_0 were to continue for a long time. Of course the excitation current increases under saturation phenomena, which means that the excitation impedance for zero-sequence voltage $\overline{Z}_{ex0}$ as a part of the equivalent circuit would have smaller values.

The transformers of Figure D have the auxiliary fourth magnetic pass of the laminated steel core by which zero-sequence flux saturation can be prevented. The transformers of Figure A (one bank of three single phase transformers) do not have such a limitation in nature.

Although excitation impedance $\overline{Z}_{ex}$ can be neglected for most cases, $\overline{Z}_{ex0}$ may rarely have to be taken into account in the zero-sequence circuit as shown in Figure 5.3c.

5.3.4 Various winding methods and the effect of delta windings

Figures a–e of Table 5.1 show typical transformer winding constructions and the symmetrical equivalent circuits. Due to the existence of the delta windings, the zero-sequence circuit of the delta winding side is isolated from the Y winding side circuits. Moreover, $_{\Delta}\overline{Z}$ is earth grounded in the zero-sequence equivalent circuit. These are two very important reasons why the role of the delta windings is explained below.

Table 5.1 The equivalent circuit of various windings of the transformer

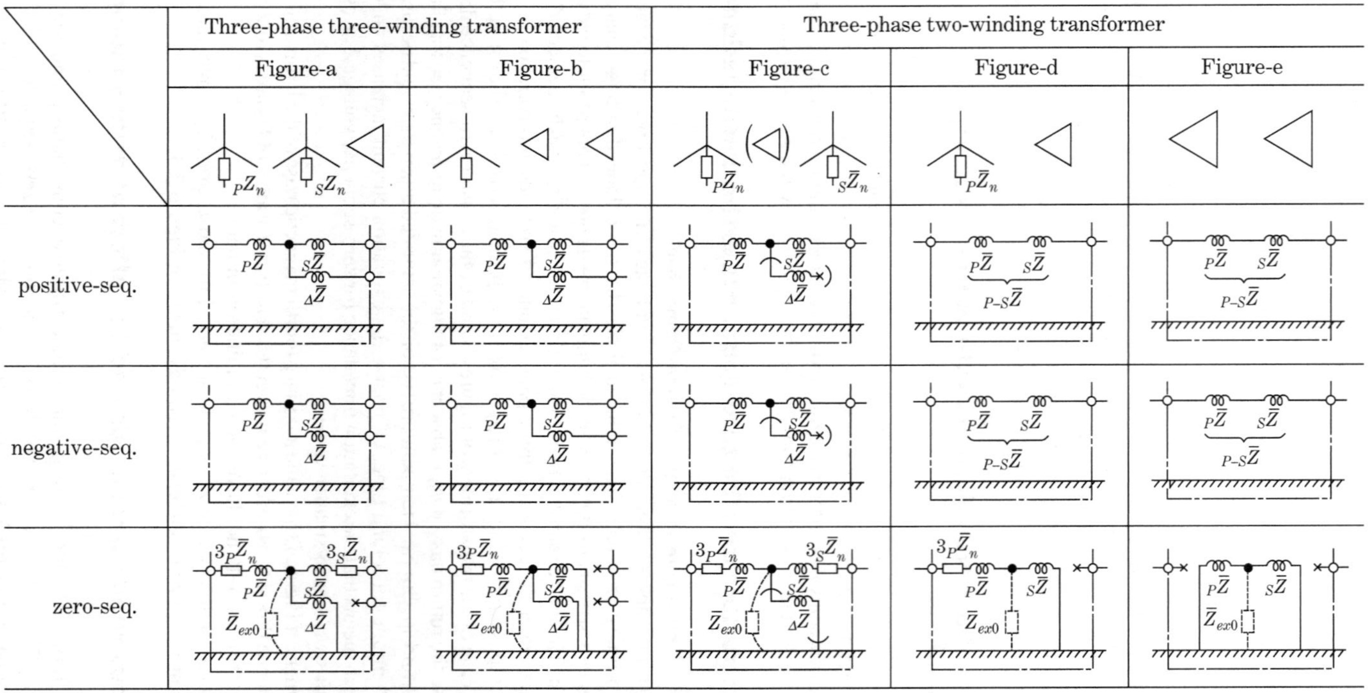

Note: The reactances $_{P}X$, $_{S}X$, $_{\Delta}X$ are calculated by Eq.(5.21), in which $_{P\text{-}S}X$, $_{P\text{-}\Delta}X$, $_{S\text{-}\Delta}X$ are given on the name-plate of the transformer as %IZ.

MVA base of $_{P\text{-}S}X$, $_{P\text{-}\Delta}X$, $_{S\text{-}\Delta}X$ should be unified before the calculation by the equation.

Table 5.2 Typical structures of thee-phase transformers

3 × Single phase transformers	Shell type	Core type(three-poles)	Core type(four poles)
Figure-A 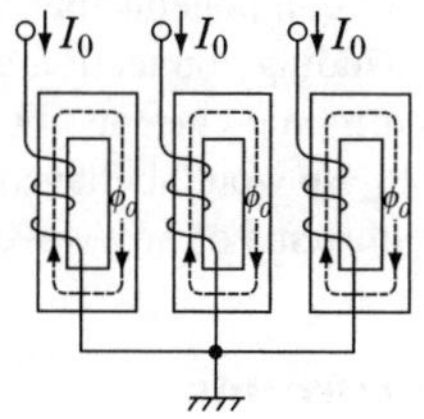 *1	Figure-B 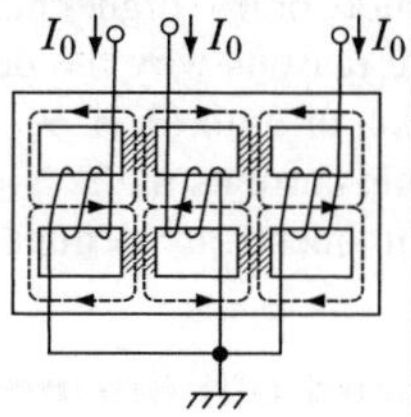 *2	Figure-C 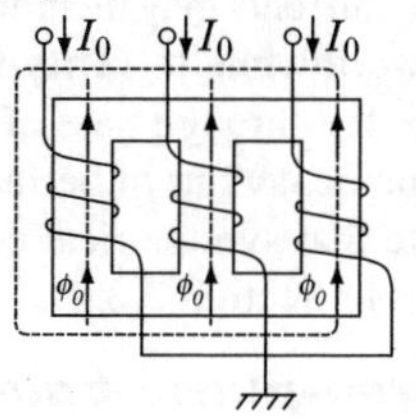 *3	Figure-D 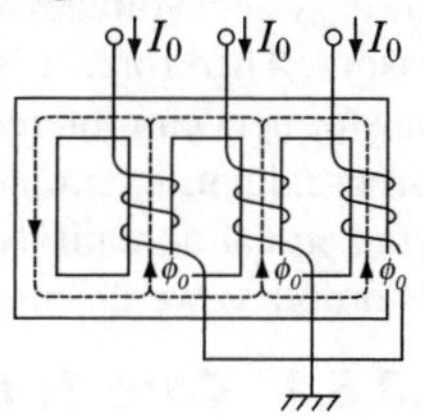 *1

Note: The zero-sequence flux ϕ_0 induced by the zero-sequence current I_0 passes only through the core. Accordingly, zero-sequence excitation impedance $\overline{Z}_{ex0}$ is usually neglected as of quite large values.

*2: The hatched part is apt to be saturated by zero-sequence current I_0. Accordingly, the zero-sequence excitation impedance $\overline{Z}_{ex0}$ is relatively small, say $\overline{Z}_{ex0} \fallingdotseq 1.0-5.0$ per unit.

*3: When the zero-sequence current I_0 is forced to flow into the transformer windings, the induced flux return pass could be through clamps, the inner wall of the tank, air gaps, rather than the laminated core. The counter electromotive force and the flux ϕ_0 induced by the zero-sequence current I_0 are rather small (because the magnetic reluctance of the return pass in particular air gaps is large). Accordingly, $\overline{Z}_{ex0}$ in the zero-sequence equivalent circuit is rather small, say $\overline{Z}_{ex0} \fallingdotseq 0.3-1.0$ per unit.

Main (step-up) transformers for large power stations contain the windings of Figure b or d and generators as well as local power station circuits are connected to the delta windings. High-voltage substation transformers would have the windings of Figure a, b, c or d. The transformer in Figure c has 人 − 人 connected windings and the third internal delta windings, although these windings are not fed out by the bushing terminals.

In total, all the transformers installed at stations in power networks probably have set-in delta windings today. The advantages gained by adopting delta windings as general practice are summarized as follows and all of them can be explained by the above-mentioned unique characteristics of the zero-sequence domain:

- To isolate the zero-sequence circuit of the delta winding side from that of the 人 winding side. In other words, to intercept zero-sequence current I_0 flowing across the different winding side circuits.
- To reduce largely zero-sequence reactance of the transformer, thus to reduce and stabilize neutral point voltages, and then to reduce phase overvoltages under normal conditions or temporary overvoltage (TOV) during faults or any other power system disturbance.
- To intercept the through pass of d.c. or harmonic currents from the high-tension (HT) to medium-tension (MT) side equipment, in particular to protect generators or motors against abnormal operation or damage.
- To protect the transformer itself from damage which may be caused by zero-sequence in-flow current or by d.c. or 3nth (3, 6, 9, . . .) harmonic currents (overheating, vibration, overvoltages, waveform distortion, etc.) in the cores, insulated windings, yokes, clamps or any other structural part.

- To reduce zero-sequence impedance and to stabilize the neutral (zero-sequence) voltages under normal condition or during system faults.

5.3.5 Harmonic frequency voltages/currents in the 0–1–2 domain

Power system voltages and currents may include more or less higher harmonic components for various reasons. Therefore, it is worthwhile to verify the reasons why the delta winding connection of the transformers can intercept the through pass of d.c. or 3*n*th (3, 6, 9,...) harmonic currents. For this purpose, let us examine the behaviour of harmonic currents in the a–b–c phase would behave in the symmetrical domain. Note that symmetrical components are defined for currents of any waveform distortion as we discussed in Section 2.2.

5.3.5.1 *Case 1: three-phase-balanced nth harmonic currents*

Three-phase-balanced *n*th harmonic currents are

$$\left.\begin{aligned} I_a &= I\mathrm{e}^{jn\omega t} \\ I_b &= I\mathrm{e}^{jn(\omega t-120^\circ)} = (\mathrm{e}^{-j120^\circ})^n \cdot I_a = a^{2n} \cdot I_a \\ I_c &= I\mathrm{e}^{jn(\omega t+120^\circ)} = (\mathrm{e}^{j120^\circ})^n \cdot I_a = a^n \cdot I_a \end{aligned}\right\} \tag{5.22}$$

Stationary harmonics caused by solid-state power conditioners (inverters, converters, rectifiers, etc.) or very little waveform distortion from generators may be classified into this category. Transforming Equation 5.22 into the 1–2–0 domain,

$$\begin{bmatrix} I_0 \\ I_1 \\ I_2 \end{bmatrix} = \frac{1}{3}\begin{bmatrix} 1 & 1 & 1 \\ 1 & a & a^2 \\ 1 & a^2 & a \end{bmatrix}\begin{bmatrix} I_a \\ a^{2n}I_a \\ a^n I_a \end{bmatrix} = \frac{1}{3}\begin{bmatrix} 1+a^{2n}+a^n \\ 1+a^{2n+1}+a^{n+2} \\ 1+a^{2n+2}+a^{n+1} \end{bmatrix} \cdot I_a \tag{5.23a}$$

Then, for the case of $n = 3m$ (0, 3, 6, 9, ...)

$$\begin{bmatrix} I_0 \\ I_1 \\ I_2 \end{bmatrix} = \begin{bmatrix} I_a \\ 0 \\ 0 \end{bmatrix} \tag{5.23b}$$

and of $n = 3m + 1$ (1, 4, 7, ...)

$$\begin{bmatrix} I_0 \\ I_1 \\ I_2 \end{bmatrix} = \begin{bmatrix} 0 \\ I_a \\ 0 \end{bmatrix} \tag{5.23c}$$

and of $n = 3m + 2\,(2, 5, 8, \ldots)$

$$\begin{bmatrix} I_0 \\ I_1 \\ I_2 \end{bmatrix} = \begin{bmatrix} 0 \\ 0 \\ I_a \end{bmatrix} \tag{5.23d}$$

That is, the behaviour of three-phase-balanced *n*th harmonic currents has characteristics of:

$n = 1, 4, 7, \ldots$: behaviour as positive-sequence currents

$n = 2, 5, 8, \ldots$: behaviour as negative-sequence currents

$n = 0, 3, 6, 9, \ldots$: behaviour as zero-sequence currents

This is the reason why transformers with delta windings can intercept 0(d.c.), 3rd, 6th, ... harmonic currents.

5.3.5.2 Case 2: nth harmonic current flow in phase a

Here

$$\left.\begin{aligned} I_a &= I\mathrm{e}^{jn\omega t} \\ I_b &= I_c = 0 \end{aligned}\right\} \tag{5.24}$$

A waveform-distorted single phase load may cause harmonic currents of this sort of category:

$$I_0 = I_1 = I_2 = \frac{1}{3} I\mathrm{e}^{jn\omega t} \tag{5.25}$$

In this case, negative- and zero-sequence currents of nth order flow through the circuit.

5.3.5.3 Case 3: nth harmonic current of synchronized delay with power frequency

Here

$$\left.\begin{aligned} I_a &= I\mathrm{e}^{jn\omega t} \\ I_b &= I\mathrm{e}^{j(n\omega t - 120^\circ)} = a^2 \cdot I_a \\ I_c &= I\mathrm{e}^{j(n\omega t + 120^\circ)} = a \cdot I_a \end{aligned}\right\} \tag{5.26}$$

Then

$$\begin{bmatrix} I_0 \\ I_1 \\ I_2 \end{bmatrix} = \frac{1}{3} \begin{bmatrix} 1 & 1 & 1 \\ 1 & a & a^2 \\ 1 & a^2 & a \end{bmatrix} \begin{bmatrix} I_a \\ a^2 \cdot I_a \\ a \cdot I_a \end{bmatrix} = \begin{bmatrix} 0 \\ I_a \\ 0 \end{bmatrix} \tag{5.27}$$

The harmonic current of Equation 5.26 consists of only positive-sequence components. In other words, the current including harmonics of Equation 5.26 is the positive-sequence current (120° phase-balanced current) with harmonic distortion.

Transient phenomena with harmonics or waveform distortion will be examined again in Chapters 7 and 22.

5.4 Base Quantity Modification of Unitized Impedance

In practical engineering, the quantities (MVA capacity, voltage, current, impedance, etc.) of individual members of a power system network (generators, transformers, transmission lines, etc.) are probably dictated by ohmic values or by PU values with different individual PU bases. On the other hand, in order to obtain a total combined system circuit for these members, MVA_{bases} have to be unified for all of the system first, and, furthermore, voltage bases for each section have to be selected to satisfy the turn ratio of the transformers (or typically to adopt rated voltages for each section). Accordingly, impedance bases of individual equipment may often have to be changed to another base value. Therefore we need to examine how to change the base quantities of impedances.

There is an impedance element of Z [Ω], which can be written as unitized impedance by two different base quantities:

$$\left.\begin{aligned} &Z\,[\Omega] = \overline{Z}_{\text{old}} \cdot Z_{\text{old base}} = \overline{Z}_{\text{new}} \cdot Z_{\text{new base}} \\ &\text{where} \\ &Z_{\text{olds base}}\,[\Omega] = \frac{(V_{\text{old}\,l-l\text{base}})^2}{VA_{\text{old}\,3\phi\text{base}}} = \frac{(kV_{\text{old}\,l-l\text{base}})^2}{MVA_{\text{old}\,3\phi\text{base}}} \\ &Z_{\text{new base}}\,[\Omega] = \frac{(V_{\text{new}\,l-l\text{base}})^2}{VA_{\text{new}\,3\phi\text{base}}} = \frac{(kV_{\text{new}\,l-l\text{base}})^2}{MVA_{\text{new}\,3\phi\text{base}}} \end{aligned}\right\} \tag{5.28}$$

Accordingly, the formula to change both the capacity base and voltage base is

$$\begin{aligned}\overline{Z}_{\text{new}} = \overline{Z}_{\text{old}} \cdot \frac{Z_{\text{old base}}}{Z_{\text{new base}}} &= \overline{Z}_{\text{old}} \cdot \left(\frac{VA_{\text{new }3\phi\text{base}}}{VA_{\text{old }3\phi\text{base}}}\right) \cdot \left(\frac{V_{\text{old }l-l\text{base}}}{V_{\text{new }l-l\text{base}}}\right)^2 \\ &= \overline{Z}_{\text{old}} \cdot \left(\frac{MVA_{\text{new }3\phi\text{base}}}{MVA_{\text{old }3\phi\text{base}}}\right) \cdot \left(\frac{kV_{\text{old }l-l\text{base}}}{kV_{\text{new }l-l\text{base}}}\right)^2\end{aligned} \tag{5.29}$$

the formula to change only the capacity base is

$$\overline{Z}_{\text{new}} = \overline{Z}_{\text{old}} \cdot \left(\frac{VA_{\text{new }3\phi\text{base}}}{VA_{\text{old }3\phi\text{base}}}\right) = \overline{Z}_{\text{old}} \cdot \left(\frac{MVA_{\text{new }3\phi\text{base}}}{MVA_{\text{old }3\phi\text{base}}}\right) \tag{5.30}$$

and the formula to change only the voltage base is

$$\overline{Z}_{\text{new}} = \overline{Z}_{\text{old}} \cdot \left(\frac{V_{\text{old }l-l\text{base}}}{V_{\text{new }l-l\text{base}}}\right)^2 = \overline{Z}_{\text{old}} \cdot \left(\frac{kV_{\text{old }l-l\text{base}}}{kV_{\text{new }l-l\text{base}}}\right)^2 \tag{5.31}$$

As general practice, one unified value of MVA_{base} has to be selected, and then kV_{base} values of individual sections across each transformer have to be decided. Through these processes, Z_{base} as well as I_{base} are dependently determined for each section. Then, the derived impedance base is adopted as the $Z_{\text{new base}}$ to obtain the unified circuit.

5.4.1 Note on % IZ of three-winding transformer

Primary and secondary MVA ratings of typical substations using three-winding transformers are usually the same, while tertiary MVA ratings may be smaller (say 30 or 35%). The %IZ described on the name-plate is usually given by different MVA bases as follows:

- Between the primary and secondary ${}_{P-S}\overline{Z}$: by primary and secondary MVA base
- Between the primary and tertiary ${}_{P-T}\overline{Z}$: by tertiary MVA base
- Between the secondary and tertiary ${}_{S-T}\overline{Z}$: by tertiary MVA base.

In order to find the equivalent circuit of the transformer, the above PU impedance values have to be modified into new PU values based on a single, common MVA base. In other words, ${}_{P-T}\overline{Z}, {}_{S-T}\overline{Z}$ are probably modified to new PU values based on primary and secondary MVA capacity and then the new ${}_{P-T}\overline{Z}, {}_{S-T}\overline{Z}$ obtained, as well as ${}_{P-S}\overline{Z}$, are put into Equation 5.21 to find ${}_{P}\overline{Z}, {}_{S}\overline{Z}, {}_{T}\overline{Z}$. (The calculation is demonstrated in the next section.)

5.5 Autotransformer

Figure 5.4 shows a single-phase three-winding transformer, in which one terminal of the primary and one terminal of the secondary winding are connected. A transformer with this type of connection is called an **autotransformer**. The related equations of this transformer connection are written as follows:

$$\left.\begin{aligned}&\begin{bmatrix} {}_{P}v \\ {}_{S}v \\ {}_{T}v \end{bmatrix} = \begin{bmatrix} Z_{PP} & Z_{PS} & Z_{PT} \\ Z_{SP} & Z_{SS} & Z_{ST} \\ Z_{TP} & Z_{TS} & Z_{TT} \end{bmatrix} \cdot \begin{bmatrix} {}_{P}i \\ {}_{S}i \\ {}_{T}i \end{bmatrix} \quad ① \\ &\text{where} \\ &\left.\begin{aligned}{}_{P}v &= {}_{P}V - {}_{S}V \\ {}_{S}v &= {}_{S}V \\ {}_{T}v &= {}_{T}V\end{aligned}\right\} ② \quad \left.\begin{aligned}{}_{P}i &= {}_{P}I \\ {}_{S}i &= {}_{P}I + {}_{S}I \\ {}_{T}i &= {}_{T}I\end{aligned}\right\} ③\end{aligned}\right\} \tag{5.32}$$

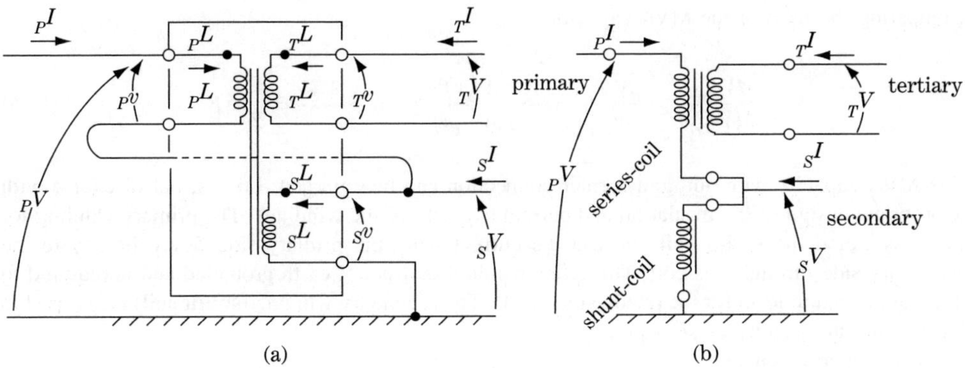

Figure 5.4 Autotransformer

Equation 5.32 ① corresponds to Equation 5.6 except that the symbols of variables V, I are replaced by v, i.

By substitution of Equation 5.32 ②③ into ① and by modification, the following equation is derived:

$$\begin{bmatrix} {}_{P}V \\ {}_{S}V \\ {}_{T}V \end{bmatrix} = \begin{bmatrix} Z_{PP}+Z_{SS}+Z_{PS}+Z_{SP} & Z_{PS}+Z_{SS} & Z_{PT}+Z_{ST} \\ Z_{SP}+Z_{SS} & Z_{SS} & Z_{ST} \\ Z_{TP}+Z_{TS} & Z_{TS} & Z_{TT} \end{bmatrix} \cdot \begin{bmatrix} {}_{P}I \\ {}_{S}I \\ {}_{T}I \end{bmatrix} \quad (5.33)$$

Equation 5.33 is of the same form as Equation 5.6, which means that a transformer of rated value ${}_{P}v, {}_{S}v, {}_{T}v$ and ${}_{P}i, {}_{S}i, {}_{T}i$ can be applied as the transformer of new rated value ${}_{P}V, {}_{S}V, {}_{T}V$ and ${}_{P}I, {}_{S}I, {}_{T}I$ under the condition that the voltage insulation of the primary winding can withstand the voltage of the primary side network ${}_{P}V = {}_{P}v + {}_{S}v$ and the current capacity of the secondary winding is for the secondary side ${}_{S}I = {}_{P}i + {}_{S}i$.

In other words, Equation 5.33 and Equation 5.6 are equivalent only when the impedance matrix of Equation 5.6 is replaced by that of Equation 5.33. Accordingly, the explanation from Equation 5.6 to Equation 5.31 in this chapter (including the application to the three-phase winding transformer and the unitization) can also be adopted for the autotransformer of Equation 5.33.

The following equations in regard to MVA capacity are derived for the transformer whose MVA capacity on the primary and secondary sides are the same:

self-winding capacity

$$\left.\begin{aligned} &MVA_{\text{self}} = {}_{P}v \cdot {}_{P}i = {}_{S}v \cdot {}_{S}i \\ &\text{where} \\ &\frac{{}_{P}v}{{}_{P}N} = \frac{{}_{S}v}{{}_{S}N} = \frac{{}_{T}v}{{}_{T}N} = k \end{aligned}\right\} ①$$

autotransformer capacity

$$\left.\begin{aligned} &MVA_{\text{auto}} = {}_{P}V \cdot {}_{P}I = ({}_{P}v + {}_{S}v) \cdot {}_{S}i = {}_{S}v \cdot ({}_{P}i + {}_{S}i) = {}_{S}V \cdot {}_{S}I \\ &\text{where} \\ &\text{co-ratio } \alpha \equiv \frac{{}_{P}V - {}_{S}V}{{}_{P}V} = \frac{{}_{P}v}{{}_{P}v + {}_{S}v} \end{aligned}\right\} ② \quad (5.34)$$

Comparing the ratio of the MVA capacities,

$$\frac{MVA_{\text{auto}}}{MVA_{\text{self}}} = \frac{{}_pV \cdot {}_pI}{{}_pv \cdot {}_pi} = \frac{({}_pv + {}_Sv) \cdot {}_pi}{{}_pv \cdot {}_pi} = \frac{{}_pv + {}_Sv}{{}_pv} = \frac{1}{\alpha} \equiv \beta \tag{5.35}$$

The MVA capacity with autotransformer connection can be enlarged β times, but of course with appropriate design of the insulation and current capacity of the windings. The primary winding for ${}_pv, {}_pi$ is called the **series coil** because the current from the primary side flows directly to the secondary side through this coil. The series winding coil is not earth grounded and is required to have an insulation level for the rated value of ${}_pV$. The secondary winding (**shunt coil**) is required to have a current capacity of ${}_SI = {}_pi + {}_Si$.

As a numerical check,

$$\text{Autotransformer } 500\,\text{kV}/275\,\text{kV}/66\,\text{kV}$$

$$\alpha = \frac{500 - 275}{500} = 0.45$$

$$\beta = 1/0.45 = 2.2$$

The weight of an autotransformer can generally be reduced in comparison with an ordinal transformer of the same MVA capacity. However, it must be noted that the percentage impedance ${}_pZ, {}_SZ, {}_TZ$ would become quite small without appropriate countermeasures in design work. It can be determined by recalculating the equivalent impedance of Equation 5.12a based on the new impedance matrix of Equation 5.33 instead of Equation 5.32, although a description is omitted here.

5.6 Numerical Example to Find the Unitized Symmetrical Equivalent Circuit

The basic theory of power system circuit analysis has been examined in Chapters 1–4 and the sections above. Now, a numerical calculation is demonstrated where a symmetrical equivalent circuit of a model system will be derived by utilizing the previously studied theories.

Table 5.3 contains the diagram of a model system including overhead lines, cable lines, generating station, substations and some loads. Although the model system may not be very realistic, it is suitable for reviewing the theoretical process and for obtaining a unified power system network circuit.

The subject of exercise is to derive the symmetrical equivalent circuit of the model system given in Table 5.3 under the condition that 1000 MVA is assigned as the MVA base quantity and the rated voltages of each section are to be selected as the voltage bases of each section.

The derived answer for this exercise is shown in Table 5.4. The calculation process is demonstrated item by item below.

1. Determination of PU base quantities

$MVA_{\text{base}} = 1000\,\text{MVA}$ is given, while the rated voltages of 22, 500, 66, 154, 66 kV of Sections A, B, C, D, E respectively are selected as the V_{base} for each section. Then, all the base quantities at each section of the model system are calculated by applying Equations 5.13a and b, and the result is summarized in Table 5.4.

Table 5.3 The model power system

Theme : To obtain symmetrical equivalent circuit by assigned base quantities of 1000MVA and rated voltage bases of each section

power frequency: 50Hz

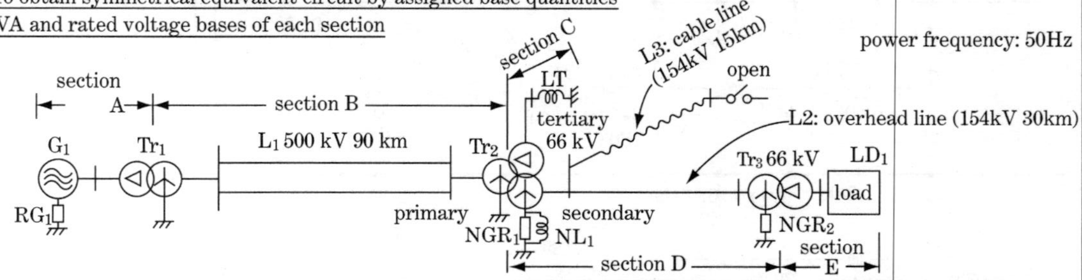

G_1 ; generator
rating 625 MVA, 22 kV
$x_d = 156\%$
$x'_d = 29\%$
$x''_d = 25\%$
$x_2 = 23\%$
$x_0 = 12\%$

RG_1 ; neutral-ground resistance 100A

Tr_1 ; transformer
1000 MVA
22 kV/500 kV
%impedance 13%

L1 overhead line (90km double circuits)
nominal square 810 mm²
(diameter 38.4 mm) × 4 bundle-conductors (averaged distance 20 cm)
grounding wire OGW 410 mm²
(diameter 28.6 mm)

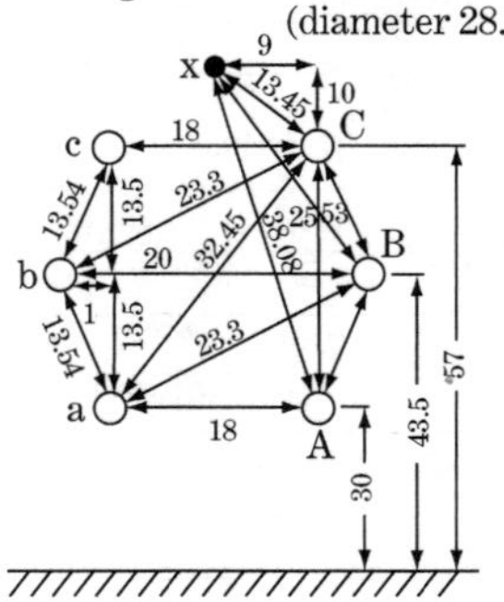

assuming $H_e = 300$ m

Stray capacitances are to be taken into account.

Tr_2 ; transformer
750 MVA
500/154/66 kV
% impedance
P-to-S ; 23%
(750 MVA base)
P-to-T ; 18%
(250 MVA base)
S-to-T ; 9%
(250 MVA base)

NGR_1 ; grounding resistance rating 200A

NL_1 ; neutral compensating reactor
rating 15000 kva

L_2 ; 154 kV 30 km
self-inductance
$L_s = 2.4$ [mH/km]
mutual-inductance
$L_m = 1.1$ [mH/km]
self-capacitance
$C_s = 0.0052$ [μF/km]
mutual-capacitance
$C_m = 0.0013$ [μF/km]

LD_1 ; 66 kV load
200 MW pf = 0.8
where $Z_1 \fallingdotseq Z_2$

TR_3 ; transformer
250 MVA 154/66 kV
$\%X_{P\text{-}s} = 12\%$
$\%X_0 = 10\%$

NGR_2 ; grounding resistance rating 100A

L_3 ; cable line 15 km
leakage current
210A per phase
single-core cable
(inductance neglected)

LT;66kV voltage regulator(reactor)
80MVA (neutral ungrounded)

Table 5.4 Equivalent circuit in symmetrical components

Base quantities for each section

		section A	section B	section C	section D	section E
capacity-base	$MVA_{3\phi base}$	$MVA_{3\phi base}$ = 1000 MVA, $MVA_{1\phi base}$ = 333.3 MVA				
voltage-base	$kV_{l\text{-}l base}$	22 kV	500 kV	66 kV	154 kV	66 kV
	$kV_{l\text{-}g base}$	$22/\sqrt{3}$ kV = 12.7 kV	$500/\sqrt{3}$ kV = 289 kV	$66/\sqrt{3}$ kV = 38 kV	$154/\sqrt{3}$ kV = 89 kV	$66/\sqrt{3}$ kV = 38 kV
current-base	$I_{l\text{-}g base} = \frac{MVA_{3\phi base}\times 10^3}{\sqrt{3}kV_{l\text{-}l base}}$	$\frac{1000\times 10^3}{\sqrt{3}\times 22} = 26240$ A	$\frac{1000\times 10^3}{\sqrt{3}\times 500} = 1155$ A	$\frac{1000\times 10^3}{\sqrt{3}\times 66} = 8748$ A	$\frac{1000\times 10^3}{\sqrt{3}\times 154} = 3749$ A	$\frac{1000\times 10^3}{\sqrt{3}\times 66} = 8748$ A
impedance-base	$Z_{l\text{-}g base} = \frac{(kV_{l\text{-}l base})^2}{MVA_{3\phi base}}$	$\frac{22^2}{1000} = 0.484\ \Omega$	$\frac{500^2}{1000} = 250\ \Omega$	$\frac{66^2}{1000} = 4.36\ \Omega$	$\frac{154^2}{1000} = 23.7\ \Omega$	$\frac{66^2}{1000} = 4.36\ \Omega$

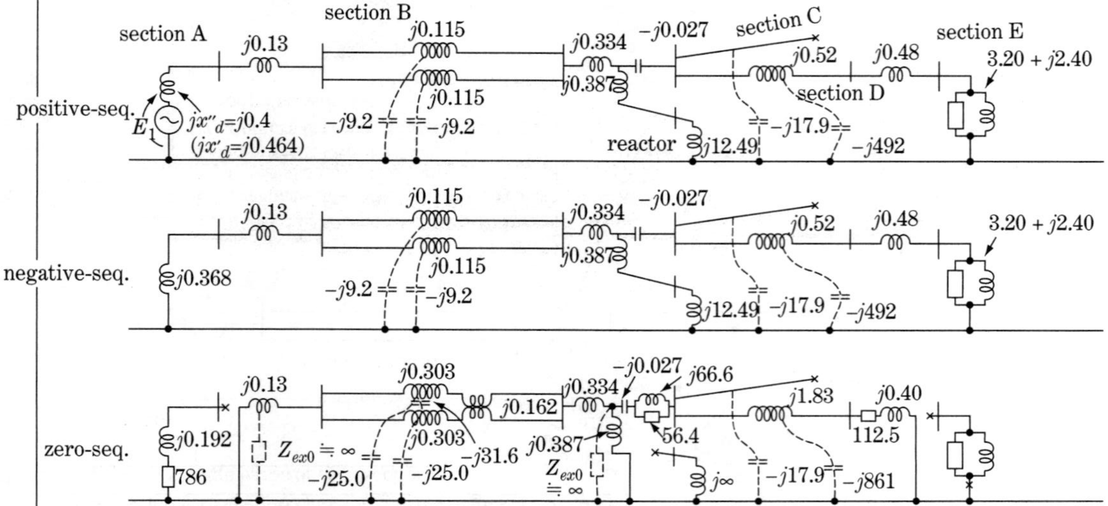

2. Generator: G_1

The given reactances are the percentage reactances based on the 625 MVA rated capacity of the generator. Accordingly, they should be modified into PU values (i.e. [pu]) based on 1000 MVA:

$$jx_1 = \begin{cases} jx''_d = j0.25 \times \dfrac{1000}{625} = j0.400[\text{pu}] \\ jx'_d = j0.29 \times \dfrac{1000}{625} = j0.464[\text{pu}] \\ jx_d = j1.56 \times \dfrac{1000}{625} = j2.495[\text{pu}] \end{cases}$$

$$jx_2 = j0.23 \times \frac{1\,000}{625} = j0.368[\text{pu}]$$

$$x_0 = j0.12 \times \frac{1\,000}{625} = j0.192[\text{pu}]$$

3. The neutral resistance of the generator R_{G_1}

The neutral earthing by 100 A resistance means the resistance value for which a current of 100 A would flow when the phase voltage $22/\sqrt{3}$ kV is charged to the generator neutral point:

$$22/\sqrt{3}\,\text{kV} \times 10^3 = R \times 100$$
$$\therefore R = 127[\Omega] = \frac{127}{0.484}[\text{pu}] = 262\ [\text{pu}] \qquad 3R = 786\ [\text{pu}]$$

4. The transformer Tr_1

The impedance of 13% is based on the rated 1000 MVA, 22/500 kV base so that base quantities need not be changed. The transformer is solidly grounded at the high-tension neutral point. Accordingly, the neutral resistance in Figure d of Table 5.1 would be zero. Then

$$jx_1 = jx_2 = j0.13\ [\text{pu}]$$
$$jx_0 = j0.13\ [\text{pu}]$$
$${}_P\overline{Z}_n = 0 \qquad \overline{Z}_{ex0} = \infty$$

The high-tension side of the zero-sequence circuit is earth grounded through jx_0 while the low-tension side is open.

5. 500 kV double circuit transmission line L_1

Each phase of this line consists of four bundled conductors. As all the conductor sizes and allocations are given, the inductances L and capacitances C can be calculated by applying the equations of Section 1.1 in Chapter 1.

The equivalent radius of the four bundled conductors is (refer to Equations 1.14a–c)

$$r = 0.0192\,\text{m}, \quad w = 0.2\,\text{m}$$
$$r_{\text{eff}} = r^{\frac{1}{n}} \cdot w^{\frac{n-1}{n}} = 0.0192^{\frac{1}{4}} \times 0.20^{\frac{3}{4}} = 0.1113\ [\text{m}]$$

The radius of the overhead grounding wire is

$$r_x = 0.0143\ [\text{m}]$$

The height of the conductors from the imaginary datum plane (see Figure 1.3) is

$$H_e = 300\,[\mathrm{m}]$$
$$h_a + H_a \risingdotseq h_b + H_b \risingdotseq \cdots \risingdotseq h_A + H_A \risingdotseq \cdots \risingdotseq H_x + h_x = 2H = 600\,[\mathrm{m}]$$

The averaged phase-to-phase distance within the same circuit is

$$S_{ll} = (S_{ab} \cdot S_{bc} \cdot S_{ca})^{\frac{1}{3}} = (13.54 \times 13.54 \times 27)^{\frac{1}{3}} = 17.04\,[\mathrm{m}]$$

The averaged distance between one phase of circuit 1 and one phase of circuit 2, S_{lL}, is

$$\begin{aligned} S_{lL} &= \{(S_{aA} \cdot S_{aB} \cdot S_{aC})^{\frac{1}{3}} \cdot (S_{bA} \cdot S_{bB} \cdot S_{bC})^{\frac{1}{3}} \cdot (S_{cA} \cdot S_{cB} \cdot S_{cC})^{\frac{1}{3}}\}^{\frac{1}{3}} \\ &= \{(18 \times 23.3 \times 32.45)^{\frac{1}{3}} \cdot (23.3 \times 20 \times 23.3)^{\frac{1}{3}} \cdot (18 \times 23.3 \times 32.45)^{\frac{1}{3}}\}^{\frac{1}{3}} \\ &= \{23.87 \times 22.14 \times 23.87)^{\frac{1}{3}} = 23.28\,[\mathrm{m}] \end{aligned}$$

The averaged distance between the phase conductor and OGW, S, is

$$S_{lx} = (S_{ax} \cdot S_{bx} \cdot S_{cx})^{\frac{1}{3}} = (13.45 \times 25.53 \times 38.08)^{\frac{1}{3}} = 23.55\,[\mathrm{m}]$$

Finally, the height of the OGW is

$$h_x = 67\,[\mathrm{m}]$$

5(a) Calculation of inductances, impedances

(a1) Self-inductances and impedances including earth grounding effect (but before modification by OGW effect)

Referring to Equation 1.14b, and putting $r_{\mathrm{eff}} = 0.1113$ and $h_a + H_a = 600$ m,

$$\begin{aligned} L_S &= 0.4605 \log_{10} \frac{h_a + H_a}{r_{\mathrm{eff}}} + 0.05\left(1 + \frac{1}{n}\right) = 0.4605 \log_{10} \frac{600}{0.1113} + 0.05\left(1 + \frac{1}{4}\right) \\ &= 1.781\,[\mathrm{mH/km}] \\ \therefore\quad & Z_S \risingdotseq Z_{aa} \risingdotseq Z_{bb} \risingdotseq Z_{AA} \risingdotseq Z_{BB} \risingdotseq \cdots \\ &= jX_S = j2\pi \cdot 50 \cdot 1.781 \times 10^{-3} = j0.559\,[\Omega/\mathrm{km}] \qquad \text{where} \quad f = 50\,\mathrm{Hz} \end{aligned}$$

(a2) Mutual inductance between phases of the same circuit including earth grounding effect (but before modification by OGW effect)

Referring to Equation 1.12a, and putting $S_{ll} = 17.04$ [m],

$$\begin{aligned} L_m &= 0.4605 \log_{10} \frac{h_a + H_a}{S_{ll}} + 0.05 = 0.4605 \log_{10} \frac{600}{17.04} + 0.05 = 0.762\,[\mathrm{mH/km}] \\ \therefore\ Z_m &\risingdotseq Z_{ab} \risingdotseq Z_{bc} \risingdotseq Z_{AB} \risingdotseq \cdots = jX_m = j2\pi \cdot 50 \times 0.762 \times 10^{-3} = j0.239\,[\Omega/\mathrm{km}] \end{aligned}$$

(a3) Mutual inductance between one phase of circuit 1 and one phase of circuit 2 including earth grounding effect (but before modification by OGW effect)

Referring to Equation 1.12a, and putting $S_{lL} = 23.28$,

$$\begin{aligned} L'_m &= 0.4605 \log_{10} \frac{h_a + H_a}{S_{lL}} + 0.05 = 0.4605 \log_{10} \frac{600}{23.28} + 0.05 = 0.700\,[\mathrm{mH/km}] \\ \therefore\ Z'_m &\risingdotseq Z_{aA} \risingdotseq Z_{aB} \risingdotseq Z_{aC} \risingdotseq Z_{bA} \risingdotseq \cdots = jX_{m'} = j2\pi \cdot 50 \times 0.700 \times 10^{-3} = j0.220\,[\Omega/\mathrm{km}] \end{aligned}$$

(a4) Correction of impedances Z_S, Z_m, Z'_m by OGW effect
Referring to Equations 1.15a,b and 1.16, the correction factor δ_{ax} by the OGW effect has to be subtracted from each impedance matrix element in the case of a single circuit line. But for double circuits, the equation before the correction by the OGW effect is given by Equation 1.17. Furthermore, the following equation is obtained as the OGW effect:

$$I_x = -\frac{1}{Z_{xx}}(Z_{xa}I_a + Z_{xb}I_b + Z_{xc}I_c + Z_{xA}I_A + Z_{xB}I_B + Z_{xC}I_C)$$

Thus, the correction factor δ_{ax} is subtracted from all the elements of the 6×6 impedance matrix, in the same way as that of the single circuit line.

Referring to Equation 1.11, and putting $r_x = 0.0143$ m and $h_x + H_x = 600$ m,

$$L_{xx} = 0.4605 \log_{10}\frac{h_x + H_z}{r_x} + 0.1 \fallingdotseq 0.4605 \log_{10}\frac{600}{0.0143} + 0.1 = 2.22 \text{ [mH/km]}$$

$$\therefore\ Z_{xx} = j2\pi \cdot 50 \times 2.18 \times 10^{-3} = j0.697 \text{ [}\Omega\text{/km]}$$

Referring to Equation 1.12b, and putting $h_x + H_x = 600$ m and $S_{lL} = 23.55$ m,

$$L_{lx} = L_{ax} \fallingdotseq L_{bx} \fallingdotseq L_{cx} \fallingdotseq L_{Ax} \fallingdotseq \cdots \fallingdotseq L_{xa} \fallingdotseq \cdots$$

$$= 0.4605 \log_{10}\frac{h_x + H_x}{S_{lx}} + 0.05 = 0.4605 \log_{10}\frac{600}{23.55} + 0.05 = 0.698 \text{ [mH/km]}$$

$$\therefore\ Z_{lx} = j2\pi \cdot 50 \times 0.698 \times 10^{-3} = j0.219 \text{ [}\Omega\text{/km]}$$

The correction factor by OGW is

$$\delta = \frac{Z_{lx} \cdot Z_{lx}}{Z_{xx}} = \frac{(j0.219)^2}{j0.697} = j0.069 \text{ [}\Omega\text{/km]}$$

Accordingly, all the self- and mutual impedance elements have to be subtracted by the same correction factor, $\delta = j0.069\ \Omega$/km:

Z_s, Z_m, Z'_m after correction by OGW effect

$$Z_s = j0.559 - j0.069 = j0.490 \text{ [}\Omega\text{/km]}$$

$$Z_m = j0.239 - j0.069 = j0.170 \text{ [}\Omega\text{/km]}$$

$$Z'_m = j0.220 - j0.069 = j0.151 \text{ [}\Omega\text{/km]}$$

This is the final calculated result for the transmission line self-/mutual impedance matrix.

(a5) Symmetrical impedance: Z_1, Z_2, Z_0, Z_{0M}
Referring to Equation 2.20a,

$$Z_1 = Z_2 = Z_s - Z_m = j0.489 - j0.169 = j0.320 \text{ [}\Omega\text{/km]}$$

$$Z_0 = Z_S + 2Z_m = j0.492 + 2 \times j0.169 = j0.830 \text{ [}\Omega\text{/km]}$$

$$Z_{0M} = 3Z'_m = 3 \times j0.151 = j0.450 \text{ [}\Omega\text{/km]}$$

(a6) Unitization of symmetrical impedance
$Z_{l-\text{gbase}} = 250\,\Omega$, line length $l = 90\,\text{km}$:

$$\overline{Z}_1 = \overline{Z}_2 = \frac{j0.320}{250} \times 90 = j0.115\,[\text{pu}]$$

$$\overline{Z}_0 = \frac{j0.830}{250} \times 90 = j0.303\,[\text{pu}]$$

$$\overline{Z}_{0M} = \frac{j0.453}{250} \times 90 = j0.162\,[\text{pu}]$$

5(b) Calculation of stray capacitance

(b1) Capacitances C_s, C_m, C'_m before correction by OGW effect
The averaged height of phase conductors is

$$h = (h_a \cdot h_b \cdot h_c)^{\frac{1}{3}} = (30 \times 43.5 \times 57)^{\frac{1}{3}} = 42.05\,[\text{m}]$$

The averaged distance between phases of the same circuit is

$$S_{ll} = 17.04\,[\text{m}]$$

The averaged distance between one phase of circuit 1 and one phase of circuit 2 is

$$S_{lL} = 23.28\,[\text{m}]$$

The equivalent conductor radius is

$$r_{\text{eff}} = 0.1113\,[\text{m}]$$

Referring to Equation 1.33,

$$C_s = \frac{0.02413}{\log_{10}\dfrac{8\,h^3}{r_{\text{eff}} \cdot S_{ll}^2}} = \frac{0.02413}{\log_{10}\dfrac{8 \times 42.05^3}{0.1113 \times 17.04^2}} = 0.00566\,[\mu\text{F/km}]$$

$$C_m = C_s \cdot \frac{\log_{10}\dfrac{2h}{S_{ll}}}{\log_{10}\dfrac{S_{ll}}{r_{\text{eff}}}} = 0.00566 \times \frac{\log_{10}\dfrac{2 \times 42.05}{17.04}}{\log_{10}\dfrac{17.04}{0.1113}} = 0.00180\,[\mu\text{F/km}]$$

$$C'_m = \frac{0.02413}{\log_{10}\dfrac{8\,h^3}{r_{\text{eff}} \cdot S_{lL}^2}} \cdot \frac{\log_{10}\dfrac{2h}{S_{lL}}}{\log_{10}\dfrac{S_{lL}}{r_{\text{eff}}}} = \frac{0.02413}{\log_{10}\dfrac{8 \times 42.05^3}{0.1113 \times 23.28^2}} \cdot \frac{\log_{10}\dfrac{2 \times 42.05}{23.28}}{\log_{10}\dfrac{23.28}{0.1113}} = 0.00145\,[\mu\text{F/km}]$$

Then, the capacitance matrix elements of Equation 1.38 are

$$C_s + 2\,C_m + 3\,C'_m = 0.0129\,[\mu\text{F/km}]$$

$$-C_m = -0.00180\,[\mu\text{F/km}]$$

$$-C'_m = -0.00145\,[\mu\text{F/km}]$$

(b2) The effect of OGW on the capacitances

Referring to Figure 1.9, one OGW (symbol x below) is to be added to the situation in the figure. Under this condition, Equation 1.37a for I_a is to be modified by addition of the term $C_{ax}(V_a - V_x)$, where $V_x = 0$. The modified new equation is

$$\begin{aligned} I_a &= j\omega[C_{aa}V_a + C_{ab}(V_a - V_b) + C_{ac}(V_a - V_c) + C_{aA}(V_a - V_A) \\ &+C_{aB}(V_a - V_B) + C_{aC}(V_a - V_C) + \underline{C_{ax}(V_a - V_x)}] \\ &= j\omega[(C_{aa} + C_{ab} + C_{ac} + C_{aA} + C_{aB} + C_{aC} + \underline{C_{ax}})\, V_a - C_{ab}V_b - C_{ac}V_c - C_{aA}V_A - C_{aB}V_B - C_{aC}V_c] \\ &\fallingdotseq j\omega[(C_s + \underline{C_{ax}} + 2\,C_m + 3\,C'_m)\, V_a - C_mV_b - C_mV_c - C'_mV_A - C'_mV_B - C'_mV_C] \end{aligned}$$

Comparing both equations, the resulting modification is $C_{aa} \to C_{aa} + C_{ax}$. That is, it can be said that the self-capacitance C_{aa} becomes a little larger by addition of OGW. In other words, referring to Equations 1.37b and 1.38 and Equations 2.24b and 2.24c, C_s would probably become a few per cent larger value by modification of $C_s \to C_s + C_{lx}$ in comparison with that for the line with the same conductor allocation and without OGW. (Details of this calculation are omitted in this book.)

(b3) Symmetrical capacitance

The modification of the value of C_s by addition of OGW is ignored below ($C_{lx} \fallingdotseq 0$) because it is only a few per cent.

Referring to Equation 2.24c and the Figures 2.8 and 4.4,

$$\begin{aligned} C_1 &= C_2 = C_s + 3C_m + 3C'_m = 0.00566 + 3 \times 0.00180 + 3 \times 0.00145 = 0.0154\ [\mu\text{F/km}] \\ C_0 &= C_s = 0.00566\ [\mu\text{F/km}] \\ C'_0 &= 3C'_m = 3 \times 0.00145 = 0.00435\ [\mu\text{F/km}] \\ \therefore\quad -jX_{c1} &= \frac{1}{jY_1} = \frac{-j}{2\pi \cdot 50 \times 0.0154 \times 10^{-6}} = -j206 \times 10^3\ [\Omega/\text{km}] \\ &= -jX_{c2} = \frac{1}{jY_2} \\ -jX_{c0} &= \frac{1}{jY_0} = \frac{-j}{2\pi \cdot 50 \times 0.00566 \times 10^{-6}} = -j562 \times 10^3\ [\Omega/\text{km}] \\ -jX'_{c0} &= \frac{1}{jY'_0} = \frac{1}{2\pi \cdot 50 \times 0.00435 \times 10^{-6}} = -j712 \times 10^3\ [\Omega/\text{km}] \end{aligned}$$

Incidentally, assuming a modification effect of 3% of C_s by OGW in the above calculation, $C_0 = C_s$ is also modified by 1.05 times, whereas $C_1 = C_2 = \overline{C_s + C_{lx}} + 3C_m + 3C'_m$ is modified 1.01 times. In other words, positive-/negative-sequence capacitances C_1, C_2 are almost not affected by the existence of the OGW, whereas the zero-sequence capacitance C_0 has a little larger value.

(b4) Unitization of symmetrical capacitance

Unitizing the ohmic values by the 250 Ω base,

$$\begin{aligned} -j\overline{X}_{c1} &= -j206 \times 10^3/250 = -j824\ [\text{pu/km}] \\ -j\overline{X}_{c0} &= -j562 \times 10^3/250 = -j2248\ [\text{pu/km}] \\ -j\overline{X}'_{c0} &= -j712 \times 10^3/250 = -j2848\ [\text{pu/km}] \end{aligned}$$

From the concentrated constants of 90 km length,

$$\left.\begin{aligned} -j\overline{X}_{c1} &= -j864 \times 90 = -j9.2\ [\text{pu}] \\ -j\overline{X}_{c0} &= -j2248/90 = -j25.0\ [\text{pu}] \\ -j\overline{X}'_{c0} &= -j2848/90 = -j31.6\ [\text{pu}] \end{aligned}\right\}$$

Accordingly, the charging current per circuit per phase under normal condition is

$$I = 1/9.2 = 0.109\ [\text{pu}] = 0.109 \times 1155\ [\text{A}]/90\ [\text{km}] = 126\ [\text{A}]/90\,\text{km}$$

6. The transformer Tr_2

$$\begin{aligned} {}_{P-S}Z &= j0.23\,[\text{pu}]\,(750\,\text{MVA base}) \\ {}_{P-T}Z &= j0.18\,[\text{pu}]\,(250\,\text{MVA base}) \\ {}_{S-T}Z &= j0.09\,[\text{pu}]\,(250\,\text{MVA base}) \end{aligned}$$

The above %IZ values are written in two different bases where 750 MVA is the rated capacity of the primary and secondary windings, while 250 MVA is that of the tertiary winding. The PU values have to be modified into those for the 1000 MVA base:

$$\begin{aligned} {}_{P-S}Z &= j0.23 \times \frac{1000}{750} = j0.307[\text{pu}](1000\,\text{MVA base}) \\ {}_{P-\Delta}Z &= j0.18 \times \frac{1000}{250} = j0.72[\text{pu}](1000\,\text{MVA base}) \\ {}_{S-\Delta}Z &= j0.09 \times \frac{1000}{250} = j0.36[\text{pu}](1000\,\text{MVA base}) \end{aligned}$$

Then

$$\begin{aligned} {}_{P}Z &= \frac{{}_{P-S}Z + {}_{P-\Delta}Z - {}_{S-\Delta}Z}{2} = \frac{j(0.307 + 0.72 - 0.36)}{2} = j0.334[\text{pu}](1000\,\text{MVA base}) \\ {}_{S}Z &= \frac{{}_{P-S}Z + {}_{S-\Delta}Z - {}_{P-\Delta}Z}{2} = \frac{j(0.307 + 0.36 - 0.72)}{2} = -j0.027[\text{pu}](1000\,\text{MVA base}) \\ {}_{\Delta}Z &= \frac{{}_{P-\Delta}Z + {}_{S-\Delta}Z - {}_{P-S}Z}{2} = \frac{j(0.72 + 0.36 - 0.307)}{2} = j0.387[\text{pu}](1000\,\text{MVA base}) \end{aligned}$$

The equivalent circuit in Figure a of Table 5.1 has been obtained. The derived impedance element ${}_{S}Z$ has a minus sign, so that the element in the equivalent circuit is a series capacitive unit (condenser).

7. The neutral resistance NGR_1 and the neutral reactor NL_1

For NGR_1 the resistance value for $154/\sqrt{3}$ kV and 200 A is

$$r_0 = \frac{154\sqrt{3}}{200} \times 10^3 = 445\,[\Omega]$$

For NL_1 the reactance value for $154/\sqrt{3}$ kV and 15 000 kVA is

$$jx_0 = j\frac{(154/\sqrt{3})^2 \times 10^3}{15\,000} = j527\,[\Omega]$$

Unitizing by the impedance base 23.7 Ω,

$$\begin{aligned} \text{NGR}_1 : r_0 &= \frac{445}{23.7} = 18.8\,[\text{pu}],\ 3r_0 = 56.4[\text{pu}](1000\,\text{MVA base}) \\ \text{NL}_1 : jx_0 &= \frac{j527}{23.7} = j22.2\,[\text{pu}],\ j3x_0 = j66.6[\text{pu}](1000\,\text{MVA base}) \end{aligned}$$

8. 66 kV, 80 MVA reactor bank for voltage regulation LT

The single phase capacity of the reactor is 80/3 MVA. Then

$$jx_1 = jx_2 = j\frac{(V_{l-g})^2}{VA_{1\phi}} = j\frac{\left(\dfrac{66}{\sqrt{3}} \times 10^3\right)^2}{\dfrac{80}{3} \times 10^6} = j54.45\,[\Omega]$$

Unitizing by the impedance base 4.36 Ω,

$$jx_1 = jx_2 = \frac{j54.45}{4.36} = j12.49\,[\text{pu}](1000\,\text{MVA base})$$
$$jx_0 = \infty$$

The zero-sequence impedance is infinitely large because the neutral reactor is open.

9. 154 kV single circuit transmission line L_2

The constants L_s, L_m, C_s, C_m are given. Referring to Equations 2.15 and 2.22b, the impedance for 30 km length is

$$jx_1 = jx_2 = j2\pi f(L_s - L_m) = j2\pi \times 50(2.4 - 1.1) \times 10^{-3} \times 30$$
$$= j12.3\,[\Omega] = j\frac{12.3}{23.7} = j0.52 \quad [\text{pu}]$$
$$jx_0 = j2\pi f(L_s + 2L_m) = j2\pi \times 50(2.4 + 2 \times 1.1) \times 10^{-3} \times 30$$
$$= j43.3\,[\Omega] = j\frac{43.3}{23.7} = j1.83 \quad [\text{pu}]$$

and the capacitance for 30 km is

$$jy_{c1} = jy_{c2} = j2\pi f(C_s + 3C_m) = j2\pi \times 50(0.0052 + 3 \times 0.0013) \times 10^{-6} \times 30 = j85.8 \times 10^{-6}\,[\Omega^{-1}]$$
$$jy_{c0} = j2\pi f C_s = j2\pi \times 50(0.0052) \times 10^{-6} \times 30 = j49.0 \times 10^{-6}\,[\Omega^{-1}]$$
$$\therefore\ -jx_{c1} = -jx_{c2} = \frac{1}{j85.8 \times 10^{-6}} = -j11\,655\,[\Omega] = \frac{-j11\,655}{23.7} = -j492\,[\text{pu}]$$
$$-jx_{c0} = \frac{1}{j49.0 \times 10^{-6}} = -j20\,408\,[\Omega] = \frac{-j20\,408}{23.7} = -j861\,[\text{pu}]$$

The leakage current per phase for 30 km is

$$\frac{1}{492} = 0.002\,[\text{pu}] = 0.002 \times 3749\,[\text{A}] \fallingdotseq 7.5\,[\text{A}]$$

10. 154 kV, 15 km power cable line (three single core cables) L_3

The line consists of three single core cables, so that $C_0 = C_1 = C_2$.

The leakage current for 15 km is 210 A per phase. Then

$$-jx_{c1} = -jx_{c2} = -jx_{c0} = \frac{\frac{154}{\sqrt{3}} \times 10^3}{210} = -j423\,[\Omega]$$
$$\therefore \quad -jx_{c1} = -jx_{c2} = -jx_{c0} = -j\frac{423}{23.7} = -j17.9\,[\text{pu}]$$

Inductance is neglected.

11. The transformer Tr_3 and NGR_2

Modifying the capacity base from 250 MVA to 1000 MVA,

$$j_{P-S}\overline{x} = j0.12 \times \frac{1\,000}{250} = j0.48[\text{pu}](1000\,\text{MVA base})$$
$$j\overline{x}_0 = j0.10 \times \frac{1\,000}{250} = j0.4[\text{pu}](1000\,\text{MVA base})$$

There is also the relation ${}_{P-S}x > x_0$, because jx_0 includes the parallel effect of $\overline{Z}_{ex0}$ in the zero-sequence equivalent circuit of Figure d in Table 5.1.

For NGR_2 at 100 A

$$r_0 = \frac{154/\sqrt{3}}{100} \times 10^3 = 889\,[\Omega]$$

$$\overline{r}_0 = \frac{889}{23.7} = 37.5\,[\text{pu}] \quad \therefore\ 3r_0 = 112.5\,[\text{pu}]$$

12. 66 kV, 200 MW load

At 200 MW with power factor $\cos\varphi = 0.8$,

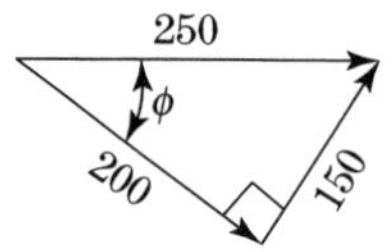

$$\frac{1.0^2}{Z_1} = P - jQ = 200 - j150[\text{MVA}] = \frac{200 - j150}{1000}[\text{pu}]$$

$$= 0.20 - j0.15[\text{pu}](1000\,\text{MVA base})$$

$$\therefore Z_1 = \frac{1}{0.20 - j0.15} = 3.20 + j2.40[\text{pu}]$$

$$= 4.0\angle 37°[\text{pu}](1000\,\text{MVA base})$$

$Z_1 \fallingdotseq Z_2$ is assumed for the load, although the impedance of a rotating load would generally be $Z_1 \neq Z_2$.

Table 5.4 shows the symmetrical equivalent circuits of the power system in Table 5.3 which has been obtained by combining all the results in 1–12 above. This is the circuit, instead of connection diagram, of the given power system which includes visually all the effective *LRC* constant elements. This is of course the first step of any system analysis.

5.7 Supplement: Transformation from Equation 5.18 to Equation 5.19

a) Transformation from Equation 5.18 ① to Equation 5.19 ①

$$\left.\begin{aligned} &\begin{bmatrix} {}_P\overline{V}_a \\ {}_P\overline{V}_b \\ {}_P\overline{V}_c \end{bmatrix} - \begin{bmatrix} {}_P\overline{V}_n \\ {}_P\overline{V}_n \\ {}_P\overline{V}_n \end{bmatrix} = \overline{Z}_{PP} \begin{bmatrix} {}_P\overline{I}_a \\ {}_P\overline{I}_b \\ {}_P\overline{I}_c \end{bmatrix} + \overline{Z}_{PS} \begin{bmatrix} {}_S\overline{I}_a \\ {}_S\overline{I}_b \\ {}_S\overline{I}_c \end{bmatrix} + \overline{Z}_{P\Delta} \begin{bmatrix} {}_\Delta\overline{I}_a \\ {}_\Delta\overline{I}_b \\ {}_\Delta\overline{I}_c \end{bmatrix} \\ \text{or} \quad &{}_P\overline{\boldsymbol{V}}_{abc} - {}_P\overline{\boldsymbol{V}}_n = \overline{Z}_{PP} \cdot {}_P\boldsymbol{I}_{abc} + \overline{Z}_{PS} \cdot {}_S\overline{\boldsymbol{I}}_{abc} + \overline{Z}_{P\Delta} \cdot {}_\Delta\overline{\boldsymbol{I}}_{abc} \end{aligned}\right\} \quad (1)$$

Multiplying the left-hand side by a,

$$\boldsymbol{a} \cdot {}_P\overline{\boldsymbol{V}}_{abc} - \boldsymbol{a} \cdot {}_P\overline{\boldsymbol{V}}_n = \overline{\overline{Z}}_{PP} \cdot \boldsymbol{a} \cdot {}_P\overline{\boldsymbol{I}}_{abc} + \overline{Z}_{PS} \cdot \boldsymbol{a} \cdot {}_S\overline{\boldsymbol{I}}_{abc} + \overline{Z}_{P\Delta} \cdot \boldsymbol{a} \cdot {}_\Delta\overline{\boldsymbol{I}}_{abc}$$

$$\uparrow \qquad \uparrow \qquad \uparrow \qquad \uparrow \qquad \uparrow$$

$${}_P\overline{V}_{012} \qquad \begin{bmatrix} {}_P\overline{V}_n \\ 0 \\ 0 \end{bmatrix} \qquad {}_P\overline{I}_{012} \qquad {}_S\overline{I}_{012} \qquad {}_\Delta\overline{I}_{012} \quad (2)$$

$$\therefore \begin{bmatrix} {}_P\overline{V}_0 \\ {}_P\overline{V}_1 \\ {}_P\overline{V}_2 \end{bmatrix} - \begin{bmatrix} {}_P\overline{V}_n \\ 0 \\ 0 \end{bmatrix} = \overline{Z}_{PP} \begin{bmatrix} {}_P\overline{I}_0 \\ {}_P\overline{I}_1 \\ {}_P\overline{I}_2 \end{bmatrix} = \overline{Z}_{PS} \begin{bmatrix} {}_S\overline{I}_0 \\ {}_S\overline{I}_1 \\ {}_S\overline{I}_2 \end{bmatrix} = \overline{Z}_{P\Delta} \begin{bmatrix} {}_\Delta\overline{I}_0 \\ {}_\Delta\overline{I}_1 \\ {}_\Delta\overline{I}_2 \end{bmatrix}$$

This is the symmetrical equation in regard to primary voltages. The secondary and tertiary side equations are derived analogously.

b) Transformation from Equation 5.18 ③④ to Equation 5.19 ②③
   This is self-explanatory.

c) Transformation from Equation 5.18⑤ to Equation 5.19 ④
   Equation 5.18⑤ is

$$\sqrt{3}\,{}_T\bar{I}_{abc} = \begin{bmatrix} 0 & -1 & 1 \\ 1 & 0 & -1 \\ -1 & 1 & 0 \end{bmatrix} \cdot {}_\Delta\bar{I}_{abc} \tag{3}$$

Then

$$\begin{aligned} {}_T\bar{\boldsymbol{I}}_{012} = a \cdot {}_T\bar{\boldsymbol{I}}_{abc} = \frac{1}{\sqrt{3}} \cdot a \cdot \begin{bmatrix} 0 & -1 & 1 \\ 1 & 0 & -1 \\ -1 & 1 & 0 \end{bmatrix} \cdot a^{-1} \cdot {}_\Delta\bar{\boldsymbol{I}}_{012} = \frac{1}{\sqrt{3}} \begin{bmatrix} 0 \cdot {}_\Delta\bar{I}_0 \\ (a - a^2) \cdot {}_\Delta\bar{I}_1 \\ (a^2 - a) \cdot {}_\Delta\bar{I}_2 \end{bmatrix} \\ = \begin{bmatrix} 0 \cdot {}_\Delta I_0 \\ j \cdot {}_\Delta I_1 \\ -j \cdot {}_\Delta I_2 \end{bmatrix} \end{aligned} \tag{4}$$

This is Equation 5.19 ④. Equation 5.19 ⑤ can be derived from Equation 5.18 ⑥ analogously.

Coffee break 3: Faraday and Henry, the discoverers of the principle of electric energy application

Michael Faraday (1791–1867), probably the greatest experimentalist in electricity and magnetism in the nineteenth century, was not only a famous contributor to chemistry as the discoverer of various new organic compounds, among them benzene, but also the first to liquefy a 'permanent' gas.

Michael Faraday (1791–1867)

In 1821, the year following Oersted's discovery, Faraday had already presumed his idea that magnetic flux should be able to be changed to electric current in contrast to Volta's current-produced magnetic flux. Faraday worked persistently trying to prove that magnetism could induce electricity.

In 1831, after 10 years, Faraday finally succeeded in producing an electric current from a magnetic field. He prepared two coils of wire wound around opposite sides of a ring of soft iron. The first coil was switched on and off by a battery, so the iron ring was magnetized, while a wire from the second coil extended to a compass needle a metre away, far enough so that it was not directly affected by any current in the first coil. When the first coil was turned on, Faraday observed a momentary deflection of the compass needle and its immediate return to its original position. When the primary current was switched off, a similar defection of the compass needle occurred but in the opposite direction. Faraday showed that changes in the magnetic field around the first coil are responsible for inducing the current in the second coil. He also showed that an electric current can be induced by moving a magnet, by turning an electromagnet on and off.

This achievement is Faraday's crowning discovery, because **magnetic induction** means that 'electricity can be steadily produced by a moving magnet which can be driven by mechanical power'. In effect, this became the foundation of the electric dynamo or generator or motor. It was the dawn of a new source of cheap and plentiful energy that was to outpace the conventional **steam engine** very quickly and revolutionize the world. Until Faraday came along, electricity and magnetism were seen as interesting but useless. However, after the invention of magnetic induction, electromagnetism became the subject of industrial energy application.

Joseph Henry (1797–1878) discovered the same principle of electromagnetic induction quite independently in 1830, but his results were not published until after he had received news of Faraday's 1831 work, nor did he develop the discovery as fully as Faraday. Henry wrote in his paper 'self-induction' which he showed by producing large electric arcs from a long helical conductor when it was disconnected from a battery. It is said that Henry met Faraday in 1837, when Henry presented a demonstration of self-induction to both Faraday and Charles Wheatstone (1802–1875). Faraday clapped his hands in delight and exclaimed, 'Hurrah for the Yankee experiment!'

Faraday wrote that the electromotive force induced by magnetic induction is proportional to the ratio of linking flux. He also introduced the concepts of **electric field** and **magnetic field** in which the **forces of electricity and magnetism** respectively exist, although they were not yet derived as mathematical equations. Furthermore, he believed in and wrote about the mutual relation between electricity and heat, between electricity and gravity, between magnetism and gravitation, as analogous to the relation between electricity and magnetism. It is believed that he almost reached the concept of the **law of energy conservation**.

Joseph Henry (1797–1878)

Faraday's electro-magnetic induction ring

6

The α–β–0 Coordinate Method (Clarke Components) and its Application

The $\alpha-\beta-0$ coordinate method ($\alpha-\beta-0$ components or Clarke components) is a useful analytical tool of almost comparable rank with the symmetrical coordinate method (0–1–2 components). Although 0–1–2 components is a very powerful approach for most phenomena, there are some cases where we encounter obstacles. The $\alpha-\beta-0$ coordinate method is another useful analytical approach which can often supplement 0–1–2 components, and by which some limitations of the symmetrical coordinate method are overcome. As a matter of fact, there are some circuits which can be solved only by $\alpha-\beta-0$ components instead of 0–1–2 components. A typical example will be introduced in Chapter 19.

The $\alpha-\beta-0$ components method is also mathematically a kind of variable transformation by the 3×3 matrix operators $\boldsymbol{\alpha}, \boldsymbol{\alpha}^{-1}$, and the important characteristic of $\alpha-\beta-0$ components is that the transformation operators contain only real-number matrix elements, while the symmetrical components method contains matrix operators $\boldsymbol{a}, \boldsymbol{a}^{-1}$ based on complex numbers a, a^{-1}. Voltage or current waveforms observed on an oscillograph, for example, have time-dependent scalar values as a matter of course, so they can be handled as vector (or complex-number) values only when the real-number equations are given for the waveforms. In other words, badly distorted waveform quantities with harmonics would usually be observed on an oscillograph as phenomena of unknown equations and so cannot be transformed into symmetrical components. Conversely, the observed real-time quantities on the oscillograph can be transformed into $\alpha-\beta-0$ components by time sequential composition, regardless of whether the equations are known or unknown.

In this chapter we study the definition and conceptual meanings of the $\alpha-\beta-0$ coordinate method first, and then study the mutual relationship of a–b–c phase quantities, $\alpha-\beta-0$ quantities and 0–1–2 quantities. Finally, we study system modelling and fault analysis by $\alpha-\beta-0$ components.

6.1 Definition of $\alpha-\beta-0$ Coordinate Method ($\alpha-\beta-0$ Components)

The $\alpha-\beta-0$ voltage and current quantities in the $\alpha-\beta-0$ coordinate method ($\alpha-\beta-0$ components) are defined by the following equations.

Handbook of Power System Engineering Yoshihide Hase
© 2007 John Wiley & Sons, Ltd

The transformation is

$$\left.\begin{aligned}
&\underbrace{\begin{bmatrix}V_\alpha\\V_\beta\\V_0\end{bmatrix}}_{\boldsymbol{V}_{\alpha\beta0}} = \underbrace{\frac{1}{3}\begin{bmatrix}2&-1&-1\\0&\sqrt{3}&-\sqrt{3}\\1&1&1\end{bmatrix}}_{\boldsymbol{\alpha}}\cdot\underbrace{\begin{bmatrix}V_a\\V_b\\V_c\end{bmatrix}}_{\boldsymbol{V}_{abc}} \qquad \therefore\ \boldsymbol{V}_{\alpha\beta0}=\boldsymbol{\alpha}\cdot\boldsymbol{V}_{abc}\\
&\underbrace{\begin{bmatrix}I_\alpha\\I_\beta\\I_0\end{bmatrix}}_{\boldsymbol{I}_{\alpha\beta0}} = \underbrace{\frac{1}{3}\begin{bmatrix}2&-1&-1\\0&\sqrt{3}&-\sqrt{3}\\1&1&1\end{bmatrix}}_{\boldsymbol{\alpha}}\cdot\underbrace{\begin{bmatrix}I_a\\I_b\\I_c\end{bmatrix}}_{\boldsymbol{I}_{abc}} \qquad \boldsymbol{I}_{\alpha\beta0}=\boldsymbol{\alpha}\cdot\boldsymbol{I}_{abc}
\end{aligned}\right\} \tag{6.1}$$

and the inverse transformation is

$$\left.\begin{aligned}
&\underbrace{\begin{bmatrix}V_a\\V_b\\V_c\end{bmatrix}}_{\boldsymbol{V}_{\alpha\beta0}} = \underbrace{\begin{bmatrix}1&0&1\\-\frac{1}{2}&\frac{\sqrt{3}}{2}&1\\-\frac{1}{2}&-\frac{\sqrt{3}}{2}&1\end{bmatrix}}_{\boldsymbol{\alpha}^{-1}}\cdot\underbrace{\begin{bmatrix}V_\alpha\\V_\beta\\V_0\end{bmatrix}}_{\boldsymbol{V}_{\alpha\beta0}} \qquad \boldsymbol{V}_{abc}=\boldsymbol{\alpha}^{-1}\cdot\boldsymbol{V}_{\alpha\beta0}\\
&\underbrace{\begin{bmatrix}I_a\\I_b\\I_c\end{bmatrix}}_{\boldsymbol{I}_{abc}} = \underbrace{\begin{bmatrix}1&0&1\\-\frac{1}{2}&\frac{\sqrt{3}}{2}&1\\-\frac{1}{2}&-\frac{\sqrt{3}}{2}&1\end{bmatrix}}_{\boldsymbol{\alpha}^{-1}}\cdot\underbrace{\begin{bmatrix}I_\alpha\\I_\beta\\I_0\end{bmatrix}}_{\boldsymbol{I}_{\alpha\beta0}} \qquad \boldsymbol{I}_{abc}=\boldsymbol{\alpha}^{-1}\cdot\boldsymbol{I}_{\alpha\beta0}
\end{aligned}\right\} \tag{6.2}$$

where

$$\boldsymbol{\alpha}\times\boldsymbol{\alpha}^{-1}=\boldsymbol{\alpha}^{1}\cdot\boldsymbol{\alpha}=\boldsymbol{1}\quad(\boldsymbol{1}\,\text{is the unit matrix})$$
$$\boldsymbol{\alpha},\ \boldsymbol{\alpha}^{-1}:\ \text{inverse matrices of each other}$$

The transform operator matrices $\boldsymbol{\alpha}, \boldsymbol{\alpha}^{-1}$ are real-number matrices in the α–β–0 coordinate method. Therefore if V_a, V_b, V_c are given as real-number quantities, V_α, V_β, V_0 are also real-number quantities, while if V_a, V_b, V_c are given as complex-number quantities, V_α, V_β, V_0 are also complex-number quantities.

The relation **a–b–c phase quantities** $\Leftrightarrow$ α–β–0 **quantities** is shown in Figure 6.1 a and b, where both quantities are demonstrated as vector values.

6.2 Interrelation Between α–β–0 Components and Symmetrical Components

The symmetrical components method is a kind of one-to-one transformation between the domains **a,b,c** $\Leftrightarrow$ **0,1,2**. The α–β–0 components method is also of one-to-one transformation between the domains **a,b,c** $\Leftrightarrow$ $\boldsymbol{\alpha}, \boldsymbol{\beta}$, 0. Therefore symmetrical components and α–β–0 components should also be the same kind of one-to-one transformation between the domains **0,1,2** $\Leftrightarrow$ $\boldsymbol{\alpha}, \boldsymbol{\beta}, \boldsymbol{0}$ for each other.

Let us examine the relation between 0, 1, 2 quantities and $\alpha, \beta, 0$ quantities in detail. We will examine the relation by using voltage symbols here, where lower case letter v means instantaneous real-number voltages, and upper case letter V means instantaneous complex-number voltages.

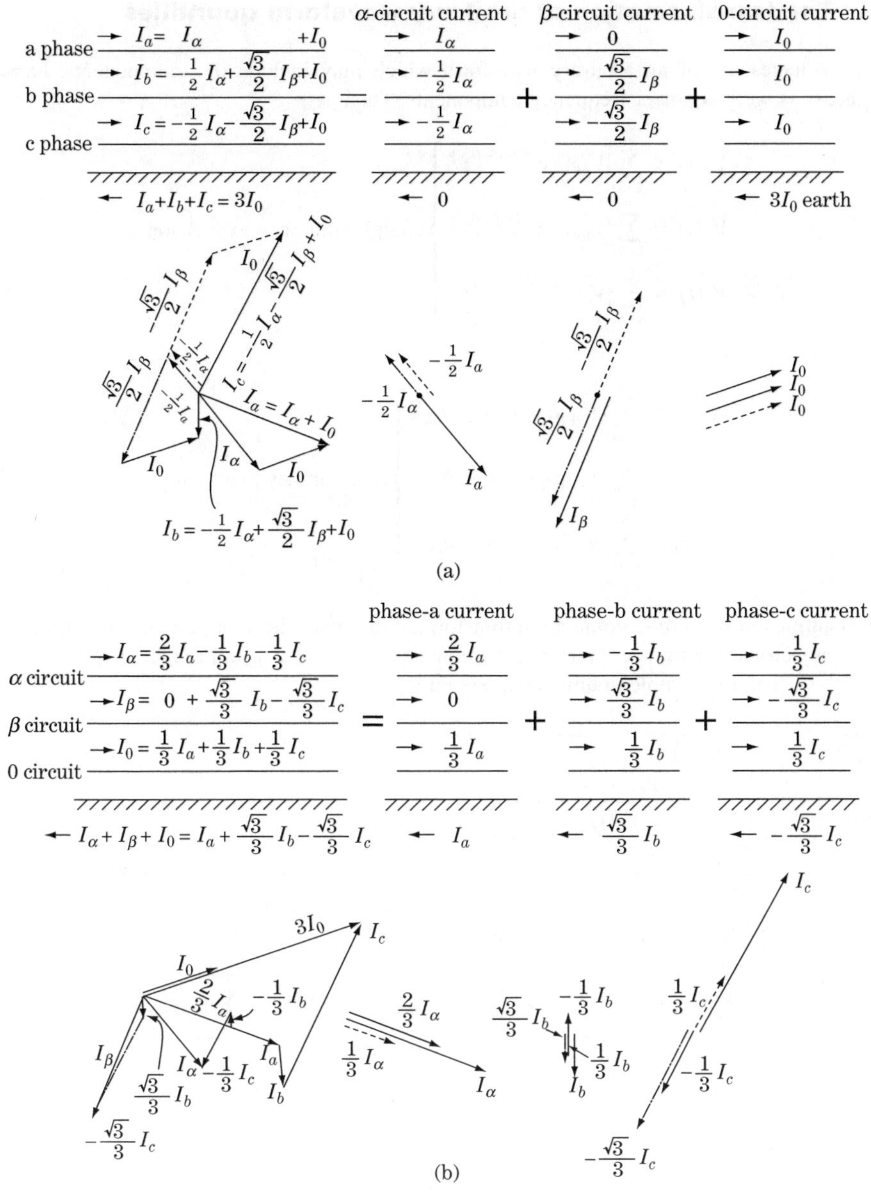

α-quantity: The currents I_α, $-1/2\ I_\alpha$, $-1/2\ I_\alpha$ flow out on the phase a, b, c circuits respectively in the same direction. In other words, the current I_α goes out from phase a and the half current $1/2\ I_\alpha$ comes back from phase b and another half current $1/2\ I_\alpha$ from phase c. As a result, the current through the ground pass is zero.

β-quantity: The currents $+\sqrt{3}/2\ I_\beta$, $-\sqrt{3}/2 I_\beta$, flow in phase b, c respectively in the same direction. In other words, current $\sqrt{3}/2\ I_\beta$ goes out from phase b and comes back from phase c. As a result, the current through phase a and the ground pass are zero.

0-quantity: The currents of the same value I_0 flow out from phase a, phase b, phase c respectively in the same direction and the current $3I_0$ comes back from the ground pass. In other words, the definition of 0-quantitiy is exactly the same as the zero-sequence quantity in symmetrical components.

Figure 6.1 Correlation of vectors I_a, I_b, I_c and I_α, I_β, I_0

6.2.1 The transformation of arbitrary waveform quantities

Suppose voltages are of an arbitrary waveform which may include d.c. components, harmonic components as well as power frequency components. Then

$$\left.\begin{aligned} V_a(t) &= \sum_{k=0}^{n} |V_{ak}| \cdot e^{j(k\omega t+\theta_{ak})} \\ V_b(t) &= \sum_{k=0}^{n} |V_{bk}| \cdot e^{j(k\omega t+\theta_{bk})} \\ V_c(t) &= \sum_{k=0}^{n} |V_{ck}| \cdot e^{j(k\omega t+\theta_{ck})} \end{aligned}\right\} \text{complex-number expression} \tag{6.3a}$$

$$\left.\begin{aligned} v_a(t) &= \sum_{k=0}^{n} |V_{ak}|\cos(k\omega t+\theta_{ak}) \\ v_b(t) &= \sum_{k=0}^{n} |V_{bk}|\cos(k\omega t+\theta_{bk}) \\ v_c(t) &= \sum_{k=0}^{n} |V_{ck}|\cos(k\omega t+\theta_{ck}) \end{aligned}\right\} \text{real-number expression} \tag{6.3b}$$

These quantities in the a–b–c domain are transformed into two different domains as follows.

For symmetrical coordinates, referring to Equations 2.1 and 2.4 and recalling that $a = e^{j120^\circ}$, $a^2 = e^{-j120^\circ}$, then the complex-number expression is

$$\begin{aligned} \begin{bmatrix} V_0(t) \\ V_1(t) \\ V_2(t) \end{bmatrix} &= \frac{1}{3} \begin{bmatrix} V_a(t)+V_b(t)+V_c(t) \\ V_a(t)+aV_b(t)+a^2V_c(t) \\ V_a(t)+a^2V_b(t)+aV_c(t) \end{bmatrix} \\ &= \frac{1}{3} \begin{bmatrix} \sum_{k=0}^{n} |V_{ak}| \cdot e^{j(k\omega t+\theta_{ak})} + \sum_{k=0}^{n} |V_{bk}| \cdot e^{j(k\omega t+\theta_{bk})} + \sum_{k=0}^{n} |V_{ck}| \cdot e^{j(k\omega t+\theta_{ck})} \\ \sum_{k=0}^{n} |V_{ak}| \cdot e^{j(k\omega t+\theta_{ak})} + \sum_{k=0}^{n} |V_{bk}| \cdot e^{j(k\omega t+\theta_{bk}+120^\circ)} + \sum_{k=0}^{n} |V_{ck}| \cdot e^{j(k\omega t+\theta_{ck}-120^\circ)} \\ \sum_{k=0}^{n} |V_{ak}| \cdot e^{j(k\omega t+\theta_{ak})} + \sum_{k=0}^{n} |V_{bk}| \cdot e^{j(k\omega t+\theta_{bk}-120^\circ)} + \sum_{k=0}^{n} |V_{ck}| \cdot e^{j(k\omega t+\theta_{ck}+120^\circ)} \end{bmatrix} \end{aligned} \tag{6.4a}$$

For the real-number expression, taking the real part of the above equation,

$$\begin{aligned} \begin{bmatrix} v_0(t) \\ v_1(t) \\ v_2(t) \end{bmatrix} &= \begin{bmatrix} \text{Re}\{V_0(t)\} \\ \text{Re}\{V_1(t)\} \\ \text{Re}\{V_2(t)\} \end{bmatrix} \\ &= \frac{1}{3} \begin{bmatrix} \sum_{k=0}^{n} |V_{ak}|\cos(k\omega t+\theta_{ak}) + \sum_{k=0}^{n} |V_{bk}|\cos(k\omega t+\theta_{bk}) + \sum_{k=0}^{n} |V_{ck}|\cos(k\omega t+\theta_{ck}) \\ \sum_{k=0}^{n} |V_{ak}|\cos(k\omega t+\theta_{ak}) + \sum_{k=0}^{n} |V_{bk}|\cos(k\omega t+\theta_{bk}+120^\circ) + \sum_{k=0}^{n} |V_{ck}|\cos(k\omega t+\theta_{ck}-120^\circ) \\ \sum_{k=0}^{n} |V_{ak}|\cos(k\omega t+\theta_{ak}) + \sum_{k=0}^{n} |V_{bk}|\cos(k\omega t+\theta_{bk}-120^\circ) + \sum_{k=0}^{n} |V_{ck}|\cos(k\omega t+\theta_{ck}+120^\circ) \end{bmatrix} \end{aligned} \tag{6.4b}$$

The α–β–0 components are defined for complex-number and real-number expressions by the same equations, 6.1 and 6.2. Then, for the complex-number expression,

$$\begin{bmatrix} V_\alpha(t) \\ V_\beta(t) \\ V_0(t) \end{bmatrix} = \frac{1}{3}\begin{bmatrix} 2V_a(t) - V_b(t) - V_c(t) \\ \sqrt{3}\{V_b(t) - V_c(t)\} \\ V_a(t) + V_b(t) + V_c(t) \end{bmatrix}$$

$$= \frac{1}{3}\begin{bmatrix} 2\sum_{k=0}^{n} |V_{ak}| e^{j(k\omega t+\theta_{ak})} - \sum_{k=0}^{n} |V_{bk}| e^{j(k\omega t+\theta_{bk})} - \sum_{k=0}^{n} |V_{ck}| e^{j(k\omega t+\theta_{ck})} \\ \sqrt{3}\left\{\sum_{k=0}^{n} |V_{bk}| e^{j(k\omega t+\theta_{bk})} - \sum_{k=0}^{n} |V_{ck}| e^{j(k\omega t+\theta_{ck})}\right\} \\ \sum_{k=0}^{n} |V_{ak}| e^{j(k\omega t+\theta_{ak})} + \sum_{k=0}^{n} |V_{bk}| e^{j(k\omega t+\theta_{bk})} + \sum_{k=0}^{n} |V_{ck}| e^{j(k\omega t+\theta_{ck})} \end{bmatrix} \quad (6.5a)$$

and for the real-number expression

$$\begin{bmatrix} v_\alpha(t) \\ v_\beta(t) \\ v_0(t) \end{bmatrix} = \frac{1}{3}\begin{bmatrix} 2v_a(t) - v_b(t) - v_c(t) \\ \sqrt{3}\{v_b(t) - v_c(t)\} \\ v_a(t) + v_b(t) + v_c(t) \end{bmatrix}$$

$$= \frac{1}{3}\begin{bmatrix} 2\sum_{k=0}^{n} |V_{ak}|\cos(k\omega t + \theta_{ak}) - \sum_{k=0}^{n} |V_{bk}|\cos(k\omega t + \theta_{bk}) - \sum_{k=0}^{n} |V_{ck}|\cos(k\omega t + \theta_{ck}) \\ \sqrt{3}\left\{\sum_{k=0}^{n} |V_{bk}|\cos(k\omega t + \theta_{bk}) - \sum_{k=0}^{n} |V_{ck}|\cos(k\omega t + \theta_{ck})\right\} \\ \sum_{k=0}^{n} |V_{ak}|\cos(k\omega t + \theta_{ak}) + \sum_{k=0}^{n} |V_{bk}|\cos(k\omega t + \theta_{bk}) + \sum_{k=0}^{n} |V_{ck}|\cos(k\omega t + \theta_{ck}) \end{bmatrix}$$

(6.5b)

6.2.2 Interrelation between α–β–0 and symmetrical components

Now we need to examine the mutual relation of both domains. Applying the complex-number expression,

$$\left.\begin{array}{l} \text{The equations of symmetrical components by Equations 2.1 and 2.4:} \\ \qquad \boldsymbol{V}_{012} = \boldsymbol{a} \cdot \boldsymbol{V}_{abc} \quad ① \qquad \boldsymbol{V}_{abc} = \boldsymbol{a}^{-1} \cdot \boldsymbol{V}_{012} \quad ② \\ \text{The equations of } \alpha\text{–}\beta\text{–}0 \text{ components by Equations 6.1 and 6.2:} \\ \qquad \boldsymbol{V}_{\alpha\beta 0} = \boldsymbol{\alpha} \cdot \boldsymbol{V}_{abc} \quad ③ \qquad \boldsymbol{V}_{abc} = \boldsymbol{\alpha}^{-1} \cdot \boldsymbol{V}_{\alpha\beta 0} \quad ④ \end{array}\right\} \quad (6.6)$$

Accordingly,

$$\left.\begin{array}{l} \boldsymbol{V}_{\alpha\beta 0} = \boldsymbol{\alpha} \cdot \boldsymbol{V}_{abc} = \boldsymbol{\alpha} \cdot (\boldsymbol{a}^{-1} \cdot \boldsymbol{V}_{012}) = (\boldsymbol{\alpha} \cdot \boldsymbol{a}^{-1}) \cdot \boldsymbol{V}_{012} \\ \boldsymbol{V}_{012} = \boldsymbol{a} \cdot \boldsymbol{V}_{abc} = \boldsymbol{a} \cdot (\boldsymbol{\alpha}^{-1} \cdot \boldsymbol{V}_{\alpha\beta 0}) = (\boldsymbol{a} \cdot \boldsymbol{\alpha}^{-1}) \cdot \boldsymbol{V}_{\alpha\beta 0} \end{array}\right\} \quad (6.7)$$

where $\boldsymbol{\alpha}\cdot\boldsymbol{a}^{-1}$ and $\boldsymbol{a}\cdot\boldsymbol{\alpha}^{-1}$ are calculated below:

$$\left.\begin{aligned}
\boldsymbol{\alpha}\cdot\boldsymbol{a}^{-1} &= \frac{1}{3}\begin{bmatrix} 2 & -1 & -1 \\ 0 & \sqrt{3} & -\sqrt{3} \\ 1 & 1 & 1 \end{bmatrix}\cdot\begin{bmatrix} 1 & 1 & 1 \\ 1 & a^2 & a \\ 1 & a & a^2 \end{bmatrix} \\
&= \frac{1}{3}\begin{bmatrix} 0 & 2-(a^2+a) & 2-(a+a^2) \\ 0 & \sqrt{3}(a^2-a) & \sqrt{3}(a-a^2) \\ 3 & 1+a^2+a & 1+a+a^2 \end{bmatrix} = \begin{bmatrix} 0 & 1 & 1 \\ 0 & -j & j \\ 1 & 0 & 0 \end{bmatrix} \\
\boldsymbol{a}\cdot\boldsymbol{\alpha}^{-1} &= \frac{1}{3}\begin{bmatrix} 1 & 1 & 1 \\ 1 & a & a^2 \\ 1 & a^2 & a \end{bmatrix}\cdot\begin{bmatrix} 1 & 0 & 1 \\ -\frac{1}{2} & \frac{\sqrt{3}}{2} & 1 \\ -\frac{1}{2} & -\frac{\sqrt{3}}{2} & 1 \end{bmatrix} = \frac{1}{2}\begin{bmatrix} 0 & 0 & 2 \\ 1 & j & 0 \\ 1 & -j & 0 \end{bmatrix}
\end{aligned}\right\} \quad (6.8)$$

Therefore

$$\underset{\boldsymbol{V}_{\alpha\beta0}}{\begin{bmatrix} V_\alpha \\ V_\beta \\ V_0 \end{bmatrix}} = \underset{(\boldsymbol{\alpha}\cdot\boldsymbol{a}^{-1})}{\begin{bmatrix} 0 & 1 & 1 \\ 0 & -j & j \\ 1 & 0 & 0 \end{bmatrix}} \cdot \underset{\cdot\boldsymbol{V}_{012}}{\begin{bmatrix} V_0 \\ V_1 \\ V_2 \end{bmatrix}} \quad \text{or} \quad \left.\begin{aligned} V_\alpha(t) &= V_1(t)+V_2(t) \\ V_\beta(t) &= -j\{V_1(t)-V_2(t)\} \\ V_0(t) &= V_0(t) \end{aligned}\right\} \quad (6.9)$$

$$\underset{\boldsymbol{V}_{012}}{\begin{bmatrix} V_0 \\ V_1 \\ V_2 \end{bmatrix}} = \frac{1}{2}\underset{(\boldsymbol{a}\cdot\boldsymbol{\alpha}^{-1})}{\begin{bmatrix} 0 & 0 & 2 \\ 1 & j & 0 \\ 1 & -j & 0 \end{bmatrix}} \cdot \underset{\cdot\boldsymbol{V}_{\alpha\beta0}}{\begin{bmatrix} V_\alpha \\ V_\beta \\ V_0 \end{bmatrix}} \quad \text{or} \quad \left.\begin{aligned} V_0(t) &= V_0(t) \\ V_1(t) &= \tfrac{1}{2}\{V_\alpha(t)+jV_\beta(t)\} \\ V_2(t) &= \tfrac{1}{2}\{V_\alpha(t)-jV_\beta(t)\} \end{aligned}\right\} \quad (6.10)$$

Equations 6.9 and 6.10 show the relation between α–β–0 components and symmetrical components that are written in complex-number quantities.

In words, V_α is the vector sum of positive-sequence voltage V_1 and negative-sequence voltage V_2, namely V_1+V_2. V_β is the product of ($-j$) and (V_1-V_2), or the vector which is obtained by a 90° clockwise rotation of subtracted vector (V_1-V_2).

For the relation for power frequency components, the symmetrical components are

$$\left.\begin{aligned}
&\text{By complex number expression} && \text{By real number expression} \\
&V_0(t) = |V_0|\,e^{j(\omega t+\theta_0)} && v_0(t) = |V_0|\cos(\omega t+\theta_0) \\
&V_1(t) = |V_1|\,e^{j(\omega t+\theta_1)} && v_1(t) = |V_1|\cos(\omega t+\theta_1) \\
&V_2(t) = |V_2|\,e^{j(\omega t+\theta_2)} && v_2(t) = |V_2|\cos(\omega t+\theta_2)
\end{aligned}\right\} \quad (6.11)$$

and the α–β–0 components are, for the complex-number expression,

$$\left.\begin{aligned}
V_\alpha(t) &= V_1(t)+V_2(t) = |V_1|e^{j(\omega t+\theta_1)} + |V_2|e^{j(\omega t+\theta_2)} \\
V_\beta(t) &= -j\{V_1(t)-V_2(t)\} = e^{-j90^\circ}\{|V_1|e^{j(\omega t+\theta_1)} - |V_2|e^{j(\omega t+\theta_2)}\} \\
V_0(t) &= |V_0|e^{j(\omega t+\theta_0)}
\end{aligned}\right\} \quad ①$$

and for the real-number expression

$$\left.\begin{aligned}
v_\alpha(t) &= |V_1|\cos(\omega t+\theta_1) + |V_2|\cos(\omega t+\theta_2) \\
v_\beta(t) &= |V_1|\cos(\omega t+\theta_1-90^\circ) - |V_2|\cos(\omega t+\theta_2-90^\circ) \\
&= |V_1|\sin(\omega t+\theta_1) - |V_2|\sin(\omega t+\theta_2) \\
v_0(t) &= |V_0|\cos(\omega t+\theta_0)
\end{aligned}\right\} \quad ②$$

(6.12)

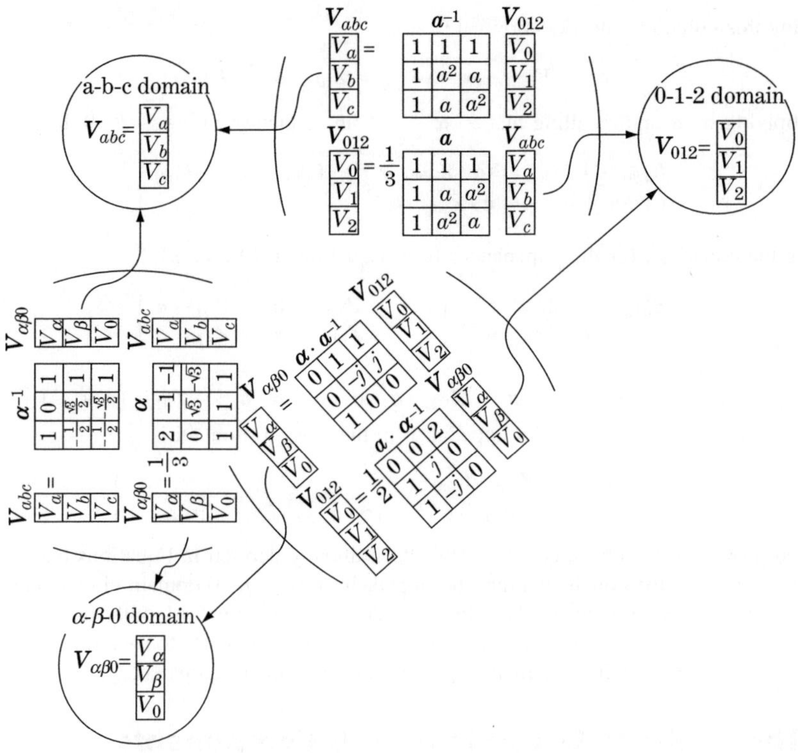

Figure 6.2 Correlation of a–b–c, 0–1–2 and $\alpha-\beta-0$ domains

Figure 6.2 is a summary of the concept showing mutual relations of the a–b–c, 0–1–2 and $\alpha-\beta-0$ domains.

6.3 Circuit Equation and Impedance by the $\alpha-\beta-0$ Coordinate Method

The general equation for a three-phase circuit is expressed by

$$\boldsymbol{E}_{abc} - \boldsymbol{V}_{abc} = \boldsymbol{Z}_{abc} \cdot \boldsymbol{I}_{abc} \tag{6.13}$$

The equation is transformed into the 0–1–2 domain as follows:

$$\left.\begin{aligned} &\boldsymbol{E}_{abc} = \boldsymbol{a}^{-1} \cdot \boldsymbol{E}_{012}, \quad \boldsymbol{V}_{abc} = \boldsymbol{a}^{-1} \cdot \boldsymbol{V}_{012}, \quad \boldsymbol{I}_{abc} = \boldsymbol{a}^{-1} \cdot \boldsymbol{I}_{012} \\ &\therefore \boldsymbol{E}_{012} - \boldsymbol{V}_{012} = (\boldsymbol{a} \cdot \boldsymbol{Z}_{abc} \cdot \boldsymbol{a}^{-1}) \cdot \boldsymbol{I}_{012} \equiv \boldsymbol{Z}_{012} \cdot \boldsymbol{I}_{012} \\ &\qquad \text{where } \boldsymbol{Z}_{012} = (\boldsymbol{a} \cdot \boldsymbol{Z}_{abc} \cdot \boldsymbol{a}^{-1}) \end{aligned}\right\} \tag{6.14}$$

Equation 6.13 can be transformed into the $\alpha-\beta-0$ domain in a similar way.

The definition of $\alpha-\beta-0$ components is

$$\left.\begin{aligned} \boldsymbol{E}_{\alpha\beta0} &= \boldsymbol{\alpha} \cdot \boldsymbol{E}_{abc}, & \boldsymbol{E}_{abc} &= \boldsymbol{\alpha}^{-1} \cdot \boldsymbol{E}_{\alpha\beta0} \\ \boldsymbol{V}_{\alpha\beta0} &= \boldsymbol{\alpha} \cdot \boldsymbol{V}_{abc}, & \boldsymbol{V}_{abc} &= \boldsymbol{\alpha}^{-1} \cdot \boldsymbol{V}_{\alpha\beta0} \\ \boldsymbol{I}_{\alpha\beta0} &= \boldsymbol{\alpha} \cdot \boldsymbol{I}_{abc}, & \boldsymbol{I}_{abc} &= \boldsymbol{\alpha}^{-1} \cdot \boldsymbol{I}_{\alpha\beta0} \end{aligned}\right\} \tag{6.15}$$

Substituting this equation into Equation 6.13,

$$\boldsymbol{\alpha}^{-1}\cdot\boldsymbol{E}_{\alpha\beta 0}-\boldsymbol{\alpha}^{-1}\cdot\boldsymbol{V}_{\alpha\beta 0}=\boldsymbol{Z}_{abc}\cdot\boldsymbol{\alpha}^{-1}\cdot\boldsymbol{I}_{\alpha\beta 0}$$

Left-multiplying by $\boldsymbol{\alpha}$ and recalling that $\boldsymbol{\alpha}\cdot\boldsymbol{\alpha}^{-1}=\mathbf{1}$, the equation in the α–β–0 domain is

$$\left.\begin{aligned}\boldsymbol{E}_{\alpha\beta 0}-\boldsymbol{V}_{\alpha\beta 0}&=(\boldsymbol{\alpha}\cdot\boldsymbol{Z}_{abc}\cdot\boldsymbol{\alpha}^{-1})\cdot\boldsymbol{I}_{\alpha\beta 0}\equiv\boldsymbol{Z}_{\alpha\beta 0}\cdot\boldsymbol{I}_{\alpha\beta 0}\\ \text{where }\boldsymbol{Z}_{\alpha\beta 0}&=\boldsymbol{\alpha}\cdot\boldsymbol{Z}_{abc}\cdot\boldsymbol{\alpha}^{-1}\end{aligned}\right\}\quad(6.16)$$

Extracting the equations for the impedances from Equations 6.14 and 6.16,

$$\left.\begin{aligned}\boldsymbol{Z}_{012}&=\boldsymbol{a}\cdot\boldsymbol{Z}_{abc}\cdot\boldsymbol{a}^{-1}\\ \boldsymbol{Z}_{\alpha\beta 0}&=\boldsymbol{\alpha}\cdot\boldsymbol{Z}_{abc}\cdot\boldsymbol{\alpha}^{-1}\end{aligned}\right\}\ \text{or}\ \left.\begin{aligned}\boldsymbol{Z}_{abc}&=\boldsymbol{a}^{-1}\cdot\boldsymbol{Z}_{012}\cdot\boldsymbol{a}\\ \boldsymbol{Z}_{abc}&=\boldsymbol{\alpha}^{-1}\cdot\boldsymbol{Z}_{\alpha\beta 0}\cdot\boldsymbol{\alpha}\end{aligned}\right\}\quad(6.17a)$$

$$\therefore\boldsymbol{Z}_{\alpha\beta 0}=\boldsymbol{\alpha}\cdot\boldsymbol{Z}_{abc}\cdot\boldsymbol{\alpha}^{-1}=\boldsymbol{\alpha}\cdot(\boldsymbol{a}^{-1}\cdot\boldsymbol{Z}_{012}\cdot\boldsymbol{a})\cdot\boldsymbol{\alpha}^{-1}=(\boldsymbol{\alpha}\cdot\boldsymbol{a}^{-1})\cdot\boldsymbol{Z}_{012}\cdot(\boldsymbol{a}\cdot\boldsymbol{\alpha}^{-1})\quad(6.17b)$$

Accordingly, the circuit equation and impedance in the α–β–0 domain is

$$\left.\begin{aligned}&\boldsymbol{E}_{\alpha\beta 0}-\boldsymbol{V}_{\alpha\beta 0}=\boldsymbol{Z}_{\alpha\beta 0}\cdot\boldsymbol{I}_{\alpha\beta 0}\\ &\text{where }\boldsymbol{Z}_{\alpha\beta 0}=\boldsymbol{\alpha}\cdot\boldsymbol{Z}_{abc}\cdot\boldsymbol{\alpha}^{-1}=(\boldsymbol{\alpha}\cdot\boldsymbol{a}^{-1})\cdot\boldsymbol{Z}_{012}\cdot(\boldsymbol{a}\cdot\boldsymbol{\alpha}^{-1})\end{aligned}\right\}\quad(6.18)$$

$\boldsymbol{\alpha}\cdot\boldsymbol{a}^{-1}$ and $\boldsymbol{a}\cdot\boldsymbol{\alpha}^{-1}$ in the above equation have been already derived in Equations 6.8.

Now we can draw the conclusion that the impedances in α–β–0 domain circuits are given by Equations 6.18. Further, Equation 6.13 for the a–b–c domain, Equation 6.14 for the 0–1–2 domain and equation 6.18 for the α–β–0 domain are in one-to-one correspondence to each other.

In the next section we will investigate $Z_{\alpha\beta 0}$ for lines and other equipment.

6.4 Three-phase Circuit in α–β–0 Components

6.4.1 Single circuit transmission line

A well-balanced three-phase single circuit transmission line between points m and n as shown in Figure 1.1 has its impedance matrix $\boldsymbol{Z}_{012}$ given by Equation 2.15 and Figure 2.5, and is again quoted here:

$$\boldsymbol{Z}_{012}=\begin{bmatrix}Z_0&0&0\\0&Z_1&0\\0&0&Z_2\end{bmatrix}\quad\left.\begin{aligned}Z_1&=Z_2=Z_s-Z_m\\Z_0&=Z_s+2Z_m\end{aligned}\right\}\quad(6.19)$$

Accordingly,

$$\begin{aligned}\boldsymbol{Z}_{\alpha\beta 0}&=\begin{bmatrix}Z_{\alpha\alpha}&Z_{\alpha\beta}&Z_{\alpha 0}\\Z_{\beta\alpha}&Z_{\beta\beta}&Z_{\beta 0}\\Z_{0\alpha}&Z_{0\beta}&Z_{00}\end{bmatrix}=(\boldsymbol{\alpha}\cdot\boldsymbol{a}^{-1})\cdot\boldsymbol{Z}_{012}\cdot(\boldsymbol{a}\cdot\boldsymbol{\alpha}^{-1})\\ &=\begin{bmatrix}0&1&1\\0&-j&j\\1&0&0\end{bmatrix}\cdot\begin{bmatrix}Z_0&0&0\\0&Z_1&0\\0&0&Z_2\end{bmatrix}\cdot\frac{1}{2}\begin{bmatrix}0&0&2\\1&j&0\\1&-j&0\end{bmatrix}\\ &=\begin{bmatrix}\frac{1}{2}(Z_1+Z_2)&\frac{1}{2}j(Z_1-Z_2)&0\\-\frac{1}{2}j(Z_1-Z_2)&\frac{1}{2}(Z_1+Z_2)&0\\0&0&Z_0\end{bmatrix}\end{aligned}\quad(6.20)$$

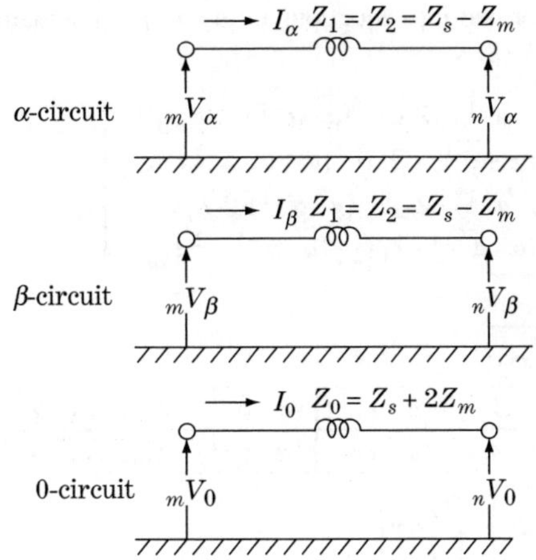

Figure 6.3 The equivalent circuit of a transmission line in α–β–0 components (single circuit line)

As $Z_1 = Z_2$ is always correct for transmission lines, then

$$\underbrace{\begin{bmatrix} {}_mV_\alpha \\ {}_mV_\beta \\ {}_mV_0 \end{bmatrix}}_{{}_mV_{\alpha\beta0}} - \underbrace{\begin{bmatrix} {}_nV_\alpha \\ {}_nV_\beta \\ {}_nV_0 \end{bmatrix}}_{{}_nV_{\alpha\beta0}} = \underbrace{\begin{bmatrix} Z_1 & 0 & 0 \\ 0 & Z_1 & 0 \\ 0 & 0 & Z_0 \end{bmatrix}}_{\boldsymbol{Z}_{\alpha\beta0}} \cdot \underbrace{\begin{bmatrix} I_\alpha \\ I_\beta \\ I_0 \end{bmatrix}}_{\boldsymbol{I}_{\alpha\beta0}} \qquad \left.\begin{aligned} Z_1 &= Z_2 = Z_s - Z_m \\ Z_0 &= Z_s + 2Z_m \end{aligned}\right\} \tag{6.21}$$

This is the fundamental equation of the transmission line in α–β–0 components and the equivalent circuit of the equation is given in Figure 6.3. The impedance matrix $\boldsymbol{Z}_{\alpha\beta0}$ is the same as $\boldsymbol{Z}_{012}$, namely $\boldsymbol{Z}_{\alpha\beta0} = \boldsymbol{Z}_{012}$. In other words, the major feature of α–β–0 components is that a well-balanced transmission line can be expressed by the same impedance matrix and the same equivalent circuits in the α–β–0 domain as well as in the 0–1–2 domain. The strange coefficient $\pm\sqrt{3/2}$ in the definition of α–β–0 components was a device to obtain this advantage.

6.4.2 Double circuit transmission line

A well-balanced double circuit line is shown by Equations 2.20a and b in the 0–1–2 domain, and is again quoted here:

$$\begin{bmatrix} {}^1_m\boldsymbol{V}_{012} \\ {}^2_m\boldsymbol{V}_{012} \end{bmatrix} - \begin{bmatrix} {}^1_n\boldsymbol{V}_{012} \\ {}^2_n\boldsymbol{V}_{012} \end{bmatrix} = \begin{bmatrix} \boldsymbol{Z}_{012} & \boldsymbol{Z}_{0M} \\ \boldsymbol{Z}_{0M} & \boldsymbol{Z}_{012} \end{bmatrix} \cdot \begin{bmatrix} {}^1\boldsymbol{I}_{012} \\ {}^2\boldsymbol{I}_{012} \end{bmatrix}$$

or

$$\left.\begin{aligned} {}^1_m\boldsymbol{V}_{012} - {}^1_n\boldsymbol{V}_{012} &= \boldsymbol{Z}_{012} \cdot {}^1\boldsymbol{I}_{012} + \boldsymbol{Z}_{0M} \cdot {}^2\boldsymbol{I}_{012} \\ {}^2_m\boldsymbol{V}_{012} - {}^2_n\boldsymbol{V}_{012} &= \boldsymbol{Z}_{0M} \cdot {}^1\boldsymbol{I}_{012} + \boldsymbol{Z}_{012} \cdot {}^2\boldsymbol{I}_{012} \end{aligned}\right\}$$

$$\text{where} \quad \boldsymbol{Z}_{012} = \begin{bmatrix} Z_0 & 0 & 0 \\ 0 & Z_1 & 0 \\ 0 & 0 & Z_1 \end{bmatrix} \quad \boldsymbol{Z}_{0M} = \begin{bmatrix} Z_{0M} & 0 & 0 \\ 0 & 0 & 0 \\ 0 & 0 & 0 \end{bmatrix} \tag{6.22}$$

Equation 6.22 can be transformed into equations in the α–β–0 domain in the same way as that of Equation 6.14 to Equation 6.18:

$$\left.\begin{aligned}{}^1_m\boldsymbol{V}_{\alpha\beta0}-{}^1_n\boldsymbol{V}_{\alpha\beta0}&=(\boldsymbol{\alpha}\cdot\boldsymbol{a}^{-1})\cdot\boldsymbol{Z}_{012}\cdot(\boldsymbol{a}\cdot\boldsymbol{\alpha}^{-1})\cdot{}^1\boldsymbol{I}_{\alpha\beta0}\\&\quad+(\boldsymbol{\alpha}\cdot\boldsymbol{a}^{-1})\cdot\boldsymbol{Z}_{0M}\cdot(\boldsymbol{a}\cdot\boldsymbol{\alpha}^{-1})\cdot{}^2\boldsymbol{I}_{\alpha\beta0}\\{}^2_m\boldsymbol{V}_{\alpha\beta0}-{}^2_n\boldsymbol{V}_{\alpha\beta0}&=(\boldsymbol{\alpha}\cdot\boldsymbol{a}^{-1})\cdot\boldsymbol{Z}_{0M}\cdot(\boldsymbol{a}\cdot\boldsymbol{\alpha}^{-1})\cdot{}^1\boldsymbol{I}_{\alpha\beta0}\\&\quad+(\boldsymbol{\alpha}\cdot\boldsymbol{a}^{-1})\cdot\boldsymbol{Z}_{012}\cdot(\boldsymbol{a}\cdot\boldsymbol{\alpha}^{-1})\cdot{}^2\boldsymbol{I}_{\alpha\beta0}\end{aligned}\right\}$$

or

$$\begin{bmatrix}{}^1_m\boldsymbol{V}_{\alpha\beta0}\\{}^2_m\boldsymbol{V}_{\alpha\beta0}\end{bmatrix}-\begin{bmatrix}{}^1_n\boldsymbol{V}_{\alpha\beta0}\\{}^2_n\boldsymbol{V}_{\alpha\beta0}\end{bmatrix}=\begin{bmatrix}(\boldsymbol{\alpha}\cdot\boldsymbol{a}^{-1})\cdot\boldsymbol{Z}_{012}\cdot(\boldsymbol{a}\cdot\boldsymbol{\alpha}^{-1}) & (\boldsymbol{\alpha}\cdot\boldsymbol{a}^{-1})\cdot\boldsymbol{Z}_{0M}\cdot(\boldsymbol{a}\cdot\boldsymbol{\alpha}^{-1})\\(\boldsymbol{\alpha}\cdot\boldsymbol{a}^{-1})\cdot\boldsymbol{Z}_{0M}\cdot(\boldsymbol{a}\cdot\boldsymbol{\alpha}^{-1}) & (\boldsymbol{\alpha}\cdot\boldsymbol{a}^{-1})\cdot\boldsymbol{Z}_{012}\cdot(\boldsymbol{a}\cdot\boldsymbol{\alpha}^{-1})\end{bmatrix}\cdot\begin{bmatrix}{}^1\boldsymbol{I}_{\alpha\beta0}\\{}^2\boldsymbol{I}_{\alpha\beta0}\end{bmatrix}$$
$$\equiv\begin{bmatrix}\boldsymbol{Z}_{\alpha\beta0} & \boldsymbol{Z}'_{\alpha\beta0}\\\boldsymbol{Z}'_{\alpha\beta0} & \boldsymbol{Z}_{\alpha\beta0}\end{bmatrix}\cdot\begin{bmatrix}{}^1\boldsymbol{I}_{\alpha\beta0}\\{}^2\boldsymbol{I}_{\alpha\beta0}\end{bmatrix}\qquad(6.23)$$

$\boldsymbol{Z}_{\alpha\beta0}$ is in the same form as Equation 6.15

$$\boldsymbol{Z}_{\alpha\beta0}=(\boldsymbol{\alpha}\cdot\boldsymbol{a}^{-1})\cdot\boldsymbol{Z}_{012}\cdot(\boldsymbol{a}\cdot\boldsymbol{\alpha}^{-1})=\begin{bmatrix}Z_1&0&0\\0&Z_1&0\\0&0&Z_0\end{bmatrix}$$

and also $\boldsymbol{Z}'_{\alpha\beta0}$ is

$$\begin{aligned}\boldsymbol{Z}'_{\alpha\beta0}&=(\boldsymbol{\alpha}\cdot\boldsymbol{a}^{-1})\cdot\boldsymbol{Z}_{0M}\cdot(\boldsymbol{a}\cdot\boldsymbol{\alpha}^{-1})\\&=\begin{bmatrix}0&1&1\\0&-j&j\\1&0&0\end{bmatrix}\cdot\begin{bmatrix}Z_{0M}&0&0\\0&0&0\\0&0&0\end{bmatrix}\cdot\frac{1}{2}\begin{bmatrix}0&0&2\\1&j&0\\1&-j&0\end{bmatrix}=\begin{bmatrix}0&0&0\\0&0&0\\0&0&Z_{0M}\end{bmatrix}\end{aligned}$$

Therefore the fundamental equation of the double circuit line in α–β–0 components is

$$\begin{bmatrix}{}^1_mV_\alpha\\{}^1_mV_\beta\\{}^1_mV_0\\{}^2_mV_\alpha\\{}^2_mV_\beta\\{}^2_mV_0\end{bmatrix}-\begin{bmatrix}{}^1_nV_\alpha\\{}^1_nV_\beta\\{}^1_nV_0\\{}^2_nV_\alpha\\{}^2_nV_\beta\\{}^2_nV_0\end{bmatrix}=\begin{bmatrix}Z_1&0&0&0&0&0\\0&Z_1&0&0&0&0\\0&0&Z_0&0&0&Z_{0M}\\0&0&0&Z_1&0&0\\0&0&0&0&Z_1&0\\0&0&Z_{0M}&0&0&Z_0\end{bmatrix}\cdot\begin{bmatrix}{}^1I_\alpha\\{}^1I_\beta\\{}^1I_0\\{}^2I_\alpha\\{}^2I_\beta\\{}^2I_0\end{bmatrix}\qquad(6.24a)$$

or

$$\left.\begin{aligned}&\alpha\text{-circuit}\quad\begin{bmatrix}{}^1_mV_\alpha\\{}^2_mV_\alpha\end{bmatrix}-\begin{bmatrix}{}^1_nV_\alpha\\{}^2_nV_\alpha\end{bmatrix}=\begin{bmatrix}Z_1&0\\0&Z_1\end{bmatrix}\cdot\begin{bmatrix}{}^1I_\alpha\\{}^2I_\alpha\end{bmatrix}\\&\beta\text{-circuit}\quad\begin{bmatrix}{}^1_mV_\beta\\{}^2_mV_\beta\end{bmatrix}-\begin{bmatrix}{}^1_nV_\beta\\{}^2_nV_\beta\end{bmatrix}=\begin{bmatrix}Z_1&0\\0&Z_1\end{bmatrix}\cdot\begin{bmatrix}{}^1I_\beta\\{}^2I_\beta\end{bmatrix}\\&0\text{-circuit}\quad\begin{bmatrix}{}^1_mV_0\\{}^2_mV_0\end{bmatrix}-\begin{bmatrix}{}^1_nV_0\\{}^2_nV_0\end{bmatrix}=\begin{bmatrix}Z_0&Z_{0M}\\Z_{0M}&Z_0\end{bmatrix}\cdot\begin{bmatrix}{}^1I_0\\{}^2I_0\end{bmatrix}\end{aligned}\right\}\qquad(6.24b)$$

The equivalent circuit corresponding to the above equation is given in Figure 6.4. Now we can conclude that the α-circuit and β-circuit of the double circuit transmission line can be expressed by the positive-sequence equivalent circuit, and the 0-circuit of course by the zero-sequence equivalent circuit.

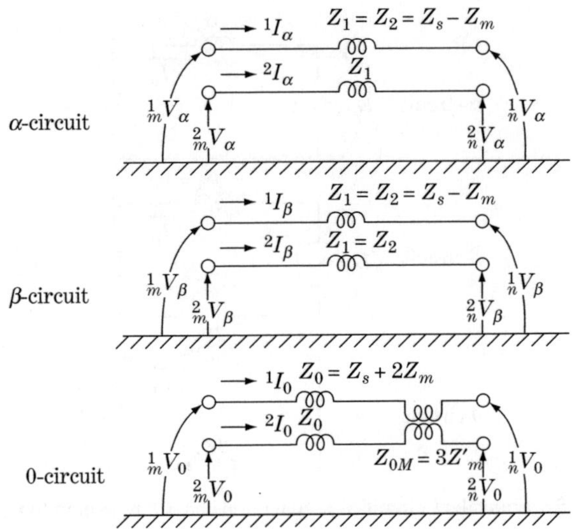

Figure 6.4 The equivalent circuit of a transmission line in α–β–0 components (double circuit line)

6.4.3 Generator

The generator circuit is described by Equation 2.27b and Figure 2.11 in the 0–1–2 domain, and is again quoted here:

$$\underbrace{\begin{bmatrix}0\\E_a\\0\end{bmatrix}}_{\boldsymbol{E}_{012}} - \underbrace{\begin{bmatrix}V_0\\V_1\\V_2\end{bmatrix}}_{\boldsymbol{V}_{012}} = \underbrace{\begin{bmatrix}Z_0 & 0 & 0\\0 & Z_1 & 0\\0 & 0 & Z_1\end{bmatrix}}_{\boldsymbol{Z}_{012}} \cdot \underbrace{\begin{bmatrix}I_0\\I_1\\I_2\end{bmatrix}}_{\boldsymbol{I}_{012}} + \underbrace{\begin{bmatrix}3Z_n \cdot I_0\\0\\0\end{bmatrix}}_{3\boldsymbol{Z}_n \cdot \boldsymbol{I}_0} \tag{6.25}$$

Left-multiplying by $\boldsymbol{\alpha} \cdot \boldsymbol{a}^{-1}$,

$$\therefore\ \boldsymbol{E}_{\alpha\beta 0} - \boldsymbol{V}_{\alpha\beta 0} = \boldsymbol{Z}_{\alpha\beta 0} \cdot \boldsymbol{I}_{\alpha\beta 0} + (\boldsymbol{\alpha} \cdot \boldsymbol{a}^{-1}) \cdot 3\boldsymbol{Z}_n \cdot \boldsymbol{I}_0$$

where $\boldsymbol{Z}_{\alpha\beta 0}$ is in the same form as Equation 6.20, and

$$\boldsymbol{E}_{\alpha\beta 0} = (\boldsymbol{\alpha} \cdot \boldsymbol{a}^{-1}) \cdot \boldsymbol{E}_{012} = \begin{bmatrix}0 & 1 & 1\\0 & -j & j\\1 & 0 & 0\end{bmatrix} \cdot \begin{bmatrix}0\\E_a\\0\end{bmatrix} = \begin{bmatrix}E_a\\-jE_a\\0\end{bmatrix}$$

$$(\boldsymbol{\alpha} \cdot \boldsymbol{a}^{-1}) \cdot 3Z_n\boldsymbol{I}_0 = \begin{bmatrix}0 & 1 & 1\\0 & -j & j\\1 & 0 & 0\end{bmatrix} \cdot \begin{bmatrix}3Z_nI_0\\0\\0\end{bmatrix} = \begin{bmatrix}0\\0\\3Z_nI_0\end{bmatrix}$$

Therefore the generator equation in the α–β–0 domain is

$$\left.\begin{aligned}&\begin{bmatrix}E_a\\-jE_a\\0\end{bmatrix} - \begin{bmatrix}V_\alpha\\V_\beta\\V_0\end{bmatrix} = \begin{bmatrix}\frac{1}{2}(Z_1+Z_2) & \frac{1}{2}j(Z_1-Z_2) & 0\\-\frac{1}{2}j(Z_1-Z_2) & \frac{1}{2}(Z_1+Z_2) & 0\\0 & 0 & Z_0\end{bmatrix} \cdot \begin{bmatrix}I_\alpha\\I_\beta\\I_0\end{bmatrix} + \begin{bmatrix}0\\0\\3Z_nI_0\end{bmatrix}\\&\text{where}\quad Z_1 = jx_1,\quad Z_2 = jx_2,\quad Z_0 = jx_0\end{aligned}\right\} \tag{6.26}$$

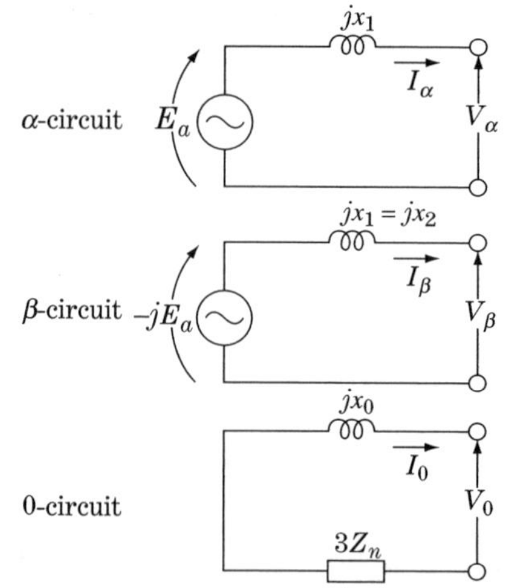

Figure 6.5 Equivalent circuit of a generator under the assumption $jx_1 \fallingdotseq jx_2$

Furthermore, the equation becomes simpler, as follows, if the assumption $Z_1 \fallingdotseq Z_2 (jx_1 \fallingdotseq jx_2)$ is justified for fault analysis distant from the generator terminal, for example:

$$\begin{bmatrix} E_a \\ -jE_a \\ 0 \end{bmatrix} - \begin{bmatrix} V_\alpha \\ V_\beta \\ V_0 \end{bmatrix} = \begin{bmatrix} Z_1 & 0 & 0 \\ 0 & Z_1 & 0 \\ 0 & 0 & Z_0 \end{bmatrix} \cdot \begin{bmatrix} I_\alpha \\ I_\beta \\ I_0 \end{bmatrix} + \begin{bmatrix} 0 \\ 0 \\ 3Z_n I_0 \end{bmatrix} \quad (6.27a)$$

or

$$\left.\begin{aligned} E_a - V_\alpha &= Z_1 \cdot I_\alpha \\ -jE_a - V_\beta &= Z_1 \cdot I_\beta \\ -V_0 &= (Z_0 + 3Z_n) \cdot I_0 \end{aligned}\right\} \quad (6.27b)$$

Figure 6.5 is the equivalent circuit of Equations 6.27a and b. As they are based on the assumption of $jx_1 = jx_2$, some errors may appear if they are adopted for analysis of phenomena around the generator terminal. However, Equation 6.26 before the assumption is the precise equation of the generator by the α–β–0 method where the circuit is described by the known symmetrical reactances, although the equation cannot be replaced by a simple equivalent circuit. Besides, it must be remembered that the generator source voltages $E_a, -jE_a$ exist on the α- and β-circuits, respectively, in the α–β–0 domain.

6.4.4 Transformer impedances and load impedances in the α–β–0 domain

Transformers do not include mutual impedances in the 0–1–2 domain as shown in Table 5.1, so Equations 6.19 and 6.20 can be applied. Moreover $Z_1 = Z_2$ is always correct, so Equations 6.21 and Figure 6.3 can be applied for the transformer. In other words, positive-sequence impedance $Z_1 = jX_1$ is applied for the α- and β-circuits, and $Z_0 = jX_0$ is applied for the 0-circuit.

Load circuit equations assumed in Equations 2.28 and 2.29 are in the same form as Equations 2.15 and 6.19 for a single circuit line. Therefore Equation 6.21 shows the load equations in the α–β–0 domain under the approximation by $Z_1 \fallingdotseq Z_2$.

6.5 Fault Analysis by $\alpha-\beta-0$ Components

As the transformed equations and the equivalent circuit of three-phase circuits in the $\alpha-\beta-0$ domain have been completed, we can begin fault analysis by the $\alpha-\beta-0$ method which can be executed using the process in Figure 2.1 or the similar one in Figure 3.1.

6.5.1 The b–c phase line to ground fault

Suppose the b–c phase l–g fault at point f is as shown in Figure 6.6. The fault condition at f is

$$ {}_fV_b = {}_fV_c = 0, \quad {}_fI_a = 0 \tag{6.28}$$

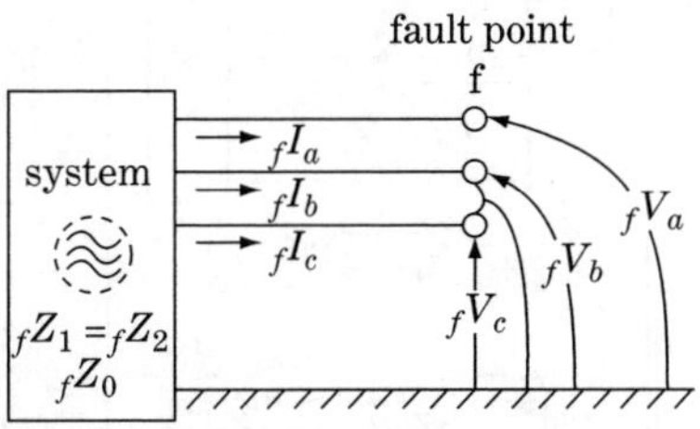

Figure 6.6 b-c phase l-g fault

Substituting Equation 6.2 into Equation 6.28, the fault terminal equations become

$$\left.\begin{array}{l} \left.\begin{array}{l} -\frac{1}{2}{}_fV_\alpha + \frac{\sqrt{3}}{2}{}_fV_\beta + {}_fV_0 = -\frac{1}{2}{}_fV_a - \frac{\sqrt{3}}{2}{}_fV_\beta + {}_fV_0 = 0 \\ {}_fI_\alpha + {}_fI_0 = 0 \end{array}\right\} \\ \left.\begin{array}{l} \therefore\ {}_fV_\alpha = 2\,{}_fV_0 \quad ①, \quad {}_fV_\beta = 0 \quad ② \\ {}_fI_\alpha + {}_fI_0 = 0 \quad ③ \end{array}\right\} \end{array}\right\} \tag{6.29}$$

the network equations:

$$\left.\begin{array}{rrcll} & E_a - {}_fV_\alpha &=& {}_fZ_1 \cdot {}_fI_\alpha & ④ \\ & -jE_a - {}_fV_\beta &=& {}_fZ_1 \cdot {}_fI_\beta & ⑤ \\ & -{}_fV_0 &=& {}_fZ_0 \cdot {}_fI_0 & ⑥ \\ \text{or} & -(2\,{}_fV_0) &=& (2\,{}_fZ_0) \cdot {}_fI_0 & ⑥' \end{array}\right\} \tag{6.30}$$

The equivalent circuit which satisfies both Equations 6.29 and 6.30 is shown in Table 6.1 #1A.

In order to satisfy Equations ① and ⑥ together, the zero-sequence circuit is expressed by ⑥′ (instead of ⑥) with terminal voltage $2\,{}_fV_0$ and impedance $2\,{}_fZ_0$.

From the equivalent circuit, the following equations are obtained:

$$\left.\begin{array}{l} {}_fI_\alpha = -{}_fI_0 = \dfrac{E_a}{{}_fZ_1 + 2\,{}_fZ_0} \\ {}_fI_\beta = \dfrac{-jE_a}{{}_fZ_1} \end{array}\right\} ① \quad \left.\begin{array}{l} {}_fV_\alpha = 2\,{}_fV_0 = E_a \cdot \dfrac{2\,{}_fZ_0}{{}_fZ_1 + 2\,{}_fZ_0} \\ {}_fV_\beta = 0 \end{array}\right\} ② \tag{6.31}$$

These are the solutions $({}_fI_\alpha, {}_fI_\beta, {}_fI_0), ({}_fV_\alpha, {}_fV_\beta, {}_fV_0)$ in the $\alpha-\beta-0$ domain. Then these solutions can be inverse transformed into the a–b–c domain by Equation 6.2, or into the 0–1–2 domain by Equation 6.10 as our final solution. The obtained final solution coincides with Equations 5C and 5D in Table 3.1.

Table 6.1 The equivalent circuits for various faults in the α–β–0 coordinates domain

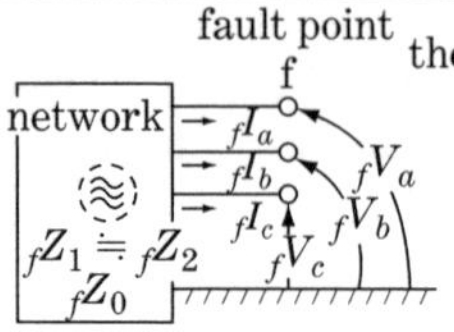

the system equations in the α-β-0 domain

$$\left.\begin{aligned} E_a - {_fV_\alpha} &= {_fZ_1}\cdot{_fI_\alpha} \\ -jE_a - {_fV_\beta} &= {_fZ_1}\cdot{_fI_\beta} \\ -{_fV_0} &= {_fZ_0}\cdot{_fI_0} \end{aligned}\right\} \quad \text{Eq.}(6\cdot 27\text{b})$$

$_fE_a$: the voltage at point f before the fault

$_fZ_1 \fallingdotseq {_fZ_2}, {_fZ_0}$: the system impedances at point f

#1 phase b-c line-to-ground fault

#1A

$$\left.\begin{aligned} {_fV_b} &= {_fV_c} = 0 \\ {_fV_a} &= 0 \end{aligned}\right\}$$

$$\left.\begin{aligned} {_fV_\alpha} &= 2{_fV_0} \\ {_fV_\beta} &= 0 \\ {_fI_\alpha} + {_fI_0} &= 0 \end{aligned}\right\}$$

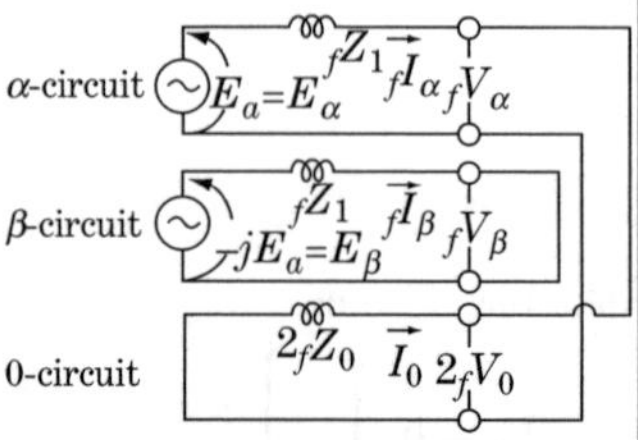

Zero-sequence circuit is from the modified equation $-2{_fV_0} = (2Z_0)\bullet{_fI_0}$

#1B

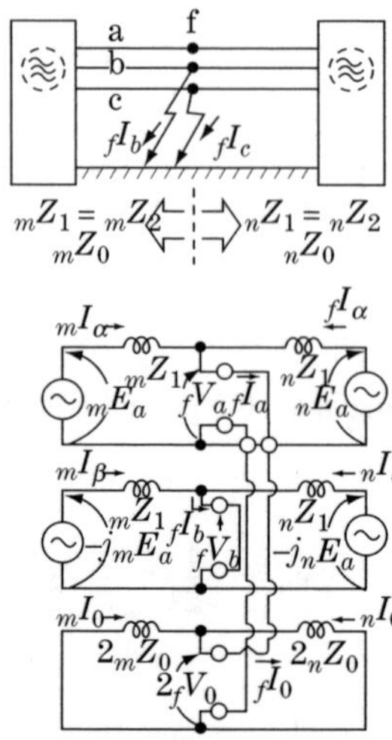

#1C

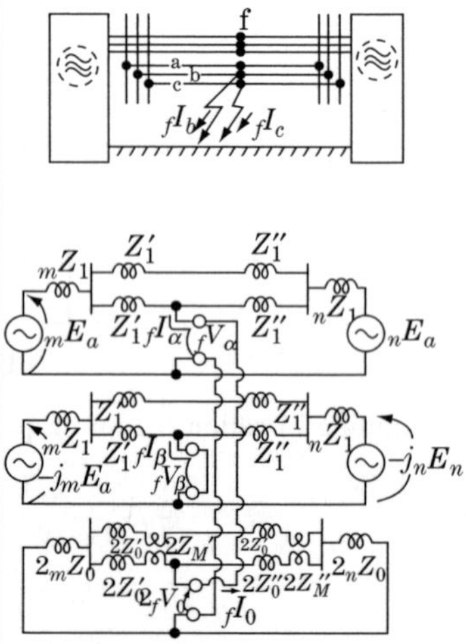

#2 phase a line-to-ground fault

$$\left.\begin{aligned} {_fV_a} &= 0 \\ {_fI_b} &= {_fI_c} = 0 \end{aligned}\right\} \quad \left.\begin{aligned} {_fV_\alpha} + {_fV_0} &= 0 \\ {_fI_\beta} &= 0 \\ {_fI_\alpha} &= 2{_fI_0} \end{aligned}\right\}$$

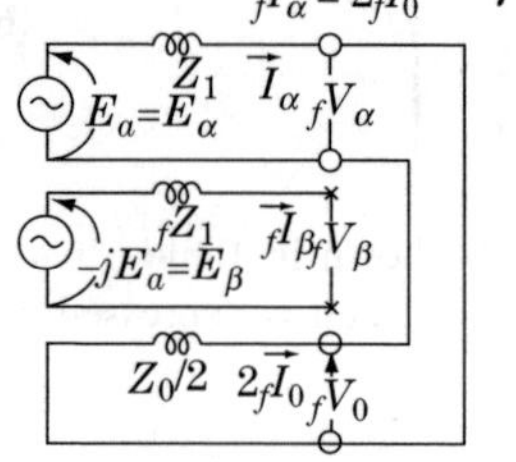

Zero-sequence circuit is from the modified equation $-{_fV_0} = \left(\frac{{_fZ_0}}{2}\right)\cdot 2{_fI_0}$

#3 phase b to c line-to-line fault

$$\left.\begin{aligned} {_fV_b} &= {_fV_c} \\ {_fI_a} &= 0 \\ {_fI_b} &= {_fI_c} = 0 \end{aligned}\right\} \quad \left.\begin{aligned} {_fV_\beta} &= 0 \\ {_fV_\alpha} &= 0 \\ {_fV_0} &= 0 \end{aligned}\right\}$$

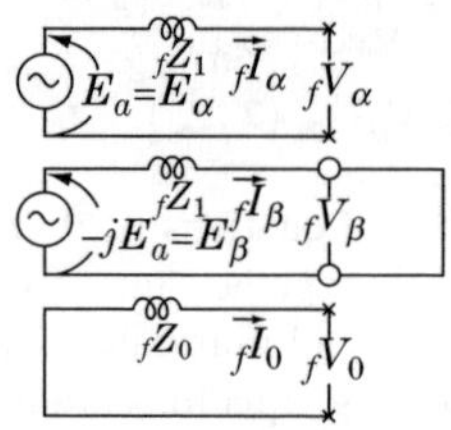

#4 three-phase fault

$$\left.\begin{aligned} {_fV_a} &= {_fV_b} = {_fV_c} = 0 \\ {_fI_a} &= {_fI_b} = {_fI_c} \end{aligned}\right\} \quad \left.\begin{aligned} {_fV_\alpha} &= 0 \\ {_fV_\beta} &= 0 \\ {_fI_0} &= 0 \end{aligned}\right\}$$

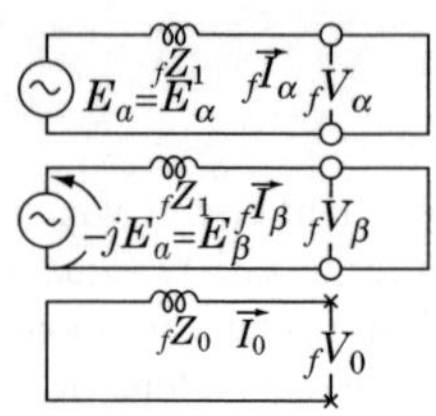

inverse transform equation to a-b-c domain

$$\left.\begin{aligned} V_a &= V_\alpha \\ V_b &= -\tfrac{1}{2}V_\alpha + \tfrac{\sqrt{3}}{2}V_\beta + V_0 \\ V_c &= -\tfrac{1}{2}V_\alpha - \tfrac{\sqrt{3}}{2}V_\beta + V_0 \end{aligned}\right\} \quad \text{Eq. (6.2)}$$

Figures # 1B and # 1C in Table 6.1 show the same equivalent circuit but for a circuit with a double source line and double circuit line.

6.5.2 Other mode short-circuit faults

The equations and equivalent circuits for the other mode faults are shown in Table 6.1, cases # 2, # 3 and # 4. In case # 2, the phase a l–g fault, the figure for the zero-sequence circuit shows voltage ${}_fV_0$, current $2\,{}_fI_0$ and impedance $Z_0/2$, because the related zero-sequence circuit equation is modified as $-{}_fV_0 = \left(\frac{{}_fZ_0}{2}\right)\cdot(2\,{}_fI_0)$.

6.5.3 Open-conductor mode faults

The equivalent circuits of open-conductor mode faults in α–β–0 components can be obtained analogously. Table 6.2 shows the equations and the equivalent circuits.

Let us try here to compare the equivalent circuits in symmetrical components, Tables 3.1 and 3.2, with those in α–β–0 components, Tables 6.1 and 6.2. The following conclusion may be derived.

In the cases of Tables 3.1 and 3.2 in the symmetrical method, there is one complicate equivalent circuit in which positive-, negative- and zero-sequence impedances are connected in series and/or parallel. On the other hand, in the cases of Tables 6.1 and 6.2 there are two or three simple and independent circuits. To solve three simple equations would generally be easier than solving one complicated equation, whether or not a computer was used. This is one of the reasons why the α–β–0 method is a worthy approach as a complement of the symmetrical method.

Again it must be stressed that the α–β–0 method is as precise as the symmetrical method, at least for Equation 6.26, instead of Equation 6.27a, which is adopted for generator equations.

Table 6.2 Equivalent circuit of conductor opening in the α–β–0 domain

#5 phase a conductor opening	#6 phase b,c conductors opening
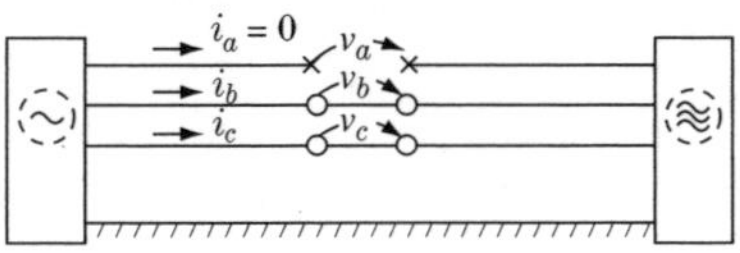	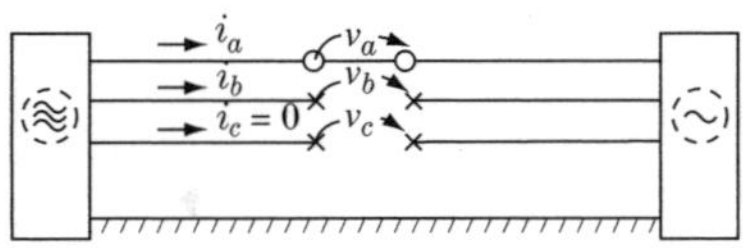
$v_b = v_c = 0$, $i_a = 0$ } $\to$ $v_\alpha = 2v_0$, $v_\beta = 0$, $I_\alpha + I_0 = 0$	$v_a = v_c = 0$, $i_b = i_c = 0$ } $\to$ $v_\alpha + v_0 = 0$, $i_\alpha = 2i_0$, $i_\beta = 0$
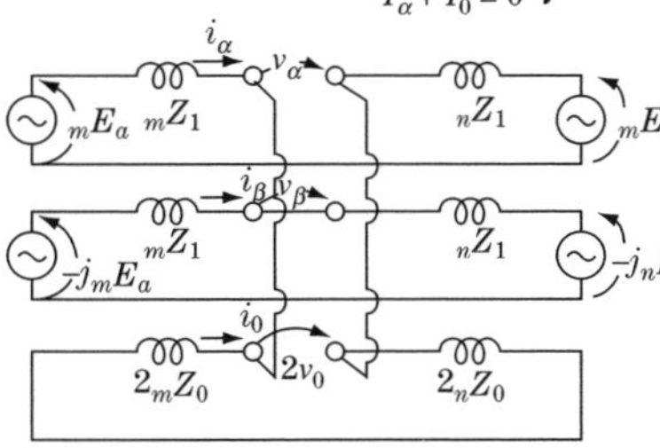	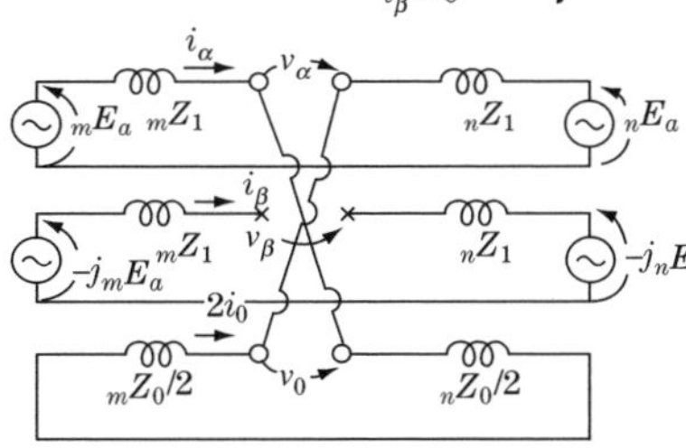
Zero-sequence circuit is from modified equation $2v_G = (2Z_0)i_0$, instead of $v_0 = Z_0 i_0$	Zero-sequence circuit is from modified equation $v_0 = \left(\frac{Z_0}{2}\right)(2i_0)$, instead of $v_0 = Z_0 i_0$

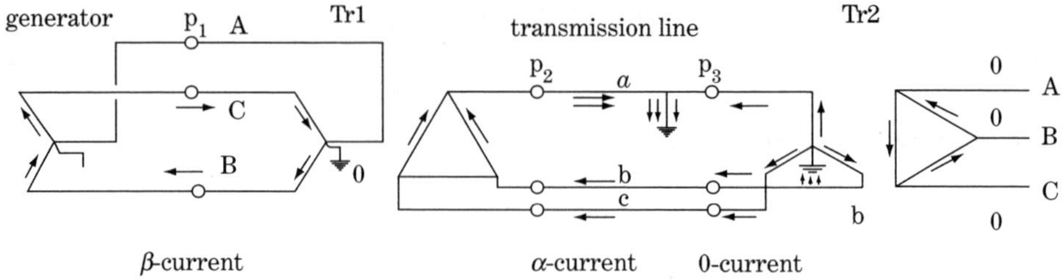

Figure 6.7 The α–β–0 component currents through transformers under *l*-g fault

Finally, Figure 6.7 shows current flow under the condition of a phase a *l*–g fault. The phase currents are represented by arrows indicating the relative magnitudes of currents in each circuit. This is the original figure for α–β–0 components first developed by W. W. Lewis. In the figure, the currents through points p_1, p_2, p_3 are the β-, α- and 0-currents, respectively. Physical current flow based on Kirchhoff's law and the cancellation law of ampere-turns can be imagined from this figure as either real-number currents or complex-number currents.

7

Symmetrical and $\alpha-\beta-0$ Components as Analytical Tools for Transient Phenomena

Most analytical engineering tasks may concern transient phenomena of three-phase circuits (whether of large networks or small partial circuits), typically system stability analysis, dynamic analysis of generators, fault analysis, switching and lightning surge analysis, harmonic resonance analysis, insulation design, analysis of factory testing of equipment, and so on. Inevitably we have to apply symmetrical and α–β–0 components as essential analytical tools for practical engineering management of these phenomena.

From such a viewpoint, it is strange that most textbooks on symmetrical components cover applications only for power frequency phenomena, while $\alpha-\beta-0$ components are seldom explained. Perhaps this is the reason why confusion is apt to arise often in applying symmetrical or $\alpha-\beta-0$ components, especially for transient phenomena. Typical misunderstanding may arise in the application of symmetrical components to transient phenomena which may often involve complex-number vector operators a, a^{-1}.

Symmetrical and $\alpha-\beta-0$ components have proved their merit, especially for transient phenomena. In this chapter, we look back briefly to the origin of transient analysis for a single-phase circuit (a review of **the complex-number symbolic method**) first, and then demonstrate short-circuit transient analysis for a three-phase network using symmetrical components and $\alpha-\beta-0$ components.

7.1 The Symbolic Method and its Application to Transient Phenomena

First, let us review the switching transient phenomena of a single phase circuit with series *LCR* elements and an a.c. power source as shown in Table 7.1.

The table shows the solution for the transient current caused just after closing the switch at time $t = 0$. The three different expressions in cases 1, 2 and 3 are demonstrated for the same phenomena for comparison.

In case 1, the power source with a sinusoidal waveform-induced voltage is written as $E\cos(\omega t + \theta)$, and the same source is written as $E\sin(\omega t + \theta)$ in case 2. In case 3, by contrast, the same power source is written as the complex number $Ee^{j(\omega t+\theta)} = E\{\cos(\omega t + \theta) + j\sin(\omega t + \theta)\}$.

Handbook of Power System Engineering Yoshihide Hase
© 2007 John Wiley & Sons, Ltd

Table 7.1 Transient current by switching of the *LCR* series circuit

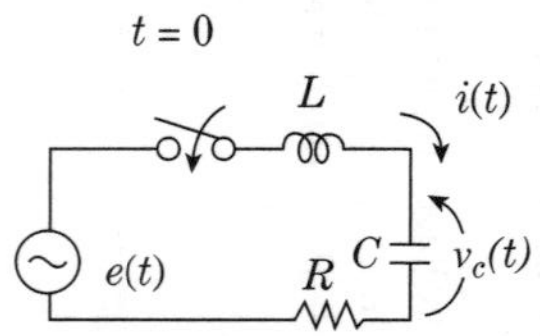

To solve the transient current equation
where $R^2 < 4L/C$ with initial conditions $i = 0$, $v_c = 0$ at $t = 0$, $L\frac{di}{dt} + Ri + \frac{1}{c}\int idt = 0$

The solution for the transient current:

		Steady-state term of the current	Transient term of the current	Circuit current
#1	Source voltage by Cosine function $e(t)' = \|E\|\cos(\omega t + \theta)$	$i(t)'s = \frac{\|E\|}{\|Z\|}\cos(\omega t + \theta - \varphi)$	$i(t)'_T = \frac{\|E\|}{\|Z\|}.e^{-\alpha t}[\cos(\theta - \varphi).\left(\frac{\alpha}{\omega_d}\sin\omega_d t - \cos\omega_d t\right) - \frac{\alpha^2 + \omega_d^2}{\omega\omega_d}\cos(\theta - \varphi + 90°)\sin\omega_d t]$	$i(t)' = i(t)'_S + i(t)'_T$
#2	Source voltage by Sine function $e(t)'' = \|E\|\sin(\omega t + \theta)$	$i(t)''s = \frac{\|E\|}{\|Z\|}\sin(\omega t + \theta - \varphi)$	$i(t)''_T = \frac{\|E\|}{\|Z\|}\cdot e^{-at}[\sin(\theta - \varphi)\cdot\left(\frac{\alpha}{\omega_d}\sin\omega_d t - \cos\omega_d t\right) - \frac{\alpha^2 + \omega_d^2}{\omega\omega_d}\sin(\theta - \varphi + 90°)\sin\omega_d t]$	$i(t)'' = i(t)''_S + i(t)''_T$
#3	Source voltage by Complex-number function **(symbolic method)** $e(t) = e(t)' + je(t)'' = \|E\|e^{j(\omega t + \theta)}$	$i(t)s = i(t)'s + ji(t)''s = \frac{\|E\|}{\|Z\|}e^{j(\omega t + \theta - \varphi)}$	$i(t)_T = i(t)'_T + ji(t)''_T = \frac{\|E\|}{\|Z\|}.e^{-\alpha t}[e^{j(\theta - \varphi)}\cdot\left(\frac{\alpha}{\omega_d}\sin\omega_d t - \cos\omega_d t\right) - \frac{\alpha^2 + \omega_d^2}{\omega\omega_d}e^{j(\theta - \varphi + 90°)}\cdot\sin\omega_d t]$	$i(t) = i(t)_S + i(t)_T = i(t)' + ji(t)''$

$\omega = 2\pi f$: angular velocity of power frequency f

$$\alpha = \frac{R}{2L},\ \omega_d = \sqrt{\omega_0^2 - \alpha^2} = \sqrt{\frac{1}{LC} - \left(\frac{R}{2L}\right)^2},\ \omega_0 = \frac{1}{\sqrt{LC}}$$

$$|Z| = \sqrt{R^2 + \left(\omega L - \frac{1}{\omega C}\right)^2},\quad \varphi = \tan^{-1}\frac{\omega L - \frac{1}{\omega C}}{R}$$

Notes: Euler's formula: $e^{\pm j\varphi} = \cos\varphi \pm \sin\varphi$.
The symbolic relation (#1) + j(#2) = (#3) is preserved for all the quantities of the circuit.
See Chapter 19 for the calculation method by Laplace transforms.

In other words, the power source is written in three different ways by the symbolic relation of (#1) + j(#2) = (#3). Note especially that, for these three different approaches for the same circuit analysis, the symbolic relation is perfectly preserved for the voltage and current quantities at any arbitrary point in the circuit and for arbitrary timing under any condition, regardless of steady-state or transient phenomena.

This review is expanded to include transient phenomena. The symbolic method using complex-number notation for electric circuit analysis is mathematically an application of Euler's formula $e^{j\varphi} = \cos\varphi + j\sin\varphi$, which was first introduced in this field by A. E. Kennelly and C. P. Steinmetz separately in 1893.

In Table 7.1, the symbolic relation of (#1) + j(#2) = (#3) is obviously preserved for steady-state terms as well as for transient terms of the current equations obtained in cases 1, 2 and 3.

The above explanation is always effective for cases where multiple numbers of transient terms are included, although this example contains a single transient term. Furthermore, the same symbolic relation is preserved not only for single phase circuit phenomena, but also for three-phase circuit phenomena.

7.2 Transient Analysis by Symmetrical and α–β–0 Components

Now let us examine the equations of a three-phase circuit for transient phenomena.

Equations 1.3 and 1.4 in Chapter 1 are the steady-state equations for the transmission line shown in the Figure 1.1b. The original equation in regard to a phase a conductor covering transient phenomena is a differential equation

$$\left.\begin{aligned} {}_mV_a(t) - {}_nV_a(t) = \left\{(r_a + r_g) + (L_{aag} + L_g)\frac{d}{dt}\right\}I_a(t) + \left\{r_g + (L_{abg} + L_g)\frac{d}{dt}\right\}I_b(t) \\ + \left\{r_g + (L_{acg} + L_g)\frac{d}{dt}\right\}I_c(t) \end{aligned}\right\} \quad (7.1)$$

where $V(t)$, $I(t)$ are complex-number expressions.

Equation 7.1 and Equations 1.3 and 1.4 are in the same form as each other except for the displacement of $j\omega \Leftrightarrow d/dt$. In exact terms, the general form of the equation was originally a differential equation in d/dt, which can be replaced by $j\omega$ for limited applications of steady-state analysis.

The matrix Equation 1.3 using d/dt (instead of $j\omega$) can be transformed into the symmetrical domain by the same procedure explained in Chapter 2, resulting in the equation

$$\left.\begin{aligned} &\begin{bmatrix} {}_mV_0(t) \\ {}_mV_1(t) \\ {}_mV_2(t) \end{bmatrix} - \begin{bmatrix} {}_nV_0(t) \\ {}_nV_1(t) \\ {}_nV_2(t) \end{bmatrix} = \begin{bmatrix} r_0 + L_0\frac{d}{dt} & 0 & 0 \\ 0 & r_1 + L_1\frac{d}{dt} & 0 \\ 0 & 0 & r_1 + L_1\frac{d}{dt} \end{bmatrix} \cdot \begin{bmatrix} I_0(t) \\ I_1(t) \\ I_2(t) \end{bmatrix} \\ &\text{where} \quad r_0 = r_s + 2r_m, \quad L_0 = L_s + 2L_m \\ &\qquad\qquad r_1 = r_s - 2r_m, \quad L_1 = L_s - L_m \\ &\text{and } V(t), I(t) \text{ are complex-number quantities} \end{aligned}\right\} \quad (7.2)$$

Extracting the real part of the above equation,

$$\left.\begin{bmatrix} {}_m v_0(t) \\ {}_m v_1(t) \\ {}_m v_2(t) \end{bmatrix} - \begin{bmatrix} {}_n v_0(t) \\ {}_n v_1(t) \\ {}_n v_2(t) \end{bmatrix} = \begin{bmatrix} r_0 + L_0 \frac{d}{dt} & 0 & 0 \\ 0 & r_1 + L_1 \frac{d}{dt} & 0 \\ 0 & 0 & r_1 + L_1 \frac{d}{dt} \end{bmatrix} \cdot \begin{bmatrix} i_0(t) \\ i_1(t) \\ i_2(t) \end{bmatrix} \right\} \tag{7.3}$$

where $v(t)$, $i(t)$ are real-number quantities. Equations 7.2 and 7.3 correspond to Equation 2.15.

Now we can recognize by the same analogy that all the equations and the transformation procedures described in the previous chapters are perfectly preserved for transient phenomena with symbolic replacement of $j\omega \Leftrightarrow d/dt$ (or the symbol $s = d/dt$ of the Laplace transform). Of course, correlations among the a–b–c, 0–1–2 and α–β–0 domains are also preserved by the same transform/inverse transform equations.

7.3 Comparison of Transient Analysis by Symmetrical and α–β–0 Components

The transient analysis for a phase b–c l–l short-circuit fault (2ϕS) is demonstrated in Table 7.2 in which the following four approaches are compared:

Case A1: symmetrical component method (by the symbolic method with complex numbers)

Case A2: symmetrical component method (by real-number expressions)

Case α1: α–β–0 component method (by the symbolic method with complex numbers)

Case α1: α–β–0 component method (by real-number expressions).

These different approaches are equivalent to each other from a mathematical viewpoint, but ease of use is actually quite different.

In the demonstrated fault analysis, the α–β–0 component method provides the solution with or without the symbolic method. In contrast, the symmetrical components method only provides the solution together with the symbolic method. In practical engineering, good senses is required to select the most appropriate method for the individual occasion from the approaches indicated in Table 7.2.

Reviewing Tables 7.1 and 7.2 overall, the following comments may be made.

The symmetrical coordinates method and the α–β–0 coordinates method are vital basic analytical methods which enable three-phase circuit analysis as practically very effective approaches whether for steady-state or transient phenomena. However, the powerful analytical capability of these methods is especially displayed when they are used in combination with **the symbolic method** of complex-number variables.

Table 7.2 Transient fault analysis by symmetrical components and α-β-0 method

	Calculation by complex-number variables (symbolic method)	Calculation by real-number variables
Three-phase circuit	$E_a(t) = \lvert E_{a1}\rvert \cdot e^{j(\omega t+\theta)}$ $E_b(t) = a^2 \cdot E_a(t) = \lvert E_{a1}\rvert \cdot e^{j(\omega t+\theta-120^\circ)}$ $E_c(t) = a \cdot E_a(t) = \lvert E_{a1}\rvert \cdot e^{j(\omega t+\theta+120^\circ)}$ 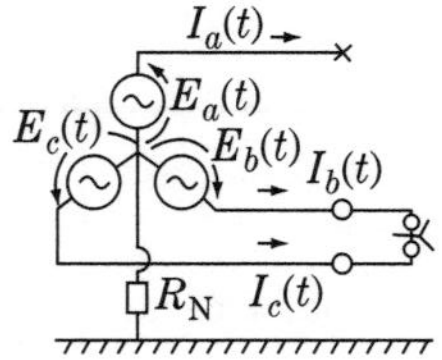	$e_a(t) = \lvert E_{a1}\rvert \cdot \cos(\omega t+\theta)$ $e_b(t) = \lvert E_{a1}\rvert \cdot \cos(\omega t+\theta-120^\circ)$ $e_c(t) = \lvert E_{a1}\rvert \cdot \cos(\omega t+\theta+120^\circ)$ 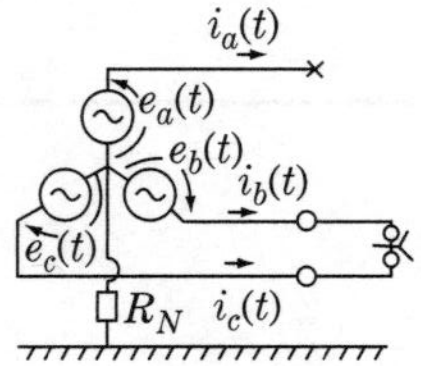
Calculation by symmetrical coordinates	Calculation A1: $E_a(t) = \lvert E_{a1}\rvert \cdot e^{j(\omega t+\theta)}$ positive seq. negative seq. 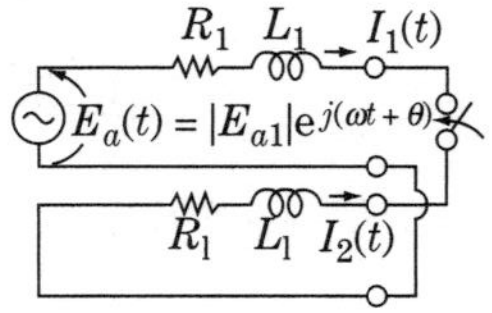 $I_0(t)=0$ $I_1(t)=-I_2(t)=\frac{\lvert E_{a1}\rvert}{\sqrt{(2R_1)^2+\omega^2(2L_1)^2}}\cdot\left\{e^{j(\omega t+\theta-\varphi)}-e^{-\frac{2R_1}{2L_1}t}\cdot e^{j(\theta-\varphi)}\right\}$ where $\varphi=\tan^{-1}\frac{2R_1}{2L_1}=\tan^{-1}\frac{L_1}{R_1}$ ① transforming into phase a, b, c current $I_a(t) = I_0(t)+I_1(t)+I_2(t)=0$ $I_b(t) = I_0(t)+a^2I_1(t)+aI_2(t)=(a^2-a)\cdot I_1(t)=\sqrt{3}\cdot e^{-j90^\circ}\cdot I_1(t)$ $= \frac{\sqrt{3}}{2}\cdot\frac{\lvert E_{a1}\rvert}{\sqrt{R_1^2+\omega^2L_1^2}}\cdot\left\{e^{j(\omega t+\theta-\varphi-90^\circ)}-e^{-\frac{R_1}{L_1}t}\cdot e^{j(\theta-\varphi-90^\circ)}\right\}$ $I_c(t) = I_0(t)+aI_1(t)+a^2I_2(t)=-(a^2-a)\cdot I_1(t)=-I_b(t)$ $= -\frac{\sqrt{3}}{2}\cdot\frac{\lvert E_{a1}\rvert}{\sqrt{R_1^2+\omega^2L_1^2}}\cdot\left\{e^{j(\omega t+\theta-\varphi-90^\circ)}-e^{-\frac{R_1}{L_1}t}\cdot e^{j(\theta-\varphi-90^\circ)}\right\}$ ②	Calculation A2: $e_a(t) = \lvert E_{a1}\rvert \cdot \cos(\omega t+\theta)$ positive seq. negative seq. 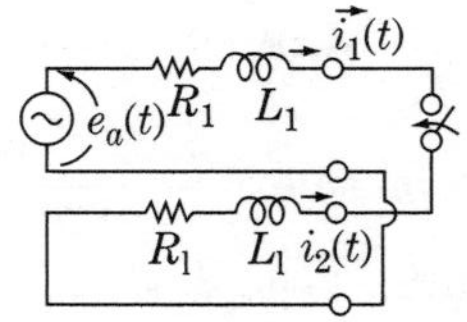 $i_0(t) = 0$ $i_1(t) = -i_2(t) = \frac{\lvert E_{a1}\rvert}{\sqrt{(2R_1)^2+\omega^2(2L_1)^2}}$ $\cdot\left\{\cos(\omega t+\theta-\varphi)-e^{-\frac{2R_1}{2L_1}t}\cdot\cos(\theta-\varphi)\right\}$ $=\frac{\lvert E_{a1}\rvert}{2\sqrt{R_1^2+\omega^2L_1^2}}\{\cos(\omega t+\theta-\varphi)$ $-e^{-\frac{R_1}{L_1}t}\cdot\cos(\theta-\varphi)\}$ ③ Solution ③ cannot be inverse transformed into the a–b–c domain. Symmetrical components can be inverse transformed into this domain only when they are given as solutions of complex-number equations
The real number part extracted from equation ② is the same with Equation ⑥ and ⑧.		

(Continued)

Table 7.2 *(Continued)*

	Calculation by complex-number variables (symbolic method)	Calculation by real-number variables
Calculation by α–β–0 coordinates	Calculation $\alpha 1$: β-circuit 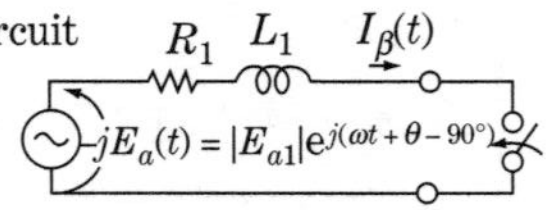 $E_a(t) = \lvert E_{a1}\rvert e^{j(\omega t+\theta)}$ $-jE_a(t) = -j\cdot\lvert E_{a1}\rvert e^{j(\omega t+\theta)} = \lvert E_{a1}\rvert e^{j(\omega t+\theta-90^\circ)}$ $I_\alpha(t) = 0, \quad I_0(t) = 0$ $I_\beta(t) = \frac{\lvert E_{a1}\rvert}{\sqrt{R_1^2+\omega^2 L_1^2}}\cdot\left\{e^{j(\omega t+\theta-\varphi-90^\circ)} - e^{-\frac{R_1}{L_1}t}\cdot e^{j(\theta-\varphi-90^\circ)}\right\}$ ④ transforming into phase a, b, c current $I_a(t) = I_\alpha(t)+I_0(t)=0$ $I_b(t) = -\frac{1}{2}I_\alpha(t)+\frac{\sqrt{3}}{2}I_\beta(t)+I_0(t)=\frac{\sqrt{3}}{2}I_\beta(t)$ $= \frac{\sqrt{3}}{2}\cdot\frac{\lvert E_{a1}\rvert}{\sqrt{R_1^2+\omega^2 L_1^2}}\cdot\left\{e^{j(\omega t+\theta-\varphi-90^\circ)} - e^{-\frac{R_1}{L_1}t}\cdot e^{j(\theta-\varphi-90^\circ)}\right\}$ $I_c(t) = -\frac{1}{2}I_\alpha(t)-\frac{\sqrt{3}}{2}I_\beta(t)+I_0(t)=-\frac{\sqrt{3}}{2}I_\beta(t)=-I_b(t)$ $= -\frac{\sqrt{3}}{2}\cdot\frac{\lvert E_{a1}\rvert}{\sqrt{R_1^2+\omega^2 L_1^2}}\cdot\left\{e^{j(\omega t+\theta-\varphi-90^\circ)} - e^{-\frac{R_1}{L_1}t}\cdot e^{j(\theta-\varphi-90^\circ)}\right\}$ ⑤ taking the real-number part $i_a(t) = 0$ $i_b(t) = \frac{\sqrt{3}}{2}\cdot\frac{\lvert E_{a1}\rvert}{\sqrt{R_1^2+\omega^2 L_1^2}}\cdot\{\cos(\omega t+\theta-\varphi-90^\circ) - e^{-\frac{R_1}{L_1}t}\cos(\theta-\varphi-90^\circ)\}$ $i_c(t) = -i_b(t)$ the solution ⑥	Calculation $\alpha 2$: β-circuit 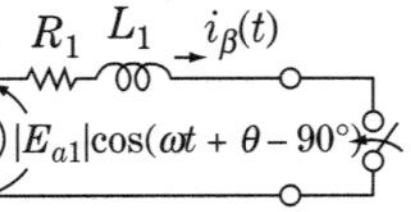 $e_a(t)=\lvert E_{a1}\rvert\cos(\omega t+\theta)$ $-je_a(t)=\lvert E_{a1}\rvert\cos(\omega t+\theta-90^\circ) =\lvert E_{a1}\rvert\sin(\omega t+\theta)$ $i_\alpha(t) = 0, \; i_0(t)=0$ $i_\beta(t) = \frac{\lvert E_{a1}\rvert}{\sqrt{R_1^2+\omega^2 L_1^2}}\cdot\{\cos(\omega t+\theta-\varphi-90^\circ) - e^{-\frac{R_1}{L_1}t}\cos(\theta-\varphi-90^\circ)\}$ ⑦ transforming into phase a, b, c current $i_a(t)=i_\alpha(t)+i_0(t)=0$ $i_b(t)=-\frac{1}{2}i_\alpha(t)+\frac{\sqrt{3}}{2}i_\beta(t)+i_0(t)=\frac{\sqrt{3}}{2}i_\beta(t)$ $=\frac{\sqrt{3}}{2}\cdot\frac{\lvert E_{a1}\rvert}{\sqrt{R_1^2+\omega^2 L_1^2}}\cdot\{\cos(\omega t+\theta-\varphi-90^\circ) - e^{-\frac{R_1}{L_1}t}\cdot\cos(\theta-\varphi-90^\circ)\}$ $i_c(t)=-\frac{1}{2}i_\alpha(t)-\frac{\sqrt{3}}{2}i_\beta(t)+i_0(t) =-\frac{\sqrt{3}}{2}i_\beta(t)=-i_b(t)$ the solution ⑧ where the source voltage is $e_a(t) = \lvert E_{a1}\rvert\cos(\omega t+\theta)$ $e_b(t) = \lvert E_{a1}\rvert\cos(\omega t+\theta-120^\circ)$ $ec(t) = \lvert E_{a1}\rvert\cos(\omega t+\theta+120^\circ)$

Note: Equation ⑥ and ⑧ are the same.

Coffee break 4: Weber and other pioneers

In the years following Faraday's great discovery, scientists made mathematical connections between electricity, magnetism and optics.

For example, **Heinrich Lenz** (1804–1865) formulated **Lenz's law** in 1833, which stated that an induced electric current flows in a direction such that the current opposes the change that induced it; this was later explained as a special case of the law of conservation of energy. Besides the law named after him, Lenz also independently discovered **Joules's law** in 1842.

Franz Ernst Neumann (1798–1895) derived the equation $U = d\phi/d\mathrm{t}$ (where U is electromotive force (emf) and ϕ the flux density) in 1841, which was actually a mathematical formulation of Faraday's law explained as 'electrical induction induced on one circuit is proportional to the decreasing rate of linking flux'.

Hermann Ludwig Helmholtz (1821–1894), **William Thomson (Lord Kelvin)** (1824–1907) and other scientists clarified the relationship between electricity and other forms of energy. **James Prescott Joule** (1818–1889) investigated the quantitative relationship between electric currents and heat during the 1840s and formulated the theory of heating effects that accompany the flow of electricity in conductors. **Gustav Kirchhoff** (1824–1887), Kelvin, Henry and **George Gabriel Stokes** (1819–1903) also extended the theory of the conduction and propagation of electricity.

Wilhelm Eduard Weber (1804–1891) also has to be specially mentioned as a great physicist. As a young assistant to **Karl Friedrich Gauss** (1777–1855), Weber started working on the experimental validation of the ampere-force. He needed to devise a new apparatus, an 'electrodynamometer', which could directly measure the angular displacement produced in a multiply wound electric coil by another coil perpendicular to it. His investigation of the force in relation to electricity and magnetism was continued over the period of 1832–1846 by a theoretical deductive approach as well as by experiment. Finally, in 1846, he published his book *Electrodynamical measurement* in which he hypothesized the existence of *positive and negative electrical charged particles* within the conductor and presented a **force law** which was dependent on velocity and acceleration. In 1846, this was at least 50 years before the concepts of the proton and electron were advocated.

Wilhelm Eduard Weber (1804–1891)

Prior to 1846, there existed three seemingly valid descriptions of the electrical interaction:
(1) Coulomb's law, describing the interaction of two electrical masses; (2) Ampre's law, describing the interaction of elements of moving electricity; and (3) a description of the laws of induction, elaborated by Lenz and Neumann. Weber achieved unification of these various phenomena under a single concept, the **fundamental electrical law**. All the above laws were well explained by the assumption that the presence of 'an electrical tension caused the positive and negative particles to move at equal velocities but in opposite directions'. For example, two

parallel conducting wires attract each other when the current in the two wires flows in the same direction, but repel each other when the opposite is the case. Coulomb's electrostatic law is of course a special case of Weber's general law, when the particles are at relative rest.

Considering all the phenomena of induction, Weber was able to formulate a general statement of the fundamental electrical law. This showed that the general law describing the force of interaction of two electrical particles depends upon the relative velocities and the relative accelerations of the positive and negative (plus and minus) particles.

Incidentally, in his work of 1855 he showed that there is a relative velocity, corresponding to constant c in his formula, at which the force between a pair of electrical positive and negative particles becomes zero. He gave the value 3.1074×10^8 m/s to the constant c, but failed to notice the fact that this was closely related to the speed of light. However, this unexpected link between electricity and optics became quite important and crucial to **Maxwell** for his theory of electromagnetic waves.

Weber put forward in 1871 the view that atoms contain positive charges that are surrounded by rotating negative particles and that the application of an electric potential to a conductor causes the negative particles to migrate from one atom to another. This is yet another prediction of the proton and electron.

Weber's discovery in regard to electricity had important revolutionary meanings in physics, because it led to the construction of a strict **scientific unit system** based on the **theory of conservation of energy**. First, it led to systematic approaches to combine various different phenomena, then to modern physics based on the concept of the **proton and electron**, and then further to today's **quantum physics**.

8

Neutral Grounding Methods

Neutral grounding methods can be classified into the **effective neutral grounding (or solidly neutral grounding) method** and the **non-effective neutral grounding method**. The difference between the two practices is the difference of the zero-sequence circuit from the viewpoint of power network theory. Therefore all power system behaviour characterized by the neutral grounding method can be explained as phenomena caused by the characteristics of the zero-sequence circuit.

Accordingly, neutral grounding methods have a wide effect on the actual practices of various engineering fields, for example in planning or operational engineering of short-circuit capacity, insulation coordination, surge protection, structure of transmission lines and towers, transformer insulation, breaker capability, protective relaying, noise interference, etc. In this section, some typical features of different neutral grounding methods are presented and their bases set out.

8.1 Comparison of Neutral Grounding Methods

The neutral grounding method of power systems can be classified as follows:

a) Effective neutral grounded system:
   - **Solidly grounded system**

b) Non-effective neutral grounded system:
   - **Resistive neutral grounded system**
   - **Arc-suppression coil (Peterson coil) neutral grounded system**
   - **Neutral ungrounded system** (may be called neutral minute-grounded system), but only adopted for distribution systems.

Table 8.1 explains in some detail the above classified neutral grounding methods.

The features of each method can be explained as features based on the zero-sequence circuit. By using a plain expression for the non-effective grounded system, grounding fault currents can be reduced considerably, but on the contrary higher temporary overvoltages would be caused during faults. The effective neutral grounded systems (solidly grounded system) has the opposite features.

Table 8.2 provides typical features of the two different grounding methods from various viewpoints.

Today's power systems are the result of continuous growth of networks since they were first constructed many years ago. Individual power systems have their own history, which has led to the applied practices of the neutral grounding method as well as the applied power frequency and nominal

Handbook of Power System Engineering Yoshihide Hase
© 2007 John Wiley & Sons, Ltd

Table 8.1 Various neutral grounded systems

A	Solidly neutral grounded system (effectively neutral grounded system)	All the transformers installed at substations belonging to the same rated voltage are solidly neutral grounded
B	Resistive neutral grounded system (impedance neutral grounded system) 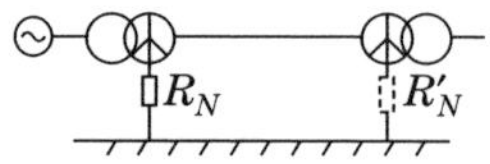 	All or some selected key transformers installed at substations belonging to the same rated voltage section are neutral grounded through a neutral grounding resistor (NGR). The resistive value [Ω] of NGR is determined mostly so that the grounding current through the NGR in one phase to ground fault is limited to 100 A or within 1000 A
C	Arc-suppression coil neutral grounded system (resonant neutral grounded system) 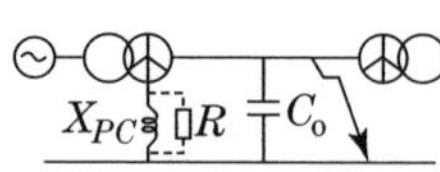 	Some key transformers are neutral grounded through tap-changeable reactors (inductance L_{pc}), whose taps are selectively controlled so that the inductive reactances $(j2\pi f \cdot L_{pc})$ are well tuned with the capacitive reactances $(-j/2\pi f \cdot C_0)$ of transmission lines over time. The zero-sequence circuit is kept under parallel quasi-resonant conditions and the zero-sequence impedance ${}_fZ_0$ of the systems have quite large values; therefore quite effective arc extinction can be expected during one phase to ground lightning faults: $${}_fZ_0 = \frac{1}{\frac{1}{j3\omega L_{PC}} + j\omega C_s} = \frac{1}{j\left(-\frac{1}{3\omega L_{PC}} + \omega C_0\right)}$$ where $$\omega L_{PC} \doteqdot \frac{1}{3\omega C_0}$$ $$\therefore \quad {}_fZ_0 \rightarrow \infty \quad {}_fI_0 \doteqdot 0$$
D	Resistive neutral grounded system with neutral compensation reactor 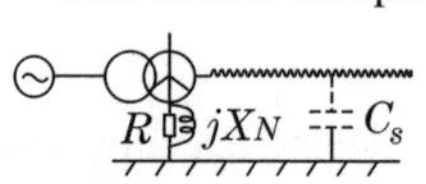 	This is essentially the same system as that of B, except the neutral compensation reactors are equipped to compensate for stray capacitances C_0 of the transmission line, in particular of long transmission lines or cable lines
E	Neutral ungrounded (isolated neutral ungrounded) systems 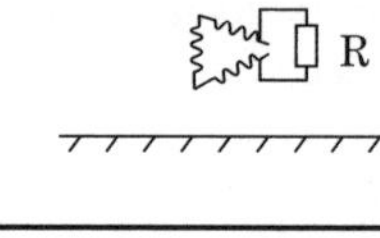 	This is a typical practice which is adopted only for distribution networks. In this system intentional neutral grounding connections do not exist, except through potential-indicating or measuring devices or other very high-impedance devices. The grounding current caused by one phase to ground fault would be limited to values of 10 mA to 1 A by a large neutral impedance ${}_fZ_0$ (the order of a few thousand ohms or more; actually ${}_fZ_0 \doteqdot \infty$ from an analytical viewpoint)

voltages of today. Therefore, the various applied neutral grounding methods may differ somewhat, in particular for lower voltage classes with an older history. However, EHV (say, over 200 kV) and UHV (say, 500 kVor higher) trunk line networks with a younger history have actually been unified by solidly grounded systems all over the world, mainly to realize EHV/UHV networks with reduced insulation levels.

In contrast, non-effective (high-impedance) neutral grounding methods have still been widely adopted for lower voltage lines and distribution networks by several countries for reasons of traditional

Table 8.2 Comparison of neutral grounding methods

	Solidly neutral grounding method (S-ngm)	High-resistive (impedance) neutral grounding method (R-ngm)
[1] Temporary overvoltages of the sound phases during line-to-ground fault ($1\phi G$)	[1S] ○ The temporary overvoltage V_b, V_c would be 0.8–1.3E The calculation check from Equation 3.10 is $\left\| {}_fV_b \right\| = \left\| \frac{j\sqrt{3}\left(a_f Z_0 - {}_fZ_2\right)}{{}_fZ_0 + {}_fZ_1 + {}_fZ_2} \cdot {}_fE_a \right\|$ If the following typical impedance is substituted ${}_fZ_1 = {}_fZ_2 = jX_1 \fallingdotseq j\frac{X_0}{3} = j\frac{{}_fX_0}{3}$, namely $jX_0 \fallingdotseq j3X_1$, then $\left\| {}_fV_b \right\| \fallingdotseq \left\| \frac{j\sqrt{3}(3a-1)}{3+1+1} \cdot {}_fE_a \right\| = 1.25 \cdot \left\| {}_fE_a \right\|$	[1R] ▼ The temporary overvoltage V_b, V_c would be 1.5–1.9E The calculation check from Equation 3.10 is $\| {}_fV_b \| = \left\| \frac{j\sqrt{3}(a_f Z_0 - {}_fZ_2)}{{}_fZ_0 + {}_fZ_1 + {}_fZ_2} \cdot {}_fE_a \right\| = \left\| \frac{j\sqrt{3}\left(a - \frac{{}_fZ_2}{{}_fZ_0}\right)}{1 + \frac{{}_fZ_1}{{}_fZ_0} + \frac{{}_fZ_2}{{}_fZ_0}} \cdot {}_fE_a \right\|$ if the following equation is assumed: $\| {}_fZ_0 \| \gg \| {}_fZ_1 \|,\ \| {}_fZ_2 \|$ $\therefore\ \| {}_fV_b \| \fallingdotseq \| j\sqrt{3}a \cdot {}_fE_a \| = \sqrt{3}\, \| {}_fE_a \|$
[2] The headway from simple faults to double or triple faults (multi-phases, or multi-point faults)	[2S] ◆ The temporary overvoltage of the sound phase under one phase to ground fault would be approximately 1.25E, so that the probability that the sound phases are forced to be time sequentially faulted would be relatively lower	[2R] ▼ The temporary overvoltage of the sound phase under one phase to ground fault would be approximately $\sqrt{3}E$ so that the probability that the sound phases are forced to be time sequentially faulted would be relatively higher
[3] Required insulation level of station equipment and transmission lines	[3S] ○ The a.c. insulation level can be reduced. This is an essential advantage of S-ngm, particularly for higher voltage systems, typically for EHV, UHV	[3R] ▼ Relatively higher insulation levels are required for R-gnm, because temporary overvoltage (TOV) becomes higher. However, this may not be necessarily disadvantageous for lower voltage systems (especially for distribution systems), because insulation levels may not be the most sensitive factor in cost reduction (see Chapter 21)
[4] Arresters	[4S] ◆ Arresters with relatively lower U_r ratings (PU values of the duty cycle voltage rating) for lower MCOV (Maximum Continuous Operating Voltage) ratings are adopted because insulation levels are relatively lower (see Chapter 21)	[4R]▼ Arresters with relatively higher U_r ratings for MCOV are adopted because insulation levels are relatively higher
[5] Transformers	[5S] ○ Stepping-reduced insulation design of the windings can be adopted Autotransformers can be adopted	[5R] ▼ Full insulation design of the windings is required Autotransformers cannot be applied

(Continued)

Table 8.2 *(Continued)*

	Solidly neutral grounding method (S-ngm)	High-resistive (impedance) neutral grounding method (R-ngm)
[6] The magnitudes of fault current passing through the earth–ground and OGW ($3f_{10}$)	[6S] ▼ The earth grounding currents during the 1ϕG or 2ϕG faults are quite large (say, 5–50 kA) The earth resistance R of substations has to be designed to fall in order to reduce $V = I_g \cdot R_d$ The sheath current of the cable lines may be disadvantageous for cable's withstanding temperature The calculation check for the system of the figure is 3ϕG $\lvert {}_fI_a \rvert = \lvert {}_fI_b \rvert = \lvert {}_fI_c \rvert = \left\lvert \frac{275\sqrt{3}}{j10} \right\rvert = 15.9\,[\text{kA}]$ $3\,{}_fI_0 = 0$ $1\phi\text{G} \lvert 3\,{}_fI_0 \rvert = \lvert {}_fI_a \rvert = \left\lvert \frac{3(275\sqrt{3})}{j(10+10+4)} \right\rvert = 19.9\,[\text{kA}]$ 2ϕG ${}_fI_1 = \frac{275\sqrt{3}}{j10 + (j10//j4)} = -j12.3\,[\text{kA}]$ $\lvert 3\,{}_fI_0 \rvert = \left\lvert 3 \times \frac{-j10}{j10 + j4} \cdot (-j12.3) \right\rvert = 26.3\,[\text{kA}]$ $\lvert {}_fI_b \rvert = \left\lvert \frac{(a^2 - a)j4 + (a^2 - 1)j10}{j4 + j10} \cdot (-j12.3) \right\rvert$ $= 19.0\,[\text{kA}]$ For 1ϕG, 2ϕG, the phase fault current may be larger than that for 3ϕG, and the earth current 3_fI_0 may be larger than the phase fault current	[6R] ○ The earth grounding currents during the 1ϕG or 2ϕG faults are limited only within 100–400 A The calculation check for the system of the figure is, assuming a 154 kV system with 200 A NGR, $R_{\text{NGR}} = \frac{154\sqrt{3}}{200} \cdot 10^3 = 445\,[\Omega]$ $\lvert {}_fZ_0 \rvert = \lvert j_fX_0 + 3R_{NGR} \rvert \gg \lvert {}_fZ_1 \rvert = \lvert j_fX_1 \rvert \doteqdot \lvert {}_fZ_2 \rvert = \lvert j_fX_2 \rvert$ 1ϕG $\lvert 3\,{}_fI_0 \rvert = \lvert {}_fI_a \rvert = \lvert 3\,{}_fE_m/\Delta \rvert \doteqdot 3 \times \frac{{}_fE_a}{{}_fZ_0} \doteqdot \frac{{}_fE_a}{R_{NGR}} = 200\,[\text{A}]$ The grounding current capacity in the stations can be reduced Communication interference can be limited 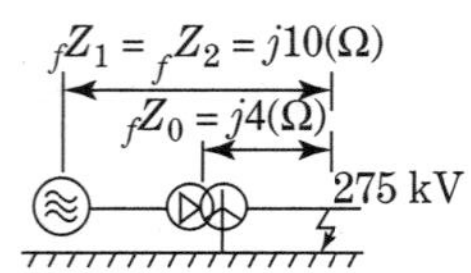
[7] Continuous zero-sequence current through the ground	[7S] ▼ Continuous zero-sequence current would flow through the ground via the neutral point of transformers, because of phase imbalance of the transmission lines Continuous sheath current would flow through cable sheaths because of the phase unbalanced allocation	[7R] ◆ The continuous zero-sequence current is actually zero. This is a very important advantage of R-ngm

[8] Radio noise or communication interference	[8S] ▼ Special countermeasures would be required against disturbances caused by zero-sequence flow-through current (including the 3rd, 6th,…, 3*n*th harmonics, especially in highly populated areas	[8R] ○ Serious problems would not arise, because zero-sequence flow-through current is limited
[9] High-speed protection against *l*–g faults	[9S] ○ High-speed selective protection scheme by differential relays as well as by directional distance relays can be adopted for all fault mode detection	[9R] ▼ High-speed selective protection scheme for line-to-ground fault detection is relatively difficult because of limited grounding fault current I_0. In particular, directional distance relays and differential relays cannot be applied for line-to-ground fault detection of transmission lines
[10] Power flow limit (*P*–δ curve characteristics) in duration of 1ϕG or 2ϕG faults	[10S]▼	[10R] ◆
	Referring to Figure 14.2, Table 14.1, Equation 14.13 and Table 3.1, the parallel-inserted reactance $jx_f = Z_2 + Z_0$ is larger in the R-ngm rather than S-ngm. Accordingly, *P*–δ curve for the stability limit during the fault would be lower in the S-ngm rather the R-ngm case	
[11] Power flow limit (*P*–δ curve characteristics) during phase opening modes	[11S] ◆	[11R] ▼
	Referring to Figure 14.2, Table 14.1, Equation 14.13 and Table 3.2, the *P*-δ curve for the stability limit during the one-phase opening mode (no-voltage time of single phase reclosing) is lower in the R-ngm. Furthermore, the power flow limit during the two-phases opening mode is actually zero in the R-ngm	
[12] Breaker's required breaking capacity (kA)	[12S] ▼ The fault current for 1ϕG could be larger than that for 3ϕG, as is demonstrated in [S6]	[12R] ◆ The fault current for 3ϕG is larger than that in any other fault modes
[13] Voltage resonance phenomena	[13S] ▼ Refer to Section 20.3. Resonance phenomena should be studied carefully although they seldom occur	[13R] ◆ Unrealistic

○: very advantageous.
◆: advantageous.
▼: disadvantageous.

history on one hand, and engineering viewpoints on the other hand. The latter may be summarized as follows:

- The largest feature of the non-effective neutral grounding method is that the continuous/temporary earth–ground flowing current ($3I_0$) is considerably reduced under normal or fault conditions. Its greatest advantages in particular for distribution systems concern human security and suppression of noise interference. These are quite important matters, especially in residential areas covered by distribution networks.
- Remarkable reductions in the system insulation level or cost by adopting the solidly grounded method cannot be expected in the lower voltage or distribution networks.
- Changing the neutral grounding method of existing networks is practically almost impossible, because major modifications or reform of existing engineering practices would be required. For example, the basic design of substation earth grounding practices (grounding varied mats, counter-poise, etc.) would have to be revised. Most arresters, protective relays and some other substation equipment would have to be replaced and so on.

8.2 Overvoltages on the Unfaulted Phases Caused by a Line-to-ground fault

If a phase a l–g fault ($1\phi G$, $V_a = 0$) occurs, the power frequency voltages on the unfaulted phases V_b, V_c are given by Equation 3.10. Accordingly, the phase c power frequency voltage V_c would become the value of the following equation during the phase a fault:

$$\left.\begin{aligned} k &= \frac{V_c}{{}_fE_a} = \frac{(a-1)_fZ_0 + (a-a^2)_fZ_1}{{}_fZ_0 + 2\,{}_fZ_1} = \frac{(a-1)\dfrac{{}_fZ_0}{{}_fZ_1} + (a-a^2)}{\dfrac{{}_fZ_0}{{}_fZ_1} + 2} \\ &= \frac{-a^2 j\sqrt{3}\cdot\dfrac{\delta + jv}{\sigma + j} + j\sqrt{3}}{\dfrac{\delta + jv}{\sigma + j} + 2} \\ &\text{where} \quad {}_fZ_0 = {}_fR_0 + j\,{}_fX_0, \quad {}_fZ_1 = {}_fR_1 + j\,{}_fX_1 \\ &\delta = \frac{{}_fR_0}{{}_fX_1}, \quad v = \frac{{}_fX_0}{{}_fX_1}, \quad \sigma = \frac{{}_fR_1}{{}_fX_1}, \quad \frac{{}_fZ_0}{{}_fZ_1} = \frac{\delta + jv}{\sigma + j} \\ &{}_fZ_1 = {}_fZ_2 \end{aligned}\right\} \quad (8.1)$$

In this equation, voltage V_c is expressed as a ratio of normal line-to neutral voltage (operating voltage) ${}_fE_a \cdot k = V_c/{}_fE_a$ is the ratio of temporary overvoltage with power frequency caused on unfaulted phase c lines during the phase a to ground fault. The absolute values of k for unfaulted phase voltages V_b, V_c are the same.

Equation 8.1 for the ratio can be expressed as curves with parameters δ, v and σ on a coordinated graph. Figure 8.1 is a typical example under the parametric conditions of $\delta = {}_fR_0/{}_fX_1 = 0 \sim +\infty$, $v = -10$ to $+10$, where $\sigma = {}_fR_1/{}_fX_1 \fallingdotseq 0$ is assumed. Also see Figure 21.2 in Chapter 21 for local detail of the same curve.

The term $v = {}_fX_0/{}_fX_1$ should have a positive value of probably 0–4 so that the zone $v < 0$ is of course unrealistic under the practical conditions of a power system. The condition $\delta \fallingdotseq 0$ to $+1$ corresponds to a solidly grounded system and $\delta \fallingdotseq 5$ to $+\infty$ to a non-effective neutral grounded system.

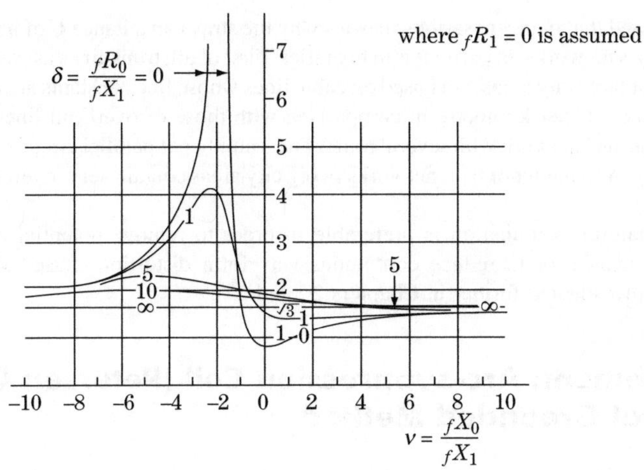

Figure 8.1 Overvoltage ratio of sound phase (phase c) under phase a *l*–g fault

Figure 8.1 as well as Figure 21.2 indicate the following.

For the non-effective neutral grounded system ($\delta \fallingdotseq 5$ to $+\infty$, $v = {}_fX_0/{}_fX_1 = 0$ to $+4$), whenever the phase a *l*-g fault occurs, the unfaulted (sound) phase voltages V_b, V_c increase and the temporary phase voltages become approximately $k = \sqrt{3}$ times nominal voltages.

For the solidly neutral grounded system ($\delta \fallingdotseq 0$ to $+1$, $v = {}_fX_0/{}_fX_1 = 0$ to $+4$), assuming $\delta \fallingdotseq 1.0$, the temporary overvoltages caused by the same fault would increase by $k = 1.2-1.3$ times, and furthermore k would be around 1.0–0.8 for the range $0 < \delta < 1$.

The above-described **temporary overvoltages** caused on the sound phase conductors by 1ϕG is actually one of the very important bases affecting the concept of insulation coordination of individual power system networks. In other words, the required **insulation level** against continuous/temporary power frequency overvoltages of individual power systems would be decided from the overvoltage coefficient k. This will be discussed in detail in Chapters 20 and 21.

Incidentally, the overvoltages caused by a double *l*–g fault (2ϕG) would be generally lower than those caused by 1ϕG. However, the overvoltage ratio by 2ϕG as well as 1ϕG should preferably be investigated in the same way. In contrast, three-phase faults (3ϕS, 3ϕG) and line-line faults (2ϕS) are of no interest.

8.3 Possibility of Voltage Resonance

We have learned that the solidly neutral grounding method is advantageous from the viewpoint of temporary overvoltages and consequently from that of the required insulation levels. However, one potential weak spot of the solidly neutral grounded system has to be considered, that is the potential possibility of series resonance (or quasi-resonance) phenomena.

Figure 8.1 indicates the existence of a serious series resonant area in $v < {}_fX_0/{}_fX_1 < 0$. In Figure 3.2b, showing a single phase-to-ground fault, or in Figures 1b and 2b, showing conductor opening, we can imagine that for the cases where C_1, C_2 and C_0 exist in positive-, negative- and zero-sequence circuits, *LC* series resonant local loops would arise in the circuits. If $j\,{}_fX_0$ or $j\,{}_fX_1$ become negative (capacitive value), regardless of the time interval, serious abnormal overvoltages would be caused. Although such resonance conditions seldom occur, engineers would still have to examine several irregular conditions including unbalanced short-circuit modes and open-conductor modes under different network connections.

It must be stressed that there are serious reasons why the stray capacitance C of networks has been increasing in today's networks, in particular in big cities. First of all, trunk lines as well as low-voltage distribution lines of large city areas, are based on cable lines whose line constants are one-fifth smaller L and 20 times larger C per kilometre in comparison with those of overhead lines. Moreover, the system may be a meshed network with several routes and a number of parallel circuits per route to meet large-load capacity. As a matter of fact, networks in big city areas contain very 'overcrowded' L and C constants.

Accordingly, careful examination is preferable in order to remove potential reasons for such possible local resonance, or to reduce continuous waveform distortion caused similarly. (These problems will be investigated further in Chapters 20 and 22).

8.4 Supplement: Arc-suppression Coil (Petersen Coil) Neutral Grounded Method

The principle of the arc-suppression coil (Petersen coil, PC coil) neutral grounded method is shown in Table 8.1(C). The actual transmission line has stray capacitances C_1, C_2, C_0 so that the zero-sequence circuit is a parallel circuit of the impedance at neutral grounding part Z_{pc} and zero-sequence line capacitance C_0. Now we recall Equation 3.10 and the equivalent circuit in Figure 3.2 of the fault $1\phi G$. If the zero-sequence impedance of the neutral point Z_{pc} is tuned with $-jX_{co} = 1/j\omega C$, this means that ${}_fZ_0 \rightarrow \infty$, ${}_fZ_{total} \rightarrow \infty$ and ${}_fI_a = 3\,I_0 \rightarrow 0$ in Equation 3.10, so we can expect easy extinction of the grounding current whenever the $1\phi G$ fault occurs.

This practice was developed in Germany around 1918 and then spread to several countries as a good instance of the solidly grounding method. However, the practice had some weak points as follows:

- Tuning of Z_{pc} to $-jX_{co}$ in the zero-sequence circuit may be easy for smaller power systems with radial feeder connections. However, it is not so easy for large power systems which include several substations that must be neutral grounded, and/or for loop-connected power systems.
- High-speed detection of the $1\phi G$ fault by protective relay is not necessarily easy because ${}_fI_a = 3\,{}_fI_0 \rightarrow 0$.
- The practice is useless against double phase faults.
- If a transformer with a suppression coil is tripped for any reason, the tuning condition of the system would be at least broken, or the power system might lose its neutral grounding point as the worst case and suffer unstable overvoltages.

Consequently, most of the power systems where the PC coil used to be adopted have been switched to a resistive grounded system or solidly neutral grounded system. It may be said that the PC coil neutral grounding method has actually become a historical feature which can be accommodated in smaller power systems based on mainly radial connections.

9

Visual Vector Diagrams of Voltages and Currents under Fault Conditions

In this chapter, diagrammatic solution of voltages and currents under various fault conditions is introduced. Simple and plain knowledge in regard to the behaviour of three-phase voltages and currents under various fault conditions and the easily derived method for that are quite important in various practical engineering activities.

9.1 Three-phase Fault: 3ϕS, 3ϕG (Solidly Neutral Grounding System, High-resistive Neutral Grounding System)

The equivalent circuit of a three-phase line-to-line fault (3ϕS, 3ϕG) and the voltage distribution by distance between the generator and the faulting point f are shown in Figures 9.1a and b, where x_1 is the positive sequence reactance of total line length. The voltage at arbitrary midpoint m can be derived as a function of $k_1 (k_1 = 0 \sim 1)$ by the equation

$$\left.\begin{array}{ll}
k_1 = \dfrac{{}_q x_1}{x_1}, \quad x_1 = {}_p x_1 + {}_q x_1 = (1-k_1)x_1 + k_1 x_1, \quad \Delta = jx_1 & ① \\
I_1 = \dfrac{E}{\Delta} = -j\dfrac{E}{x_1}, \quad I_2 = I_0 = 0 & ② \\
{}_f V_1 = 0, \quad {}_f V_2 = {}_f V_0 = 0 & ③ \\
\dfrac{{}_m V_1}{E} = j_q x_1 \cdot \dfrac{I_1}{E} = \dfrac{{}_q x_1}{x_1} = k_1, \quad \dfrac{{}_m V_2}{E} = \dfrac{{}_m V_0}{E} = 0 & ④ \\
{}_f V_a = {}_f V_b = {}_f V_c = 0 & ⑤ \\
\dfrac{{}_m V_a}{E} = k_1, \quad \dfrac{{}_m V_b}{E} = a^2 k_1, \quad \dfrac{{}_m V_c}{E} = a k_1 & ⑥
\end{array}\right\} \quad (9.1)$$

Accordingly, the voltage vectors at point m can be drawn as shown in Figure 9.1c by the parameter of k_1. Finally we can draw the three-dimensional vector diagram, Figure 9.1d, where the current vectors are the same at any point because leakage current is neglected, and the phase angle is approximately 90° (say 85°, considering line resistances) lagged from the voltages.

Handbook of Power System Engineering Yoshihide Hase
© 2007 John Wiley & Sons, Ltd

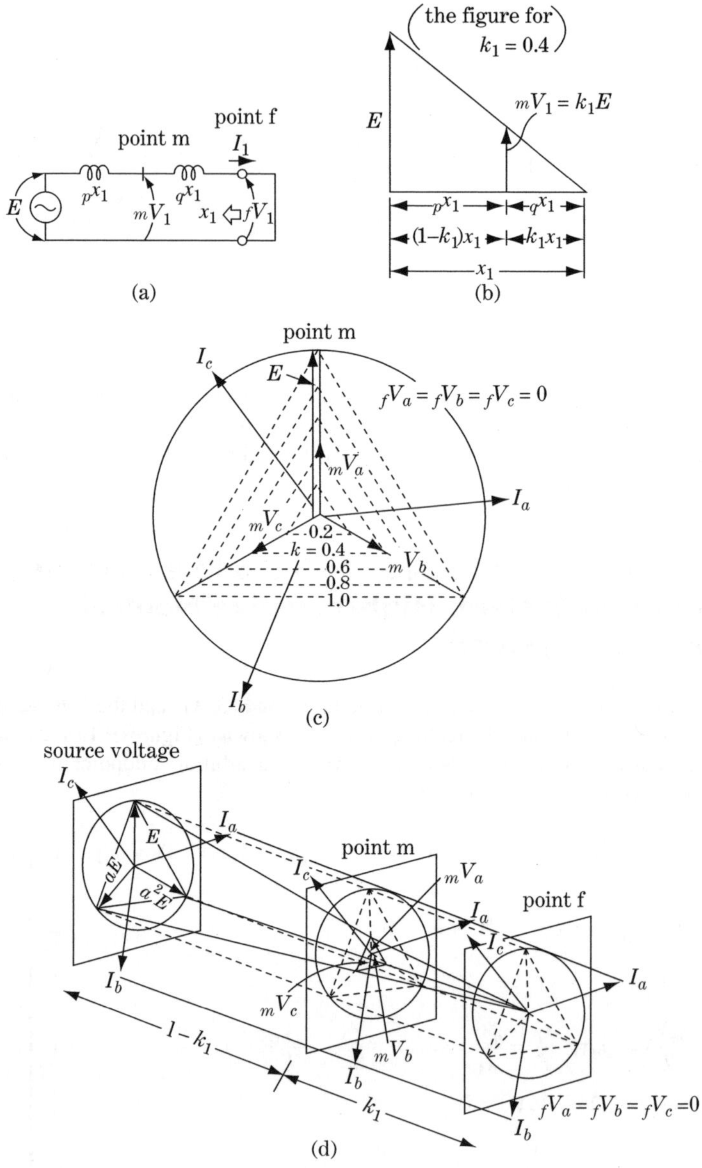

Figure 9.1 Three-phase fault: $3\phi S, 3\phi G$ (solidly neutral grounding system, high-resistive neutral grounding system)

9.2 Phase b–c Fault: $2\phi S$ (for Solidly Neutral Grounding System, High-resistive Neutral Grounding System)

The equivalent circuit and the related voltage distribution through the series circuit of positive- and negative-sequence reactances are shown in Figures 9.2a and b.

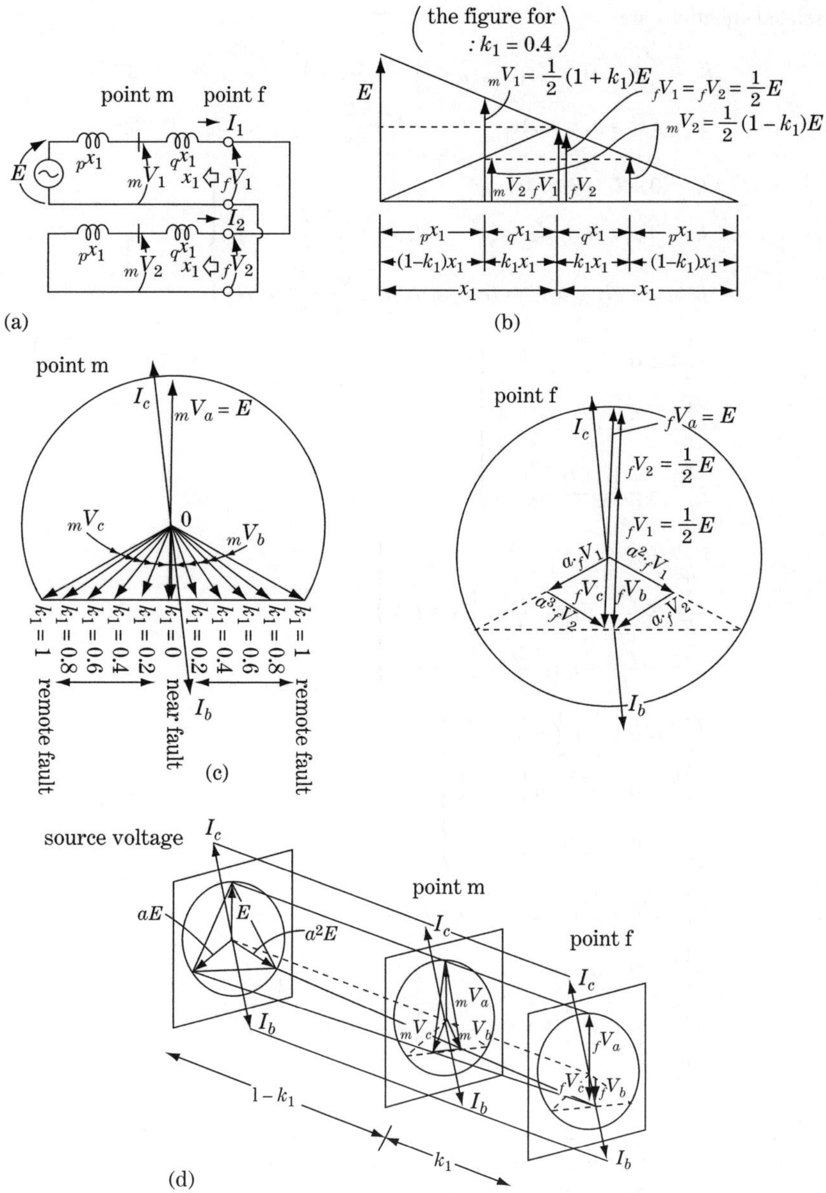

Figure 9.2 Phase b–c fault: 2ϕS (for solidly neutral grounding system, high-resistive neutral grounding system)

The related equations are

$$\left.\begin{array}{ll}
k_1 = \dfrac{{}_q x_1}{x_1}, \quad x_1 = {}_p x_1 + {}_q x_1 = (1-k_1)x_1 + k_1 x_1 & \text{①} \\
\Delta = j(x_1 + x_1) = j2x_1, \quad I_1 = -I_2 = \dfrac{E}{\Delta} = -j\dfrac{E}{2x_1}, \quad I_0 = 0 & \text{②} \\
\left.\begin{aligned}
I_a &= 0 + I_1 + I_2 = 0 \\
I_b &= 0 + a^2 I_1 + a I_2 = (a^2 - a) I_1 = -\frac{\sqrt{3}}{2}\cdot\frac{E}{x_1} \\
I_c &= 0 + a I_1 + a^2 I_2 = (a^2 - a) I_2 = +\frac{\sqrt{3}}{2}\cdot\frac{E}{x_1} = -I_b
\end{aligned}\right\} & \text{③} \\
\left.\begin{aligned}
\frac{{}_f V_0}{E} &= 0 \\
\frac{{}_f V_1}{E} &= \frac{E - jx_1 \cdot I_1}{E} = \frac{1}{2} \\
\frac{{}_f V_2}{E} &= \frac{-jx_1 \cdot I_2}{E} = \frac{1}{2}
\end{aligned}\right\} & \text{④} \\
\left.\begin{aligned}
\frac{{}_m V_0}{E} &= 0 \\
\frac{{}_m V_1}{E} &= \frac{{}_f V_1}{E} + j_q x_1 \cdot \frac{I_1}{E} = \frac{1}{2}(1 + k_1) \\
\frac{{}_m V_2}{E} &= \frac{{}_f V_2}{E} + j_q x_1 \cdot \frac{I_2}{E} = \frac{1}{2}(1 - k_1)
\end{aligned}\right\} & \text{⑤} \\
\left.\begin{aligned}
\frac{{}_f V_a}{E} &= 0 + \frac{1}{2} + \frac{1}{2} = 1 \\
\frac{{}_f V_b}{E} &= 0 + a^2 \frac{1}{2} + a\frac{1}{2} = -\frac{1}{2} \\
\frac{{}_f V_c}{E} &= 0 + a\frac{1}{2} + a^2 \frac{1}{2} = -\frac{1}{2}
\end{aligned}\right\} & \text{⑥} \\
\left.\begin{aligned}
\frac{{}_m V_a}{E} &= 0 + \frac{1}{2}(1 + k_1) + \frac{1}{2}(1 - k_1) = 1 \\
\frac{{}_m V_b}{E} &= 0 + a^2 \cdot \frac{1}{2}(1 + k_1) + a \cdot \frac{1}{2}(1 - k_1) = -\frac{1}{2} - j\frac{\sqrt{3}}{2}k_1 \\
\frac{{}_m V_c}{E} &= 0 + a \cdot \frac{1}{2}(1 + k_1) + a^2 \cdot \frac{1}{2}(1 - k_1) = -\frac{1}{2} + j\frac{\sqrt{3}}{2}k_1 \\
\frac{{}_m V_{bc}}{E} &= -j\sqrt{3}k_1
\end{aligned}\right\} & \text{⑦}
\end{array}\right\} \quad (9.2)$$

The vector diagram in Figure 9.2c for arbitrary point m is derived from Equation 9.2⑦, and the diagram in Figure 9.2d for fault point f is a special case with $k = 0$.

The fault currents in the case of $2\phi S$ become

$$|I_b| = |I_c| = \frac{\sqrt{3}}{2}\left|\frac{E}{x_1}\right|$$

which is 0.87 times the current in the case of

$$3\phi S \left|\frac{E}{x_1}\right|$$

Zero-sequence voltage and current are zero in this case, so that neutral voltage is zero potential at any point. Accordingly, the equation $E = {}_m V_a = {}_f V_a$ (where $I_a = 0$) is found and the phase a to ground voltage ${}_m V_a$ at an arbitrary point m is almost not affected by the distance from point f.

9.3 Phase a to Ground Fault: 1ϕG (Solidly Neutral Grounding System)

The equivalent circuit of this case is given by Figure 9.3a, from which the related Equation 9.3 as well as the diagrams in Figures 9.3b1 and b2 are derived in a similar way to that of Section 9.2.

Figure 9.3b1 is drawn to satisfy precisely the circuit condition of Figure 9.3a, that is the straight line $\overline{\mathrm{ad}}$ is divided by points b and c which are determined to satisfy the ratio

$$\overline{\mathrm{ab}} : \overline{\mathrm{bc}} : \overline{\mathrm{cd}} = x_1 : x_1 : x_0 = 1 : 1 : v$$

${}_fV_1$, ${}_fV_2$ are drawn at point b and ${}_fV_0$ at point c. The arrows for ${}_fV_2$, ${}_fV_0$ are in the opposite direction to that of ${}_fV_1$ because ${}_fV_1 = -({}_fV_2 + {}_fV_0)$.

${}_mV_1$ is drawn at the point b′, which divides $\overline{\mathrm{ab}}$ by $\overline{\mathrm{ab'}} : \overline{\mathrm{b'b}} = {}_p\mathrm{x}_1 : {}_q\mathrm{x}_1$. ${}_mV_2$, ${}_mV_0$ are derived analogously.

Figure 9.3b2 is obtained by folding Figure 9.3b1. The related equations are

$$\left.\begin{aligned}
&v = \frac{x_0}{x_1} = \frac{x_0}{x_2}, \quad x_1 = x_2 = \frac{x_0}{v}\\
&k_1 = \frac{{}_qx_1}{x_1}, \quad x_1 = {}_px_1 + {}_qx_1 = (1-k_1)x_1 + k_1x_1\\
&k_0 = \frac{{}_qx_0}{x_0}, \quad x_0 = {}_px_0 + {}_qx_0 = (1-k_0)x_0 + k_0x_0\\
&\Delta = j(x_0 + x_1 + x_2) = j(v+2)x_1
\end{aligned}\right\} \quad ①$$

$$I_1 = I_2 = I_0 = \frac{E}{\Delta} = -j\frac{1}{v+2}\cdot\frac{E}{x_1} \quad ②$$

$$\left.\begin{aligned}
&\frac{{}_fV_0}{E} = -jx_0\cdot\frac{I_0}{E} = \frac{-v}{v+2}\\
&\frac{{}_fV_1}{E} = -\left(\frac{{}_fV_0}{E} + \frac{{}_fV_2}{E}\right) = \frac{v+1}{v+2}\\
&\frac{{}_fV_2}{E} = -jx_1\cdot\frac{I_2}{E} = \frac{-1}{v+2}
\end{aligned}\right\} \quad ③$$

$$\left.\begin{aligned}
&\frac{{}_mV_0}{E} = \frac{{}_fV_0}{E} + j_qx_0\cdot\frac{I_0}{E} = \frac{{}_fV_0}{E} + \frac{k_0v}{v+2} = \frac{-v}{v+2} + \frac{k_0v}{v+2}\\
&\frac{{}_mV_1}{E} = \frac{{}_fV_1}{E} + j_qx_1\cdot\frac{I_1}{E} = \frac{{}_fV_1}{E} + \frac{k_1}{v+2} = \frac{v+1}{v+2} + \frac{k_1}{v+2}\\
&\frac{{}_mV_2}{E} = \frac{{}_fV_2}{E} + j_qx_1\cdot\frac{I_2}{E} = \frac{{}_fV_2}{E} + \frac{k_1}{v+2} = \frac{-1}{v+2} + \frac{k_1}{v+2}
\end{aligned}\right\} \quad ④ \qquad (9.3)$$

$$\left.\begin{aligned}
&{}_fV_a = 0\\
&\frac{{}_fV_b}{E} = \frac{{}_fV_0}{E} + a^2\frac{{}_fV_1}{E} + a\frac{{}_fV_2}{E} = \frac{(a^2-1)v + (a^2-a)}{v+2} = \frac{-3v}{2(v+2)} - j\frac{\sqrt{3}}{2}\\
&\frac{{}_fV_c}{E} = \frac{{}_fV_0}{E} + a\frac{{}_fV_1}{E} + a^2\frac{{}_fV_2}{E} = \frac{(a-1)v + (a-a^2)}{v+2} = \frac{-3v}{2(v+2)} + j\frac{\sqrt{3}}{2}\\
&\frac{{}_fV_{bc}}{E} = \frac{{}_fV_b}{E} - \frac{{}_fV_c}{E} = -j\sqrt{3}
\end{aligned}\right\} \quad ⑤$$

$$\left.\begin{aligned}
&\frac{{}_mV_a}{E} = \frac{{}_fV_a}{E} + \frac{k_0v + k_1 + k_1}{v+2} = \frac{k_0v + 2k_1}{v+2}\\
&\frac{{}_mV_b}{E} = \frac{{}_fV_b}{E} + \frac{k_0v + a^2k_1 + ak_1}{v+2} = \frac{{}_fV_b}{E} + \frac{k_0v - k_1}{v+2}\\
&\frac{{}_mV_c}{E} = \frac{{}_fV_c}{E} + \frac{k_1v + ak_1 + a^2k_1}{v+2} = \frac{{}_fV_c}{E} + \frac{k_0v - k_1}{v+2}\\
&\frac{{}_mV_{bc}}{E} = \frac{{}_mV_b}{E} - \frac{{}_mV_c}{E} = \frac{{}_fV_{bc}}{E} = -j\sqrt{3}
\end{aligned}\right\} \quad ⑥$$

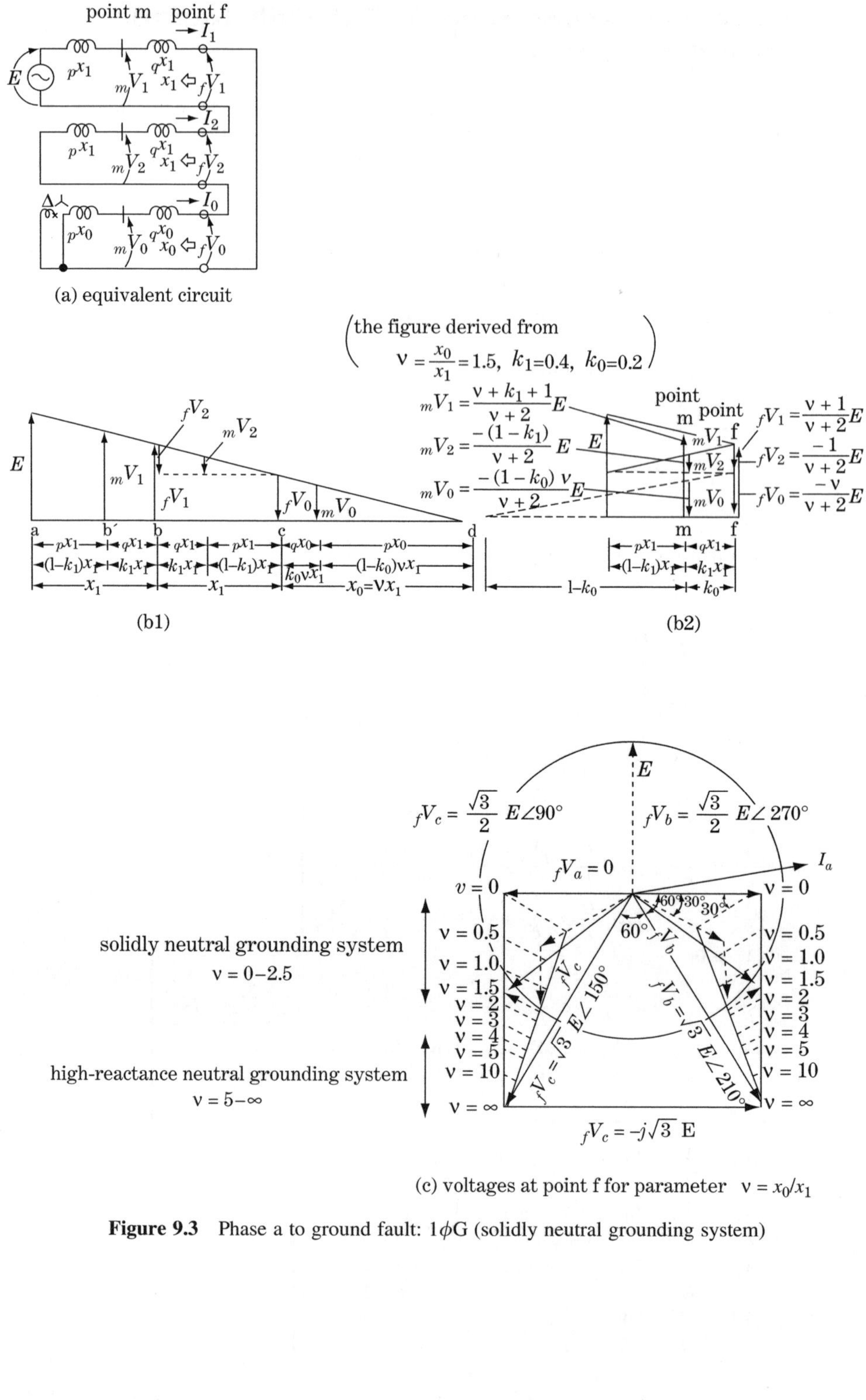

Figure 9.3 Phase a to ground fault: $1\phi G$ (solidly neutral grounding system)

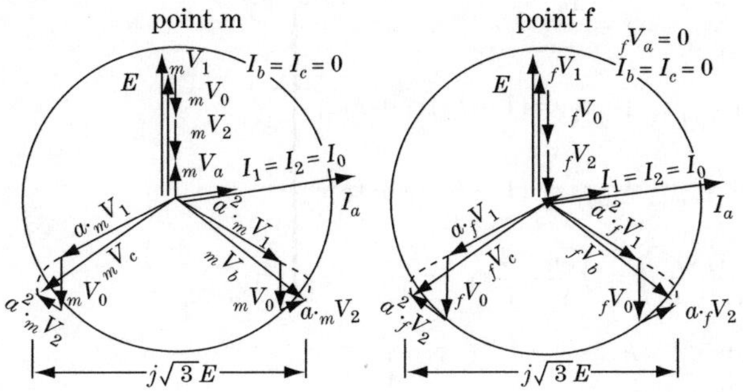

(d) voltages, currents at points f, m

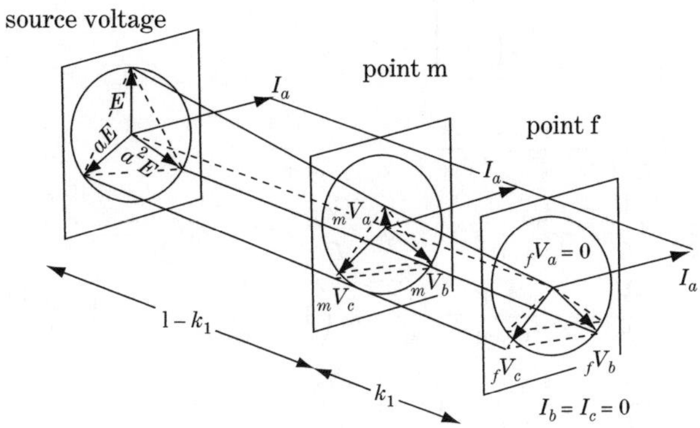

(e) voltages and currents overview

Figure 9.3 (*Continued*)

Accordingly, $_fV_1$, $_fV_2$, $_fV_0$ and $_mV_1$, $_mV_2$, $_mV_0$ are easily calculated by referring to Equation 9.3①–④ or Figure 9.3b2. Then, referring to Equation 9.3⑤, the phase voltages at fault point f, $_fV_a$, $_fV_b$, $_fV_c$, can be drawn with the parameter $v = x_0/x_1$ as is shown in Figure 9.3c. The parameter v is $v = x_0/x_1 = 0$ to 2 or 3 at most for the solidly neutral grounding system, so that the magnitudes of unfaulted phase voltages $_fV_b$, $_fV_c$ are around 0.8–1.1. In other words, jumping overvoltage phenomena of unfaulted phases do not occur for 1ϕG in the solidly neutral grounding system, which is obviously a great advantage from the viewpoint of reducing the level of insulation coordination. This is the essential reason why modern EHV and UHV systems over 200 or 300 kV utilize the solidly neutral grounding system without exception (see Chapters 20 and 21). Figure 9.3d shows the vector diagrams at points f and m and Figure 9.3e is the three-dimensional overview of the vector diagram of the total system.

Incidentally, Figure 9.3c corresponds to the case which was studied in Section 8.2 and the resulting Figure 8.1 (where $\delta = 0$). If a high-reactance neutral grounded system of large V (say, $v \geqq 8$ or 10) is assumed, unfaulted phase voltages $_fV_b$, $_fV_c$ would become $\sqrt{3}$ times E as shown in Figures 9.3c–e.

9.4 Double Line-to-ground (Phases b and c) Fault: 2ϕG (Solidly Neutral Grounding System)

Equation 9.4 and Figure 9.4 can be derived in the same manner as in the previous section. The related equations are

$$\left.\begin{aligned}
&v = \frac{x_0}{x_1} = \frac{x_0}{x_2}, \quad x_1 = x_2 = \frac{x_0}{v}\\
&k_1 = \frac{{}_q x_1}{x_1}, \quad x_1 = {}_p x_1 + {}_q x_1 = (1-k_1)x_1 + k_1 x_1\\
&k_0 = \frac{{}_q x_0}{x_0}, \quad x_0 = {}_p x_0 + {}_q x_0 = (1-k_0)x_0 + k_0 x_0\\
&\Delta = j\left(x_1 + \frac{x_1 x_0}{x_1 + x_0}\right) = j\frac{2v+1}{v+1}x_1
\end{aligned}\right\} \quad ①$$

$$\left.\begin{aligned}
&I_1 = \frac{E}{\Delta} = -j\frac{v+1}{2v+1}\cdot\frac{E}{x_1}\\
&I_2 = \frac{-x_0}{x_1+x_0}I_1 = \frac{-v}{v+1}I_1 = j\frac{v}{2v+1}\cdot\frac{E}{x_1}\\
&I_0 = \frac{-x_1}{x_1+x_0}I_1 = \frac{-1}{v+1}I_1 = j\frac{1}{2v+1}\cdot\frac{E}{x_1}
\end{aligned}\right\} \quad ②$$

$$\frac{{}_f V_0}{E} = \frac{{}_f V_1}{E} = \frac{{}_f V_2}{E} = j\frac{x_1 x_0}{x_1+x_0}\cdot\frac{I_1}{E} = j\frac{v x_1}{v+1}\cdot\frac{I_1}{E} = \frac{v}{2v+1} \quad ③$$

$$\left.\begin{aligned}
&\frac{{}_m V_0}{E} = \frac{{}_f V_0}{E} + j_q x_0\frac{I_0}{E} = \frac{{}_f V_0}{E} - \frac{k_0 v}{2v+1} = (1-k_0)\frac{{}_f V_0}{E} = (1-k_0)\frac{v}{2v+1}\\
&\frac{{}_m V_1}{E} = \frac{{}_f V_1}{E} + j_q x_1\frac{I_1}{E} = \frac{{}_f V_1}{E} + k_1\frac{v+1}{2v+1} = \frac{{}_f V_1}{E} + k_1\left(1 - \frac{{}_f V_1}{E}\right) = \frac{v + k_1(v+1)}{2v+1}\\
&\frac{{}_m V_2}{E} = \frac{{}_f V_2}{E} + j_q x_1\frac{I_2}{E} = \frac{{}_f V_2}{E} - \frac{k_1 v}{2v+1} = (1-k_1)\frac{{}_f V_2}{E} = (1-k_1)\frac{v}{2v+1}
\end{aligned}\right\} \quad ④ \qquad (9.4)$$

$$\left.\begin{aligned}
&I_a = 0\\
&I_b = j\frac{1 - a^2(v+1) + av}{2v+1}\cdot\frac{E}{x_1} = \frac{\sqrt{3}(a-v)}{2v+1}\cdot\frac{E}{x_1}\\
&I_c = j\frac{1 - a(v+1) + a^2 v}{2v+1}\cdot\frac{E}{x_1} = \frac{\sqrt{3}(-a^2+v)}{2v+1}\cdot\frac{E}{x_1}
\end{aligned}\right\} \quad ⑤$$

$$\left.\begin{aligned}
&\frac{{}_f V_a}{E} = \frac{{}_f V_0}{E} + \frac{{}_f V_1}{E} + \frac{{}_f V_2}{E} = 3\frac{{}_f V_1}{E} = \frac{3v}{2v+1}\\
&\frac{{}_f V_b}{E} = 0\\
&\frac{{}_f V_c}{E} = 0
\end{aligned}\right\} \quad ⑥$$

$$\left.\begin{aligned}
\frac{{}_m V_a}{E} &= \frac{{}_f V_a}{E} + \frac{(-k_0 v) + k_1(v+1) + (-k_1 v)}{2v+1} = \frac{{}_f V_a}{E} + \frac{k_1 - k_0 v}{2v+1}\\
&= \frac{3v}{2v+1} + \frac{k_1 - k_0 v}{2v+1}\\
\frac{{}_m V_b}{E} &= \frac{{}_f V_b}{E} + \frac{(-k_0 v) + a^2 k_1(v+1) + a(-k_1 v)}{2v+1} = \frac{(a^2 - j\sqrt{3}v)k_1 - k_0 v}{2v+1}\\
\frac{{}_m V_c}{E} &= \frac{{}_f V_c}{E} = \frac{(-k_0 v) + a k_1(v+1) + a^2(-k_1 v)}{2v+1} = \frac{(a^2 + j\sqrt{3}v)k_1 - k_0 v}{2v+1}
\end{aligned}\right\}$$

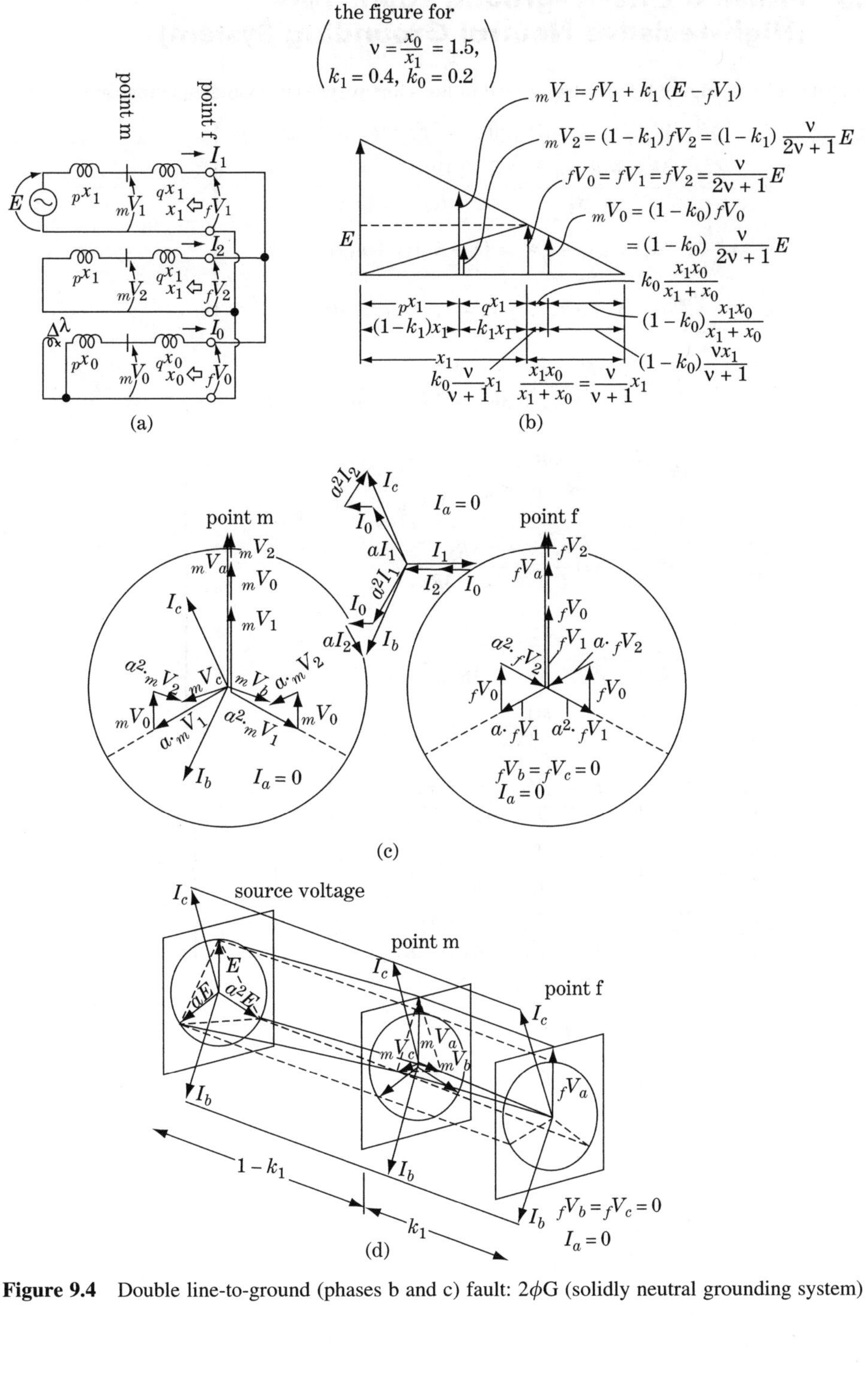

Figure 9.4 Double line-to-ground (phases b and c) fault: $2\phi G$ (solidly neutral grounding system)

9.5 Phase a Line-to-ground Fault: $1\phi G$ (High-resistive Neutral Grounding System)

Equation 9.5 and Figure 9.5 can be derived in the same way. The related equations are

$$\left.\begin{aligned}
&Z_0 = 3R_N + j({}_px_0 + {}_qx_0) \fallingdotseq 3R_N, \quad 3R_N \gg {}_px_0 + {}_qx_0 = x_0 \\
&v = \frac{Z_0}{jx_1} = \frac{3R_N + j({}_px_0 + {}_qx_0)}{jx_1} \fallingdotseq \frac{3R_N}{jx_1} \\
&k_1 = \frac{{}_qx_1}{x_1}, \quad x_1 = {}_px_1 + {}_qx_1 = (1-k_1)x_1 + k_1x_1 \\
&k_{arc} = \frac{3R_{arc}}{3R_N}, \quad \Delta = 3R_N + j(x_0 + x_1 + x_2) \fallingdotseq 3R_N
\end{aligned}\right\} \quad ①$$

$$I_1 = I_2 = I_0 = \frac{E}{\Delta + 3R_{arc}} \fallingdotseq \frac{E}{3R_N + 3R_{arc}} = \frac{E}{3R_N} \cdot \frac{1}{1+k_{arc}} \quad ②$$

$$\left.\begin{aligned}
&\frac{{}_fV_{arc}}{E} = 3R_{arc}\frac{I_0}{E} \fallingdotseq \frac{3R_{arc}}{3R_N + 3R_{arc}} = \frac{1}{1 + \dfrac{R_N}{R_{arc}}} = \frac{1}{1 + \dfrac{1}{k_{arc}}} \\
&\frac{{}_fV_0}{E} = -(3R_N + jx_0)\frac{I_0}{E} \fallingdotseq \frac{-R_N}{3R_N + 3R_{arc}} = \frac{-1}{1 + \dfrac{R_{arc}}{R_N}} = \frac{-1}{1+k_{arc}} \\
&\frac{{}_fV_1}{E} = \frac{E - jx_1I_1}{E} \fallingdotseq 1 - \frac{jx_1}{3R_N + 3R_{arc}} \fallingdotseq 1 \\
&\frac{{}_fV_2}{E} = -jx_1\frac{I_2}{E} \fallingdotseq \frac{-jx_1}{3R_N + 3R_{arc}} \fallingdotseq 0
\end{aligned}\right\} \quad ③$$

$$\left.\begin{aligned}
&\frac{{}_mV_0}{E} = \frac{{}_fV_0}{E} + j_qx_0\frac{I_0}{E} = \frac{{}_fV_0}{E} + \frac{j_qx_0}{3R_N + 3R_{arc}} \fallingdotseq \frac{{}_fV_0}{E} = \frac{-1}{1+k_{arc}} \\
&\frac{{}_mV_1}{E} = \frac{{}_fV_1}{E} + j_qx_1\frac{I_1}{E} = \frac{{}_fV_1}{E} + \frac{j_qx_1}{3R_N + 3R_{arc}} \fallingdotseq \frac{{}_fV_1}{E} \fallingdotseq 1 \\
&\frac{{}_mV_2}{E_m} = \frac{{}_fV_2}{E} + j_qx_1\frac{I_2}{E} = \frac{{}_fV_2}{E} + \frac{j_qx_1}{3R_N + 3R_{arc}} \fallingdotseq \frac{{}_fV_2}{E} \fallingdotseq 0
\end{aligned}\right\} \quad ④$$

$$\left.\begin{aligned}
&\frac{{}_fV_a}{E} = \frac{{}_fV_0}{E} + \frac{{}_fV_1}{E} + \frac{{}_fV_2}{E} \fallingdotseq \frac{-1}{1+k_{arc}} + 1 \\
&\frac{{}_fV_b}{E} = \frac{{}_fV_0}{E} + a^2\frac{{}_fV_1}{E} + a\frac{{}_fV_2}{E} \fallingdotseq \frac{-1}{1+k_{arc}} + a^2 \\
&\frac{{}_fV_c}{E} = \frac{{}_fV_0}{E} + a\frac{{}_fV_1}{E} + a^2\frac{{}_fV_2}{E} \fallingdotseq \frac{-1}{1+k_{arc}} + a
\end{aligned}\right\} \quad ⑤$$

$$\left.\begin{aligned}
&\frac{{}_mV_a}{E} = \frac{{}_mV_0}{E} + \frac{{}_mV_1}{E} + \frac{{}_mV_2}{E} \fallingdotseq \frac{{}_fV_0}{E} + 1 \fallingdotseq \frac{{}_fV_a}{E} \\
&\frac{{}_mV_b}{E} = \frac{{}_mV_0}{E} + a^2\frac{{}_mV_1}{E} + a\frac{{}_mV_2}{E} \fallingdotseq \frac{{}_fV_0}{E} + a^2 \fallingdotseq \frac{{}_fV_b}{E} \\
&\frac{{}_mV_c}{E} = \frac{{}_mV_0}{E} + a\frac{{}_mV_1}{E} + a^2\frac{{}_mV_2}{E} \fallingdotseq \frac{{}_fV_0}{E} + a \fallingdotseq \frac{{}_fV_c}{E}
\end{aligned}\right\} \text{where } \frac{{}_fV_0}{E} = \frac{-1}{1+k_{arc}} \quad ⑥$$

(9.5)

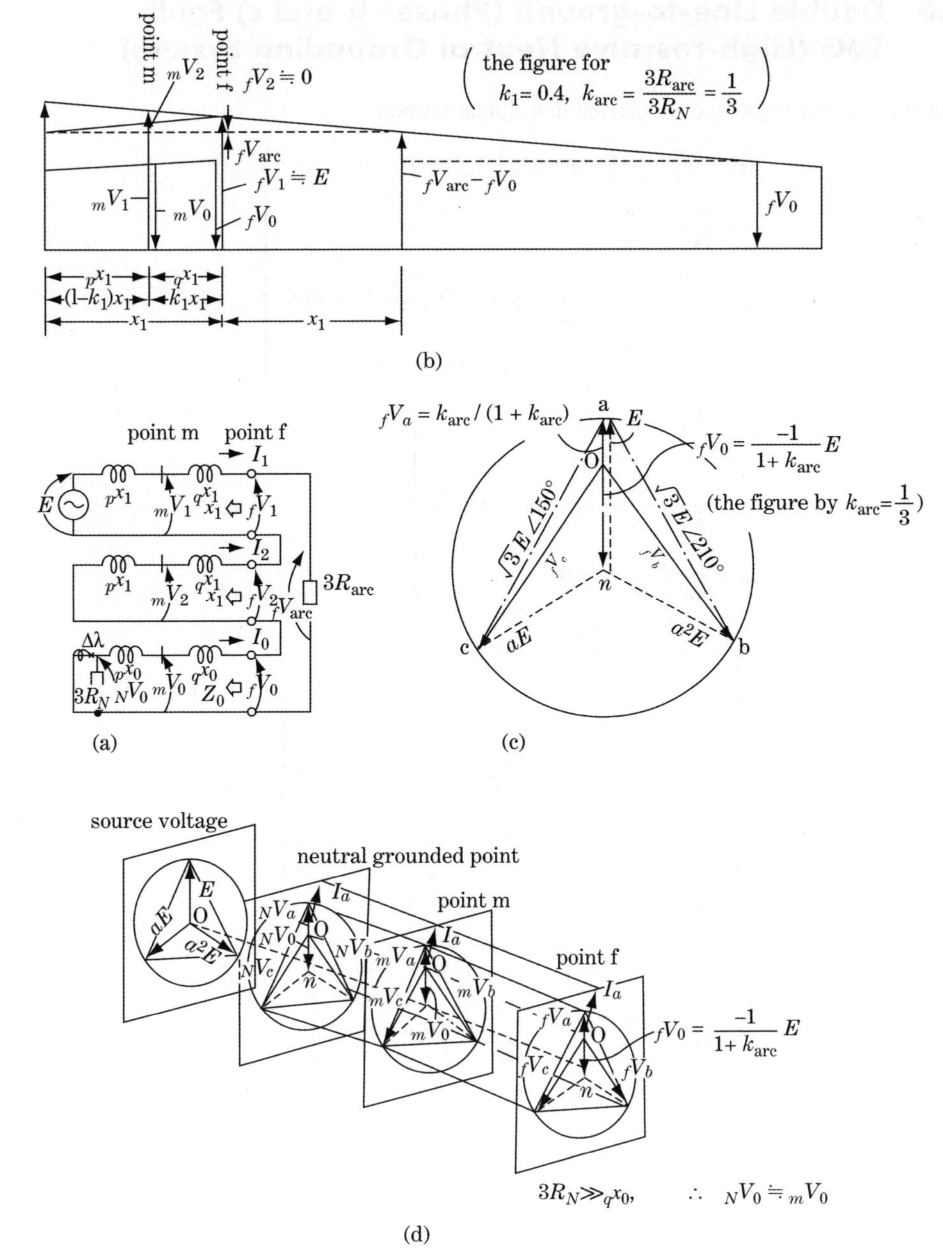

Figure 9.5 Phase a to ground fault: $1\phi G$ (high-resistive neutral grounding system)

9.6 Double Line-to-ground (Phases b and c) Fault: 2ϕG (High-resistive Neutral Grounding System)

Equation 9.6 and Figure 9.6 are derived in a similar fashion:

$$\left.\begin{aligned}
&Z_0 = 3R_N + j\left({}_p x_0 + {}_q x_0\right) \fallingdotseq 3R_N \gg jx_1 \\
&v = \frac{Z_0}{jx_1} \fallingdotseq \frac{3R_N}{jx_1} \\
&k_1 = \frac{{}_q x_1}{x_1}, \quad x_1 = {}_p x_1 + {}_q x_1 = (1-k_1)x_1 + k_1 x_1 \\
&\Delta = jx_1 + \frac{Z_0 jx_1}{Z_0 + jx_1} \fallingdotseq jx_1 + jx_1 = j2x_1
\end{aligned}\right\} \text{①}$$

$$\left.\begin{aligned}
&I_1 = \frac{E}{\Delta} \fallingdotseq -j\frac{1}{2}\cdot\frac{E}{x_1} \\
&I_2 = \frac{-Z_0}{Z_0 + jx_1} I_1 \fallingdotseq -I_1 = j\frac{1}{2}\cdot\frac{E}{x_1} \\
&I_0 = \frac{-jx_1}{Z_0 + jx_1} I_1 \fallingdotseq 0
\end{aligned}\right\} \text{②}$$

$$\frac{{}_f V_1}{E} = \frac{{}_f V_2}{E} = \frac{{}_f V_0}{E} = -jx_1\frac{I_2}{E} \fallingdotseq \frac{1}{2} \quad \text{③}$$

$$\left.\begin{aligned}
&\frac{{}_m V_0}{E} \fallingdotseq \frac{{}_f V_0}{E} \fallingdotseq \frac{1}{2} \\
&\frac{{}_m V_1}{E} = \frac{{}_f V_1}{E} + j_q x_1\frac{I_1}{E} = \frac{{}_f V_1}{E} + k_1\frac{1}{2} \fallingdotseq (1+k_1)\frac{1}{2} \\
&\frac{{}_m V_2}{E} = \frac{{}_f V_2}{E} + j_q x_1\frac{I_2}{E} = \frac{{}_f V_2}{E} - k_1\frac{1}{2} \fallingdotseq (1-k_1)\frac{1}{2}
\end{aligned}\right\} \text{④} \qquad (9.6)$$

$$\left.\begin{aligned}
&I_a = 0 \\
&I_b = 0 + a^2 I_1 + aI_2 = -j\sqrt{3}I_1 = -\frac{\sqrt{3}}{2}\cdot\frac{E}{x_1} \\
&I_c = 0 + aI_1 + a^2 I_2 = -I_b = \frac{\sqrt{3}}{2}\cdot\frac{E}{x_1}
\end{aligned}\right\} \text{⑤}$$

$$\left.\begin{aligned}
&\frac{{}_f V_a}{E} = 3\frac{{}_f V_1}{E} = \frac{3}{2} \\
&\frac{{}_f V_b}{E} = 0 \\
&\frac{{}_f V_c}{E} = 0
\end{aligned}\right\} \text{⑥}$$

$$\left.\begin{aligned}
&\frac{{}_m V_a}{E} \fallingdotseq \frac{1}{2} + \frac{1+k_1}{2} + \frac{1-k_1}{2} = \frac{3}{2} \\
&\frac{{}_m V_b}{E} \fallingdotseq \frac{1}{2} + a^2\frac{1+k_1}{2} + a\frac{1-k_1}{2} = -j\frac{\sqrt{3}}{2}k_1 \\
&\frac{{}_m V_c}{E} \fallingdotseq \frac{1}{2} + a\frac{1+k_1}{2} + a^2\frac{1-k_1}{2} = j\frac{\sqrt{3}}{2}k_1
\end{aligned}\right\} \text{⑦}$$

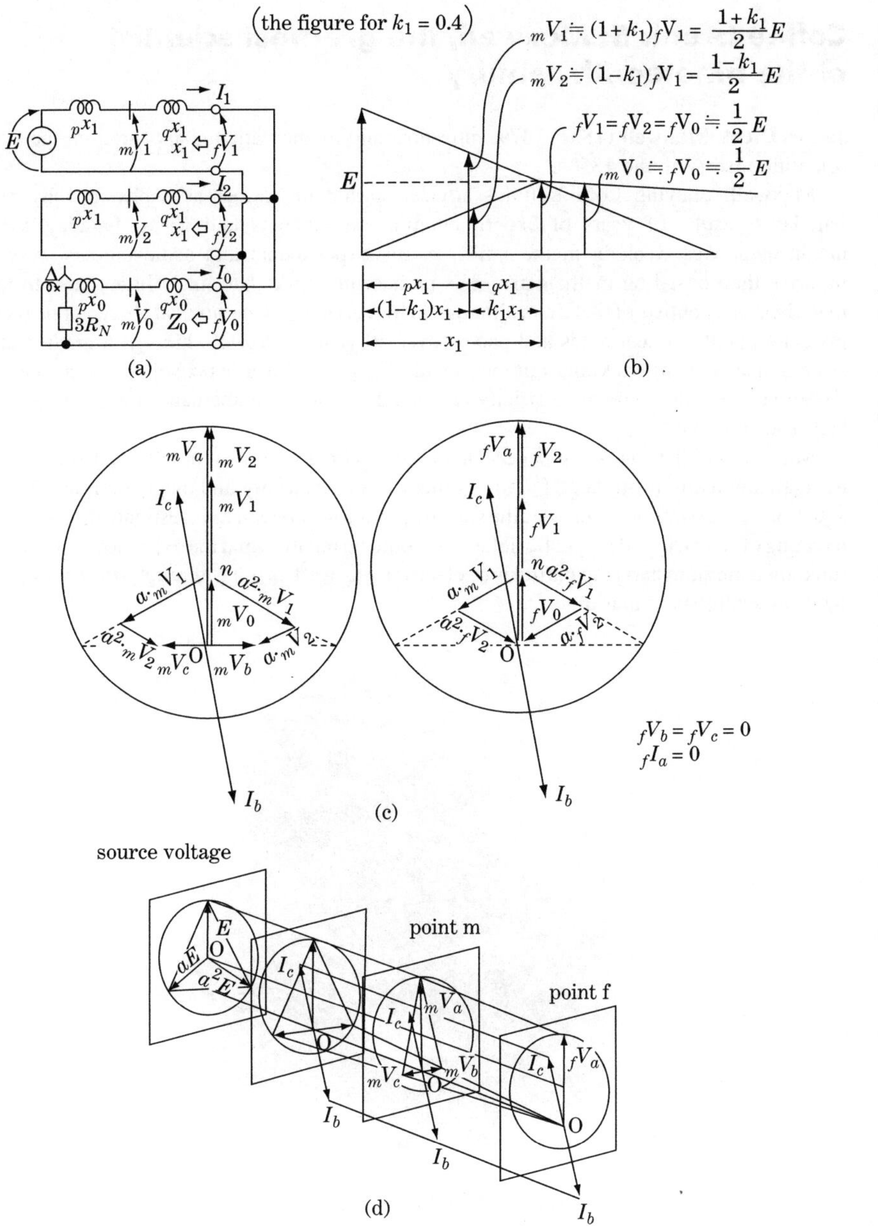

Figure 9.6 Double line-to-ground (phases b and c) fault: $2\phi G$ (high-resistive neutral grounding system)

Coffee break 5: Maxwell, the greatest scientist of the nineteenth century

James Clerk Maxwell (1831–1879), physicist and mathematician, appeared on the stage of scientific history in the 1850s.

Maxwell, obeying Lord Kelvin's advice, started his career as a physicist by reading Faraday's results of years of experimentation on magnets and wires. Faraday, virtually uneducated, was weak in mathematics, so his experiments and achievements were more intuitive than based on mathematics. His pioneering works had made little sense to mathematicians or scientists of that time, although in his earnest and methodical character he recorded his experiments in neat notes and papers over 30 years. Maxwell, himself a mathematician, systematically went back and climbed inside Faraday's head. Maxwell wrote at the outset, 'Before I began the study of electricity I resolved to read no mathematics on the subject until I had read [Faraday].'

Maxwell published his first paper 'On Faraday's lines of forces' in 1855 and 1856, in which he mathematically formulated Faraday's theories of electricity and magnetic lines of force as equations of streamlines. This was the starting point for Maxwell to investigate the true physical meaning of 'electric and magnetic fields'. It should be recalled that the existence of 'ether' as the 'unknown medium carrying particles of electricity as well as light' through space was believed by most scientists at that time.

James Clerk Maxwell (1831–1879)

Maxwell published his second paper 'On physical line of forces' in 1862 and the third, 'Dynamic theory of the electromagnetic fields', in 1864, in which he introduced his new concept that a dielectric material receives **tensile stress** or **compressive stress** in the space of a dielectric field and also introduced **displacement current** as his logical result.

In 1873, he published his corpus *Treatise on Electricity and Magnetism* in which his great presumption of **electromagnetic waves** in the form of partial differential equations was disclosed in fully developed form, known today as **'Maxwell's fundamental equations of the**

electromagnetic field' or **'Maxwell's four equations'**. Maxwell's original equations were later reformed by Heaviside into four partial differential equations without impairment, namely

$$\left.\begin{array}{ll}\operatorname{div}\boldsymbol{D}=\rho\left(=\dfrac{\partial\boldsymbol{D}}{\partial x}+\dfrac{\partial\boldsymbol{D}}{\partial y}+\dfrac{\partial\boldsymbol{D}}{\partial z}\right) & \text{(a)}\\ \operatorname{div}\boldsymbol{B}=0 & \text{(b)}\\ \operatorname{rot}\boldsymbol{H}=\boldsymbol{I}+\dfrac{\partial\boldsymbol{D}}{\partial t} & \text{(c)}\\ \operatorname{rot}\boldsymbol{E}+\dfrac{\partial\boldsymbol{B}}{\partial t}=0 & \text{(d)}\end{array}\right\} \quad (1)$$

where $\boldsymbol{D}$: electric displacement or electric flux density (in uniform free space $\boldsymbol{D}=\varepsilon\boldsymbol{E}$, where ε is the dielectric constant)
ρ: density of charge
$\boldsymbol{H}$: intensity of magnetic field
$\boldsymbol{E}$: intensity of electric field ($\boldsymbol{E}=-\operatorname{grad}V$[V/m], definition of voltage)
$\boldsymbol{I}$: conduction current or current in a conductor
($\operatorname{div}\boldsymbol{I}=-\partial\rho/\partial t$, definition of current; $\boldsymbol{I}=0$ in an insulated space)
$\boldsymbol{B}$: magnetic flux density (in uniform free space $\boldsymbol{B}=\mu\boldsymbol{H}$, where μ is permeability)

Maxwell showed that the above equations could express the behaviour of electric and magnetic fields and their interrelation, and that all the already discovered phenomena or equations by other scientists in regard to electricity and magnetism, from Coulomb to Neumann, could be explained as special cases of his equations.

Maxwell also showed in this paper that the above equations can be reformed into new equations having the form of wave properties.

These are

$$\left.\begin{array}{ll}\dfrac{\partial^2 E}{\partial x^2}=\varepsilon\mu\dfrac{\partial^2 E}{\partial t^2} & \text{(a)}\\ \dfrac{\partial^2 H}{\partial x^2}=\varepsilon\mu\dfrac{\partial^2 H}{\partial t^2} & \text{(b)}\end{array}\right\} \quad (2)$$

$$\left.\begin{array}{ll}E=E_1(x-ct)+E_2(x+ct) & \text{(a)}\\ H=H_1(x-ct)+H_2(x+ct) & \text{(b)}\\ \text{where } c=1/\sqrt{\varepsilon\mu}\,[\text{m/sec}] & \end{array}\right\} \quad (3)$$

Equations 2 and 3 are the wave equation and its solution respectively, and are of course the same as Equations 18.5 and 18.6 in Chapter 18 of this book for travelling waves on overhead transmission lines.

The constant c in Equation 3 is the speed of the electromagnetic wave travelling in uniform space of dielectric constant ε and permeability μ, which was calculated by Maxwell as the travelling velocity of the wave in a vacuum, $c=1/\sqrt{\varepsilon\mu}\fallingdotseq 30\times10^8$ m/sec.

Electromagnetic waves can travel in a **pure vacuum** without 'ether'. Maxwell of course presumed from this result that **light** would also be an electromagnetic wave having the same properties as in electricity and magnetism, and presumed that light as well as electromagnetic fields would have the same velocity in free space. He wrote, 'We can scarcely avoid the conclusion that light consists in transverse undulations of the same medium which is the cause of electric and magnetic phenomena.'

However, it was said that most scientists of that time could not understand Maxwell's presumptions and the 'ether' was slow to disappear, although **Hermann Helmholtz** (1821–1894) and **Ludwig Boltzmann** (1844–1906) soon made important discoveries justifying Maxwell's theory.

Maxwell also continued his work in the kinetic theory of gases. By treating gases statistically in 1866, he and Boltzmann formulated independently the 'Maxwell–Boltzmann kinetic theory of gases', which showed that temperature and heat involved only molecular movement, instead of any particle, under statistical conditions.

Maxwell's great discovery of electromagnetic waves in 1873 was 15 years before experimental proof of his theory by **Heinrich Hertz** in 1888 and 22 years before the success of wireless communication by **Guglielmo Marconi** (1874–1937) in 1895. **Einstein's theory of relativity** was disclosed over the period 1905–1916. It is well known that **Albert Einstein** (1879–1955) examined deeply the scientific meanings of the 'essence of light' and 'rest and motion', so Maxwell's theory was always central to Einstein. It is interesting that Maxwell's equations needed no revision when Einstein disclosed his theories some 40 years later, although **Newton's** laws had to be revised. **Richard Feynmann**, Noble laureate and influential twentieth-century physicist, paid his respects in this way: 'From a long view of the history of mankind, seen from, say, 1000 years from now, there can be little doubt that the most significant event of the nineteenth century will be judged as Maxwell's discovery of the laws of electrodynamics.'

To derive Equation 2 from Equation 1 above, in uniform free space

$$\boldsymbol{B} = \mu \boldsymbol{H}, \text{ where } \mu \text{ is constant(the permeability)} \tag{a}$$

$$\boldsymbol{D} = \varepsilon \boldsymbol{E}, \text{ where } \varepsilon \text{ is constant(the dielectric constant)} \tag{b}$$

From Equations 1d

$$\text{rot}\,\boldsymbol{E} = -\mu \frac{\partial \boldsymbol{H}}{\partial \text{t}} \tag{c}$$

From Equations 1c

$$\text{rot}\,\boldsymbol{H} = \boldsymbol{I} + \varepsilon \cdot \frac{\partial \boldsymbol{E}}{\partial \text{t}} \tag{d}$$

Thus

$$\begin{aligned} \text{rot}(\text{rot}\,\boldsymbol{E}) &= -\text{rot}\left(\mu \frac{\partial \boldsymbol{H}}{\partial \text{t}}\right) = -\mu \frac{\partial(\text{rot}\,\boldsymbol{H})}{\partial t} \\ &= -\mu \frac{\partial(\boldsymbol{I} + \partial \boldsymbol{D}/\partial t)}{\partial t} = -\left\{\mu \frac{\partial \boldsymbol{I}}{\partial \text{t}} + \mu\varepsilon \frac{\partial^2 \boldsymbol{E}}{\partial t^2}\right\} \end{aligned} \tag{e}$$

From vector analysis

$$\text{rot}(\text{rot}\,\boldsymbol{E}) = \text{grad}(\text{div}\,\boldsymbol{E}) - \nabla^2 \boldsymbol{E} = -\nabla^2 \boldsymbol{E} \tag{f}$$

$$\text{div}\,\boldsymbol{E} = \left(\frac{\partial \boldsymbol{E}}{\partial x} + \frac{\partial \boldsymbol{E}}{\partial y} + \frac{\partial \boldsymbol{E}}{\partial z}\right) = \rho = 0 \tag{g}$$

in uniform space, where the Laplacian is

$$\nabla^2 = \left(\frac{\partial^2}{\partial x^2} + \frac{\partial^2}{\partial y^2} + \frac{\partial^2}{\partial z^2}\right) \tag{h}$$

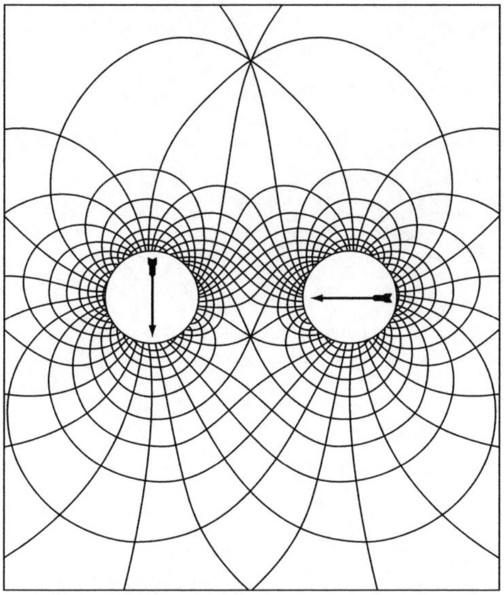

Maxwell's graph of a magnetic field

From (e), (f) and (g)

$$\nabla^2 \boldsymbol{E} = \left(\frac{\partial^2}{\partial x^2} + \frac{\partial^2}{\partial y^2} + \frac{\partial^2}{\partial z^2}\right)\boldsymbol{E} = \mu\frac{\partial \boldsymbol{I}}{\partial t} + \mu\varepsilon\frac{\partial^2 \boldsymbol{E}}{\partial t^2} \tag{i}$$

We now imagine a conductor laid in a uniform medium of conductivity $k(= 1/\sigma)$ along the direction of the x-axis in orthogonal coordinates.

Ohm's law is justified for any type of medium, regardless of its conductivity. Then

$$\boldsymbol{I} = \sigma \boldsymbol{E} \tag{j}$$

However, the air space (or vacuum) of a transmission line is actually an insulated medium of $k = 1/\sigma = \infty$, so the current flowing through the insulation free space is zero. Then

$$\boldsymbol{I} = 0, \quad \text{so} \quad \frac{\partial \boldsymbol{I}}{\partial t} = 0 \tag{k}$$

and

$$E_x = E,\ E_y = E_z = 0 \tag{l}$$

Accordingly, substituting (k) and (l) into (i) gives the differential Equation 2(a). By a similar process, Equation 2(b) can be derived by calculating rot (rot H) using Equations (c) (d) (h) (k) (l). The derivation of Equation 3 from Equation 2 is given in Chapter 18.

Maxwell's graph of a magnetic field surrounding two cylindrical magnets is shown in the above.

10

Theory of Generators

The generator model in Figure 2.10 and its equivalent circuit by symmetrical components in Figure 2.11 are not strictly correct in the strict sense stated in Chapter 2.

The generator's appropriate mathematical model is an essential concept with which most engineers in the field of power system engineering are required to be familiar, because any kind of behaviour analysis of power systems cannot be conducted without appropriate generator model circuits, regardless of the analysis for large or small systems and for total or partial systems. It must be remembered that the so-called 'connection diagram' and 'the circuit' of generators or any power system networks are important but quite different, and analytical works are always based on 'the circuit' instead of 'the connection diagram'.

Our generator model has to be accurate primarily and then easily connected to other power system networks (such as transmission lines, transformers, other generators, loads, and so on) in the symmetrical sequence domain. Park's generator model and equations and the resulting equations in the d–q–0 domain gives us a satisfactory answer for the above requirement. Due to these theoretical results, generators can be connected with other network facilities in the symmetrical coordinate domain, and we acquire the circuit of the power system by which we can analyse the power system accurately whether by hand or by computer or by analogue simulation.

In this chapter, starting from the basic concept of a three-phase rotating machine, we introduce Park's theory for generators based on d–q–0 transformation first, and then examine the generator's equations and equivalent circuits in relation to the 0–1–2 domain as well as the d–q–0 domain, and furthermore demonstrate transient analysis of the generator for short-circuit faults.

10.1 Mathematical Description of a Synchronous Generator

A generator is an electromechanical machine composed of a static part (the stator) and a rotating part (the rotor) whose relative position is changed periodically by rotating angle ωt. In other words, a generator is a three-phase electromagnetic machine composed of $l(t)$ and resistance r of stator and rotor windings (we neglect leakage capacitances of the stator), while the inductance $l(t)$ should be periodically changed depending on the relative angular position $\theta = \omega t$ between the stator and rotor windings. The voltages $v(t)$ [V], currents $i(t)$ [A], flux linkage $\psi(t)$ [weber-turn] and the magnetic reluctances of the related magnetic passes should also be functions of ωt.

10.1.1 The fundamental model

A generator's fundamental electrical structure can be expressed as shown in Figure 10.1. Although a two-pole machine is shown, a multi-pole machine with any number of pairs of poles can be treated as a

© 2007 John Wiley & Sons, Ltd

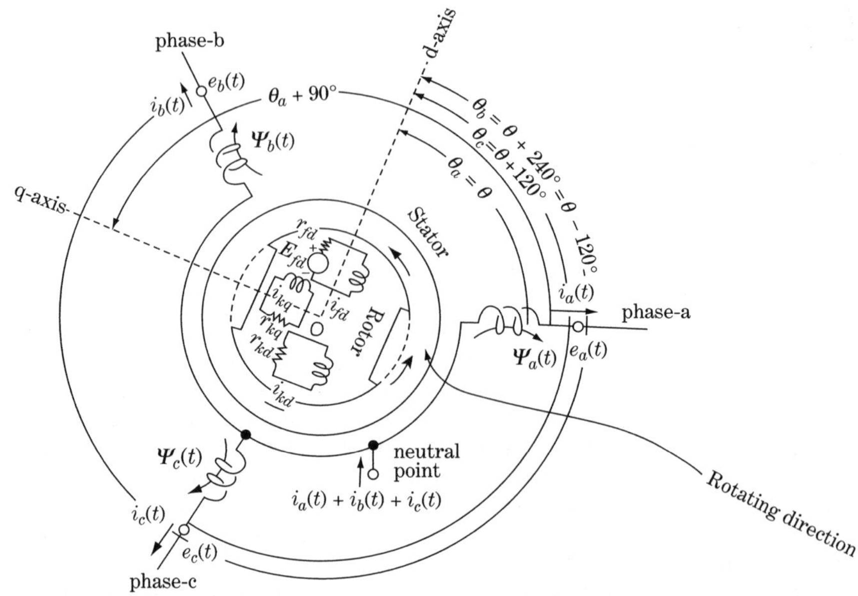

Figure 10.1 Electrical concept of a synchronous generator

two-pole machine electrically, because armature (stator) windings are identically arranged with respect to each pair of poles.

We have to investigate Figure 10.1 in detail because it is the starting point for developing the mathematical model.

10.1.1.1 Rotor

As the rotor has two axes of mechanical rectangular symmetry, we call them the 'd-axis' and 'q-axis'. Namely:

- *d-axis or direct-axis*: the axis from the axial centre point o in the pole direction
- *q-axis or quadrature-axis*: the axis from the axial centre point o in the direction 90° ahead (leading) of the d-axis.

As the rotor is designed symmetrically for the d-axis and q-axis, we can assume the rotor is as follows.

10.1.1.1.1 Field windings The field winding (named 'd-axis filed coil') is a closed circuit connected to a source of d.c. voltage E_{fd}, and with an inductance to produce flux only in the direction of the d-axis.

The flux may flow into the left-hand side (i.e. the +q-axis component) and right-hand side (i.e. the −q-axis component) of the d-axis, but such ±q-axis components would be balanced because of the design symmetry and must actually be cancelled in total. Therefore we can assume that the field windings produce flux only in the d-axis direction and not in the q-axis direction. This is the reason why we can justify the 'd-axis field coil' as a very good approximation that the field winding circuit exists only in the d-axis direction (d-axis field coil.).

10.1.1.1.2 Damper windings Typical hydraulic-turbine-driven generators (vertical type with salient poles) have amortisseur or squirrel-cage windings in the pole face (damper windings), which consists of copper bars through the pole connected at their ends as a closed circuit.

Thermal or nuclear-turbine-driven generators (horizontal type with cylindrical non-salient poles) have field windings only and do not have such damper windings. However, eddy currents are forced to flow into the pass of the rotor solid steel for the duration of transient or currrent-unbalanced conditions. Therefore we have to assume that steam-driven generators also have damper windings.

The currents in these damper windings may be assumed to flow in only two closed circuits, one in the d-axis and the other in the q-axis as an approximate electrical model again because of the rotor's symmetrical design for the d-axis and q-axis.

With all these bases, the rotor circuit model consists of one field coil and one damper coil in the d-axis and one damper coil in the q-axis, as shown in Figure 10.1.

10.1.1.2 Stator (armature)

The stator has three stator windings in name for phases a, b and c connected at their ends commonly as the neutral terminal. The three windings are arranged electrically by 120° symmetrically to each other. We can justify the assumption that the stator windings are sinusoidally distributed along with the air gap as far as all the mutual effects of the rotor are concerned, because the generator windings are distributed so as to minimize harmonics in its design.

On these bases, the stator circuit model consists of three star-connected phase a, b, c coils, each of which has its own self-inductance and resistance as well as mutual inductances between all other stator coils and rotor coils, as shown in Figure 10.1.

10.1.1.3 Relative angular position between rotor and stator

The stator is immovable and the rotor is rotating counterclockwise at an angular speed of $\omega = d\theta/dt$; therefore the relative position between the rotor and the stator is measured by the rotating angle of the rotor d-axis. That is, the rotating position of each coil in time can be written as follows on a d-axis basis:

$$\begin{array}{lll} \text{Stator} & \text{a phase coil} & : \ \theta_a = \theta = \omega t = 2\pi f t \\ & \text{b phase coil} & : \ \theta_b = \theta + 240^\circ = \theta - 120^\circ = \omega t - 120^\circ \\ & \text{c phase coil} & : \ \theta_c = \theta + 120^\circ = \omega t + 120^\circ \\ \text{Rotor} & \text{field coil} & : \ 0^\circ \\ & \text{damper d-axis coil} & : \ 0^\circ \\ & \text{damper q-axis coil} & : \ +90^\circ \end{array} \qquad (10.1)$$

The position of each stator coil is a function of time t on the d-axis basis.

Now, refer to Figure 10.1 which has been prepared as the electrical model of the generator in the phase a–b–c domain in order to introduce the equations of the generator.

10.1.2 Fundamental three-phase circuit equations

We define the quantities of each coil in Figure 10.1 as follows:

$\psi_a(t), \psi_b(t), \psi_c(t)$	:	total flux linkage of phase a, b, c coil, respectively [Wb-turn]
$e_a(t), e_b(t), e_c(t)$	:	terminal voltage of phase a, b, c coil, respectively [V]
$i_a(t), i_b(t), i_c(t)$	:	terminal current of phase a, b, c coil, respectively [A]
$\psi_{fd}(t)$	:	total flux linkage of d-axis field coil [Wb-turn]
$\psi_{kd}(t)$	:	total flux linkage of d-axis damper coil [Wb-turn]
$\psi_{kq}(t)$	:	total flux linkage of q-axis damper coil [Wb-turn]
E_{fd}	:	field excitation voltage [V]
$i_{fd}(t)$	:	current in d-axis field coil [A]
$i_{kd}(t)$	:	current in d-axis damper coil [A]
$i_{kq}(t)$	:	current in q-axis damper coil [A]

With the above definitions, the following equations can be derived.
The equation of the stator (armature) coils voltage is

$$\underbrace{\begin{bmatrix} e_a(t) \\ e_b(t) \\ e_c(t) \end{bmatrix}}_{\boldsymbol{e}_{abc}(t)} = \underbrace{\begin{bmatrix} \frac{d}{dt}\psi_a(t) \\ \frac{d}{dt}\psi_b(t) \\ \frac{d}{dt}\psi_c(t) \end{bmatrix}}_{\frac{d}{dt}\boldsymbol{\psi}_{abc}(t)} - r \underbrace{\begin{bmatrix} i_a(t) \\ i_b(t) \\ i_c(t) \end{bmatrix}}_{\boldsymbol{i}_{abc}(t)} \tag{10.2}$$

the equation of the rotor coils voltage is

$$\underbrace{\begin{bmatrix} E_{fd} \\ 0 \\ 0 \end{bmatrix}}_{\boldsymbol{E}_F(t)} = \underbrace{\begin{bmatrix} \frac{d}{dt}\psi_{fd}(t) \\ \frac{d}{dt}\psi_{kd}(t) \\ \frac{d}{dt}\psi_{kq}(t) \end{bmatrix}}_{\frac{d}{dt}\boldsymbol{\psi}_F(t)} + \underbrace{\begin{bmatrix} r_{fd}\cdot i_{fd}(t) \\ r_{kd}\cdot i_{kd}(t) \\ r_{kq}\cdot i_{kq}(t) \end{bmatrix}}_{\text{voltage drop}} \tag{10.3}$$

the equation of the stator (armature) coils flux linkage is

$$\underbrace{\begin{bmatrix} \psi_a(t) \\ \psi_b(t) \\ \psi_c(t) \end{bmatrix}}_{\boldsymbol{\psi}_{abc}(t)} = \underbrace{\begin{bmatrix} -l_{aa}(t) & l_{ab}(t) & l_{ac}(t) \\ l_{ba}(t) & -l_{bb}(t) & l_{bc}(t) \\ l_{ca}(t) & l_{cb}(t) & -l_{cc}(t) \end{bmatrix}}_{\boldsymbol{l}_{abc}(t)} \cdot \underbrace{\begin{bmatrix} i_a(t) \\ i_b(t) \\ i_c(t) \end{bmatrix}}_{\boldsymbol{i}_{abc}(t)} + \underbrace{\begin{bmatrix} l_{afd}(t) & l_{akd}(t) & l_{akq}(t) \\ l_{bfd}(t) & l_{bkd}(t) & l_{bkq}(t) \\ l_{cfd}(t) & l_{ckd}(t) & l_{ckq}(t) \end{bmatrix}}_{\boldsymbol{l}_{abc-F}(t)} \cdot \underbrace{\begin{bmatrix} i_{fd}(t) \\ i_{kd}(t) \\ i_{kq}(t) \end{bmatrix}}_{\boldsymbol{i}_F(t)} \tag{10.4}$$

and the equation of the rotor coils flux linkage is

$$\underbrace{\begin{bmatrix} \psi_{fd}(t) \\ \psi_{kd}(t) \\ \psi_{kq}(t) \end{bmatrix}}_{\boldsymbol{\psi}_F(t)} = -\underbrace{\begin{bmatrix} l_{fad}(t) & l_{fbd}(t) & l_{fcd}(t) \\ l_{kad}(t) & l_{kbd}(t) & l_{kcd}(t) \\ l_{kaq}(t) & l_{kbq}(t) & l_{kcq}(t) \end{bmatrix}}_{\boldsymbol{l}_{F-abc}(t)} \cdot \underbrace{\begin{bmatrix} i_a(t) \\ i_b(t) \\ i_c(t) \end{bmatrix}}_{\boldsymbol{i}_{abc}(t)} + \underbrace{\begin{bmatrix} L_{ffd} & L_{fkd} & 0 \\ L_{fkd} & L_{kkd} & 0 \\ 0 & 0 & L_{kkq} \end{bmatrix}}_{\boldsymbol{L}_F} \cdot \underbrace{\begin{bmatrix} i_{fd}(t) \\ i_{kd}(t) \\ i_{kq}(t) \end{bmatrix}}_{\boldsymbol{i}_F(t)} \tag{10.5}$$

where

r : resistance of each stator coil [Ω]
r_{fd}, r_{kd}, r_{kq} : resistance of field d-axis, damper d-axis coil, damper q-axis coil
$\boldsymbol{l}_{abc}(t)$ $\begin{cases} l_{aa}(t), l_{bb}(t), l_{cc}(t) \\ l_{ab}(t), l_{bc}(t), \text{etc.} \end{cases}$: self-inductances of stator coils [H]; : mutual inductances among stator coils [H]
$\boldsymbol{l}_{abc-F}(t)$: mutual inductance matrix between stator coils and rotor coils. [H]
$\boldsymbol{l}_{F-abc}(t)$: the same as the above. [H]
$\boldsymbol{L}_F$: mutual inductance matrix among three rotor coils. [H]

The matrices $\boldsymbol{l}_{abc-F}(t)$ and $\boldsymbol{l}_{F-abc}(t)$ are obviously dependent on time t, because all the mutual inductances between stator phase coils and rotor coils ($l_{afd}(t)$, $l_{akd}(t)$, $l_{akq}(t)$) are affected by the changing relative position over time between the stator and the rotor. Therefore $l_{afd}(t), l_{akd}(t), l_{akq}(t)$, etc., include the symbol (t) to emphasize their time dependency by rotation. Matrices $\boldsymbol{l}_{abc-F}(t)$ and $\boldsymbol{l}_{F-abc}(t)$ also include (t) for the same reason.

The relative position between the phase a stator coil and phase b stator coil does not change all the time. However, the self-inductance $l_{aa}(t)$ of the phase a coil and the mutual inductance $l_{ab}(t)$ between phase a and b coils may be affected by the changing rotor position over time, because the flux passes of

ψ_{aa} as well as ψ_{ab} may flow partly through the periodically changing air gap and rotor structure. Therefore self- and mutual inductances $l_{aa}(t)$, $l_{ab}(t)$ must be time dependent. By such reasoning the matrix $\boldsymbol{l}_{abc}(t)$ also has time-dependent inductances and thus symbol (t).

The self-inductances of each rotor coil L_{ffd}, L_{kkd}, L_{kkq} in the incident matrix $\boldsymbol{l}_F$ are not affected by ωt. In other words, they are time independent because the field d-axis, damper d-axis and damper q-axis coils are fixed on the d-axis or q-axis, and all the linking flux of these coils in the rotor are not affected by the relative position of stator phase coils from the rotor.

The mutual inductance $L_{fkd} = L_{kfd}$ between the rotor d-axis coil and the damper d-axis coil also exists as time-independent mutual inductance. On the other hand, the mutual inductance between the rotor d-axis coil and the rotor q-axis does not exist physically. Therefore the matrix $\boldsymbol{L}_F$ is not time dependent and includes some zero elements as seen in Equation10.5.

10.1.3 Characteristics of inductances in the equations

Now, we have to examine how $\boldsymbol{l}_{abc}(t)$, $\boldsymbol{l}_{abc-F}(t)$, $\boldsymbol{l}_{F-abc}(t)$ can be written as time-dependent (ωt) inductance matrices.

The conclusive equations are shown first by the equations below, followed by our reasons for justifying these equations.

10.1.3.1 Inductance matrix of stator coils

$$\boldsymbol{l}_{abc}(t) = \begin{bmatrix} -l_{aa}(t) & l_{ab}(t) & l_{ac}(t) \\ l_{ba}(t) & -l_{bb}(t) & l_{bc}(t) \\ l_{ca}(t) & l_{cb}(t) & -l_{cc}(t) \end{bmatrix}$$

$$= \begin{bmatrix} -\{L_{aa0} + L_{aa2}\cos 2\theta_a\} & L_{ab0} - L_{aa2}\cos(\theta_a + \theta_b) & L_{ab0} - L_{aa2}\cos(\theta_a + \theta_c) \\ L_{ab0} - L_{aa2}\cos(\theta_a + \theta_b) & -\{L_{aa0} + L_{aa2}\cos 2\theta_b\} & L_{ab0} - L_{aa2}\cos(\theta_b + \theta_c) \\ L_{ab0} - L_{aa2}\cos(\theta_a + \theta_c) & L_{ab0} - L_{aa2}\cos(\theta_b + \theta_c) & -\{L_{aa0} + L_{aa2}\cos 2\theta_c\} \end{bmatrix}$$

$$= \begin{bmatrix} -\{L_{aa0} + L_{aa2}\cos 2\theta\} & L_{ab0} - L_{aa2}\cos(2\theta - 120°) & L_{ab0} - L_{aa2}\cdot\cos(2\theta + 120°) \\ L_{ab0} - L_{aa2}\cdot\cos(2\theta - 120°) & -\{L_{aa0} + L_{aa2}\cdot\cos(2\theta + 120°)\} & L_{ab0} - L_{aa2}\cos 2\theta \\ L_{ab0} - L_{aa2}\cdot\cos(2\theta + 120°) & L_{ab0} - L_{aa2}\cos 2\theta & -\{L_{aa0} + L_{aa2}\cdot\cos(2\theta - 120°)\} \end{bmatrix}$$

$$\theta_a = \theta, \quad \theta_b = \theta - 120°, \quad \theta_c = \theta + 120° \tag{10.6}$$

10.1.3.2 Mutual inductance matrix between stator coils and rotor coils

$$\boldsymbol{l}_{abc-F}(t) = \begin{bmatrix} l_{afd}(t) & l_{akd}(t) & l_{akq}(t) \\ l_{bfd}(t) & l_{bkd}(t) & l_{bkq}(t) \\ l_{cfd}(t) & l_{ckd}(t) & l_{ckq}(t) \end{bmatrix}$$

$$= \begin{bmatrix} L_{afd}\cos\theta_a & L_{akd}\cos\theta_a & -L_{akq}\sin\theta_a \\ L_{afd}\cos\theta_b & L_{akd}\cos\theta_b & -L_{akq}\sin\theta_b \\ L_{afd}\cos\theta_c & L_{akd}\cos\theta_c & -L_{akq}\sin\theta_c \end{bmatrix} \tag{10.7}$$

$$= \begin{bmatrix} L_{afd}\cos\theta & L_{akd}\cos\theta & -L_{akq}\sin\theta \\ L_{afd}\cos(\theta - 120°) & L_{akd}\cos(\theta - 120°) & -L_{akq}\sin(\theta - 120°) \\ L_{afd}\cos(\theta + 120°) & L_{akd}\cos(\theta + 120°) & -L_{akq}\sin(\theta + 120°) \end{bmatrix}$$

$$
\boldsymbol{l}_{F-abc}(t) = \begin{array}{|c|c|c|}\hline l_{fad}(t) & l_{fbd}(t) & l_{fcd}(t) \\ \hline l_{kad}(t) & l_{kbd}(t) & l_{kcd}(t) \\ \hline l_{kaq}(t) & l_{kbq}(t) & l_{kcq}(t) \\ \hline \end{array}
$$

$$
= \begin{array}{|c|c|c|}\hline L_{afd}\cos\theta_a & L_{afd}\cos\theta_b & L_{afd}\cos\theta_c \\ \hline L_{akd}\cos\theta_a & L_{akd}\cos\theta_b & L_{akd}\cos\theta_c \\ \hline -L_{akq}\sin\theta_a & -L_{akq}\sin\theta_b & -L_{akq}\sin\theta_c \\ \hline \end{array} \qquad (10.8)
$$

$$
= \begin{array}{|c|c|c|}\hline L_{afd}\cos\theta & L_{afd}\cos(\theta-120°) & L_{afd}\cos(\theta+120°) \\ \hline L_{akd}\cos\theta & L_{akd}\cos(\theta-120°) & L_{akd}\cos(\theta+120°) \\ \hline -L_{akq}\sin\theta & -L_{akq}\sin(\theta-120°) & -L_{akq}\sin(\theta+120°) \\ \hline \end{array} = [\boldsymbol{l}_{abc-F}(t)]^t
$$

where the matrix $[\,]^t$ is the transposed matrix of [], that is the rows and columns are interchanged.

Now let us examine how Equations 10.6, 10.7 and 10.8 are derived.
The total flux linkage of the phase a armature coil can be described as follows:

$$
\begin{aligned}
\psi_a(t) &= -\psi_{aa}(t) + \psi_{ab}(t) + \psi_{ac}(t) + \psi_{afd}(t) + \psi_{akd}(t) + \psi_{akq}(t) \\
&= -l_{aa}(t)\cdot i_a(t) + l_{ab}(t)\cdot i_b(t) + l_{ac}(t)\cdot i_c(t) \\
&\quad + l_{afd}(t)\cdot i_{fd}(t) + l_{akd}(t)\cdot i_{kd}(t) + l_{akq}(t)\cdot i_{kq}(t)
\end{aligned} \qquad (10.9)
$$

Here $l_{aa}(t) = \psi_{aa}(t)/i_a(t)$ is the self-inductance of the phase a stator coil (the flux linkage of phase a coil induced by unit current of phase a coil); $l_{ab}(t) = \psi_{ab}(t)/i_b(t)$ is the mutual inductance between the phase a stator coil and phase b coil (the flux linkage of phase a coil induced by unit current of phase b coil); $l_{afd}(t) = \psi_{afd}(t)/i_{fd}(t)$ is the mutual inductance between the phase a stator coil and rotor d-axis coil (the flux linkage of phase a coil induced by unit current of rotor d-axis coil); and so on.

10.1.3.2.1 Introduction of Equation 10.6: $\boldsymbol{l}_{abc}(t)$ The surface of any rotor is uneven and the air-gap length between the stator and rotor varies depending on the relative position of the rotor to the stator. In other words, any rotor is not a uniform cylinder from the viewpoint of magnetic passes. This means that the magnetic reluctance $\boldsymbol{l}_{abc}(t)$ varies depending on the relative angular position from the d-axis.

Therefore, the self-inductance of any armature winding varies periodically as a function of ωt, and it must become a maximum when the pole (d-axis) is in line with the phase axis, and a minimum when the interpole (q-axis) is in line with the phase axis. That is, $l_{aa}(t)$ must be a periodic function by electrical angle 180° and an even function by θ_a. In other words, $l_{aa}(t)$ can be written as an equation in the Fourier series expansion

$$
l_{aa}(t) = L_{aa0} + L_{aa2}\cos 2\theta_a + L_{aa4}\cos 4\theta_a + L_{aa6}\cos 6\theta_a + \cdots
$$

$$
\left.\begin{aligned} \therefore\ l_{aa}(t) &= L_{aa0} + L_{aa2}\cos 2\theta_a \\ \theta_a &= \omega t \end{aligned}\right\} \qquad (1)
$$

The actual armature coil is designed as sinusoidal distribution windings, so that the third term and other smaller terms on the right-hand side can be neglected. Figure 10.2 shows the state.
The equation of $l_{ab}(t) = l_{ba}(t)$ is presumed below in the same way:

$$
\left.\begin{aligned} & l_{ab}(t) = l_{ba}(t) = L_{ab0} - L_{ab2}\cos(\theta_a + \theta_b) \\ & \text{here } \theta_a = \omega t, \quad \theta_b = \omega t - 120° \end{aligned}\right\} \qquad (2)
$$

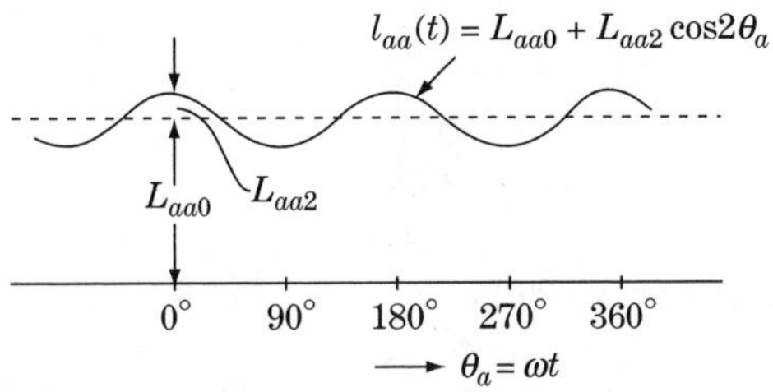

Figure 10.2 Self-inductance of phase a coil

However, the reasoning behind Equation 2 is not still clear, and moreover the physical images and values of L_{aa0}, L_{aa2}, L_{ab0}, L_{ab2} cannot be obtained by the above explanation. Let us examine this further from the viewpoint of electromagnetism.

The equation of a magnetic circuit is

$$\text{magnetic motive force } (mmf) = \text{flux } (\phi) \times \text{magnetic reluctance } (\Re)$$

Using symbols

${}_a mmf$, ${}_a\phi$: magnetic motive force and flux induced by phase a current $\mathrm{i}_a(\mathrm{t}) = 1.0$ (unit value).

${}_a\phi_d(t)$, ${}_a\phi_q(t)$: d-axis and q-axix component of ${}_a\phi$

$\Re_d$, $\Re_q$: magnetic reluctance of flux pass for ${}_a\phi_d(t)$ in d-axis direction and for ${}_a\phi_q(t)$ in q-axis direction

The flux ${}_a\phi$ induced by current $i_a(t) = 1.0$ can be into the d-axix component ${}_a\phi_d(t)$ and q-axix component ${}_a\phi_q(t)$ as shown in Figure 10.3, where the reluctance $\Re_d$ for ${}_a\phi_d(t)$ and $\Re_q$ for ${}_a\phi_q(t)$ are of time-independent constants because ${}_a\phi_d(t)$, ${}_a\phi_q(t)$as well as $i_a(t)$ are in synchronization with the rotating d-axix and q-axix.

then

$$\left.\begin{aligned} {}_a\phi_d(t) &= \frac{{}_a mmf}{\Re_d}\cos\theta_a \\ {}_a\phi_q(t) &= -\frac{{}_a mmf}{\Re_q}\cos(\theta_a + 90°) = -\frac{{}_a mmf}{\Re_q}\sin\theta_a \end{aligned}\right\} \tag{3}$$

The self inductance $l_{aa}(t)$ is defined as the number of linking flux induced by $i(t) = 1.0$ and links with the phase a coil. Referring the Figure 4.3, the stator phase a coil links to ${}_a\phi_d(t)$ by angle θ_a and link to ${}_a\phi_q(t)$ by angle $\theta_a + 90°$

$$\begin{aligned} l_{aa}(t) &= {}_a\phi_d(t)\cos\theta_a + {}_a\phi_q(t)\cos(90° + \theta_a) \\ &= \frac{{}_a mmf}{\Re_d}\cos^2\theta_a + \frac{{}_a mmf}{\Re_q}\sin^2\theta_a = \frac{{}_a mmf}{\Re_d}\cdot\frac{1+\cos 2\theta_a}{2} + \frac{{}_a mmf}{\Re_q}\cdot\frac{1-\cos 2\theta_a}{2} \\ &= \frac{{}_a mmf}{2}\left(\frac{1}{\Re_d} + \frac{1}{\Re_q}\right) + \frac{{}_a mmf}{2}\left(\frac{1}{\Re_d} - \frac{1}{\Re_q}\right)\cos 2\theta_a \\ &= A + B\cos 2\theta_a = A + B\cos 2\theta \end{aligned} \tag{4}$$

Here

$$A = \frac{{}_a mmf}{2}\left(\frac{1}{\Re_d} + \frac{1}{\Re_q}\right), \quad B = \frac{{}_a mmf}{2}\left(\frac{1}{\Re_d} - \frac{1}{\Re_q}\right) \tag{5}$$

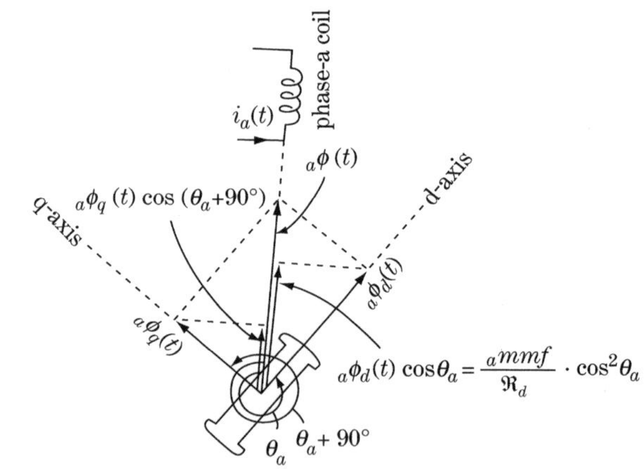

Figure 10.3 Linking flux of phase a winding coil

The mutual inductance $l_{ba}(t)$ is defined as the number of linking flux induced by $i_a(t) = 1.0$ and links with the phase b coil. While the stator phase b coil links to ${}_a\phi_d$ but by angle θ_b, and links to ${}_a\phi_q(t)$ by angle $(\theta_b + 90°)$.
then

$$
\left.\begin{aligned}
l_{ba}(t) &= {}_a\phi_d(t)\cos\theta_b + {}_a\phi_q(t)\,\cos(\theta_b + 90°)\\
&= \frac{{}_a mmf}{\Re_d}\cos\theta_a\,\cos\theta_b + \frac{{}_a mmf}{\Re_q}\sin\theta_a\,\sin\theta_b\\
&= \frac{{}_a mmf}{\Re_d}\cdot\frac{\cos(\theta_a - \theta_b) + \cos(\theta_a + \theta_b)}{2} + \frac{{}_a mmf}{\Re_q}\cdot\frac{\cos(\theta_a - \theta_b) - \cos(\theta_a + \theta_b)}{2}\\
&= \frac{{}_a mmf}{2}\left(\frac{1}{\Re_d} + \frac{1}{\Re_q}\right)\cos(\theta_a - \theta_b) + \frac{{}_a mmf}{2}\left(\frac{1}{\Re_d} - \frac{1}{\Re_q}\right)\cos(\theta_a + \theta_b)\\
&= A\,\cos(\theta_a - \theta_b) + B\cos(\theta_a + \theta_b) = A\cos 120° + B\cos(\theta_a + \theta_b)\\
&= -\left\{\frac{1}{2}A - B\cos(\theta_a + \theta_b)\right\}
\end{aligned}\right\} \quad (6)
$$

Therefore, arranging Equations 4, 5 and 6,

$$
\left.\begin{aligned}
&l_{aa}(t) = \psi_{aa}(t)/i_a(t) = L_{aa0} + L_{aa2}\,\cos 2\theta_a && ①\\
&l_{ba}(t) = \psi_{ba}(t)/i_a(t) = L_{ab0} - L_{aa2}\,\cos(\theta_a + \theta_b) && ②\\
&\text{where}\\
&L_{aa0} = \frac{A}{i_a(t)} = \frac{{}_a mmf}{i_a(t)}\cdot\frac{1}{2}\left(\frac{1}{\Re_d} + \frac{1}{\Re_q}\right) = 2L_{ab0} && ③\\
&L_{aa2} = \frac{B}{i_a(t)} = \frac{{}_a mmf}{i_a(t)}\cdot\frac{1}{2}\left(\frac{1}{\Re_d} - \frac{1}{\Re_q}\right) ll && ④
\end{aligned}\right\} \quad (7)
$$

Equation 7 proves that the intuitively written Equations 1 and 2 are correct. Also, L_{ab2} in Equation 2 is the same as L_{aa2} in Equation 7②, so it can be replaced by L_{aa2}. Furthermore, Equation 7③④ indicates $L_{aa0} > L_{aa2}$. Accordingly, $l_{aa}(t)$ of Equation 1 can be represented as the curve drawn in Figure 10.2.

The time-dependent inductances $l_{bb}(t), l_{cc}(t), l_{ac}(t), l_{bc}(t)$, etc., can be derived analogously. Now, arranging all these inductances, the inductance matrix of the armature coil, Equation 10.6, is obtained.

10.1.3.2.2 Introduction of Equations 10.7 and 10.8: $l_{abc-F}(t)$, $l_{F-abc}(t)$

For $l_{afd} = l_{fad}(t)$, $l_{afd}(t)$ is the mutual inductance by the linkage of field flux ϕ_{fd} (induced by $i_{fd}(t)$) to the stator phase a coil. From Figure 10.1, $l_{afd}(t)$ reaches its peak value when the field d-axis coil and the stator phase-a coil are in line, and have a negative peak value at the 180° reverse position.

Accordingly,

$$l_{afd}(t) = l_{fad}(t) = L_{afd}\cos\theta_a \tag{8}$$

The mutual inductance is obtained as only the term of fundamental sinusoidal form $\cos\theta_a$ without including higher harmonic components, because generators are mechanically designed so that the stator mmf is sinusoidally distributed. For $l_{akd}(t) = l_{kad}(t)$, the damper d-axis coil is in line with the field d-axis coil, so $l_{afd}(t) = l_{fad}(t)$ is in the same form as Equation 8, namely

$$l_{akd}(t) = l_{kad}(t) = L_{akd}\cos\theta_a \tag{9}$$

For $l_{akq}(t) = l_{kaq}(t)$, the damper q-axis coil is in the position of 90° ahead of the d-axis. Therefore

$$l_{akq}(t) = l_{kaq}(t) = L_{akq}\cos(\theta_a + 90°) = -L_{akq}\sin\theta_a \tag{10}$$

In the same way, the mutual inductances $l_{bfd}(t), l_{cfd}(t)$, etc., in regard to the phase b coil and phase c coil can be obtained by replacing θ_a in Equations 8, 9 and 10 with θ_b and θ_c.

From all the explanation above, the mutual inductance matrices between the stator and the rotor, Equations 10.7 and 10.8, are obtained.

In conclusion, Equations 10.1–10.8 are the equations for the generator in the a–b–c domain.

10.2 Introduction of d–q–0 Method (d–q–0 Components)

The derived equations are of no use as they are, because inductances are periodically time-dependent variables and, secondly, the equations cannot be connected with transmission lines and other equipment. Consequently, we now need to introduce the d–q–0 method.

10.2.1 Definition of d–q–0 method

10.2.1.1 Mathematical definition

The d–q–0 method is a transformation from three variables in the a–b–c domain to the another three variables in the d–q–0 domain from a mathematical viewpoint. The d–q–0 method is defined by the following equations including ωt in its transformation matrix and $\boldsymbol{e}_{abc}(t)$ by real number crest value expression

$$\underbrace{\begin{bmatrix} e_d(t) \\ e_q(t) \\ e_0(t) \end{bmatrix}}_{\boldsymbol{e}_{dq0}(t)} = \underbrace{\frac{2}{3}\begin{bmatrix} \cos\theta_a & \cos\theta_b & \cos\theta_c \\ -\sin\theta_a & -\sin\theta_b & -\sin\theta_c \\ \frac{1}{2} & \frac{1}{2} & \frac{1}{2} \end{bmatrix}}_{\boldsymbol{D}(t)} \cdot \underbrace{\begin{bmatrix} e_a(t) \\ e_b(t) \\ e_c(t) \end{bmatrix}}_{\boldsymbol{e}_{abc}(t)} \tag{10.10a}$$

$$\underbrace{\begin{bmatrix} i_d(t) \\ i_q(t) \\ i_0(t) \end{bmatrix}}_{\boldsymbol{i}_{dq0}(t)} = \underbrace{\frac{2}{3}\begin{bmatrix} \cos\theta_a & \cos\theta_b & \cos\theta_c \\ -\sin\theta_a & -\sin\theta_b & -\sin\theta_c \\ \frac{1}{2} & \frac{1}{2} & \frac{1}{2} \end{bmatrix}}_{\boldsymbol{D}(t)} \cdot \underbrace{\begin{bmatrix} i_a(t) \\ i_b(t) \\ i_c(t) \end{bmatrix}}_{\boldsymbol{i}_{abc}(t)} \tag{10.10b}$$

$$\underbrace{\begin{bmatrix} \psi_d(t) \\ \psi_q(t) \\ \psi_0(t) \end{bmatrix}}_{\boldsymbol{\psi}_{dq0}(t)} = \underbrace{\frac{2}{3}\begin{bmatrix} \cos\theta_a & \cos\theta_b & \cos\theta_c \\ -\sin\theta_a & -\sin\theta_b & -\sin\theta_c \\ \frac{1}{2} & \frac{1}{2} & \frac{1}{2} \end{bmatrix}}_{\boldsymbol{D}(t)} \cdot \underbrace{\begin{bmatrix} \psi_a(t) \\ \psi_b(t) \\ \psi_c(t) \end{bmatrix}}_{\boldsymbol{\psi}_{abc}(t)} \tag{10.10c}$$

$$\underbrace{\begin{bmatrix} e_a(t) \\ e_b(t) \\ e_c(t) \end{bmatrix}}_{\boldsymbol{e}_{abc}(t)} = \underbrace{\begin{bmatrix} \cos\theta_a & -\sin\theta_a & 1 \\ \cos\theta_b & -\sin\theta_b & 1 \\ \cos\theta_c & -\sin\theta_c & 1 \end{bmatrix}}_{\boldsymbol{D}^{-1}(t)} \cdot \underbrace{\begin{bmatrix} e_d(t) \\ e_q(t) \\ e_0(t) \end{bmatrix}}_{\boldsymbol{e}_{dq0}(t)} \tag{10.11a}$$

$$\underbrace{\begin{bmatrix} i_a(t) \\ i_b(t) \\ i_c(t) \end{bmatrix}}_{\boldsymbol{i}_{abc}(t)} = \underbrace{\begin{bmatrix} \cos\theta_a & -\sin\theta_a & 1 \\ \cos\theta_b & -\sin\theta_b & 1 \\ \cos\theta_c & -\sin\theta_c & 1 \end{bmatrix}}_{\boldsymbol{D}^{-1}(t)} \cdot \underbrace{\begin{bmatrix} i_d(t) \\ i_q(t) \\ i_0(t) \end{bmatrix}}_{\boldsymbol{i}_{dq0}(t)} \tag{10.11b}$$

$$\underbrace{\begin{bmatrix} \psi_a(t) \\ \psi_b(t) \\ \psi_c(t) \end{bmatrix}}_{\boldsymbol{\psi}_{abc}(t)} = \underbrace{\begin{bmatrix} \cos\theta_a & -\sin\theta_a & 1 \\ \cos\theta_b & -\sin\theta_b & 1 \\ \cos\theta_c & -\sin\theta_c & 1 \end{bmatrix}}_{\boldsymbol{D}^{-1}(t)} \cdot \underbrace{\begin{bmatrix} \psi_d(t) \\ \psi_q(t) \\ \psi_0(t) \end{bmatrix}}_{\boldsymbol{\psi}_{dq0}(t)} \tag{10.11c}$$

where

$$\theta_a = \theta = \omega t, \quad \theta_b = \theta - 120^\circ = \omega t - 120^\circ, \quad \theta_c = \theta + 120^\circ = \omega t + 120^\circ \tag{10.12}$$

The transformation matrices $\boldsymbol{D}(t)$, $\boldsymbol{D}^{-1}(t)$ are functions of time t because they include θ_a, θ_b, θ_c.

10.2.1.2 Physical meaning of d–q–0 transformation method

Now we examine the physical meaning of the d–q–0 transformation by using stator flux linkage $\psi_a(t)$, $\psi_b(t)$, $\psi_c(t)$ and their transformed flux linkages $\psi_d(t)$, $\psi_q(t)$, $\psi_0(t)$.

The positions of stator phase a, b, c coils to the d-axis are θ_a, θ_b, θ_c, respectively, so that the components in the d-axis direction of $\psi_a(t)$, $\psi_b(t)$, $\psi_c(t)$ are $\psi_a(t)\cos\theta_a$, $\psi_b(t)\cos\theta_b$, $\psi_c(t)\cos\theta_c$, respectively. Then the definition of $\psi_d(t)$ by Equation 10.10c is

$$\begin{aligned} \psi_d(t) &= \frac{2}{3}\{\psi_a(t)\cos\theta_a + \psi_b(t)\cos\theta_b + \psi_c(t)\cos\theta_c\} \\ &= \frac{2}{3}\{(\text{d-axis component of } \psi_a(t)) + (\text{d-axis component of } \psi_b(t)) \\ &\quad + (\text{d-axis component of } \psi_c(t))\} \end{aligned} \tag{10.13a}$$

The components in the q-axis direction of $\psi_a(t)$, $\psi_b(t)$, $\psi_c(t)$ are $\psi_a(t)\cos(\theta_a + 90^\circ) = -\psi_a(t)\sin\theta_a$, $\psi_b(t)\cos(\theta_b + 90^\circ) = -\psi_b(t)\sin\theta_b$, $\psi_c(t)\cos(\theta_c + 90^\circ) = -\psi_c(t)\sin\theta_c$, respectively. Then, the equation for $\psi_q(t)$ from Equation 10.10c is

$$\begin{aligned} \psi_q(t) &= \frac{2}{3}\{-\psi_a(t)\sin\theta_a - \psi_b(t)\sin\theta_b - \psi_c(t)\sin\theta_c\} \\ &= \frac{2}{3}\{(\text{q-axis component of } \psi_a(t)) + (\text{q-axis component of } \psi_b(t)) \\ &\quad + (\text{q-axis component of } \psi_c(t))\} \end{aligned} \tag{10.13b}$$

In other words, the physical meaning of $\psi_d(t)$ is 2/3 times the values of the total sum of the d-axis components of $\psi_a(t)$, $\psi_b(t)$, $\psi_c(t)$. The meaning of $\psi_q(t)$ is the same but for the q-axis.

$\psi_0(t)$ from Equation 10.10c is the same as that of the zero-sequence symmetrical component:

$$\psi_0(t) = \frac{1}{3}\{\psi_a(t) + \psi_b(t) + \psi_c(t)\} \tag{10.13c}$$

The voltages and currents in the d–q–0 domain are defined analogously.

10.2.2 Mutual relation of d–q–0, a–b–c and 0–1–2 domains

The theory of the double axes armature reaction, which appears in books on synchronous machine design, can be said to be the same as the d–q–0 transformation but eliminating zero-sequence quantities in principle. Although the theory is worthy for generator design, it may be imperfect and useless as a theoretical tool because it cannot explain unbalanced or transient phenomena, neither can it provide the method to connect the generator to other network equipment. On the other hand, the d–q–0 method is a mathematical transformation from the a–b–c domain to the d–q–0 domain by three variables, covering precisely any phenomenon, and furthermore provides a 'generator circuit' to be connected to other network equipments.

Let us examine the mutual relations of the a–b–c, d–q–0 and 0–1–2 domains, put together. The relations of quantities in the three different domains are defined by the equations below.

For the 0–1–2 ⇔ a–b–c domain:

$$\left.\begin{aligned} \boldsymbol{e}_{012}(t) &= \boldsymbol{a}\cdot\boldsymbol{e}_{abc}(t)\\ \boldsymbol{e}_{abc}(t) &= \boldsymbol{a}^{-1}\cdot\boldsymbol{e}_{012}(t)\end{aligned}\right\} \tag{10.14}$$

For the d–q–0 ⇔ a–b–c domain:

$$\left.\begin{aligned} \boldsymbol{e}_{dq0}(t) &= \boldsymbol{D}(t)\cdot\mathrm{Re}\left[\boldsymbol{e}_{abc}(t)\right]\\ \mathrm{Re}\left[\boldsymbol{e}_{abc}(t)\right] &= \boldsymbol{D}^{-1}(t)\cdot\boldsymbol{e}_{dq0}(t)\end{aligned}\right\} \tag{10.15}$$

For the d–q–0 ⇔ 0–1–2 domain:

$$\left.\begin{aligned} \boldsymbol{e}_{dq0}(t) &= \boldsymbol{D}(t)\cdot\mathrm{Re}\left[\boldsymbol{a}^{-1}\boldsymbol{e}_{012}(t)\right]\\ \mathrm{Re}\left[\boldsymbol{e}_{012}(t)\right] &= \mathrm{Re}\left[\boldsymbol{a}\cdot\boldsymbol{D}^{-1}(t)\cdot\boldsymbol{e}_{dq0}(t)\right]\end{aligned}\right\} \tag{10.16}$$

It is obvious that the transformation between the d–q–0 and 0–1–2 domains is defined by Equation 10.16, which is to be derived from Equations 10.14 and 10.15.

However it is to be reminded that $e_d(t)$ and $e_q(t)$ by Equation 10.10 and Equation 10.11 are defined as real number quantities, because they are d.c. values under three phase balanced condition as is soon mathematically confirmed. Therefore, the three phase quantities $\boldsymbol{e}_{abc}(t)$, $\boldsymbol{i}_{abc}(t)$, $\boldsymbol{\psi}_{abc}(t)$ in the Equation 10.10 and Equation 10.11 should be of real number expression.

10.2.3 Characteristics of d–q–0 domain quantities

Now we examine how arbitrary power frequency quantities having positive; negative, zero sequence components are transferred into d-q-0 domain quantities.

$$\underbrace{\begin{bmatrix} e_d(t)\\ e_q(t)\\ e_0(t)\end{bmatrix}}_{\boldsymbol{e}_{dq0}(t)} = \underbrace{\frac{2}{3}\begin{bmatrix} \cos\omega t & \cos(\omega t-120°) & \cos(\omega t+120°)\\ -\sin\omega t & -\sin(\omega t-120°) & -\sin(\omega t+120°)\\ \frac{1}{2} & \frac{1}{2} & \frac{1}{2}\end{bmatrix}}_{D(t)}$$

$$\times\left\{\underbrace{\begin{bmatrix} E_{a1}\cos(\omega t+\alpha_1)\\ E_{a1}\cos(\omega t+\alpha_1-120°)\\ E_{a1}\cos(\omega t+\alpha_1+120°)\end{bmatrix}}_{\text{Positive-seq}} + \underbrace{\begin{bmatrix} E_{a2}\cos(\omega t+\alpha_2)\\ E_{a2}\cos(\omega t+\alpha_2+120°)\\ E_{a2}\cos(\omega t+\alpha_2-120°)\end{bmatrix}}_{\text{negative-seq}} + \underbrace{\begin{bmatrix} E_{a0}\cos(\omega t+\alpha_0)\\ E_{a0}\cos(\omega t+\alpha_0)\\ E_{a0}\cos(\omega t+\alpha_0)\end{bmatrix}}_{\text{zero-seq}}\right\}$$

$$= \underbrace{\begin{bmatrix} E_{a1}\cos\alpha_1\\ E_{a1}\sin\alpha_1\\ 0\end{bmatrix}}_{\text{p-seq}} + \underbrace{\begin{bmatrix} E_{a2}\cos(2\omega t+\alpha_2)\\ -E_{a2}\sin(2\omega t+\alpha_2)\\ 0\end{bmatrix}}_{\text{n-seq}} + \underbrace{\begin{bmatrix} 0\\ 0\\ E_{a0}\cos(\omega t+\alpha_0)\end{bmatrix}}_{\text{zero-seq}} \tag{10.17}$$

namely

$$\left.\begin{aligned} e_d(t) &= E_{a1}\cos\alpha_1 + E_{a2}\cos(2\omega t + \alpha_2) \\ e_q(t) &= E_{a1}\sin\alpha_1 - E_{a2}\sin(2\omega t + \alpha_2) \\ e_0(t) &= E_{a0}\cos(\omega t + \alpha_0) \end{aligned}\right\} \tag{10.18}$$

Equation 10.17 and 10.18 explain that positive sequence component appears as d.c. component on d.q.0 domain which is time independent, while negative sequence component appears as double frequency component.

Equation 10.18 can be recasted as

$$\left.\begin{aligned} \{e_d(t) + je_q(t)\} &= E_{a1}e^{j\alpha_1} + E_{a1}e^{-j(2\omega t+\alpha_2)} && \text{(a)} \\ \{e_d(t) + je_q(t)\}e^{j\omega t} &= \underline{E_{a1}e^{j(\omega t+\alpha_1)}} + E_{a2}e^{-j(\omega t+\alpha_2)} && \text{(b)} \end{aligned}\right\} \tag{10.19}$$

or

$$\left.\begin{aligned} \{e_d(t) - je_q(t)\} &= E_{a2}e^{-j\alpha_1} + E_{a2}e^{j(2\omega t+\alpha_2)\}} && \text{(a)} \\ \{e_d(t) - je_q(t)\}e^{-j\omega t} &= E_{a1}e^{-j(\omega t+\alpha_1)} + \underline{E_{a2}e^{j(\omega t+\alpha_2)}} && \text{(b)} \end{aligned}\right\} \tag{10.20}$$

Therefore, positive and negative sequence quantities by complex number expression and real number expression can be derived from Equation 10.19(b) and Equation 10.20(b) respectively. Then symmetrical component quantities by complex number expression as well as real number expression are written as follows in terms of d- and q-axes quantities.

Positive sequence quantities

$$e_1(t) = E_{a1}e^{j(\omega t+\alpha_1)} = \{e_d(t) + je_q(t)\}e^{j\omega t} \tag{10.21}$$

$$\begin{aligned} e_1(t) &= E_{a1}\cos(\omega t + \alpha_1) = \mathrm{Re}[\{e_d(t) + je_q(t)\}(\cos\omega t + j\sin\omega t)] \\ &= e_d(t)\cos\omega t - e_q(t)\sin\omega t \end{aligned} \tag{10.22}$$

Negative sequence quantities

$$e_2(t) = E_{a2}e^{j(\omega t+\alpha_2)} = \{e_d(t) - je_q(t)\}e^{-j\omega t} \tag{10.23}$$

$$\begin{aligned} e_2(t) &= E_{a2}\cos(\omega t + \alpha_2) = \mathrm{Re}[\{e_d(t) - je_q(t)\}(\cos\omega t - j\sin\omega t)] \\ &= e_d(t)\cos\omega t - e_q(t)\sin\omega t \end{aligned} \tag{10.24}$$

Zero sequence quantities

$$e_0(t) = E_{a0}e^{j(\omega t+\alpha_0)} \tag{10.25}$$

$$e_0(t) = E_{a0}\cos(\omega t + \alpha_0) \tag{10.26}$$

Now, we have found that:

- **Positive-sequence voltage** is rotating by angular velocity ω in synchronization with the rotor or d-axis and q-axis, so that it is at a standstill from the d- and q-axes viewpoints. Therefore positive sequence voltage appears as d-c variables $E_{a1}e^{j\alpha 1}$ in $e_d(t)$ and $e_q(t)$.
- **Negative-sequence voltage** is rotating on a-b-c domain by inverse rotation (angular velocity $-\omega$) to the rotor or d- and q-axis, so that it appears as voltage components of angular velocity 2ω in $e_d(t)$, $e_q(t)$.
- **Zero-sequence voltage** is the same as that of symmetrical components.

Equation 10.19(a), 20(a) is just the symmetrical components voltages looking from the stator, and Equation 10.19(b), 20(b) may be said to be the same but looking from the rotor.

10.3 Transformation of Generator Equations from a–b–c to d–q–0 Domain

10.3.1 Transformation of generator equations to d–q–0 domain

We examine the transformation of a generator's equations, Equations 10.2–10.8, to equations in the d–q–0 domain using the definitions in Equations 10.10 and 10.11.

10.3.1.1 Transformation of equation 10.2

Now

$$\boldsymbol{e}_{abc}(t) = \frac{d}{dt}\boldsymbol{\psi}_{abc}(t) - r\boldsymbol{i}_{abc}(t) \tag{10.27}$$

Reminding $\boldsymbol{D}(t)$, $\boldsymbol{D}^{-1}(t)$ are the matrices of real number elements,

Left-multiplying both sides by $\boldsymbol{D}(t)$,

$$\begin{aligned}\boldsymbol{e}_{dq0}(t) &= \mathrm{Re}\left[\boldsymbol{D}(t)\boldsymbol{e}_{abc}(t)\right] = \mathrm{Re}\left[\boldsymbol{D}(t)\left\{\frac{d}{dt}\boldsymbol{\psi}_{abc}(t)\right\} - \boldsymbol{D}(t)r\boldsymbol{i}_{abc}(t)\right]\\ &= \boldsymbol{D}(t)\frac{d}{dt}\{\boldsymbol{D}^{-1}(t)\boldsymbol{\psi}_{dq0}(t)\} - \boldsymbol{D}(t)r\boldsymbol{D}^{-1}(t)\boldsymbol{i}_{dq0}(t)\\ &= \boldsymbol{D}(t)\left\{\frac{d}{dt}\boldsymbol{D}^{-1}(t)\right\}\boldsymbol{\psi}_{dq0}(t) + \boldsymbol{D}(t)\boldsymbol{D}^{-1}(t)\frac{d}{dt}\boldsymbol{\psi}_{dq0}(t) - r\boldsymbol{i}_{dq0}(t)\end{aligned}$$

$\boldsymbol{D}^{-1}(t)$ and $\boldsymbol{\psi}_{abc}(t)$ are functions of time t, so the first term on the right-hand side in the above equation should be expressed as follows by applying a differential equation formula: (Appendix A.3)

$$\frac{d}{dt}\{\boldsymbol{D}^{-1}(t)\cdot\boldsymbol{\psi}_{dq0}(t)\} = \left\{\frac{d}{dt}\boldsymbol{D}^{-1}(t)\right\}\cdot\boldsymbol{\psi}_{dq0}(t) + \boldsymbol{D}^{-1}(t)\cdot\frac{d}{dt}\boldsymbol{\psi}_{dq0}(t)$$

$$\therefore\ \boldsymbol{e}_{dq0}(t) = \boldsymbol{D}(t)\left\{\frac{d}{dt}\boldsymbol{D}^{-1}(t)\right\}\boldsymbol{\psi}_{dq0}(t) + \frac{d}{dt}\boldsymbol{\psi}_{dq0}(t) - r\boldsymbol{i}_{dq0}(t) \tag{10.28}$$

where

$\theta_a = \omega t, \quad \theta_b = \omega t - 120°, \quad \theta_c = \omega t + 120°$

$$\frac{d}{dt}\boldsymbol{D}^{-1}(t) = \frac{d}{dt}\begin{bmatrix}\cos\theta_a & -\sin\theta_a & 1\\ \cos\theta_b & -\sin\theta_b & 1\\ \cos\theta_c & -\sin\theta_c & 1\end{bmatrix} = \begin{bmatrix}-\sin\theta_a\dfrac{d\theta_a}{dt} & -\cos\theta_a\dfrac{d\theta_a}{dt} & 0\\ -\sin\theta_b\dfrac{d\theta_b}{dt} & -\cos\theta_b\dfrac{d\theta_b}{dt} & 0\\ -\sin\theta_c\dfrac{d\theta_c}{dt} & -\cos\theta_c\dfrac{d\theta_c}{dt} & 0\end{bmatrix} = \begin{bmatrix}-\sin\theta_a & -\cos\theta_a & 0\\ -\sin\theta_b & -\cos\theta_b & 0\\ -\sin\theta_c & -\cos\theta_c & 0\end{bmatrix}\cdot\frac{d\theta}{dt}$$

$$\frac{d\theta_a}{dt} = \frac{d\theta_b}{dt} = \frac{d\theta_c}{dt} = \frac{d\theta}{dt} = \omega$$

Then

$$
\boldsymbol{D}(t)\left\{\frac{d}{dt}\boldsymbol{D}^{-1}(t)\right\} = \underbrace{\frac{2}{3}\begin{bmatrix} \cos\theta_a & \cos\theta_b & \cos\theta_c \\ -\sin\theta_a & -\sin\theta_b & -\sin\theta_c \\ \frac{1}{2} & \frac{1}{2} & \frac{1}{2} \end{bmatrix}}_{\boldsymbol{D}(t)} \cdot \underbrace{\begin{bmatrix} -\sin\theta_a & -\cos\theta_a & 0 \\ -\sin\theta_b & -\cos\theta_b & 0 \\ -\sin\theta_c & -\cos\theta_c & 0 \end{bmatrix}}_{\frac{d}{dt}\boldsymbol{D}^{-1}(t)} \cdot \frac{d\theta}{dt} = \begin{bmatrix} 0 & -\frac{d\theta}{dt} & 0 \\ \frac{d\theta}{dt} & 0 & 0 \\ 0 & 0 & 0 \end{bmatrix}
$$

That is, the d–q–0 domain equation transformed from Equation 10.2 is

$$
\left.\begin{bmatrix} e_d(t) \\ e_q(t) \\ e_0(t) \end{bmatrix} = \begin{bmatrix} 0 & -\frac{d\theta}{dt} & 0 \\ \frac{d\theta}{dt} & 0 & 0 \\ 0 & 0 & 0 \end{bmatrix} \cdot \begin{bmatrix} \psi_d(t) \\ \psi_q(t) \\ \psi_0(t) \end{bmatrix} + \begin{bmatrix} \frac{d}{dt}\psi_d(t) \\ \frac{d}{dt}\psi_q(t) \\ \frac{d}{dt}\psi_0(t) \end{bmatrix} - r\begin{bmatrix} i_d(t) \\ i_q(t) \\ i_0(t) \end{bmatrix}\right\} \quad \textbf{Park's equation} \qquad (\mathbf{10.29})
$$

where

$$
\frac{d\theta}{dt} = \omega = 2\pi f
$$

10.3.1.2 Transformation of equation 10.3

The equations include only d–q–0 domain quantities, so we do not need the transformation. It is again written below:

$$
\begin{bmatrix} E_{fd} \\ 0 \\ 0 \end{bmatrix} = \begin{bmatrix} \frac{d}{dt}\psi_{fd}(t) \\ \frac{d}{dt}\psi_{kd}(t) \\ \frac{d}{dt}\psi_{kq}(t) \end{bmatrix} + \begin{bmatrix} r_{fd}i_{fd}(t) \\ r_{kd}i_{kd}(t) \\ r_{kq}i_{kq}(t) \end{bmatrix} \qquad (\mathbf{10.30})
$$

10.3.1.3 Transformation of equation 10.4

Now

$$
\boldsymbol{\psi}_{abc}(t) = \boldsymbol{l}_{abc}(t)\boldsymbol{i}_{abc}(t) + \boldsymbol{l}_{abc-F}(t)\boldsymbol{i}_F(t) \qquad (10.31)
$$

Then

$$
\boldsymbol{\psi}_{dq0}(t) = \{D(t)\boldsymbol{l}_{abc}(t)D^{-1}(t)\}\boldsymbol{i}_{dq0}(t) + \{D(t)\boldsymbol{l}_{abc-F}(t)\}\boldsymbol{i}_F \qquad (10.32)
$$

$\boldsymbol{D}(t)\boldsymbol{l}_{abc}(t)\boldsymbol{D}^{-1}(t)\}$ and $\{D(t)\boldsymbol{l}_{abc-F}(t)\}$ in this equation can be calculated from $\boldsymbol{l}_{abc}(t)$ and $\boldsymbol{l}_{abc-F}(t)$ in Equations 10.6 and 10.7, and the equations below are derived. The result can be proved manually, although the process is rather time consuming and the demonstration is omitted in this book:

$$
\boldsymbol{D}(t)\boldsymbol{l}_{abc}(t)\boldsymbol{D}^{-1}(t) = -\begin{bmatrix} L_{aa0} + L_{ab0} + \frac{3}{2}L_{aa2} & 0 & 0 \\ 0 & L_{aa0} + L_{ab0} - \frac{3}{2}L_{aa2} & 0 \\ 0 & 0 & L_{aa0} - 2L_{ab0} \end{bmatrix}
$$

$$
\equiv -\begin{bmatrix} L_d & 0 & 0 \\ 0 & L_q & 0 \\ 0 & 0 & L_0 \end{bmatrix}
$$

$$\boldsymbol{D}(t)\boldsymbol{l}_{abc-F}(t) = \begin{bmatrix} L_{afd} & L_{akd} & 0 \\ 0 & 0 & L_{akq} \\ 0 & 0 & 0 \end{bmatrix}$$

Accordingly, the following equations in the d–q–0 domain have been derived as the transformation of the original Equation 10.4 from the a–b–c to the d–q–0 domain:

$$\left.\begin{array}{l} \begin{bmatrix} \psi_d(t) \\ \psi_q(t) \\ \psi_0(t) \end{bmatrix} = -\begin{bmatrix} L_d & 0 & 0 \\ 0 & L_q & 0 \\ 0 & 0 & L_0 \end{bmatrix} \cdot \begin{bmatrix} i_d(t) \\ i_q(t) \\ i_0(t) \end{bmatrix} + \begin{bmatrix} L_{afd} & L_{akd} & 0 \\ 0 & 0 & L_{akq} \\ 0 & 0 & 0 \end{bmatrix} \cdot \begin{bmatrix} i_{fd}(t) \\ i_{kd}(t) \\ i_{kq}(t) \end{bmatrix} \\ \text{where} \\ \quad \text{self-inductance of stator d-axis coil} \quad L_d = L_{aa0} + L_{ab0} + \frac{3}{2}L_{aa2} \\ \quad \text{self-inductance of stator q-axis coil} \quad L_q = L_{aa0} + L_{ab0} - \frac{3}{2}L_{aa2} \\ \quad \text{self-inductance of stator zero-sequence coil} \quad L_0 = L_{aa0} - 2L_{ab0} \end{array}\right\} \quad \textbf{(10.33)}$$

10.3.1.4 Transformation of equation 10.5

Here

$$\boldsymbol{\psi}_F(t) = -\boldsymbol{l}_{F-abc}(t)\boldsymbol{i}_{abc}(t) + \boldsymbol{L}_F\boldsymbol{i}_F(t) \tag{10.34}$$

Accordingly,

$$\boldsymbol{\psi}_F(t) = -\{\boldsymbol{l}_{F-abc}(t)\boldsymbol{D}^{-1}(t)\}\boldsymbol{i}_{dq0}(t) + \boldsymbol{L}_F\boldsymbol{i}_F(t) \tag{10.35}$$

and

$$\{\boldsymbol{l}_{F-abc}(t)\boldsymbol{D}^{-1}(t)\}$$

$$= \begin{bmatrix} L_{afd}\cos\theta_a & L_{afd}\cos\theta_b & L_{afd}\cos\theta_c \\ L_{akd}\cos\theta_a & L_{akd}\cos\theta_b & L_{akd}\cos\theta_c \\ -L_{akq}\sin\theta_a & -L_{akq}\sin\theta_b & -L_{akq}\sin\theta_c \end{bmatrix} \cdot \begin{bmatrix} \cos\theta_a & -\sin\theta_a & 1 \\ \cos\theta_b & -\sin\theta_b & 1 \\ \cos\theta_c & -\sin\theta_c & 1 \end{bmatrix} = \frac{3}{2}\begin{bmatrix} L_{afd} & 0 & 0 \\ L_{akd} & 0 & 0 \\ 0 & L_{akq} & 0 \end{bmatrix}$$

This equation is the same as the transposed matrix of $\boldsymbol{D}(t)\,\boldsymbol{l}_{abc-F}(t)$ multiplied by 3/2. Then

$$\{\boldsymbol{l}_{F-abc}(t)\boldsymbol{D}^{-1}(t)\} = \frac{3}{2}\cdot\{\boldsymbol{D}(t)\boldsymbol{l}_{abc-F}(t)\}^t$$

That is, the d–q–0 domain equation transformed from Equation 10.5 is

$$\begin{bmatrix} \psi_{fd}(t) \\ \psi_{kd}(t) \\ \psi_{kq}(t) \end{bmatrix} = -\frac{3}{2}\begin{bmatrix} L_{afd} & 0 & 0 \\ L_{akd} & 0 & 0 \\ 0 & L_{akq} & 0 \end{bmatrix} \cdot \begin{bmatrix} i_d(t) \\ i_q(t) \\ i_0(t) \end{bmatrix} + \begin{bmatrix} L_{ffd} & L_{fkd} & 0 \\ L_{fkd} & L_{kkd} & 0 \\ 0 & 0 & L_{kkq} \end{bmatrix} \cdot \begin{bmatrix} i_{fd}(t) \\ i_{kd}(t) \\ i_{kq}(t) \end{bmatrix} \quad \textbf{(10.36)}$$

In conclusion, a generator's equations in the d–q–0 domain are shown by Equations 10.29, 10.30, 10.33 and 10.36. Note that all these equations are described using only fixed inductances L (independent of ωt), and all the ωt-dependent inductances have disappeared.

Equation 10.29 is Park's equation, named after R. H. Park, which is simply written as

$$\left.\begin{aligned} e_d(t) &= \frac{d}{dt}\psi_d(t) - ri_d(t) - \psi_q(t)\frac{d\theta}{dt} \\ e_q(t) &= \frac{d}{dt}\psi_q(t) - ri_q(t) + \psi_d(t)\frac{d\theta}{dt} \\ e_0(t) &= \frac{d}{dt}\psi_0(t) - ri_0(t) \end{aligned}\right\} \textbf{Park's equation} \qquad \textbf{(10.37)}$$

10.3.2 Unitization of generator d–q–0 domain equations

Next, Equations 10.37 or 10.29 and 10.36 have to be unitized to obtain mathematical models and to enable generators to be connected to transmission lines and other equipment.

10.3.2.1 Examination of self-inductances of stator L_d and L_q

On the unitization procedure, first we need to examine what the inductances L_d and L_q mean physically.

The inductances L_d and L_q are the self-inductances of the stator coils, which can be presumed as below, from a physical viewpoint.

L_d and L_q are made up of two parts:

L_d = {the stator d-axis inductance due to flux which does not link any rotor circuits (the leakage inductance) (*1)}

+ {the inductance due to flux which does link the rotor d-axis circuit (the mutual inductance)}

L_q = {the stator q-axis inductance due to flux which does not link any rotor circuits (the leakage inductance) (*2)}

+ {the inductance due to flux which does link the rotor q-axis circuit (the mutual inductance)}

where

(the stator d-axis leakage inductance) (*1)
= d-axis component of {(slot leakage) + (air-gap leakage) + (end-coil leakage)}

(the stator q-axis leakage inductance) (*2)
= q-axis component of {(slot leakage) + (air-gap leakage) + (end-coil leakage)}

The structure of stator windings, air gaps and end-coil parts (except the rotor) is designed almost uniformly in any round section. Accordingly, we can justify the assumption $*1 = *2 = L_l$.

Therefore

$$\left.\begin{aligned} L_d &= L_l + L_{ad} \\ L_q &= L_l + L_{aq} \end{aligned}\right\} \qquad (10.38)$$

where L_l: the stator d-axis and q-axis leakage inductances caused by leakage flux in the stator which does not link the rotor coils

L_{ad}: the mutual inductances between the stator and rotor in the d-axis

L_{aq}: the mutual inductances between the stator and rotor in the q-axis

Now we replace $L_d \rightarrow L_l + L_{ad}$ and $L_q \rightarrow L_l + L_{aq}$ in Equation 10.33. Recall that the base rotor currents for the unitization are defined in terms of inductances L_{ad} and L_{aq} (instead of L_d and L_q), as shown later in Equation 10.41, where L_{ad} and L_{aq} are based on so-called 'effective interlinking flux between stator and rotor coils'.

10.3.2.2 Setting the base quantities for unitization

10.3.2.2.1 Notation of base quantities for unitization The base quantities for the stator quantities (suffix s means stator) are ${}_se_{\text{base}}$, ${}_si_{\text{base}}$, ${}_s\psi_{\text{base}}$, ${}_sZ_{\text{base}}$, ${}_sL_{\text{base}}$, etc., where

$$\left.\begin{aligned}
{}_sZ_{\text{base}} &= 2\pi f_{\text{base}} \cdot {}_sL_{\text{base}} && ① \\
{}_s\psi_{\text{base}} &= {}_sL_{\text{base}} \cdot {}_si_{\text{base}} && ② \\
{}_se_{\text{base}} &= {}_sZ_{\text{base}} \cdot {}_si_{\text{base}} = 2\pi f_{\text{base}} \cdot {}_sL_{\text{base}} \cdot {}_si_{\text{base}} \\
&= 2\pi f_{\text{base}} \cdot {}_s\psi_{\text{base}} && ③
\end{aligned}\right\} \quad (10.39\text{a})$$

The base quantities for the rotor field coil quantities (suffix f means field) are ${}_fe_{\text{base}}$, ${}_fi_{\text{base}}$, ${}_f\psi_{\text{base}}$, ${}_fZ_{\text{base}}$, ${}_fL_{\text{base}}$, where

$$\left.\begin{aligned}
{}_fZ_{\text{base}} &= 2\pi f_{\text{base}} \cdot {}_fL_{\text{base}} && ④ \\
{}_f\psi_{\text{base}} &= {}_fL_{\text{base}} \cdot {}_fi_{\text{base}} && ⑤ \\
{}_fe_{\text{base}} &= {}_fZ_{\text{base}} \cdot {}_fi_{\text{base}} = 2\pi f_{\text{base}} \cdot {}_fL_{\text{base}} \cdot {}_fi_{\text{base}} \\
&= 2\pi f_{\text{base}} \cdot {}_f\psi_{\text{base}} && ⑥
\end{aligned}\right\} \quad (10.39\text{b})$$

The base quantities for the rotor damper coil quantities (suffix k means damper coil) are ${}_ke_{\text{base}}$, ${}_ki_{\text{base}}$, ${}_k\psi_{\text{base}}$, ${}_kZ_{\text{base}}$, ${}_kL_{\text{base}}$, where

$$\left.\begin{aligned}
{}_kZ_{\text{base}} &= 2\pi f_{\text{base}} \cdot {}_kL_{\text{base}} && ⑦ \\
{}_k\psi_{\text{base}} &= {}_kL_{\text{base}} \cdot {}_ki_{\text{base}} && ⑧ \\
{}_ke_{\text{base}} &= {}_kZ_{\text{base}} \cdot {}_ki_{\text{base}} = 2\pi f_{\text{base}} \cdot {}_kL_{\text{base}} \cdot {}_ki_{\text{base}} \\
&= 2\pi f_{\text{base}} \cdot {}_k\psi_{\text{base}} && ⑨ \\
\omega_{\text{base}} &= 2\pi f_{\text{base}}[\text{rad/s}],\ f_{\text{base}} = 50\ \text{Hz} \quad \text{or} \quad 60\ \text{Hz} && ⑩
\end{aligned}\right\} \quad (10.39\text{c})$$

10.3.2.2.2 Definition of base quantities for s-coils, f-coil and k-coils

1. Capacity bases (VA_{base} or MVA_{base}) for s-coils, f-coil and k-coils are unified into the rated stator winding capacity of the generator. Namely,

$$VA_{3\phi\text{base}} = 3\left(\frac{{}_se_{\text{base}}}{\sqrt{2}}\right)\left(\frac{{}_si_{\text{base}}}{\sqrt{2}}\right) = \frac{3}{2}\,{}_se_{\text{base}} \cdot {}_si_{\text{base}} = {}_fe_{\text{base}} \cdot {}_fi_{\text{base}} = {}_ke_{\text{base}} \cdot {}_ki_{\text{base}}$$

where

${}_se_{\text{base}}, {}_si_{\text{base}}$: voltage and current bases by crest value (10.40)

${}_se_{\text{base}}/\sqrt{2}, {}_si_{\text{base}}/\sqrt{2}$: voltage and current bases by rms value

$VA_{3\phi\text{base}}$: VA capacity as multiplication of rms voltage and current

Recall that the voltage and current bases of f-coils and k-coils as well as s-coils are defined so that the VA capacity base of each coil coincides with that of the stator three-phase-rated VA capacity base. In other words, the base capacity for the rotor should be selected with the same value as the rated capacity of stator coils, even though the actual rated capacity of the excitor is far smaller (about 10%) than that of the stator.

2. Current bases for s-coils, f-coil and k-coils are defined as

$$\left.\begin{aligned}
L_{ad} \cdot {}_si_{\text{base}} &= L_{afd} \cdot {}_fi_{\text{base}} = L_{akd} \cdot {}_ki_{\text{base}} \\
L_{aq} \cdot {}_si_{\text{base}} &= L_{akq} \cdot {}_ki_{\text{base}}
\end{aligned}\right\} \quad (10.41)$$

Base rotor currents are then defined in terms of inductances L_{ad} and L_{aq}. In other words, base rotor current and base stator current are defined so that the base of effective flux linkage $\psi_{\text{base}} = (L_{\text{base}} \cdot i_{\text{base}})$ of the s-, f- and k-coils coincide, instead of L_d, L_q as was explained in Equation 10.38.

3. Voltage bases for s-coils, f-coil and k-coils are subordinately defined from Equations 10.40 and 10.41.

4. Time t is unitized below from seconds to radians by the base quantity $\omega_{\text{base}} = 2\pi f_{\text{base}}$ [rad/sec] for simplicity, although it does not necessarily have to be unitized:

$$\bar{t} = \omega_{\text{base}} \cdot t = 2\pi f_{\text{base}} \cdot t \,[\text{rad/sec}], \quad d\bar{t} = \omega_{\text{base}} \cdot t = 2\pi f_{\text{base}} \cdot dt \,[\text{rad/sec}] \tag{10.42}$$

Unitized t is counted in electrical angle $\bar{t}$ radians, instead of t seconds. Accordingly, the written symbols are changed as $e_a(t) \to e_a(\bar{t})$, $i_{fd}(t) \to i_{fd}(\bar{t})$.

10.3.2.3 Unitization of equations 10.29, 10.30, 10.33 and 10.36

10.3.2.3.1 Unitization of equation 10.29 Recalling $d\bar{t} = 2\pi f_{\text{base}} \cdot dt$, the first row of Equation 10.29 can be unitized as

$$e_d(\bar{t}) = -2\pi f_{\text{base}} \frac{d\theta}{d\bar{t}} \psi_q(\bar{t}) + 2\pi f_{\text{base}} \frac{d}{d\bar{t}} \psi_d(\bar{t}) - r i_d(\bar{t})$$

Dividing both sides by Equation 10.39a⑦,

$$\frac{e_d(\bar{t})}{{}_s e_{\text{base}}} = -2\pi f_{\text{base}} \frac{d\theta}{d\bar{t}} \left(\frac{1}{2\pi f_{\text{base}}} \cdot \frac{\psi_q(\bar{t})}{{}_s\psi_{\text{base}}} \right) + 2\pi f_{\text{base}} \frac{d}{d\bar{t}} \left(\frac{1}{2\pi f_{\text{base}}} \cdot \frac{\psi_d(\bar{t})}{{}_s\psi_{\text{base}}} \right) - \frac{r}{{}_s Z_{\text{base}}} \cdot \frac{i_d(\bar{t})}{{}_s i_{\text{base}}}$$

Then
$$\bar{e}_d(\bar{t}) = -\frac{d\theta}{d\bar{t}} \bar{\psi}_q(\bar{t}) + \frac{d}{d\bar{t}} \bar{\psi}_d(\bar{t}) - \bar{r}\bar{i}_d(\bar{t})$$

$e_q(t)$ and $e_0(t)$ are unitized in the same way.

Unitized Park's equation is

$$\begin{bmatrix} \bar{e}_d(\bar{t}) \\ \bar{e}_q(\bar{t}) \\ \bar{e}_0(\bar{t}) \end{bmatrix} = \begin{bmatrix} 0 & -\dfrac{d\theta}{d\bar{t}} & 0 \\ \dfrac{d\theta}{d\bar{t}} & 0 & 0 \\ 0 & 0 & 0 \end{bmatrix} \cdot \begin{bmatrix} \bar{\psi}_d(\bar{t}) \\ \bar{\psi}_q(\bar{t}) \\ \bar{\psi}_0(\bar{t}) \end{bmatrix} + \begin{bmatrix} \dfrac{d}{d\bar{t}} \bar{\psi}_d(\bar{t}) \\ \dfrac{d}{d\bar{t}} \bar{\psi}_q(\bar{t}) \\ \dfrac{d}{d\bar{t}} \bar{\psi}_0(\bar{t}) \end{bmatrix} - \bar{r} \begin{bmatrix} \bar{i}_d(\bar{t}) \\ \bar{i}_q(\bar{t}) \\ i_0(\bar{t}) \end{bmatrix} \tag{10.43}$$

d-and q-axix quantities are not independent for each other, because e_d is related with ψ_q and e_q is related with ψ_d.

10.3.2.3.2 Unitization of equation 10.30 Recalling $d\bar{t} = 2\pi f_{\text{base}} \cdot dt$, and dividing the first equation of Equation 10.30 by Equation 10.39b⑥,

$$\frac{E_{fd}}{{}_f e_{\text{base}}} = 2\pi f_{\text{base}} \frac{d}{d\bar{t}} \left(\frac{1}{2\pi f_{\text{base}}} \cdot \frac{\psi_{fd}(\bar{t})}{{}_f\psi_{\text{base}}} \right) + \frac{r_{fd}}{{}_f Z_{\text{base}}} \cdot \frac{i_{fd}(\bar{t})}{{}_f i_{\text{base}}}$$

Therefore
$$\bar{E}_{fd} = \frac{d}{d\bar{t}} \bar{\psi}_{fd}(\bar{t}) + \bar{r}_{fd} \cdot \bar{i}_{fd}(\bar{t})$$

The second and third equations are divided by Equation 10.39⑨:

$$\therefore \quad \begin{bmatrix} \bar{E}_{fd} \\ 0 \\ 0 \end{bmatrix} = \begin{bmatrix} \dfrac{d}{d\bar{t}} \bar{\psi}_{fd}(\bar{t}) \\ \dfrac{d}{dt} \bar{\psi}_{kd}(\bar{t}) \\ \dfrac{d}{d\bar{t}} \bar{\psi}_{kq}(\bar{t}) \end{bmatrix} + \begin{bmatrix} \bar{r}_{fd} \cdot \bar{i}_{fd}(\bar{t}) \\ \bar{r}_{kd} \cdot \bar{i}_{kd}(\bar{t}) \\ \bar{r}_{kq} \cdot \bar{i}_{kq}(\bar{t}) \end{bmatrix} \tag{10.44}$$

10.3.2.3.3 Unitization of equation 10.33 From Equations 10.33 and 10.39②, recalling $L_d = L_l + L_{ad}$,

$$\frac{\psi_d(\bar{t})}{{}_s\psi_{\text{base}}} = -\left(\frac{L_l}{{}_sL_{\text{base}}} + \frac{L_{ad}}{{}_sL_{\text{base}}}\right) \cdot \frac{i_d(\bar{t})}{{}_si_{\text{base}}} + \frac{L_{afd}(t)}{{}_sL_{\text{base}} \cdot \dfrac{{}_si_{\text{base}}}{{}_fi_{\text{base}}}} \cdot \frac{i_{fd}(\bar{t})}{{}_fi_{\text{base}}} + \frac{L_{akd}}{{}_sL_{\text{base}} \cdot \dfrac{{}_si_{\text{base}}}{{}_ki_{\text{base}}}} \cdot \frac{i_{kd}(\bar{t})}{{}_ki_{\text{base}}}$$

$$\therefore\ \bar{\psi}_d(\bar{t}) = -(\bar{L}_l + \bar{L}_{ad}) \cdot \bar{i}_d(\bar{t}) + \left(\frac{L_{afd}}{{}_sL_{\text{base}} \cdot {}_si_{\text{base}}/{}_fi_{\text{base}}}\right) \cdot \bar{i}_{fd}(\bar{t}) + \left(\frac{L_{akd}}{{}_sL_{\text{base}} \cdot {}_si_{\text{base}}/{}_ki_{\text{base}}}\right) \cdot \bar{i}_{kd}(\bar{t})$$

The terms in the large parentheses of the above equation are rearranged below:

$$\bar{L}_{afd} \equiv \left(\frac{L_{afd}}{{}_sL_{\text{base}} \cdot {}_si_{\text{base}}/{}_fi_{\text{base}}}\right) = \frac{L_{afd} \cdot {}_fi_{\text{base}}}{{}_sL_{\text{base}} \cdot {}_si_{\text{base}}} = \frac{L_{ad} \cdot {}_si_{\text{base}}}{{}_sL_{\text{base}} \cdot {}_si_{\text{base}}} = \frac{L_{ad}}{{}_sL_{\text{base}}} = \bar{L}_{ad}$$

$$\bar{L}_{akd} \equiv \left(\frac{L_{akd}}{{}_sL_{\text{base}} \cdot {}_si_{\text{base}}/{}_ki_{\text{base}}}\right) = \frac{L_{akd} \cdot {}_ki_{\text{base}}}{{}_sL_{\text{base}} \cdot {}_si_{\text{base}}} = \frac{L_{ad} \cdot {}_si_{\text{base}}}{{}_sL_{\text{base}} \cdot {}_si_{\text{base}}} = \frac{L_{ad}}{{}_sL_{\text{base}}} = \bar{L}_{ad}$$

Therefore, the unitized first equation is

$$\bar{\psi}_d(\bar{t}) = -(\bar{L}_l + \bar{L}_{ad}) \cdot \bar{i}_d(\bar{t}) + \bar{L}_{ad} \cdot \bar{i}_{fd}(\bar{t}) + \bar{L}_{ad} \cdot \bar{i}_{kd}(\bar{t})$$

and the second equation can be unitized in the same way, namely

$$\bar{\psi}_q(\bar{t}) = -(\bar{L}_l + \bar{L}_{aq}) \cdot \bar{i}_q(\bar{t}) + \left(\frac{L_{akq}}{{}_sL_{\text{base}} \cdot {}_si_{\text{base}}/{}_ki_{\text{base}}}\right) \cdot \bar{i}_{kq}(\bar{t})$$

where

$$\bar{L}_{akq} \equiv \left(\frac{L_{akq}}{{}_sL_{\text{base}} \cdot {}_si_{\text{base}}/{}_ki_{\text{base}}}\right) = \frac{L_{akq} \cdot {}_ki_{\text{base}}}{{}_sL_{\text{base}} \cdot {}_si_{\text{base}}} = \frac{L_{aq} \cdot {}_si_{\text{base}}}{{}_sL_{\text{base}} \cdot {}_si_{\text{base}}} = \frac{L_{aq}}{{}_sL_{\text{base}}} = \bar{L}_{aq}$$

The unitized first equation is thus

$$\bar{\psi}_q(\bar{t}) = -(\bar{L}_l + \bar{L}_{aq}) \cdot \bar{i}_q(\bar{t}) + \bar{L}_{aq} \cdot \bar{i}_{kq}(\bar{t})$$

The third equation can be unitized by dividing by Equation 10.39a②:

$$\bar{\psi}_0(\bar{t}) = -\bar{L}_0(\bar{t}) \cdot \bar{i}_0(\bar{t})$$

Accordingly, the unitized equation of Equation 10.33 is

$$\left.\begin{aligned} \begin{bmatrix} \bar{\psi}_d(\bar{t}) \\ \bar{\psi}_q(\bar{t}) \\ \bar{\psi}_0(\bar{t}) \end{bmatrix} &= -\begin{bmatrix} \bar{L}_d & 0 & 0 \\ 0 & \bar{L}_q & 0 \\ 0 & 0 & \bar{L}_0 \end{bmatrix} \cdot \begin{bmatrix} \bar{i}_d(\bar{t}) \\ \bar{i}_q(\bar{t}) \\ \bar{i}_0(\bar{t}) \end{bmatrix} \\ &\quad + \begin{bmatrix} \bar{L}_{ad} & \bar{L}_{ad} & 0 \\ 0 & 0 & \bar{L}_{aq} \\ 0 & 0 & 0 \end{bmatrix} \cdot \begin{bmatrix} \bar{i}_{fd}(\bar{t}) \\ \bar{i}_{kd}(\bar{t}) \\ \bar{i}_{kq}(\bar{t}) \end{bmatrix} \\ &\text{where } \bar{L}_d = \bar{L}_l + \bar{L}_{ad}, \quad \bar{L}_q = \bar{L}_l + \bar{L}_{aq} \end{aligned}\right\} \tag{10.45}$$

10.3.2.3.4 Unitization of equation 10.36

From Equations 10.36 and 10.39

$$\bar\psi_{fd}(\bar t) = -\left(\frac{\frac{3}{2}L_{afd}}{{}_fL_{base}\cdot\frac{{}_fi_{base}}{{}_si_{base}}}\right)^{*1}\cdot\bar i_d(\bar t) + \bar L_{ffd}\cdot\bar i_{fd}(\bar t) + \left(\frac{L_{fkd}}{{}_fL_{base}\cdot\frac{{}_fi_{base}}{{}_ki_{base}}}\right)^{*2}\cdot\bar i_{kd}(\bar t)$$

$$\bar\psi_{kd}(\bar t) = -\left(\frac{\frac{3}{2}L_{akd}}{{}_kL_{base}\cdot\frac{{}_ki_{base}}{{}_si_{base}}}\right)^{*3}\cdot\bar i_d(\bar t) + \left(\frac{L_{fkd}}{{}_kL_{base}\cdot\frac{{}_ki_{base}}{{}_fi_{base}}}\right)^{*4}\cdot\bar i_{fd}(\bar t) + \bar L_{kkd}\cdot\bar i_{kd}(\bar t)$$

$$\bar\psi_{kq}(\bar t) = -\left(\frac{\frac{3}{2}L_{akq}}{{}_kL_{base}\cdot\frac{{}_ki_{base}}{{}_si_{base}}}\right)^{*5}\cdot\bar i_q(\bar t) + \bar L_{kkq}\,.\,\bar i_{kq}(\bar t)$$

Substituting Equation 10.39③⑥⑨ into Equation 10.40,

$$\frac{3}{2}2\pi f_{base}\cdot{}_sL_{base}\cdot{}_si_{base}^2 = 2\pi f_{base}\cdot{}_fL_{base}\cdot{}_fi_{base}^2 = 2\pi f_{base}\cdot{}_kL_{base}\cdot{}_ki_{base}^2$$

$$\therefore\ \frac{3}{2}\,{}_sL_{base}\cdot{}_si_{base}^2 = {}_fL_{base}\cdot{}_fi_{base}^2 = {}_kL_{base}\cdot{}_ki_{base}^2$$

Then

$$*1 = \frac{\frac{3}{2}L_{afd}}{{}_fL_{base}\cdot\frac{{}_fi_{base}}{{}_si_{base}}} = \frac{\frac{3}{2}L_{afd}}{\frac{3}{2}\,{}_sL_{base}\cdot\frac{{}_si_{base}}{{}_fi_{base}}} = \frac{L_{afd}\cdot{}_fi_{base}}{{}_sL_{base}\cdot{}_si_{base}} = \frac{L_{ad}\cdot{}_si_{base}}{{}_sL_{base}\cdot{}_si_{base}} = \frac{L_{ad}}{{}_sL_{base}} = \bar L_{ad}$$

$$*2 = \frac{L_{fkd}}{{}_fL_{base}\cdot\frac{{}_fi_{base}}{{}_ki_{base}}} = \frac{L_{fkd}}{{}_kL_{base}\cdot\frac{{}_ki_{base}}{{}_fi_{base}}} = *4 \equiv \bar L_{fkd}$$

$$*3 = \frac{\frac{3}{2}L_{akd}}{{}_kL_{base}\cdot\frac{{}_ki_{base}}{{}_si_{base}}} = \frac{\frac{3}{2}L_{akd}}{\frac{3}{2}\,{}_sL_{base}\cdot\frac{{}_si_{base}}{{}_ki_{base}}} = \frac{L_{akd}\cdot{}_ki_{base}}{{}_sL_{base}\cdot{}_si_{base}} = \frac{L_{ad}\cdot{}_si_{base}}{{}_sL_{base}\cdot{}_si_{base}} = \frac{L_{ad}}{{}_sL_{base}} = \bar L_{ad}$$

$$*5 = \frac{\frac{3}{2}L_{akq}}{{}_kL_{base}\cdot\frac{{}_ki_{base}}{{}_si_{base}}} = \frac{\frac{3}{2}L_{akq}}{\frac{3}{2}\,{}_sL_{base}\cdot\frac{{}_si_{base}}{{}_ki_{base}}} = \frac{L_{akq}\cdot{}_ki_{base}}{{}_sL_{base}\cdot{}_si_{base}} = \frac{L_{aq}\cdot{}_si_{base}}{{}_sL_{base}\cdot{}_si_{base}} = \frac{L_{aq}}{{}_sL_{base}} = \bar L_{aq}$$

Accordingly, the unitized equation of Equation 10.36 is

$$\begin{bmatrix}\bar\psi_{fd}(\bar t)\\ \bar\psi_{kd}(\bar t)\\ \bar\psi_{kq}(\bar t)\end{bmatrix} = -\begin{bmatrix}\bar L_{ad} & 0 & 0\\ \bar L_{ad} & 0 & 0\\ 0 & \bar L_{aq} & 0\end{bmatrix}\cdot\begin{bmatrix}\bar i_d(\bar t)\\ \bar i_q(\bar t)\\ \bar i_0(\bar t)\end{bmatrix} + \begin{bmatrix}\bar L_{ffd} & \bar L_{fkd} & 0\\ \bar L_{fkd} & \bar L_{kkd} & 0\\ 0 & 0 & \bar L_{kkq}\end{bmatrix}\cdot\begin{bmatrix}\bar i_{fd}(\bar t)\\ \bar i_{kd}(\bar t)\\ \bar i_{kq}(\bar t)\end{bmatrix} \quad \textbf{(10.46)}$$

In total, the unitized equations of the generator are Equations 10.43–10.46.

L_{afd}, L_{akd} in Equation 10.33 and $\frac{3}{2}L_{afd}, \frac{3}{2}L_{akd}$ in Equation 10.36 are unified into $\bar L_{ad}$. Also, L_{akq} in Equation 10.33 and $\frac{3}{2}L_{akq}$ in Equation 10.36 are unified into $\bar L_{aq}$. Furthermore, the stator inductance matrices in Equations 10.45 and 10.46 are the transposed matrices of each other, which represents the reciprocal mutual inductances between the stator and the rotor.

10.3.3 Introduction of d–q–0 domain equivalent circuits

Now let us introduce the equivalent circuits for Equations 10.43–10.46. Below we omit $(\bar t)$, and replace $d\theta/d\bar t$ by $s\bar\theta(s \to d/dt)$ to avoid symbolic complications.

Differentiating Equation 10.45 and substituting the result into the second term of Equation 10.43 so that, eliminating the flux variables,

$$\left.\begin{aligned}-(\bar{e}_d + s\bar{\theta}\cdot\bar{\psi}_q) &= \bar{L}_l\cdot s\bar{i}_d + \bar{r}\cdot\bar{i}_d + \bar{L}_{ad}\cdot s(\bar{i}_d - \bar{i}_{fd} - \bar{i}_{kd})\\ -(\bar{e}_q - s\bar{\theta}\cdot\bar{\psi}_d) &= \bar{L}_l\cdot s\bar{i}_q + \bar{r}\cdot\bar{i}_q + \bar{L}_{aq}\cdot s(\bar{i}_q - \bar{i}_{kq})\\ -\bar{e}_0 &= \bar{L}_0\cdot s\bar{i}_0 + \bar{r}\cdot\bar{i}_0\end{aligned}\right\} \quad \textbf{(10.47)}$$

Differentiating Equation 10.46 and substituting into the second term of Equation 10.44, then by clever modification

$$\left.\begin{aligned}\bar{E}_{fd} &= -\bar{L}_{ad}\cdot s(\bar{i}_d - \bar{i}_{fd} - \bar{i}_{kd}) + (\bar{L}_{fkd} - \bar{L}_{ad})\cdot s(\bar{i}_{fd} + \bar{i}_{kd})\\ &\quad + (\bar{L}_{ffd} - \bar{L}_{fkd})\cdot s\bar{i}_{fd} + \bar{r}_{fd}\cdot\bar{i}_{fd}\\ 0 &= -\bar{L}_{ad}\cdot s(\bar{i}_d - \bar{i}_{fd} - \bar{i}_{kd}) + (\bar{L}_{fkd} - \bar{L}_{ad})\cdot s(\bar{i}_{fd} + \bar{i}_{kd})\\ &\quad + (\bar{L}_{kkd} - \bar{L}_{fkd})\cdot s\bar{i}_{kd} + \bar{r}_{kd}\cdot\bar{i}_{kd}\\ 0 &= -\bar{L}_{aq}\cdot s(\bar{i}_q - \bar{i}_{kq}) + (\bar{L}_{kkq} - \bar{L}_{aq})\cdot s\bar{i}_{kq} + \bar{r}_{kq}\cdot\bar{i}_{kq}\end{aligned}\right\} \quad \textbf{(10.48)}$$

Equations 10.47 and 10.48 are the unitized equations of a generator in the d–q–0 domain, which were derived from the physical concept of Figure 10.1 and were introduced through a very accurate mathematical procedure. Furthermore, the quite accurate equivalent circuit of these derived equations can be drawn as shown in Figure 10.4, which corresponds one-to-one correspondent with the equations. The related flux linkage can be added to the figure in the form of $\bar{\psi} = \bar{L}\cdot\bar{i}$ if necessary.

Incidentally, we need to examine the inductive element $\bar{L}_{fkd} - \bar{L}_{ad}$ in Figure 10.4. The damper windings and the stator winding are very close across the narrow air gap so that most of the flux induced by field current $\bar{i}_{fd}$ reaches not only to the damper coil but also to the stator coils (see Figure 16.7c). In other words, the flux linkage induced by $\bar{i}_{fd}$ on the damper d-axis coil ($\bar{L}_{fkd}\times\bar{i}_{fd}$) and the flux linkage induced by $\bar{i}_{fd}$ on the stator winding ($\bar{L}_{afd}\times\bar{i}_{fd}$) can be considered to have similar magnitudes, which then means $\bar{L}_{fkd} \fallingdotseq \bar{L}_{afd}$. Next, $\bar{L}_{afd} = \bar{L}_{ad}$ is found by comparing Equations 10.45 and 10.33.

Then

$$\bar{L}_{fkd} \fallingdotseq \bar{L}_{ad} \qquad \text{or} \qquad \bar{L}_{fkd} - \bar{L}_{ad} \fallingdotseq 0 \qquad (10.49\text{a})$$

Accordingly, the inductance ($\bar{L}_{fkd} - \bar{L}_{ad}$) in the d-axis circuit can actually be neglected.

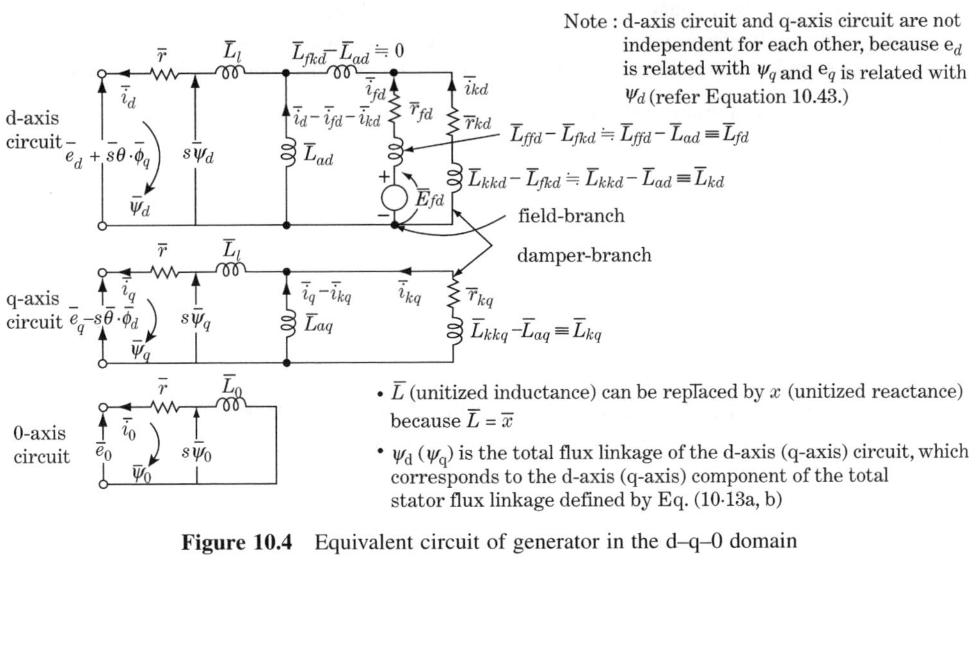

Figure 10.4 Equivalent circuit of generator in the d–q–0 domain

Here, we introduce new symbols for inductances $\bar{L}_{fd}, \bar{L}_{kd}, \bar{L}_{kq}$ as are defined by the equations below and shown in the equivalent circuits:

$$\left.\begin{aligned} \bar{L}_{fd} &\equiv \bar{L}_{ffd} - \bar{L}_{ad} \doteqdot \bar{L}_{ffd} - \bar{L}_{fkd} \\ \bar{L}_{kd} &\equiv \bar{L}_{kkd} - \bar{L}_{ad} \doteqdot \bar{L}_{kkd} - \bar{L}_{fkd} \\ \bar{L}_{kq} &\equiv \bar{L}_{kkq} - \bar{L}_{aq} \end{aligned}\right\} \tag{10.49b}$$

The relations between inductances $\bar{L}$ and reactances $\bar{x}$ in unitized equations are the same because

$$\bar{L} = \frac{L}{L_{\text{base}}} = \frac{2\pi f_{\text{base}} \cdot L}{2\pi f_{\text{base}} \cdot L_{\text{base}}} = \frac{x}{x_{\text{base}}} = \bar{x} \tag{10.50}$$

Therefore, all the inductances of symbol $\bar{L}$ can be replaced by reactances of symbol $\bar{x}$.
Next, we have the relation below in regard to the resistances.

For the comparison of $\bar{r}_{fd}$ and $\bar{r}_{kd}$, $r_{fd} \ll r_{kd}$ is obvious by nature, so the per unit magnitudes of the common rotor impedance base are also $\bar{r}_{fd} \ll \bar{r}_{kd}$.

For the comparison of $\bar{r}$ and $\bar{r}_{fd}$, the rated capacity of the field winding is of the order of 10% in comparison with the rated capacity of the stator windings in general. However, the per unit base capacity of the rotor is selected to coincide with the rated capacity of the stator windings by Equation 10.40. This means that the selected current bases of the rotor ${}_fI_{\text{base}}, {}_kI_{\text{base}}$ are approximately 10 times larger than the actual rated current of the field winding. Accordingly, the impedance bases ${}_fZ_{\text{base}}, {}_kZ_{\text{base}}$ are very small in comparison with the stator impedance base Z_{base}. Then we can justify $\bar{r} \ll \bar{r}_{fd}$ as the per unit value.

In total,

$$\bar{r} \ll \bar{r}_{fd} \ll \bar{r}_{kd}, \quad \bar{r} \ll \bar{r}_{kq} \tag{10.51}$$

Summarizing the above sections:

1. We introduced the fundamental equations of the generator based on the physical model shown in Figure 10.1,which included ωt-dependent inductances.
2. The d–q–0 transformation was introduced and the relations of the quantities in the d–q–0, 0–1–2 and a–b–c domains were examined. Then, the generator equations were transformed into the d–q–0 domain.
3. The unitized generator equations in the d–q–0 domain and the equivalent circuit in Figure 10.4 were introduced by clever unitization, in that all the inductances are written as ωt-independent constants L, and furthermore with the reciprocal mutual inductances between the stator and rotor.

The relation between the above-derived d–q–0 circuit and the 0–1–2 domain equivalent circuit will be examined later.

Further to the summary above, some comments on the linearity of a generator were introduced. Typical non-linear characteristics of a generator are hysteresis of the core, skin effects of the windings, and so on. Such non-linearities can be neglected by good design, and any generator can usually be treated as a linear machine for normal operating voltage range.

Special non-linear saturation phenomena which may be caused by unbalanced currents, higher harmonic currents and/or overvoltages will be discussed in Chapters 16, 20 and 22.

10.4 Generator operating characteristics and it's Vector Diagrams on d-and q-axes plain

In this section we examine a generator's dynamic operating characteristics and derive the vector diagrams on d-and q-axes plain, which enables us visualization of the dynamic behavior.

We start with the following equation of three-phase-balanced quantities as the initial condition of generator dynamic behavior (note that all the quantities here are unitized):

$$\left.\begin{array}{ll} \bar{e}_a(\bar{t}) = \bar{E}_{a1}\cos(\bar{t}+\bar{\alpha}_1) = \bar{e}_1(\bar{t}) & \bar{i}_a(\bar{t}) = \bar{I}_{a1}\cos(\bar{t}+\bar{\beta}_1) = \bar{i}_1(t) \\ \bar{e}_b(\bar{t}) = \bar{E}_{a1}\cos\left(\bar{t}+\bar{\alpha}_1-\dfrac{2\pi}{3}\right) & \bar{i}_b(\bar{t}) = \bar{I}_{a1}\cos\left(\bar{t}+\bar{\beta}_1-\dfrac{2\pi}{3}\right) \\ \bar{e}_c(\bar{t}) = \bar{E}_{a1}\cos\left(\bar{t}+\bar{\alpha}_1+\dfrac{2\pi}{3}\right) & \bar{i}_c(\bar{t}) = \bar{I}_{a1}\cos\left(\bar{t}+\bar{\beta}_1+\dfrac{2\pi}{3}\right) \end{array}\right\} \tag{10.52}$$

$\bar{E}_{a1}$, $\bar{I}_{a1}$ are the sinusoidal real-number crest values. The transformed equations in the d–q–0 domain are written below, referring to Equation 10.17:

$$\left.\begin{array}{ll} \bar{e}_d(\bar{t}) = \bar{E}_{a1}\cos\bar{\alpha}_1 & \bar{i}_d(\bar{t}) = \bar{I}_{a1}\cos\bar{\beta}_1 \\ \bar{e}_q(\bar{t}) = \bar{E}_{a1}\sin\bar{\alpha}_1 & \bar{i}_q(\bar{t}) = \bar{I}_{a1}\sin\bar{\beta}_1 \\ \bar{e}_0(\bar{t}) = 0 & \bar{i}_0(\bar{t}) = 0 \end{array}\right\} \tag{10.53}$$

The quantities under the three phase balanced initial condition on the d-axis and q-axis are d.c. values (time independent). The situation is the same for flux quantities $\bar{\psi}_d(\bar{t})$, $\bar{\psi}_q(\bar{t})$. Equation 10.53 can be modified into the equations

$$\begin{array}{l} \bar{e}_d(\bar{t}) + je_q(\bar{t}) = \overline{E}_{a1}\cdot e^{j\bar{\alpha}_1} \\ \bar{i}_d(\bar{t}) + ji_q(\bar{t}) = \bar{I}_{a1}\cdot e^{j\bar{\beta}_1} \end{array} \tag{10.54}$$

or

$$\begin{array}{ll} \bar{e}_1(\bar{t}) = \{\bar{e}_d(\bar{t}) + j\bar{e}_q(\bar{t})\}e^{j\bar{t}} = \overline{E}_{a1}\cdot e^{j(\bar{t}+\bar{\alpha}_1)}: & \text{positive-sequence voltage by complex number} \\ \bar{i}_1(\bar{t}) = \{\bar{i}_d(\bar{t}) + j\bar{i}_q(\bar{t})\}e^{j\bar{t}} = \bar{I}_{a1}\cdot e^{j(\bar{t}+\bar{\beta}_1)}: & \text{positive-sequence current by complex number} \end{array} \tag{10.55}$$

The following conditions are justified under the initial condition.

$$\left.\begin{array}{ll} s\bar{\theta} = \dfrac{d\theta}{d\bar{t}} = \dfrac{\omega}{\omega_{\text{base}}} = 1.0 & ① \\ \dfrac{d}{d\bar{t}}\bar{\psi}_d(\bar{t}) = \dfrac{d}{d\bar{t}}\bar{\psi}_q(\bar{t}) = \dfrac{d}{d\bar{t}}\bar{\psi}_0(\bar{t}) = 0 & ② \\ \bar{i}_{kd}(\bar{t}) = \bar{i}_{kq}(\bar{t}) = 0 & ③ \end{array}\right\} \tag{10.56}$$

Referring to Equation 10.54, $\bar{\psi}_d(\bar{t})$, $\bar{\psi}_q(\bar{t})$, $\bar{\psi}_0(\bar{t})$ are the d.c. components under the three-phase-balanced condition so that their derivatives are obviously zero (Equation ②).

Referring to Equation 10.44, ψ_{kd} are also d.c. quantities so that $s\bar{\psi}_{kd} = -\bar{r}_{kd}\cdot\bar{i}_{kd} = 0$. In other words, damper currents $\bar{i}_{kd}$ and $\bar{i}_{kq}$ are zero under the three-phase-balanced condition (Equation ③).

Equation ③ may be explained in another way referring to the d-axis and q-axis circuits of Figure 10.4. Recalling that $L\cdot di/dt = 0$ for d.c. current, all the inductance elements in the equivalent circuit are actually short-circuited for d.c. currents $\bar{i}_d, \bar{i}_q$. Accordingly, the d.c. current distribution of the three branched passes of the d-axis circuit is dominated by only the inverse ratio of resistances. Therefore, d.c. current cannot flow in the damper branch at extremely large $\bar{r}_{kd}$ (see Equation 10.51), so $\bar{i}_{kd} = 0$ under steady-state operation.

Now, substituting Equation 10.56 into Equations 10.43 and 10.45, and replacing $\bar{L}$ by unitized reactances $\bar{x}$ (because $\bar{L} = \bar{x}$), then

$$\left.\begin{array}{l} \bar{e}_d = -\bar{\psi}_q - \bar{r}\cdot\bar{i}_d \\ \bar{e}_q = +\bar{\psi}_d - \bar{r}\cdot\bar{i}_q \\ \bar{e}_0 = -\bar{r}\cdot\bar{i}_0 \end{array}\right\} \tag{10.57}$$

$$\left.\begin{aligned}\bar{\psi}_d &= -\bar{x}_d \cdot \bar{i}_d + \bar{x}_{ad} \cdot \bar{i}_{fd} = -\bar{x}_d \cdot \bar{i}_d + \overline{E}_f \\ \bar{\psi}_q &= -\bar{x}_q \cdot \bar{i}_q \\ \bar{\psi}_0 &= -\bar{x}_0 \cdot i_0 \quad \text{where} \quad \overline{E}_f \equiv \bar{x}_{ad} \cdot \bar{i}_{fd}\end{aligned}\right\} \tag{10.58}$$

Substituting Equation 10.58 into Equation 10.57,

$$\left.\begin{aligned}\bar{e}_d &= \bar{x}_q \cdot \bar{i}_q - \bar{r} \cdot \bar{i}_d \\ \bar{e}_q &= \overline{E}_f - \bar{x}_d \cdot \bar{i}_d - \bar{r} \cdot \bar{i}_q \\ &\text{where} \quad \overline{E}_f \equiv \bar{x}_{ad} \cdot \bar{i}_{fd}\end{aligned}\right\} \tag{10.59}$$

Rearranging the above equations in the form of $\bar{e}_d + j\bar{e}_q$,

$$\begin{aligned}\bar{e}_d + j\bar{e}_q &= j(\bar{\psi}_d + j\bar{\psi}_q) - \bar{r}(\bar{i}_d + j\bar{i}_q) \\ &= (\bar{x}_q \cdot \bar{i}_q - \bar{r} \cdot \bar{i}_d) + j(-\bar{x}_d \cdot \bar{i}_d + \bar{x}_{ad} \cdot \bar{i}_{fd} - \bar{r} \cdot \bar{i}_q) \\ &= -(\bar{r} + j\bar{x}_q)(\bar{i}_d + j\bar{i}_q) - j(\bar{x}_d - \bar{x}_q) \cdot \bar{i}_d + j\bar{x}_{ad} \cdot \bar{i}_{fd}\end{aligned}$$

then

$$\left.\begin{aligned}& j(\bar{\psi}_d + j\bar{\psi}_q) = (\bar{e}_d + j\bar{e}_q) + \bar{r}(\bar{i}_d + j\bar{i}_q) \\ & (\bar{e}_d + j\bar{e}_q) + (\bar{r} + j\bar{x}_q)(\bar{i}_d + j\bar{i}_q) + j(\bar{x}_d - \bar{x}_q) \cdot \bar{i}_d = j\bar{x}_{ad} \cdot \bar{i}_{fd} \equiv j\bar{E}_f \\ \therefore \quad & \bar{E}_{a1} \cdot e^{j\bar{\alpha}_1} + (\bar{r} + j\bar{x}_q) \cdot \bar{I}_{a1} \cdot e^{j\bar{\beta}_1} + j(\bar{x}_d - \bar{x}_q) \cdot \bar{i}_d = j\bar{x}_{ad} \cdot \bar{i}_{fd} \equiv j\bar{E}_f\end{aligned}\right\} \tag{10.60}$$

In Equation 10.60, we define $j\bar{E}_f$, which is the voltage proportional to the amount of excitation current $\bar{i}_{fd}$.

Equation 10.60 is the equation of voltage and current in the d–q–0 domain under three-phase-balanced conditions. The equations can also be written as the vector diagram of Figure 10.5, in which the stator voltage and current quantities are drawn as time-independent vector (complex-number) quantities in d–q–0 coordinates.

Incidentally, as shown in Table 10.1, the magnitudes of $\bar{x}_d$ and $\bar{x}_q$ for the cylindrical-rotor-type machine (two poles for thermal, four poles for nuclear) are equal:

$$\bar{x}_d \doteqdot \bar{x}_q \tag{10.61}$$

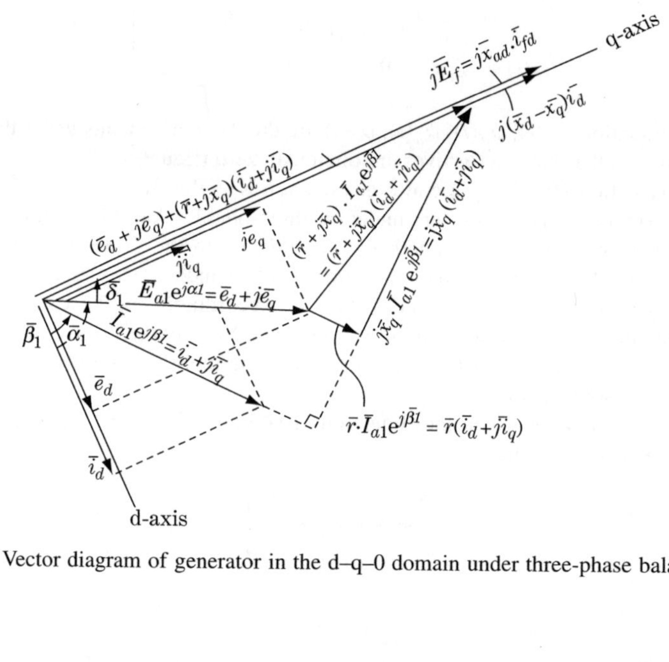

Figure 10.5 Vector diagram of generator in the d–q–0 domain under three-phase balanced condition

Thus the term $j(\bar{x}_d - \bar{x}_q)$ can be neglected, and Equation 10.60 is further simplified as

$$\bar{E}_{a1} \cdot e^{j\bar{\alpha}_1} + (\bar{r} + j\bar{x}_q) \cdot \bar{I}_{a1} \cdot e^{j\bar{\beta}_1} = j\bar{x}_{ad} \cdot \bar{i}_{fd} \equiv j\bar{E}_f \quad ①$$

Multiplying by $e^{j\bar{t}}$, (10.62)

$$\bar{E}_{a1} \cdot e^{j(\bar{t}+\bar{\alpha}_1)} + (\bar{r} + j\bar{x}_q) \cdot \bar{I}_{a1} \cdot e^{j(\bar{t}+\bar{\beta}_1)} = j\bar{x}_{ad} \cdot \bar{i}_{fd} \cdot e^{j\bar{t}} = j\bar{E}_f \cdot e^{j\bar{t}} \quad ②$$

It may be said that Equation 10.62① explains the generator behaviour which an observer riding on the rotor can see on the generator, while ② explains the same behaviour which an observer, standing on the floor looking into the generator, can see. Equation 10.62 can be written as follows. For the generator terminal voltage and current equation,

$$\dot{\bar{e}}_{a1}(\bar{t}) + (\bar{r} + j\bar{x}_q) \cdot \dot{\bar{i}}_{a1}(\bar{t}) = j\bar{E}_f \cdot e^{j\bar{t}} = \dot{\bar{e}}_f(\bar{t})$$

where (10.63)

positive-sequence voltage : $\dot{\bar{e}}_{a1}(\bar{t}) = \bar{E}_{a1} \cdot e^{j(\bar{t}+\bar{\alpha}_1)} = (\bar{e}_d + j\bar{e}_q) \cdot e^{j\bar{t}}$

positive-sequence current : $\dot{\bar{i}}_{a1}(\bar{t}) = \bar{I}_{a1} \cdot e^{j(\bar{t}+\bar{\beta}_1)} = (\bar{i}_d + j\bar{i}_q) \cdot e^{j\bar{t}}$

generator back-source voltage : $\dot{\bar{e}}_f(\bar{t}) \equiv j\bar{E}_f \cdot e^{j\bar{t}} = j\bar{x}_{ad} \cdot \bar{i}_{fd} \cdot e^{j\bar{t}}$

From this equation, we can derive Figure 10.6, which is the positive-sequence equivalent circuit for a synchronous machine under steady-state conditions, and with constant speed and constant field excitation approximated by $\bar{x}_d \doteq \bar{x}_q$.

The positive-sequence equivalent circuit of the simple generator model shown in Figure 2.11 is in the same form as Figure 10.6, although the figure was drawn intuitively without theoretical explanation.

In case of a salient-pole machine (for hydro-generators of multi-poles), $\bar{x}_d$ and $\bar{x}_q$ differ to some extent as shown in Table 10.1. If the term $j(\bar{x}_d - \bar{x}_q)\bar{i}_d$ cannot be neglected, there is no simple equivalent circuit, so Equation 10.61 should be used for detailed analysis.

Table 10.1 Generator's typical reactances, time constants (non-saturable values)

	Ratings				Reactances [%]							Time constants [s]			
	Capacity [MVA]	Frequency [Hz]	Pole number		x_d	x_q	x'_d	x''_d	x''_q	x_2	x_0	T'_{d0}	T'_d	T''_d	T_a
Turbine generator	1300	50	4	W	185	185	38	29	29	29	19	6.9	1.5	0.03	0.25
	800	60	2	W	179	177	34	26	25	25	12	6.4	1.2	0.02	0.40
	585	50	4	W	180	175	36	27	27	27	13	8.0	2.3	0.03	0.22
	556	60	2	HD	174	172	29	25	24	24	10	5.2	0.9	0.02	0.55
	270	60	2	HI	183	183	31	24	24	24	13	6.0	0.9	0.03	0.40
	53	60	2	A	205	194	22	17	17	17	9	6.3	0.7	0.03	0.25
Hydro-generator	280	60	24	A	110	78	34	22	24	23	17	7.6	2.3	0.04	0.31
	26	60	72	A	112	76	42	33	41	37	15	3.3	1.2	0.03	0.16
	21	50	12	A	123	71	33	23	21	22	14	4.9	1.3	0.06	0.17
Generator motor for pumped storage	390	50	14	A	135	84	27	16	17	17	14	11.0	2.3	0.06	0.35
DG	6	50	8	A	190	102	35	22	19	20	13	4.9	0.9	0.05	0.08

DG: diegel generator. W: hydro-cooling. HD: hydrogen gas direct cooling. HI: hydrogen gas indirect cooling. A: air cooling.

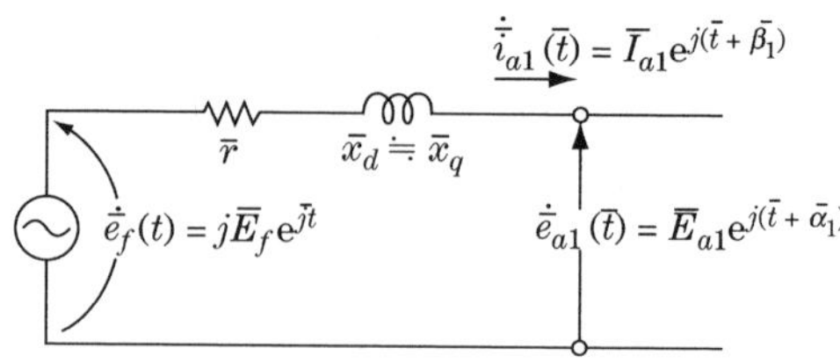

Figure 10.6 Positive-sequence equivalent circuit under approximation $\bar{x}_d \doteqdot \bar{x}_q$

10.5 Transient Phenomena and the Generator's Transient Reactances

Now we examine the transient behaviour of a generator caused by sudden disturbances in the power system (phase faults, feeder switching, sudden load changes, rapid dynamic instability, etc.).

10.5.1 Initial condition just before sudden change

A generator's equivalent circuit in the d–q–0 domain is given by Figure 10.4, in which the d-axis circuit is an active circuit including a d.c. source, while the q-axis and 0-axis circuits are passive circuits composed only of reactances and resistances. However, recall that these circuits are not electrically independent. In other words, they are mutually coupled through the flux linkage terms $\bar{\psi}_d, \bar{\psi}_q$. The reason can be explained by Equations 10.43 and 10.57, where $\bar{e}_d$ is related to the term $\bar{\psi}_q$, and $\bar{e}_q$ is related to $\bar{\psi}_d$.

Next, our generator is connected to the outer part of the power system network (let us call it the 'outer system'), which usually includes other generators. The outer system can also be transformed into the 'outer circuit in the d–q–0 domain', and the transformed d-axis circuit includes d.c. sources if other generators exist. The equivalent circuits of the total power system can be obtained by connecting each d-, q- and 0-axis equivalent circuits of our generator and the 'outer circuit' as shown in Figure 10.8. If we assume three-phase-balanced steady-state operation of this total system, all the flux/voltage/current quantities flowing through our generator circuit as well as the outer circuit are of d.c. components in the d–q–0 domain.

Now, referring to Equations 10.57 and 10.58, the three-phase-balanced initial conditions before system disturbance are

$$\left.\begin{aligned} \bar{e}_d(0) &= -\bar{\psi}_q(0) - \bar{r}\cdot\bar{i}_d(0) \\ \bar{e}_q(0) &= \bar{\psi}_d(0) - \bar{r}\cdot\bar{i}_q(0) \\ \bar{e}_0(0) &= 0, \;\; \bar{i}_0(0) = 0 \end{aligned}\right\} \tag{10.64}$$

$$\left.\begin{aligned} \bar{\psi}_d(0) &= -\bar{x}_d\cdot\bar{i}_d(0) + \bar{E}_f \\ \bar{\psi}_q(0) &= -\bar{x}_q\cdot\bar{i}_q(0) \\ \bar{\psi}_0(0) &= 0, \\ \text{where}\quad \bar{E}_f &= \bar{x}_{ad}\cdot\bar{i}_{fd} \end{aligned}\right\} \tag{10.65}$$

Referring to Equation 10.60,

$$\left.\begin{array}{l}\{\bar{e}_d(0)+j\bar{e}_q(0)\}+(\bar{r}+j\bar{x}_q)\cdot\{\bar{i}_d(0)+j\bar{i}_q(0)\}+j(\bar{x}_d-\bar{x}_q)\cdot\bar{i}_d(0)=j\bar{E}_f\\ \{\bar{e}_d(0)+j\bar{e}_q(0)\}+\bar{r}\{\bar{i}_d(0)+j\bar{i}_q(0)\}=j\{\bar{\psi}_d(0)+j\bar{\psi}_q(0)\}\\ \bar{e}_0(0)=0,\quad \bar{i}_0(0)=0,\quad \bar{\psi}_0(0)=0\end{array}\right\} \tag{10.66}$$

This is the initial condition at $t = 0$, the time just before sudden system disturbance.

10.5.2 Assorted d-axis and q-axis reactances for transient phenomena

Now we examine how our generator circuit in the d–q–0 domain behaves over time when the 'outer circuit' is suddenly changed by the system disturbance. The initial conditions at $t = 0-$ are given by Equation 10.64, 10.65, 10.66.

Just after system disturbance at $t = 0+$, the circuit quantities in our generator circuits would become the situation described by

$$\left.\begin{array}{lll}\bar{\psi}_d(\bar{t})=\bar{\psi}_d(0)+\Delta\bar{\psi}_d & \bar{e}_d(\bar{t})=\bar{e}_d(0)+\Delta\bar{e}_d & \bar{i}_d(\bar{t})=\bar{i}_d(0)+\Delta\bar{i}_d\\ \bar{\psi}_d(\bar{t})=\bar{\psi}_q(0)+\Delta\bar{\psi}_q & \bar{e}_q(\bar{t})=\bar{e}_q(0)+\Delta\bar{e}_q & \bar{i}_q(\bar{t})=\bar{i}_q(0)+\Delta\bar{i}_q\\ \bar{\psi}_0(\bar{t})=\Delta\bar{\psi}_0 & \bar{e}_0(\bar{t})=\Delta\bar{e}_0 & \bar{i}_0(\bar{t})=\Delta\bar{i}_0\end{array}\right\} \tag{10.67}$$

10.5.2.1 *Time interval t = 0–3 cycles (0–45 or 60 ms)*

It should be remembered that the voltage across any L and r series-connected circuit is $v = Ldi/dt + ri$, and ri dominates for constant d.c. current because $di/dt = 0$, while Ldi/dt dominates for sudden large changes of i because di/dt is large.

Now, we observe the transient phenomena caused at $t = 0$ on the d-axis equivalent circuit of Figure 10.4.

In the first interval just after the disturbance ($t = 0$–3 cycles), all the branches of the d-axis circuit are dominated by each inductance instead of resistance. Therefore it is obvious that the terminal voltages of the d-axis are dominated by the composite total inductance looking into the d-axis circuit at the terminal. In other words, the inductance $\bar{L}''_d$ in this time interval becomes the composite value of L_l and the three branched inductances $\bar{L}_{ad}$, $\bar{L}_{fd}$, $\bar{L}_{kd}$. Of course, the q-axis circuit can be derived similarly. Therefore the generator reactances in this period are

d-axis sub-transient reactance

$$\bar{x}''_d=\bar{L}''_d=-\frac{\Delta\bar{\psi}_d}{\Delta\bar{i}_d}=\bar{L}_l+\frac{1}{\dfrac{1}{\bar{L}_{ad}}+\dfrac{1}{\bar{L}_{fd}}+\dfrac{1}{\bar{L}_{kd}}}$$

q-axis sub-transient reactance

$$\bar{x}''_q=\bar{L}''_q=-\frac{\Delta\bar{\psi}_q}{\Delta\bar{i}_q}=\bar{L}_l+\frac{1}{\dfrac{1}{\bar{L}_{aq}}+\dfrac{1}{\bar{L}_{kq}}} \tag{10.68}$$

The flux linkage of the stator d-and q-axis coils in this period are

$$\left.\begin{array}{l}\bar{\psi}_d(\bar{t})=\bar{\psi}_d(0)+\Delta\bar{\psi}_d=\bar{\psi}_d(0)-\bar{x}''_d\cdot\Delta\bar{i}_d=\bar{\psi}_d(0)+\bar{x}''_d\cdot\bar{i}_d(0)-\bar{x}''_d\cdot\bar{i}_d(\bar{t})\\ \bar{\psi}_q(\bar{t})=\bar{\psi}_q(0)+\Delta\bar{\psi}_q=\bar{\psi}_q(0)-\bar{x}''_q\cdot\Delta\bar{i}_q=\bar{\psi}_q(0)+\bar{x}''_q\cdot\bar{i}_q(0)-\bar{x}''_q\cdot\bar{i}_q(\bar{t})\end{array}\right\} \tag{10.69}$$

The transient terms $\Delta\bar{i}_d$, $\Delta\bar{i}_q$ as well as $\Delta\bar{i}_{fd}$, $\Delta\bar{i}_{kd}$, $\Delta\bar{i}_{kq}$ appear in this period.

10.5.2.2 Time interval t = 3 to approximately 60 cycles (45 or 60 ms to 1 sec)

As transient phenomena have already been attenuated to some extent, domination of the damper coil branch by $\bar{r}_{kd}$ instead of $s\bar{L}_{kd}$ will begin in this period. As $\bar{r}_{kd}$ is much larger than $\bar{r}_{fd}$ in the d-axis circuit, the magnitudes of the three branched pass impedances become the relative orders of magnitude of $s\bar{L}_{ad}, s\bar{L}_{fd}+\bar{r}_{fd} \ll \bar{r}_{kd} \fallingdotseq (s\bar{L}_{kd}+\bar{r}_{kd})$ (where $s = d/d\bar{t}$) so that the damper current is extinguished at all ($\bar{i}_{kd}=0$). In other words, the damper's effect has already disappeared in this time. Therefore the generator reactances in this period are obtained as

$$\left.\begin{aligned}
&\text{d-axis transient reactance} \quad \bar{x}'_d = \bar{L}'_d = -\frac{\Delta\bar{\psi}_d}{\Delta\bar{i}_d} = \bar{L}_l + \cfrac{1}{\cfrac{1}{\bar{L}_{ad}}+\cfrac{1}{\bar{L}_{fd}}}\\
&\text{q-axis transient reactance} \quad \bar{x}'_q = \bar{L}'_q = -\frac{\Delta\bar{\psi}_q}{\Delta\bar{i}_q} = \bar{L}_l + \bar{L}_{aq} = \bar{x}_q
\end{aligned}\right\} \tag{10.70}$$

The inductance of the q-axis circuit can be derived analogously.

Here, the q-axis reactance $\bar{x}'_q = \bar{x}_q$ can be justified because the q-axis circuit will not be changed in the third period described below. The flux linkages of the stator d- and q- axis coils in this period are

$$\left.\begin{aligned}
\bar{\psi}_d(\bar{t}) &= \bar{\psi}_d(0)+\Delta\bar{\psi}_d = \bar{\psi}_d(0) - \bar{x}'_d\cdot\Delta\bar{i}_d = \bar{\psi}_d(0)+\bar{x}'_d\cdot\bar{i}_d(0) - \bar{x}'_d\cdot\bar{i}_d(\bar{t})\\
\bar{\psi}_q(\bar{t}) &= \bar{\psi}_q(0)+\Delta\bar{\psi}_q = \bar{\psi}_q(0) - \bar{x}_q\cdot\Delta\bar{i}_q = \bar{\psi}_q(0)+\bar{x}_q\cdot\bar{i}_q(0) - \bar{x}_q\cdot\bar{i}_q(\bar{t})
\end{aligned}\right\} \tag{10.71}$$

The damper flux terms $\Delta\bar{i}_{kd}$, $\Delta\bar{i}_{kq}$ disappear in this period.

10.5.2.3 Time interval t = 1 sec to steady-state condition

In this period, the field coil branch in the d-axis circuit is already dominated by $\bar{r}_{fd}$ instead of $s\bar{L}_{fd}$. As a result, the impedance magnitudes of the two branches reach the relative condition of $s\bar{L}_{ad} \ll \bar{r}_{fd} \fallingdotseq (s\bar{L}_{fd}+\bar{r}_{fd})$, so that the d-axis circuit terminal reactances are obtained from the following equation (the reactance $\bar{x}_q$ in this period is obviously the same as the already derived $\bar{x}'_q$):

$$\left.\begin{aligned}
&\text{d-axis steady-state reactance} \quad \bar{x}_d = \bar{L}_d = -\frac{\Delta\bar{\psi}_d}{\Delta\bar{i}_d} = \bar{L}_l + \bar{L}_{ad}\\
&\text{q-axis steady-state reactance} \quad \bar{x}_q = \bar{L}_q = -\frac{\Delta\bar{\psi}_q}{\Delta\bar{i}_q} = \bar{L}_l + \bar{L}_{aq}
\end{aligned}\right\} \tag{10.72}$$

The flux linkage of the stator d- and q-axis coils in this period are

$$\left.\begin{aligned}
\bar{\psi}_d(\bar{t}) &= \bar{\psi}_d(0)+\Delta\bar{\psi}_d = \bar{\psi}_d(0) - \bar{x}_d\cdot\Delta\bar{i}_d = \bar{\psi}_d(0)+\bar{x}_d\cdot\bar{i}_d(0) - \bar{x}_d\cdot\bar{i}_d(\bar{t})\\
\bar{\psi}_q(\bar{t}) &= \bar{\psi}_q(0)+\Delta\bar{\psi}_q = \bar{\psi}_q(0) - \bar{x}_q\cdot\Delta\bar{i}_q = \bar{\psi}_q(0)+\bar{x}_q\cdot\bar{i}_q(0) - \bar{x}_q\cdot\bar{i}_q(\bar{t})
\end{aligned}\right\} \tag{10.73}$$

The transient term $\Delta\bar{i}_{fd}$ disappears in this period.

10.6 Symmetrical Equivalent Circuits of Generators

The generator's equations and the equivalent circuits in the d–q–0 domain have been derived. Now we need to find the equations and the equivalent circuits in the 0–1–2 domain.

10.6.1 Positive-sequence circuit

The positive-sequence circuit of a generator under transient conditions can be checked by studying the three-phase short-circuit fault at the generator terminal.

The following conditions are given as the generator's initial conditions just before a sudden change at $\bar{t} = 0+$:

$$\left.\begin{aligned} s\bar{\theta} &= 1.0 \text{ (the rotor is rotating at constant speed)} \quad ① \\ s\bar{\psi}_d(\bar{t}) &= s\bar{\psi}_q(\bar{t}) = 0 \quad ② \end{aligned}\right\} \tag{10.74}$$

Equation 10.74② does not mean that $\bar{\psi}_d(\bar{t}), \bar{\psi}_q(\bar{t})$ are unchangeable constants after the disturbance at $\bar{t} = 0+$. It means that only the fundamental frequency components of the fault current can be calculated after $\bar{t} = 0$, and the d.c. offset does not appear in the solution.

The flux $\bar{\psi}_d(\bar{t})$ at $\bar{t} = 0+$ cannot be changed suddenly from the magnitude of the d.c. quantity $\bar{\psi}_d(0)$, because any flux can be changed continuously over time. Accordingly, the voltage quantities $s\bar{\psi}_d(\bar{t})$, $s\bar{\psi}_q(\bar{t})$ in Equation 10.43 must be terms corresponding to the transient components. Therefore, if only the fundamental currents are to be calculated, this can be accomplished by putting $s\bar{\psi}_d(\bar{t}) = s\bar{\psi}_q(\bar{t}) = 0$. Applying this initial condition in the calculation does not mean assuming that $\bar{\psi}_d(\bar{t}), \bar{\psi}_q(\bar{t})$ are constants, which is why the fundamental frequency components can be calculated accurately.

Now, Equation 10.43 is simplified under the conditions of Equations 10.74 and 10.75:

$$\left.\begin{aligned} \bar{e}_d(\bar{t}) &= -\bar{\psi}_q(\bar{t}) - \bar{r} \cdot \bar{i}_d(\bar{t}) \\ \bar{e}_q(\bar{t}) &= \bar{\psi}_d(\bar{t}) - \bar{r} \cdot \bar{i}_q(\bar{t}) \end{aligned}\right\} \tag{10.75}$$

We start our examination of the three-phase short-circuit fault at the generator terminal at $\bar{t} = 0$ thus:

$$\text{for } \bar{t} \geqq 0 \qquad \bar{e}_a(\bar{t}) = \bar{e}_b(\bar{t}) = \bar{e}_c(\bar{t}) = 0 \tag{10.76}$$

Then

$$\text{for } \bar{t} \geqq 0 \qquad \bar{e}_d(\bar{t}) = \bar{e}_q(\bar{t}) = \bar{e}_0(\bar{t}) = 0 \tag{10.77}$$

From Equations 10.75 and 10.77, for $\bar{t} \geqq 0$

$$\left.\begin{aligned} 0 = \bar{e}_d(\bar{t}) &= -\bar{\psi}_q(\bar{t}) - \bar{r} \cdot \bar{i}_d(\bar{t}) \\ 0 = \bar{e}_q(\bar{t}) &= \bar{\psi}_d(\bar{t}) - \bar{r} \cdot \bar{i}_q(\bar{t}) \end{aligned}\right\} \quad \text{or} \quad \left.\begin{aligned} \bar{\psi}_q(\bar{t}) &= -\bar{r} \cdot \bar{i}_d(\bar{t}) \\ \bar{\psi}_d(\bar{t}) &= \bar{r} \cdot \bar{i}_q(\bar{t}) \end{aligned}\right\} \tag{10.78}$$

10.6.1.1 Sub-transient period: t = 0–3 cycles (0–45 or 60 ms)

From Equations 10.78 and 10.69, eliminating $\bar{\psi}_d(\bar{t}), \bar{\psi}_q(\bar{t})$,

$$\left.\begin{aligned} \bar{\psi}_d(0) + \bar{x}_d'' \cdot \bar{i}_d(0) &= \bar{r} \cdot \bar{i}_q(\bar{t}) + \bar{x}_d'' \cdot \bar{i}_d(\bar{t}) \quad ① \\ \bar{\psi}_q(0) + \bar{x}_q'' \cdot \bar{i}_q(0) &= -\bar{r} \cdot \bar{i}_d(\bar{t}) + \bar{x}_q'' \cdot \bar{i}_q(\bar{t}) \quad ② \end{aligned}\right\} \tag{10.79}$$

Putting the above equations in the form of $j\{① + j②\}$ and modifying,

$$\begin{aligned} &(\bar{r} + j\bar{x}_q'')\{\bar{i}_d(\bar{t}) + j\bar{i}_q(\bar{t})\} + j(\bar{x}_d'' - \bar{x}_q'') \cdot \bar{i}_d(\bar{t}) \\ &= j\{\bar{\psi}_d(0) + j\bar{\psi}_q(0)\} + j\bar{x}_q''\{\bar{i}_d(0) + j\bar{i}_q(0)\} + j(\bar{x}_d'' - \bar{x}_q'') \cdot \bar{i}_d(0) \end{aligned} \tag{10.80}$$

From this equation and the second equation of Equation 10.66,

$$\begin{aligned} &(\bar{r} + j\bar{x}_q'')\{\bar{i}_d(\bar{t}) + j\bar{i}_q(\bar{t})\} + j(\bar{x}_d'' - \bar{x}_q'') \cdot \bar{i}_d(\bar{t}) \\ &= \{\bar{e}_d(0) + j\bar{e}_q(0)\} + (\bar{r} + j\bar{x}_q'')\{\bar{i}_d(0) + j\bar{i}_q(0)\} + j(\bar{x}_d'' - \bar{x}_q'') \cdot \bar{i}_d(0) \equiv \dot{\bar{E}}'' \end{aligned} \tag{10.81}$$

Now, let us simplify the equation by imagining a cylindrical type of machine:

$$\bar{x}''_d \doteqdot \bar{x}''_q \tag{10.82}$$

Then

$$\left.\begin{aligned}&(\bar{r}+j\bar{x}''_d)\{\bar{i}_d(\bar{t})+j\bar{i}_q(\bar{t})\}=\dot{\bar{E}}''\\&\text{or}\\&(\bar{r}+j\bar{x}''_d)\{\bar{i}_d(\bar{t})+j\bar{i}_q(\bar{t})\}\cdot e^{j\bar{t}}=\dot{\bar{E}}''\cdot e^{j\bar{t}}\end{aligned}\right\} \tag{10.83a}$$

Referring to Equation 10.55, the positive-sequence complex-number current of fundamental frequency can be written as

$$\dot{\bar{i}}_1(\bar{t})=\{\bar{i}_d(\bar{t})+j\bar{i}_q(\bar{t})\}\cdot e^{j\bar{t}}$$

Then

$$\left.\begin{aligned}&(\bar{r}+j\bar{x}''_d)\cdot\dot{\bar{i}}_1(\bar{t})=\dot{\bar{E}}''\cdot e^{j\bar{t}}\\&\text{where}\quad \bar{E}''=\{\bar{e}_d(0)+j\bar{e}_q(0)\}+(\bar{r}+j\bar{x}''_q)\{\bar{i}_d(0)+j\bar{i}_q(0)\}+j(\bar{x}''_d-\bar{x}''_q)\cdot\bar{i}_d(0)\end{aligned}\right\} \tag{10.83b}$$

Recall that Equation 10.81 is quite a precise equation only under the condition of Equation 10.74①②, although the equivalent circuit cannot be written. Under the additional condition of $\bar{x}''_d=\bar{x}''_q$, Equation 10.83b and the equivalent circuit in Figure 10.7a have also been obtained.

10.6.1.2 Transient period: t = 3 to approximately 60 cycles (45 or 60 ms to 1 sec)

Eliminating $\bar{\psi}_d(\bar{t})$, $\bar{\psi}_q(\bar{t})$ from Equations 10.79 and 10.71,

$$\left.\begin{aligned}\bar{\psi}_d(0)+\bar{x}'_d\cdot\bar{i}_d(0)&=\bar{r}\cdot\bar{i}_q(\bar{t})+\bar{x}'_d\cdot\bar{i}_d(\bar{t}) \quad ①\\ \bar{\psi}_q(0)+\bar{x}_q\cdot\bar{i}_q(0)&=-\bar{r}\cdot\bar{i}_d(\bar{t})+\bar{x}_q\cdot\bar{i}_q(\bar{t}) \quad ②\end{aligned}\right\} \tag{10.84}$$

Using the form $j\{①+j②\}$ and referring to Equation 10.66,

$$\begin{aligned}&(\bar{r}+j\bar{x}_q)\{\bar{i}_d(\bar{t})+j\bar{i}_q(\bar{t})\}+j(\bar{x}'_d-\bar{x}_q)\cdot\bar{i}_d(\bar{t})\\&=\{\bar{e}_d(0)+j\bar{e}_q(0)\}+(\bar{r}+j\bar{x}_q)\{\bar{i}_d(0)+j\bar{i}_q(0)\}+j(\bar{x}'_d-\bar{x}_q)\cdot\bar{i}_d(0)\equiv\bar{E}'\end{aligned} \tag{10.85}$$

We assume the equation below in order to find the equivalent circuit, although this is a bold as seen in Table 10.1:

$$\bar{x}'_d \doteqdot \bar{x}_q \tag{10.86}$$

Then

$$\left.\begin{aligned}&(\bar{r}+j\bar{x}'_d)\{i_d(t)+ji_q(t)\}=\dot{\bar{E}}'\\&\text{or}\\&(\bar{r}+j\bar{x}'_d)(i_d(\bar{t})+j\bar{i}_q(\bar{t})\}e^{j\bar{t}}=\dot{\bar{E}}'\cdot e^{j\bar{t}}\end{aligned}\right\} \tag{10.87a}$$

In the form of $\dot{\bar{i}}_1(\bar{t})=\{\bar{i}_d(\bar{t})+j\bar{i}_q(\bar{t})\}\cdot e^{j\bar{t}}$

$$\left.\begin{aligned}&(\bar{r}+j\bar{x}'_d)\cdot\dot{\bar{i}}_1(\bar{t})=\bar{E}'\cdot e^{j\bar{t}}\\&\text{where}\quad \bar{E}'=\{\bar{e}_d(0)+j\bar{e}_q(0)\}+(\bar{r}+j\bar{x}_q)\{\bar{i}_d(0)+j\bar{i}_q(0)\}+j(\bar{x}'_d-\bar{x}_q)\cdot\bar{i}_d(0)\end{aligned}\right\} \tag{10.87b}$$

This is the positive-sequence equation of a generator for the period of 3–60 cycles and the corresponding equivalent circuit is shown in Figure 10.7a.

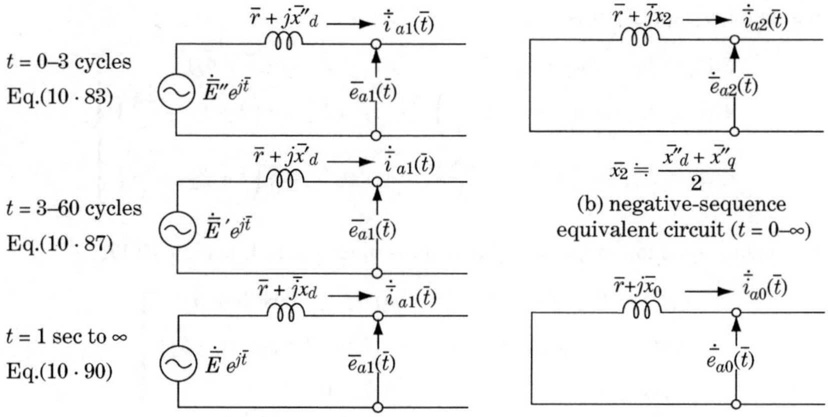

Figure 10.7 Symmetrical equivalent circuit of generator

Recall that Equation 10.85 is quite precise, while Equations 10.87a and b are under bold approximations of $\bar{x}'_d \fallingdotseq \bar{x}_q$.

10.6.1.3 Steady-state period: t = 1 sec

Eliminating $\bar{\psi}_d(\bar{t})$, $\bar{\psi}_q(\bar{t})$ from Equations 10.78 and 10.73,

$$\left.\begin{aligned}\bar{\psi}_d(0) + \bar{x}_d \cdot \bar{i}_d(0) &= \bar{r} \cdot \bar{i}_q(\bar{t}) + \bar{x}_d \cdot \bar{i}_d(\bar{t}) \quad ① \\ \bar{\psi}_q(0) + \bar{x}_q \cdot \bar{i}_q(0) &= -\bar{r} \cdot \bar{i}_d(\bar{t}) + \bar{x}_q \cdot \bar{i}_q(\bar{t}) \quad ②\end{aligned}\right\} \tag{10.88}$$

Using the form $j\{① + j②\}$ and referring to Equation 10.66,

$$\begin{aligned}&(\bar{r} + j\bar{x}_q)\{\bar{i}_d(\bar{t}) + j\bar{i}_q(\bar{t})\} + j(\bar{x}_d - \bar{x}_q) \cdot \bar{i}_d(\bar{t}) \\ &= \{\bar{e}_d(0) + j\bar{e}_q(0)\} + (\bar{r} + j\bar{x}_q)\{\bar{i}_d(0) + j\bar{i}_q(0)\} + j(\bar{x}_d - \bar{x}_q) \cdot \bar{i}_d(0) \equiv \dot{\bar{E}}\end{aligned} \tag{10.89}$$

This is quite a precise equation, while the equivalent circuit cannot be drawn. Then, we assume $\bar{x}_d \fallingdotseq \bar{x}_q$ in order to obtain the equivalent circuit, although this is a little bold for hydro-generators as seen in Table 10.1:

$$\left.\begin{aligned}&(\bar{r} + j\bar{x}_d)\{i_d(\bar{i}) + j\bar{i}_q(\bar{t})\} = \dot{\bar{E}} \\ &\text{or} \\ &(\bar{r} + j\bar{x}_d)\{\bar{i}_d(\bar{t}) + j\bar{i}_q(\bar{t})\}e^{j\bar{t}} = \dot{\bar{E}} \cdot e^{j\bar{t}}\end{aligned}\right\} \tag{10.90a}$$

In the form of $\dot{\bar{i}}_1(\bar{t}) = \{\bar{i}_d(\bar{t}) + j\bar{i}_q(\bar{t})\} \cdot e^{j\bar{t}}$

$$\left.\begin{aligned}&(\bar{r} + j\bar{x}_d) \cdot \dot{\bar{i}}_1(\bar{t}) = \dot{\bar{E}} \cdot e^{j\bar{t}} = \dot{\bar{e}}_1(\bar{t}) \\ &\dot{\bar{E}} = \{\bar{e}_d(0) + j\bar{e}_q(0)\} + (\bar{r} + j\bar{x}_q)\{\bar{i}_d(0) + j\bar{i}_q(0)\} + j(\bar{x}_d - \bar{x}_q) \cdot \bar{i}_q(0)\end{aligned}\right\} \tag{10.90b}$$

10.6.2 Negative-sequence circuit

We start from the equations below for the negative-sequence fundamental frequency voltage and current.

The real-number expression is

$$\left.\begin{aligned}\bar{e}_a(t) &= \bar{E}_{a2}\cos(\bar{t}+\bar{\alpha}_2) & \bar{i}_a(t) &= \bar{I}_{a2}\cos(\bar{t}+\bar{\beta}_2)\\ \bar{e}_b(t) &= \bar{E}_{a2}\cos\left(\bar{t}+\bar{\alpha}_2+\frac{2\pi}{3}\right) & \bar{i}_b(t) &= \bar{I}_{a2}\cos\left(\bar{t}+\bar{\beta}_2+\frac{2\pi}{3}\right)\\ \bar{e}_c(t) &= \bar{E}_{a2}\cos\left(\bar{t}+\bar{\alpha}_2-\frac{2\pi}{3}\right) & \bar{i}_c(t) &= \bar{I}_{a2}\cos\left(\bar{t}+\bar{\beta}_2-\frac{2\pi}{3}\right)\end{aligned}\right\} \tag{10.91}$$

The corresponding equation in the d–q–0 domain is, referring to Equation 10.17,

$$\left.\begin{aligned}\bar{e}_d(t) &= \bar{E}_{a2}\cos(2\bar{t}+\bar{\alpha}_2) & \bar{i}_d(t) &= \bar{I}_{a2}\cos(2\bar{t}+\bar{\beta}_2)\\ \bar{e}_q(t) &= -\bar{E}_{a2}\sin(2\bar{t}+\bar{\alpha}_2) & \bar{i}_q(t) &= -\bar{I}_{a2}\sin(2\bar{t}+\bar{\beta}_2)\\ \bar{e}_0(t) &= 0 & \bar{i}_0(t) &= 0\end{aligned}\right\} \tag{10.92}$$

The complex-number expression is

$$\left.\begin{aligned}&\left.\begin{aligned}\bar{E}_{a2}e^{j(2\bar{t}+\bar{\alpha}_2)} &= \bar{e}_d(\bar{t})-j\bar{e}_q(\bar{t})\\ \bar{I}_{a2}e^{j(2\bar{t}+\bar{\beta}_2)} &= \bar{i}_d(\bar{t})-j\bar{i}_q(\bar{t})\end{aligned}\right\}① \\ &\text{or}\\ &\left.\begin{aligned}\dot{\bar{e}}_2(\bar{t}) &= \bar{E}_{a2}e^{j(\bar{t}+\bar{\alpha}_2)} = \{\bar{e}_d(\bar{t})-j\bar{e}_q(\bar{t})\}\cdot e^{-j\bar{t}}\\ \dot{\bar{i}}_2(\bar{t}) &= \bar{I}_{a2}e^{j(\bar{t}+\bar{\beta}_2)} = \{\bar{i}_d(\bar{t})-j\bar{i}_q(\bar{t})\}\cdot e^{-j\bar{t}}\end{aligned}\right\}②\end{aligned}\right\} \tag{10.93}$$

Equation 10.93 explains that the negative-sequence current of Equation 10.91 in the a–b–c domain flowing through a generator is equivalent to the double frequency current of Equation 10.92 or 10.93 flowing through a generator circuit in the d–q–0 domain of Figure 10.4.

Now we examine the difference in behaviour of positive- and negative-sequence quantities in the d-axis circuit of Figure 10.4.

10.6.2.1 Case: positive-sequence current behaviour on d-axis circuit

Positive-sequence quantities are d.c. quantities in the d–q–0 domain. Accordingly, in the d-axis equivalent circuit, just after a sudden change in generator condition, $\bar{r}_{kd}$ soon dominates $s\bar{L}_{kd}$ in the damper branch, and presently $\bar{r}_{fd}$ dominates $s\bar{L}_{fd}$ in the field branch. Finally the current from the outer circuit flows only through $s\bar{L}_{ad}$. By this reason, the d-axis terminal reactance of the generator will change from time to time. The behaviour in the q-axis circuit is similar.

10.6.2.2 Case: negative-sequence current behaviour on d-axis circuit

In this case, all the quantities are of double frequency in the d-axis equivalent circuit. Therefore for the time just after the sudden change of generator condition, $s\bar{L}_{kd}$ continues to dominate $\bar{r}_{kd}$ in the damper branch, and $s\bar{L}_{fd}$ continues to dominate $\bar{r}_{fd}$ in the field branch. By this reason, the d-axis terminal reactance of the generator for the negative-sequence quantities has the same magnitude as $\bar{x}_d''$ (given by Equation 10.68) and will not change over time. The q-axis terminal reactance of the generator for the negative-sequence quantities $\bar{x}_q''$ (given by Equation 10.68) will similarly not be changed over time.

Accordingly, for the negative-sequence quantities,

$$\left.\begin{aligned} \frac{-\bar{\psi}_d(\bar{t})}{\bar{i}_d(\bar{t})} &= \bar{x}''_d \qquad \bar{\psi}_d(\bar{t}) = -\bar{x}''_d \cdot \bar{i}_d(\bar{t}) \\ -\frac{\bar{\psi}_q(\bar{t})}{\bar{i}_q(\bar{t})} &= \bar{x}''_q \qquad \bar{\psi}_q(\bar{t}) = -\bar{x}''_q \cdot \bar{i}_q(\bar{t}) \end{aligned}\right\} \tag{10.94}$$

Substituting this equation into Equation 10.43,

$$\left.\begin{aligned} \bar{e}_d(\bar{t}) &= -\bar{\psi}_q(\bar{t}) + s\bar{\psi}_d(\bar{t}) - \bar{r}\cdot\bar{i}_d(\bar{t}) = \bar{x}''_q\cdot\bar{i}_q(\bar{t}) - \bar{x}''_d\cdot s\bar{i}_d(\bar{t}) - \bar{r}\cdot\bar{i}_d(\bar{t}) \\ \bar{e}_q(\bar{t}) &= \bar{\psi}_d(\bar{t}) + s\bar{\psi}_q(\bar{t}) - \bar{r}\cdot\bar{i}_q(\bar{t}) = -\bar{x}''_d\cdot\bar{i}_d(\bar{t}) - \bar{x}''_q\cdot s\bar{i}_q(\bar{t}) - \bar{r}\cdot\bar{i}_q(\bar{t}) \end{aligned}\right\} \tag{10.95}$$

$\bar{i}_d(t)$ and $\bar{i}_q(t)$ are given by Equation 10.92, and the derivatives are

$$\left.\begin{aligned} s\bar{i}_d(\bar{t}) &= \frac{d}{d\bar{t}}\bar{i}_d(\bar{t}) = -2\bar{I}_{a2}\sin(2\bar{t}+\bar{\beta}_2) \\ s\bar{i}_q(\bar{t}) &= -2\bar{I}_{a2}\cos(2\bar{t}+\bar{\beta}_2) \end{aligned}\right\} \tag{10.96}$$

Substituting the current terms of Equations 10.92 and 10.96 into Equation 10.95, and neglecting the stator resistance ($\bar{r}=0$),

$$\left.\begin{aligned} \bar{e}_d(\bar{t}) &= (2\bar{x}''_d - \bar{x}''_q)\bar{I}_{a2}\sin(2\bar{t}+\bar{\beta}_2) \\ \bar{e}_q(\bar{t}) &= (2\bar{x}''_q - \bar{x}''_d)\bar{I}_{a2}\cos(2\bar{t}+\bar{\beta}_2) \end{aligned}\right\} \tag{10.97}$$

Now, refering Equation 10.93 the relation between the negative-sequence voltage $\bar{e}_2$ and the d–q–0 sequence voltages $\bar{e}_d$, $\bar{e}_q$ is, for the real-number expression,

$$\begin{aligned} \bar{e}_2(\bar{t}) &= \bar{E}_{a2}\cos(\bar{t}+\bar{\alpha}_2) = \mathrm{Re}[\dot{\bar{e}}_2(\bar{t})] = \mathrm{Re}[\bar{E}_{a2}e^{j(\bar{t}+\bar{\alpha}_2)}] \\ &= \mathrm{Re}[\bar{E}_{a2}e^{j(2\bar{t}+\bar{\alpha}_2)}\cdot e^{-j\bar{t}}] \\ &= \mathrm{Re}[\{\bar{e}_d(t) - j\bar{e}_q(t)\}(\cos\bar{t} - j\sin\bar{t})] \\ &= \bar{e}_d(\bar{t})\cos\bar{t} - \bar{e}_q(\bar{t})\sin\bar{t} \end{aligned} \tag{10.98}$$

Substituting Equation 10.97 into Equation 10.98,

$$\left.\begin{aligned} e_2(\bar{t}) &= (2\bar{x}''_d - \bar{x}''_q)\bar{I}_{a2}\sin(2\bar{t}+\bar{\beta}_2)\cos\bar{t} - (2\bar{x}''_q - \bar{x}''_d)\bar{I}_{a2}\cos(2\bar{t}+\bar{\beta}_2)\sin\bar{t} \\ &= \frac{1}{2}(2\bar{x}''_d - \bar{x}''_q)\bar{I}_{a2}\{(\sin(3\bar{t}+\bar{\beta}_2) + \sin(\bar{t}+\bar{\beta}_2)\} \\ &\quad - \frac{1}{2}(2\bar{x}''_q - \bar{x}''_d)\bar{I}_{a2}\{\sin(3\bar{t}+\bar{\beta}_2) - \sin(\bar{t}+\bar{\beta}_2)\} \\ \therefore\quad \bar{e}_2(\bar{t}) &= \frac{\bar{x}''_d + \bar{x}''_q}{2}\bar{I}_{a2}\sin(\bar{t}+\bar{\beta}_2) + \frac{3}{2}(\bar{x}''_d - \bar{x}''_q)\bar{I}_{a2}\sin(3\bar{t}+\bar{\beta}_2) \\ &\equiv \bar{x}_2\bar{I}_{a2}\sin(\bar{t}+\bar{\beta}_2) + (\text{the third - harmonic term}) \\ \text{where}\quad \bar{x}_2 &\equiv \frac{\bar{x}''_d + \bar{x}''_q}{2} \end{aligned}\right\} \tag{10.99}$$

This is the equation of negative-sequence circuit quantities. The negative-sequence reactance $\bar{x}_2$ is given by $(\bar{x}''_d + \bar{x}''_q)/2$ and will not obviously be changed over time.

Equation 10.99 shows that the negative-sequence current consists of the fundamental frequency component and the third-frequency term. As a matter of fact, the third-frequency current will appear, although the magnitude is very small because the term is proportional to $(\bar{x}''_d - \bar{x}''_q)$. The negative-sequence equivalent circuit is shown in Figure 10.7b under the condition of neglecting the third-frequency term.

Incidentally, if negative-sequence current flows into a generator for some duration (which means the second-harmonic current continues to flow in the equivalent circuit of Figure 10.4), serious problems will arise at the generator. This scheme will be discussed in Chapter 16.

10.6.3 Zero-sequence circuit

The generator's zero-sequence circuit is shown as the third equation of Equation 10.47 and the equivalent circuit is shown in Figure 10.4.

The examination of the zero-sequence fundamental frequency current flowing into the generator is quite simple. It is given by (real-number expression)

$$\bar{i}_0(\bar{t}) = \frac{1}{3}\{\bar{i}_a(\bar{t}) + \bar{i}_b(\bar{t}) + \bar{i}_c(\bar{t})\} = \bar{I}_{a0}e^{j(\bar{t}+\bar{\alpha}_0)} \tag{10.100}$$

The generator's zero-sequence circuit equation has already been derived as the third equation of Equation 10.47 and the equivalent circuit in Figure 10.4, namely

$$\dot{\bar{e}}_0(\bar{t}) = -(\bar{r} + j\bar{x}_0) \cdot \dot{\bar{i}}_0(\bar{t}) \tag{10.101}$$

The equivalent circuit is again shown in Figure 10.7c.

Now we have found the generator's symmetrical equivalent circuit in Figure 10.7, we need to recollect that the equivalent circuit was drawn with the assumption of $\bar{x}''_d \doteqdot \bar{x}''_q$ for Equation 10.83, $\bar{x}'_d \doteqdot \bar{x}_q$ for Equation 10.87 and $\bar{x}_d \doteqdot \bar{x}_q$ for Equation 10.90. These assumptions would be justified for most of the analysis of power system phenomena (with one exceptional case, namely the terminal fault of the hydro-generator) because the inductances of transmission lines and transformers connected to the generator mitigate the possible error in the assumptions. For accurate analysis of generator terminal faults, we have to go back to the equations before the assumptions of Equations 10.83, 10.87 and 10.90.

Typical reactance values of generators are shown in Table 10.1.

10.7 Laplace-transformed Generator Equations and the Time Constants

10.7.1 Laplace-transformed equations

A generator's equivalent circuits in the 0–1–2 domain (Figure 10.7) as well as the α–β-0 domain (Figure 6.5) include some small assumptions as already discussed. For accurate fault transient analysis at the generator terminal or a closer point, we need to go back to Equations 10.43–10.46 before the assumptions and solve them by applying Laplace transforms.

Our generator is rotating at constant speed. The equations of the stator quantities are given by Equation 10.43 where $d\theta/d\bar{t} = s\bar{\theta} = 1.0$. Accordingly, Equation 10.43 can be simplified to

$$\left.\begin{aligned} \bar{e}_d(\bar{t}) &= -\bar{\psi}_q(\bar{t}) + \frac{d}{d\bar{t}}\bar{\psi}_d(\bar{t}) - \bar{r}\cdot\bar{i}_d(\bar{t}) \\ \bar{e}_q(\bar{t}) &= \bar{\psi}_d(\bar{t}) + \frac{d}{d\bar{t}}\bar{\psi}_q(\bar{t}) - \bar{r}\cdot\bar{i}_q(\bar{t}) \\ \bar{e}_0(\bar{t}) &= \frac{d}{d\bar{t}}\bar{\psi}_0(\bar{t}) - \bar{r}\cdot\bar{i}_0(\bar{t}) \end{aligned}\right\} \tag{10.102}$$

The Laplace-transformed equations are

$$\left.\begin{aligned}\bar{e}_d(s) &= -\bar{\psi}_q(s) + s\bar{\psi}_d(s) - \bar{r}\cdot\bar{i}_d(s)\\ \bar{e}_q(s) &= \bar{\psi}_d(s) + s\bar{\psi}_q(s) - \bar{r}\cdot\bar{i}_q(s)\\ \bar{e}_0(s) &= s\bar{\psi}_0(s) - \bar{r}\cdot\bar{i}_0(s)\end{aligned}\right\} \qquad (10.103)$$

Obviously from Figure 10.4

$$\left.\begin{aligned}s\bar{\psi}_d(s) &= -\bar{x}_d(s)\cdot s\bar{i}_d(s) \quad \text{or} \quad \bar{\psi}_d(s) = -\bar{x}_d(s)\cdot\bar{i}_d(s)\\ s\bar{\psi}_q(s) &= -\bar{x}_q(s)\cdot s\bar{i}_q(s) \qquad\quad\; \bar{\psi}_q(s) = -\bar{x}_q(s)\cdot\bar{i}_q(s)\\ s\bar{\psi}_0(s) &= -\bar{x}_0(s)\cdot s\bar{i}_0(s) \qquad\quad\; \bar{\psi}_0(s) = -\bar{x}_0(s)\cdot\bar{i}_0(s)\end{aligned}\right\} \qquad (10.104)$$

$\bar{x}_d(s), \bar{x}_q(s), \bar{x}_0(s)$ are the generator terminal impedances of Figure 10.4 in terms of the Laplace-transformed s functions and are given by the equations below, named **operational reactances**:

$$\left.\begin{aligned}\bar{x}_d(s) &= \bar{x}_l + \frac{1}{\dfrac{1}{\bar{x}_{ad}} + \dfrac{1}{\bar{x}_{fd} + \dfrac{\bar{r}_{fd}}{s}} + \dfrac{1}{\bar{x}_{kd} + \dfrac{\bar{r}_{kd}}{s}}} = \left\{\bar{x}_l + \frac{1}{\dfrac{1}{\bar{x}_{ad}} + \dfrac{1}{\bar{x}_{fd}} + \dfrac{1}{\bar{x}_{kd}}}\right\}\cdot\frac{\left(s+\dfrac{1}{\bar{T}'_d}\right)\left(s+\dfrac{1}{\bar{T}''_d}\right)}{\left(s+\dfrac{1}{\bar{T}'_{d0}}\right)\left(s+\dfrac{1}{\bar{T}''_{d0}}\right)}\\ &= \bar{x}''_d\cdot\frac{\left(s+\dfrac{1}{\bar{T}'_d}\right)\left(s+\dfrac{1}{\bar{T}''_d}\right)}{\left(s+\dfrac{1}{\bar{T}'_{d0}}\right)\left(s+\dfrac{1}{\bar{T}''_{d0}}\right)}\\ \bar{x}_q(s) &= \bar{x}_l + \frac{1}{\dfrac{1}{\bar{x}_{aq}} + \dfrac{1}{\bar{x}_{kq} + \dfrac{\bar{r}_{kq}}{s}}} = \left\{\bar{x}_l + \frac{1}{\dfrac{1}{\bar{x}_{aq}} + \dfrac{1}{\bar{x}_{kq}}}\right\}\cdot\frac{\left(s+\dfrac{1}{\bar{T}''_q}\right)}{\left(s+\dfrac{1}{\bar{T}''_{q0}}\right)} = \bar{x}''_q\cdot\frac{\left(s+\dfrac{1}{\bar{T}''_q}\right)}{\left(s+\dfrac{1}{\bar{T}''_{q0}}\right)}\\ \bar{x}_0(s) &= \bar{x}_0\end{aligned}\right\} \qquad (10.105)$$

Eliminating flux linkage ψ from Equations 10.103 and 10.104,

$$\left.\begin{aligned}\bar{e}_d(s) &= -\{\bar{r} + s\cdot\bar{x}_d(s)\}\cdot\bar{i}_d(s) + \bar{x}_q(s)\cdot\bar{i}_q(s)\\ \bar{e}_q(s) &= -\bar{x}_d(s)\cdot\bar{i}_d(s) - \{\bar{r} + s\cdot\bar{x}_q(s)\}\cdot\bar{i}_q(s)\\ \bar{e}_0(s) &= -\{\bar{r} + s\cdot\bar{x}_0(s)\}\cdot\bar{i}_0(s)\end{aligned}\right\} \qquad (10.106)$$

Thus the Laplace-transformed generator equations and the reactances have now been derived.

10.7.1.1 Open-circuit transient time constants $\bar{T}'_{d0}, \bar{T}''_{d0}, \bar{T}'_{q0}, \bar{T}''_{q0}$

In Figures 10.8a–c, the generator is under operation in connection with the three-phase-balanced outer circuit, where $\bar{X}_{\text{outd}}, \bar{X}_{\text{outq}}$ are the d-, q-axis reactances of the outer network circuit between the generator terminal and arbitrary point F in the network.

If a generator is suddenly tripped from the network, transient phenomena in the d-, q- axis would be explained by Figure 10.8, in that the switches Sw1, Sw2 are simultaneously opened. Accordingly, the time constants $\bar{T}'_{d0}, \bar{T}''_{d0}, \bar{T}'_{q0}, \bar{T}''_{q0}$ can be derived as follows (the suffix 0 means opening mode).

Note: Generator tripping: Sw1, Sw2 opening (Sw3, Sw4 are in the open condition)
Generator short-circuit fault: Sw3, Sw4 closing (Sw1, Sw2 are in the closed condition)

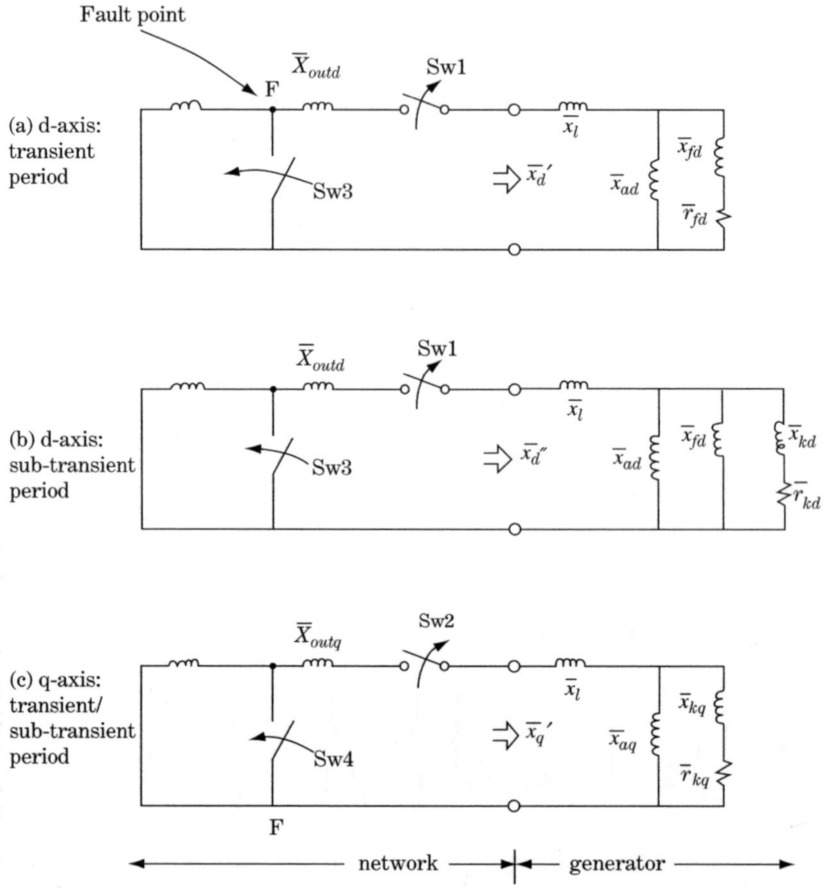

Figure 10.8 Generator's transient analysis (tripping/short circuit) in d–q–0 domain

10.7.1.1.1 $\bar{T}'_{d0}$: Open-circuit transient d-axis time constant

Sw1 is opened in Figure 10.8a. Then the time constant is

$$\bar{T}'_{d0} = \frac{\bar{x}_{ad}+\bar{x}_{fd}}{\bar{r}_{fd}}[\text{rad}] = \frac{\bar{x}_{ad}+\bar{x}_{fd}}{\bar{r}_{fd}}\cdot\frac{1}{2\pi f}[\text{sec}] \quad \left(=\frac{\bar{x}_{ffd}}{\bar{r}_{fd}}[\text{rad}]\right) \tag{10.107a}$$

10.7.1.1.2 $\bar{T}''_{d0}$: Open-circuit sub-transient d-axis time constant

Sw1 is opened in Figure 10.8b. Then

$$\bar{T}''_{d0} = \frac{(\bar{x}_{ad}//\bar{x}_{fd})+\bar{x}_{kd}}{\bar{r}_{kd}} = \frac{\dfrac{\bar{x}_{ad}\bar{x}_{fd}}{\bar{x}_{ad}+\bar{x}_{fd}}+\bar{x}_{kd}}{\bar{r}_{kd}}[\text{rad}] \quad \left(=\frac{\bar{x}_{kkd}-\dfrac{\bar{x}_{ad}^2}{\bar{x}_{ffd}}}{\bar{r}_{kd}}[\text{rad}]\right) \tag{10.107b}$$

10.7.1.1.3 $\overline{T}'_{q0} = \overline{T}''_{q0}$: Open-circuit transient q-axis time constant Sw2 is opened in Figure 10.8c. Then

$$\overline{T}'_{q0} = \overline{T}''_{q0} = \frac{\overline{x}_{aq} + \overline{x}_{kq}}{\overline{r}_{kq}}\,[\text{rad}] \left(= \frac{\overline{x}_{kkq}}{\overline{r}_{kq}}\,[\text{rad}] \right) \tag{10.107c}$$

10.7.1.2 Short-circuit transient time constants $\overline{T}'_d$, $\overline{T}''_d$, $\overline{T}'_q$, $\overline{T}''_q$

If a short-circuit fault occurs at the point F (Sw3, Sw4 are simultaneously closed) in Figure 10.8, the time constants of the transient and sub-transient periods would be as follows:

$$\left.\begin{aligned} \overline{T}'_d &= \overline{T}'_{d0}\frac{\overline{x}'_d + \overline{X}_{outd}}{\overline{x}_d + \overline{X}_{outd}} \\ \overline{T}''_d &= \overline{T}''_{d0}\frac{\overline{x}''_d + \overline{X}_{outd}}{\overline{x}'_d + \overline{X}_{outd}} \\ \overline{T}'_q &= \overline{T}''_q = \overline{T}'_{q0}\frac{\overline{x}'_q + \overline{X}_{outq}}{\overline{x}_q + \overline{X}_{outq}} \end{aligned}\right\} \tag{10.108}$$

The generator terminal fault is of course a special case of $\overline{X}_{outd} = \overline{X}_{outq} = 0$ that gives the generator transient/sub-transient short-circuit time constants

$$\left.\begin{aligned} \overline{T}'_d &= \overline{T}'_{d0}\frac{\overline{x}'_d}{\overline{x}_d} && \text{short-circuit transient d-axis time constant} \\ \overline{T}''_d &= \overline{T}''_{d0}\frac{\overline{x}''_d}{\overline{x}'_d} && \text{short-circuit sub-transient d-axis time constant} \\ \overline{T}'_q &= \overline{T}''_q = \overline{T}'_{d0}\frac{\overline{x}'_q}{\overline{x}_q} && \text{short-circuit transientq-axis time constant} \end{aligned}\right\} \tag{10.109}$$

The time constants would become a little larger if the fault point were located far from the generator.

The short-circuit transient current can be decomposed to the three current terms of {steady-state}, {transient}, {sub-transient} and written in the following form:

$$\left.\begin{aligned} \overline{i}_d(t) &= \{\text{steady-state term}\} + \{\text{transient term}\} \cdot e^{-\overline{t}/\overline{T}'_d} + \{\text{sub-transient term}\} \cdot e^{-\overline{t}/\overline{T}''_d} \\ \overline{i}_q(t) &= \{\text{steady-state term}\} + \{\text{transient term}\} \cdot e^{-\overline{t}/\overline{T}'_q} \end{aligned}\right\} \tag{10.110}$$

10.7.1.3 Short-circuit time constant of armature winding $\overline{T}_a$

$\overline{T}_a$ is the time constant of the stator required for the d.c. component of the armature current under the transient condition of the generator terminals (typically the short-circuit fault there).

The armature phase coil is operating in a position in line with the d-axis under normal conditions before the short circuit. Whenever a short circuit occur, the rotor d-axis and q-axis alternate rapidly in position, so that the phase coil is forced rapidly to align with the q-axis.

Accordingly, the reactance of the short-circuited armature winding, which depends upon the mutual coupling between the armature and rotor circuits, changes rapidly from $\overline{x}''_d$ to $\overline{x}''_q$, and is approximately $\frac{1}{2}(\overline{x}''_d + \overline{x}''_q) = \overline{x}_2$. Accordingly, $\overline{T}_a$ is given by the following equation as an approximate value:

$$\overline{T}_a \fallingdotseq \frac{\overline{x}''_d + \overline{x}''_q}{2\overline{r}} = \frac{\overline{x}_2}{\overline{r}}\,[\text{rad}] = \frac{\overline{x}_2}{\overline{r}} \cdot \frac{1}{2\pi f}\,[\text{sec}] \tag{10.111a}$$

$\bar{T}_a$ may be written as follows under the condition of $\bar{x}''_d \doteq \bar{x}''_q$:

$$\bar{T}_a \doteq \frac{2}{\bar{r}\left(\frac{1}{\bar{x}''_d}+\frac{1}{\bar{x}''_q}\right)} = \frac{2\bar{x}''_d \cdot \bar{x}''_q}{\bar{r}(\bar{x}''_d+\bar{x}''_q)} \tag{10.111b}$$

(Subtraction of both approximations in Equations 10.111a and 10.111b gives $(\bar{x}''_d - \bar{x}''_q)^2/\{2\bar{r}(\bar{x}''_d+\bar{x}''_q)\}$. This is also a supplementary explanation of Equation 10.118.)

Note that all the time constants above are unitized [radian] values by base value $2\pi f_{base}$:

$$\langle 1 \text{ sec} \Leftrightarrow 2\pi\, f_{\text{base}} = 2\pi \times 50 = 314 \text{ rad}\rangle$$

Accordingly, the above radian value can be transformed into [sec] by multiplying by $1/(314) = 0.0032$ for 50 Hz, or $1/(2\pi \times 60) = 0.0027$ for 60 Hz.

10.8 Relations Between the d–q–0 and α–β–0 Domains

The relations between the d–q–0 and α–β–0 domains are determined here.

Referring to Equations 10.15 and 6.6, and reminding $\boldsymbol{D}(t)$ and $\boldsymbol{\alpha}^{-1}$ are matrices of real number elements,

$$\boldsymbol{e}_{dq0}(t) = \boldsymbol{D}(t)\mathrm{Re}\left[\boldsymbol{e}_{abc}(t)\right] = (\boldsymbol{D}(t)\cdot\boldsymbol{\alpha}^{-1})\cdot\mathrm{Re}\left[\boldsymbol{e}_{\alpha\beta0}\right] \tag{10.112}$$

where $\boldsymbol{D}(t)$ is given by Equation 10.10a and $\boldsymbol{\alpha}^{-1}$ is given by Equation 6.2.

Then $\boldsymbol{D}(t)\cdot\boldsymbol{\alpha}^{-1}$ can be calculated and the following equations are obtained:

$$\underset{\boldsymbol{e}_{dq0}(t)}{\begin{bmatrix} e_d(t) \\ e_q(t) \\ e_0(t) \end{bmatrix}} = \underset{\boldsymbol{D}(t)\cdot\boldsymbol{\alpha}^{-1}}{\begin{bmatrix} \cos\omega t & \sin\omega t & 0 \\ -\sin\omega t & \cos\omega t & 0 \\ 0 & 0 & 1 \end{bmatrix}} \cdot \underset{\boldsymbol{e}_{\alpha\beta0}(t)}{\begin{bmatrix} e_\alpha(t) \\ e_\beta(t) \\ e_0(t) \end{bmatrix}} \tag{10.113a}$$

Namely

$$\left.\begin{aligned} e_d(t) &= \cos\omega t\, e_\alpha(t) + \sin\omega t\, e_\beta(t) \\ e_q(t) &= -\sin\omega t\, e_\alpha(t) + \cos\omega t\, e_\beta(t) \\ e_0(t) &= e_0(t) \\ &\text{with inverse solution} \\ e_\alpha(t) &= \cos\omega t\, e_d(t) - \sin\omega t\, e_q(t) \\ e_\beta(t) &= \sin\omega t\, e_d(t) + \cos\omega t\, e_q(t) \\ e_0(t) &= e_0(t) \end{aligned}\right\} \tag{10.113b}$$

10.9 Detailed Calculation of Generator Short-circuit Transient Current under Load Operation

The transient analysis of a sudden three-phase short-circuit fault at the generator terminal end is demonstrated in this section.

The generator is operating under the three-phase-balanced on-load condition and with constant speed ($d\theta/d\bar{t} = 1.0$) before $t = 0$. Referring to Equations 10.52 and 10.53 or Figure 10.5, the initial condition at $t = 0$ is

$$\left.\begin{aligned} &\bar{e}_d(0-) = \bar{E}_{a1}\cos\bar{\alpha}_1 = \bar{E}_{a1}\sin\bar{\delta}_1, \quad \bar{i}_d(0-) = \bar{I}_{a1}\cos\bar{\beta}_1 \\ &\bar{e}_q(0-) = \bar{E}_{a1}\sin\bar{\alpha}_1 = \bar{E}_{a1}\cos\bar{\delta}_1, \quad \bar{i}_q(0-) = \bar{I}_{a1}\sin\bar{\beta}_1 \\ &\bar{e}_0(0-) = 0, \quad \bar{i}_0(0-) = 0 \end{aligned}\right\} \tag{10.114}$$

As shown in Figure 10.5, δ_1 is the generator's inner operating angle and is the complementary angle of α_1 (i.e. $\delta_1 + \alpha_1 = 90°$).

Now, the three-phase short-circuit fault suddenly occurs at the generator terminal at $\bar{t} = 0$ under the above initial condition, which means, by equations, to obtain the a–b–c terminal voltages $\bar{e}_a(t) = \bar{e}_b(t) = \bar{e}_c(t) = 0$ at $\bar{t} = 0+$. It also means, in the d–q–0 domain, to obtain the voltages $\bar{e}_d(\bar{t}) = \bar{e}_q(\bar{t}) = \bar{e}_0(\bar{t}) = 0$ at $\bar{t} = 0+$, or, in other words, to shorten simultaneously the d-, q-, 0-axis terminals in Figure 10.4 or Figure 10.8. Accordingly, the transient current $\bar{i}_d(\bar{t})$ in the d-axis circuit after $\bar{t} = 0$ can be calculated with the initial voltages $\bar{e}_d(0-)$ but with opposite sign (based on Thévenin's theorem). Calculation of the q- and 0-axis circuits is done similarly. That is,

$$\left.\begin{aligned} &\bar{e}_d(\bar{t}) = -\bar{e}_d(0-)1(t) \\ &\bar{e}_q(\bar{t}) = -\bar{e}_q(0-)1(t) \\ &\bar{e}_0(\bar{t}) = 0 \\ &\bar{e}_0(\bar{t}) = 0 \end{aligned} \quad \text{where} \quad \mathbf{1}(t) = \begin{cases} 0(t<0) \\ 1(t \geqq 0) \end{cases} \right\} \tag{10.115}$$

$\mathbf{1}(t)$ is the stepping function, and its Laplace transform is $\mathcal{L}[\mathbf{1}(t)] = 1/s$. Then

$$\left.\begin{aligned} \bar{e}_d(s) &= \mathcal{L}[\bar{e}_d(\bar{t})] = -\bar{e}_d(0-) \cdot \frac{1}{s} \\ \bar{e}_q(s) &= \mathcal{L}[\bar{e}_q(\bar{t})] = -\bar{e}_q(0-) \cdot \frac{1}{s} \\ \bar{e}_0(s) &= \mathcal{L}[\bar{e}_0(\bar{t})] = 0 \end{aligned}\right\} \tag{10.116}$$

The transient calculation can be executed by applying Equation 10.106 for the generator circuit and Equation 10.116 for the initial condition at the generator terminals in the d–q–0 Laplace domain.

Eliminating $e_d(s)$, $e_q(s)$, $e_0(s)$ from both equations,

$$\left.\begin{aligned} &-\{\bar{r} + s \cdot \bar{x}_d(s)\} \cdot \bar{i}_d(s) + \bar{x}_q(s) \cdot \bar{i}_q(s) = -\bar{e}_d(0-) \cdot \frac{1}{s} \\ &-\bar{x}_d(s) \cdot \bar{i}_d(s) - \{\bar{r} + s \cdot \bar{x}_q(s)\} \cdot \bar{i}_q(s) = -\bar{e}_q(0-) \cdot \frac{1}{s} \\ &-\{\bar{r} + s \cdot \bar{x}_0(s)\} \cdot \bar{i}_0(s) = 0 \end{aligned}\right\} \tag{10.117a}$$

This is a set of simultaneous equations in $\bar{i}_d(s)$, $\bar{i}_q(s)$ which can be solved as follows:

$$\left.\begin{aligned} \bar{i}_d(s) &= \frac{\bar{e}_d(0-) \cdot \left(s + \dfrac{\bar{r}}{\bar{x}_q(s)}\right) + \bar{e}_q(0-)}{s \cdot \bar{x}_d(s)\left\{s^2 + \bar{r}\left(\dfrac{1}{\bar{x}_d(s)} + \dfrac{1}{\bar{x}_q(s)}\right)s + 1 + \dfrac{\bar{r}^2}{\bar{x}_d(s) \cdot \bar{x}_q(s)}\right\}} & ① \\ \bar{i}_q(s) &= \frac{-\bar{e}_d(0-) + \bar{e}_q(0-) \cdot \left(s + \dfrac{\bar{r}}{\bar{x}_d(s)}\right)}{s \cdot \bar{x}_q(s)\left\{s^2 + \bar{r}\left(\dfrac{1}{\bar{x}_d(s)} + \dfrac{1}{\bar{x}_q(s)}\right)s + 1 + \dfrac{\bar{r}^2}{\bar{x}_d(s) \cdot \bar{x}_q(s)}\right\}} & ② \\ i_0(s) &= 0 & ③ \end{aligned}\right\} \tag{10.117b}$$

Here $\bar{x}_d(s)$, $\bar{x}_q(s)$ are given by Equation 10.105.

Now, Equation 10.117b with Equation 10.105 is the answer to our problem in the d–q–0 and time s domains, and accordingly what we have to do is to inverse-transform the answer into the d–q–0 and time t domains, and finally into the a–b–c and time t domains, at least from a mathematical viewpoint.

However, the inverse transform is very hard and we have to simplify somewhat by introducing justifiable assumptions as follows:

a) We neglect the third term of { } including $\bar{r}^2$ in the denominator. This is not a problem because $\bar{r}$ is very small.

b) We apply the simplification as follows:

$$\left.\begin{aligned} &\bar{r}\left(\frac{1}{\bar{x}_d(s)}+\frac{1}{\bar{x}_q(s)}\right)s \fallingdotseq \bar{r}\left(\frac{1}{\bar{x}''_d}+\frac{1}{\bar{x}''_q}\right)s=\frac{2}{\bar{T}_a}s \\ &\text{where}\quad \bar{T}_a=\frac{2}{\bar{r}\left(\dfrac{1}{\bar{x}''_d}+\dfrac{1}{\bar{x}''_q}\right)} \end{aligned}\right\} \qquad (10.118)$$

This simplification can be justified from an actual engineering viewpoint by considering assorted reactance values.

c) We neglect the term including $\bar{r}$ in the numerators of Equation 10.117b①②. This is also not a problem because $\bar{r}$ is very small.

Now, referring Equation 10.105, Equation 10.117b is simplified somewhat and recast into the equations

$$\left.\begin{aligned} &\bar{i}_d(s)=\frac{\{\bar{e}_d(0-)\cdot s+\bar{e}_q(0-)\}}{s\cdot\bar{x}_d(s)\cdot\left\{s^2+\dfrac{2}{\bar{T}_a}s+1\right\}}=\frac{\{\bar{e}_d(0-)\cdot s+\bar{e}_q(0-)\}\cdot\left(s+\dfrac{1}{\bar{T}'_{d0}}\right)\left(s+\dfrac{1}{\bar{T}''_{d0}}\right)}{\bar{x}''_d\cdot s\left(s+\dfrac{1}{\bar{T}'_d}\right)\left(s+\dfrac{1}{\bar{T}''_d}\right)\left(s^2+\dfrac{2}{\bar{T}_a}s+1\right)} \quad ① \\ &\bar{i}_q(s)=\frac{\{-\bar{e}_d(0-)+\bar{e}_q(0-)\cdot s\}}{s\cdot\bar{x}_q(s)\cdot\left\{s^2+\dfrac{2}{\bar{T}_a}s+1\right\}}=\frac{\{-\bar{e}_d(0-)+\bar{e}_q(0-)\cdot s\}\cdot\left(s+\dfrac{1}{\bar{T}''_{q0}}\right)}{\bar{x}''_q\cdot s\left(s+\dfrac{1}{\bar{T}''_q}\right)\left(s^2+\dfrac{2}{\bar{T}_a}s+1\right)} \quad ② \\ &i_0(s)=0 \quad ③ \\ &\text{where} \\ &\left(s^2+\frac{2}{\bar{T}_a}s+1\right)=\left(s+\frac{1}{\bar{T}_a}-j\sqrt{1-\frac{1}{\bar{T}_a^2}}\right)\left(s+\frac{1}{\bar{T}_a}+j\sqrt{1-\frac{1}{\bar{T}_a^2}}\right) \\ &\qquad\fallingdotseq\left(s+\frac{1}{\bar{T}_a}-j\right)\left(s+\frac{1}{\bar{T}_a}+j\right) \end{aligned}\right\} \qquad (10.119)$$

Although Equation 10.119 is still complicated, it can be solved relatively easily because the denominators are already in factorized form. The equations can be modified into expanded equations and the resulting equations are as follows (the process of the expansion is shown in Supplement 1 and Supplement 2):

$$
\left.
\begin{aligned}
&\bar{i}_d(s) = \frac{k_1}{s} + \frac{k_2}{s+\frac{1}{\bar{T}'_d}} + \frac{k_3}{s+\frac{1}{\bar{T}''_d}} + \left\{ \frac{k_4\angle\theta_4}{s+\frac{1}{\bar{T}_a}-j} + \frac{k_4\angle-\theta_4}{s+\frac{1}{\bar{T}_a}+j} \right\} \quad ① \\
&\text{where } k_1 = \frac{\bar{e}_q(0-)}{\bar{x}_d}, \qquad k_2 \doteqdot \bar{e}_q(0-)\cdot\left(\frac{1}{\bar{x}'_d}-\frac{1}{\bar{x}_d}\right) \\
&k_3 \doteqdot e_q(0-)\cdot\left(\frac{1}{\bar{x}''_d}-\frac{1}{\bar{x}'_d}\right) \\
&k_4\angle\theta_4 \doteqdot \frac{\bar{E}_{a1}}{2\bar{x}''_d}\angle-\left(\bar{\alpha}_1+\frac{\pi}{2}\right) = \frac{\bar{E}_{a1}}{2\bar{x}''_d}\angle(\bar{\delta}_1-\pi) \\
&k_4\angle-\theta_4 \doteqdot \frac{\bar{E}_{a1}}{2\bar{x}''_d}\angle\left(\bar{\alpha}_1+\frac{\pi}{2}\right) = \frac{\bar{E}_{a1}}{2\bar{x}''_d}\angle-(\bar{\delta}_1-\pi) \\
&\bar{i}_q(s) = \frac{k_5}{s} + \frac{k_6}{s+\frac{1}{\bar{T}''_q}} + \left\{ \frac{k_7\angle\theta_7}{s+\frac{1}{\bar{T}_a}-j} + \frac{k_7\angle-\theta_7}{s+\frac{1}{\bar{T}_a}+j} \right\} \quad ② \\
&\text{here } k_5 = -\frac{\bar{e}_d(0-)}{\bar{x}_q}, \qquad k_6 \doteqdot -e_d(0-)\cdot\left(\frac{1}{\bar{x}''_q}-\frac{1}{\bar{x}_q}\right) \\
&k_7\angle\theta_7 \doteqdot \frac{\bar{E}_{a1}}{2\bar{x}''_q}\angle-\bar{\alpha}_1 = \frac{\bar{E}_{a1}}{2\bar{x}''_q}\angle-\left(\frac{\pi}{2}-\bar{\delta}_1\right) \\
&\bar{E}_{a1} = \sqrt{\bar{e}_d(0-)^2+\bar{e}_q(0-)^2} \\
&\bar{i}_0(s) = 0 \quad ③
\end{aligned}
\right\} \quad (10.120)
$$

Each term on the right-hand side of Equation ① can be inverse transformed into the t domain by applying Laplace transforms (see Supplement 3):

$$
\left.
\begin{aligned}
\bar{i}_d(\bar{t}) &= \left[k_1 + k_2\cdot e^{-\frac{1}{\bar{T}'_d}\bar{t}} + k_3\cdot e^{-\frac{1}{\bar{T}''_d}\bar{t}} + k_4\angle\theta_4\cdot e^{-\left(\frac{1}{\bar{T}_a}-j\right)\bar{t}} + k_4\angle-\theta_4\cdot e^{-\left(\frac{1}{\bar{T}_a}+j\right)\bar{t}}\right]\mathbf{1}(t) \\
&= \left[\bar{e}_q(0-)\cdot\left\{\frac{1}{\bar{x}_d}+\left(\frac{1}{\bar{x}'_d}-\frac{1}{\bar{x}_d}\right)e^{-\bar{t}/\bar{T}'_d}+\left(\frac{1}{\bar{T}''_d}-\frac{1}{\bar{x}'_d}\right)e^{-\bar{t}/\bar{T}''_d}\right\} - \frac{\bar{E}_{a1}}{\bar{x}''_d}\cdot e^{-\bar{t}/\bar{T}_a}\cos(\bar{t}+\bar{\delta}_1)\right]\mathbf{1}(\bar{t}) \quad ① \\
\bar{i}_q(\bar{t}) &= \left[k_5 + k_6\cdot e^{-\frac{1}{\bar{T}''_q}\bar{t}} + k_7\angle\theta_7\cdot e^{-\left(\frac{1}{\bar{T}_a}-j\right)\bar{t}} + k_7\angle-\theta_7\cdot e^{-\left(\frac{1}{\bar{T}_a}+j\right)\bar{t}}\right]\mathbf{1}(t) \quad ② \\
&= \left[-\bar{e}_d(0-)\cdot\left\{\frac{1}{\bar{x}_q}+\left(\frac{1}{\bar{x}''_q}-\frac{1}{\bar{x}_q}\right)e^{-\bar{t}/\bar{T}''_q}\right\} + \frac{\bar{E}_{a1}}{\bar{x}''_q}\cdot e^{-\bar{t}/\bar{T}_a}\sin(\bar{t}+\bar{\delta}_1)\right]\mathbf{1}(\bar{t}) \\
\bar{i}_0(\bar{t}) &= 0 \quad ③ \\
&\text{where } \bar{e}_d(0-) = \bar{E}_{a1}\cos\bar{\alpha}_1 = \bar{E}_{a1}\sin\bar{\delta}_1 \\
&\bar{e}_q(0-) = \bar{E}_{a1}\sin\bar{\alpha}_1 = \bar{E}_{a1}\cos\bar{\delta}_1 \\
&\sqrt{e_d(0-)^2+e_q(0-)^2} = \bar{E}_{a1} \\
&\sin(\bar{t}-\bar{\alpha}_1) = -\cos(\bar{t}+\bar{\delta}_1), \qquad \cos(\bar{t}-\bar{\alpha}_1) = \sin(\bar{t}+\bar{\delta}_1)
\end{aligned}
\right\} \quad (10.121)
$$

This is the solution of the transient current in the d–q–0 and time t domains but without load current. Referring to Equation 10.64, the load currents are (putting $\bar{r}=0$ for simplicity)

$$\left.\begin{aligned} \bar{e}_d(0-) &= \bar{x}_q \cdot \bar{i}_q(0-) \\ \bar{e}_q(0-) &= \bar{E}_f - \bar{x}_d \cdot \bar{i}_d(0-) \\ \bar{e}_0(0-) &= 0 \end{aligned}\right\} \quad \text{or} \quad \left.\begin{aligned} \bar{i}_d(0-) &= \frac{\bar{E}_f - \bar{e}_q(0-)}{\bar{x}_d} \\ \bar{i}_q(0-) &= \frac{\bar{e}_d(0-)}{\bar{x}_q} \\ \bar{i}_0(0-) &= 0 \end{aligned}\right\} \tag{10.122}$$

The total fault currents are derived by addition of Equations 10.121 and 10.122, namely

$$\left.\begin{aligned} \bar{i}_d(\bar{t}) &= \bar{e}_q(0-)\cdot\left\{\left(\frac{1}{\bar{x}'_d}-\frac{1}{\bar{x}_d}\right)\mathrm{e}^{-\bar{t}/\bar{T}'_d}+\left(\frac{1}{\bar{x}''_d}-\frac{1}{\bar{x}'_d}\right)\mathrm{e}^{-\bar{t}/\bar{T}''_d}\right\} \\ &\quad +\frac{\bar{E}_f}{\bar{x}_d}-\frac{\bar{E}_{a1}}{\bar{x}''_d}\mathrm{e}^{-\bar{t}/\bar{T}_a}\cos(\bar{t}+\bar{\delta}_1) \\ &= \left\{\left(\frac{1}{\bar{x}'_d}-\frac{1}{\bar{x}_d}\right)\mathrm{e}^{-\bar{t}/\bar{T}'_d}+\left(\frac{1}{\bar{x}''_d}-\frac{1}{\bar{x}'_d}\right)\mathrm{e}^{-\bar{t}/\bar{T}''_d}\right\}\cdot\bar{E}_{a1}\cos\bar{\delta}_1 \\ &\quad +\frac{\bar{E}_f}{\bar{x}_d}-\frac{\bar{E}_{a1}}{\bar{x}''_d}\mathrm{e}^{-\bar{t}/\bar{T}_a}\cos(\bar{t}+\bar{\delta}_1) \\ \bar{i}_q(\bar{t}) &= -\bar{e}_d(0-)\cdot\left(\frac{1}{\bar{x}''_q}-\frac{1}{\bar{x}_q}\right)\mathrm{e}^{-\bar{t}/\bar{T}''_q}+\frac{\bar{E}_{a1}}{\bar{x}''_q}\cdot\mathrm{e}^{-\bar{t}/\bar{T}_a}\sin(\bar{t}+\bar{\delta}_1) \\ &= -\left(\frac{1}{\bar{x}''_q}-\frac{1}{\bar{x}_q}\right)\mathrm{e}^{-\bar{t}/\bar{T}''_q}\cdot\bar{E}_{a1}\sin\bar{\delta}_1+\frac{\bar{E}_{a1}}{\bar{x}''_q}\cdot\mathrm{e}^{-\bar{t}/\bar{T}_a}\sin(\bar{t}+\bar{\delta}_1) \\ \bar{i}_0 &= 0 \qquad \bar{t}\geqq 0 \end{aligned}\right\} \tag{10.123}$$

Finally, the above currents are inverse transformed from the d–q–0 domain to the a-b-c domain, referring to Equation 10.11b:

$$\begin{bmatrix}\bar{i}_a(\bar{t})\\ \bar{i}_b(\bar{t})\\ \bar{i}_c(\bar{t})\end{bmatrix} = \begin{bmatrix}\bar{i}_d(\bar{t})\cos\bar{t}-\bar{i}_q(\bar{t})\sin\bar{t}+\bar{i}_0(\bar{t})\\ \bar{i}_d(\bar{t})\cos\left(\bar{t}-\frac{2\pi}{3}\right)-\bar{i}_q(\bar{t})\sin\left(\bar{t}-\frac{2\pi}{3}\right)+\bar{i}_0(\bar{t})\\ \bar{i}_d(\bar{t})\cos\left(\bar{t}+\frac{2\pi}{3}\right)-\bar{i}_q(\bar{t})\sin\left(\bar{t}+\frac{2\pi}{3}\right)+\bar{i}_0(\bar{t})\end{bmatrix}$$

$$= \frac{\bar{E}_f}{\bar{x}_d}\cdot\begin{bmatrix}\cos\bar{t}\\ \cos\left(\bar{t}-\frac{2\pi}{3}\right)\\ \cos\left(\bar{t}+\frac{2\pi}{3}\right)\end{bmatrix} + \left\{\left(\frac{1}{\bar{x}'_d}-\frac{1}{\bar{x}_d}\right)\mathrm{e}^{-\bar{t}/\bar{T}'_d}+\left(\frac{1}{\bar{x}''_d}-\frac{1}{\bar{x}'_d}\right)\mathrm{e}^{-\bar{t}/\bar{T}''_d}\right\}\times\bar{E}_{a1}\cos\bar{\delta}_1\cdot\begin{bmatrix}\cos\bar{t}\\ \cos\left(\bar{t}-\frac{2\pi}{3}\right)\\ \cos\left(\bar{t}+\frac{2\pi}{3}\right)\end{bmatrix}$$

$$+\left(\frac{1}{\bar{x}''_q}-\frac{1}{\bar{x}_q}\right)\mathrm{e}^{-\bar{t}/\bar{T}''_q}\cdot\bar{E}_{a1}\sin\bar{\delta}_1\cdot\begin{bmatrix}\sin\bar{t}\\ \sin\left(\bar{t}-\frac{2\pi}{3}\right)\\ \sin\left(\bar{t}+\frac{2\pi}{3}\right)\end{bmatrix}$$

$$-\bar{E}_{a1}\mathrm{e}^{-\bar{t}/\bar{T}_a}\cdot\left\{\frac{1}{2}\left(\frac{1}{\bar{x}''_d}+\frac{1}{\bar{x}''_q}\right)\cdot\begin{bmatrix}\cos\bar{\delta}_1\\ \cos\left(\bar{\delta}_1+\frac{2\pi}{3}\right)\\ \cos\left(\bar{\delta}_1-\frac{2\pi}{3}\right)\end{bmatrix}+\frac{1}{2}\left(\frac{1}{\bar{x}''_d}-\frac{1}{\bar{x}''_q}\right)\cdot\begin{bmatrix}\cos(2\bar{t}+\bar{\delta}_1)\\ \cos\left(2\bar{t}+\bar{\delta}_1-\frac{2\pi}{3}\right)\\ \cos\left(2\bar{t}+\bar{\delta}_1+\frac{2\pi}{3}\right)\end{bmatrix}\right\} \tag{10.124}$$

This is our final solution, where:

- The first term is the steady-state term.
- The second term is the oscillatory term with attenuation factors of time constants $\bar{T}'_d$ and $\bar{T}''_d$.
- The third term is the oscillatory term with attenuation factors of time constant $\bar{T}''_q$.
- The fourth term is the d.c. attenuation term with attenuation factor of time constant $\bar{T}_a$.
- The fifth term is the second-harmonic term with attenuation factor of time constant $\bar{T}_a$.

Besides the fundamental frequency terms and d.c. term, the second-harmonic transient term has unexpectedly appeared. This term is obviously negligible for the thermal generator ($\because \bar{x}''_d \doteq \bar{x}''_q$).

As a numerical check, consider a hydro-generator with $\bar{x}''_d = 0.33$, $\bar{x}''_q = 0.41$ as one example. In this case

$$\{\ \}\text{ of the last term} = \left\{\frac{1}{2}\left(\frac{1}{\bar{x}''_d}+\frac{1}{\bar{x}''_q}\right)\cos\bar{\delta}_1 + \frac{1}{2}\left(\frac{1}{\bar{x}''_d}-\frac{1}{\bar{x}'_q}\right)\cos(2\bar{t}+\bar{\delta}_1)\right\}$$
$$= \{2.74\cos\delta_1 + 0.29\cos(2t+\delta_1)\}$$

Accordingly, the second-harmonic transient term is rather small in comparison with the d.c. term even for hydro-generators.

10.9.1 Transient fault current by sudden three-phase terminal fault under no-load condition

This is a special case of the above derived equations, where the initial condition of Equation 10.122 is replaced by the simple equations

$$\left.\begin{aligned} 0 = \bar{i}_q(0-) &= \frac{\bar{e}_d(0-)}{\bar{x}_q} \\ 0 = \bar{i}_d(0-) &= \frac{\bar{E}_f - e_q(0-)}{\bar{x}_d} \end{aligned}\right\}$$

That is,

$$\left.\begin{aligned} e_d(0-) = \bar{E}_{a1}\sin\bar{\delta}_1 &= 0 \qquad \text{or} \quad \bar{\delta}_1 = 0 \\ e_q(0-) = \bar{E}_{a1}\cos\bar{\delta}_1 &= \bar{E}_f \qquad\qquad \bar{E}_{a1} = \bar{E}_f \end{aligned}\right\} \tag{10.125}$$

and in Equations 10.123 and 10.124 replacement of $e_d(0-) = 0$, $e_q(0-) = \bar{E}_f$, $\bar{\delta}_1 = 0$, $\bar{E}_{a1}\sin\bar{\delta}_1 = \bar{E}_f$ is carried out.

It must be emphasized at the end of this chapter that a power system network, regardless of its size, can be described as 'one combined electrical equational circuit' by virtue of the mathematically elegant Park's equation explained here, together with the symmetrical components.

10.10 Supplement 1: The Equations of the Rational Function and Their Transformation into Expanded Sub-sequential Fractional Equations

A rational function $F(s)$ whose numerator $N(s)$ and denominator $M(s)$ are of polynomial form with n and m order respectively (where $m \geqq n$) can be expressed as

$$F(s) = \frac{N(s)}{M(s)} = \frac{s^n + d_1 s^{n-1} + \cdots + d_{n-1}s + d_n}{s^m + c_1 s^{m-1} + \cdots + c_{m-1}s + c_m} = \frac{s^n + b_1 s^{n-1} + \cdots + b_{n-1}s + b_n}{(s-s_1)\cdot(s-s_2)\cdot\cdots\cdot(s-s_m)} \tag{1}$$

$F(s)$ can be expanded into the sub-sequential fractional equation

$$F(s) = \frac{A_1}{s - s_1} + \frac{A_2}{s - s_2} + \cdots + \frac{A_m}{s - s_m} \tag{2}$$

$s_1, s_2, \ldots, s_m$ are the roots of $M(s)$. Here, $A_k(k = 1, 2, 3, \ldots, m)$ is a real number if s_k is a real-number root. On the contrary, A_k is a complex-number if s_k is a complex-number root.

We examine the next equation as a typical example, whose denominator is a polynomial of fifth order with three real-number roots $-a_1$, $-a_2$, $-a_3$ and two complex-number roots $-\alpha \pm j\beta$:

$$\left.\begin{aligned} F(s) &= \frac{N(s)}{(s + a_1)(s + a_2)(s + a_3)(s + \overline{\alpha - j\beta})(s + \overline{\alpha + j\beta})} \\ &= \frac{k_1}{s + a_1} + \frac{k_2}{s + a_2} + \frac{k_3}{s + a_3} + \left\{ \frac{k_4 \angle \delta_4}{s + \overline{\alpha - j\beta}} + \frac{k_4 \angle -\delta_4}{s + \overline{\alpha + j\beta}} \right\} \end{aligned}\right\} \tag{3}$$

k_1, k_2, k_3, k_4 are real numbers, and the fourth and the fifth terms on the right-hand side are the conjugates of each other.

Now we try to find k_1. Multiplying both sides of Equation 3 by $(s + a_1)$,

$$\begin{aligned} (s + a_1) \cdot F(s) &= \frac{N(s)}{(s + a_2)(s + a_3)(s + \overline{\alpha - j\beta})(s + \overline{\alpha + j\beta})} \\ &= k_1 + (s + a_1) \left[\frac{k_2}{s + a_2} + \frac{k_3}{s + a_3} + \left\{ \frac{k_4 \angle \delta_4}{s + \overline{\alpha - j\beta}} + (\text{conjugate}) \right\} \right] \end{aligned} \tag{4}$$

Putting $s = -a_1$,

$$k_1 = (s + a_1) \cdot F(s)|_{s=-a_1} = \frac{N(-a_1)}{(-a_1 + a_2)(-a_1 + a_3)(-a_1 + \overline{\alpha - j\beta})(-a_1 + \overline{\alpha + j\beta})} \tag{5}$$

Thus k_1 is found. All other coefficients k_2, k_3, $k_4 \angle \delta_4$, $k_4 \angle -\delta_4$ can be found similarly. For example,

$$\begin{aligned} k_4 \angle \delta_4 &= (s + \overline{\alpha - j\beta}) \cdot F(s)|_{s=-\alpha+j\beta} \\ &= \frac{N(-\alpha + j\beta)}{(\overline{-\alpha + j\beta} + a_1)(\overline{-\alpha + j\beta} + a_2)(\overline{-\alpha + j\beta} + a_3)(2j\beta)} \\ k_4 \angle -\delta_4 &= (s + \overline{\alpha + j\beta}) \cdot F(s)|_{s=\alpha-j\beta} \\ &= \frac{N(-\alpha - j\beta)}{(\overline{-\alpha - j\beta} + a_1)(\overline{-\alpha - j\beta} + a_2)(\overline{-\alpha - j\beta} + a_3)(-2j\beta)} = \{k_4 \angle \delta_4\}^* \end{aligned}$$

10.11 Supplement 2: Calculation of the Coefficients of Equation 10.120

10.11.1 Calculation of Equation ① k_1, k_2, k_3, $k_4 \angle \delta$, $k_4 \angle -\delta$

$$\begin{aligned} \bar{i}_d(s) &= \frac{\{\bar{e}_d(0-) \cdot s + \bar{e}_q(0-)\} \left(s + \frac{1}{\bar{T}'_{d0}}\right) \left(s + \frac{1}{\bar{T}''_{d0}}\right)}{\bar{x}''_d \cdot s \left(s + \frac{1}{\bar{T}'_d}\right) \left(s + \frac{1}{\bar{T}''_d}\right) \left(s^2 + \frac{2}{\bar{T}_a} s + 1\right)} = \frac{k_1}{s} + \frac{k_2}{s + \frac{1}{\bar{T}'_d}} + \frac{k_3}{s + \frac{1}{\bar{T}''_d}} \\ &\quad + \left\{ \frac{k_4 \angle \theta_4}{s + \frac{1}{\bar{T}_a} - j} + \frac{k_4 \angle -\theta_4}{s + \frac{1}{\bar{T}_a} + j} \right\} \end{aligned}$$

Then

$$k_1 = s \cdot \bar{i}_d(s)|_{s=0} = \frac{\{\bar{e}_d(0-) \cdot 0 + \bar{e}_q(0-)\}\left(0 + \frac{1}{\bar{T}'_{d0}}\right)\left(0 + \frac{1}{\bar{T}''_{d0}}\right)}{\bar{x}''_d \cdot \left(0 + \frac{1}{\bar{T}'_d}\right)\left(0 + \frac{1}{\bar{T}''_d}\right)\left(0^2 + \frac{2}{\bar{T}_a} \cdot 0 + 1\right)}$$

$$= \frac{\bar{e}_q(0-)}{\bar{x}''_d} \cdot \frac{\bar{T}'_d}{\bar{T}'_{d0}} \cdot \frac{\bar{T}''_d}{\bar{T}''_{d0}} = \frac{\bar{e}_q(0-)}{\bar{x}''_d} \cdot \frac{\bar{x}'_d}{\bar{x}_d} \cdot \frac{\bar{x}''_d}{\bar{x}'_d} = \frac{\bar{e}_q(0-)}{\bar{x}_d}$$

$$k_2 = \left(s + \frac{1}{\bar{T}'_d}\right) \cdot \bar{i}_d(s)|_{s=-\frac{1}{\bar{T}'_d}} = \frac{\left\{-\frac{\bar{e}_d(0-)}{\bar{T}'_d} + \bar{e}_q(0-)\right\}\left(-\frac{1}{\bar{T}'_d} + \frac{1}{\bar{T}''_{d0}}\right)\left(-\frac{1}{\bar{T}'_d} + \frac{1}{\bar{T}'_{d0}}\right)}{\bar{x}''_d \cdot \left(-\frac{1}{\bar{T}'_d}\right)\left(-\frac{1}{\bar{T}'_d} + \frac{1}{\bar{T}''_d}\right)\left(\frac{1}{\bar{T}'^2_d} - \frac{2}{\bar{T}'_d \cdot \bar{T}_a} + 1\right)}$$

The conversion ratio between [s] and [radian] is 1 s $\Leftrightarrow 2\pi \times 60 = 377$ rad for 60 Hz (314 rad for 50 Hz base) so that all the associated time constants in Table 10.1 are quite larger than 1 in radians. Namely, $\bar{T}'_d \gg 1$, $\bar{T}_a \gg 1$, $\bar{T}'_d \gg \bar{T}''_d$, $\bar{T}'_d \gg \bar{T}''_{d0}$, $\bar{T}'_{d0} \gg \bar{T}''_{d0} \gg 1$, $\bar{T}''_d 2 \gg 1$, etc.. Then

$$k_2 \doteqdot \frac{\bar{e}_q(0-)}{\bar{x}''_d}\left(1 - \frac{\bar{T}'_d}{\bar{T}'_d}\right) \cdot \frac{\bar{T}''_d}{\bar{T}''_{d0}} = \frac{\bar{e}_q(0-)}{\bar{x}''_d}\left(1 - \frac{\bar{x}'_d}{\bar{x}_d}\right) \cdot \frac{\bar{x}''_d}{\bar{x}'_d} = \bar{e}_q(0-) \cdot \left(\frac{1}{\bar{x}'_d} - \frac{1}{\bar{x}_d}\right)$$

$$k_3 = \left(s + \frac{1}{\bar{T}''_d}\right) \cdot \bar{i}_d(s)|_{s=-\frac{1}{\bar{T}''_d}} = \frac{\left\{-\frac{\bar{e}_d(0-)}{\bar{T}''_d} + \bar{e}_q(0-)\right\}\left(-\frac{1}{\bar{T}''_d} + \frac{1}{\bar{T}''_{d0}}\right)\left(-\frac{1}{\bar{T}''_d} + \frac{1}{\bar{T}'_{d0}}\right)}{\bar{x}''_d\left(-\frac{1}{\bar{T}''_d}\right)\left(-\frac{1}{\bar{T}''_d} + \frac{1}{\bar{T}'_d}\right)\left(\frac{1}{\bar{T}''^2_d} - \frac{2}{\bar{T}''_d \cdot \bar{T}_a} + 1\right)}$$

$$\therefore k_3 \doteqdot \frac{\bar{e}_q(0-)}{\bar{x}''_d} \cdot \left(1 - \frac{\bar{T}''_d}{\bar{T}''_{d0}}\right) = \frac{\bar{e}_q(0-)}{\bar{x}''_d} \cdot \left(1 - \frac{\bar{x}''_d}{\bar{x}'_d}\right) = \bar{e}_q(0-) \cdot \left(\frac{1}{\bar{x}''_d} - \frac{1}{\bar{x}'_d}\right)$$

Next, for $k_4 \angle \delta_4$, recalling that

$$\left(s^2 + \frac{2}{\bar{T}_a} s + 1\right) \doteqdot \left(s + \frac{1}{\bar{T}_a} - j\right)\left(s + \frac{1}{\bar{T}_a} + j\right)$$

then

$$k_4 \angle \delta_4 = \left(s + \frac{1}{\bar{T}_a} - j\right) \cdot \bar{i}_d(s)|_{s=-\frac{1}{\bar{T}_a}+j}$$

$$= \frac{\left\{\bar{e}_d(0-) \cdot \left(-\frac{1}{\bar{T}_a} + j\right) + \bar{e}_q(0-)\right\}\left(j - \frac{1}{\bar{T}_a} + \frac{1}{\bar{T}'_{d0}}\right)\left(j - \frac{1}{\bar{T}_a} + \frac{1}{\bar{T}''_{d0}}\right)}{\bar{x}'_d \cdot \left(j - \frac{1}{\bar{T}_a}\right)\left(j - \frac{1}{\bar{T}_a} + \frac{1}{\bar{T}'_d}\right)\left(j - \frac{1}{\bar{T}_a} + \frac{1}{\bar{T}''_d}\right)(2j)}$$

Here $\bar{T}'_{d0} \gg \bar{T}_a > \bar{T}''_{d0} \gg 1$, $\bar{T}'_d \gg 1$, $\bar{T}''_d \gg 1$, etc., so

$$k_4 \angle \delta_4 \doteqdot \frac{\{j \cdot \bar{e}_d(0-) + \bar{e}_q(0-)\} j^2}{\bar{x}''_d \cdot 2j^4} = \frac{-j}{2\bar{x}''_d} \cdot \{\bar{e}_d(0-) - j\bar{e}_q(0-)\}$$

$$= \frac{\sqrt{\bar{e}_d(0-)^2 + \bar{e}_q(0-)^2}}{2\bar{x}''_d} \cdot \angle \tan^{-1}\frac{-\bar{e}_q(0-)}{\bar{e}_d(0-)} \cdot \angle - \frac{\pi}{2} = \frac{\bar{E}_{a1}}{2\bar{x}''_d} \angle - \bar{\alpha}_1 \cdot \angle - \frac{\pi}{2}$$

$$\therefore k_4\angle\delta_4 \doteqdot \frac{\bar{E}_{a1}}{2\bar{x}_d''}\angle\left(\bar{\alpha}_1+\frac{\pi}{2}\right)=\frac{\bar{E}_{a1}}{2\bar{x}_d''}\angle(\bar{\delta}_1-\pi), \quad \text{where } \bar{\alpha}_1+\bar{\delta}_1=\frac{\pi}{2}$$

In the same way

$$k_4\angle-\delta_4 \doteqdot \frac{\bar{E}_{a1}}{2\bar{x}_d''}\angle\left(\bar{\alpha}_1+\frac{\pi}{2}\right)=\frac{\bar{E}_{a1}}{2\bar{x}_d''}\angle-(\bar{\delta}_1-\pi)$$

Accordingly, the summation of the third and fourth terms of Equation 10.115② is

$$\begin{aligned}&\{k_4\angle\delta_4\cdot \mathrm{e}^{-\left(\frac{1}{T_a}-j\right)\bar{t}}+k_4\angle-\theta_4\cdot \mathrm{e}^{-\left(\frac{1}{T_a}+j\right)\bar{t}}\}\\&=\frac{\bar{E}_{a1}}{2\bar{x}_d''}\cdot \mathrm{e}^{-\bar{t}/\bar{T}_a}\cdot\{\mathrm{e}^{j(\bar{\delta}_1-\pi+\bar{t})}+\mathrm{e}^{-j(\bar{\delta}_1-\pi+\bar{t})}\}\\&=-\frac{\bar{E}_{a1}}{\bar{x}_d''}\cdot \mathrm{e}^{-\bar{t}/\bar{T}_a}\cdot\cos(\bar{t}+\bar{\delta}_1)\end{aligned}$$

10.11.2 Calculation of equation 10.120② k_5, k_6, $k_7\angle\pm\delta_7$

In the same way,

$$k_5 \doteqdot s\cdot\bar{i}_q(s)|_{s=0}=\frac{-\bar{e}_d(0-)}{\bar{x}_q''}\cdot\frac{\bar{T}_q''}{\bar{T}_{q0}''}=\frac{-\bar{e}_d(0-)}{\bar{x}_q''}\cdot\frac{\bar{x}_q''}{\bar{x}_q}=-\frac{\bar{e}_d(0-)}{\bar{x}_q}$$

$$k_6 \doteqdot \frac{-\bar{e}_d(0-)}{\bar{x}_q''}\cdot\left(1-\frac{\bar{T}_q''}{\bar{T}_{q0}''}\right)=-\bar{e}_d(0-)\cdot\left(\frac{1}{\bar{x}_q''}-\frac{1}{\bar{x}_q}\right)$$

$$k_7\angle\delta_7 \doteqdot \frac{\bar{e}_d(0-)-j\bar{e}_q(0-)}{2\bar{x}_q''}=\frac{\bar{E}_{a1}}{2\bar{x}_q''}\angle-\bar{\alpha}_1=\frac{\bar{E}_{a1}}{2\bar{x}_q''}\angle\left(\bar{\delta}_1-\frac{\pi}{2}\right)$$

Readers might like to find $k_7\angle-\delta$ by themselves.

10.12 Supplement 3: The Formulae of the Laplace Transform (see also Appendix A)

$$\mathcal{L}^{-1}\left[\frac{1}{s}\right]=\mathbf{1}(t), \qquad \mathcal{L}^{-1}\left[\frac{1}{s\pm a}\right]=\mathrm{e}^{\mp at}\cdot\mathbf{1}(t)$$

$$\mathcal{L}\left[\frac{s+\alpha}{(s+\alpha)^2+\beta^2}\right]=\mathrm{e}^{-at}\cos\beta t\cdot\mathbf{1}(t), \quad \mathcal{L}^{-1}\left[\frac{\beta}{(s\pm\alpha)^2+\beta^2}\right]=\mathrm{e}^{-at}\sin\beta t\cdot\mathbf{1}(t)$$

Also, recalling $(s+\overline{\alpha-j\beta})(s+\overline{\alpha+j\beta})=(s+\alpha^2)+\beta^2$, then

$$\mathcal{L}^{-1}\left[\frac{A\angle\theta}{s+\alpha-j\beta}+\frac{A\angle-\theta}{s+\alpha+j\beta}\right]=2A\mathrm{e}^{-at}\cos(\beta t+\theta)\cdot\mathbf{1}(t)$$

11

Apparent Power and its Expression in the 0–1–2 and d–q–0 Domains

11.1 Apparent Power and its Symbolic Expression for Arbitrary Waveform Voltages and Currents

11.1.1 Definition of apparent power

According to custom, effective power P, reactive power Q and apparent power $\dot{S} = P + jQ$ are defined and applied in relation to fundamental frequency (or power frequency, as the more practical name) voltages and currents. So, let us examine first the power for fundamental frequency voltage and current given by the equations

$$\left.\begin{aligned} v(t) &= V\cos(\omega t+\alpha) = \sqrt{2}\,V_e\cos(\omega t+\alpha)\\ i(t) &= I\cos(\omega t+\beta) = \sqrt{2}\,I_e\cos(\omega t+\beta)\end{aligned}\right\} \tag{11.1}$$

where

V, I: the crest values
V_e, I_e: the rms values (suffix e means 'effective')

Here

$$V_e = \frac{V}{\sqrt{2}}, \quad I_e = \frac{I}{\sqrt{2}}$$

The instantaneous power $\tilde{P}(t)$ for the above voltage and current is

$$\begin{aligned}\tilde{P}(t) &= v(t)\cdot i(t) = VI\cos(\omega t+\alpha)\cos(\omega t+\beta)\\ &= \frac{VI}{2}\cos(\alpha-\beta) + \frac{VI}{2}\cos(2\omega t+\alpha+\beta)\end{aligned} \tag{11.2}$$

The first term on the right-hand side is the d.c. component without including time t, and the second term is the double frequency component. The instantaneous power $\tilde{P}(t)$ in this case is the offset-biased double frequency alternating power as shown in Figure 11.1.

Handbook of Power System Engineering Yoshihide Hase
© 2007 John Wiley & Sons, Ltd

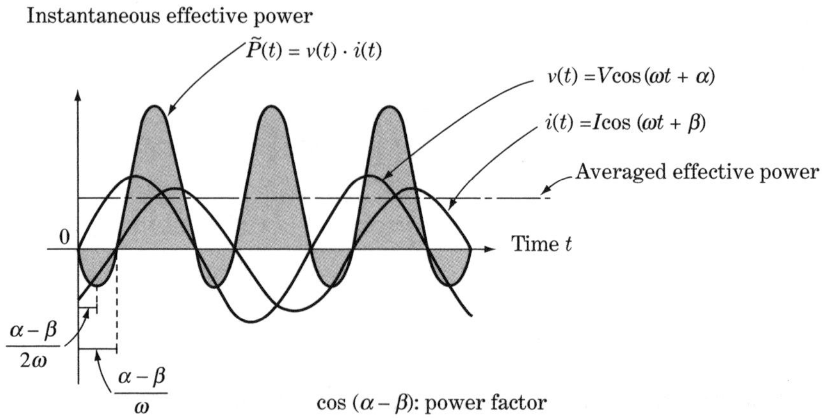

Figure 11.1 Instantaneous power $\tilde{P}(t)$ and apparent power P

The effective power P is defined as the time-averaged power of $\tilde{P}(t)$. Since the averaged value of the second term is zero, then

$$P = \frac{1}{T}\int_0^T \tilde{P}(t)dt = \frac{VI}{2}\cos(\alpha - \beta) = \frac{V}{\sqrt{2}} \cdot \frac{I}{\sqrt{2}}\cos(\alpha - \beta) = V_e I_e \cos(\alpha - \beta) \qquad (11.3)$$

where $T = 2\pi/\omega$ is the time for one periodic cycle.

As an additional comment, if the voltage and current in Equation 11.1 are expressed by the sine instead of cosine, only the sign of the second term on the right-hand side of Equation 11.2 is replaced from + to −. Therefore the definition of P by Equation 11.3 is not affected by the selection of the cosine or sine.

The reactive power Q is defined by the following equation in response to Equation 11.3:

$$Q = V_e I_e \sin(\alpha - \beta) \qquad (11.4)$$

The effective power P and reactive power Q have been defined for the real-number voltage and current of observable waveforms. The defined P and Q and the apparent power $\dot{S}$ can be defined as below in the symbolic method based on complex-number expressions. (In this chapter, all the variables with the form of a complex number will be expressed by letters with the 'dot' notation just to emphasize that they are complex-number variables.)

For the voltage and current in the symbolic method $v(t)$, $i(t)$, and for the apparent power, respectively,

$$\left.\begin{array}{l} \left.\begin{array}{l} \dot{v}(t) \equiv \frac{V}{\sqrt{2}} e^{j(\omega t + \alpha)} = V_e e^{j(\omega t + \alpha)} \\ \dot{i}(t) \equiv \frac{I}{\sqrt{2}} e^{j(\omega t + \beta)} = I_e e^{j(\omega t + \beta)} \end{array}\right\} \quad ① \\ \dot{S} = P + jQ \equiv \dot{v}(t) \cdot \dot{i}(t)^* = V_e I_e e^{j(\alpha - \beta)} \\ \qquad = V_e I_e \cos(\alpha - \beta) + jV_e I_e \sin(\alpha - \beta) \quad ② \\ \text{where } V_e I_e = (V/\sqrt{2}) \cdot (I/\sqrt{2}) = VI/2. \end{array}\right\} \qquad (11.5)$$

That is, the apparent power $\dot{S} = P + jQ$ is defined as the product of $\dot{v}(t)$ and $\dot{i}(t)^*$ ($\dot{i}(t)^*$ is the conjugate of $\dot{i}(t)$). The sign $\equiv$ signifies that the equation is a definition. The definition of P, Q in Equation 11.5 is the same as that in Equations 11.3 and 11.4. Further, we need to distinguish carefully the rms values V_e, I_e and the crest values V, I in the above equations.

11.1.2 Expansion of apparent power for arbitrary waveform voltages and currents

Let us assume a voltage of angular velocity ω_1 and a current of angular velocity ω_2 at an arbitrary point in a circuit. Then

$$\left.\begin{aligned} \dot{v}(t) &= V_e \mathrm{e}^{j(\omega_1 t+\alpha)} \\ \dot{i}(t) &= I_e \mathrm{e}^{j(\omega_2 t+\beta)} \\ \dot{\tilde{S}}(t) &= \dot{v}(t)\cdot\dot{i}(t)^* = V_e I_e \mathrm{e}^{j\{(\omega_1-\omega_2)t+(\alpha-\beta)\}} \end{aligned}\right\} \tag{11.6}$$

Therefore, $\dot{S} = P + jQ = \{\text{the averaged value of } \dot{v}(t)\cdot\dot{i}(t)^*\}$, that is

$$\dot{S} = \begin{cases} V_e I_e \mathrm{e}^{j(\alpha-\beta)} = V_e I_e \cos(\alpha-\beta) + jV_e I_e \sin(\alpha-\beta) & \text{for } \omega_1 = \omega_2 \\ 0 \quad \text{for } \omega_1 \neq \omega_2 \end{cases} \tag{11.7}$$

This equation means that the apparent power appears only as the product of voltage and current of the same frequency. Bearing this result in mind, we examine the power for voltage and current with an arbitrary waveform.

For the instantaneous voltage and current with an arbitrary waveform $\dot{v}(t)$, $\dot{i}(t)$

$$\left.\begin{aligned} \dot{v}(t) &= V_{e0}\mathrm{e}^{j\alpha_0} + V_{e1}\mathrm{e}^{j(\omega t+\alpha_1)} + V_{e2}\mathrm{e}^{j(2\omega t+\alpha_2)} + V_{e3}\mathrm{e}^{j(3\omega t+\alpha_3)} + \ldots \\ \dot{i}(t) &= I_{e0}\mathrm{e}^{j\beta_0} + I_{e1}\mathrm{e}^{j(\omega t+\beta_1)} + I_{e2}\mathrm{e}^{j(2\omega t+\beta_2)} + I_{e3}\mathrm{e}^{j(3\omega t+\beta_3)} + \ldots \end{aligned}\right\} \tag{11.8}$$

for the instantaneous apparent power $\dot{\tilde{S}}(t) = \tilde{P}(t) + j\tilde{Q}(t)$

$$\begin{aligned} \dot{\tilde{S}}(t) &= \tilde{P}(t) + j\tilde{Q}(t) = \{\dot{v}(t)\cdot\dot{i}(t)^*\} \\ &= \left\{\sum_{k=0} V_{ek}\mathrm{e}^{j(k\omega t+\alpha_k)}\right\}\cdot\left\{\sum_{k=0} I_{ek}\mathrm{e}^{-j(k\omega t+\beta_k)}\right\} \end{aligned} \tag{11.9}$$

and for the apparent power $\dot{S} = P + jQ$

$$\begin{aligned} \dot{S} &= P + jQ = \{\text{the averaged value of } \dot{v}(t)\cdot\dot{i}(t)^*\} \\ &= V_{e0}I_{e0}\mathrm{e}^{j(\alpha_0-\beta_0)} + V_{e1}I_{e1}\mathrm{e}^{j(\alpha_1-\beta_1)} + V_{e2}I_{e2}\mathrm{e}^{j(\alpha_2-\beta_2)} + V_{e3}I_{e3}\mathrm{e}^{j(\alpha_3-\beta_3)} + \cdots \\ &= \left\{\sum_{k=0} V_{ek}I_{ek}\cos(\alpha_k-\beta_k)\right\} + j\left\{\sum_{k=0} V_{ek}I_{ek}\sin(\alpha_k-\beta_k)\right\} \end{aligned} \tag{11.10}$$

The apparent power of a single phase circuit for the voltage and current of an arbitrary waveform has thus been introduced.

11.2 Apparent Power of a Three-phase Circuit in the 0–1–2 Domain

Applying the results of Equations 11.8–11.10 for a single phase circuit, the apparent power of a three-phase circuit with the voltages and currents of an arbitrary waveform are introduced as follows.

For the instantaneous apparent power per phase

$$\left.\begin{array}{l}\begin{bmatrix}\dot{\tilde{S}}_a(t)\\ \dot{\tilde{S}}_b(t)\\ \dot{\tilde{S}}_c(t)\end{bmatrix}=\begin{bmatrix}\tilde{P}_a(t)\\ \tilde{P}_b(t)\\ \tilde{P}_c(t)\end{bmatrix}+j\begin{bmatrix}\tilde{Q}_a(t)\\ \tilde{Q}_b(t)\\ \tilde{Q}_c(t)\end{bmatrix}=\begin{bmatrix}\dot{v}_a(t)\cdot\dot{i}_a(t)^*\\ \dot{v}_b(t)\cdot\dot{i}_b(t)^*\\ \dot{v}_c(t)\cdot\dot{i}_c(t)^*\end{bmatrix}\\ \text{or}\\ \dot{\tilde{\boldsymbol{S}}}_{abc}(t)=\tilde{\boldsymbol{P}}_{abc}(t)+j\tilde{\boldsymbol{Q}}_{abc}(t)=\dot{\boldsymbol{v}}_{abc}(t)\cdot\dot{\boldsymbol{i}}_{abc^*}\end{array}\right\}\quad(11.11\text{a})$$

and for the instantaneous apparent power for the total three phases

$$\left.\begin{array}{rl}\dot{\tilde{S}}_{3\phi}(t)= & \dot{\tilde{S}}_a(t)+\dot{\tilde{S}}_b(t)+\dot{\tilde{S}}_c(t)=\{\tilde{P}_a(t)+\tilde{P}_b(t)+\tilde{P}_c(t)\}+j\{\tilde{Q}_a(t)+\tilde{Q}_b(t)+\tilde{Q}_c(t)\}\\ = & \dot{v}_a(t)\cdot\dot{i}_a(t)^*+\dot{v}_b(t)\cdot\dot{i}_b(t)^*+\dot{v}_c(t)\cdot\dot{i}_c(t)^*\\ = & \underbrace{\begin{bmatrix}\dot{v}_a(t) & \dot{v}_b(t) & \dot{v}_c(t)\end{bmatrix}}_{{}^t\dot{\boldsymbol{v}}_{abc}(t)}\cdot\underbrace{\begin{bmatrix}(\dot{i}_a(t)*\\ \dot{i}_b(t)*\\ \dot{i}_c(t)*\end{bmatrix}}_{\dot{\boldsymbol{i}}_{abc}(t)*}\\ \text{or} & \\ & \dot{\tilde{\boldsymbol{S}}}_{3\phi}(t)={}^t\dot{\boldsymbol{v}}_{abc}(t)\cdot\dot{\boldsymbol{i}}_{abc}(t)^*\end{array}\right\}\quad(11.11\text{b})$$

Here, the voltages and currents with arbitrary waveforms are

$$\left.\begin{array}{l}\left.\begin{array}{l}\dot{v}_a(t)=\sum_k V_{ak}e^{j(k\omega t+\alpha_{ak})}\\ \dot{v}_b(t)=\sum_k V_{bk}e^{j(k\omega t+\alpha_{bk})}\\ \dot{v}_c(t)=\sum_k V_{ck}e^{j(k\omega t+\alpha_{ck})}\end{array}\right\}\text{①}\\ \left.\begin{array}{l}\dot{i}_a(t)=\sum_k I_{ak}e^{j(k\omega t+\beta_{ak})}\\ \dot{i}_b(t)=\sum_k I_{bk}e^{j(k\omega t+\beta_{bk})}\\ \dot{i}_c(t)=\sum_k I_{ck}e^{j(k\omega t+\beta_{ck})}\end{array}\right\}\text{②}\end{array}\right.\quad(11.12)$$

The dimensions of voltages and currents are volts [V] and amperes [A], while those for the apparent power P, Q, $\dot{S}$ are volt amperes [VA] with P also in watts [W].

Now we transform Equation 11.11 into the equation of power in symmetrical components:

$$\left.\begin{array}{l}\dot{\boldsymbol{v}}_{abc}(t)=\boldsymbol{a}^{-1}\dot{\boldsymbol{v}}_{012}(t)\\ \dot{\boldsymbol{i}}_{abc}(t)=\boldsymbol{a}^{-1}\dot{\boldsymbol{i}}_{012}(t)\end{array}\right\}\quad(11.13)$$

Recalling the following matrix formulae,

$${}^t\boldsymbol{a}^{-1}=\boldsymbol{a}^{-1},\quad {}^t\{\mathbf{A}\cdot\mathbf{B}\}={}^t\mathbf{B}\cdot{}^t\mathbf{A}\quad(11.14)$$

then

$$\left.\begin{array}{rl}\dot{\boldsymbol{S}}_{3\phi}(t) & ={}^t\boldsymbol{v}_{abc}(t)\cdot\boldsymbol{i}_{abc}(t)^*\\ {}^t\boldsymbol{v}_{abc}(t) & ={}^t\{\boldsymbol{a}^{-1}\cdot\boldsymbol{v}_{012}\}={}^t\boldsymbol{v}_{012}(t)\cdot{}^t\boldsymbol{a}^{-1}={}^t\boldsymbol{v}_{012}(t)\cdot\boldsymbol{a}^{-1}\\ \boldsymbol{i}_{abc}(t)^* & =\{\boldsymbol{a}^{-1}\cdot\boldsymbol{i}_{012}(t)\}^*=\boldsymbol{a}^{-1*}\cdot\boldsymbol{i}_{012}(t)^*=3\boldsymbol{a}\cdot\boldsymbol{i}_{012}(t)^*\end{array}\right\}\quad(11.15)$$

while

$$\left.\begin{array}{l} \boldsymbol{a}^{-1*} = \begin{array}{|c|c|c|}\hline 1 & 1 & 1 \\ \hline 1 & (a^2)^* & (a)^* \\ \hline 1 & (a)^* & (a^2)^* \\ \hline \end{array} = \begin{array}{|c|c|c|}\hline 1 & 1 & 1 \\ \hline 1 & a & a^2 \\ \hline 1 & a^2 & a \\ \hline \end{array} = 3\boldsymbol{a} \quad \begin{pmatrix} \boldsymbol{a}^2 = \mathrm{e}^{-j120^0} = \boldsymbol{a}^* \\ \boldsymbol{a} = \mathrm{e}^{j120^0} = \boldsymbol{a}^{2*} \end{pmatrix} \\ \therefore \quad \dot{\boldsymbol{i}}_{012}(t)^* = \frac{1}{3}\boldsymbol{a}^{-1} \cdot \dot{\boldsymbol{i}}_{abc}(t)^* \\ \qquad \dot{\boldsymbol{i}}_{abc}(t)^* = 3\boldsymbol{a} \cdot \dot{\boldsymbol{i}}_{012}(t)^* \end{array}\right\} \tag{11.16}$$

Accordingly,

$$\left.\begin{array}{l} \left.\begin{array}{l} \dot{\tilde{\mathbf{S}}}_{3\phi}(t) = {}^t\dot{\boldsymbol{v}}_{abc}(t) \cdot \dot{\boldsymbol{i}}_{abc}(t)^* = \{{}^t\dot{\boldsymbol{v}}_{012}(t).\boldsymbol{a}^{-1}\}\{3\boldsymbol{a} \cdot \dot{\boldsymbol{i}}_{012}(t)^*\} \\ \quad = 3{}^t\dot{v}_{012}(t) \cdot \dot{i}_{012}(t)^* = 3 \begin{array}{|c|c|c|}\hline \dot{v}_0(t) & \dot{v}_1(t) & \dot{v}_2(t) \\ \hline \end{array} \begin{array}{|c|}\hline \dot{i}_0(t)^* \\ \hline \dot{i}_1(t)^* \\ \hline \dot{i}_2(t)^* \\ \hline \end{array} \\ \quad = 3\{\dot{v}_0(t) \cdot \dot{i}_0(t)^* + \dot{v}_1(t) \cdot \dot{i}_1(t)^* + \dot{v}_2(t) \cdot \dot{i}_2(t)^*\} \\ \quad = 3\{\dot{\tilde{S}}_0(t) + \dot{\tilde{S}}_1(t) + \dot{\tilde{S}}_2(t)\} \end{array}\right\} \text{①} \\ \text{or} \\ \frac{\dot{\tilde{S}}_{3\phi}(t)}{3} = \dot{\tilde{S}}_{1\phi}(t) = \frac{1}{3}\{\dot{S}_a(t) + \dot{S}_b(t) + \dot{S}_c(t)\} \\ \quad = \frac{1}{3}\{\dot{v}_a(t)\dot{i}_a(t)^* + \dot{v}_b(t) \cdot \dot{i}_b(t)^* + \dot{v}_c(t) \cdot \dot{i}_c(t)^*\} \\ \quad = \dot{\tilde{S}}_0(t) + \dot{\tilde{S}}_1(t) + \dot{\tilde{S}}_2(t) \\ \quad = \dot{v}_0(t) \cdot \dot{i}_0(t)^* + \dot{v}_1(t) \cdot \dot{i}_1(t)^* + \dot{v}_2(t) \cdot \dot{i}_2(t)^* \qquad \text{②} \\ \text{where} \\ \left.\begin{array}{l} \dot{\tilde{S}}_0(t) = \tilde{P}_0(t) + j\tilde{Q}_0(t) = \dot{v}_0(t) \cdot \dot{i}_0(t)^* \\ \dot{\tilde{S}}_1(t) = \tilde{P}_1(t) + j\tilde{Q}_1(t) = \dot{v}_1(t) \cdot \dot{i}_1(t)^* \\ \dot{\tilde{S}}_2(t) = \tilde{P}_2(t) + j\tilde{Q}_2(t) = \dot{v}_2(t) \cdot \dot{i}_2(t)^* \end{array}\right\} \text{③} \end{array}\right\} \tag{11.17}$$

We can conclude as follows from Equation 11.17. For voltages $\dot{v}_{abc}(t)$ and currents $\dot{i}_{abc}(t)$ with arbitrary waveforms in a three-phase circuit:

- The apparent power of positive-, negative-, zero-sequence circuits is derived independently as $\dot{\tilde{S}}_0(t)$, $\dot{\tilde{S}}_1(t)$, $\dot{\tilde{S}}_2(t)$ of Equation 11.17 ③ in the 0–1–2 domain.
- The summation $\dot{\tilde{S}}_0(t) + \dot{\tilde{S}}_1(t) + \dot{\tilde{S}}_2(t)$ is equal to $\dot{\tilde{S}}_{1\phi}(t)$, the averaged power of $\dot{\tilde{S}}_a(t), \dot{\tilde{S}}_b(t), \dot{\tilde{S}}_c(t)$.

Equation 11.17 is unitized as follows.

Recalling $VA_{3\phi\text{base}} = 3VA_{1\phi\text{base}}$,

$$\begin{aligned} \frac{\dot{\tilde{S}}_{3\phi}/3}{\dot{S}_{3\phi\text{base}}/3} &= \frac{\left(\dot{\tilde{S}}_a + \dot{\tilde{S}}_b + \dot{\tilde{S}}_c\right)/3}{\dot{S}_{1\phi\text{base}}} = \frac{\dot{\tilde{S}}_{1\phi}}{\dot{S}_{1\phi\text{base}}} = \frac{\dot{\tilde{S}}_0}{\dot{S}_{1\phi\text{base}}} + \frac{\dot{\tilde{S}}_1}{\dot{S}_{1\phi\text{base}}} + \frac{\dot{\tilde{S}}_2}{\dot{S}_{1\phi\text{base}}} \\ &= \frac{\dot{v}_0(t)}{V_{1\phi\text{base}}} \cdot \frac{\dot{i}_0(t)^*}{I_{1\phi\text{base}}} + \frac{\dot{v}_1(t)}{V_{1\phi\text{base}}} \cdot \frac{\dot{i}_1(t)^*}{I_{1\phi\text{base}}} + \frac{\dot{v}_2(t)}{V_{1\phi\text{base}}} \cdot \frac{\dot{i}_2(t)^*}{I_{1\phi\text{base}}} \end{aligned} \tag{11.18}$$

namely

$$\dot{\bar{S}}_{3\phi} = \dot{\bar{S}}_{1\phi} = \left(\dot{\bar{S}}_a + \dot{\bar{S}}_b + \dot{\bar{S}}_c\right)/3$$
$$= \dot{\bar{S}}_0 + \dot{\bar{S}}_1 + \dot{\bar{S}}_2 = \bar{v}_0(t) \cdot \dot{\bar{i}}_0(t)^* + \bar{v}_1 \cdot (t) \cdot \dot{\bar{i}}_2 \cdot (t)^* + \bar{v}_2(t) \cdot \dot{\bar{i}}_2(t)^* \tag{11.19}$$

$\bar{S}_{3\phi}$ (total three-phase power) and $\bar{S}_{1\phi}$ (averaged single phase power) have the same value by unitization.

11.3 Apparent Power in the d–q–0 Domain

The relation between the apparent power in the 0–1–2 domain and that in the d–q–0 domain will be investigated now.

Effective power P_a of phase-a can be calculated directly in terms of d-and q-axes quantities $e_d(t)$, $e_q(t)$ and $i_d(t),\ i_q(t)$ that are defined by Equation 10.11a and 10.11b as of real-number expression.

$$P_a = e_a(t) \cdot i_a(t) = \{\cos\theta_a \cdot e_d(t) - \sin\theta_a \cdot e_q(t) + \mathrm{Re}[e_0(t)]\} \cdot \{\cos\theta_a \cdot i_d(t) - \sin\theta_a \cdot i_q(t) + \mathrm{Re}[i_0(t)]\}$$
$$= [\cos^2\theta_a] \cdot e_d(t) \cdot i_d(t) + [\sin^2\theta_a] \cdot e_q(t) \cdot i_q(t) - [\cos\theta_a \sin\theta_a]\{e_q(t) \cdot i_d(t) + \cdot e_d(t) \cdot i_q(t)\}$$
$$+ \mathrm{Re}[e_0(t) \cdot i_0^*(t)]$$
$$\text{where } \theta_a = \omega t - 120^\circ, \theta_b = \omega t - 120^\circ, \theta_c = \omega t + 120^\circ. \tag{11.20}$$

$e_d(t), e_q(t)$ and $i_d(t), i_q(t)$ are d.c values under three-phase balanced condition, while may not be time-independent whenever phase-unbalanced.

P_b and P_c are derived as equations of the same form, then reminding the formulae $\sin\theta_a + \sin\theta_b + \sin\theta_c = 0,\ \cos^2\theta_a + \cos^2\theta_b + \cos^2\theta_c = 3/2$ etc.,

$$P_a + P_b + P_c = [\cos^2\theta_a + \cos^2\theta_b + \cos^2\theta_c] \cdot e_d(t) \cdot i_d(t) + [\sin^2\theta_a + \sin^2\theta_b + \sin^2\theta c] \cdot e_q(t) \cdot i_q(t)$$
$$- [\cos\theta a \sin\theta a + \cos\theta_b \sin\theta_b + \cos\theta_c \sin\theta_c]\{(e_d(t) \cdot i_q(t) + e_q(t) \cdot i_d(t)\}$$
$$+ 3\mathrm{Re}[e_0(t) \cdot i_0^*(t)] = (3/2)e_d(t) \cdot i_d(t) + (3/2)e_q(t) \cdot i_q(t) + 3\mathrm{Re}[e_0(t) \cdot i_0^*(t)]$$

$\therefore$

$$(P_a + P_b + P_c)/3 = (1/2)\{e_d(t) \cdot i_d(t) + e_q(t) \cdot i_q(t)\} + \mathrm{Re}[e_0(t) \cdot i_0^*(t)]$$
$$= (1/2)\{P_d(t) + P_q(t)\} + \mathrm{Re}[e_0(t) \cdot i_0^*(t)] \tag{11.21}$$

In next, reactive power Q_a of phase-a can be calculated directly as the equation of the same form with Equation 10.20 but by replacing θ_a of current $i_a(t)$ to $\theta_a + 90^\circ$.

$$Q_a = e_a(t) \cdot i_a(t) = \{\cos\theta_a \cdot e_d(t) - \sin\theta_a \cdot e_q(t) + \mathrm{Im}[e_0(t)]\} \cdot \{\cos(\theta_a + 90) \cdot i_d(t)$$
$$- \sin(\theta_a + 90) \cdot i_q(t) + \mathrm{Im}[i_0^*(t)]\}$$
$$= [\cos\theta_a \sin\theta_a] \cdot \{(-e_d(t) \cdot i_d(t) + e_q(t) \cdot i_q(t))\} + [\sin^2\theta_a](e_q(t) \cdot i_d(t) - [\cos^2\theta_a]e_d(t) \cdot i_q(t)$$
$$+ \mathrm{Im}[e_0(t) \cdot i_0^*(t)]$$
$$\therefore (Q_a + Q_b + Q_c) = (3/2)e_q(t) \cdot i_d(t) - (3/2)e_d(t) \cdot i_q(t) + 3\mathrm{Im}[e_0(t) \cdot i_0^*(t)]$$

and then

$$\therefore\ (Q_a + Q_b + Q_c)/3 = (1/2)\{e_q(t) \cdot i_d(t) - e_d(t) \cdot i_q(t)\} + \mathrm{Im}[e_0(t) \cdot i_0^*(t)]$$
$$= (1/2)\{Q_d(t) + Q_q(t)\} + \mathrm{Im}[e_0(t) \cdot i_0^*(t)] \tag{11.22}$$

Furthermore, Equation 11.21 and 11.22 can be recasted as a complex number equation of P + jQ..

$$
\begin{aligned}
S_{3\phi/3} = S_{1\phi} &= (P_a + P_b + P_c)/3 + j(Q_a + Q_b + Q_c)/3 \\
&= (1/2)\{e_d(t)\cdot i_d(t) + e_q(t)\cdot i_q(t)\} + (1/2)j\{e_q(t)\cdot i_d(t) - e_d(t)\cdot i_q(t)\} + e_0(t)\cdot i_0^*(t) \\
&= (1/2)\{e_d(t) + j\,e_q(t)\}\cdot\{i_d(t) - ji_q(t))\} + e_0(t)\cdot i_0{}^*(t) \\
&= (1/2)\{(P_d + j\,Q_d) + (P_q + j\,Q_q)\} + e_0(t)\cdot i_0{}^*(t) \\
&= (1/2)\{S_d + S_q\} + S_0
\end{aligned}
\tag{11.23}
$$

The equations can be unitized by the same capacity base as adopted in Equation 10.40. Accordingly, whenever the generator's apparent power is concerned, the rated rms capacity of the generator by [VA] or [MVA] is usually adopted as the base quantity for the unitization of the apparent power. Thus

$$
\frac{VA_{3\phi\text{base}}}{3} = \left(\frac{{}_s e_{\text{base}}}{\sqrt{2}}\right)\cdot\left(\frac{{}_s i_{\text{base}}}{\sqrt{2}}\right) = \frac{1}{2}\cdot {}_s e_{\text{base}}\cdot {}_s i_{\text{base}} \tag{11.24}
$$

${}_s e_{\text{base}}/\sqrt{2}$: rms phase voltage base
${}_s i_{\text{base}}/\sqrt{2}$: rms phase current base

Dividing Equation 11.23 by Equation 11.24, the coefficient 1/2 disappears and we have

$$
\begin{aligned}
\dot{\bar{S}}_{3\phi} = \dot{\bar{S}}_{1\phi} &= \bar{P}_{3\phi} + j\bar{Q}_{3\phi} = \bar{P}_{1\phi} + j\bar{Q}_{1\phi} \\
&= \dot{\bar{e}}_d(t)\cdot \dot{\bar{i}}_d(t)^* + \dot{\bar{e}}_q(t)\cdot \dot{\bar{i}}_q(t)^* + 2\dot{\bar{e}}_0(t)\cdot \dot{\bar{i}}_0(t)^* \quad ① \\
&= \dot{\bar{S}}_d(t) + \dot{\bar{S}}_q(t) + 2\dot{\bar{S}}_0(t)
\end{aligned}
\tag{11.25}
$$

or referring to Equation (11.19)

$$
\dot{\bar{S}}_{3\phi} = \dot{\bar{S}}_{1\phi} = \dot{\bar{S}}_1 + \dot{\bar{S}}_2 + \dot{\bar{S}}_0 = \dot{\bar{S}}_d + \dot{\bar{S}}_q + 2\dot{\bar{S}}_0 \quad ②
$$

In the unitized equation, fortunately the coefficient 1/2 disappears. However, the zero-sequence power of $2\dot{\bar{S}}_0$ has unexpectedly appeared as a term of unitized total apparent power $\dot{\bar{S}}_{3\phi} = \dot{\bar{S}}_{1\phi}$.

Again in the d–q–0 domain, the apparent powers of d-axis, q-axis and zero-axis circuits $\dot{\bar{S}}_d, \dot{\bar{S}}_q, \dot{\bar{S}}_0$ can be independently treated in the form of $\dot{\bar{e}}(t)\cdot\dot{\bar{i}}(t)^*$. The summation value $(\dot{\bar{S}}_d + \dot{\bar{S}}_q + 2\dot{\bar{S}}_0)$ is equal to the unitized three-phase power $\dot{\bar{S}}_{3\phi} = \dot{\bar{S}}_{1\phi}$.

The above derived equations of apparent power are the expanded equations which can be applied for arbitrary unbalanced phase voltages and currents. The equations shows that apparent power S, effective power P, reactive/capacitive power Q can be interchangeable on a-b-c domain, symmetrical domain as well as d-q-0 domain.

Coffee break 6: Hertz, the discoverer and inventor of radio waves

In the 1870s and 1880s, most scientists doubted Maxwell's theory and still continued their study or discussion of the 'existence of ether', because the profound meaning of Maxwell's equations was not easily understood at first, and in particular any definite proof that electricity and light were electromagnetic waves did not exist. However, Hertz's famous experiment in 1888 and his profound theory became very important turning points for proving Maxwell's predictions.

Heinrich Rudolf Hertz (1857–1894) carried out his famous experiment in his laboratory in 1888, where radio waves were sent by one coil and detected by another coil (antenna coil) some

Heinrich Rudolf Hertz (1857–1894)

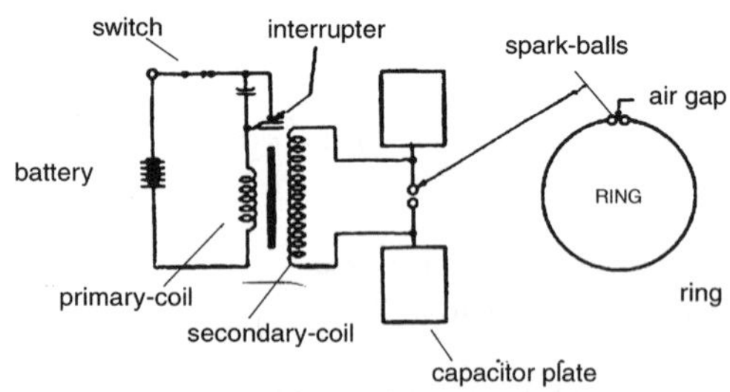

Hertz's first transmitter (1986)

distance from each other. The first coil contained a metal rod with a small gap at its midpoint, and when sparks crossed this gap, violent oscillations at high frequency were set up in the rod and the noise signal was detected by the second coil circuit. Hertz proved that these waves were transmitted through air from one coil to the other. This is the first instance of wave radiation by humankind.

Hertz also showed that, like light waves, these waves were reflected and refracted and, in particular, they travelled at the same speed as light but had a much longer wavelength. He also noted that electrical conductors reflected the waves and that they can be focused by concave reflectors. He found that non-conductors (including air and vacuum spaces) allow most of the waves to pass through. These waves, of course just **radio waves** to us today, conclusively confirmed Maxwell's prediction of the existence of electromagnetic waves, in the form of both light and radio waves, and further of electricity in telegraph cables. The **Hertz dipole** or **Hertz's oscillator** originated from the above experiment and became fundamentals of **radio wave applications** of the twentieth century.

It is said that in 1891 Oliver Heaviside quoted: 'Three years ago electromagnetic waves were nowhere. Shortly afterwards, they were everywhere.'

12

Generating Power and Steady-state Stability

We studied generator theory in Chapter 10 as the characteristics of voltage v, current i and flux linkage ψ without using P, Q. However, in Chapter 11, we discussed what apparent power $\dot{S} = P + jQ$ means in the a–b–c, 0–1–2 and d–q–0 domain. Now we need to extend generator theory to network theory as the characteristics of v, i, P, Q, f.

In this chapter, we begin by examining the generator's fundamental characteristics in combination with the connected network. As the first step, the equations of generator operation by v, i, P, Q under three-phase-balanced conditions are introduced, and then steady-state stability of a power system is discussed.

Incidentally, we will study the behaviour of various power systems at fundamental frequency as well as higher harmonics and surge frequency using simple power system models in the latter part of the book. We would like to emphasize that the essence of power system behaviour can be clearly understood only by studying simple models.

12.1 Generating Power and the *P*–δ and *Q*–δ Curves

A generator is operating under three-phase-balanced conditions and rotating at constant speed. This case coincides with that of Equations 10.52–10.56 and of the vector diagram in Figure 10.5. That is,

$$\left.\begin{aligned} \bar{e}_d + j\bar{e}_q &= \bar{E}_1 e^{j\bar{\alpha}_1} \\ \bar{i}_d + j\bar{i}_q &= \bar{I}_1 e^{j\bar{\beta}_1} \end{aligned}\right\} \tag{12.1a}$$

$$\left.\begin{aligned} (\bar{e}_d + j\bar{e}_q)e^{j\bar{t}} &= \bar{E}_1 e^{j(\bar{t}+\bar{\alpha}_1)} = \dot{\bar{e}}_1(\bar{t}) \\ (\bar{i}_d + j\bar{i}_q)e^{j\bar{t}} &= \bar{I}_1 e^{j(\bar{t}+\bar{\beta}_1)} = \dot{\bar{i}}_1(\bar{t}) \end{aligned}\right\} \tag{12.1b}$$

$$\left.\begin{aligned} \bar{e}_d &= \bar{E}_1 \cos\bar{\alpha}_1 = \bar{E}_1 \sin\delta \\ \bar{e}_q &= \bar{E}_1 \sin\bar{\alpha}_1 = \bar{E}_1 \cos\delta \end{aligned}\right\} \tag{12.1c}$$

where $\delta = (\pi/2) - \alpha_1$ (the suffix and overbar on δ are omitted for simplicity).

The symbol (t) in $\bar{e}_d, \bar{e}_q, \bar{i}_d, \bar{i}_q$ has been omitted because all the quantities are for time-independent d.c. components under three-phase-balanced conditions. $\bar{t}$ is the unitized value [rad] based on Equation 10.42. Referring to Figure 10.5, δ is the angular difference between the generator's induced voltage $j\bar{E}_f (\equiv j\bar{x}_{ad} \cdot \bar{i}_{fd}$, proportional to the field excitation) and the terminal voltage $\bar{E}_1 e^{j\bar{\alpha}_1}$ (namely, the generator's inner angular difference) and $\delta = (\pi/2) - \bar{\alpha}_1$.

© 2007 John Wiley & Sons, Ltd

The apparent power is, referring to Equations 12.1a and b,

$$\left.\begin{aligned} \dot{\bar{S}}_{3\phi} = \dot{\bar{S}}_{1\phi} = \dot{\bar{e}}_1(\bar{t}) \cdot \dot{\bar{i}}_1(t)^* &= \left(\bar{E}_1 e^{j\bar{\alpha}_1}\right) \cdot \left(\bar{I}_1 e^{j\beta_1}\right)^* \\ &= (\bar{e}_d + j\bar{e}_q)e^{j\bar{t}} \cdot \left(\bar{i}_d - j\bar{i}_q\right)e^{-j\bar{t}} \\ &= (\bar{e}_d\bar{i}_d + \bar{e}_q\bar{i}_q) + j(\bar{e}_q\bar{i}_d - \bar{e}_d\bar{i}_q) \\ \therefore \quad \bar{P}_{3\phi} = \bar{P}_{1\phi} = \bar{e}_d\bar{i}_d + \bar{e}_q\bar{i}_q & \\ \bar{Q}_{3\phi} = \bar{Q}_{1\phi} = \bar{e}_q\bar{i}_d - \bar{e}_d\bar{i}_q & \end{aligned}\right\} \tag{12.2}$$

This is the apparent power of the generator under three-phase-balanced conditions, and all the d- and q-domain quantities are for d.c. components.

Next, Equation 10.59 is utilized under three-phase-balanced conditions. Rewriting the equation as

$$\left.\begin{aligned} \bar{i}_d &= \frac{\bar{E}_f - \bar{e}_q - \bar{r}\bar{i}_q}{\bar{x}_d} \\ \bar{i}_q &= \frac{\bar{e}_d + \bar{r}\bar{i}_d}{\bar{x}_q} \end{aligned}\right\} \tag{12.3}$$

neglecting terms in $\bar{r}$ (because $\bar{x}_d, \bar{x}_q \gg \bar{r}$) and substituting Equation 12.3 into Equation 12.2,

$$\left.\begin{aligned} \bar{P}_{\text{gen}} \equiv \bar{P}_{3\phi} = \bar{P}_{1\phi} &= \frac{\bar{E}_f\bar{e}_d}{\bar{x}_d} + \bar{e}_d\bar{e}_q\left(\frac{1}{\bar{x}_q} - \frac{1}{\bar{x}_d}\right) & ① \\ \bar{Q}_{\text{gen}} = \bar{Q}_{3\phi} = \bar{Q}_{1\phi} &= \frac{\bar{E}_f\bar{e}_q}{\bar{x}_d} - \left(\frac{\bar{e}_d^2}{\bar{x}_q} + \frac{\bar{e}_q^2}{\bar{x}_d}\right) & ② \end{aligned}\right\} \tag{12.4}$$

On substituting Equation 12.1c into 12.4, the following very important equations are derived.

For the P–δ and Q–δ curve characteristics

$$\left.\begin{aligned} &\boldsymbol{P-\delta}\ \textbf{curve} \\ &\bar{P}_{\text{gen}} = \bar{P}_{3\phi} = \frac{\bar{E}_f\bar{E}_1}{\bar{x}_d}\sin\delta + \frac{\bar{E}_1^2}{2}\left(\frac{1}{\bar{x}_q} - \frac{1}{\bar{x}_d}\right)\sin 2\delta_1 & ① \\ &\boldsymbol{Q-\delta}\ \textbf{curve} \\ &\bar{Q}_{\text{gen}} = \bar{Q}_{3\phi} = \left\{\frac{\bar{E}_f\bar{E}_1}{\bar{x}_d}\cos\delta - \frac{\bar{E}_1^2}{2}\left(\frac{1}{\bar{x}_q} + \frac{1}{\bar{x}_d}\right)\right\} + \frac{\bar{E}_1^2}{2}\left(\frac{1}{\bar{x}_q} - \frac{1}{\bar{x}_d}\right)\cos 2\delta_1 & ② \end{aligned}\right\} \tag{12.5}$$

where the second terms on the right-hand sides in the Equations ①② are called the 'saliency effect' terms.

These are very important equations which explain the generator's essential characteristics in terms of the ***P–δ* curve** and ***Q–δ* curve**. Figure 12.1a shows the vector diagram of the generator quantities under three-phase-balanced operation. Figures 12.1b and c are the sets of the P–δ curve and Q–δ curve derived from Equation 12.5 which show effective power P and reactive power Q as parameters of δ. The saliency effect appears under the condition of $\bar{x}_d \neq \bar{x}_q$.

Equation 12.5 ① ② can be written as the combined equation

$$\left.\begin{aligned} &\dot{\bar{S}}_{\text{gen}} = \bar{P}_{\text{gen}} + j\bar{Q}_{\text{gen}} = \frac{\bar{E}_1^2}{2}\left(\frac{1}{j\bar{x}_q} + \frac{1}{j\bar{x}_d}\right) - \frac{\bar{E}_f\bar{E}_1}{j\bar{x}_d}e^{-j\delta} - \frac{\bar{E}_1^2}{2}\left(\frac{1}{j\bar{x}_q} - \frac{1}{j\bar{x}_d}\right)e^{-j2\delta} & ① \\ &\text{for the non-salient-pole machine } (\bar{x}_d = \bar{x}_q) \\ &\dot{\bar{S}}_{\text{gen}} = \bar{P}_{\text{gen}} + j\bar{Q}_{\text{gen}} = \bar{E}_1 \cdot \frac{\bar{E}_1 - \bar{E}_f e^{-j\delta}}{j\bar{x}_d} & ② \end{aligned}\right\} \tag{12.6}$$

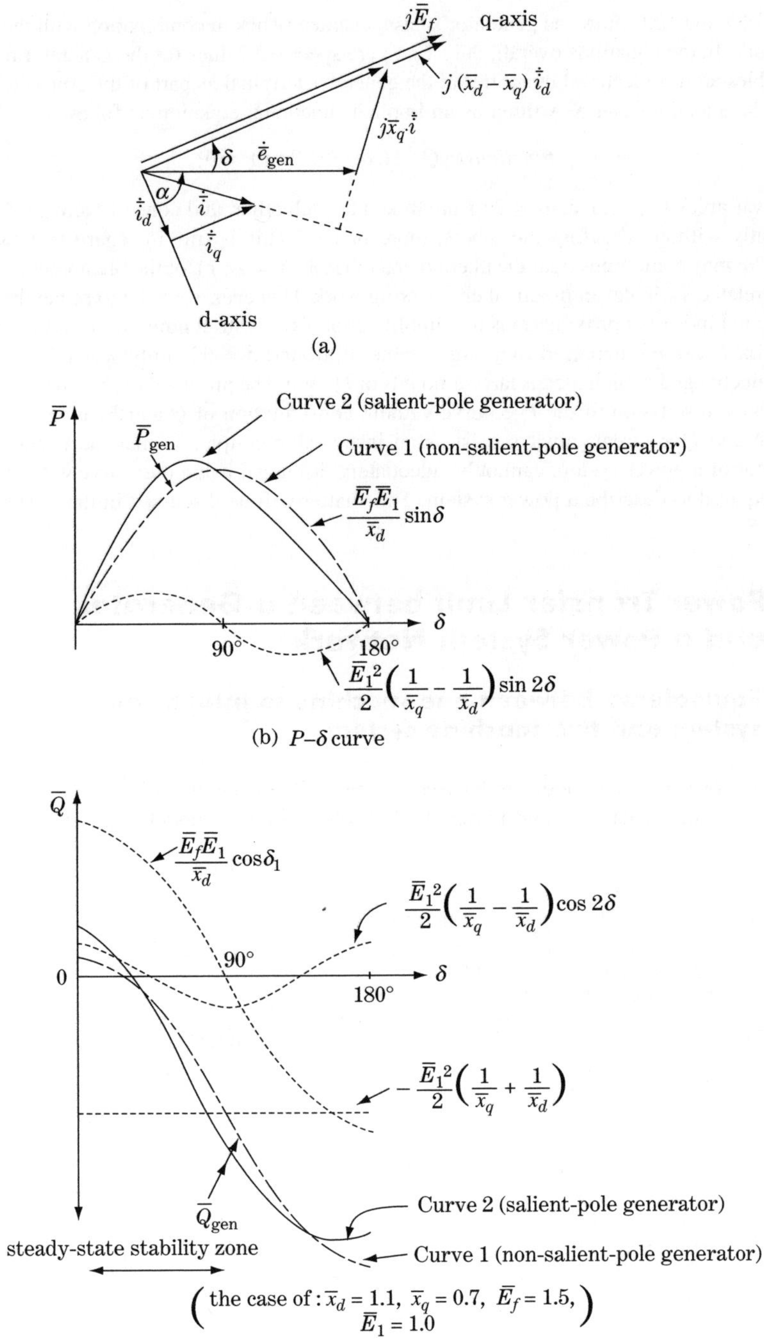

(a)

(b) P–δ curve

(c) Q–δ curve

Figure 12.1 *P*–δ and *Q*–δ curves of a generator

Equations 12.5 and 12.6 show the generator's basic characteristics in combination with the connected outer network. In the equations overall, only $\bar{x}_d, \bar{x}_q$ are specified values for the generator machine; all other variables are for electrical quantities at the generator terminal as part of the connected network. Therefore the equations can be written as an implicit functional equation as follows:

$$function(P, Q, v, i, \delta, \omega = 2\pi f) = 0 \tag{12.7}$$

All the variables P, Q, v, i, δ, $\omega = 2\pi f$ are linked to each other, and none of them can be changed independently without affecting the others, more or less. This is true in regard to power system networks. We may sometimes treat the phenomena of (P, δ, $\omega = 2\pi f$), or the phenomena of (Q, V) as closely correlated variables in practical engineering work. However, we need to remember that such treatments are kinds of approximations for simplification from a strict point of view. For example, in Figure 12.1b, P can be increased to a larger value of around $\delta = 90°$ only when E_1 (the terminal voltage) is unchanged, which means large amounts of Q should be provided as shown in Figure 12.1c. In other words, discussion of the P–δ curve without consideration of Q and E_1 is meaningless.

The P–δ and Q–δ curves always exist as an inseparable couple, so that the characteristics or the behaviour of a power system cannot be adequately discussed using one curve without the other. Both are required to describe a power system. This matter will be discussed in the next section and Section 14.5.

12.2 Power Transfer Limit between a Generator and a Power System Network

12.2.1 Equivalency between one-machine to infinite-bus system and two-machine system

Figure 12.2a shows a system model with two machines in that generator 1 is connected to generator 2 through a network reactance, and Figure 12.3a shows a system model with one machine to an

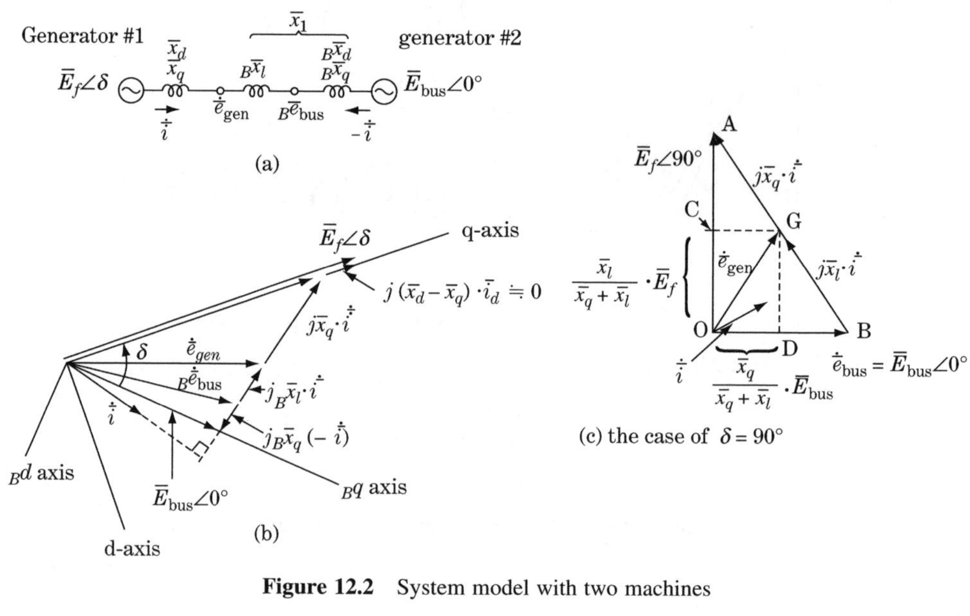

Figure 12.2 System model with two machines

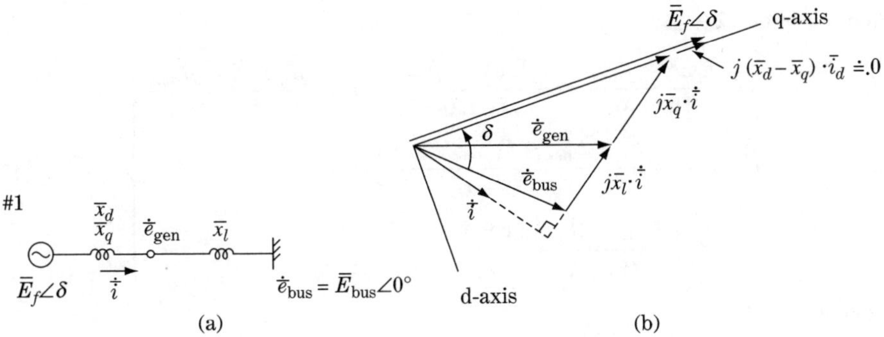

Figure 12.3 System model with one machine to infinite bus

infinite bus, in which generator 1 is connected to the infinite bus. The vector diagram for each case is also shown in the figures.

First of all, let us examine the equivalence of both figures. The electrical condition of generator 1 becomes equivalent in both figures if the condition in the following equation below is satisfied:

$$ {}_B\bar{x}_l + {}_B\bar{x}_d \doteqdot {}_B\bar{x}_l + {}_B\bar{x}_q \equiv \bar{x}_l \tag{12.8} $$

Equation 12.8 means that the d-axis circuit and the q-axis circuit of the outer system can be considered as the same as each other in Figure 12.2a. (The difference for the zero-sequence circuit is neglected as we are studying three-phase-balanced normal operating conditions.)

The assumption of Equation 12.8 can be justified if generator 2 is of cylindrical type ($\bar{x}_d = \bar{x}_q$), or even if generator 2 is of salient type (${}_B\bar{x}_d \neq {}_B\bar{x}_q$). Equation 12.8 may be approximately justified by insertion of some line reactance ${}_B\bar{x}_l$ between both generators.

In other words, at least the behaviour of generator 1 can be examined equally by a two-machine system or by a one-machine to infinite-bus system. In an actual power system with a total number n of generators, we can imagine that our generator 1 is connected to another equivalent generator which actually consists of $n - 1$ parallel connected generators through the network.

12.2.2 Apparent power of a generator

We assume that Equation 12.8 is justified and that generator 1 is a thermal driven generator ($\bar{x}_d = \bar{x}_q$) in Figures 12.2 and 12.3. Then

$$
\left.
\begin{array}{ll}
\dot{e}_{bus} = \bar{E}_{bus}\angle 0° = \bar{E}_{bus} & ① \\
\dot{e}_{gen} = \bar{E}_{bus} + j\bar{x}_l\dot{\bar{i}} & ② \\
\bar{E}_f = \bar{E}_f\angle\delta = \bar{E}_{bus} + j(\dot{\bar{x}}_q + \dot{\bar{x}}_l)\dot{\bar{i}}, \quad j(\bar{x}_d - \bar{x}_q)\dot{\bar{i}}_d \doteqdot 0 & ③ \\
\dot{S}_{bus} = \bar{P}_{bus} + j\bar{Q}_{bus} = \bar{E}_{bus}\dot{\bar{i}}^* & ④ \\
\dot{S}_{gen} = \bar{P}_{gen} + j\bar{Q}_{gen} = (\bar{E}_{bus} + j\bar{x}_l\dot{\bar{i}})\dot{\bar{i}}^* & ⑤
\end{array}
\right\}
\tag{12.9}
$$

where the suffix gen refers to quantities of generator 1.

From Equation 12.9 ④ ⑤

$$ \bar{P}_{gen} + j\bar{Q}_{gen} = (\bar{P}_{bus} + j\bar{Q}_{bus}) + j\bar{x}_l\bar{i}^2 \tag{12.10} $$

From Equation 12.9 ③

$$\left.\begin{aligned} \dot{i} &= \frac{\overline{E}_f e^{j\delta} - \overline{E}_{\text{bus}}}{j(\overline{x}_q + \overline{x}_l)} = \frac{-j\overline{E}_f e^{j\delta} + j\overline{E}_{\text{bus}}}{\overline{x}_q + \overline{x}_l}, \quad \dot{i}^* = \frac{j\overline{E}_f e^{-j\delta} - j\overline{E}_{\text{bus}}}{\overline{x}_q + \overline{x}_l} \\ \dot{i}^2 &= \dot{i} \cdot \dot{i}^* = \frac{\overline{E}_f^2 + \overline{E}_{\text{bus}}^2 - \overline{E}_f \overline{E}_{\text{bus}}(e^{j\delta} + e^{-j\delta})}{(\overline{x}_q + \overline{x}_l)^2} \\ &= \frac{\overline{E}_f^2 + \overline{E}_{\text{bus}}^2 - 2\overline{E}_f \overline{E}_{\text{bus}} \cos\delta}{(\overline{x}_q + \overline{x}_l)^2} \end{aligned}\right\} \tag{12.11}$$

Accordingly,

$$\left.\begin{aligned} \dot{S}_{\text{bus}} &= \overline{P}_{\text{bus}} + j\overline{Q}_{\text{bus}} = \overline{E}_{\text{bus}} \cdot \dot{i}^* = \overline{E}_{\text{bus}} \cdot \frac{j\overline{E}_f e^{-j\delta} - j\overline{E}_{\text{bus}}}{\overline{x}_q + \overline{x}_l} \\ &= \frac{\overline{E}_f \overline{E}_{\text{bus}}}{\overline{x}_q + \overline{x}_l} \sin\delta + j\left\{\frac{\overline{E}_{\text{bus}}(\overline{E}_f \cos\delta - \overline{E}_{\text{bus}})}{\overline{x}_q + \overline{x}_l}\right\} & ① \\ \dot{S}_{\text{gen}} &= \overline{P}_{\text{gen}} + j\overline{Q}_{\text{gen}} = \frac{\overline{E}_f \overline{E}_{\text{bus}}}{\overline{x}_q + \overline{x}_l} \sin\delta + j\left\{\frac{\overline{E}_f^2 \overline{x}_l - \overline{E}_{\text{bus}}^2 \overline{x}_q + \overline{E}_f \overline{E}_{\text{bus}}(\overline{x}_q - \overline{x}_l)\cos\delta}{(\overline{x}_q + \overline{x}_l)^2}\right\} & ② \\ \overline{P}_{\text{gen}} &= \frac{\overline{E}_f \overline{E}_{\text{bus}}}{\overline{x}_q + \overline{x}_l} \sin\delta \quad (\boldsymbol{P - \delta}\ \textbf{curve}) & ③ \\ \overline{Q}_{\text{gen}} &= \frac{\overline{E}_f^2 \overline{x}_l - \overline{E}_{\text{bus}}^2 \overline{x}_q + \overline{E}_f \overline{E}_{\text{bus}}(\overline{x}_q - \overline{x}_l)\cos\delta}{(\overline{x}_q + \overline{x}_l)^2} \quad (\boldsymbol{Q - \delta}\ \textbf{curve}) \end{aligned}\right\} \tag{12.12}$$

$\overline{P}_{\text{gen}}, \overline{Q}_{\text{gen}}$ have been derived as P–δ and Q–δ characteristics with the parameters of $\overline{E}_f, \overline{E}_{\text{bus}}$, where usually $\overline{E}_{\text{bus}} = 1.0\angle 0°$.

The equation shows that $\overline{Q}_{\text{gen}}$ is of positive (+) magnitude (lagging operation) for large $\overline{E}_f$, while it is of negative (−) magnitude (leading operation) for smaller $\overline{E}_f$ (weak excitation). Also $\overline{P}_{\text{gen}} = \overline{P}_{\text{bus}}$ is recognized because line resistance is neglected, while $\overline{Q}_{\text{gen}} \neq \overline{Q}_{\text{bus}}$ because reactive power consumption on the transmission line exists. The special case of $\overline{x}_l \to 0$, $\overline{E}_{\text{bus}} \to \overline{E}_1$, $\overline{x}_d = \overline{x}_q$ in Equation 12.12 coincides with Equation 12.6 ②.

12.2.3 Power transfer limit of a generator (steady-state stability)

A generator normally operates in synchronization with the connected power system. This means that $\overline{E}_f\angle\delta$ and $\overline{E}_{\text{bus}}\angle 0°$ are running at the same speed and the angular $\angle\delta$ displacement is always within some upper limit ($\delta < 90°$).

Suppose generators 1 and 2 are operating in synchronization, and generator 2 is accelerated a little for some reason and leads to $\delta \to \delta + \Delta\delta$. Then generator 1 automatically tries to recover the delay of $\Delta\delta$ by releasing the kinetic energy stored in its rotor, so that the electrical output of generator 1 is immediately increased by $\overline{P} \to \overline{P} + \Delta\overline{P}$. Such inherent recovering characteristics of the generator are called **synchronizing power**. The synchronizing power of the generator is effective so long as the recovering power $\Delta\overline{P}$ is of positive sign for some disturbance $\Delta\delta$.

In other words, the critical condition in which the generator can be operated with the power system in synchronization is

$$\text{Synchronizing power} \quad \frac{\partial \overline{P}_{\text{gen}}}{\partial\delta} = \frac{\Delta\overline{P}}{\Delta\delta} \geq 0 \tag{12.13}$$

Applying Equation 12.12 ③ to Equation 12.13,

$$\frac{\partial \overline{P}_{\text{gen}}}{\partial \delta} = \frac{E_f \overline{E}_{\text{bus}}}{\overline{x}_g + \overline{x}_l} \cos\delta \geqq 0 \quad \therefore \delta \leqq 90^\circ \tag{12.14}$$

The generator reaches the power transfer upper limit at large δ of approximately 90°. Such an upper limit condition is called the **steady-state stability limit**. Using the suffix max for such conditions,

$$\left.\begin{aligned} \dot{\overline{S}}_{g\max} &= \overline{P}_{g\max} + j\overline{Q}_{g\max} & ① \\ \overline{P}_{g\max} &= \frac{\overline{E}_f \overline{E}_{\text{bus}}}{\overline{x}_q + \overline{x}_l} & ② \\ \overline{Q}_{g\max} &= \frac{\overline{E}_f^2 \overline{x}_l - \overline{E}_{\text{bus}}^2\ \overline{x}_q}{(\overline{x}_q + \overline{x}_l)^2} & ③ \\ &\text{where } \delta = 90^\circ \text{(with the condition } \overline{x}_d = \overline{x}_q) \end{aligned}\right\} \tag{12.15}$$

In Figure 12.1b, the steady-state stability limit is at $\delta = 90^\circ$ for a generator of non-salient poles ($\overline{x}_d = \overline{x}_q$, curve 1) and is at an angle of less than 90° (say, 70°) for a generator of salient poles ($\overline{x}_d \neq \overline{x}_q$, curve 2).

It should be noted that in order to transfer such maximum power $P_{g\max}$ through the line, the corresponding large value of Q_{gen} in Figure 12.1c must be supplied to the system so that the terminal voltage E_1 is kept as the normal voltage. Otherwise, E_1 decreases and the P–δ curve shrinks. The necessary value for Q_{gen} becomes quite large for the operation with δ exceeding 50° as shown in Figure 12.1c. This is why we always need to treat P and Q together as coupled quantities.

12.2.4 Visual description of generator's apparent power transfer limit

$\overline{P}_{g\max}, \overline{Q}_{g\max}$ are functions of $\overline{E}_f, \overline{E}_{\text{bus}}$ and so cannot be independent of each other. Therefore we want to find an equation for $\overline{P}_{g\max}$ and $\overline{Q}_{g\max}$ by elimination of $\overline{E}_f$ and $\overline{E}_{\text{bus}}$. For this purpose we need one more equation besides Equation 12.15 ② ③, which can be obtained from Figure 12.2c. Under the condition of $\delta = 90^\circ$, $\overline{\text{AO}} : \overline{\text{CO}} = \overline{\text{AB}} : \overline{\text{GB}} = (\overline{x}_l + \overline{x}_q) : \overline{x}_l$, so that Pythagoras's theorem can be applied:

$$\dot{e}_{\text{gen}}^2 = \left(\frac{\overline{x}_l}{\overline{x}_q + \overline{x}_l}\right)^2 \overline{E}_f^2 + \left(\frac{\overline{x}_q}{\overline{x}_q + \overline{x}_l}\right)^2 \overline{E}_{\text{bus}}^2 \tag{12.16}$$

Eliminating $\overline{E}_f, \overline{E}_{\text{bus}}$ from Equations 12.15 ② ③ and 12.16, we obtain the following equation (refer to the supplement at the end of the chapter for the process):

$$\overline{P}_{g\max}^2 + \left\{\overline{Q}_{g\max} - \frac{1}{2}\left(\frac{1}{\overline{x}_l} - \frac{1}{\overline{x}_q}\right)\overline{e}_{\text{gen}}^2\right\}^2 = \left\{\frac{1}{2}\left(\frac{1}{\overline{x}_l} + \frac{1}{\overline{x}_q}\right)\overline{e}_{\text{gen}}^2\right\}^2 \tag{12.17}$$

Unitizing $\overline{P}_{\max}, \overline{Q}_{\max}$ by $\overline{e}_{\text{gen}}^2$ and writing them as $\overline{p}, \overline{q}$,

$$\left.\begin{aligned}
&\overline{p}^2 + \left\{\overline{q} - \frac{1}{2}\left(\frac{1}{\overline{x}_l} - \frac{1}{\overline{x}_q}\right)\right\}^2 = \left\{\frac{1}{2}\left(\frac{1}{\overline{x}_l} + \frac{1}{\overline{x}_q}\right)\right\}^2 \\
&\qquad \text{where} \quad \overline{p} = \frac{\overline{P}_{\text{gmax}}}{\overline{e}_{\text{gen}}^2}, \quad \overline{q} = \frac{\overline{Q}_{\text{gmax}}}{\overline{e}_{\text{gen}}^2} \\
&\quad \underline{\text{the circular locus in } p - q \text{ coordinates :}} \\
&\quad \text{centre:} \quad \left(0, \frac{1}{2}\left(\frac{1}{\overline{x}_l} - \frac{1}{\overline{x}_q}\right)\right) \\
&\quad \text{radius:} \quad \frac{1}{2}\left(\frac{1}{\overline{x}_l} + \frac{1}{\overline{x}_q}\right) \\
&\quad \text{diameter:} \quad \text{the straight line connecting the points } \left(0, \frac{1}{\overline{x}_l}\right) \text{ and } \left(0, -\frac{1}{\overline{x}_q}\right)
\end{aligned}\right\} \qquad (12.18)$$

Figure 12.4 can be drawn from Equation 12.18. The circle in p–q coordinates gives the steady-state stability limit and the generator cannot be operated outside of the circle.

Equation 12.18 can be modified to the equation below:

$$\left.\begin{aligned}
&(\overline{p}\,\overline{x}_q)^2 + \left\{(\overline{q}\,\overline{x}_q) - \frac{1}{2}\left(\frac{\overline{x}_q}{\overline{x}_l} - 1\right)\right\}^2 = \left\{\frac{1}{2}\left(\frac{\overline{x}_q}{\overline{x}_l} + 1\right)\right\}^2 \\
&\underline{\text{For the circular locus in } (\overline{p}\,\overline{x}_q) - (\overline{q}\,\overline{x}_q) \text{ coordinates:}} \\
&\quad \text{centre:} \quad \left(0, \frac{1}{2}\left(\frac{\overline{x}_q}{\overline{x}_l} - 1\right)\right) \\
&\quad \text{radius:} \quad \frac{1}{2}\left(\frac{\overline{x}_q}{\overline{x}_l} + 1\right) \\
&\quad \text{diameter:} \quad \text{the straight line connecting points } \left(0, \frac{\overline{x}_q}{\overline{x}_l}\right) \text{ and } (0, -1)
\end{aligned}\right\} \qquad (12.19)$$

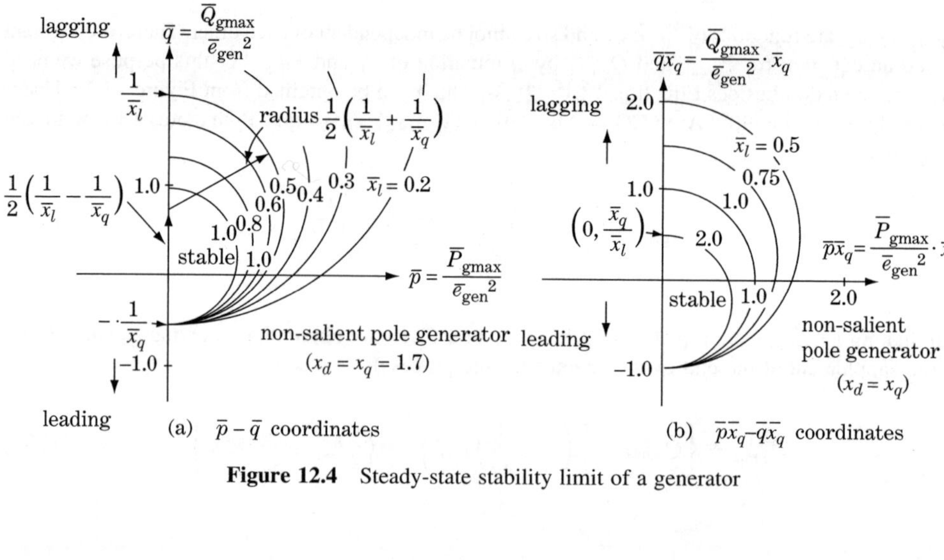

(a) $\overline{p} - \overline{q}$ coordinates (b) $\overline{p}\overline{x}_q$–$\overline{q}\overline{x}_q$ coordinates

Figure 12.4 Steady-state stability limit of a generator

Figure 12.4b can be drawn from the equation, that is circles based on the parameter $\bar{x}_q/\bar{x}_l$. If the power system capacity is relatively smaller ($\bar{x}_l \to$ large), the generator's stable operating zone shrinks. In other words, if the generator's output power exceeds the limit, the generator has to be stepped out in deceleration mode.

12.3 Supplement: Derivation of Equation 12.17 from Equations 12.15 ② ③ and 12.16

Putting

$$A = \frac{\bar{E}_f}{\bar{x}_q + \bar{x}_l} \qquad B = \frac{\bar{E}_{\text{bus}}}{\bar{x}_q + \bar{x}_l} \tag{1}$$

we obtain

$$\bar{P}_{\text{gmax}} = AB(\bar{x}_q + \bar{x}_l) \tag{2}$$

$$\bar{Q}_{\text{gmax}} = A^2\bar{x}_l - B^2\bar{x}_q \tag{3}$$

$$\bar{e}^2_{\text{gen}} = A^2\bar{x}^2_l + B^2\bar{x}^2_q \tag{4}$$

From (3) (4)

$$A^2 = \frac{\bar{e}^2_{\text{gen}} + \bar{Q}_{\text{gmax}}\bar{x}_q}{\bar{x}_l(\bar{x}_q + \bar{x}_l)} \qquad B^2 = \frac{\bar{e}^2_{\text{gen}} - \bar{Q}_{\text{gmax}}\bar{x}_l}{\bar{x}_q(\bar{x}_q + \bar{x}_l)} \tag{5}$$

Substituting (5) into (3),

$$\begin{aligned} \bar{P}^2_{\text{gmax}} &= \frac{\left(\bar{e}^2_{\text{gen}} + \bar{Q}_{\text{gmax}}\bar{x}_q\right) \cdot \left(\bar{e}^2_{\text{gen}} - Q_{\text{gmax}}\bar{x}_l\right)}{\bar{x}_q \cdot \bar{x}_l} \\ &= -\left\{Q_{\text{gmax}} - \frac{1}{2}\left(\frac{1}{\bar{x}_l} - \frac{1}{\bar{x}_q}\right)\bar{e}^2_{\text{gen}}\right\}^2 + \left\{\frac{1}{2}\left(\frac{1}{\bar{x}_l} + \frac{1}{\bar{x}_q}\right)\bar{e}^2_{\text{gen}}\right\}^2 \end{aligned} \tag{12.17}$$

This is Equation 12.17.

13

The Generator as Rotating Machinery

We have studied the generator's characteristics from the viewpoint of electrical theory in the previous chapters. In this chapter we examine the generator as a rotating mechanical machine. This base knowledge is essential in order to understand the dynamic characteristics of a generator and of the power system as a congregation of generators.

Note that we omit the overbar symbol for the per unit value below and for all subsequent chapters, even though most of all the quantities are in per unit values.

13.1 Mechanical (Kinetic) Power and Generating (Electrical) Power

We examine first the relation between the generator's mechanical input (driving force by steam turbine or by water wheel) and electrical output (generating power).

13.1.1 Mutual relation between mechanical input power and electrical output power

A generator connected to a power system network is operating under three-phase-balanced conditions. The generator's electrical quantities are given by Equation 10.43 in the d–q–0 domain, and the apparent power is given by Equation 11.25, all in per unit values.

Substituting Equation 10.43 into Equation 11.25 to eliminate voltage variables, the following equation is derived:

$$\begin{aligned}\dot{S}_{3\phi} &= \dot{S}_{1\phi} = P_{1\phi} + jQ_{1\phi} = \dot{S}_d(t) + \dot{S}_q(t) + 2\dot{S}_0(t)\\ &= \left\{-\omega_m(t)\dot{\psi}_q(t) + \frac{d}{dt}\dot{\psi}_d(t) - r\dot{i}_d(t)\right\}\dot{i}_d^* + \left\{\omega_m(t)\dot{\psi}_d(t) + \frac{d}{dt}\dot{\psi}_q(t) - r\dot{i}_q(t)\right\}\dot{i}_q^*\\ &\quad + 2\left\{\frac{d}{dt}\dot{\psi}_0(t) - r\dot{i}_0(t)\right\}\dot{i}_0(t)^*\end{aligned} \tag{13.1a}$$

$$\left.\begin{aligned}&\therefore\ \dot{S}_{3\phi} = \dot{S}_{1\phi} = \left\{\dot{\psi}_d(t)\dot{i}_q(t)^* - \dot{\psi}_q(t)\dot{i}_d(t)^*\right\}\omega_m(t)\\ &\quad + \left\{\frac{d}{dt}\dot{\psi}_d(t)\cdot \dot{i}_d(t)^* + \frac{d}{dt}\dot{\psi}_q(t)\cdot \dot{i}_q(t)^* + 2\frac{d}{dt}\dot{\psi}_0(t)\cdot \dot{i}_0(t)^*\right\} - r\left\{i_d^2(t) + i_q^2(t) + 2i_0^2(t)\right\} \quad ①\\ &\text{where } \omega_m(t) = \frac{d}{dt}\theta_m(t) \text{ is the mechanical angular velocity of the rotor} \quad ②\\ &\qquad\text{(suffix } m \text{ means mechanical quantities)}\end{aligned}\right\}$$

(13.1b)

© 2007 John Wiley & Sons, Ltd

The angular velocity ω_m in the equation is the instantaneous value of the rotor shaft, which fluctuates and does not necessarily coincide with the electrical angular velocity of the power network $\omega = 2\pi f$.

Equation 13.1b is a very important equation which combines the concepts of mechanical power and electrical power. Now, we investigate each term on the right-hand side of this equation.

13.1.1.1 The first term

From the dynamic theory of rotating devices in physics,

$$(\text{mechanical power}\, P_m) = (\text{torque}\, T_m) \times (\text{revolving velocity}\, \omega_m) \tag{13.2}$$

Comparing the first term on the right-hand side of Equation 13.1b with Equation 13.2, the part in { } corresponds to the mechanical torque T_m, and the first term itself corresponds to mechanical power P_m which is given from the prime-mover. This mechanical power P_m intervenes from flux linkages and finally is transferred from the rotor to the stator armature windings in the form of electrical power across the air gap. In other words, the first term on the right-hand side is the mechanical power P_m which is given from the prime-mover to the rotor and transferred to the stator coil windings as electrical power by the form of flux linkage through the air gap. Or, we should adopt the above explanation for the real-number part of the equation, because mechanical power is not directly comparable with reactive power in electrical theory.

Accordingly, the part in { } in the equation can be treated as the term for electrical torque with the symbol $\dot{T}_e(t)$. Then

$$\left.\begin{array}{lll}\text{electrical torque} & \dot{T}_e(t) & = \dot{\psi}_d(t)\dot{i}_q(t)^* - \dot{\psi}_q(t)\dot{i}_d(t)^* \\ \text{mechanical torque} & T_m & = Re\{\dot{T}_e(t)\}\end{array}\right\} \tag{13.3}$$

13.1.1.2 The second term

This is the term for the rate of decrease of armature magnetic energy. It is zero under three-phase-balanced conditions because ψ_d, ψ_q, ψ_0 are d.c. quantities under steady-state conditions and accordingly their derivatives are zero. In other words, the second term corresponds to the transient term which appears when the armature fluxes in the d–q–0 domain are being changed.

13.1.1.3 The third term

This is the term for Joule losses, caused by the resistance of armature windings. The zero-sequence component includes the coefficient 2 as in Equation 11.25.

In total, Equation 13.1 can be understood as follows:

$$\dot{S}_{3\phi}(t) = T_e(t)\omega_m(t) + \underbrace{\left\{\begin{array}{c}\text{rate of change of armature magnetic}\\ \text{energy in d-q-0 domain}\end{array}\right\}}_{\text{transient term}} - \left\{\begin{array}{c}\text{Joule losses of armature}\\ \text{windings}\end{array}\right\} \tag{13.4}$$

The real-number part of Equation 13.4 is, for the effective electrical output power,

$$P_{3\phi}(t) = \underset{*1}{T_m(t)\omega_m(t)} + \underset{*2}{\left\{\begin{array}{c}\text{rate of change of armature magnetic}\\ \text{energy in d-q-0 domain}\end{array}\right\}} - \underset{*3}{\left\{\begin{array}{c}\text{Joule losses of the}\\ \text{armature winding}\end{array}\right\}} \tag{13.5}$$

*1 mechanical power the rotor receives from the prime-mover

*2 transient electric power caused by discharge (or charge) of the rotor's kinetic energy in the transient condition of deceleration (or acceleration) of the rotor

*3 Joule losses of the armature winding

13.1.1.4 Steady-state condition

In this case, T_m, ω_m, $P_{3\phi}$, etc., are invariable, so the second term *2 of Equation 13.5 or 13.1b① is zero. The energy of revolution is also invariable because $\omega_m(t)$ is invariable in this condition.

13.1.1.5 Transient condition by sudden disturbance

Suppose that the load demand of the generator is suddenly increased from $P_{3\phi}$ to $P_{3\phi} + \Delta P_{3\phi}$ for some reason on the outer network. This means a sudden increase on the left-hand side of Equation 13.5, while term *1 does not vary because the term will be changed only by the prime-mover. The change in term *3 is negligible. Accordingly, term *2 has to be increased instantaneously. Therefore, the rotor begins to release its stored kinetic energy ($k = (1/2)I\omega_m^2$, see the next section) to compensate for the transient power imbalance, while slowing down of the rotor (or electrical angular slip) continues until the prime-mover begins to increase the mechanical input power P_m to the rotor (term *1); consequently, the rotor begins to recover the energy imbalance caused. If the increased electrical angular displacement of the rotor exceeds the stability limit of approximately 90°, the generator will be forced to lose synchronization with the outer power system. In other words, capricious load fluctuations have to be followed after mechanical input power (P_m) increase/decrease control by all the generators within the power system from time to time as a function of the individual speed governor operation, and, furthermore, as a function of the total automatic frequency control (AFC) of the power system.

Incidentally, the **'simultaneity and equality of the demanding power and the supplying power'** are often referred to as the fatal characteristics of a power system. These result from the above described generator characteristics.

Returning to Equation 13.2, the operating frequency of most power utility systems is kept typically at 50/60 ± 0.05 Hz by AFC system operation from the central dispatching centre. Then the angular velocity of the power system and of all the operating generators is $\omega(t) = \omega_m(t) = 1.0 \pm 0.01 \fallingdotseq 1.0$ on a per unit basis. Accordingly, putting $\omega_m(t) = 1.0$ into Equation 13.2, we find that the power and torque actually have the same per unit values as each other (namely, $P_m = T_m, S_{3\phi} = T_e$ by PU expression), although they are quantities with different dimensional units electrically as well as mechanically. It is also clear that the electrical and mechanical quantities are equivalent under synchronized operation.

13.2 Kinetic Equation of the Generator

13.2.1 Dynamic characteristics of the generator (kinetic motion equation)

A generator's rotor can be assumed to be a homogeneous rigid cylindrical body as shown in Figure 13.1.

We introduce here the kinetic motion equation of the rotor.

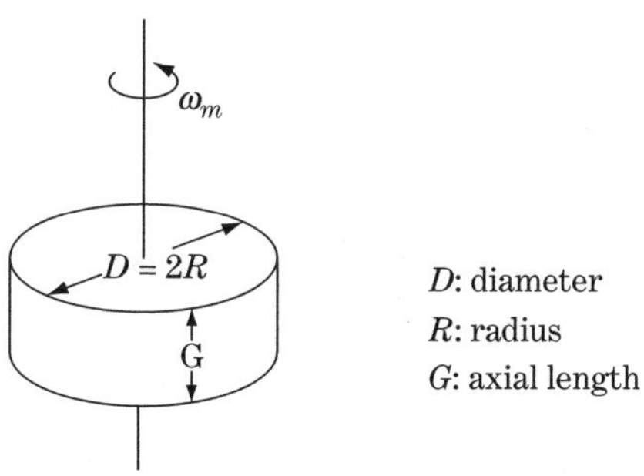

Figure 13.1 Cylindrical revolving body

The mechanical acceleration equation (the kinetic motion equations) is

$$\left.\begin{aligned} &\Delta P_m = \frac{M_0}{\omega_m}\cdot\frac{d\omega_m}{dt} = \omega_m \Delta T_m [\mathrm{kg\cdot m/s}] && \text{①} \\ &\frac{d^2\theta_m}{dt^2} = \frac{d\omega_m}{dt} = \frac{\omega_m}{M_0}\Delta P_m = \frac{\omega_m^2}{M_0}\Delta T_m = \frac{\Delta T_m}{I} && \text{②} \\ &M_0 = I\omega_m^2 = 2\times\left(\frac{1}{2}I\omega_m^2\right) = 2K && \text{③} \\ &K = \frac{1}{2}I\omega_m^2 && \text{④} \end{aligned}\right\} \quad (13.6)$$

where I : moment of inertia [kg · m/s] $I = \sum_i m_i r_i^2$

m_i is the weight of the material point and is r_i is its distance from the rotating axis

M_0 : inertia constant, $M_0 = I\omega_m^2$

θ_m, ω_m : mechanical phase angle [radian], angular velocity [rad/s]

P_m : mechanical power transferred from the prime-mover to the generator rotor.

T_m : mechanical torque of the rotor

K : kinetic energy of the rotor $K = \frac{1}{2}I\omega^2$

P_e : effective power transferred from the rotor to the armature windings across the air gap [W]

T_e : electrical torque of the rotor [W · s/rad]

I and ω_m can be replaced by the practical scale units below:

$$\begin{aligned} I &= \frac{GR^2}{g} = \frac{GD^2}{4g} \\ \omega_m &= 2\pi\cdot\frac{N_{\mathrm{rate}}}{60[\mathrm{sec}]}[\mathrm{rad}] \end{aligned} \quad (13.7)$$

where g : gravity accelerator

G : effective axial length [m]

R, D : radius and diameter of rotor [m]

GD^2 : flywheel effect [kg · m²]

N_{rate} : rated rotating speed per minute [rpm]

$N_{\mathrm{rate}}/60$: rated rotating speed per second [s^{-1}]

Accordingly,

$$\left.\begin{aligned} &\text{motion energy} \quad K = \frac{M_0}{2} = \frac{1}{2}I\omega_m^2 = \frac{1}{2}\left(\frac{GD^2}{4g}\right)\cdot\left(2\pi\frac{N_{\mathrm{rate}}}{60}\right)^2 [\mathrm{kg\cdot m}] \\ &\qquad = 1.37\times GD^2\left(\frac{N_{\mathrm{rate}}}{1000}\right)^2 [\mathrm{kW\cdot sec}] \\ &\text{where} \quad 1\,\mathrm{kg\cdot m} = 9\cdot 8\,[\mathrm{W\cdot sec}] \end{aligned}\right\} \quad (13.8)$$

G, D are the axial length and diameter of the rotor, so that GD^2 is the volume of the cylinder. All the values in the equation are specific mechanical values for the generator rotor.

Now we introduce M (**unit inertia constant**), which is defined by $M = M_0/P_{\text{rate}}$:

$$\left.\begin{aligned} M &\equiv \frac{M_0[\text{kW} \cdot \text{sec}]}{P_{\text{rate}}[\text{kW}]} = \frac{2.74 \times GD^2 \left(\frac{N_{\text{rate}}}{1000}\right)^2}{P_{\text{rate}}}[\text{sec}] \\ M_0 &= M \cdot P_{\text{rate}} \\ &\text{where } P_{\text{rate}} : \text{the rated output capacity [kW]} \end{aligned}\right\} \tag{13.9}$$

GD^2 as well as P_{rate} (the rated capacity) and N_{rate} (rated rotating speed) are given on the name-plate of each generator, as essential specifications, so the unit inertia constant M is also a specified value for each generator. Generators with larger M obviously have larger synchronizing power, because M is proportional to GD^2 and N_{rate}^2. Typical values of M are shown in Table 13.1.

GD^2 for thermal generators is generally smaller than that for hydro-generators of the same reted capacity, while the rotating speed N_{rate} of thermal generators (3000/3600 rpm) is 3–10 ($= n$, where $2n$ is the number of poles of hydro-generators) times faster than that of hydro-generators. Accordingly, M for thermal generators is typically a little larger than that for hydro-generators despite the fact that GD^2 is relatively smaller.

13.2.2 Dynamic equation of generator as an electrical expression

In Equation 13.6, the symbols of variable quantities can be replaced as follows:

$$\left.\begin{aligned} \Delta P_m(t) &\to P_m(t) - P_e(t) \\ \theta_m(t) &\to \theta_e(t)/n \\ \omega_m(t) &\to \omega_e(t)/n \end{aligned}\right\} \tag{13.10}$$

where

$P_m(t)$: mechanical input power given by the prime-mover (friction loss, windage loss are already subtracted)
$P_e(t)$: electrical output power (including armature resistance loss)
$2n$: pole numbers: thermal generator ($2n = 2$), hydro-generator ($2n = 6$ to 20), nuclear generator ($2n = 4$)
ω_m, ω_e : mechanical angular velocity, electrical angular velocity [rad]
ω_0 : rated angular velocity
$\omega_e \fallingdotseq \omega_0 = 2\pi f_0 (f_0 = 50 \text{ or } 60\,\text{Hz})$

The mechanical acceleration Equation 13.6, can be modified as follows.
For the dynamic equation of a generator

$$\frac{d^2\theta_e(t)}{dt^2} = \frac{d\omega_e(t)}{dt} = \frac{\omega_0}{M_0}(P_m(t) - P_e(t)) \tag{13.11}$$

Table 13.1 Unit inertia constants of generators ($M = M/P_{\text{rate}}$)

	***M* value**
Hydro-generator	6–8
Thermal generators (forced cooling type)	7–10
Thermal generators (natural cooling type)	10–15
Synchronous motors	3–5

P_e, θ_e, ω_e in this equation are the same as the variables P_e, θ_e, ω_e in Chapter 10 for generator electrical theory. The symbol (t) is added in order to emphasize that this is a dynamic equation applicable to dynamic transient phenomena.

The equation can be unitized by ω_0/M_0 so that the coefficient on the right-hand side disappears:

$$\frac{d^2\overline{\theta}_e(t)}{dt^2} = \frac{d\,\overline{\omega}_e(t)}{dt} = \overline{P}_m(t) - \overline{P}_e(t) \tag{13.12}$$

Equation 13.11 or 13.12 shows that the rotating speed is constant ($d\omega/dt = 0$) under balanced conditions of mechanical input and electrical output ($\overline{P}_m = \overline{P}_e$). The generator rotor is decelerated if $\overline{P}_m < \overline{P}_e$, and is accelerated if $\overline{P}_m > \overline{P}_e$. Therefore, whenever mechanical input power $\overline{P}_m$ from the prime-mover or electrical output power $\overline{P}_e$ to the outer power system fluctuates, the generator rotor is forced to swing towards or slip from the power system angular velocity $\omega_0 = 2\pi f_0$.

13.2.3 Speed governors, the rotating speed control equipment for generators

The mechanical input of the hydro/thermal prime-mover must always be automatically controlled so that the rotor speed ω_m always meets the angular velocity of the power system $\omega = 2\pi f$. Every generator is equipped with a **speed governor**, by which the prime-mover is controlled so as immediately to increase (decrease) the mechanical input whenever the rotating speed is decelerated (accelerated) towards power system angular velocity $\omega = 2\pi f$. In other words, the governor has the function of speed restoration which tends to keep the present rotating speed $\{\Delta\omega = (\omega_m - 2\pi f)\} \to 0$ within the power capacity range of 3–5% of the rated MW values of the TG unit.

Figure 13.2 shows the operating mechanism of the speed governor (mechanical type) for hydro/thermal turbines. Although the principle is the same, the governors for thermal turbines have quick response characteristics in comparison with that for water-wheel turbines.

Figure 13.3 shows photographs and artwork of a typical water turbine generating unit for a pumped storage station. The photographs show the stator–rotor unit, wheel casing with guide vane and Francis-type runner unit for a high-head station.

In the case of a hydro-generating station, the amount of water in the long penstock system has large kinetic energy, so rapid changes of water flow cause severe water hammer or sudden vacuum phenomena on the penstock or on the guide-vane system, in spite of the installation of a surge tank. Therefore, rapid water flow control cannot be expected in order to avoid damage to the penstock or turbine systems. Accordingly, the speed governor for the hydro-unit is equipped with a dash-pot mechanism to prevent quite rapid mechanical vane (water flow) control.

In the case of the thermal turbine unit, the control valve can be operated quickly by the speed governor (say, order of 0.1 s), because pressure variation phenomena like hammering need not be taken into account for the high pressure dry steam gas. Accordingly, the speed governor for this unit is not equipped with mechanisms like dash-pots. This unit is, incidentally, equipped with an **emergency governor**, which takes emergency action and closes the main stop valve whenever the rotor speed reaches the physical upper limit.

In conclusion, the function of the speed governor is to control the rotor speed to follow the fluctuating power system frequency. Thus the governor has an important role in maintaining synchronization of the generator with the network. In addition to the above, it carries out a major role in limiting fringing-fluctuated system frequency f as a supporting function of AFC (from the central dispatching centre).

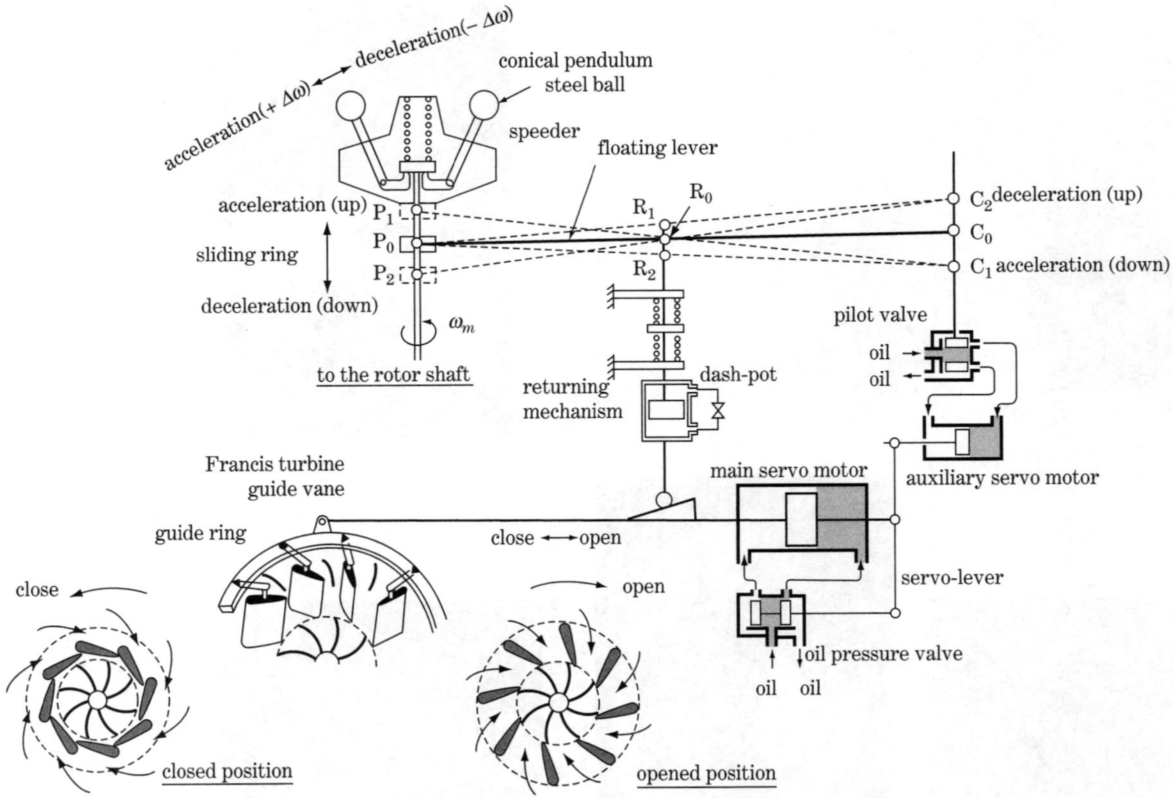

Figure 13.2 Speed governor for water wheel (mechanical type)

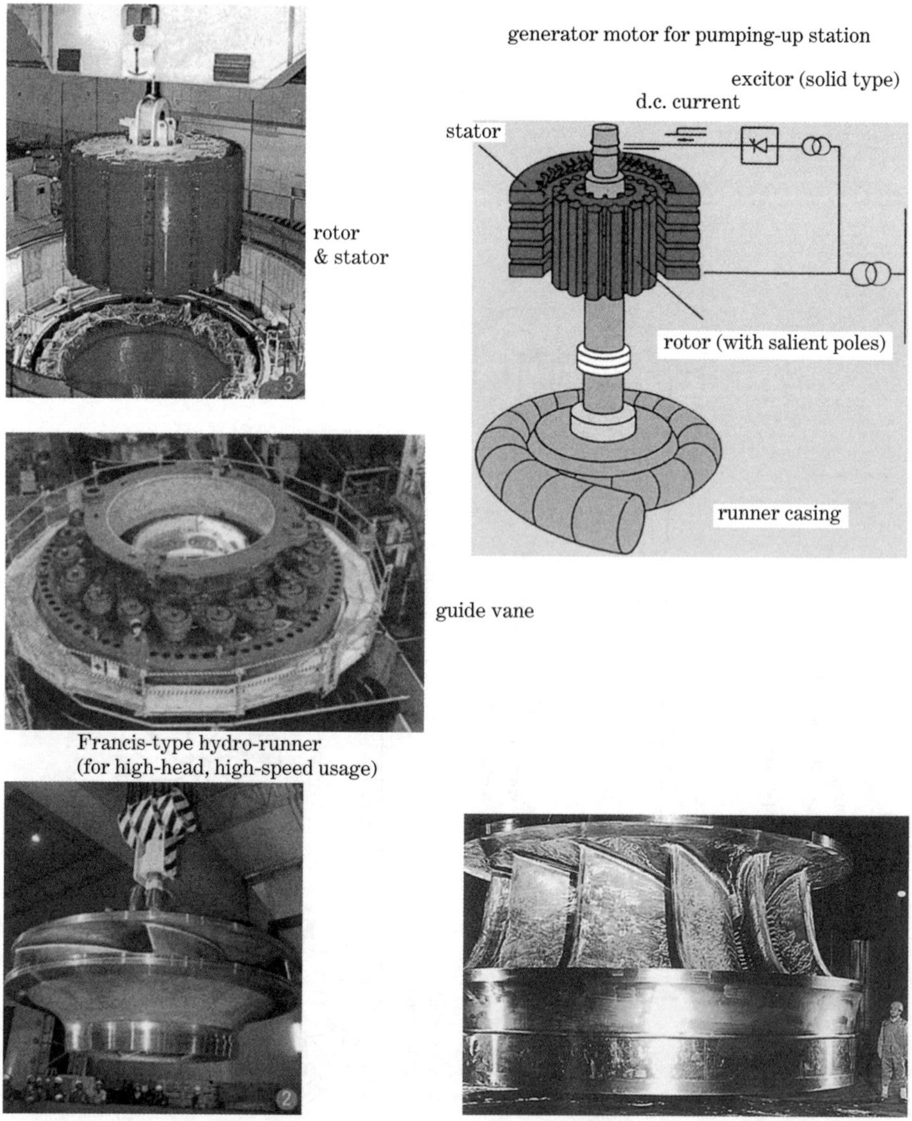

(a) Water TG unit for high-head pumped, storage power station (360 MW, 50 Hz, 12 poles, 500 rpm)

(b) Francis-type hydro-runner (for low-head, low-speed usage)

Courtesy of Toshiba

Figure 13.3 Hydro-turbine and generator

Coffee break 7: Heaviside, the great benefactor of electrical engineering

Hertz's experiment became the great turning point when the 'ether' vanished and the door of modern science and technology of electricity was opened. The history of electricity in the nineteenth century seemed to have branched into two streams by the twentieth century. The first concerned the science of physics in the fields of, for example, atomic and quantum physics, radio waves, astronomy, and Einstein's relativity. The second stream concerned, of course, engineering for social and industrial applications of electricity, perhaps divided into three categories, namely (a) radio wave and electron devices, (b) electric power generation and transmission, (c) various uses of energy applied typically in lighting and motors.

Let us look at some of the 'electricians' who made great contributions to the analytical methods of electricity from the engineering viewpoint, in particular in categories (b) and (c), and closely related to power system engineering.

The first person is **Oliver Heaviside** (1850–1925). His contribution in the various engineering fields as the developer of analytical methods for electricity is so direct and great that he is the equal of Maxwell and Hertz in the nineteenth century. Heaviside was the first physicist to recognize the worth of Maxwell's equations and popularized them. His greatest achievements are described below.

He was able to simplify greatly Maxwell's quite complicated 20-odd equations and introduce a simpler treatment of the 'four equations' without impairing accuracy. Maxwell's four equations quoted in a previous chapter may even be called the **Maxwell–Heaviside equations** on this view.

In 1874, he designed the **symbolic method** or **operational form calculus** for analysing differential equations. The **Heaviside operator** p was experimentally developed by him to reduce complicated differential equations to simple algebraic ones. The stepping function $\mathbf{1}(t)$ is

Oliver Heaviside (1850–1925)

also due to Heaviside's originality. His methods were said to be 'imperfect' and of 'no consequence' by most mathematicians, despite the fact that they gave correct answers for most of the cases. Later, **T. J. Bromwich** (1916), **J. R. Carson** (1918) and **K. W. Wagner** (1925) vindicated Heaviside and his operational calculus was justified by the proof that it was mathematically a functional transformation to the domain p. 'Heaviside's operator p' is of course equivalent (one-to-one correspondence, although the definition is different) to **Laplace's operator** s of the Laplace transform, which is an essential technique for every electrical engineer today for solving transient phenomena (that is, for solving partial differential equations) or as ***s*-functions in the theory of automatic control** or **automation**.

Equations 1 and 2 given in the first supplement of Chapter 10 Section 10.10 (page 211) are called **Heaviside's expansion theorem**, even though the Laplace operator s is adopted, and the solution demonstrated in all three supplements to that chapter is Heaviside's method of solution. Although it is likely that the operator p-domain has been dominated by s-domain today in practical engineering, s may be said a kind of the revival of p.

Heaviside made great contributions to establish modern scientific 'units' based on length L, weight M and time T systematically, a great work that could only be done by someone who fully understood modern physics based on theories of energy preservation in a wide sense. Interestingly, the values of electric flux initiated from the surface of a ball of radius r (surface area $4\pi r^2$) may be counted as multiples of units of r^2, or of $4\pi r^2$, and of course the latter is a rational unit. It was Heaviside who originated the **rational unit system** by eliminating formally the coefficients 4π or $2\sqrt{\pi}$ that were originally included in Maxwell's equations. His unit is called the Heaviside–Lorentz unit because it was Hendrik Anton Lorentz (1853–1928) who adopted it first. The unit is actually the **CGS unit** or the equivalent **MKS rational unit**, which are practical scale units today.

Telephony by wire was invented by **Alexander Graham Bell** (1847–1922) in 1877, in addition to the already utilized **telegram**, and the application was expanded year by year. Soon the problems of signal attenuation or distortion for long-distance communication became more serious. In 1881, Heaviside originated the **telegram equation with four constants *L, C, R, G***, which enabled the analysis of attenuation and waveform distortion phenomena. He also showed that the condition $R/L = G/C$ could satisfy distortion-less transmission of a signal wave (see Equation 18.20 in Chapter 18). This work by him was called the **principle of loading cable**. His analytical conclusion was soon utilized practically under the more familiar name of the **pupin-coil**, which adds inductance L intentionally to cable wire in order to obtain a matching balance of $R/L = G/C$ and to improve balance with the cable capacitance C. Incidentally, the conventional telegram equation at that time was the famous **Thomson equation by *KR* law**. The equation was, so to speak, two constant equations with R, K (K is the symbol of capacitance, today C), which was advocated by Kelvin in 1855 in his famous paper 'On the theory of the electrical telegraph' and had become quite important for the engineering basis of wired telegram theory at that time. William Thomson's famous KR law is

$$\frac{\partial^2 v}{\partial x^2} = KR\frac{\partial v}{\partial t}$$

However, the equation cannot obviously explain attenuation or distortion of waveforms (see Chapter 18). Heaviside overcame this problem, (refer section 18. 1.1), which led to quite important advances in wire communication engineering at the time. He emphasized the role of metallic circuits as 'guides' rather than 'conductors of a.c. currents'.

The idea of matching R/L and G/C is a basic concept of every type of communication or signal transfer technology today. Heaviside is obviously the originator of the travelling wave theory of transmission lines as well as telecommunications, although it was five years before the first practical a.c. transmission line of 2000 V, 27 km was installed in 1886, and fifteen years before the first radio communication by Marconi, who sent radio signals across the Atlantic. However, no one could explain why the signals were not stopped by the curvature of the Earth.

In 1902 Heaviside's famous prediction of an ionized layer in the atmosphere which would deflect radio waves was published in an article titled 'telegraphy'. The idea arose when he was considering the analogy between the movement of electric waves along a pair of conducting wires and a conducting earth. He thought that waves travelling around the Earth might accommodate themselves to the surface of the sea in the same way as waves follow wires: 'There may be a sufficiently conducting layer in the upper air. If so, the waves will, so to speak, catch on to it more or less. Then guidance will be by the sea on one side and the upper layer on the other.' This is of course the **Kennelly–Heaviside layer**.

His great works can be seen in his famous book *Electromagnetic Theory* published in 1893, 1889 and 1912. He was a nephew of Wheatstone.

Although Heaviside was a bachelor who spent much of his time studying and writing scientific papers in complete solitude, his contribution to modern advanced electrical engineering is without doubt one of the greatest.

Note: d/dt, d^2/dt^2, d^3/dt^3, ... are replaced by p, p^2, p^3, ... respectively in the Heaviside transform, and by s, s^2, s^3, ... respectively in the Laplace transform. However, the definition in the replaced domain is different by one order. Below are some typical examples in both transformations:

t domain	p domain	s domain
$\lim_{\tau \to 0} \frac{1}{\tau}[1(t) - 1(t-\tau)]$	p	1
$1(t)$	1	$\frac{1}{s}$
$e^{\mp \alpha t}$	$\frac{p}{p \pm \alpha}$	$\frac{1}{s \pm \alpha}$
$\cos \omega t$	$\frac{p^2}{p^2 + \omega^2}$	$\frac{s}{s^2 + \omega^2}$

14

Transient/Dynamic Stability, *P–Q–V* Characteristics and Voltage Stability of a Power System

The dynamic characteristics of a power system deeply depend on the characteristics of generators, which we studied in Chapters 10–13. In this chapter, transient and dynamic stability are examined first, then the *P–Q–V* characteristics of a power system and voltage stability phenomena are examined.

14.1 Steady-state Stability, Transient Stability, Dynamic Stability

Power system stability is typically defined as the property of the power system that it will remain in operating equilibrium through normal and abnormal conditions. In terms of interconnected synchronous machines, in order to be stable, the machines must maintain synchronism through normal and abnormal conditions. Instability is expediently classified into three categories which will be introduced here first.

14.1.1 Steady-state stability

Steady-state stability is defined as the operating state of a power system which is characterized by slow and gradual changes. The steady-state stability limit is actually explained by Equations 12.5, 12.12 and 12.17 and Figures 12.1 and 12.4.

14.1.2 Transient-state stability

The transient state is defined here as the operating state of a power system which is characterized by a sudden change in load or circuit conditions. Transient-state stability is defined as stability under such transient states. Short-circuit fault and fault tripping/reclosing, switching of circuits, abrupt significant load changes, sudden tripping of generators, etc., are typical disturbances. A sudden change in excitation of generators which may be caused by some irregular conditions in automatic voltage regulator (AVR) equipment (a sudden change in the AVR set value, for example), or in the mechanical power of prime-movers, or a change of power flow in the network caused by changes of power

Handbook of Power System Engineering Yoshihide Hase
© 2007 John Wiley & Sons, Ltd

distribution among generators or by changes in network connection, have to be included as kinds of disturbances. Also, hunting phenomena among plural generators may be another kind of disturbance.

14.1.3 Dynamic stability

The power system stability limit can be improved far beyond the steady-state stability limit by the use of appropriately designed AVR equipment. Also, automatic speed-governor control (automatically controlling the mechanical power of the prime-movers) of each generator by detecting a sudden significant frequency change under fault or no-fault conditions may also improve the stability limit.

Dynamic stability can be defined as the concept of improved stability by applying appropriately quick excitation control (jE_f control by AVR) as well as appropriately quick speed-governor control (P_m control by frequency detection) at each generating station.

Dynamic stability might have been so named originally in contrast to ordinary steady-state stability. However, it is obvious that appropriate AVR control and speed-governor control at each generating plant improve not only steady-state stability but also transient-state stability caused by various cascades of sudden changes in power system conditions.

The time constants of the 'AVR + field excitation circuit' are very small (say, 0.1–0.5 s), while those of the 'speed governor + prime-mover' must be larger (say, a few seconds). Accordingly, AVR must be more effective for an initial rapid response to serious disturbances.

14.2 Mechanical Acceleration Equation for the Two-generator System, and Disturbance Response

The generator's mechanical acceleration equation was derived in Equation 13.11 or 13.12. Now we examine the power system shown in Figure 14.1, which contains generators G and B connected by a double circuit line:

$$\begin{array}{ll} \text{generator G} & \text{generator B} \\ \dfrac{d^2\theta_G(t)}{dt^2} = \dfrac{\omega_G(t)}{M_G}(P_{Gm} - P_{Ge}) & \dfrac{d^2\theta_B(t)}{dt^2} = \dfrac{\omega_B(t)}{M_B}(P_{Bm} - P_{Be}) \end{array} \tag{14.1}$$

There are mutual relations between generators G and B, as follows:

$$\begin{array}{ll} P_{Ge} = -P_{Be} & \text{(line resistance neglected)} \\ \delta(t) = \theta_G(t) - \theta_B(t) = \int \{\omega_G(t) - \omega_B(t)\} dt < 90^\circ & \left(\begin{array}{l}\text{angular difference of induced voltages} \\ \text{of both generators}\end{array}\right) \\ \omega_G(t) \fallingdotseq \omega_B(t) \fallingdotseq 2\pi f_0 \equiv \omega_0, & f_0\text{:power frequency} \end{array} \tag{14.2}$$

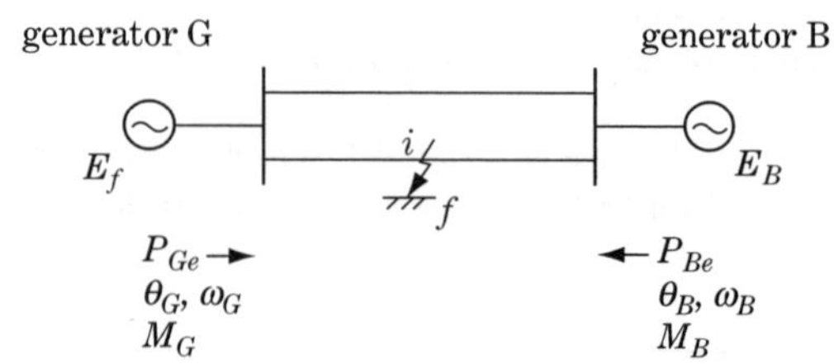

Figure 14.1 Two-generator system

$\delta(t)$ is the angular difference between the two generators which must be within $\pm 90^{\circ}$ under synchronization. From Equation 14.2,

$$\frac{d^2\delta}{dt^2} = \omega_0 \left\{ \frac{P_{Gm} - P_{Ge}}{M_G} - \frac{P_{Bm} - P_{Be}}{M_B} \right\} \tag{14.3}$$

Now we assume a non-salient-pole machine for simplicity and take the effective power from Equation 12.12:

$$P_{Ge} = -P_{Be} = \frac{E_f E_B}{x_q + x_l} \sin \delta \tag{14.4}$$

where E_f, E_B: internal induced voltages of generators G and B, respectively.

The mechanical input P_{Gm}, P_{Bm} cannot be changed for 0–3 s from the magnitudes just before system disturbance (because of the inertia of the prime-mover system). Then

$$P_{Gm} = P_{Ge} = -P_{Be} = -P_{Bm} \tag{14.5}$$

From Equations 14.3 and 14.5, for the mechanical acceleration equation of a generator G,

$$\left.\begin{aligned} \frac{d^2\delta}{dt^2} &= \omega_0 \left(\frac{1}{M_G} + \frac{1}{M_B} \right) \cdot (P_{Gm} - P_{Ge}) \\ &= \frac{\omega_0}{M_0} \left(P_{Gm} - \frac{E_f E_B}{x_q + x_l} \sin \delta \right) \quad \text{where } M_0 = \frac{M_G M_B}{M_G + M_B} \end{aligned}\right\} \tag{14.6a}$$

where $\delta(t)$: phase angular difference between the induced voltages of both generators ($-90^\circ < \delta(t) < 90^\circ$ under normal conditions).

If generator B is of quite a large capacity in comparison with generator G, this means that $M_B \to \infty$ and $M_0 \to M_G$, which correspond to the one machine to infinite bus.

In Equation 14.6, $\omega_0 = 2\pi f_0$ (where $f_0 = 50/60\,\text{Hz}$) is of fixed value. M_g, M_B, are the specific machine constants. Reactance x_l is the network reactance connected to the generator terminal, which would suddenly take on a large value if a fault were to occur in the network.

Therefore Equation 14.6 can be written as $\delta(t) = \mathit{function}(P_{Gm}, E_f)$, from the viewpoint of controllable quantities. In other words, this indicates that we can control at the generating station only the excitation of the generator and the mechanical input power from the prime-mover.

Incidentally, the equivalent inertia constant M_0 for two machines is written as the weighted average value of each generator's inertia constants. Analogously, the equivalent inertia constant M_0 for multiple-machine systems can be written as the weighted average value of each generator's inertia constants. That is,

$$\frac{1}{M_0} = \sum_{k}^{n} \frac{1}{M_k} \tag{14.6b}$$

14.3 Transient Stability and Dynamic Stability (Case Study)

Let us assume a power system with a parallel circuit transmission line as shown in Figure 14.1, and where a short-circuit fault occurs at point f of the first circuit. A proper protective relay would detect the fault immediately and the associated breakers of the circuit would trip to remove the fault successfully within 3 or 6 cycles so that the remaining system could continue ordinary operation without causing instability.

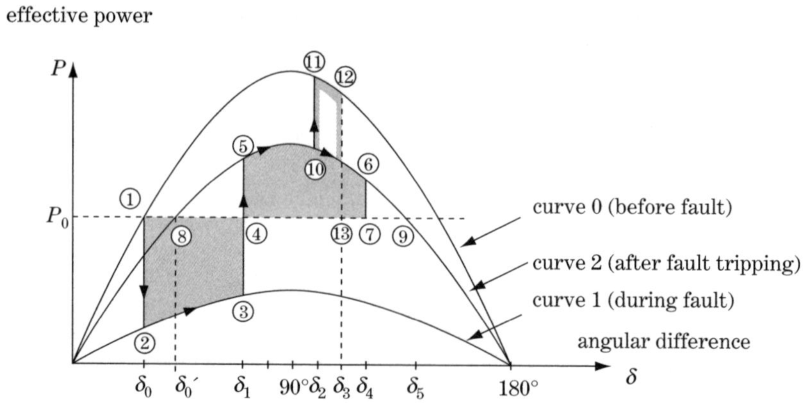

Figure 14.2 Transient stability (equal-area method)

We will study the concepts of transient-state stability and dynamic stability for the typical cascade disturbances described above. The system behaviour in the face of such disturbances will be explained step by step in Figure 14.2, which is a $P-\delta$ curve explaining stability by the **equal-quadrant method**.

Figure 14.2 is the $P-\delta$ curve for generator G. The condition before the fault is shown by curve 0 whose peak value is specified by Equation 12.5① or Equation 12.12③. When the fault occurs at point f, reactance x_l suddenly becomes large (see the next section) so that the generator condition has to be stepped down from curve 0 to curve 1. Next, immediately after fault tripping, the generator condition will jump from curve 1 to curve 2, because reactance x_l returns to a smaller value very close to the original value (curve 2 is a little lower than curve 0 because of the difference due to removal of the faulted line in this timing).

14.3.1 Transient stability

14.3.1.1 Case 1: The transient stability is successfully maintained

In this case transient stability is successfully maintained and the system continues stable operation after removing the fault.

The generator G is operating with power output $P_0(\delta_0)$ at point ① on curve 0 before the fault (P_0 is the power from the prime-mover). When the fault occurs at point f, the operating point moves suddenly from point ① $P_0(\delta_0)$ to ② $P_1(\delta_0)$ on curve 1. As the result, the generator causes an excess power of $P_0(\delta_0) - P_1(\delta_0)$ (say, accelerating mode) so that δ begins to increase on curve 1 from ② $P_1(\delta_0)$ to ③ $P_1(\delta_1)$. Then the fault is removed by the associated breakers tripping at the timing of point ③ $P_1(\delta_1)$, and the generator condition jumps from ③ to ⑤ $P_2(\delta_1)$. In this condition, the generator suddenly causes a power shortage of $P_2(\delta_1) - P_0(\delta_1)$ (say, decelerating mode). However, δ still continues to increase up to ⑥ $P_2(\delta_4)$ because of the rotor inertia, and then in turn decreases towards the new stable point ⑧ $P_0(\delta_0')$, though δ may repeatedly over-swing a little across the new stable angular difference δ_0'.

In the above process, the maximum angle δ_4 at point ⑥ satisfies the following relation:

$$\overbrace{\int_{\delta_0}^{\delta_1}\{P_0(\delta_0) - P_1(\delta)\}d\delta}^{\text{accelerating energy}} = \overbrace{\int_{\delta_1}^{\delta_4}\{P_2(\delta) - P_0(\delta)\}d\delta}^{\text{decelerating energy}}$$

or

{the area ①②③④} = {the area ④⑤⑥⑦} (14.7a)

It should be noted that δ exceeds 90° at point ⑥ for a short period.

14.3.1.2 Case 2: The system condition exceeds the transient stability limit

This is the case when the system condition unfortunately exceeds the transient stability limit and the system fails to continue stable operation after the fault.

After reaching point ⑤ $P_2(\delta_1)$ by the same process as in case 1, if δ still continues to increase beyond ⑥, in spite of the decelerating mode, and exceeds point ⑨ $P_2(\delta_5)$ at last, the generator will again enter accelerating mode and the synchronizing force is lost entirely. As a result, the generator will lose synchronization.

The point ⑨ $P_2(\delta_5)$ is the critical point of the synchronization, where critical angle δ_5 has a value exceeding 90°.

The critical condition of the transient stability limit is

{area ①②③④} ≦ {area ④⑤⑥⑨} (14.7b)

It is obvious that δ_1 should be small enough (this means fast fault tripping) to satisfy the above condition.

14.3.1.3 Case 3: Reclosing is successfully executed and transient stability is maintained

After reaching point ⑤ $P_2(\delta_1)$ by the same process as in case 1, δ continues to increase along curve 2 by inertia. Next, reclosing of the fault line is executed at the timing point $P_2(\delta_2)$. If a faulted arc were extinguished and the insulation at the faulted point had been recovered, the generator condition would jump from $P_2(\delta_2)$ to ⑪ $P_0(\delta_2)$ by the successful reclosing and δ would increase up to point ⑫ $P_0(\delta_3)$ and then turn to decrease. The point ⑫ $P_0(\delta_3)$ satisfies the following equation:

{area ①②③④} = {area ⑤⑫⑪⑫⑬ (14.7c)

Returning to our main theme, let us now examine how transient stability can be improved.

First of all, the height of curve 1 is actually determined by $1/x_l$ in the fault mode, so it is an out-of-control matter (we discuss x_l in the next section). Accordingly, the most effective countermeasure to improve transient stability is to shorten the fault-tripping time, by decreasing the acceleration energy {area ①②③④}. Clearly, the time delay of fault tripping by any reason means that point ③ in the figure would be shifted to the right so that acceleration energy increases remarkably.

It must be stressed that stable power system operation can be secured by **high-speed fault tripping.** Today, due to the advanced technology of protective relays and circuit-breakers, fast fault tripping for high-voltage trunk lines has been realized, where typical operating times are

{fault-detecting time by the relays 1 – 3 cycles} + {tripping time by the breakers 1 – 3 cycles}
= {total tripping time 2 – 6 cycles}

14.3.2 Dynamic stability

Besides high-speed fault tripping, there are two other effective countermeasures to improve stability which can be realized by decreasing the acceleration energy {area ①②③④} or by increasing the deceleration energy {area ④⑤⑥⑦}.

14.3.2.1 Quick excitation control by AVR

This is the countermeasure to increase the decelerating energy {area ④⑤⑥⑦} by enlarging curve 2 immediately after the fault. The peak value of curve 2 is $E_f E_B/(X_q + X_l)$, so it can be enlarged by increasing excitation E_f of the generator.

AVR increases excitation jE_f very quickly immediately after detecting voltage drop $-\Delta V$ caused by a fault, so curve 2 would be enlarged and the decelerating energy would be increased.

Today, due to the advanced technology of AVR and excitation equipment, rapid excitation control (time constant of, say, 0.1 s) can be exercised. (This is again discussed in Chapter 15.)

14.3.2.2 Quick driving-power adjustment by speed-governor control of the prime-mover

This is the countermeasure to decrease the accelerating energy {area ①②③④} by depressing the mechanical input from the prime-mover immediately after the fault. The speed-governor, upon detecting a sudden acceleration of the rotor speed, reduces the mechanical input from the prime-mover by decreasing the water/steam flow (i.e. $P_0 \rightarrow (P_0 - \Delta P_0)$ in Figure 14.2).

However, the amount of input power the speed-governor can decrease in a short time is limited (say, $\Delta P_0/P_0 = 3 - 10\%$). In addition, especially in the case of a hydro-unit, water flow cannot be quickly changed because of the time delay characteristics of the water system (time constant, say, 1–3 s). Accordingly, the contribution of the speed-governor as the countermeasure to improve dynamic stability may be limited, especially for the initial short duration (of 0–1 s) just after disturbance.

The function of power control by speed-governor is quite important in reducing frequency fluctuations of the power system on the one hand, while on the other hand it may be a supplementary countermeasure to improve dynamic stability.

As described above, dynamic stability is the concept of greatly improved stability beyond the steady-state stability limit, which can be realized by quick excitation (AVR) control and supplementary speed-governor control of the generators.

14.4 Four-terminal Circuit and the $P-\delta$ Curve under Fault Conditions

In Figure 14.2, the peak value of the $P-\delta$ curve is given by $E_f E_B/(X_q + X_l)$. We conducted our study with the understanding that curve 0 before the fault has a large peak value, while curve 1 during the fault has a very small peak value, because the reactance x_l (the equivalent reactance) included in the denominator becomes quite large under fault conditions in comparison with that under normal conditions before the fault (say, 10 or 20 times).

Why would x_l become so large under the fault condition in comparison with the value before the fault? What is the reactance x_l which is included in the denominator of Equation 14.6a, or Equation 12.12, in particular under fault conditions? We need to clear up these simple questions.

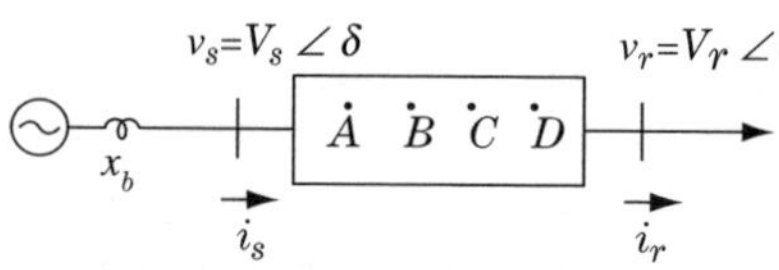

Figure 14.3 Four-terminal circuit

Let us examine Figure 14.3, which is a single phase four-terminal circuit between the sending point s and the receiving point r. The equation for the circuit is

$$\left.\begin{aligned} &\begin{bmatrix} \dot{v}_s \\ \dot{i}_s \end{bmatrix} = \begin{bmatrix} \dot{A} & \dot{B} \\ \dot{C} & \dot{D} \end{bmatrix} \cdot \begin{bmatrix} \dot{v}_r \\ \dot{i}_r \end{bmatrix} \\ &\text{where} \quad \dot{v}_s = V_s\angle\delta = V_s \cdot e^{j\delta} \\ &\qquad\qquad \dot{v}_r = V_r\angle 0 = V \end{aligned}\right\} \tag{14.8a}$$

eliminating $\dot{i}_r$

$$\dot{i}_s = \frac{\dot{D}}{\dot{B}}\dot{v}_s + \frac{\dot{B}\dot{C} - \dot{A}\dot{D}}{\dot{B}}\dot{v}_r \tag{14.8b}$$

The apparent power at the sending point s is

$$\begin{aligned} \dot{S}_s &= P_s + jQ_s = \dot{v}_s\dot{i}_s^* \\ &= \frac{\dot{D}^*}{\dot{B}^*}\dot{v}_s\dot{v}_s^* + \frac{\dot{B}^*\dot{C}^* - \dot{A}^*\dot{D}^*}{\dot{B}^*}\dot{v}_s\dot{v}_r^* \\ &= \frac{\dot{D}^*}{\dot{B}^*}V_s^2 + \frac{\dot{B}^*\dot{C}^* - \dot{A}^*\dot{D}^*}{\dot{B}^*}V_sV_re^{j\delta} \end{aligned} \tag{14.9}$$

Using this equation, we compare circuits 1 and 2 in Figure 14.4.

14.4.1 Circuit 1

This case corresponds to the case of Equation 12.12 ②③ under the relation $X_l \Leftrightarrow x + x'$. The equation is

$$\left.\begin{aligned} &\begin{bmatrix} \dot{v}_s \\ \dot{i}_s \end{bmatrix} = \begin{bmatrix} 1 & j(x+x') \\ 0 & 1 \end{bmatrix} \cdot \begin{bmatrix} \dot{v}_r \\ \dot{i}_r \end{bmatrix} = \begin{bmatrix} \dot{A} & \dot{B} \\ \dot{C} & \dot{D} \end{bmatrix} \cdot \begin{bmatrix} \dot{v}_r \\ \dot{i}_r \end{bmatrix} \\ &\dot{A} = \dot{D} = 1, \quad \dot{B} = j(x+x'), \quad \dot{C} = 0 \end{aligned}\right\} \tag{14.10}$$

Substituting the conjugates $\dot{A}^*$, $\dot{B}^*$, $\dot{C}^*$, $\dot{D}^*$ into Equation 14.9,

$$S_s = P_s + jQ_s = \frac{V_sV_r}{x+x'}\sin\delta + j\left\{\frac{V_s^2 - V_sV_r\cos\delta}{x+x'}\right\} \tag{14.11}$$

The first and second term on the right-hand side gives the $P-\delta$ curve and the $Q-\delta$ curve at point s, and the denominator is $x + x'$ (the reactance between points s and r).

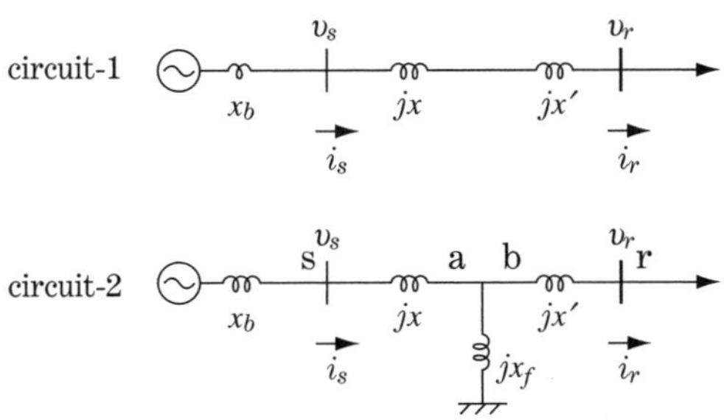

Figure 14.4 Comparison of sending power

14.4.2 Circuit 2

This is the case where reactance x_f is added in parallel at the midpoint of the line. The four terminal circuit equations can be calculated by multiplying the impedance matrices of each section s–a, a–b, b–r in the following order:

$$\begin{bmatrix} \dot{v}_s \\ \dot{i}_s \end{bmatrix} = \overset{\text{marix}}{\underset{s-a}{\begin{bmatrix} 1 & jx \\ 0 & 1 \end{bmatrix}}} \cdot \overset{a-b}{\begin{bmatrix} 1 & 0 \\ \dfrac{1}{jx_f} & 1 \end{bmatrix}} \cdot \overset{b-r}{\begin{bmatrix} 1 & jx' \\ 0 & 1 \end{bmatrix}} \cdot \begin{bmatrix} \dot{v}_r \\ \dot{i}_r \end{bmatrix}$$

$$= \begin{bmatrix} 1+\dfrac{x}{x_f} & j\left(x+x'+\dfrac{x\cdot x'}{x_f}\right) \\ -j\dfrac{1}{x_f} & 1+\dfrac{x'}{x_f} \end{bmatrix} \cdot \begin{bmatrix} \dot{v}_r \\ \dot{i}_r \end{bmatrix} \equiv \begin{bmatrix} \dot{A} & \dot{B} \\ \dot{C} & \dot{D} \end{bmatrix} \cdot \begin{bmatrix} \dot{v}_r \\ \dot{i}_r \end{bmatrix} \tag{14.12}$$

$\dot{A}, \dot{B}, \dot{C}, \dot{D}$ are found and their conjugates $\dot{A}^*, \dot{B}^*, \dot{C}^*, \dot{D}^*$ substituted into Equation 14.9:

$$\begin{aligned} S_s &= P_s + jQ_s \\ &= \frac{V_s V_r}{x+x'+\dfrac{x\cdot x'}{x_f}}\sin\delta + j\left\{\frac{V_s^2\cdot\left(1+\dfrac{x\cdot x'}{x_f}\right) - V_s V_r\cos\delta}{x+x'+\dfrac{x\cdot x'}{x_f}}\right\} \end{aligned} \tag{14.13}$$

In this case, the denominator reactance which corresponds to x_l of Equation 12.12 ②③ becomes $x+x'+(x\cdot x')/(x_f)$.

Now, comparing the results of both circuits, the equivalent reactance of circuit 2 becomes larger than that of circuit 1 as follows:

$$\begin{array}{ccc} \text{circuit 1} & & \text{circuit 2} \\ x+x' & \rightarrow & x+x'+\dfrac{x\cdot x'}{x_f} \end{array} \tag{14.14}$$

That is, Equation 14.11 under normal conditions is replaced by Equation 14.13 under short-circuit fault conditions with the reactance jx_f (or Z_f) inserted at the point f.

The above result can be applied to the positive-sequence circuit of Figure 3.2, for example, in which faults occur at the midpoint f of the line. In case of 1ϕG (phase a to ground fault), the reactance ${}_fx_2 + {}_fx_0$ is inserted at point f, which corresponds to the inserted reactance ${}_fx_2 + {}_fx_0$ of circuit 2.

In the same way, the above result can be applied to all other fault modes by referring to Tables 3.1 and 3.2 which are summarized as Table 14.1.

In cases of short-circuit mode faults, the $P-\delta$ curve obviously becomes smaller (lower) for smaller $Z_f = jx_f$. Then, referring to Table 14.1, the $P-\delta$ curve becomes smaller and of the order of 'before fault $\rightarrow$ 1ϕG $\rightarrow$ 2ϕG $\rightarrow$ 2ϕS $\rightarrow$ 3ϕS. The extreme case is 3ϕS, in that $x_f \rightarrow 0, D(x) \rightarrow \infty, P_s(\delta) \rightarrow 0$, which means the power cannot be transferred at all during the three-phase fault.

Conversely, in cases of phase-opening modes, the $P-\delta$ curves becomes smaller (lower) for larger $Z_f = jX_f$. Then, referring to Table 14.1, the $P-\delta$ curve becomes smaller and of the order of 'before fault $\rightarrow$ 1ϕOp $\rightarrow$ 2ϕOp $\rightarrow$ 3ϕOp.

Table 14.1 The $P-\delta$ curve during a fault and the positive-sequence equivalent impedance

P–δ	$P_s = \dfrac{V_s V_r}{D(x)} \cdot \sin\delta$
Before fault $D(x) = x + x'$	Impedance $Z_f = (r_f + jx_f)$ to be inserted in the positive-sequence circuit
Short-circuit modes $D(x) = x + x' + \dfrac{xx'}{x_f}$	3ϕS $Z_f = 0$ 2ϕS $Z_f = {}_fZ_2$ 1ϕG $Z_f = {}_fZ_0 + {}_fZ_2$ 2ϕG $Z_f = {}_fZ_0 \cdot {}_fZ_2/({}_fZ_0 + {}_fZ_2)$
Phase opening modes $D(x) = x + x' + x_f$	3ϕOp $Z_f = \infty$ 2ϕOp $Z_f = {}_fZ_0 + {}_fZ_2$ 1ϕOp $Z_f = {}_fZ_0 \cdot {}_fZ_2/({}_fZ_0 + {}_fZ_2)$

14.4.3 Trial calculation under assumption of $x_1 = x_2 = x$, $x'_1 = x'_2 = x'$

14.4.3.1 Case 1: 3ϕS

This case corresponds to the condition of $x_f \to 0, D(x) \to \infty$, so the height of the $P - \delta$ curve becomes almost zero. In other words, power can seldom be transformed to the outer system during the fault, and the generator would immediately begin to be accelerated quickly.

14.4.3.2 Case 2: 2ϕS

Referring to Table 14.1 and Table 3.1,

$$x_f = {}_fx_2 = (x/\!/x') = \frac{xx'}{(x + x')}$$

$$D(x) = x + x' + \frac{xx'}{x_f} = 2(x + x')$$

Then, the height of the $P - \delta$ curve under 2ϕS becomes one-half of that before the fault.

14.4.3.3 Case 3: 1ϕG

Referring to Tables 14.1 and 3.1,

$$x_f = {}_fx_2 + {}_fx_0 = (x/\!/x') + (x_0/\!/x'_0) = \frac{xx'}{(x + x')} + \frac{x_0 x'_0}{x_0 + x'_0}$$

$$D(x) = x + x' + \frac{xx'}{x_f}$$

If a solidly neutral grounded system with $k = x_0/x = x'_0/x'$ is assumed, then

$$x_f = (1 + k) \cdot xx'/(x + x')$$

$$D(x) = \{(2 + k)/(1 + k)\} \cdot (x + x')$$

Accordingly, the height of the $P-\delta$ curve under the 1ϕG fault becomes $(1 + k)/(2 + k)$ times that before the fault (then 4/5 times for $k = 3$). In the case of a high-resistive neutral grounding system, $(x_0/\!/x'_0) \to \infty, x_f \to \infty, D(x) \to x + x'$ so that the height of the $P-\delta$ curve is not greatly changed from that before the fault. (refer Table 8.2 [10]).

14.4.3.4 Case 4: 2ϕOp

Referring to Table 14.1 and Table 3.2[2B],

$$x_f = {}_f x_2 + {}_f x_0 = (x + x') + (x_0 + x'_0) = (1 + \alpha)(x + x')$$

where $(x_0 + x'_0) = \alpha(x + x')$ is assumed and

$$D(x) = x + x' + x_f = (2 + \alpha)(x + x')$$

Then, the height of the $P-\delta$ curve under 2ϕOp becomes $1/(2 + \alpha)$ times that before the fault.

14.4.3.5 Case 5: 1ϕOp

Referring to Table 14.1 and Table 3.2[1B],

$$x_f = {}_f x_2 // {}_f x_0 = (x + x') // (x_0 + x'_0) = \frac{(x + x')(x_0 + x'_0)}{x + x' + x_0 + x'_0}.$$
$$D(x) = x + x' + x_f = \left\{1 + \frac{x_0 + x'_0}{x + x' + x_0 + x'_0}\right\} \cdot (x + x')$$

Then, the height of the $P-\delta$ curve under 2ϕOp becomes $1/\{1 + (x_0 + x'_0)/(x + x' + x_0 + x'_0)\}$ times that before the fault.

14.5 P–Q–V Characteristics and Voltage Stability (Voltage Instability Phenomena)

14.5.1 Apparent power at sending terminal and receiving terminal

Referring to the four-terminal circuit of Figure 14.3, the voltages and currents at both terminals are given by Equation 14.8 and the apparent power at the sending terminal is given by Equation 14.9. We calculate the apparent power also at the receiving terminal.

From Equation 14.8a

$$\dot{i}_r = \frac{1}{\dot{B}} \dot{v}_s - \frac{\dot{A}}{\dot{B}} \dot{v}_r \tag{14.15}$$

The apparent power at the receiving terminal is given by

$$\left.\begin{aligned} \dot{S}_r = {} & P_r + jQ_r = \dot{v}_r \dot{i}_r^* \\ = {} & \frac{1}{\dot{B}^*} \dot{v}_s^* \dot{v}_r - \frac{\dot{A}^*}{\dot{B}^*} \dot{v}_r \dot{v}_r^* = \frac{1}{\dot{B}^*} V_s V_r e^{-j\delta} - \frac{\dot{A}^*}{\dot{B}^*} V_r^2 \end{aligned}\right\} \tag{14.16}$$

Next, we derive the equations of apparent power of Figure 14.5a, where the line impedance between points s and r is $z = r + jx$.

For the four-terminal circuit equation

$$\left.\begin{bmatrix} \dot{v}_s \\ \dot{i}_s \end{bmatrix} = \underset{Z = r + jx}{\begin{bmatrix} \dot{A} & \dot{B} \\ \dot{C} & \dot{D} \end{bmatrix}} \cdot \begin{bmatrix} \dot{v}_r \\ \dot{i}_r \end{bmatrix} = \begin{bmatrix} 1 & \dot{Z} \\ 0 & 1 \end{bmatrix} \cdot \begin{bmatrix} \dot{v}_r \\ \dot{i}_r \end{bmatrix} \right\} \tag{14.17}$$

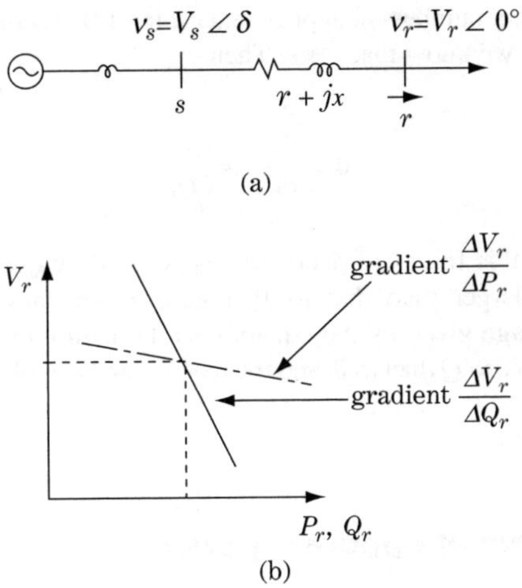

Figure 14.5 Gradient-decreasing characteristics of voltage V by P, Q

Substituting the conjugates of $A = D = 1$, $B = \dot{Z}$, $C = 0$ into Equations 14.9 and 14.16, the apparent power at the sending and receiving terminals, respectively, is derived as follows:

$$\begin{aligned} \dot{S}_s &= P_s + jQ_s = \frac{1}{r - jx}\{V_s^2 - V_s V_r e^{j\delta}\} \\ &= \frac{V_s}{r^2 + x^2}[\{xV_r \sin\delta + r(V_s - V_r\cos\delta)\} + j\{x(V_s - V_r\cos\delta) - rV_r\sin\delta\}] \quad ① \\ \dot{S}_r &= P_r + jQ_r = \frac{1}{r - jx}\{V_s V_r e^{-j\delta} - V_r^2\} \\ &= \frac{V_r}{r^2 + x^2}[\{xV_s\sin\delta + r(V_s\cos\delta - V_r)\} + j\{x(V_s\cos\delta - V_r) - rV_s\sin\delta\}] ② \end{aligned} \tag{14.18}$$

14.5.2 Voltage sensitivity by small disturbance $\Delta P, \Delta Q$

We examine voltage sensitivity caused by small disturbance of P, Q.
From Equation 14.18 ②

$$\begin{aligned} &(P_r + jQ_r)(r - jx) = V_r\{V_s e^{-j\delta} - V_r\} \\ \therefore\quad &(rP_r + xQ_r + V_r^2) + j(rQ_r - xP_r) = V_s V_r e^{-j\delta} \end{aligned} \tag{14.19a}$$

We will get two equations by separating the real and imaginary parts of the above equation, and then, eliminating δ from both equations:

$$(rP_r + xQ_r + V_r^2)^2 + (rQ_r - xP_r)^2 = V_s^2 V_r^2 \tag{14.19b}$$

Putting $P_r \to P_r + \Delta P_r$, $V_r \to V_r + \Delta V_r$, then $\Delta V_r/\Delta P_r$ can be calculated. $\Delta V_r/\Delta Q_r$ can be calculated similarly. (Refer to Supplement 1 for the derivation process.) Thus

$$\left.\begin{aligned} \frac{\partial V_r}{\partial P_r} &= \frac{\Delta V_r}{\Delta P_r} = \frac{(x^2 + r^2)P_r + rV_r^2}{V_r\{V_s^2 - 2V_r^2 - 2(rP_r + xQ_r)\}} \quad ① \\ \frac{\partial V_r}{\partial Q_r} &= \frac{\Delta V_r}{\Delta Q_r} = \frac{(x^2 + r^2)P_r + xV_r^2}{V_r\{V_s^2 - 2V_r^2 - 2(rP_r + xQ_r)\}} \quad ② \end{aligned}\right\} \tag{14.20}$$

In this equation, V_r, V_s are real numbers of approximate value 1.0. Accordingly, both equations are of negative value. In addition, we know that $x \gg r$. Then

$$0 > \frac{\partial V_r}{\partial P_r} \gg \frac{\partial V_r}{\partial Q_r} \tag{14.21}$$

This equation tells us that V is forced to decrease whenever P or Q is increased, and the voltage sensitivity to Q is much larger than that to P. Figure 14.5b shows the **gradient-decreasing characteristics of the voltage** given by the equation as the nature of the power system.

Voltage V is more sensitive to Q than to P, and the reason can be explained by the fact that $x \gg r$ for the power system.

14.5.3 Circle diagram of apparent power

We examine the system shown in Figure 14.6, where resistance of the line is neglected for simplicity. Putting $r = 0$ in Equation 14.18,

$$\left.\begin{aligned} S_r &= P_r + jQ_r = \frac{V_s V_r}{x}\sin\delta + j\frac{V_r(V_s\cos\delta - V_r)}{x} & ① \\ P_r &= \frac{V_s V_r}{x}\sin\delta & ② \\ Q_r &= \frac{V_r(V_s\cos\delta - V_r)}{x} & ③ \end{aligned}\right\} \tag{14.22}$$

The equation can be modified as follows.

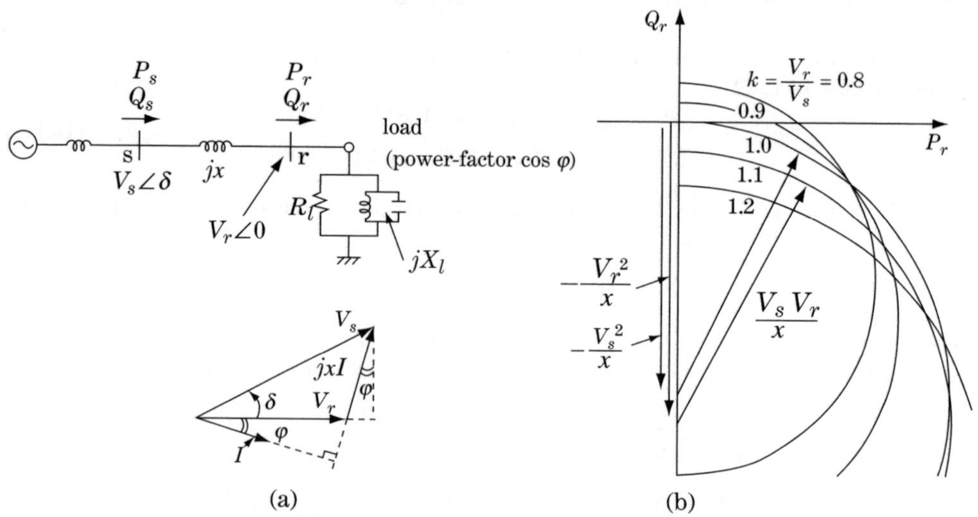

Figure 14.6 Power circle diagram at the receiving terminal (line resistance neglected)

For the equation of the circle diagram in coordinates (P_r, Q_r)

$$\left.\begin{aligned}
&P_r + j\left(Q_r + \frac{V_r^2}{x}\right) = j\frac{V_s V_r}{x} e^{-j\delta} \quad ①\\
&\text{or}\\
&P_r^2 + \left(Q_r + \frac{V_r^2}{x}\right)^2 = \left(\frac{V_s V_r}{x}\right) \quad ②\\
&\text{or}\\
&\left.\begin{aligned}
&P_r^2 + \left(Q_r + \frac{k^2 V_s^2}{x}\right)^2 = \left(\frac{k V_s^2}{x}\right)^2\\
&\text{centre of the circle } \left(0, -\frac{V_r^2}{x}\right), \text{ radius } \frac{V_s V_r}{x}\\
&\text{where } \quad k = \frac{V_r}{V_s}
\end{aligned}\right\} ③
\end{aligned}\right\} \tag{14.23}$$

14.5.4 *P–Q–V* characteristics, and *P–V* and *Q–V* curves

Here we introduce the load equation at the receiving terminal r, whose power factor is $\cos\varphi$:

$$\left.\begin{aligned}
&\dot{S}_r = P_r + jQ_r = |S_r|\cos\varphi + j|S_r|\sin\varphi \quad ①\\
&Q_r = P_r \tan\varphi, \quad \cos\varphi = P_r/\sqrt{P_r^2 + Q_r^2} \quad ②
\end{aligned}\right\} \tag{14.24}$$

Substituting P_r, Q_r from Equation 14.22 into Equation 14.24 ② and modifying it using formulae for trigonometric functions,

$$V_r = V_s\{\cos\delta - \tan\varphi \sin\delta\} = V_s \frac{\cos(\varphi + \delta)}{\cos\varphi} \tag{14.25}$$

Applying this equation to Equation 14.22, either V_s or V_r can be eliminated and we have the following equations.

For the equation of P_r, Q_r by V,

$$\left.\begin{aligned}
&\underline{P_r - V_s \text{ curve}}\\
&P_r = \frac{V_s^2}{x}\cdot\frac{\cos(\varphi+\delta)\sin\delta}{\cos\varphi} = \frac{V_s^2}{2x}\cdot\frac{\{\sin(\varphi+2\delta) - \sin\varphi\}}{\cos\varphi} \quad ①\\
&\underline{Q_r - V_s \text{ curve}}\\
&Q_r = P_r\tan\varphi = \frac{V_s^2}{2x}\cdot\frac{\{\sin(\varphi+2\delta) - \sin\varphi\}}{\cos\varphi}\tan\varphi \quad ②
\end{aligned}\right\} \tag{14.26}$$

For the equation of P_r, Q_r by V_r,

$$\left.\begin{aligned}
&\underline{P_r - V_r \text{ curve}}\\
&P_r = \frac{V_r^2}{x}\cdot\frac{\cos\varphi\sin\delta}{\cos(\varphi+\delta)} = \frac{V_r^2}{2x}\cdot\frac{\{\sin(\varphi+\delta) - \sin(\varphi-\delta)\}}{\cos(\varphi+\delta)} \quad ①\\
&\underline{Q_r - V_r \text{ curve}}\\
&Q_r = P_r\tan\varphi = \frac{V_r^2}{2x}\cdot\frac{\{\sin(\varphi+\delta) - \sin(\varphi-\delta)\}}{\cos(\varphi+\delta)}\tan\varphi \quad ②
\end{aligned}\right\} \tag{14.27}$$

The receiving point quantities V_r, P_r, Q_r of Equations 14.25 and 14.26 are expressed as functions of variables V_s, δ, φ, where $V_s = 1.0$. Accordingly, V_r, P_r, Q_r can be written only by parameters δ (the angular difference between the sending and receiving terminals) and φ (the angle of load power factor). In other words, the curved surface of P–Q–V characteristics in three-dimensional coordinates can be written by the parameters δ and φ. Also, P–V, Q–V and P–Q curves can be written as projections in two-dimensional coordinates. Figure 14.7 shows the curves of the P–Q–V characteristics.

On the P–V curve in Figure 14.7a, if P is gradually increased from zero under the load of the leading power factor (φ is negative), V tends to increase slowly. However, in turn it begins to decrease at roughly $\delta = 40°$, and P reaches the critical point of the upper limit at roughly $\delta = 70°$.

If P is gradually increased from zero under the load of the lagging power factor (φ is positive), V tends to decrease rather quickly, and P reaches the critical point of the upper limit at roughly $\delta = 45 - 70°$. The Q–V curve in Figure 14.7b shows similar characteristics to that in Figure 14.7a.

Incidentally, Figure 14.7a indicates that Equation 14.27 gives two voltage solutions for one value of P that are higher and lower solutions. In other words, the upper half of the P–V curve includes the actual operating zone, while the lower half is meaningless. Also Figure 14.7b indicates two voltage solutions for one value of Q, and the actual stable voltage is given by the higher solution.

14.5.5 P–Q–V characteristics and voltage instability phenomena

The P–Q–V characteristics of the power system having been discussed, we examine next the P–Q–V characteristics of the load system. Our load can be expressed by the parallel circuit of resistance R_l and reactance jX_l (negative value for the capacitive load) as shown in Figure 14.6a:

$$\left.\begin{aligned} &P_l + jQ_l = V_r I_l^* = V_r \frac{V_r^*}{Z_l^*} = V_r^2\left(\frac{1}{R_l} + \frac{1}{jX_l}\right) && ① \\ &\text{power factor} \cos\varphi = \frac{P_l}{|P_l + jQ_l|}, \quad Q_l = P_l \tan\varphi && ② \end{aligned}\right\} \qquad (14.28)$$

where suffix l means load, and

$$\left.\begin{aligned} &\left.\begin{aligned} P_l &= \frac{V_r^2}{R_l} \\ Q_l &= \frac{V_r^2}{X_l} \end{aligned}\right\} ① \qquad \left.\begin{aligned} V_r &= \sqrt{R_l}\,.\,\sqrt{P_l} \\ V_r &= \sqrt{X_l}\,.\,\sqrt{Q_l} \end{aligned}\right\} ② \end{aligned}\right\} \qquad (14.29)$$

The voltage V_r is a function of the square root of P_l or Q_l. The above equation gives the P–V characteristics and Q–V characteristics of the load system. The set of magnitudes P, Q, V at the intersecting point of the power system characteristics and the load characteristics gives the operating values at the receiving point. In order to visualize the total P–Q–V characteristics, let us examine the phenomena in two-dimensional coordinates.

14.5.5.1 Voltage collapse phenomena (P–V avalanche)

The P–V characteristics at the receiving point (Equation 14.27①) and those of the load (Equation 14.29) are shown together in Figure 14.7d.

P_r, V_r at the intersecting point are the actual operating power and voltage, respectively. If the load gradually increases (under nearly constant Q_r), the operating point will move from the load point ① to

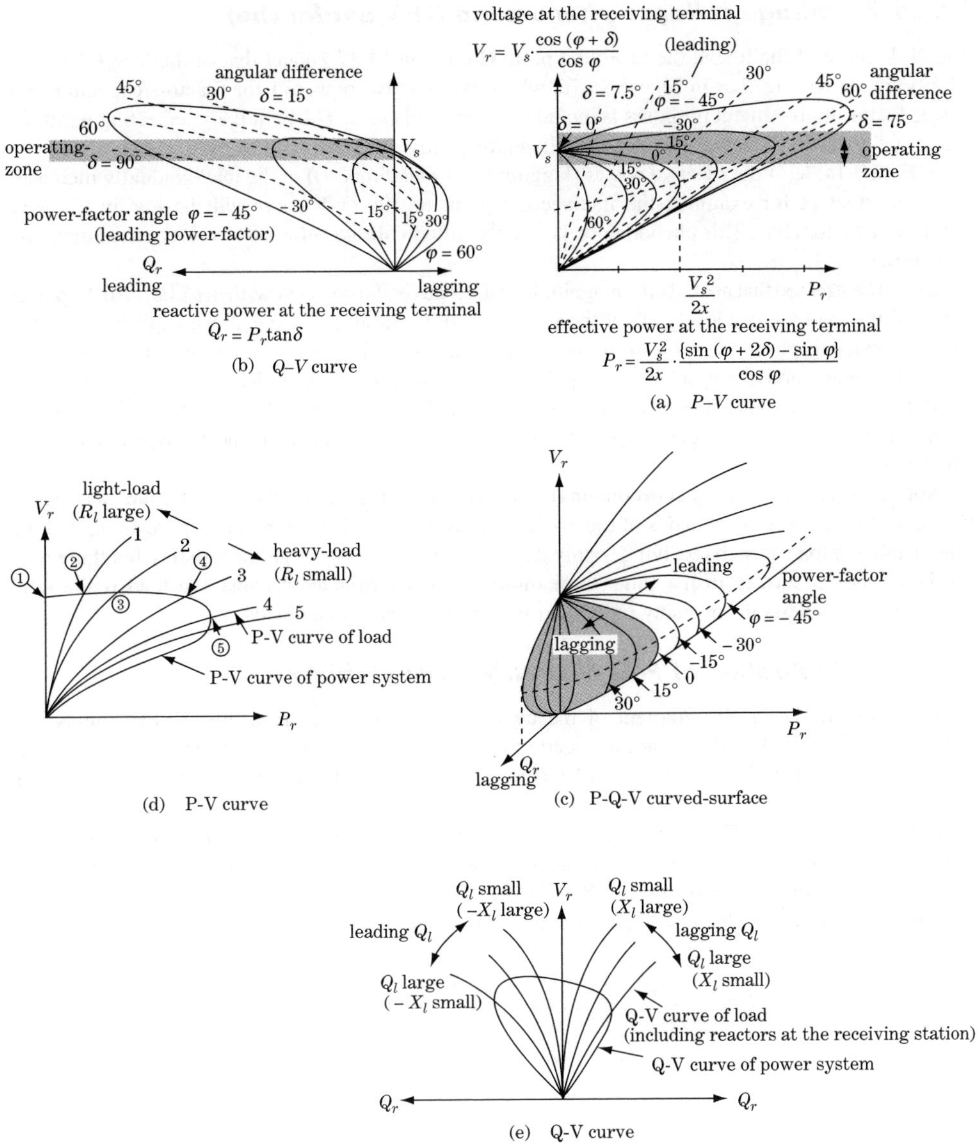

Figure 14.7 *P–Q–V* characteristics (*P–Q–V* steady-state stability)

② → ③ → ④ → ⑤. Around point ④, the voltage begins to decrease suddenly and rapidly, and before long P_r, V_r are forced to decrease or miss the critical point ⑤ (the maximum load point). In other words, if the system is operating at point ④ and the load is increased for some reason, exceeding the critical point ⑤, the stable operating point no longer exists and the voltage and the power of the network are entirely lost. This phenomenon is usually called **voltage collapse** or **voltage avalanche by *P–V* mode**.

14.5.5.2 Voltage collapse phenomena (Q–V avalanche)

The Q–V curve of the line at the receiving point (Equation 14.27②) and that of the load (Equation 14.29) are shown together in Figure 14.7e, where the reactive power of the reactors/capacitors for power-factor improvement purposes is included in the load curve. Q_r, V_r at the intersecting point are the actual operating reactive power and voltage, respectively.

In Figure 14.7e, if the reactive power (lagging (+) or leading (−)) of the load gradually increases under constant P, for example, the intersecting point on the Q–V curve will be lost in the very low-power-factor zone. This phenomenon is usually called **voltage collapse** or **voltage avalanche by** ***Q–V*** **mode**.

It must be stressed that not only the magnitude of the load (or the power flow, then δ), but also the power factor of the load $\cos\varphi$, closely affect the voltage stability limit. Also, the above explanation in the two-dimensional plane is only to make understanding easier. The actual operating condition should be understood as the intersecting point of the three-dimensional characteristics. With this in mind, the characteristics explained in Figure 14.7 should be called 'the P–Q–V steady-state stability characteristics' and accordingly 'P–Q–V collapse' or 'P–Q–V avalanche' respectively, as more exact expressions.

Note also that the power system contains the smooth curved surface of the P–Q–V characteristics in the three-dimensional coordinates of the V-axis, P-axis and Q-axis as shown in Figure 14.7c. On the other hand, the load at each receiving terminal also contains the P–Q–V characteristics, but there may not be a smooth surface in the same three-dimensional coordinates because the load is the total composite characteristics of many individual time to time capricious loads.

14.5.5.3 Evaluation of P–Q–V steady-state stability

Again in Figure 14.7a, the gradient of the curve $\partial V_r/\partial P_r$ changes from plus (+) to minus (−) around $\delta = 40°$. On the other hand, we need to recall the explanation that the steady-state stability limit was given at $\delta = 90°$ in Figure 12.1 (for a non-salient-pole machine). We must clarify the difference.

Let us look again at Equation 14.18②, which shows the apparent power at the receiving point before neglecting line resistance r. We can separate the real and imaginary parts and then eliminate δ to derive the following equation (refer to Supplement 2 for the process).

For the power circle diagram (line resistance r is considered)

$$\left.\begin{aligned}&\left(P_r+\frac{rV_r^2}{r^2+x^2}\right)^2+\left(Q_r+\frac{xV_r^2}{r^2+x^2}\right)^2=\frac{V_s^2V_r^2}{r^2+x^2}\\&\text{where}\quad V_s\fallingdotseq V_r\fallingdotseq 1.0,\quad x\gg r\end{aligned}\right\}\quad ①$$

Or neglecting r^2 in the denominators,

$$\left.\begin{aligned}&\left(P_r+\frac{rV_r^2}{x^2}\right)^2+\left(Q_r+\frac{xV_r^2}{x^2}\right)^2=\left(\frac{V_sV_r}{x}\right)^2\\&\text{with centre}\left(-\frac{rV_r^2}{x^2},-\frac{xV_r^2}{x^2}\right)\text{ and radius}\frac{V_sV_r}{x}\end{aligned}\right\}\quad ② \qquad (14.30)$$

The **power circle diagram** of the receiving terminal (P_r, Q_r) considering line resistance is derived as shown in Figure 14.8, in which the diagram of the sending terminal is also shown.

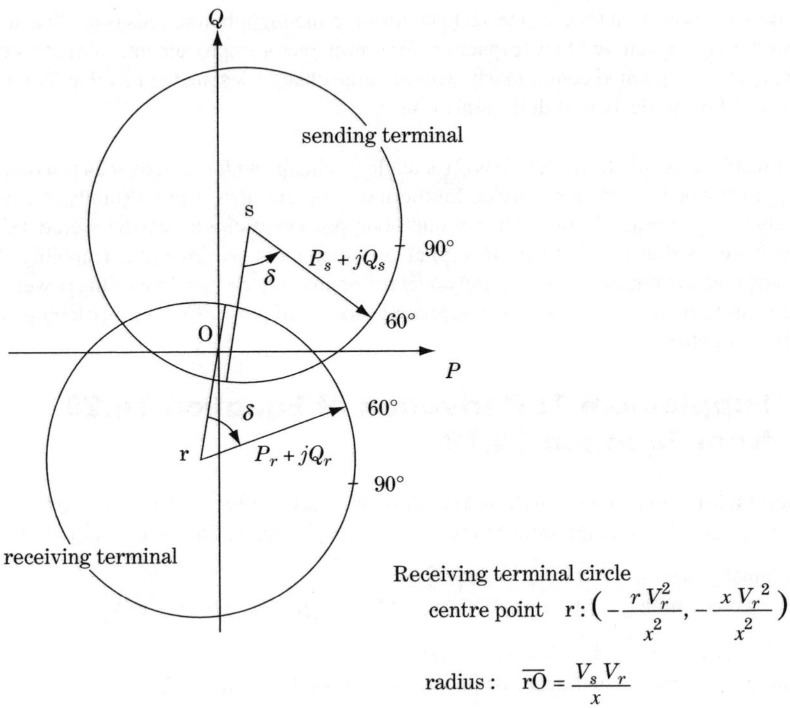

Figure 14.8 The power circle diagram (line resistance considered)

We can conclude from Equation 14.30, Figure 14.8 and Figure 14.7 as follows:

- In order to increase P_r gradually without changing the voltage ($k = V_r/V_s$: fixed, say 1.0), the angular difference δ and Q_r become larger, and before long very much larger Q_r is required around $\delta = 30 - 35°$ for only a small increase in ΔP_r.
- If P_r is gradually increased under fixed Q_r, $k = V_r/V_s$ has to become smaller; in other words, V_r has to become smaller quickly under fixed V_s, which means voltage collapse.

This is yet another explanation of the characteristics given by Figure 12.1. In actual power system practice, operation under a condition of δ exceeding 40° is unrealistic, because the capacity of Q_r is actually limited.

14.5.5.4 V–Q control (voltage and reactive power control) of power systems

The load characteristics of each load area must be quite different as they are related to the social and geographical environments of the area.

In the first place, load varies from time to time by nature, so the actual load behaviour may not easily be expressed in simple equational form. In addition, as a typical example, the load characteristics of an ordinary industrialized area may have the following features:

- In the night hours, the magnitude of real power P consumption decreases remarkably and the ratio of this load to the peak load of the same day may decrease to a value of 1/2 or even 1/3.

- In addition, the power factor cos φ tends to be lower in the night hours. This is because many of the facilities having capacitive MVA (capacitors for power-factor improvement, cable stray capacities, etc.) are apt to be operated continuously without large change despite the fact that facilities having real power MW are decreased during this time.

In other words, load quantities P, Q, cos ϕ (or angle φ) change widely and over 24 h, as conditions of the total network or local area networks. Furthermore, severe disturbances (faults etc.) may occur capriciously at any location. Under all these conditions, power systems have to be operated all the time, keeping voltages within 1.0 ± 0.1 (typically) at any point and ensuring system stability. To control voltage V and reactive power Q appropriately over time and at all the locations of the power system is a very tough engineering matter and is as important as total real power ($\sum P_i$) dispatching control and frequency (f) control.

14.6 Supplement 1: Derivation of Equation 14.20 from Equation 14.19

In Equation 14.19b, replacing $P_r \to P_r + \Delta P_r$, $V_r \to V_r + \Delta V_r$ and $P_r^2 \to P_r^2 + 2P_r \cdot \Delta P_r$, $V_r^2 \to V_r^2 + 2V_r \cdot \Delta V_r$ by neglecting second-order terms $(\Delta P_r)^2, (\Delta V_r)^2$, we get the two equations below:

The original equations $\quad A^2 + B^2 = V_s^2 V_r^2 \quad ①$

The derived equations $\quad (A + r \cdot \Delta P_r + 2V_r \cdot \Delta V_r)^2 + (B - x \cdot \Delta P_r)^2 = V_s^2 V_r^2 + V_s^2 (2V_r \cdot \Delta V_r) \quad ②$

where $A = rP_r + xQ_r + V_r^2$, $B = rQ_r - xP_r$

Substituting ① into ② and again neglecting second-order terms $(\Delta P_r)^2, (\Delta V_r)^2$,

$$\frac{\Delta V_r}{\Delta P_r} = \frac{rA - xB}{V_r(V_s^2 - 2A)} = \text{Eq.}(14.20)①$$

Equation 14.20 ② can be derived analogously.

Note that the above calculations are mathematically a partial derivation of the implicit function.

14.7 Supplement 2: Derivation of Equation 14.30 from Equation 14.18 ②

From Equation 14.18 ②

$$P_r + \frac{rV_r^2}{r^2 + x^2} = \frac{V_s V_r}{r^2 + x^2}(x \sin \delta + r \cos \delta) \quad ①$$

$$Q_r + \frac{xV_r^2}{r^2 + x^2} = \frac{V_s V_r}{r^2 + x^2}(x \cos \delta - r \sin \delta) \quad ②$$

Then, the following equation is derived from $①^2 + ②^2$:

$$\left(P_r + \frac{rV_r^2}{r^2 + x^2}\right)^2 + \left(Q_r + \frac{xV_r^2}{r^2 + x^2}\right)^2 = \frac{V_s^2 V_r^2}{r^2 + x^2} \tag{14.30}$$

15

Generator Characteristics with AVR and Stable Operation Limit

A generator's stable operation is closely related to AVR (Automatic Voltage Regulator) characteristics and the load condition. In this chapter, the operational characteristics of a generator with AVR and its visualization in *p–q* coordinates are introduced. Then, the critical conditions of the generator's stable operation will be investigated.

15.1 Theory of AVR, and Transfer Function of Generator System with AVR

Our generator is running in combination with an AVR and supplying power to the load Z as shown in Figure 15.1.

Whenever the generator terminal voltage e is decreased a little, $e \to e - \Delta e$, AVR will detect the small deviation Δe and will immediately increase the generator excitation $E_f \to E_f + \Delta E_f$ so that the terminal voltage is recovered quickly. This behaviour has to be examined as a problem of response characteristics in regard to the total system including the generator, AVR, transmission line and load. This will be investigated by applying the concept of transfer functions, which is a familiar tool in automatic control theory based on Laplace transforms. In the Laplace transformation, $s = d/dt$ is treated like an algebraic equation. We start with the generator's transfer function.

15.1.1 Inherent transfer function of generator

We begin with the results of Equations 10.43–10.46 and Figure 10.4 of Park's theory for a generator in Chapter 10. The equations are again written below as Laplace-transformed equations, while d/dt is replaced by symbol s:

$$\left.\begin{aligned}
e_d(s) &= -\psi_q(s)s\theta + s\psi_d(s) - ri_d(s) && ① \\
e_q(s) &= +\psi_d(s)s\theta + s\psi_q(s) - ri_q(s) && ② \\
E_{fd}(s) &= s\psi_{fd}(s) + r_{fd}i_{fd}(s) && ③ \\
\psi_d(s) &= -x_d i_d(s) + x_{ad}i_{fd}(s) + x_{ad}i_{kd}(s) && ④ \\
\psi_q(s) &= -x_q i_q(s) + x_{aq}i_{kq}(s) && ⑤ \\
\psi_{fd}(s) &= -x_{ad}i_d(s) + x_{ffd}i_{fd}(s) + x_{fkd}i_{kd}(s) && ⑥
\end{aligned}\right\} \qquad (15.1)$$

Handbook of Power System Engineering Yoshihide Hase
© 2007 John Wiley & Sons, Ltd

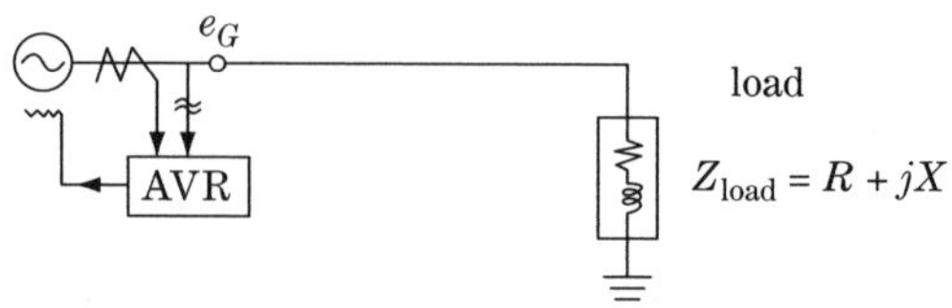

Figure 15.1 Model power system of one generator with AVR and the load

These are the general equations of a generator, which can be simplified as follows for practical reasons:

- The damper currents $i_{kd}(s)$, $i_{kq}(s)$ vanish within 0–3 cycles after the disturbance, so they can be neglected for the phenomena 3 cycles after the power system disturbance.
- The generator is operating in synchronism at a speed of $50/60 \pm 0.05$ Hz. Accordingly, the unitized angular velocity is 1.0, namely

$$s\theta = \omega = \frac{2\pi(50 \pm 0.05)}{2\pi \times 50} = 1.0 \pm 0.001 \fallingdotseq 1.0 \tag{15.2}$$

- Flux linkages ψ_d, ψ_q are of d.c. quantities at $t = 0-$, as was explained in detail in Section 10.6 and Equation 10.75, so that the derivatives $s\psi_d$, $s\psi_q$ (corresponding to transient d.c. current) can be neglected for the purposes of this chapter.

Therefore Equation 15.2 is as simplified below:

$$\left.\begin{aligned}
e_d(s) &= -\psi_q(s) - ri_d(s) && ① \\
e_q(s) &= +\psi_d(s) - ri_q(s) && ② \\
E_{fd}(s) &= s\psi_{fd}(s) + r_{fd}i_{fd}(s) && ③ \\
\psi_d(s) &= -x_d i_d(s) + x_{ad} i_{fd}(s) && ④ \\
\psi_q(s) &= -x_q i_q(s) && ⑤ \\
\psi_{fd}(s) &= -x_{ad} i_d(s) + x_{ffd} i_{fd}(s) && ⑥
\end{aligned}\right\} \tag{15.3}$$

As we are studying phenomena of excitation control, $E_{fd}(s)$ has to be treated as a function of s. Eliminating variables of flux linkage ψ,

$$\left.\begin{aligned}
&\text{from ①⑤} \\
&\quad e_d(s) = x_q i_q(s) - ri_d(s) && ① \\
&\text{from ②④} \\
&\quad e_q(s) = -x_d i_d(s) + x_d i_{fd}(s) - ri_q(s) && ② \\
&\text{from ③⑥} \\
&\quad E_{fd}(s) = -x_{ad} s i_d(s) + (x_{ffd}s + r_{fd}) i_{fd}(s) && ③
\end{aligned}\right\} \tag{15.4}$$

Substituting i_{fd} from Equation 15.4 ③ into ②,

$$\left.\begin{aligned}
e_q(s) &= \frac{x_{ad}}{x_{ffd}s + r_{fd}} E_{fd}(s) - \left\{ x_d - \frac{x_{ad}^2 s}{x_{ffd}s + r_{fd}} \right\} i_d(s) - ri_q(s) \\
&= \frac{1}{1 + T'_{d0}s} \cdot \frac{x_{ad}}{r_{fd}} E_{fd}(s) - \frac{1}{1 + T'_{d0}s} \left\{ x_d + \left(x_d T'_{d0} - \frac{x_{ad}^2}{r_{fd}} \right) s \right\} i_d(s) - ri_q(s) && ① \\
\therefore e_q(s) &= \frac{1}{1 + T'_{d0}s} e_f(s) - \frac{1}{1 + T'_{d0}s} \left\{ x_d + \left(x_d - \frac{x_{ad}^2}{x_{ffd}} \right) T'_{d0}s \right\} i_d(s) - ri_q(s) && ② \\
&\text{where } e_f(s) \equiv \frac{x_{ad}}{r_{fd}} E_{fd}(s), \quad T'_{d0} \equiv \frac{x_{ffd}}{r_{fd}}
\end{aligned}\right\} \tag{15.5}$$

The equation can be modified further. From Equation 10.70

$$\left.\begin{aligned} & x'_d = x_l + \frac{1}{\dfrac{1}{x_{ad}} + \dfrac{1}{x_{fd}}} && ① \\ & \text{from Eq. (10.72)} \\ & x_d = x_l + x_{ad} && ② \\ & \text{from Eq. (10.49b)} \\ & x_{fd} = x_{ffd} - x_{ad} && ③ \\ & \therefore\ x'_d = x_d - x_{ad} + \frac{x_{ad}x_{fd}}{x_{ad} + x_{fd}} = x_d - \frac{x_{ad}^2}{x_{ffd}} && ④ \end{aligned}\right\} \quad (15.6)$$

Then, replacing ( ) on the right of Equation 15.5 ② by x'_d, we obtain the transfer function of a generator:

$$\left.\begin{aligned} & \left.\begin{aligned} & e_d(s) = x_q i_q(s) - r i_d(s) && ① \\ & e_q(s) = \frac{1}{1 + T'_{d0}s} e_f(s) - \frac{x_d + x'_d T'_{d0}s}{1 + T'_{d0}s} i_d(s) - r\, i_q(s) && ② \end{aligned}\right\} \\ & \text{where} \\ & T'_{d0} = \frac{x_{ffd}}{r_{fd}} : \text{the time constant of excitation circuit of d-axis equivalent circuit} \\ & \qquad\qquad \text{(see Figure 10.4 and Equation 10.107a)} \\ & x'_d = x_d - \frac{x_{ad}^2}{x_{ffd}} : \text{transient reactance of d-axis circuit} \\ & e_f(s) = \frac{x_{ad}}{r_{fd}} E_{fd}(s) : \text{excitation voltage} \end{aligned}\right\} \quad (15.7)$$

The first term on the right of Equation 15.7 ② is the term which would be changed proportionally by the excitation e_f, and is a first-order time delay s function with time constant T'_{d0}.

The second term on the right of Equation 15.7 ② is the term related to the armature current, and corresponds to the voltage drop caused by armature reactance. The second term obviously becomes $x'_d i_d$ when $t \to 0+ \ (s \to \infty)$, and finally becomes $x_d i_d$ when $t \to \infty (s \to 0+)$. The result of course coincides with the explanation of Sections 10.6 and 10.7.

15.1.2 Transfer function of generator + load

Next, the generator is connected to the load impedance $Z = R + jX$ as shown in Figure 15.1, which already includes line impedance. Also referring to Equation 10.55 or Figure 10.5,

$$\left.\begin{aligned} & \left.\begin{aligned} & e_G(s) = \sqrt{e_d^2(s) + e_q^2(s)} && ① \\ & e_d(s) + je_q(s) = \{i_d(s) + ji_q(s)\}(R + jX) && ② \end{aligned}\right\} \\ & \left.\begin{aligned} & \therefore\ i_d(s) = \frac{R}{R^2 + X^2} e_d(s) + \frac{X}{R^2 + X^2} e_q(s) && ③ \\ & i_q(s) = \frac{R}{R^2 + X^2} e_q(s) - \frac{X}{R^2 + X^2} e_d(s) && ④ \end{aligned}\right\} \end{aligned}\right\} \quad (15.8)$$

By substituting $i_d(s)$, $i_q(s)$ into Equation 15.7 ①②, we obtain simultaneous equations of first order and two variables $e_d(s)$, $e_q(s)$ and the following solution (see Supplement 1 for details):

$$\left.\begin{aligned} e_d(s) &= \frac{x_q R - Xr}{\left\{(X+x_d)(X+x_q)+(R+r)^2\right\}+\left\{(X+x'_d)(X+x_q)+(R+r)^2\right\}T'_{d0}s}e_f(s) \quad ① \\ e_q(s) &= \frac{X^2+R^2+Xx_q+Rr}{\left\{(X+x_d)(X+x_q)+(R+r)^2\right\}+\left\{(X+x'_d)(X+x_q)+(R+r)^2\right\}T'_{d0}s}e_f(s) \quad ② \end{aligned}\right\} \tag{15.9}$$

Substituting the results into Equation 15.8 ① (see Supplement 2 for the solution), then

$$\left.\begin{aligned} & e_G(s) = \frac{A}{1+Ts}e_f(s) && ① \\ & G_G(s) \equiv \frac{e_G(s)}{e_f(s)} = \frac{A}{1+Ts} : \text{the transfer function of the generator terminal voltage under load condition} && ② \\ & \text{where} \\ & A = \frac{\sqrt{X^2+R^2}\cdot\sqrt{(X+x_q)^2+(R+r)^2}}{(X+x_d)(X+x_q)+(R+r)^2} : \text{the gain of generator + load} && ③ \\ & T = \frac{(X+x'_d)(X+x_q)+(R+r)^2}{(X+x_d)(X+x_q)+(R+r)^2}T'_{d0} : \text{the time constant of generator + load} && ④ \\ & e_f(s) = \frac{x_{ad}}{r_{fd}}E_{fq}(s) : \quad \text{excitation} && ⑤ \end{aligned}\right\} \tag{15.10}$$

Equations 15.9 and 15.10 explain that the total system of a generator + load has the following features. The value of the time constant T is affected by the load condition (R, jX) and is more or less smaller than the generator's specific time constant T'_{d0}, namely $T < T'_{d0}$, because $x_d > x'_d$ in Equation 15.10 ④. Incidentally, the armature resistance r can actually be neglected in the above equations, which we will now examine further.

15.1.2.1 Cases of special load conditions

15.1.2.1.1 Case 1: no-load condition

Putting $R = \infty$, $X = \infty$, then $A = 1$, $T = T'_{d0}$

$$\therefore\ G_G(s) = \frac{1}{1+T'_{d0}s} \tag{15.11}$$

15.1.2.1.2 Case 2: load with power factor cos ϕ = 1.0

Putting $X = 0$, $r = 0$,

$$\therefore \quad G_G(s) = \frac{\dfrac{R\sqrt{x_q^2+R^2}}{x_d x_q + R^2}}{1+\dfrac{x'_d x_q + R^2}{x_d x_q + R^2}T'_{d0}} \tag{15.12}$$

For a light load $R \gg x_d$, x_q, then

$$G_G(s) = \frac{1}{1+T'_{d0}s} \tag{15.13}$$

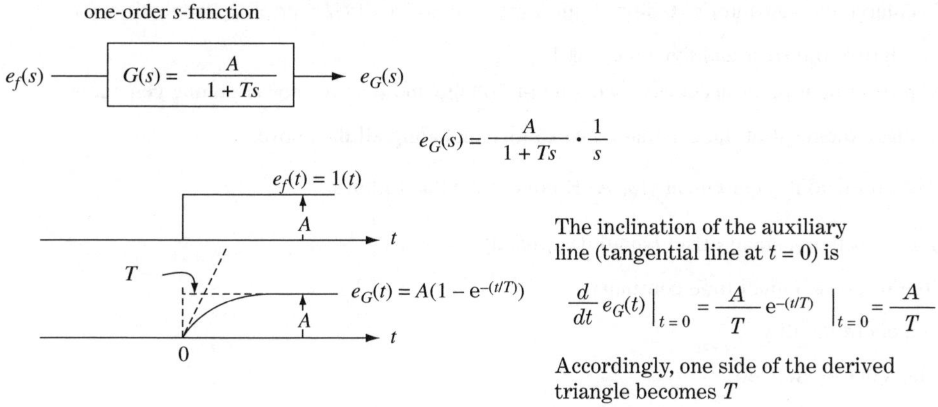

Figure 15.2 The transfer function of first-order time delay

15.1.2.1.3 Case 3: inductive load with power factor $\cos\phi = 0$

Putting $R = 0,\ r = 0,$

$$\therefore\ G_G(s) = \frac{\dfrac{X}{X + x_d}}{1 + \dfrac{X + x'_d}{X + x_d} T'_{d0} s} \tag{15.14}$$

For an inductive light load, $X \gg x_d,\ x'_d$, so $G_G(s)$ is the same as in Equation 15.13.

15.1.2.1.4 Case 4: capacitive load with power factor $\cos\phi = 0$

In this case, $G_G(s)$ is equal to Equation 15.14, but X is of negative magnitude $X = -X_c$. The typical case of this mode is where a transmission line is charged by a generator with no load condition. If the absolute value of X_c becomes close to the value of x_d, an abnormal solution of $(X + x_d) = (-X_c + x_d) \to 0,\ A \to \infty,\ T \to \infty$ would occur. Such an extraordinary large gain A means a quite unstable system condition from the viewpoint of automatic control theory. Of course, this condition physically means very unstable series resonance. This phenomenon is again discussed in Section 15.4. Figure 15.2 shows the transfer function of a generator, and also the behaviour when a step-function signal $e_f(t) = 1(t)$ (namely, $e_f(s) = 1/s$) is input.

15.2 Duties of AVR and Transfer Function of Generator + AVR

We have found the transfer function of the **generator** + **load impedance**. Now we will find the total transfer function of the **AVR** + **exciter** + **generator** + **load impedance**, by which we can investigate the dynamic behaviour of the generator system with AVR.

The duties of AVR are outlined as follows:

- To maintain generator terminal voltage within a permissible certain band around the rated voltage and to prevent generator operation in the unstable zone or in the prohibited operating zone.
- To generate appropriate Var(Q) in relation to effective power (P) generation (V–Q control and P–Q control).
- To control voltage and reactive power automatically to maintain stability during large disturbances.

- To enlarge the generator's stability limit, especially at small GD^2 or of short-circuit ratio.
- To improve transient and dynamic stability.
- To protect hunting or necessary swinging of voltage and reactive power among generators.
- To make the most of the generator's capability including all the above.

The substantial requirements for AVR are outlined as follows:

- High sensitivity, small offset (small dead-band)
- High response (small time constant)
- Fine controllability
- Wide control area, etc.

AVRs installed at every generator terminal of a power system play leading roles in power system operation, in order to protect individual generators from prohibited operation ,or to maintain stable power system operation over time by avoiding unstable conditions among other generators, such as loss of synchronization or unnecessary hunting phenomena of voltages, cross-currents and reactive power . If AVR were categorized as only accessory equipment for a generator, having the function to keep its terminal voltage at a certain value, this would be a serious underestimation.

Now we will study the response characteristics of AVR and of the exciter first, and then examine the dynamic characteristics of a generator with AVR. Incidentally, AVR is treated in this book only from a functional viewpoint, not a hardware one.

For the transfer function of the exciter, $G_f(s)$, the excitation system of a generator may be typically classified as follows:

- D.C. generator type (self-excitation method, separate excitation method)
- A.C. generator type with a rectifier circuit
- Solid-state type (by semiconductors for power use)

Regardless of the type of excitation circuit, the transfer function of the exciter can be assumed as a typical form of first-order delay for our purposes of power system behaviour, although the hardware of individual excitation circuits may be different. The time constant T_f is of order, say, 0.1 s, and is usually smaller than the time constant T'_{d0} of the paired generator.

For the transfer function of AVR, $G_{\text{avr}}(s)$, AVR is a closed-loop control system with negative feedback pass in which generator terminal voltage e_G is stabilized to a certain value. Figure 15.3 shows the block diagram of a total system including a generator, an exciter and AVR equipment written as a

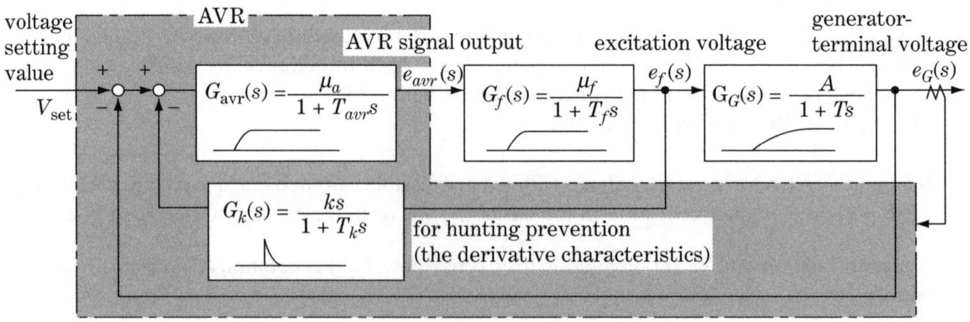

Figure 15.3 Transfer function of a generator with AVR connected to the outer load

combination of transfer functions. In the figure, V_{set} is a value set by AVR for the generator terminal voltage. The generator voltage $e_G(s)$ detected by the potential transformer (PT) is directly fed back to the AVR. The transfer function of the forward element $G_{avr}(s)$ of AVR can also be written in the form of first-order time delay with gain μ_a and time constant T_{avr}. AVR has also a duty to prevent hunting, so the supplementary negative feedback function $G_k(s)$ of a typical derivative characteristic is supplied.

Further, another supplementary function to minimize cross-currents from neighbouring generators will be examined in Chapter 16.

Advanced AVR equipment in recent years is mostly of solid-state type and equipped with outstanding characteristics of high sensitivities as well as quick response times, where the time constant T_{avr} of AVR itself is very small. Accordingly, the actual values of each time constant can be estimated as $T, T_f \gg T_{avr}$.

From Figure 15.3

$$\left.\begin{aligned} &e_G(s) = G_G(s)e_f(s) && ① \\ &e_f(s) = G_{avr}(s)G_f(s)\{V_{set} - e_G(s) - G_k(s)e_f(s)\} && ② \\ &\text{From ①②} \\ &e_f(s) = \frac{G_{avr}(s)G_f(s)}{1 + G_{avr}(s)G_f\{G_k(s) + G_G(s)\}}V_{set} && ③ \end{aligned}\right\} \quad (15.15)$$

Then the total transfer function is

$$e_G(s) = G_G(s)e_f(s) = \frac{G_{avr}(s)G_f(s)G_G(s)}{1 + G_{avr}(s)G_f(s)\{G_k(s) + G_G(s)\}}V_{set} \quad ①$$

where

$$G_G(s) = \frac{A}{1 + Ts} \quad \text{: transfer function of generator in relation to load} \quad ②$$

$$G_f(s) = \frac{\mu_f}{1 + T_f s} \doteq \mu_f \quad \text{: transfer function of 'exciter'} \quad ③$$

$$G_{avr}(s) = \frac{\mu_a}{1 + T_{avr}s} \doteq \mu_a \quad \text{: transfer function of AVR forward element} \quad ④$$

$$G_k(s) = \frac{ks}{1 + T'_A s} \doteq ks \quad \text{: transfer function of feedback circuit to prevent hunting} \quad ⑤$$

(15.16)

Simplification of $G_{avr}(s), G_f(s), G_k(s)$ in Equation 15.16 is justified by the difference in time constants, namely by $T, T_f \gg T_{avr}$.

In addition, we can put $k = 0$ because the gain of $G_k(s)$ is set by a very small value. Accordingly, for the total transfer function,

$$\left.\begin{aligned} e_G(s) &\doteq \frac{\mu_a\mu_f G_G(s)}{1 + \mu_a\mu_f\{ks + G_G(s)\}}V_{set} \doteq \frac{\mu_a\mu_f G_G(s)}{1 + \mu_a\mu_f G_G(s)}V_{set} \\ &= \underbrace{\left(\frac{A}{A + (1/\mu_a\mu_f)}\right)}_{A_{total}} \cdot \frac{1}{1 + \underbrace{\left(\frac{T}{1 + \mu_a\mu_f A}\right)}_{T_{total}} s} V_{set} \equiv A_{total}\frac{1}{1 + T_{total}\, s}V_{set} \end{aligned}\right\} \quad (15.17)$$

In conclusion, dynamic characteristics of a generator with AVR can be written as the **transfer function of a generator with AVR** given by Equation 15.17, which is in the form of a first-order delay mode with gain A_{total} and time constant T_{total}.

If we put $s \to 0$ in the equation (i.e. $t \to \infty$ in the time domain), then we find $e_G \to V_{set}$, which means that generator terminal voltage $e_G(t)$ is controlled to maintain the AVR's set value μ_a. It should be recalled that, in Equation 15.17, only the value of μ_a is controllable and all other symbols have the uncontrollable specific values of the actual machines.

We can introduce the following conclusion from the observation of A_{total} and T_{total} in Equation 15.17: if μ_a is set to a large value, the total system time constant T_{total} can be set to a much smaller value (almost to zero) in comparison with the generator's specific time constant T, while the total system gain A_{total} would not differ greatly from the generator's specific gain A. In other words, due to AVR, the time constant T can be minimized so that quick recovery of generator terminal voltages $e_G(t)$ against ordinary fluctuations as well as against large system disturbances can be realized, maintaining $e_G(t)$ at its set value V_{set} of AVR over time.

Note that generator operation without AVR (in other words, manual operation of excitation) is actually impossible. Without AVR, the generator would be forced to rush into the specifically prohibited operating zone of voltages, or $(p + jq)$ coordinate zone, or serious voltages and reactive power hunting phenomena would be caused among other generators, regardless of their physical distance from our generator, which might lead to mechanical breakdown or damage to the generators. Furthermore, various modes of power system instability might also arise. These problems will be discussed in detail in Chapter 16.

In Figure 15.3, the special case where the gain of AVR is set to zero ($\mu_a = 0$) corresponds to the case when AVR is out of service and the total transfer function consists only of $G_f(s)$ and $G_G(s)$, although such operation is unreal. Incidentally, if we do not neglect the gain k of $G_k(s)$ to prevent hunting, the equation would be written as a second-order function, instead of Equation 15.17. This also will be discussed in Chapter 16.

15.3 Response Characteristics of Total System and Generator Operational Limit

15.3.1 Introduction of s functions for AVR + exciter + generator + load

Our next task is to find conditions in which a generator with AVR can be stably operated.

Note, however, that the load condition has to be taken into account for this purpose, because we know that the operating characteristics of a generator are closely affected by the load condition. Therefore, we examine the behaviour of a generator with AVR in the power system shown in Figure 15.4a, where the load is expressed as a parallel circuit of R_l, jX_l (X_l may be negative value for capacitive loads).

The relation between the generator terminal e_G, i_G and the load impedance $Z(R_l // jX_l)$ is

$$\left.\begin{aligned} i_G &= Z^{-1} e_G & &① \\ Z^{-1} &= \frac{1}{R_l} + \frac{1}{jX_l} = R_l^{-1} - jX_l^{-1} & &② \end{aligned}\right\} \tag{15.18}$$

where the impedance of the transmission line is already contained within the load impedance.

As the system is operated under three-phase-balanced conditions, the following equation can be derived from Equations 10.55 and 15.17:

$$\left.\begin{aligned} (i_d + ji_q)e^{jt} &= Z^{-1}(e_d + je_q)e^{jt} \\ \therefore \quad i_d + ji_q &= Z^{-1}(e_d + je_q) \end{aligned}\right\} \tag{15.19}$$

During a system disturbance, the quantities would suffer some deviations, such as

$$\left.\begin{aligned} &\Delta i_G = (R_l^{-1} - jX_l^{-1})\Delta e_G \qquad \Delta e_G = \frac{R_l + jX_l}{R_l^2 + X_l^2} \cdot \Delta i_G & &① \\ &\therefore \ \Delta i_d + j\Delta i_q = (R_l^{-1} - jX_l^{-1}) \cdot (\Delta e_d + j\Delta e_q) & &② \\ &\Delta i_d = R_l^{-1}\Delta e_d + X_l^{-1}\Delta e_q & &③ \\ &\Delta i_q = -X_l^{-1}\Delta e_d + R_l^{-1}\Delta e_q & &④ \end{aligned}\right\} \tag{15.20}$$

On the other hand, we can derive the following equations from Equation 15.7 ①② and Equation 15.8 ① (putting armature resistance $r = 0$):

$$\left.\begin{aligned} &\Delta e_d = x_q \Delta i_q && \text{①} \\ &\Delta e_q = \frac{1}{1+T'_{d0}s}\Delta e_f - \frac{x_d + x'_d T'_{d0}s}{1+T'_{d0}s}\Delta i_d && \text{②} \\ &\Delta e_G = \sqrt{(\Delta e_d)^2 + (\Delta e_q)^2} && \text{③} \end{aligned}\right\} \quad (15.21)$$

Next, referring to Figure 15.3, the following equation is derived as the s function of the system including AVR (where subsidiary feedback function $G_k(s)$ is neglected for simplicity, which is to prevent hunting ($k = 0$):

$$\left.\begin{aligned} &e_f(s) = \{V_{\text{set}} - e_G(s)\}G_{\text{avr}}(s)G_f(s) && \text{①} \\ \therefore\ &\Delta e_f(s) = -\{G_{\text{avr}}(s)G_f(s)\}\Delta e_G(s) \equiv -\mu\Delta e_G(s) && \text{②} \\ &\mu \equiv \mu_a\mu_f = G_{\text{avr}}(s)\cdot G_f(s) : \text{ total gain of AVR and exciter} && \text{③} \end{aligned}\right\} \quad (15.22)$$

Eliminating Δi_q from Equations 15.20 ④ and 15.21 ①,

$$\left(x_q^{-1} + X_l^{-1}\right)\Delta e_d = R_l^{-1}\Delta e_q \quad \therefore\ \Delta e_d = \left(x_q^{-1} + X_l^{-1}\right)^{-1}\cdot R_l^{-1}\Delta e_q \quad (15.23)$$

Substituting this equation into Equations 15.20 ③④ and 15.21 ③,

$$\left.\begin{aligned} &\Delta i_d = \left\{R_l^{-2}\left(x_q^{-1}+X_l^{-1}\right)^{-1} + X_l^{-1}\right\}\Delta e_q && \text{①} \\ &\Delta i_q = \left\{-X_l^{-1}R_l^{-1}\left(x_q^{-1}+X_l^{-1}\right)^{-1} + R_l^{-1}\right\}\Delta e_q && \text{②} \\ &\Delta e_G = \sqrt{\left\{\left(x_q^{-1}+X_l^{-1}\right)^{-1}\cdot R_l^{-1}\right\}^2 + 1}\cdot\Delta e_q \\ &\quad = \left(x_q^{-1}+X_l^{-1}\right)^{-1}\sqrt{\left(R_l^{-1}\right)^2 + \left(x_q^{-1}+X_l^{-1}\right)^2}\cdot\Delta e_q && \text{③} \end{aligned}\right\} \quad (15.24)$$

Eliminating Δe_f by substituting Equation 15.22 ② into 15.21 ②,

$$(1 + T'_{d0}s)\Delta e_q + G_{\text{avr}}(s)G_f(s)\Delta e_G + (x_d + x'_d T'_{d0}s)\Delta i_d = 0 \quad (15.25)$$

Substituting Δi_d, Δe_G from Equation 15.24 ①③ into Equation 15.25, then all terms including Δe_q vanish and the following equation is obtained:

$$\left.\begin{aligned} &1 + T'_{d0}s + G_{\text{avr}}(s)G_f(s)\left(x_q^{-1}+X_l^{-1}\right)^{-1}\cdot\sqrt{\left(R_l^{-1}\right)^2 + \left(x_q^{-1}+X_l^{-1}\right)^2} \\ &\quad + (x_d + x'_d T'_{d0}s)\left\{R_l^{-2}\left(x_q^{-1}+X_l^{-1}\right)^{-1} + X_l^{-1}\right\} = 0 \\ &\therefore\ G_{avr}(s)G_f(s)\sqrt{\left(R_l^{-1}\right)^2 + \left(x_q^{-1}+X_l^{-1}\right)^2} + \left(X_q^{-1}+X_l^{-1}\right) + x_d\left\{X_l^{-1}\left(x_q^{-1}+X_l^{-1}\right) + R_l^{-2}\right\} \\ &\quad + T'_{d0}s\left[\left(x_q^{-1}+X_l^{-1}\right) + x'_d\left\{X_l^{-1}\left(x_q^{-1}+X_l^{-1}\right) + R_l^{-2}\right\}\right] = 0 \end{aligned}\right\} \quad (15.26)$$

namely

$$\begin{aligned} &\left[G_{\text{avr}}(s)G_f(s)\sqrt{\left(\frac{1}{R_l}\right)^2 + \left(\frac{1}{x_q}+\frac{1}{X_l}\right)^2} + \left(\frac{1}{x_q}+\frac{1}{X_l}\right) + \frac{x_d}{X_l}\left(\frac{1}{x_q}+\frac{1}{X_l}\right) + \frac{x_d}{R_l^2}\right] \\ &\quad + T'_{d0}s\left[\left(\frac{1}{x_q}+\frac{1}{X_l}\right) + \frac{x'_d}{X_l}\left(\frac{1}{x_q}+\frac{1}{X_l}\right) + \frac{x'_d}{R_l^2}\right] = 0 \end{aligned} \quad (15.27)$$

On the other hand, the relation between apparent power of the generator and the load impedance Z (parallel impedance $R_l // jX_l$) is

$$P + jQ = e_G i_G^* = e_G \left(\frac{e_G}{R_l} + \frac{e_G}{jX_l} \right)^* = \frac{e_G^2}{R_l} + j\frac{e_G^2}{X_l} \tag{15.28}$$

Unitizing P, Q by the base value e_G^2,

$$p + jq = \frac{P}{e_G^2} + j\frac{Q}{e_G^2} = \frac{1}{R_l} + j\frac{1}{X_l}$$

Now, Equation 15.27 can be expressed as below by replacing the symbols $1/R_l \rightarrow p$, $1/X_l \rightarrow q$.

For the s function of the total system, generator + exciter + AVR + load, under the load condition $(p + jq)$

$$\begin{aligned} &\left[G_{\text{avr}}(s) G_f(s) \sqrt{p^2 + \left(q + \frac{1}{x_q} \right)^2} + x_d \left\{ p^2 + \left(q + \frac{1}{x_d} \right) \left(q + \frac{1}{x_q} \right) \right\} \right] \\ &\quad + T'_{d0} s \cdot \left[x'_d \left\{ p^2 + \left(q + \frac{1}{x'_d} \right) \left(q + \frac{1}{x_q} \right) \right\} \right] = 0 \end{aligned} \tag{15.29}$$

This is the **Laplace-transformed equation of the operating condition for the total system**, in p–q coordinates which consists of a generator, an exciter and AVR plus load, as shown in Figure 15.4a.

The necessary condition for the above equation to be stable is found with the help of the method for determining system stability within the essential theory of automatic control.

Here we neglect the time lag of AVR and the exciter, putting $T_{\text{avr}} = 0$, $T_f = 0$ (which is of no consequence for stability analysis),

$$G_{\text{avr}}(s) G_f = \mu_a \mu_f \equiv \mu \quad \text{the total gain of AVR and the exciter} \tag{15.30}$$

Accordingly, Equation 15.29 can be modified in the form of an s function of only first order, namely

$$\left.\begin{aligned} &A + BT'_{d0} s = 0 \quad \text{with one root only and} \quad s = -A/(BT'_{d0}) && ① \\ &\text{here} \quad A = \mu \cdot \sqrt{p^2 + \left(q + \frac{1}{x_q} \right)^2} + x_d \left\{ p^2 + \left(q + \frac{1}{x_d} \right) \left(q + \frac{1}{x_q} \right) \right\} && ② \\ &B = T'_{d0} x'_d \left\{ p^2 + \left(q + \frac{1}{x'_d} \right) \left(q + \frac{1}{x_q} \right) \right\} && ③ \end{aligned}\right\} \tag{15.31}$$

15.3.2 Generator operational limit and its p–q coordinate expression

Our purpose is to find the condition from the above equation where the generator can be stably operated.

The method for determining system stability as in the theory of automation indicates that the s function of first order in the form of Equation 15.31 ① can be stable only when the root of s is a negative real number (Nyquist Stability Criterion). Accordingly,

$$\left.\begin{aligned} &\text{the condition of system stability} \\ &\qquad A \cdot B \geqq 0 \\ &\text{the boundary condition of the stable condition} \\ &\qquad A = 0, \quad B = 0 \end{aligned}\right\} \tag{15.32}$$

Consequently, the critical borderline for stable operation can be given by putting $A = 0$, $B = 0$ in Equation 15.31 ②③.

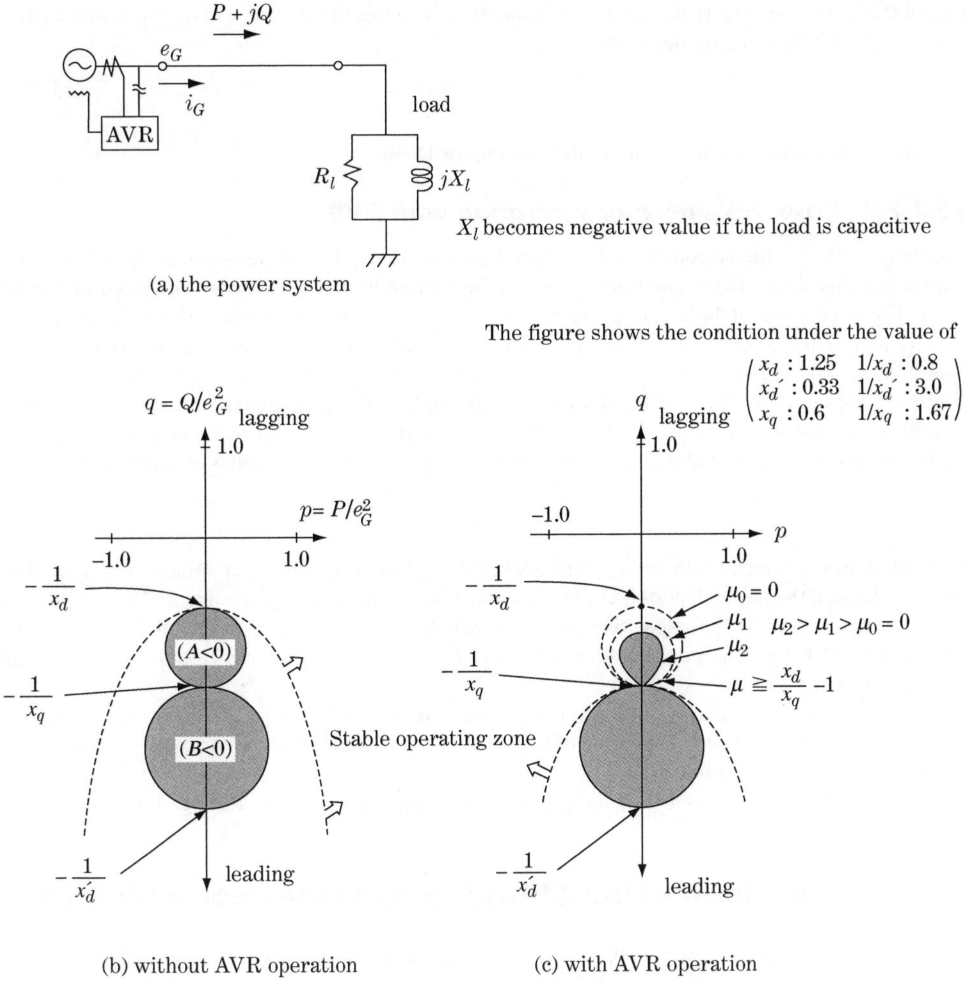

Figure 15.4 Operating limit of the generator under weak excitation (leading power-factor operation)

15.3.2.1 Case 1: generator operation without AVR

If AVR is out of service, so that $\mu_a = 0, \mu = \mu_a \cdot \mu_f = 0$ in Equation 15.22, then the equation of the critical stability limit is

$$
\begin{aligned}
&A = p^2 + \left(q + \frac{1}{x_d}\right)\left(q + \frac{1}{x_q}\right) = 0 \quad ① \\
&B = p^2 + \left(q + \frac{1}{x'_d}\right)\left(q + \frac{1}{x_q}\right) = 0 \quad ② \\
&\text{the stable condition} \\
&A \cdot B \geqq 0 \quad ③
\end{aligned}
\qquad (15.33)
$$

Equation 15.33 ①② can be written as in the $p-q$ coordinates diagram shown in Figure 15.4b.

Equation 15.33 ① is the circle whose diameter is given by the straight line connecting points $(0, -1/x_d)$ and $(0, -1/x_q)$, and the outer area of the circle is $A \geqq 0$. On the other hand, Equation 15.33 ② is the circle whose diameter is given by the straight line connecting $(0, -1/x'_d)$, $(0 - 1/x_q)$. The outer area of the two circles satisfy the stable condition $AB \geqq 0$, while the area of $-1/x'_d \geqq q$ is unrealistic. Accordingly, for the realistic stable area,

$$q > -1/x_d \tag{15.34a}$$

That is, the upper area from small circle in Figure 15.4b.

15.3.2.2 Case 2: Generator operation with AVR

Equation 15.31 ② with the condition $A = 0$ and Equation 15.31 ③ with the condition $B = 0$ give the critical stability limit. These conditions can also be written in p–q coordinates as shown in Figure 15.4c. The larger circle at the bottom is given by $B = 0$, which is the same circle as that in Figure 15.4b. The smaller distorted circle at the top is given by the equation $A = 0$ and can be written in p–q coordinates with the parameter $\mu(>0)$. The case with $\mu = 0$ coincides with case 1, that is generator operation without AVR. If the parameter μ (the total gain) is selected to be larger, the distorted circle shrinks more, and finally converges to the special point $(0, -1/x_q)$. The value of μ at this special converged point can be calculated by the conditions $p = 0, A = 0, B = 0$, and is given by the equation

$$\mu = \frac{x_d}{x_q} - 1 \tag{15.34b}$$

Now, comparing Figures 15.4b and c, if the AVR gain μ_a is adjusted to larger values ($\mu = \mu_a \mu_f$ also becomes larger), the upper distorted circle is shrinks towards the converged point, and as a result the stable operating zone in the capacitive area (q is negative) is enlarged to the point $(0, -1/x_q)$. In other words, AVR can enlarge the leading power-factor zone of the generator from the critical point $(0, -1/x_d)$ to $(0, -1/x_q)$.

From an actual engineering viewpoint, the operating zone of hydro-generators can be greatly improved (perhaps by 40–50%) by virtue of $x_d > x_q$, but that of thermal generators only moderately so because $x_d \fallingdotseq x_q$ (refer to Table 10.4).

The problem of the generator's leading power-factor operation is examined again in Chapter 16.

15.4 Transmission Line Charging by Generator with AVR

We examine the special case of $p = 0$, which means a generator is operating with pure inductive (L) or capacitive (C) load, namely by power-factor zero. A typical example of this situation is the so-called 'no-load transmission line charging by a generator', that is one terminal of the line is opened and the other terminal is charged by the generator, so that the stray capacitance of the line becomes the pure capacitive load of the generator.

Equation 15.29 becomes very simple by putting $p = 0$:

$$\left\{\mu + x_d\left(q + \frac{1}{x_d}\right)\right\} + T'_{d0}x'_d\left(q + \frac{1}{x'_d}\right)s = 0 \tag{15.35}$$

For the system to be stable, the root of s should be a negative real number, namely, for the stability condition of line charging,

$$\left.\begin{aligned} s = -\frac{\mu + x_d\left(q + \dfrac{1}{x_d}\right)}{T'_{d0}x'_d\left(q + \dfrac{1}{x'_d}\right)} \leqq 0 \quad & ① \\ \therefore \left\{\mu + x_d\left(q + \frac{1}{x_d}\right)\right\} \cdot \left\{q + \frac{1}{x'_d}\right\} \geqq 0 \quad & ② \end{aligned}\right\} \tag{15.36}$$

15.4.1 Line charging by generator without AVR

As this is the case of $\mu = 0$, the stable condition is

$$\left.\begin{aligned} q &\geqq -\frac{1}{x_d} \quad \text{(realistic stable zone)} \quad ① \\ q &\leqq -\frac{1}{x'_d} \quad \text{(unrealistic zone)} \quad ② \end{aligned}\right\} \tag{15.37}$$

Equation ① is the actual stable condition, while ② has physically no meaning. Therefore,

The case of capacitive load $(q < 0)$: critical point of stable operation is $q = -1/x_d$[pu]
The case of inductive load $(q > 0)$: stable for all the inductive zone.

15.4.2 Line charging by generator with AVR

This is the case of $\mu \neq 0$. Equation 15.36 ② is a second-order inequality of q, so it can be solved. The enlarged stable zone of line charging is

$$q \geqq \sqrt{\left\{\frac{1}{2}\left(\frac{1}{x_d} - \frac{1}{x'_d}\right)\right\}^2 - \mu\frac{1}{x_d}} - \frac{1}{2}\left(\frac{1}{x_d} + \frac{1}{x'_d}\right) \tag{15.38}$$

The absolute value of the left-hand side of Equation 15.38 is larger than $1/x_d$ of Equation 15.37, and the special case of $\mu = 0$ in the former equation gives the latter. Thus, due to the effect of the term μ/x_d by AVR, the stable limit of leading power-factor operation in *p–q* coordinates has been enlarged in the down direction in the Figure 15.4b.

15.4.2.1 Trial calculation

For the case of generator reactance $x_d = 1.8$ pu, $x'_d = 0.4$ pu, then, without AVR, from Equation 15.37 ① $q \geqq -0.556$; and with AVR, from Equation 15.37 ① $q \geqq -0.904$ (gain $\mu = 1.0$).

Obviously the stable zone of the generator's capacitive load operation is remarkably improved by AVR.

The theme of transmission line charging is again discussed in Chapter 20 as a problem of overvoltage phenomena.

15.5 Supplement 1: Derivation of Equation 15.9 from Equations 15.7 and 15.8

The calculation takes some time but can be solved.

For the first step, substituting $i_d(s), i_q(s)$ in Equation 15.8 into 15.7 ①②, we can derive simultaneous equations for $e_d(s), e_q(s)$:

$$\left.\begin{aligned} &(R^2 + X^2 + Xx_q + Rr)e_d(s) + (Xr - Rx_q)e_q(s) = 0 \\ &\{(Rx_d - Xr) + (Rx'_d - Xr)T'_d s\}e_d(s) + \{(R^2 + X^2 + Rr + Xx_d) \\ &\quad + (R^2 + X^2 + Xx'_d + Rr)T'_{d0}s\}e_q(s) = e_f(s) \cdot (R^2 + X^2) \end{aligned}\right\} \tag{1}$$

The equations can be written as follows:

$$\begin{aligned} &m_1 e_d(s) + m_2 e_q(s) = 0 \\ &\{m_3 + m_4 T'_{d0}(s)s\}e_d(s) + \{m_5 + m_6 T'_{d0}(s)\, s\}e_q(s) = m_7 e_f(s) \end{aligned}$$

The solution is

$$e_d(s) = \frac{m_2 m_7}{(m_2 m_3 - m_1 m_5) + (m_2 m_4 - m_1 m_6)T'_{d0}(s)s} \cdot e_f(s)$$

$$e_q(s) = -\frac{m_1}{m_2} \cdot e_d(s)$$

where

$$\begin{aligned}
m_1 &= R^2 + X^2 + Xx_q + Rr = X(X + x_q) + R(R + r) \\
m_2 &= Xr - Rx_q = x(R + r) - R(X + x_q) \\
m_3 &= Rx_d - Xr = -X(R + r) + R(X + x_d) \\
m_4 &= Rx'_d - Xr = -X(R + r) + R(X + x'_d) \\
m_5 &= R^2 + X^2 + Rr + Xx_d = R(R + r) + X(X + x_d) \\
m_6 &= R^2 + X^2 + Rr + Xx'_d = R(R + r) + X(X + x'_d) \\
m_7 &= R^2 + X^2
\end{aligned}$$

m_1 to m_6 have been modified into the form shown on their right-hand sides, leading to the following results:

$$\begin{aligned}
(m_2 m_3 - m_1 m_5) &= -(X^2 + R^2)\left\{(X + x_d)(X + x_q) + (R + r)^2\right\} \\
(m_2 m_4 - m_1 m_6) &= -(X^2 + R^2)\left\{(X + x'_d)(X + x_q) + (R + r)^2\right\} \\
m_2 m_7 &= -(X^2 + R^2)(x_q R - Xr)
\end{aligned}$$

Finally we obtain Equation 15.9.

15.6 Supplement 2: Derivation of Equation 15.10 from Equations 15.8 and 15.9

Equation 15.10 is derived by substituting $e_d(s)$, $e_q(s)$ from Equation 15.9 into Equation 15.8 ①. However, the following modification can be adopted in the process:

$$x_q R - Xr = -X(R + r) + R(X + x_q)$$

Accordingly,

$$\{\text{numerator of Equation 15.9 ①}\}^2 + \{\text{numerator of Equation 15.9 ②}\}^2$$
$$= (X^2 + R^2)\left\{(X + x_q)^2 + (R + r)^2\right\}$$

which leads easily to Equation 15.10.

Coffee break 8: The symbolic method by complex numbers and Arthur Kennelly, the prominent pioneer

Heaviside's **symbolic method** or **transform** and the **Laplace transform** for analysing differential equations made prominent contributions to the advancement of electrical engineering in the twentieth century. However, referring to the simple circuit in Table 7.1, for example, can we solve the transient phenomena of the circuit using only the Laplace transform without using the concept of complex numbers? Or can we solve the steady-state phenomena without using $j\omega$? The answer is actually no. Electrical engineers today are very familiar with the concept of voltages and currents in complex numbers or impedance in the complex number $R + j\omega L$, so they may understand as if 'the physical substance of electricity is the quantities of complex number!'

Arthur Edwin Kennelly (1861–1939)

In a similar sense, phenomena of the simple three-phase circuit in Table 7.2 cannot actually be solved without using the methods of **electricity description by complex numbers** and **symmetrical components**. As a matter of fact, we believe that every page of this book, in regard to power systems engineering, cannot be properly described without such outstanding symbolic methods.

We electricians today know that the substance of electricity is an electromagnetic field or wave which propagates through a vacuum or in air along wires with the same speed of light. Electricity is of course a form of energy which obeys the law of energy conservation. However, electricians working in the fields of electronic devices, communications/broadcasting, power systems, for example, usually treat electricity as 'current flowing through a circuit' just like 'water flowing through a pipe'. It is fortunate that electromagnetic waves with such profound characteristics of propagation in space can be handled so simply in most of our practical engineering works. We owe this partly to the substantial theories of electricity that have been well established, but also partly to the excellent methodology of handling electricity that has been established.

Considering all these things, we engineers are required to be familiar with such a methodology as much as possible, while not forgetting the actual substance of electricity.

We emphasize the important roles of the **mathematical method of description or symbolic methods of electricity**. The second half of this coffee break below contains short biographies of those prominent pioneers who developed such important methods of mathematical treatment.

The first is **Arthur Edwin Kennelly** (1861–1939), who developed or, better, 'invented' **the symbolic method of electricity by complex numbers**. Another independent inventor, Steinmetz, is described later. Kennelly was born in India, the son of an Irish naval officer. After his mother died he was sent to England when he was 3 years old to be educated. After working for a few years as a telegraph operator for the Eastern Telegraph Company, he left for the United States in 1887. In December of the same year 1887, **Thomas Alva Edison** (1847–1931) opened his West Orange Laboratory and Kennelly joined it as chief assistant to Edison.

Edison had half-a-dozen assistants, chosen because of their expertise in fields in which Edison felt weaker. Kennelly became Edison's closest expert, especially in mathematics and electromagnetic studies.

The following year, Edison was anxious to characterize the a.c. generating system as too dangerous, since it was being vigorously promoted by **George Westinghouse** (1846–1914) and his competing group. Edison had sufficient reason to do so, because he was the great innovator, as much as the inventor, for his enterprise in d.c. electricity containing the functions of power utility and manufacture of dynamos, electric lamps, telephones, etc. Kennelly worked for Edison for six years. He had to manage his work to justify d.c. as a member of Edison's staff, but still presented a prominent paper for treating a.c. electricity in 1893.

In 1893, he submitted a paper on 'Impedance' to the AIEE (American Institute of Electrical Engineers) in which he discussed the first **use of complex numbers as applied to Ohm's law in a.c. theory**. In this paper, he wrote that inductance and capacitance can be written as the 'resistances of $pl\sqrt{-1}$, $-1/kp\sqrt{-1}$ respectively, where $p = \omega = 2\pi f$ and k $= C$. Further, in the next year he proposed new signal $\angle\theta$ and its converse.

His AIEE paper was praised quite highly in the periodical *Electrical World*: 'admirable faculty of taking an involved and little understood subject, and by a remarkably lucid treatment placing it within the grasp of the merest tyro in electricity'. The publication of Kennelly's paper immediately allowed complex-number techniques to be applied to a.c. theory. We think that the above comment by the editor of *Electrical World* is not sufficient: we cannot imagine the substantial physical concept of reactive power Q without using complex numbers, for example.

Kennelly should also be remembered as the person who initiated the application of **complex hyperbolic functions** in the solution of transmission of radio waves, which is essentially the mathematical theory of telephone wires and later of waveguides. 'The table and chart atlas of complex hyperbolic and circular functions **(Smithsonian tables)**' is one of his monuments.

Kennelly left Edison in 1894, joining the consulting firm Houston and Kennelly in Philadelphia. In 1901, he came back again as a consulting engineer with the Edison General Electric Company in New York, and then became a professor at Harvard University from 1902 to 1930 and at the Massachusetts Institute of Technology (MIT) from 1913 to 1924.

On 12 December 1901, Marconi received the radio signal 'S' sent by Morse code from England to Newfoundland. Kennelly thought that the signal received was far better than predicted by existing radio wave theory. The following year he deduced that the reason why Marconi's radio waves were able to cross the Atlantic Ocean was because they were being reflected back to Earth from an ionized layer in the upper atmosphere. He predicted the existence of a conducting stratum at a height of 50 miles (80 km) and with a conductivity several times as large as that of sea water. This is of course the **'Kennelly–Heaviside layer'**, which is now called the 'E region of the ionosphere'.

Kennelly was the sole author of 10 books and co-author of 18 more, and wrote about 350 papers. He died in Boston in 1939 after becoming a leader of various societies of electricians and a leading promoter of the metric system of weights and measures.

16

Operating Characteristics and the Capability Limits of Generators

Besides the item of ratings, each generator has its own specified operational limits and prohibited operating conditions. A generator's capability curves specifying the operational limits and some common weak points in generators are discussed in this chapter. The subject is quite important for power station design and operational engineering as well as for network design.

16.1 General Equations of Generators in Terms of *p–q* Coordinates

We examine the operating condition of a generator which is connected to an infinite bus through a transmission line as shown in Figure 16.1a. Figure 16.1b shows the vector diagram of this system's voltages and currents. First of all we start from the definition of *p–q* coordinates here, although it has already appeared in Chapter 15:

$$\left.\begin{aligned} &P+jQ=e_G i_G^* = e_G\left(\frac{e_G}{Z_l}\right)^* = \frac{e_G^2}{Z_l^*} & ① \\ &p+jq=\frac{P}{e_G^2}+j\frac{Q}{e_G^2}, \quad p=\frac{P}{e_G^2}, \quad q=\frac{Q}{e_G^2} & ② \end{aligned}\right\} \tag{16.1}$$

where Z_l is the **characteristic impedance** looking into the network from the generator terminal.

$P-Q$ coordinates and $p-q$ coordinates are in one-to-one correspondence to each other under the parametric coefficient of $1/e_G^2$, while $e_G = 1.0 \pm 0.05$ and $1/e_G^2 \fallingdotseq 1.0 \pm 0.10$ under nominal generator operation.

The generator terminal voltage and current equations are

$$\left.\begin{aligned} &P+jQ=e_G i_G \mathrm{e}^{j\varphi} & ① \\ &p+jq=\frac{P}{e_G^2}+j\frac{Q}{e_G^2}=\frac{i_G}{e_G}\mathrm{e}^{j\varphi} & ② \\ &\therefore \quad p=\frac{i_G}{e_G}\cos\varphi, \quad q=\frac{i_G}{e_G}\sin\varphi & ③ \\ &p^2+q^2=\frac{i_G^2}{e_G^2} & ④ \end{aligned}\right\} \tag{16.2}$$

with φ: the power-factor angle

Handbook of Power System Engineering Yoshihide Hase
© 2007 John Wiley & Sons, Ltd

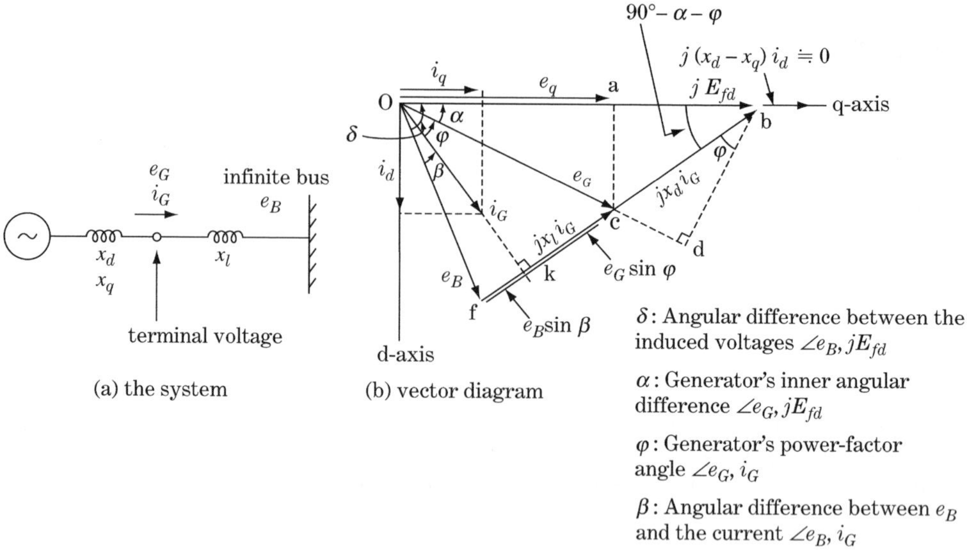

Figure 16.1 One-generator to infinite-bus system

Referring to vector diagram, where the salient-pole effect $(x_d - x_q)i_d$ is neglected,

$$\left.\begin{aligned} E_{fd} &= \overline{\mathrm{Oa}} + \overline{\mathrm{ab}} = e_q + x_d i_G \cos(90° - \alpha - \varphi) \\ &= e_q + x_d i_G \sin(\alpha + \varphi) = e_q + x_d i_d & ① \\ e_q &= e_G \cos\alpha & ② \\ i_d &= i_G \sin(\alpha + \varphi) & ③ \end{aligned}\right\} \tag{16.3}$$

Substituting Equations 16.2 ③ and 16.3 ② ③ into Equation 16.3 ①,

$$\begin{aligned} E_{fd} &= e_q + x_d i_G(\sin\alpha \cos\varphi + \cos\alpha \sin\varphi) \\ &= e_G x_d \left\{ p \sin\alpha + \left(q + \frac{1}{x_d}\right)\cos\alpha \right\} \end{aligned} \tag{16.4}$$

Then if α is expressed by p and q, E_{fd} can be written by p,q. Here, the following equations are derived from the vector diagram:

$$\left.\begin{aligned} \tan\alpha &= \frac{\overline{\mathrm{bd}}}{\overline{\mathrm{Oc}} + \overline{\mathrm{cd}}} = \frac{x_d i_G \cos\varphi}{e_G + x_d i_G \sin\varphi} = \frac{x_d(e_G i_G \cos\varphi)}{e_G^2 + x_d(e_G i_G \sin\varphi)} \\ &= \frac{x_d P}{e_G^2 + x_d Q} = \frac{x_d e_G^2 p}{e_G^2 + x_d e_G^2 q} = \frac{x_d p}{1 + x_d q} \\ \therefore \quad \cos\alpha &= \frac{1 + x_d q}{\sqrt{(x_d p)^2 + (1 + x_d q)^2}} = \frac{q + \dfrac{1}{x_d}}{\sqrt{p^2 + \left(q + \dfrac{1}{x_d}\right)^2}} \\ \sin\alpha &= \frac{x_d p}{\sqrt{(x_d p)^2 + (1 + x_d q)^2}} = \frac{p}{\sqrt{p^2 + \left(q + \dfrac{1}{x_d}\right)^2}} \end{aligned}\right\} \tag{16.5}$$

Accordingly, Equation 16.4 is

$$E_{fd} = e_G x_d \left\{ p \cdot \frac{p}{\sqrt{p^2 + \left(q + \frac{1}{x_d}\right)^2}} + \left(q + \frac{1}{x_d}\right) \cdot \frac{q + \frac{1}{x_d}}{\sqrt{p^2 + \left(q + \frac{1}{x_d}\right)^2}} \right\}$$
$$= e_G x_d \sqrt{p^2 + \left(q + \frac{1}{x_d}\right)^2} \qquad (16.6)$$

E_{fd} has simply been written only in p, q and e_G.
For the network equations, from the vector diagram

$$\left.\begin{array}{ll} e_G \sin\varphi + e_B \sin\beta = x_l i_G & ① \\ e_G \cos\varphi = e_B \cos\beta & ② \\ \therefore \quad e_B^2 = (e_G \cos\varphi)^2 + (x_l i_G - e_G \sin\varphi)^2 & \\ \quad = e_G^2 + x_l^2 i_G^2 - 2e_G x_l i_G \sin\varphi & ③ \end{array}\right\} \qquad (16.7)$$

Eliminating i_G, φ by substituting Equation 16.2 ③ ④,

$$\left.\begin{array}{l} e_B^2 = e_G^2 + e_G^2 x_l^2 (p^2 + q^2) - 2e_G x_l q e_G = e_G^2 x_l^2 \left\{ p^2 + \left(q - \frac{1}{x_l}\right)^2 \right\} \\ \text{where} \quad e_B = e_G x_l \sqrt{p^2 + \left(q - \frac{1}{x_l}\right)^2} \end{array}\right\} \qquad (16.8)$$

Accordingly, the general equation of generators in terms of p–q coordinates is

$$\left.\begin{array}{ll} e_G = \frac{e_B}{x_l} \cdot \frac{1}{\sqrt{p^2 + \left(q - \frac{1}{x_l}\right)^2}} & ① \\ i_G = e_G \sqrt{p^2 + q^2} = \frac{e_B}{x_l} \cdot \frac{\sqrt{p^2 + q^2}}{\sqrt{p^2 + \left(q - \frac{1}{x_l}\right)^2}} & ② \\ P + jQ = e_G^2 (p + jq) = \left(\frac{e_B}{x_l}\right)^2 \cdot \frac{p + jq}{p^2 + \left(q - \frac{1}{x_l}\right)^2} & ③ \\ E_{fd} = e_G x_d \sqrt{p^2 + \left(q + \frac{1}{x_d}\right)^2} = e_B \frac{x_d}{x_l} \cdot \frac{\sqrt{p^2 + \left(q + \frac{1}{x_d}\right)^2}}{\sqrt{p^2 + \left(q - \frac{1}{x_l}\right)^2}} & ④ \end{array}\right\} \qquad (16.9)$$

The generator voltage e_G has been derived as a function of p, q and e_B. The equations correspond to the vector diagram of Figure 16.1b. In the above equations, x_d is the specific constant of the generator, while x_l is the reactance of the network which may be changed by the network conditions. The voltage of the infinite bus can usually be taken as $e_B = 1.0$. Accordingly, the characteristics of a generator can be written as an implicit function $F(e_G,\ i_G,\ E_{fd},\ p,\ q) = 0$. All these variables are related to each other as a set of variables.

Equation 16.9 is very important because it is widely utilized as the basis for actual engineering of power generation and power system operation.

16.2 Rating Items and the Capability Curve of the Generator

16.2.1 Rating items and capability curve

Each generator has its own specific capability limits, most of which are given on its name-plate as ratings, or may be specified by the appropriate standards.

In addition, each generator has its own 'specified capability curve', which indicates allowable operating zone or critical operating limits of the generator, usually in the p–q coordinate plane. Generators always have to be operated within these limits.

For the principal terms of the ratings, the specification on the name-plate includes physical weight and dimensions, cooling practices, winding connections, insulation level, maximum temperature rise, excitation method, and so on. Further, electrical terms of ratings are described, which indicate the guaranteed capabilities or allowable operating zone of the generator. These are:

- **Rated capacity [MVA]:** Every part of the generator causes a temperature rise as the result of the balance between heat generation and radiation. The rated capacity is determined by the allowable maximum temperature rise of each part, including the metal structure and insulation of the stator and the rotor.
- **Rated effective power [MW]:** The maximum effective power guaranteed by the rated power factor.
- **Rated power factor:** Specified power factor by which the rated capacity [MVA] has to be guaranteed. Generators of lower power factor tend to have increased weight. Typically $\cos\varphi =$ 0.8 or 0.85 or 0.9.
- **Rated voltage [kV]:** The rated voltage is selected from specified series numbers in the appropriate standards. Typical figures are 3.3, 6.6, 13.2,. 20.0, 24.0, 33.0, 36.0 kV, etc.
- **Rated frequency f [Hz]:** 50 Hz or 60 Hz in most areas of the world.
- **Rated rotation [rpm]:** For the $2n$-pole machine (pole number $2n$): $N = 60f/n$. For the two-pole machine ($2n=2$): $N=3000$ rpm (50 Hz) or 3600 rpm (60 Hz).
- **Short-circuit ratio:** That is, {necessary excitation current to induce rated voltage under rated rotating speed}/{necessary excitation current to induce rated current under short-circuit condition}.
- **Various reactances, time constants:** Items as listed in Table 10.1.

For the allowable operating range in regard to voltage and frequency (or rpm), this is generally specified by the electrical standards for generators. Figure 16.2 shows a typical example of the standards.

For the capability curve of a generator (allowable operating range in terms of P–Q or p–q coordinate plane), Figure 16.3 shows such a curve which indicates the generator's allowable operating zone as the closed zone in the p–q coordinate plane.

Any generator should be operated strictly within the specified capability curve to prevent serious damage to the machine and/or to maintain stable power system operation. In order to ensure such operation of generators, various countermeasures, mainly by AVR and other supervising control equipment as well as by associated protective relays, are invariably adopted at any power plant.

The essential points of the capability curve are the four items below, referring to Figure 16.3:

a) The upper limit curve of apparent power $P+jQ$ or $p+jq$ (curve ②–③).

b) The upper limit curve of excitation voltage E_f (equivalent to i_f) (curve ①–②).

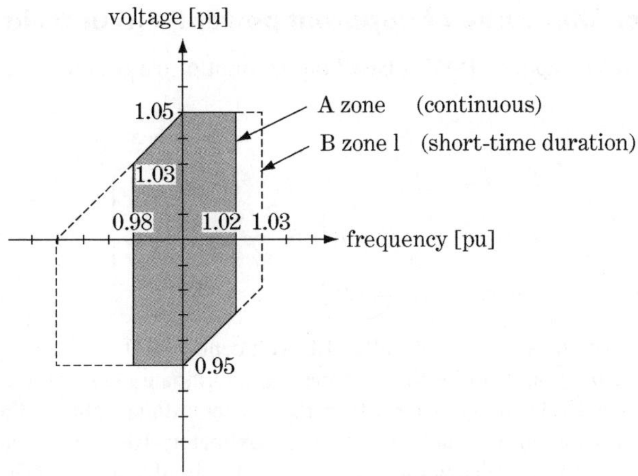

Figure 16.2 Synchronous generators: allowable operating ranges of the terminal voltage and frequency (over 10 MVA)

c) The limit curve to maintain stability (curve ⑤–⑦).

d) The limit curve to prevent extraordinary local heating mainly caused around the structure of the stator coil end (curve ③–④).

We examine these curves one by one.

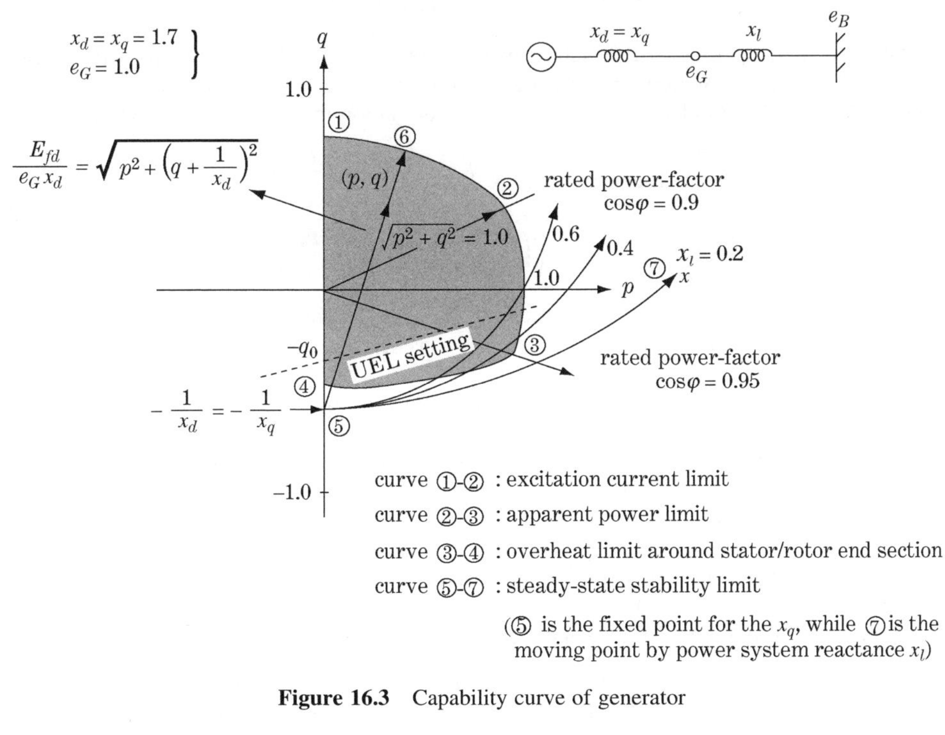

Figure 16.3 Capability curve of generator

16.2.1.1 Upper limit curve of apparent power P+jQ or p+jq (curve ②–③)

A generator has its rated capacity [MVA] based on the limit of temperature rise caused by current and flux.

From Equation 16.9 ③

$$\left.\begin{aligned} & S = P + jQ = e_G^2(p + jq) \leqq S_{\max} \quad ① \\ \therefore \quad & p + jq = \frac{1}{e_G^2}(P + jQ) \leqq \frac{1.0}{e_G^2} \quad ② \\ & \text{where} \quad S_{\max} = 1.0 \end{aligned}\right\} \tag{16.10}$$

This equation is written as a circle of radius 1.0 with centre (0, 0) in the P–Q coordinate plane, or a circle of radius $1.0/e_G^2$ and centre (0, 0) in the p–q coordinate plane. However, the curve may be written simply as a circle of diameter 1.0 in the p–q coordinate plane under the assumption $e_G = 1.0$. In such cases, the limit should be read as approximately 10% retouched curve because the voltage is actually $e_G = 1.0 \pm 0.05$ and $e_G^2 = 1.0 \pm 0.1$. The circular arc ②–③ in Figure 16.3 shows this limitation.

16.2.1.2 Upper limit curve of excitation voltage E_{fd} (equivalent to i_f) (curve ①–②)

Rewriting the left-hand side and middle of Equation 16.9 ④,

$$\left.\begin{aligned} & E_{fd} = e_G x_d \sqrt{p^2 + \left(q + \frac{1}{x_d}\right)^2} \\ \therefore \quad & p^2 + \left(q + \frac{1}{x_d}\right)^2 = \left(\frac{E_{fd}}{e_G x_d}\right)^2 \end{aligned}\right\} \tag{16.11}$$

Arbitrary point (p, q) which satisfies the above equation is on a circle whose centre is $(0, -1/x_d)$ and radius $E_{fd}/(e_G x_d)$. Accordingly, the upper limit can be drawn as a circle for the largest excitation $E_{fd\max}$. The circular arc ①–② shows this limitation in the excitation circuit:

$$\begin{aligned} & \text{the circle equation} \quad p^2 + \left(q + \frac{1}{x_d}\right)^2 \leqq \left(\frac{E_{fd\max}}{e_G x_d}\right)^2 \\ & \text{centre} \quad \left(0, -\frac{1}{x_d}\right), \quad \text{radius} \quad \frac{E_{fd\max}}{e_G x_d} \end{aligned} \tag{16.12}$$

The design value of excitation circuit capacity $E_{fd\max}$ has to guarantee the generator's operation by the rated capacity MVA under the rated voltage and power factor, which corresponds to point ② in Figure 16.3.

The distance from point $(0, -1/x_d)$ to arbitrary point (p, q) is obviously proportional to the magnitude of excitation E_{fd}.

Accordingly, operation in the lagging zone $(q > 0)$ requires strong excitation, while operation in the leading zone $(q < 0)$ requires weak excitation. The special condition of $E_{fd} = 0$ corresponds to the special point ⑤ $(0, -1/x_d)$.

For the generator whose x_d is smaller (short-circuit ratio $1/x_d$ is larger), point ⑤ goes down to the lower point in the p–q coordinate plane, so that larger excitation capacity is obviously required.

16.2.1.3 The stability limit curve (curve ⑤–⑦)

The steady-state stability limit under the condition of $\delta = 90°$ is given by Equation 12.18 and by the circle in Figure 12.4a. The circle can be transferred to Figure 16.3 as the curve of the steady-state stability limit ⑤–⑦. This curve also shares the same special point ⑤. It is interesting that this point is not only the centre of the excitation limit circle, but also one end of the diameter of the circle of steady-state stability.

16.2.1.4 The limit curve to prevent extraordinary local heating mainly caused around the structure of the stator coil end (curve ③–④)

A generator has the special problem of extraordinary local heating under operation in the leading power-factor ($q < 0$) zone (in other words, under weak field operation), which tends to be caused mainly around the structure of the stator coil end. Curve ③–④ in the figure indicates the critical curve for this problem, for the reason discussed in Section 16.3. This curve should be indicated by the generator manufacturer.

16.2.2 Generator's locus in the *p–q* coordinate plane under various operating conditions

The generator's specific capability was discussed with the parameter of generator terminal voltage e_G in the previous section. Now let us examine the generator's loci with the parameter e_B (usually $e_B = 1.0$) instead of e_G, under various operating conditions.

16.2.2.1 The locus under the condition of fixed excitation E_{fd}

Rewriting the left- and right-hand sides of Equation 16.9 ④,

$$E_{fd} = e_B \frac{x_d}{x_l} \cdot \frac{\sqrt{p^2 + \left(q + \dfrac{1}{x_d}\right)^2}}{\sqrt{p^2 + \left(q - \dfrac{1}{x_l}\right)^2}} \tag{16.13}$$

This is a second-order equation in p and q, and can be modified into the equation of a circle as shown below. (See the supplement for the calculation process.)

For the circle

$$\left.\begin{aligned}
&\text{the circle equation } p^2 + \left(q + \frac{b}{1-a^2}\right)^2 = \left\{\sqrt{\left(\frac{b}{1-a^2}\right)^2 + c}\right\}^2 \\
&\text{where} \quad a = \left(\frac{E_{fd}}{e_B x_d}\right) x_l \\
&\qquad b = \frac{1}{x_d} + \frac{a^2}{x_l} = \left(\frac{E_{fd}}{e_B x_d}\right)^2 x_l + \frac{1}{x_d} \\
&\qquad c = \frac{\dfrac{a^2}{x_l^2} - \dfrac{1}{x_d^2}}{1-a^2} = \frac{\left(\dfrac{E_{fd}}{e_B x_d}\right)^2 - \dfrac{1}{x_d^2}}{1-a^2} \\
&\text{centre}\left(0, -\frac{b}{1-a^2}\right), \quad \text{radius} \quad \sqrt{\left(\frac{b}{1-a^2}\right)^2 + c}
\end{aligned}\right\} \tag{16.14}$$

The operational locus (p, q) of the generator is drawn as a circle under the conditions of $e_B = 1.0$, fixed E_{fd} and constant x_d, x_l. This circle is shown as loci a_1, a_2 in Figure 16.4. Assuming generator

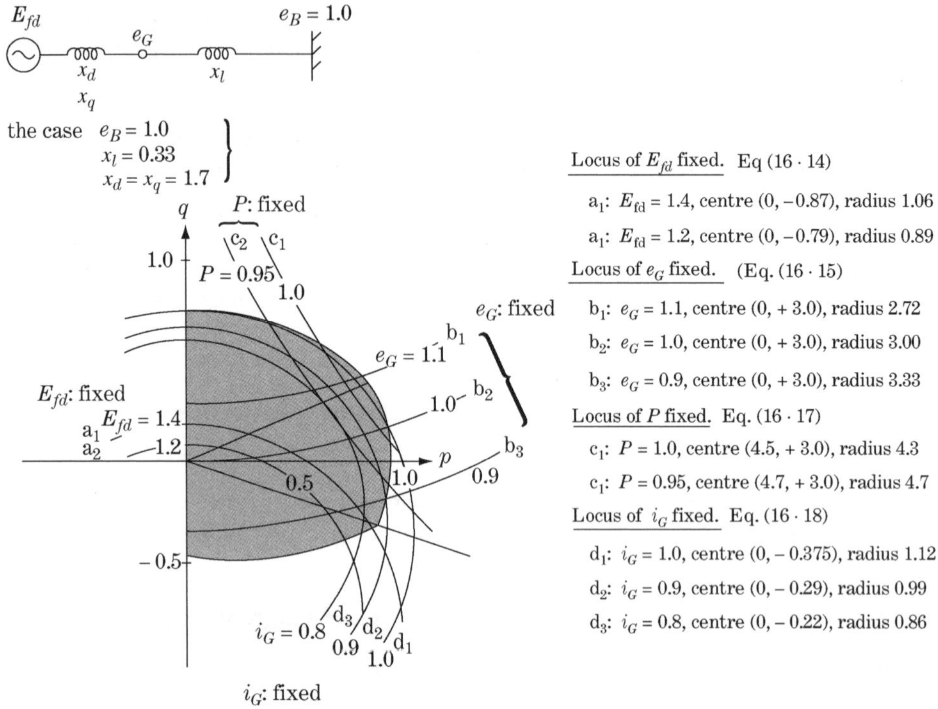

Figure 16.4 Operational locus in the p–q coordinate plane

operation under fixed E_{fd} (i.e. AVR is out of service) and if p is increased by the prime-mover, the generator locus (p, q) moves down immediately to the direction of leading power-factor operation. That is, to the prohibited operating zone.

16.2.2.2 The locus under the condition of fixed terminal voltage e_G

The terminal voltage e_G is given by Equation 16.9 ①, which can be modified as

$$\left.\begin{array}{lll} \text{the circle equation} & p^2 + \left(q - \dfrac{1}{x_l}\right)^2 = \left(\dfrac{1}{e_G} \cdot \dfrac{e_B}{x_l}\right)^2 & \\ \text{centre} & \left(0, \dfrac{1}{x_l}\right), \quad \text{radius } \dfrac{1}{e_G} \cdot \dfrac{e_B}{x_l} & \end{array}\right\} \tag{16.15}$$

This is also drawn as a circle, which is shown as loci b_1, b_2, b_3 in Figure 16.4.

16.2.2.3 The locus under the condition of fixed effective power P

From the real part of Equation 16.9 ③

$$P = e_G^2 p = \left(\frac{e_B}{x_l}\right)^2 \cdot \frac{p}{p^2 + \left(q - \dfrac{1}{x_l}\right)^2} \tag{16.16}$$

Then

$$\left.\begin{array}{lll} \text{the circle equation} & \left(p-\dfrac{e_B^2}{2Px_l^2}\right)^2+\left(q-\dfrac{1}{x_l}\right)^2=\left(\dfrac{e_B^2}{2Px_l^2}\right)^2 \\ \text{with} \quad \text{centre} \left(\dfrac{e_B^2}{2Px_l^2},\dfrac{1}{x_l}\right), & \text{radius } \dfrac{e_B^2}{2Px_l^2} \end{array}\right\} \quad (16.17)$$

The circle is shown as loci c_1, c_2 in Figure 16.4.

16.2.2.4 The locus under the condition of fixed terminal current i_G

The equation of fixed terminal current i_G is derived from Equation 16.9 ②, which can be modified as

$$\left.\begin{array}{l} \text{the circle equation} \quad p^2+\left(q-\dfrac{1}{x_l(1-A^2)}\right)^2=\left\{\dfrac{A}{x_l(1-A^2)}\right\}^2 \\ \text{where} \quad A=\dfrac{e_B}{x_l i_G} \\ \text{centre} \left(0,\dfrac{1}{x_l(1-A^2)}\right), \quad \text{radius } \dfrac{A}{x_l(1-A^2)} \end{array}\right\} \quad (16.18)$$

The circle is shown as loci d_1, d_2, d_3 in Figure 16.4.

The curve in which i_G is replaced by $i_{G\max}$ is obviously the current limit curve.

Lastly, the vector diagram for the four typical operating points in the p–q coordinate plane are shown in Figure 16.5. The magnitudes of e_B and i are drawn with a similar size for each diagram.

16.3 Leading Power-factor (Under-excitation Domain) Operation, and UEL Function by AVR

16.3.1 Generator as reactive power generator

The load of power system $\sum\{P_{\text{load}}+j(Q_l-Q_c)\}$ (where P, Q consumed on the transmission lines are included) is incidental by nature and always changing with time. On the other hand, **simultaneity and equality** of the demanding power and the supplying power are other essentials of a power system. Therefore, in regard to effective power, the total amount of generation $\sum P_{\text{gen}}$ is controlled to meet capricious load $\sum P_{\text{load}}$ and to maintain the frequency within $50/60 \pm \alpha$ Hz (α may typically be 0.05 Hz, although it may be different for utilities) over time by means of the processes of **generating power dispatching control** and **AFC** based at the control centre.

Reactive and capacitive power also have to be supplied over time to meet the load requirement $j(Q_l-Q_c)$; however so-called Var power control must be conducted not only on a total system basis but also on an individual partial network basis in order to maintain the voltages of each local area within allowable levels (say, 1.0 ± 0.05).

The generating sources of reactive power to meet reactive/capacitive load demand are as follows:

- **Reactors** (for jQ_l), **capacitors** (for jQ_c) installed at receiving substations.
- **Generators** (for jQ_l and jQ_c).
- **Synchronous phase modifiers** (rotary condensers, for jQ_l and jQ_c) installed at receiving substations. (Synchronous phase modifiers may be uncommon today because they are expensive in comparison with reactors and/or capacitors.)

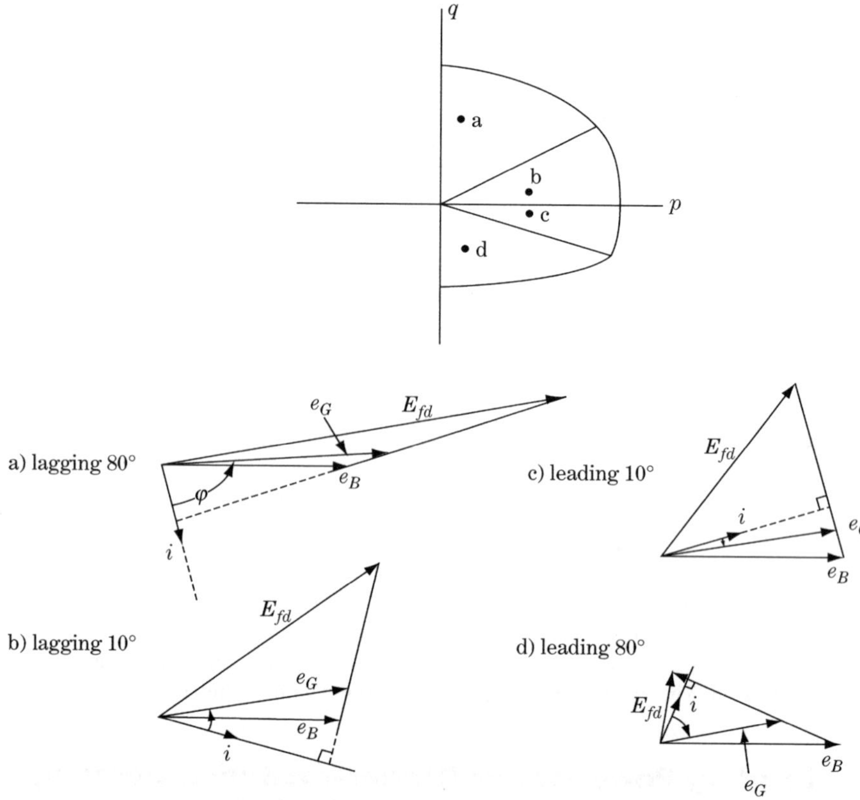

Figure 16.5 Vector diagrams for various p–q operation modes

Generators also share important roles as reactive/capacitive power generators, and have very large supplying capacity of reactive power (Var, MVar), but rather poor supplying capacity of capacitive power.

Typically, in the daytime, each generator supplies reactive power as well as effective power to the load. At night, however, capacitive load may be required, which can be partly compensated in parallel by reactor banks installed at the receiving stations, while some generators may be required partly to share the role to supply capacitive power by leading power-factor operation.

As well as such scheduled leading power-factor operation, generators may be suddenly forced into leading power-factor operation according to the special incidental conditions listed below:

- Sudden tripping of another generator under leading power-factor operation.
- Sudden increase of excitation (increasing lagging reactive power) by another generator operating in parallel.
- Sudden tripping of reactors at receiving substations, or sudden increase in capacitive loads.
- Sudden increase of effective power caused by the prime-mover.
- Sudden change of AVR setting voltage (V_{set}) to a lower value (by mistake).

Now we need to study the reason why leading power-factor operation may cause severe conditions for generators.

16.3.2 Overheating of stator core end by leading power-factor operation (low excitation)

Generators have a problem with abnormal overheating of the stator core end, caused during leading power-factor operation. Here we study the limit curve ③–④ in the leading power-factor zone of Figure 16.3.

Figure 16.6 and Figure 16.7 show the structural concepts of a generator. The rotor is a rotating electromagnet having N- and S-poles. All the flux ϕ generated by excitation current i_f flows out

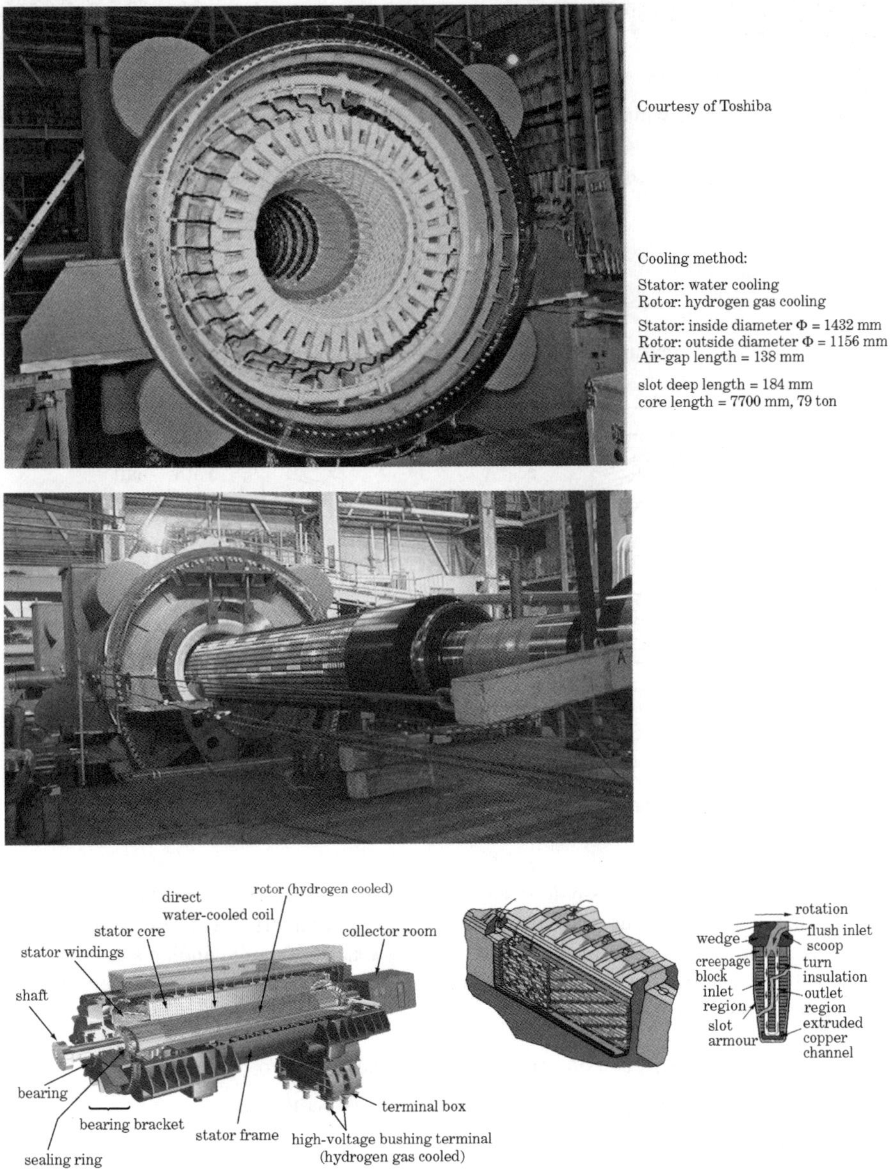

Figure 16.6 Thermal generator (1120 MVA, 1000 MW, 2-pole, 60 Hz, 360 rpm)

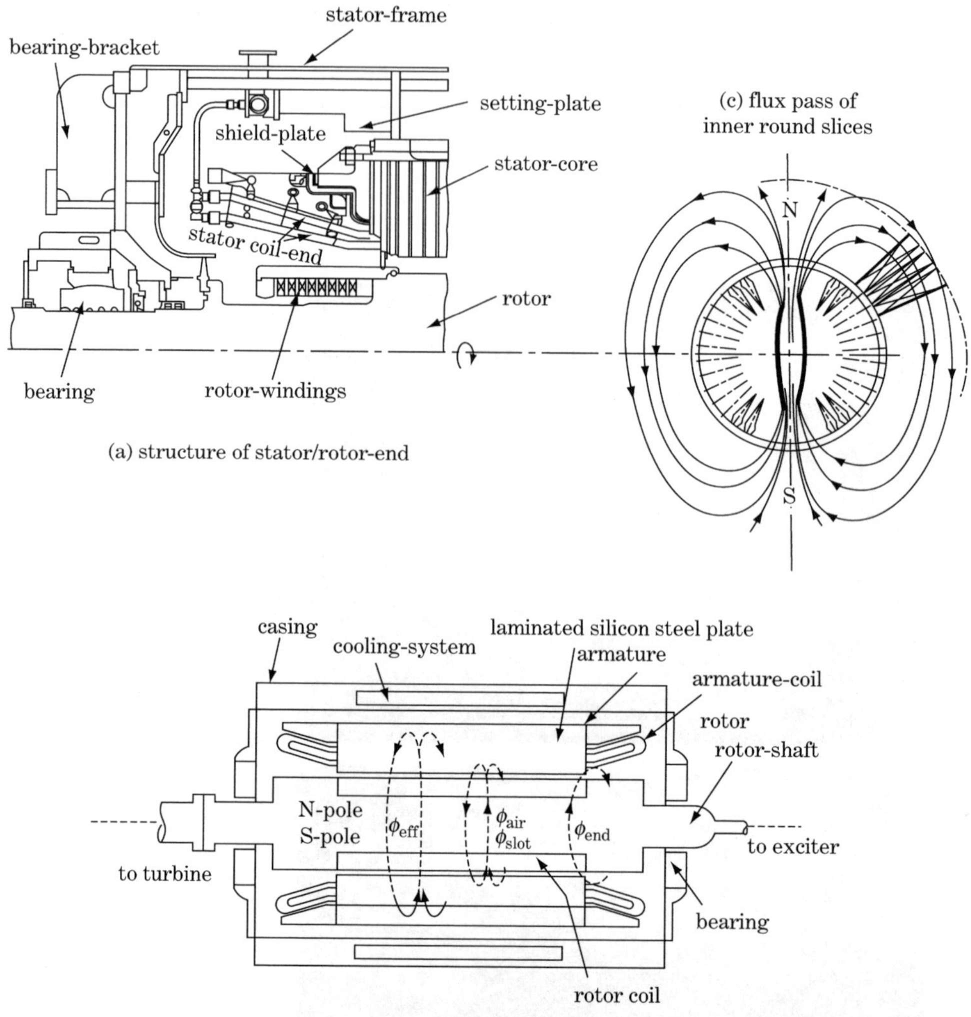

(b) concept of flux pass

Figure 16.7 Flux passes of thermal generator

from the N-pole and returns to the S-pole as shown in Figure 16.7b, c. Most of the flux from the N-pole reaches the stator winding zone and returns to the S-pole after effectively interlinking with the armature coil. However, there is some flux which does not reach the stator winding zone and returns to the S-pole without interlinking with the armature coil. That is, the total flux produced by the rotor magnet can be categorized as follows:

$$\phi_{total} = \phi_{eff}\,(\text{linking flux}) + \phi_{leak}\,(\text{leakage flux}) \fallingdotseq \phi_{eff} + \phi_{end}$$

$$\text{where}\quad \psi_{total}\,(\text{flux linkage}) = \phi_{eff} \cdot N\,(N : \text{coil turns})$$

$$\phi_{leak} = \phi_{air} + \phi_{slot} + \phi_{end} \fallingdotseq \phi_{end}$$

$$\phi_{end} \gg \phi_{air},\quad \phi_{slot}$$

- ϕ_{air} (air-gap leakage flux): Flux flowing down through the air gap to the S-pole without interlinking with the armature coil.
- ϕ_{slot} (slot leakage flux): Flux flowing down around the stator surface to the S-pole without interlinking with the armature coil.
- ϕ_{end} (end-coil leakage flux): Flux starting from the N-pole around the coil end of the rotor cylinder and flowing down to the S-pole without interlinking with the armature coil.

Considering round slices of a generator and inner slices along the axial length, most of the flux interlinks effectively with the armature coil as $\psi_{total} = \phi_{eff} \cdot N$ (where N is the turn number of the armature coil) so that ϕ_{air} and ϕ_{slot} will seldom exist.

However, at the coil end, for a round slice of the cylinder end, most of the flux flowing out from the N-pole turns down to the S-pole as ϕ_{end} without interlinking with the stator coil, through various routes of the stator end metal structure (such as parts of the core, yoke, coil support, shield plate, cooling pipe, etc.). Accordingly, eddy currents appear in the high-resistive metal structure and tend to raise the temperature. This temperature rise is not too serious under operation at a power factor of 1.0, but it is extremely serious under operation at leading power factor. The reason is explained below.

Figure 16.8 shows vector diagrams for operation under three different power factors, namely case (a) for lagging power factor, case (b) for power factor 1.0, case (c) for leading power factor. The vector magnitudes of terminal voltage e_G and current i are drawn equally for the three cases.

If we change the operating power-factor angle δ slowly from 90° (lagging) $\rightarrow$ 0°($\cos\delta =$ 1.0) $\rightarrow -90°$ (leading) under the condition of fixed values of e_G and i, the excitation voltage jE_{fd}

ϕ_{eff}: total linking flux
ϕ_{end}: total end-coil leakage flux

$\phi_{total} = \phi_{eff} + (\phi_{leak}) \fallingdotseq \phi_{eff} + (\phi_{end})$

where $(\phi_{leak}) = (\phi_{air} + \phi_{slot} + \phi_{end}) \fallingdotseq (\phi_{end}) \because \phi_{air} \fallingdotseq \phi_{slot} \fallingdotseq 0$

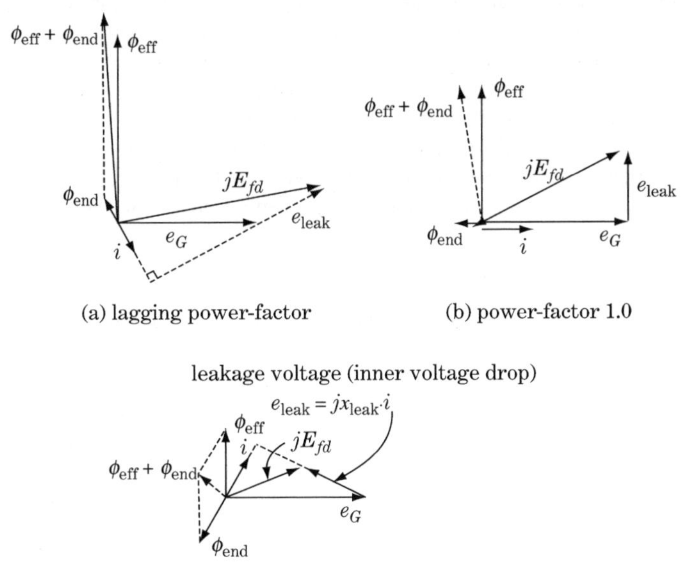

Figure 16.8 Flux vector diagrams around coil end structure

becomes smaller, as studied already. Next, effective linkage flux ϕ_{eff} is 90° leading from e_G and obviously proportional to E_{fd}. Accordingly, ϕ_{eff} is the largest in case (a).

The counter electromotive force $e_{\text{leak}} = (jx_{\text{leak}} \cdot i)$ is the same value for each case because stator current i has the same magnitude. The leakage flux ϕ_{end} is produced by e_{leak} and is 90° leading from e_{leak}. However, ϕ_{end} does not have the same magnitude for each case even though the magnitudes of e_{leak} are the same.

In case (a) (lagging power-factor operation), ϕ_{eff} is large and, in addition, the composed flux $(\phi_{\text{eff}} + \phi_{\text{end}})$ becomes larger. However, the flux density of ϕ_{eff} is so high that superposed flux ϕ_{end} cannot have a large magnitude because of magnetic saturation of the flux pass. The situation in case (b) (power factor 1.0) is similar to case (a).

On the other hand, in case (c) (leading power-factor operation), ϕ_{eff} is very small (because E_{fd} is small) and, in addition, the composed flux $(\phi_{\text{eff}} + \phi_{\text{end}})$ becomes smaller. That is, the flux density of ϕ_{eff} is very low and becomes weaker on addition of ϕ_{end}. Accordingly, superposed ϕ_{end} would have a very large magnitude, because magnetic saturation would not occur.

Now we focus on the leakage flux ϕ_{end} of the coil end round slice. Even though the magnitude of e_{leak} is the same for all three cases, the flux produced at the coil end round slice ϕ_{end} becomes extremely large only for case (c). Therefore, a relatively large current is forced to flow through the metal structure of the cylinder end so that large eddy-current loss and magnetic loss occur around the coil end structure and cause serious local heating of the coil end metal structure.

If a generator's operational mode is suddenly changed from the lagging power-factor zone to the leading power-factor zone, leakage flux around the coil end would be suddenly increased, and the temperature would rise rapidly in a very short time, and cause decay of mechanical strength by annealing, melting and/or damage to insulation. It should be stressed that wedges (for example) must have the strength to withstand the huge centrifugal force of approximately Mach 1 peripheral speed occurring on the rotor coils. This is the reason for the generator's leading power-factor operational limit (low-excitation problem).

The capability curve for low-excitation limit ③–④ in Figure 16.3 is to protect generators from such an effect.

Again, consider a general view of Figure 16.3. In regard to operation in the lagging power-factor zone, a generator can withstand such operation for some time even if the excitation limit curve ①–②: (1) is suddenly exceeded, (2) the exciter can generate current exceeding its rating, and (3) there is no stability problem. Therefore, exceeding the curve ①–② is not necessarily a major problem for the excitation limit or the generator's lagging mode operation.

On the other hand, in regard to operation in the leading power-factor zone, time becomes very important if operation exceeds the coil end overheat limit curve ③–④ and the stability limit curve ⑤–⑦, and such operation should be prevented or avoided within a few or tens of seconds.

16.3.3 UEL (under-excitation limit) protection by AVR

There are several reasons why a generator is often forced to rush into the low-excitation zone and probably tends to exceed the limit curves ③–④ and ⑤–⑦ without countermeasures being taken within a few seconds, because the time constants $T_{\text{avr}} + T_f + T$ are less than a few seconds. The countermeasure to prevent such operation is the UEL function of AVR.

This function is supplied by AVR as one of its essential functions. As shown in Figure 16.3, the UEL-setting zone is usually within some reserved margin of a straight line from $(0, -q_0)$ to the upper right. (The setting characteristics can be written as the equation $q = ap - q_0$, where a, q_0 are positive PU values.) AVR changes excitation freely within the setting area of Figure 16.3 in order to maintain the terminal voltage at the set value V_{set}. However, if the operating point in the p–q domain reaches the UEL line, AVR will no longer weaken the excitation and the voltage may consequently differ from the value of V_{set}. The UEL function of AVR is essential protection for a generator.

Loss of excitation protection will be discussed in Chapter 17.

16.3.4 Operation in the over-excitation domain

Engineering practices to stop the generator exceeding the over-excitation limit curve ①–② may be rather simple, and may involve only over-current relays for the armature winding current i_G and/or field current i_{fd}, because a generator can tolerate operation exceeding the excitation limit curve ①–② for some time with a relatively large margin.

16.4 V–Q (Voltage and Reactive Power) Control by AVR

16.4.1 Reactive power distribution for multiple generators and cross-current control

Two generators are operating in parallel as shown in Figure 16.9, and with the same operating output of $(P_1 + jQ_1) = (P_2 + jQ_2)$. Balance of the effective power distribution $P_1 = P_2$ is to be maintained by controlling the mechanical input volume from the prime-mover. Balance of the reactive power distribution $Q_1 = Q_2$ is to be maintained by controlling the generator excitation, and each AVR plays a role.

Assuming that the reactive power of generator 1 is suddenly increased from $Q_1 \rightarrow Q_1 + \Delta Q$ for some reason, generator 2 would then decrease from $Q_2 \rightarrow Q_2 - \Delta Q$. This represents unnecessary hunting phenomena. Assuming a very small error difference in the AVR set values V_{set1} and V_{set2}, one generator may run up to the lagging power-factor zone, while the other may be forced to run down to the leading power-factor zone. The countermeasure to solve these problems is also provided by AVR, namely the function of **cross-current compensation**.

In Figure 16.9a, the two generators are operating under the condition that the total power is constant, namely $(P_1 + jQ_1) + (P_2 + jQ_2) = \text{constant}$.

The currents of generators 1 and 2 are $i_1(t)$, $i_2(t)$, and the averaged current is $i_{av}(t)$. Thus

$$
\begin{aligned}
&e_G = E_{fd1} - i_1 \cdot jx_d = E_{fd2} - i_2 \cdot jx_d \\
&i_{av} = \frac{i_1 + i_2}{2} \qquad i_{av}\text{ : the averaged current of } i_1,\ i_2
\end{aligned} \tag{16.19a}
$$

$$
\begin{aligned}
&i_1 - i_{av} = \frac{i_1 - i_2}{2} \equiv \Delta i \\
&i_2 - i_{av} = \frac{i_2 - i_1}{2} \equiv -\Delta i \\
&\Delta i\text{: the difference between } i_1,\ i_{av} \text{ or between } i_2,\ i_{av}
\end{aligned} \tag{16.19b}
$$

Accordingly,

$$
\begin{aligned}
&i_1 = i_{av} + \Delta i \\
&i_2 = i_{av} - \Delta i \\
&\Delta i = \frac{i_1 - i_2}{2} = \frac{1}{2} \cdot \frac{E_{fd1} - E_{fd2}}{jx_d} = \frac{1}{2} \cdot \frac{\Delta E}{jx_d}
\end{aligned} \tag{16.19c}
$$

Δi is called the **cross-current** between the generators 1 and 2. The cross-current Δi can be explained as the current induced by the difference in the excitation voltages of 1 and 2, $\Delta E = E_{fd1} - E_{fd2}$.

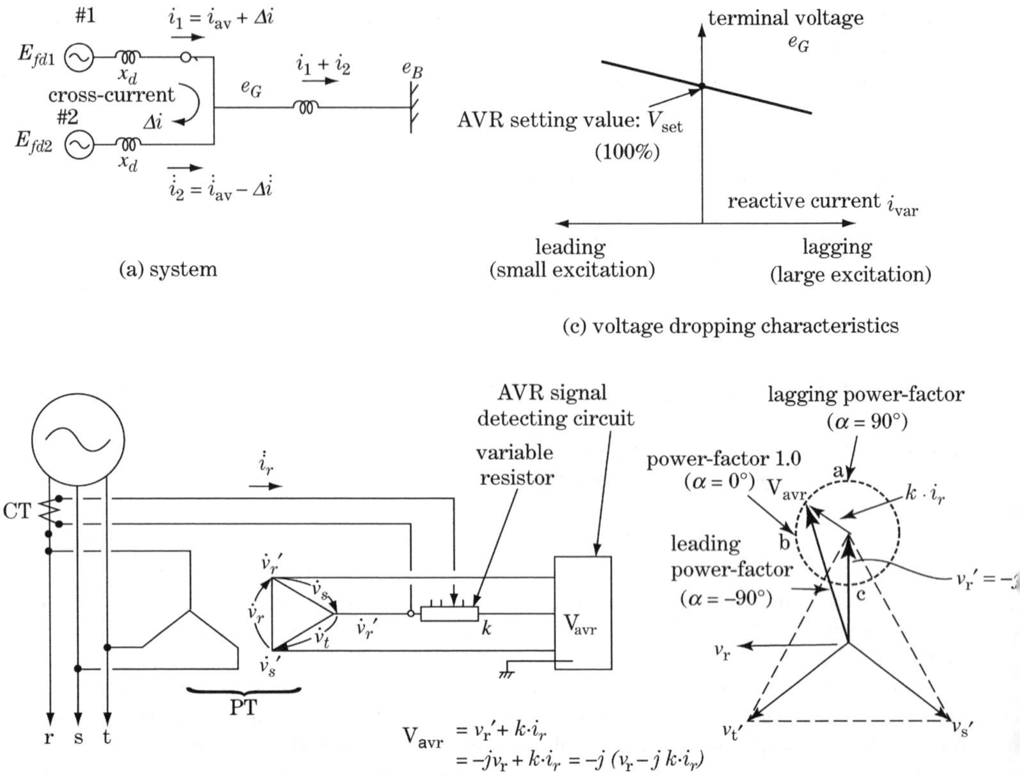

$\dot{v}_r \cdot \dot{v}_s \cdot \dot{v}_t \cdot \dot{i}_r$: generator terminal voltages and currrents (the secondary values of PT, CT)

$\dot{v}_r' \cdot \dot{v}_s' \cdot \dot{v}_t'$: potential voltage of PT secondary delta-connected points (the phases named by 90° lagging)

Vavr : signal voltage applied to the AVR signal detecting circuit

(b) signal detecting circuit (with cross-current compensation function)

Figure 16.9 AVR and the principle of voltage detection

The currents i_1, i_2 are of measurable value, so the averaged current i_{av} is also measurable. Then, the excitation of each generator can be continuously controlled, $i_1 \rightarrow i_{av}$, $i_2 \rightarrow i_{av}$ (i.e. to minimize the cross-current $\Delta i \rightarrow 0$), with high sensitivity and rapid response time, and as a result equal reactive power generation $Q_1(t) = Q_2(t)$ is realized and hunting can also be prevented. This is the principle of cross-current compensation control by AVR. The hunting prevention circuit in the block diagram of Figure 15.3 is the supplementary feedback circuit of AVR to realize a rapid response.

Figure 16.9 shows the principle of cross-current control by AVR, including a typical signal detection method. Each AVR produces positive-sequence quantities $v_1(t) - jki_1(t)$ at its signal-detecting circuit (as shown in (b)), and controls the excitation E_{fd} by the dropping characteristics of AVR as shown in (c). Accordingly, cross-current can always be minimized to zero.

This principle can be commonly applied to a system with two parallel operating generators of different BTG (boiler–turbine–generator) units in the same plant, or for two generators belonging to a common boiler unit.

Also the principle can be applied after a small modification for other operational modes, as follows.

In one case, namely operation of two generators with different reactive power shared ratios, operation by Q_1:$Q_2 \rightarrow 1$:α is realized by the excitation control of each generator, as below:

$$\begin{aligned} &\text{Generator 1} \quad i_1 \rightarrow \frac{1}{\alpha+1}\cdot i_{av} \\ &\text{Generator 2} \quad i_2 \rightarrow \frac{\alpha}{\alpha+1}\cdot i_{av} \\ &\text{where} \quad i_{av} = \frac{i_1+i_2}{2} \end{aligned} \tag{16.20a}$$

In another case, namely operation of n generators by equal reactive power generation, operation by $Q_1 : Q_2 : \ldots : Q_n = 1 : 1 : \ldots : 1$ is, for generator k,

$$i_k \rightarrow i_{av} = (i_1 + i_2 + \cdots + i_n)/n \tag{16.20b}$$

There are cases where one thermal boiler unit is combined with two or more turbines and multiple generators that are electrically operated in parallel. Of course, appropriate excitation control by the same principle can be realized for a variety of BTG units, although the actual functional circuit of AVR becomes rather complex.

The application of various BTG structures has been realized due to recent trends for large capacity and higher total efficiency as well as for various types of fuel. As a matter of fact, 7 to 10 generators comprise one thermal unit in the case of a modern ACC (Advanced Combined Cycle) thermal generation unit, which is a combined system of gas turbines and steam turbines applied for the system and fuelled by liquefied natural gas.

16.4.2 *P–f* control and *V–Q* control

Each generator has its own dynamic equation, Equation 13.11. By adding the equations of all the generators operating in synchronism, the following equation is derived for the system:

$$\left.\begin{aligned} &\sum_i P_{mi} - \sum_i P_{ei} = \sum_i \frac{M_i}{\omega_i}\cdot\frac{d\omega_i}{dt} \doteq \frac{\sum_i M_1}{2\pi f}\cdot\frac{d2\pi f}{dt} = \frac{\sum_i M_i}{f}\cdot\frac{df}{dt} \\ &\text{where} \quad \omega_1 = \omega_2 = \cdots \doteq \omega_{\text{average}} = 2\pi f \end{aligned}\right\} \tag{16.21}$$

This equation indicates that system frequency can be maintained by controlling the total mechanical input of prime-movers $\underset{i}{\Sigma} P_{mi}$ to meet the electrical load $\underset{i}{\Sigma} P_{ei}$ over time in the total power system.

This is the duty of **power dispatching** and **AFC** (Automatic Frequency Control) in that system frequency can be maintained within permissible limits. How to distribute total generating power to each generator (dispatching) does not matter in this case, although **generation power dispatching** (effective generating power distribution among numbers of generators) is quite important from the viewpoint of generation economy (**ELD**, Economic Load Dispatching) and the **best fuel combination policy** (fuel selection policy: hydro, coal, oil, natural gas, nuclear, various renewables). The central dispatching centre has a duty for all of them and commands the generation capacity to all the generation units through **ALD** (Automatic Load Dispatching).

On the other hand, V and Q have to be controlled to maintain balance not only in the total power system, but also in individual local subsystems, because voltages have to be maintained within allowable bands at each part of the network. In other words, ***V–Q* control** of the network has to be of individual locally dispersed type. Accordingly, the **AQR** (Automatic Reactive/Capacitive power control system) has to be installed for major generating plants as well as for major substations as the local area control system.

In the generating plant, the AQR system controls the reactive power (or the terminal voltage) of each generating unit by controlling V_{set} (the set value) of AVR. In the major receiving substations, the AQR system controls the operating capacity of reactor/capacitor banks as well as LTC-Trs (On-Load Tap-Changing Transformers) to maintain the regional voltages and reactive power balance.

16.5 Thermal Generators' Weak Points (Negative-sequence Current, Higher Harmonic Current, Shaft-torsional Distortion)

Besides the capability curves, we need to examine some weak points of thermal generators that are more or less common in synchronous motors/condensers as well as in large induction motors.

As the starting premise we need to recognize some basic features of modern advanced generators.

16.5.1 Features of large generators today

The unit capacity MVA of a generator is determined by the total magnitude of effective flux linkage whose practical relation with the generator volume can be explained by the equation

$$\begin{aligned} &\mathit{Rated\ MVA} \propto \{\psi_a + \psi_b + \psi_c : \text{total flux linkage}\} \propto \{C \cdot N_{\text{turn}}\} \cdot \{B\} \cdot \{RPM\} \\ &N_{\text{turn}} = \{\text{stator coil volume}\} \propto \{\pi(D_{\text{out}}/2)^2 - \pi(D_{\text{in}}/2)^2\} \cdot L \propto D^2L \end{aligned} \tag{16.22}$$

where
- B: flux density at rotor surface or air gap
- C: current density of armature
- N_{turn}: turn numbers of armature coil
- $D_{\text{out}}, D_{\text{in}}$: outer and inner diameters of stator
- L: effective axial length of stator and rotor
- D^2L: generator volume structure
- RPM: rotating speed per minute

$$\therefore \mathit{Rated\ MVA} = \{B\text{: flux density}\}\cdot\{C\text{: current density}\}\cdot\{D^2L\text{: generator volume}\}\cdot\{RPM\}$$

The unit capacity of the largest thermal generators today is are 1000–1300 MVA, while that of generators in 1950 was only 50–100 MVA. The capacity of the largest generators has increased twenty times in half a century; however, they have a relatively smaller physical size. In other words, a dramatic capacity enlargement has been realized not only by enlargement of the generator's physical size (D and L), but also by enlargement of flux density (B) and current density (C). Such enlargements have been based on advanced technology over the last half-century, which makes the following quantitative explanation possible.

For C (stator current density):

- direct stator coil cooling by water or hydrogen gas coolant flowing through the hollow piped coil
- high-temperature-withstanding stator coil insulation.

For B (linking flux density):

- direct rotor coil cooling by hydrogen gas coolant flowing through the hollow pipe coil
- advanced silicon–steel plate with high flux density and heat-withstanding capability
- high-temperature-withstanding rotor winding insulation.

A rotor of thermal generator is a horizontal cylinder which rotates at 50/60 turns per second (the peripheral rotating velocity is close to the velocity of sound, Mach 1). Furthermore. the rotor requires a very large dynamic metal strength and balancing accuracy in order to withstand various mechanical effects such as bending caused its weight, centrifugal forces, vibrations, eccentricity, mechanical resonance, heat expansion, torque, electromagnetic motive force, etc. Naturally there are manufacturing upper limits on the sizes of D and L, especially in forging technology for large-scale metal bodies with sufficient homogeneity and strength.

16.5.2 The thermal generator: smaller I_2-withstanding capability

Whenever unbalanced faults occur in the network, the negative-sequence current I_2 is forced to flow into a generator from outside the network. I_2 immediately induces large eddy currents in the rotor surface metal and consequently causes a rapid and serious temperature rise around the rotor surface metal.

These phenomena can be explained by the d- and q-axis equivalent circuits of Figure 10.4.

The currents i_d, i_q of the d- and q- axis circuit are d.c. components for positive-sequence current I_1 (under normal operating conditions), yet become second-harmonic $(2\omega t)$ currents for negative-sequence current I_2, because the rotating direction of the negative-sequence current is opposite to the direction of rotation of the rotor (see Equation 10.26). Now, in the d- and q-axis circuits of Figure 10.4, the second-harmonic currents are suddenly forced to flow into the circuits. In the d-axis circuit, it should be recalled that $Ldi/dt + ri$ is dominated by ri for d.c. and by Ldi/dt for second-order harmonics. The second-harmonic current i_d seldom flows into the L_{ad} branch or field branch and mostly is forced to flow into the damper branch. Accordingly, the damper branch quickly becomes overheated by the Joule heat losses at high resistances, r_{kd}.

Incidentally, the damper branch of the cylindrical rotor effectively means rotor surface metal. In other words, negative-sequence fault current I_2 flowing into the stator winding causes extreme eddy currents on the rotor surface so that the rotor surface metal (core, wedges, etc.) experiences a large rise in temperature. An advanced winding cooling system is only effective for conductor cooling and is useless against such temperature rise on rotor surface. The weakening of the mechanical strength of wedges by annealing is the most serious effect, because they have the important role of withstanding the centrifugal force on the rotor windings in the slots. This is the reason why the negative-sequence current I_2 withstanding capability of thermal generators is generally quite poor.

Table 16.1 shows the generator's withstanding capacity against negative-sequence current indicated by typical standards. These values are decided mainly by the critical metal strength of rotor wedges whose mechanical strength would be rapidly weakened around 200°C because of metal annealing.

The negative-sequence current I_2 forced to flow into the generator may be classified as follows:

a) Continuous negative-sequence current caused by three-phase imbalance of network and/or loads.

b) Intermittent negative-sequence current caused by special loads (rail traffic loads, electric furnace, etc.).

Table 16.1 Withstanding limit of I_2 (negative-sequence current) (JEC 2130 (2000))

Salient-pole machine	Maximum limit of I_2/I_{rate} for continuous I_2 current	Maximum limit of $(I_2/I_{rate})^2 \cdot t$ for I_2 fault current	Cylindrical machine	Maximum limit of I_2/I_{rate} for continuous I_2 current	Maximum limit of $(I_2/I_{rate})^2 \cdot t$ for fault current
1 Direct cooling			1 Indirect cooling rotor		
Motor	0.1	20	Air cooling type	0.1	15
Generator	0.08	20	Hydrogen-gas cooling type	0.1	10
Synchronous phase modifier	0.1	20			
2 Indirect cooling			2 Direct cooling rotor		
Motor	0.08	15	≤350 MVA	0.08	8
Generator	0.05	15	≤900 MVA	(*1)	(*2)
Synchronous phase modifier	0.08	15	≤1250 MVA	(*1)	5
			≤1600 MVA	0.05	5

Time t: sec.

Note: (*1) $\frac{I_2}{I_{rate}} = 0.08 - \frac{S_N - 350}{3 \times 10^4}$ S_N: rated apparent power [MVA]

(*2) $(I_2/I_{rate})^2 \cdot t = 8 - 0.00545\,(S_{rate} - 350)$

c) Large negative-sequence fault current caused by unbalanced fault (+ single phase/multi-phase reclosing).

d) Breakers fail to trip (open-mode failure) etc.

In categories (a) and (b), current I_2 caused by category (a) may not be serious for generators because the magnitudes are rather small for most cases. Current I_2 caused by (b) may be serious on occasion in that a large unbalance load source exists very close to the generator. In such cases countermeasures should be taken by the special load sides (see Section 24.3 in Chapter 24).

In category (c), for the unbalanced fault current, I_2 caused by a short-circuit fault has to be carefully investigated because I_2 could become quite large in comparison with I_{rate}, so I_2 could only be allowed for a short time.

As a check, for a 1000 MVA class cylindrical generator

$$(I_2/I_{rate})^2 \cdot t \leqq 5 \text{ (from Table 16.1)}$$

then

$$\text{for} \quad I_2/I_{rate} = 5\text{: the allowable time is } t = 0.2 \text{ sec (10–12 cycles)}$$

$$\text{for} \quad I_2/I_{rate} = 7\text{: the allowable time is } t = 0.10 \text{ sec (5–6 cycles)}$$

The allowable duration for a large fault current is obviously quite small. Although the above critical condition may have some redundancy, repetition of such a fault may cause serious metal fatigue and may badly damage the generator in the worst case.

Now we can recognize that **high-speed fault tripping** by protective relays + breakers is vitally important not only for stability or protection of transmission lines or substation equipment, but also to prevent generator fatigue caused by repeated fault current. Meanwhile, the propriety of **single phase reclosing** at the thermal station against single phase short faults on neighbouring transmission lines close to thermal/nuclear generating plants has to be carefully determined.

In the case of zero-sequence current I_0 flowing into a generator commented upon here, I_0 caused by a high-voltage line would not flow into the generator because I_0 could not flow through the delta winding of the generator-connected main transformer. In the case of a grounding fault between the generator and the step-up transformer, I_0 would flow into the generator. However, the magnitude of I_0 is restricted to a small value (say, 200 A) because generators are mostly neutral resistive grounded, so that serious situations would not occur.

Hydro-generators have relatively large withstanding capability to negative-sequence current because of the large size of the rotor, natural cooling and smaller centrifugal force.

16.5.3 Rotor overheating caused by d.c. and higher harmonic currents

We now study the phenomena that arise whenever d.c. and higher harmonic currents flow into a generator.

16.5.3.1 Current of order n flowing into the phase a coil of a generator

This case can be described by the equation below where the real-number symbolic method is applied ($n = 0$ means d.c.):

$$\left.\begin{aligned} i_a &= I\cos n\omega t \\ i_b &= i_c = 0 \end{aligned}\right\} \tag{16.23a}$$

or in the 0–1–2 domain

$$i_0 = i_1 = i_2 = \frac{1}{3}\cdot I_a = \frac{1}{3}I\cos n\omega t \tag{16.23b}$$

In the d–q–0 domain (see Equation 10.10b), substituting Equation 16.23a into Equation 10.10b.

$$\left.\begin{aligned} i_d &= \frac{2}{3}\cos\omega t\cdot I\cos n\omega t = +\frac{1}{3}I\{\cos\overline{n+1}\cdot\omega t + \cos\overline{n-1}\cdot\omega t\} \\ i_q &= -\frac{2}{3}\sin\omega t\cdot I\cos n\omega t = -\frac{1}{3}I\{\sin\overline{n+1}\cdot\omega t - \sin\overline{n-1}\cdot\omega t\} \\ i_0 &= \frac{1}{3}I\cos n\omega t \\ \therefore\quad i_d + ji_q &= \frac{1}{3}I\left\{e^{-j(n+1)\omega t} + e^{j(n-1)\omega t}\right\} \\ &= \frac{1}{3}I\{e^{-jn\omega t} + e^{jn\omega t}\}\cdot e^{-j\omega t} = \frac{2}{3}I\cos n\omega t\cdot e^{-j\omega t} \end{aligned}\right\} \tag{16.23c}$$

Equation 16.23b indicates that i_2 and i_0 appear regardless of the value of n of the phase current. Accordingly, eddy currents would be induced in the cores and wedges of the rotor surface and cause local overheating for the same reason explained in the previous Section.

Furthermore, Equation 16.23c indicates that i_d i_q cannot be d.c. regardless of the value of n. In other words, harmonic current is forced to flow into the damper branch of the d- and q-axis equivalent circuits in Figure 10.4, and cause overheating of the rotor. Note that the special case of the above equation for $n=1$ corresponds to Equations 10.92 and 10.93.

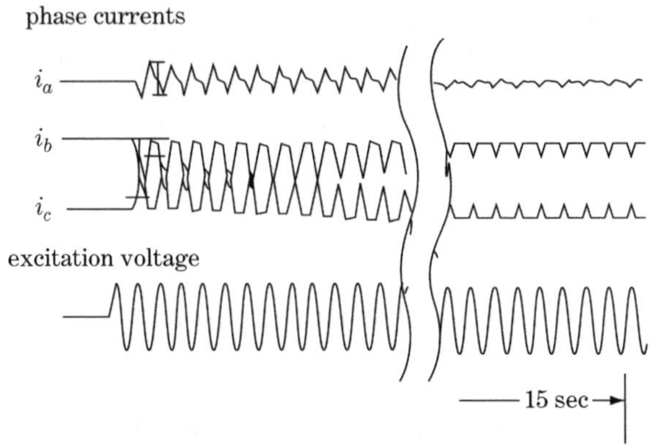

Figure 16.10 Magnetizing inrush current caused by transformer charging.

16.5.3.2 D.C. current flow

This is the case of $n=0$ in the above equations, and i_2 appears. The transient d.c. components included in the short-circuit fault would usually have a duration within 1–3 cycles or 15–60 ms (see Equation 10.124) so that it does not matter for the generator, although it may be of concern for breakers (Section 19.3, zero-miss phenomena) and high-speed protective relays (Section 22.1).

On the other hand, the **magnetizing inrush current** of a transformer is a unique transient phenomenon which appears immediately after the transformer is charged by breaker closing and which is a phase-unbalanced wave distorted by d.c. offset currents caused as magnetic flux transient phenomena with the hysteresis saturation characteristics of the transformer magnetic core. The magnitude of this inrush current could exceed by five times the value of the rated current, although this depends on the residual flux and the voltage value at the time of breaker closing. The duration depends on the resistance and reactance of the transformer, and would be around 1 second or less for smaller transformers, and longer for large transformers. It may even exceed 10 seconds, as shown in Figure 16.10, so that special attention should be paid to the generator as well as to the tripping capabilities of breakers and the fault-detecting capabilities of protective relays.

16.5.3.3 Three-phase-balanced nth-order current flow

This case can be written as the equation below where the complex-number symbolic expression is applied:

$$\left.\begin{aligned} i_a &= I\cos n(\omega t) \\ i_b &= I\cos n(\omega t - 120°) \\ i_c &= I\cos n(\omega t + 120°) \end{aligned}\right\} \quad (16.24a) \qquad \left.\begin{aligned} i_a &= I e^{jn\omega t} \\ i_b &= I e^{jn(\omega t-120°)} = a^{2n}\cdot I e^{jn\omega t} \\ i_c &= I e^{jn(\omega t+120°)} = a^{n}\cdot I e^{jn\omega t} \\ a^2 &= e^{-j120°} \quad a = e^{j120°} \end{aligned}\right\} \quad (16.24b)$$

Then, symmetrical components are

$$\left.\begin{aligned} 3i_0 &= (1+a^{2n}+a^{n})I e^{jn\omega t} \\ 3i_1 &= (1+a^{2n+1}+a^{n+2})I e^{jn\omega t} \\ 3i_2 &= (1+a^{2n+2}+a^{n+1})I e^{jn\omega t} \end{aligned}\right\} \quad (16.25)$$

i_d and i_q are derived by substituting Equation 16.24a into Equation (10.10b).

$$\left.\begin{aligned} i_d &= (2/3)I[\cos(\omega t)\cdot\cos n(\omega t)+\cos(\omega t-120°)\cdot\cos n(\omega t-120°)+\cos(\omega t+120°)\cdot\cos n(\omega t+120°)] \\ &= (1/3)\{\cos(n+1)(\omega t)+\cos(n+1)(\omega t-120°)+\cos(n+1)(\omega t+120°)\} \\ &\quad +\{\cos(n-1)(\omega t)+\cos(n-1)(\omega t-120°)+\cos(n-1)(\omega t+120°)\}] \\ i_q &= -(2/3)I[\sin(\omega t)\cdot\cos n(\omega t)+\sin(\omega t-120°)\cdot\cos n(\omega t-120°)+\sin(\omega t+120°)\cdot\cos n(\omega t+120°)] \\ &= -(1/3)I[\{\sin(n+1)(\omega t)+\sin(n+1)(\omega t-120°)+\sin(n+1)(\omega t+120°)\} \\ &\quad +\sin(n-1)(\omega t)+\sin(n-1)(\omega t-120°)+\sin(n-1)(\omega t+120°)\}] \end{aligned}\right\} \tag{16.26}$$

Then,
Fot the case n $= 3m$: d.c., 3rd, 6th, 9th order harmonic quantities

$$\left.\begin{aligned}\left.\begin{aligned} i_1&=0\\ i_2&=0\\ i_0&=I\cos n\omega t\end{aligned}\right\} \quad \left.\begin{aligned} i_d&=0\\ i_q&=0\end{aligned}\right\} \quad i_d+j_q=0\end{aligned}\right\} \tag{16.27a}$$

For the case $n = 3m+1$: 1st, 4th, 7th order balanced harmonic quantities,

$$\left.\begin{aligned}\left.\begin{aligned} i_1&=I\cos n\omega t\\ i_2&=0\\ i_d&=0\end{aligned}\right\} \quad \left.\begin{aligned} i_d&=I\cos(n+1)\omega t\\ i_q&=-I\sin(n+1)\omega t\end{aligned}\right\} \quad i_d+ji_q=Iej^{(n-1)\omega t}\}\end{aligned}\right\} \tag{16.27b}$$

For the case $n = 3m+2$: 2nd, 5th, 8th order balanced harmonic quantities,

$$\left.\begin{aligned}\left.\begin{aligned} i_1&=0\\ i_2&=I\cos n\omega t\\ i_0&=0\end{aligned}\right\} \quad \left.\begin{aligned} i_d&=I\cos(n+1)\omega t\\ i_q&=-I\sin(n+1)\omega t\end{aligned}\right\} \quad i_d+ji_q=Ie^{j(n+1)\omega t}\}\end{aligned}\right\} \tag{16.27c}$$

In the case of $n = 0$(d.c.),3rd, 6th, 9th order harmonic quantities, only zero sequence quantities of n-th order exist, so that i_l, i_2 as well as i_d, i_q are zero. In the case of n=1st, 4th, 7th order balanced harmonic quantities, only positive sequence quantities of n-th order exist, while i_d and i_q of (n−1)th order harmonics appears on the d-q-axes circuit.

In the case of n=2nd, 5th, 8th, order balanced harmonic quantities, only negative sequence quantities of n-th order exist, while i_d and i_q of (n+1)th order harmonics appears on the d-q-axes circuit.

The cases $n =$ 2nd, 5th, 8th, 11th are of nth-order negative-sequence currents and $i_d + j\,i_q$ is of $(n+1)$th-order a.c.

Recall that i_d, i_q are d.c. components under normal conditions in the d- and q-axis equivalent circuit of Figure 10.4, so the rotor surface is protected from abnormal overheating. On the contrary, the rotor surface cannot avoid serious overheating if i_d, i_q are a.c. components. Accordingly, higher harmonic currents cause abnormal conditions on the generator regardless of whether the current has positive-sequence components (n=3m + 1) or negative-sequence components (n = 3m + 2).

Higher harmonic components (or waveform distortion including higher harmonics) of the voltage and current are weak points for generators and for all other rotating machines, so the magnitude and duration of harmonics should always be treated carefully.

Waveform distortion will be discussed in Chapter 22.

16.5.4 Transient torsional twisting torque of TG coupled shaft

16.5.4.1 *Transient torsional torque arising from electrical transient by sudden network disturbance*

A generator and a turbine are mechanically coupled directly as a rigidly connected TG unit and are stationary operated under the balanced condition of mechanical input T_m and electrical output T_e. If a sudden disturbance occurs on either the connected network side or the prime-mover side, the generator or the turbine is immediately accelerated or decelerated by the torque $\Delta T = T_m - T_e$, so that transient torsional torque stress is caused on the rigidly coupled TG shaft train. Typical examples are the cases of sudden short-circuit faults and/or tripping of the on-load operating generator. Immediately after the electrical output is lost ($T_e = 0$), all the mechanical input power from the prime-mover becomes excess power ($\Delta T = T_m$) and the turbine is suddenly accelerated by the torque ΔT on both the shafts and the coupling device. This is the TG torsional shaft-twisting phenomenon. Repetition of such phenomena may cause mechanical fatigue. It may also cause mechanical vibration or pulsation.

Typical sudden disturbances causing shaft twisting are:

- Sudden tripping of the generator
- Sudden tripping of another generator operating in parallel
- Short-circuit fault at a relatively close point in the network
- Large-capacity load tripping
- Reclosing after fault tripping
- Emergency shutdown of BT (Boiler–Turbine) system.

Figure 16.11 is a diagram of a typical thermal TG unit which consists of three turbine shafts and one generator shaft (four mechanical points system) mutually and rigidly coupled as one rotating unit. Thermal TG units are generally very long and slender cylindrical structures with high-speed rotation (a 1300 MVA class tandem compound unit would be approximately 100 m in length and diameter of the 'shaft' only 1–1.5 m), so a mechanical system cannot tolerate shaft-twisting phenomena, while a hydro TG system is generally tougher.

The transient torsional torque modes can be classified into a few different modes as follows:

- D.C. offset mode torsional torque (the positive-sequence mode) ΔT_1: Three-phase-balanced disturbances (three-phase short-circuit fault, three-phase reclosing, load tripping, sudden shutdown

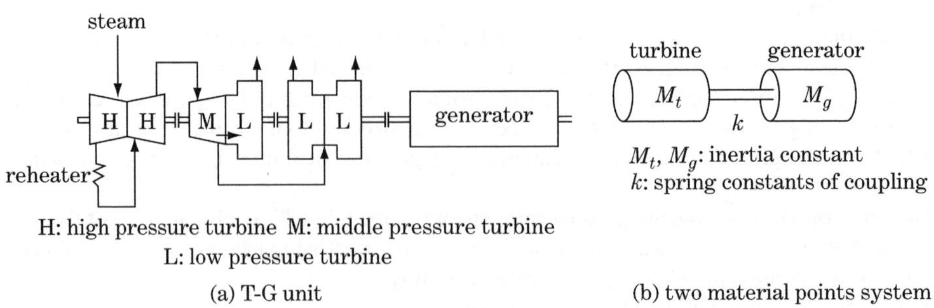

Figure 16.11 Concept of TG unit

of prime-mover, etc.) cause sudden stepping mode changes ΔP_e or ΔP_m; then the shaft torque caused must also be of stepping mode (ΔT_1).

- Double harmonic mode torsional torque (the negative-sequence mode) ΔT_2: Phase-unbalanced disturbances (phase-unbalanced short- circuit fault, single phase reclosing, etc.) cause negative-sequence current. Accordingly, the shaft twisting caused must be of double harmonic frequency pulsation mode (ΔT_2).
- Fundamental frequency mode torsional torque (the transient d.c. mode) ΔT_3: Whenever faults occur, unbalanced d.c. offset currents would appear in each phase as superposed transient components of short-circuit currents (the duration may be a few cycles, say 0.1 s), and the torsional torque caused must be of fundamental frequency pulsation mode (ΔT_3).

All of these torques cause severe stresses in the TG shafts and especially in the coupling mechanism.

16.5.4.2 Amplification of torsional torque

We examine transient torsional torque phenomena in the case of a three-phase fault and the reclosing and final tripping as shown in Figure 16.12. We assume that the TG unit is a simplified two material points system (one point for turbine and one point for generator) as s6hown in Figure 16.11b and the mechanical input is not changed for the duration of the process.

16.5.4.2.1 Time interval t_0–t_1 Referring to Figure 16.12, a three-phase short-circuit fault occurs at time t_0 at a point very close to the generator terminal. The generator output power P_e is lost and the corresponding electrical torque T_e becomes almost zero ($T_e \doteqdot 1 \rightarrow 0$) for the duration t_0–t_1, because the terminal three-phase voltages are almost zero, so that the continued transmission of power through the second circuit line is very small. Accordingly, accelerating torque of the stepping mode ($\Delta T = 0 \rightarrow 1.0$) occurs on the generator at time t_0, in that torsional torque oscillation is caused by the mechanical natural oscillating frequency ω_0 of the TG system, and continues oscillation with very slow damping by an amplitude of ± 1.0 (between $+2 \Leftrightarrow -0$).

16.5.4.2.2 Time interval t_1–t_2 The three-phase fault on the circuit#1 is cleared at t_1 (perhaps 50–100 ms after occurrence of the fault) and the power transmission P_e is recovered through the

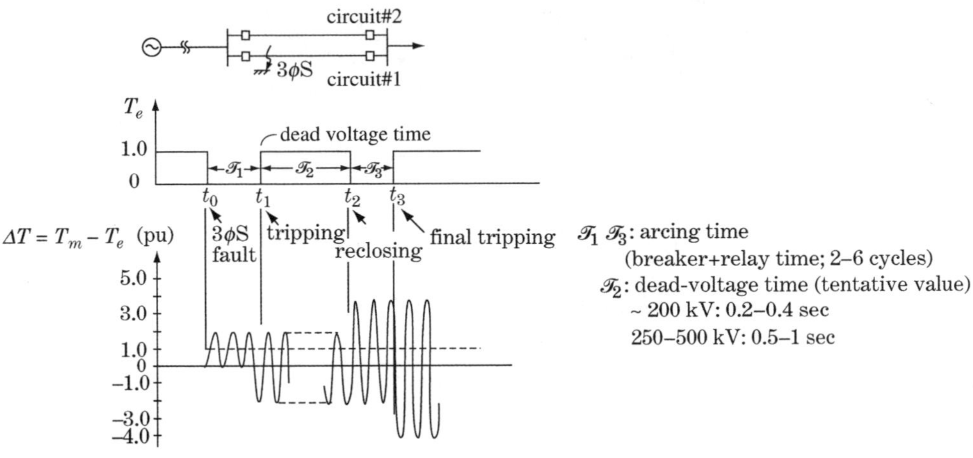

Figure 16.12 Transient torsional torque of TG coupled shaft under reclosing process

second circuit line for the interval t_1–t_2. T_e begins to recover ($T_e = 0 \rightarrow 1$) at time t_1. The instantaneous value of oscillating ΔT at time t_1 is the value between +2 and 0 and may be $\Delta T = +2$ by chance. Then ΔT begins to oscillate with an amplitude of ±2.0 (between $\Delta T = +2$ and −2) for the new interval t_1–t_2.

16.5.4.2.3 Time interval t_2–t_3 Three-phase reclosing of the first circuit#1 is conducted at time t_2 (perhaps of order 0.5 s after t_1). Accordingly, the sending power is lost again ($P_e \doteq 1 \rightarrow 0$) because the metal fault is still on the circuit#1. If $\Delta T = -2$ at t_2 by chance, ΔT begins to oscillate between $\Delta T = +4$ and −2 with an amplitude of ±3 for the interval t_2–t_3.

16.5.4.2.4 Time interval t_3–t_4 The circuit#1 line is finally tripped at time t_3. If $\Delta T = +4$ at t_3 by chance, ΔT begins to oscillate between $\Delta T = +4$ and −4 with an amplitude of ±4 for the interval of after t_3.

In the above example, four big disturbances (fault, fault tripping, reclosing, final tripping) occurred in succession. As a result, the amplitude of the shaft mechanical torsional torque oscillation could become extremely large by chance.

Furthermore, in the intervals t_0–t_1 and t_2–t_3, the shaft-torsional torque of the fundamental frequency pulsation mode (ΔT_3) should be superposed because unbalanced d.c. offset currents would appear in each phase.

In the case of phase-unbalanced faults, the shaft-torsional torque of the double frequency pulsation mode (ΔT_2) would be caused by negative-sequence current.

The damping of these mechanical oscillations is very slow (the time constants of the mechanical oscillation are quite long) in comparison with electrical phenomena so that rapid oscillation reduction cannot be expected. Figure 16.13 shows the simulation chart of a transient torsional oscillation in a four-mass model for fault tripping and reclosing.

In actual engineering work, each TG unit should be clearly assigned a specific guideline for the application of reclosing which may be specified by severity and the repeat time of fault tripping and reclosing. Also, the pros and cons regarding the application of the reclosing function of the lines' protective equipment in the neighbourhood of transmission lines from a thermal generating station have to be carefully decided from the viewpoints of generating plant and power system operation.

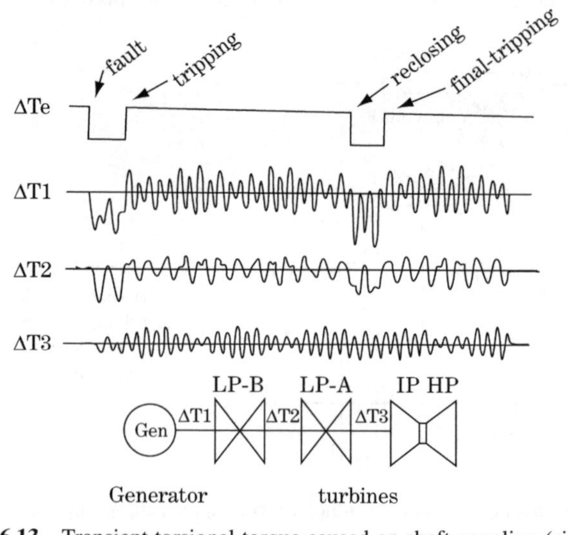

Figure 16.13 Transient torsional torque caused on shaft coupling (simulation)

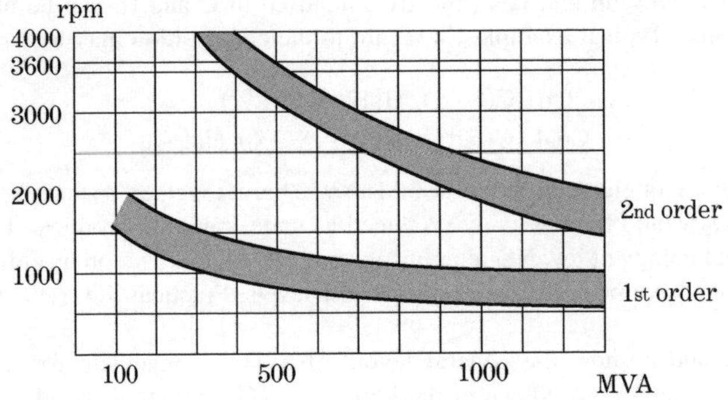

Figure 16.14 Speed–danger curve by rotational resonance

Harmonic resonance of the rotating TG unit is another matter to be considered. Any hardware has its own specific natural resonant frequency. Figure 16.14 shows the 'danger curves' of generators.

Steam turbines also have similar curves whose first-order resonant zone is generally below the rated rotating speed and second-order resonant zone over the rated speed. Whenever the turbine is started up from zero speed to the rated 3000/3600 rpm, a quick pass through the first-order resonant area is required. Needless to say, overspeed rotation close to the second-order resonant area should be strictly avoided.

Special attention has to be paid to loads in a typical **electrical furnace** because it has an intermittent repeated rushing load, and also a large unbalanced load with negative-sequence components as well as rippled harmonic components. Generators and motors installed close to such loads require special attention.

A thermal generating unit connected to the d.c. transmission line may be another concern, because the **power conditioners (converters)** of such a line produce stationary ripple harmonics, which may contribute to mechanical fatigue of the generator. Application of a power filter would be an effective countermeasure.

16.6 General Description of Modern Thermal/Nuclear TG Unit

Most electrical engineers may be required to know something about modern TG units in thermal or nuclear generating stations in the general sense described below, although a detailed description is beyond this book.

16.6.1 Steam turbine (ST) unit for thermal generation

The elements of hydrogen (H) as well as carbon (C) can be burnt to produce heat energy and the process is expressed by the following chemical equations:

$$\text{Hydrogen:}\quad H_2 + \frac{1}{2}O_2 = H_2O + 14.2 \times 10^7\ \text{J/kg(or 142 000 kW} \cdot \text{sec/kg)}$$

$$\text{Carbon:}\quad C + O_2 = CO_2 + 3.4 \times 10^7\ \text{J/kg(or 34 000 kW} \cdot \text{sec/kg)}$$

$$\text{Sulfur:}\quad S + O_2 = SO_2 + 0.92 \times 10^7\ \text{J/kg(or 9200 kW} \cdot \text{sec/kg)}$$

This is the reason why **oil** and **coal**, mostly composed of C and H, can be utilized as thermal energy sources (fuels). Typical examples in regard to the composition may be written as follows:

$$\text{Oil: C(86\%), H(12\%), S(2\%)}$$
$$\text{Coal: C(62\%), H(5\%), S(2\%), N(1\%)}$$

Incidentally, the minor elements S or N would also be burnt together in steam boilers, so that SO_x, NO_x as well as CO_2 would be inevitably produced as undesirable by-products. However, with the recent advanced technology of low NO_x combustion and NO_x neutralization by ammonia into N_2 and H_2O, NO_x emissions have been reduced to quite low levels. Practical SO_x reduction is also being widely achieved.

Figures 16.15a and b show the general layout of a TG arrangement for a 1000 MW class steam turbine and generator. As shown in the figure, the TG unit based on oil or coal combustion typically consists of three sections of turbines (HP-T/IP-T/LP-T, with different steam pressure and temperature ratings) and a generator with two poles of 3000/3600 rpm ratings, whose rotary shafts are rigidly coupled as 'one train shaft' (tandem type). Superheated steam typically at 550–600°C is fed from the oil- or coal-fuelled steam boiler to the TG area and flows through the HP-T → reheater of boiler → IP-T → LP-T → main condenser. Exhaust steam at the outlet of the LP-T is fed into the main condenser and then cooled by water (from a river water reservoir with a cooling tower or from sea water, at 0–32°C). In the condenser, most of the exhaust steam is condensed or liquidized, resulting in quite a low pressure (almost vacuum gauge pressure). The steam flow is then naturally extracted through the LP-T outlet to the main condenser. Typical examples of the dry-steam pressures and temperatures at each stage are shown in Figure 16.15b.

The **thermal efficiency** of the total BTG system with an ST unit is derived by using the **Rankine-cycle model** in thermodynamics and is limited by the temperature difference $\Delta t = t_1 - t_2$ between the superheated steam temperature t_1 and the cooling water temperature t_2 in the main condenser (0–32°C). Because t_2 is technically almost uncontrollable, the temperature t_1 is actually the main factor in determining the thermal efficiency η. The total efficiency η of a typical modern generating unit with $t_1 = 550$–$600°C$ is within the range $\eta = 38$–43%, which is the theoretical upper limit for a steam temperature of 600°C or less.

16.6.2 Combined cycle system with gas/steam turbines

LNG (Liquefied Natural Gas) has recently become a very important fuel for power generation in parallel with coal and oil, supported by the remarkably advanced technology of excavation and liquefaction for easy transportation/storage, as well as of combustion technology.

The essential quality of natural gas (**NG**) is due to hydrocarbon cyclic compounds whose chemical formation is generally given by C_nH_{2n+2} (methane (CH_4) sharing about 90% of ethane (C_2H_6), propane (C_3H_8), butane (C_4H_{10}), etc.). NG can be burnt efficiently in the presence of oxygen (O_2); in other words, under well-mixed conditions with plenty of compressed air supplied by the air compressor to the combustor. The chemical equation of burning hydrocarbon gas is written as follows:

$$C_nH_{2n+2} + \frac{3n+1}{2}O_2 = nCO_2 + (n+1)H_2O + Q \text{ J/kg}$$

For methane ($n=1$)

$$CH_4 + 2O_2 = CO_2 + 2H_2O \text{ J/kg or } 55\,700\,\text{kW} \cdot \text{sec/kg}$$

NG has the characteristic that it is liquefied into LNG at a very low temperature of −162°C (or 111 K). Liquefaction of gas obviously enables effective transportation by ship and storage.

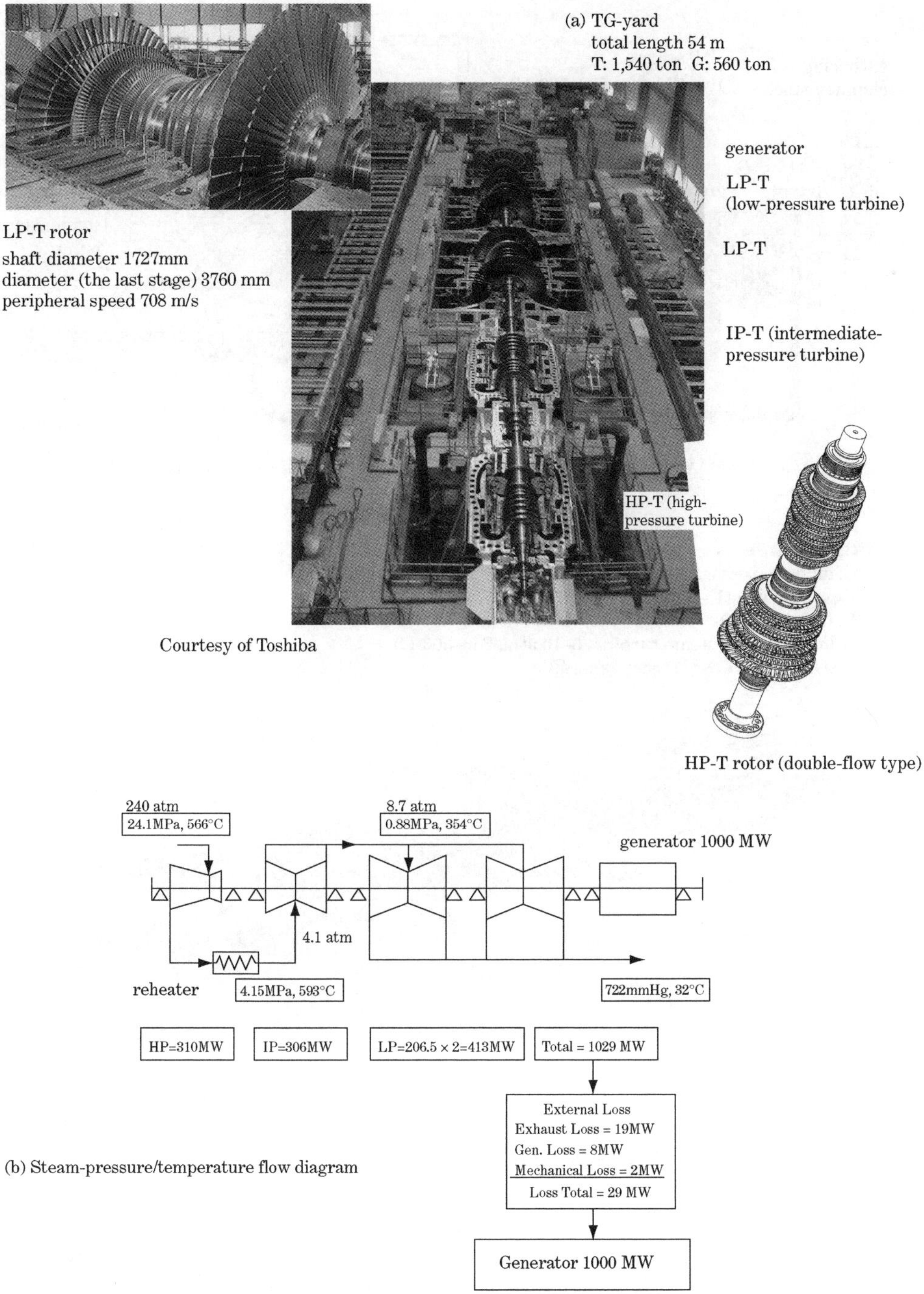

Note: 1 atm = 1.013×10^5 Pa = 760 mm Hg

Figure 16.15 Large-capacity TG unit (1000 MW)

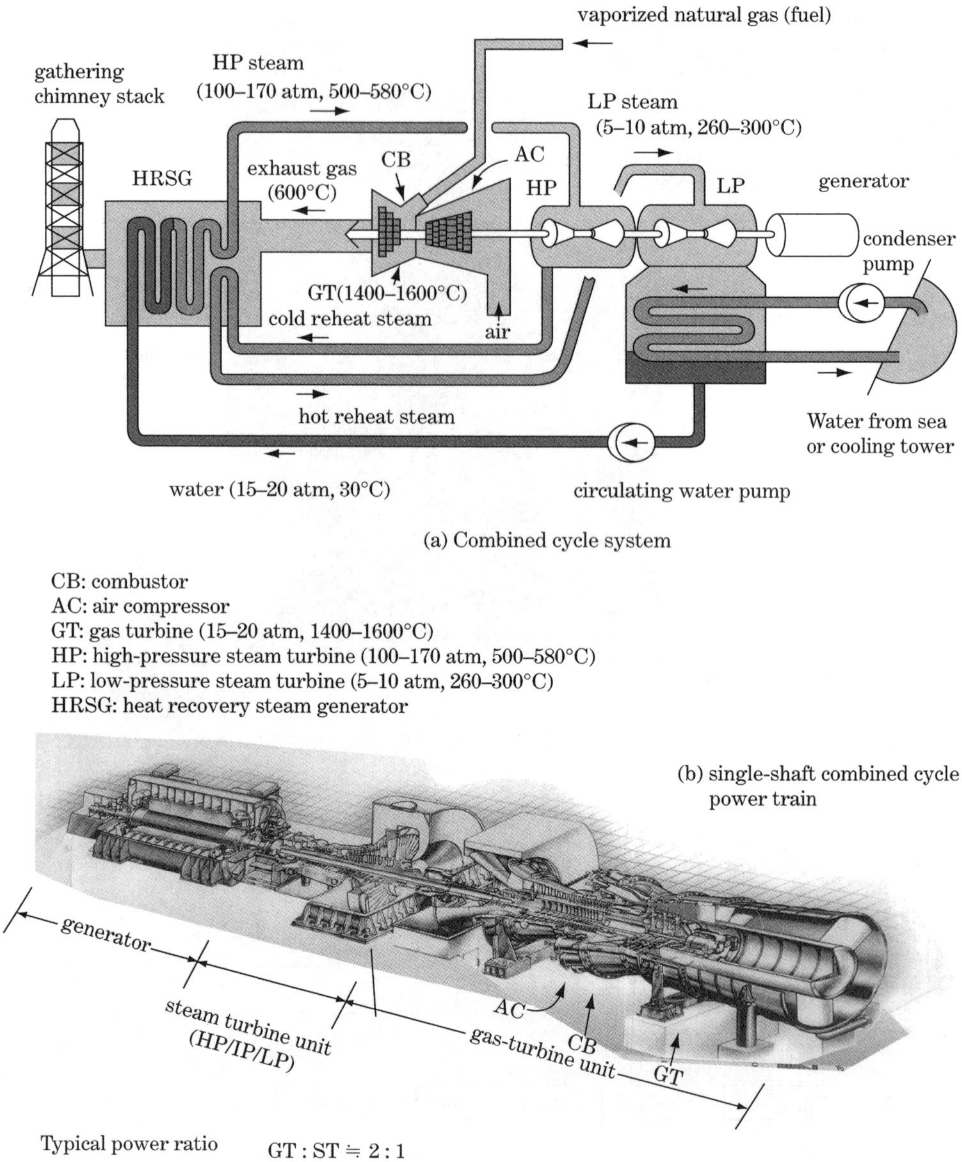

Courtesy of TEPCO/GE/Toshiba

Figure 16.16 Combined cycle system

Due to the advances in **NG excavation** and **LNG ship transportation/storage technology** achieved since about 1970, **NG resources** and large-scale, large-capacity LNG thermal power generation have become quite familiar generating practices in the succeeding years, particularly since the 1990s.

Figure 16.16a shows the configuration of **(advanced) combined cycle power generation** (ACC or CC) by **LNG fuel combustion**, based on state-of-the-art technology. As shown in the figure, the first section of the power system is the **gas turbine unit** which consists of an **air compressor, combustor** and **gas turbine**. LNG is vaporized in the **vaporizer** (LNG can be easily vaporized to gas through a warming process with water) and is led to the combustor (CB). LNG is then mixed with highly compressed air and burnt in the combustor (CB). The produced **combustible gas** at very high temperature (say, 1500°C) is led to the gas turbine chamber (GT) and drives the GT's rotary shaft. The principle and basic structure of the GT unit are similar to that of a jet engine for an aircraft. The exhaust gas at the outlet of the GT with a temperature of 600–700°C and is led to the **heat recovery steam generator (HRSG)**. This is a kind of large-scale, high-temperature heat exchanger, in which the thermal energy of the inlet exhaust gas is used to convert circulatory steam–water into superheated steam of 580–600°C. The steam is then led to the high-pressure steam-turbine (HPST) and drives the shaft. The principles of the processes after the HP ST are the same as in conventional ST generating stations.

In brief, the CC consists of a GT and ST unit. Figure 16.16b shows a configuration of a typical ACC system of 150–500 MW capacity in that GT, ST as well as generator G are mechanically directly coupled as a **single shaft power train**. Another system configuration of ACC is where the GT/G train and ST/G train are arranged separately without mechanical shaft coupling.

In contrast to the CC system, the **simple cycle system** (SC, with only GT and without ST) is also available, mainly for spinning reserve generation purposes to meet peak power demands of individual power systems, the virtue of which is a quick-start capability.

The rated firing temperature at the combustor/GT using recent technology is 1400–1500°C, although it used to be typically 1100–1300°C in the 1980s. The firing temperature at the combustor/GT is far higher than the inlet steam temperature for a conventional ST, so that the heat efficiency η of an ACC or CC generating system becomes quite high in comparison with that of the conventional ST system. The efficiency of ACC at 1500°C, for example, would be 50–54%, while that of conventional steam generation would be 40–43%.

An extremely high temperature for the combustible gas could be quite close to the annealing/melting critical temperature of the alloys and metals adopted as the construction materials of the combustor and GT (a few top stages) nozzles/buckets, so prudent thermal design is required in order to cool the hot-gas parts without relying on film cooling and to avoid causing abnormal hotspots.

In the case of an advanced single shaft power-train model with a class 1400–1600°C GT, for example, the first-stage rotational and stationary aerofoils of the GT are made of single crystal alloy and thermal barrier coated. Furthermore, a three-dimensional geometry with closed-loop steam cooling (using high-temperature steam of 400°C extracted from the HRSG as the cooling fluid) would be employed. Using today's advanced technology, GT classes over 1600°C may be achieved in the near future.

Note also that NG contains such impurities as dust, sulfur and water, but almost all of these can be removed during the process of liquefaction. Therefore, emission levels of substances responsible for air pollution (NO_x, SO_x, CO_2) would be reduced to quite low levels, which is why thermal power generation by LNG has been recognized as so-called 'clean thermal energy' in comparison with that of coal or oil.

16.6.3 ST unit for nuclear generation

Figure 16.17 shows the TG unit for nuclear generating stations, including typical steam pressures and temperatures of turbines. The principles of the steam/water circulation system including turbines/condensers are the same as that for conventional thermal units. However, the superheated steam fed from the nuclear reactor is typically at 280–300°C, which is much lower than that by thermal steam boiler. Therefore the TG unit of nuclear plant consists of two sections of turbines, HP-T/LP-T (without IP-T), and a generator with four poles of 1500/1800 rpm ratings. In other words, the TG coupled rotor

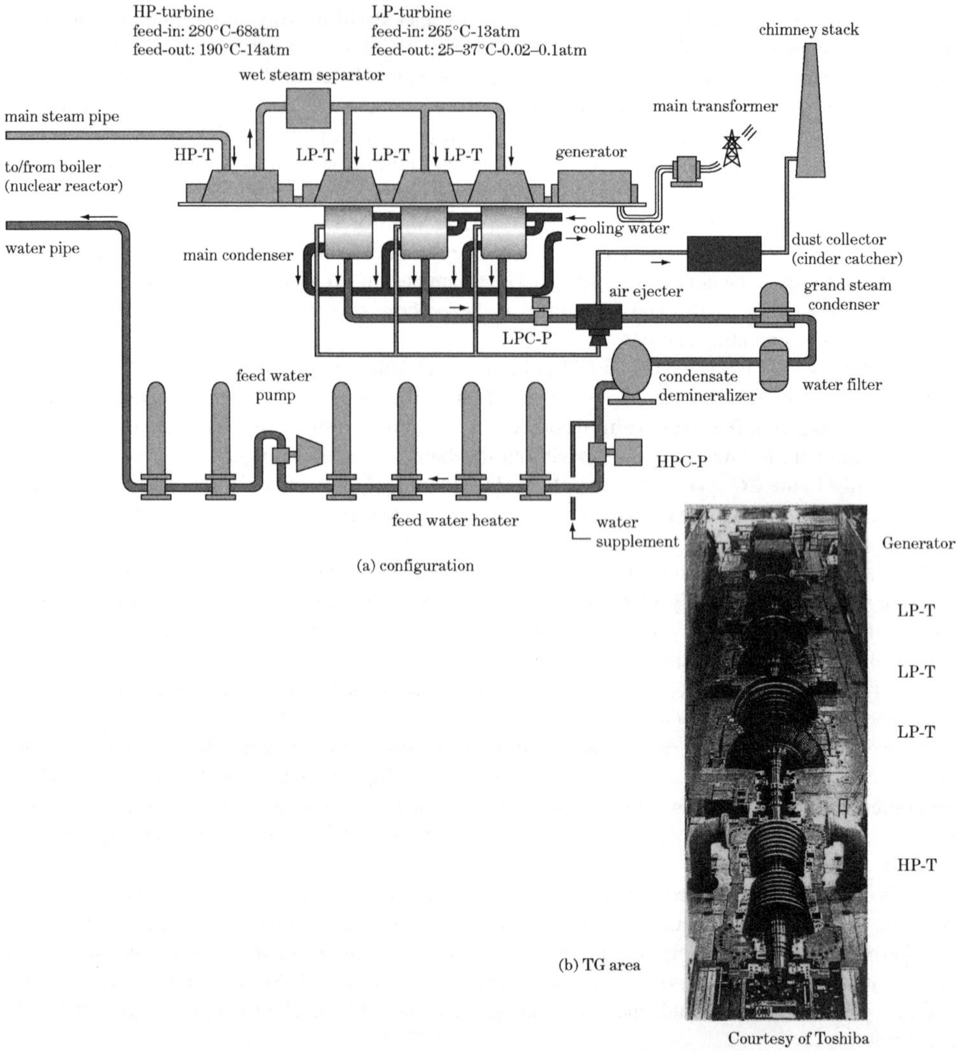

Figure 16.17 TG unit for nuclear power generating station

for the nuclear unit is half the speed and the generator rotor has a relatively larger mechanical volume in comparison with those in a thermal unit.

16.7 Supplement: Derivation of Equation 16.14 from Equation 16.9 ④

Modifying Equation 16.9 ④ and making it equivalent to a,

$$\frac{E_{fd}}{x_d}\cdot\frac{x_l}{e_B}=\frac{\sqrt{p^2+\left(q+x_d^{-1}\right)^2}}{\sqrt{p^2+\left(q-x_l^{-1}\right)^2}}\equiv a \tag{1}$$

modifying the middle and right side of the above equation,

$$p^2 + q^2 + 2q \cdot \frac{x_d^{-1} + a^2 x_l^{-1}}{1 - a^2} = \frac{-x_d^{-2} + a^2 x_l^{-2}}{1 - a^2} \tag{2}$$

Putting the numerator of the third term on the left equivalent to b and the right to c, then

$$\left.\begin{aligned} & p^2 + q^2 + 2q \cdot \frac{b}{1 - a^2} = c \\ & \therefore \quad p^2 + \left(q + \frac{b}{1 - a^2}\right)^2 = \left(\frac{b}{1 - a^2}\right)^2 + c \\ & b = \frac{1}{x_d} + \frac{a^2}{x_i} \\ & c = \frac{\frac{a^2}{x_l^2} - \frac{1}{x_d^2}}{1 - a^2} \end{aligned}\right\} \tag{3}$$

Thus Equation 16.14 is obtained.

17

R–X Coordinates and the Theory of Directional Distance Relays

Rapid fault tripping by high-speed protective relays is essential for the stable operation of modern power systems. Among various types of relays, the two principal types are directional distance relays and differential relays (including carrier differential relay equipment for transmission line protection). The study in this chapter mostly concerns directional distance relays, because they are primarily the most widely used as essential relays of various protection equipment, and their theoretical background is complicated, at least from the application viewpoint for crowded networks, so knowledge of them is important.

17.1 Protective Relays, Their Mission and Classification

Wherever a fault occurs in a network, a minimum section including the faulted point has to be removed immediately so that the power system maintains sound operation. Operating times of protective relays are typically 1.0–3 cycles and the tripping times of breakers are 1.5–3 cycles, giving a total fault tripping time of 2.5–6 cycles (50–120 ms at 50 Hz, 42–100 ms at 60 Hz). Malfunction or a slight delay of fault tripping by the associated relays may rarely become a trigger for serious black-outs by a **domino effect**, that is disruption of a power system caused typically by the following system behaviour:

- Cascade trips caused by over/undervoltages, over-currents, or by back-up tripping at the adjacent substations.
- Power stability collapse (step-out) or voltage stability collapse (instability caused by the imbalance of real power and/or voltage losing).
- Cascade trips of generators caused by abnormal frequency exceeding each machine's over-/under-frequency limits.
- Tearing (cutting) of a network caused by simultaneous lightning faults.
- Tearing (cutting) by extended cascade trippings (multiple faults, extensive damage to equipment, mal-operation or back-up trips of protective relays, etc.,) and so on.

Protective relays shoulder the vitally important duty of protecting the power system from such threats or at least of minimizing the failed tripping zone.

Handbook of Power System Engineering Yoshihide Hase
© 2007 John Wiley & Sons, Ltd

17.1.1 Duties of protective relays

The duties of protective relays can be classified as follows:

- To detect the faulted section immediately and to order tripping commands to the associated breakers of the section which includes the fault:
  –**Primary protection**, to detect the minimum section including the faulted point.
  –**Back-up protection**, to detect the faults of adjacent sections and to order tripping commands to adjacent breakers whenever primary protection fails to trip the faulted section.
- Continuous monitoring of various electrical quantities (voltages $V(t)$, currents $I(t)$, real power (P), apparent power $(P + jQ)$, frequency (f), phase angle (δ), synchronization $V(t)$, $V(t)'$, etc.) and the necessary commands to change the network connection immediately.
- Reclosing function (in the case of overhead transmission lines).
- Instability preventive control equipment (a combination of several protective relays over a wide region).

Protective relays have the responsibility to conduct 'instantaneous fault tripping or network connection changes'.

The last item above is a relatively new application of protective relay engineering. For example, a few sections may be tripped almost simultaneously by plural faults, or by back-up relays tripping (which may be caused whenever fault clearing by the associated primary relays or breakers fails). In such cases, a real power imbalance condition (over- or under-frequency phenomena) or critical stability condition (P–Q–V critical condition) are apt to cause cascade behaviours. Instability preventive control equipment continuously monitors the operation of various relays installed at plural stations and dispatches special commands (such as generator tripping, load shading, inter-tie tripping, etc.) whenever these conditions occur in order to prevent power system collapse by the domino effect.

17.1.2 Classification of major relays

Protective relays may be classified by the method described below.

17.1.2.1 Directional distance relays (DZ-relays)

The appearance of **high-speed directional distance relays** (mechanical type) around 1950 should be remembered not only as part of the history of modern protective relay equipment, but also as the realization of modern power system operating practices. In the first half of the twentieth century, relays commonly had simple functions and slow operating speeds, typically like the induction disc-type OC (over-current) or UV (undervoltage) relays. Technical improvements led to the appearance of greatly advanced relays which had the capability of judging the direction and modes of faults as well as measuring the distance from the relay to the fault points. Furthermore, faults were detected within 2–3 cycles.

These high-speed directional distance relays became leading practice in complicated, modern, high-voltage transmission line networks as essential relays for primary protection (**carrier-relaying equipment**) as well as for **back-up protection**, and for detecting **loss of excitation**, **step-outs**, etc. Today, directional distance relays still have the same important roles, though the hardware has been changed from the mechanical type to the electrostatic (solid-state analogue or digital) type.

17.1.2.2 Differential relays

These relays too have a long history and have mainly been applied for generator and transformer protection. The inflow current $i_{in}(t)$ and outflow current $i_{out}(t)$ of the associated equipment (generator or transformer) are sent to the relay through current transformers, and the relay operates whenever the differential current $\Delta i(t) = i_{\text{in}}(t) - i_{\text{out}}(t)$ exceeds a specified threshold.

Today, advanced digital communication facilities through microwave networks or OPGW (OGW with optical fibre) enable fault current waveforms to be sent from one terminal station to another. Due to the wide adoption of such facilities in recent years, differential relay practice has begun to be widely applied not only for **equipment protection** at generating plants/substations, but also for the **primary differential protection of transmission lines**.

17.1.2.3 Other relays

There are various other relays and typical examples are:

- Relays for voltages, currents, power, frequency, phase angle.
- Relays for rate of change quantities for power, frequency, and so on.

The operating principles and the application methods of the relays above are rather simple from the application point of view, although the design and manufacture are based on advanced hardware and software technology. However, the application theory of directional distance relays is more complicated, especially for applications to crowded networks, so knowledge of the appropriate utilization (relay settings, for example) is important.

This chapter concentrates on the application theory of directional distance relays for the above reasons. It can be said that directional distance relays are the essential basis of today's modern high-speed protection (as carrier-relaying equipment for primary protection, time-coordinated back-up relaying schemes, machine protective relaying, and so on).

17.2 Principle of Directional Distance Relays and *R–X* Coordinates Plane

17.2.1 Fundamental function of directional distance relays

High-speed directional distance relays (simply DZ-Rys hereafter) have the fundamental function of detecting the direction and distance of faults occurring on a transmission line.

The relays are installed on each feeder at a substation, to which the secondary voltages and currents of the PT (voltage transformers) and CT (current transformers) of the associated feeder line are applied as shown in Figure 17.1a. The fundamental theory from an application viewpoint will be briefly described below, although the actual hardware/software practices are very sophisticated.

Six DZ-Rys are utilized as a set for feeder protection at one station terminal, namely:

- phase-to-phase fault detection relays (DZ-S) for phase a–b, phase b–c, phase c–a
- phase-to-ground fault detection relays (DZ-G) for phase a–g, phase b–g, phase c–g.

Each relay produces new quantities as described below:

$$\left.\begin{aligned} &\text{DZ-S: for phase a-b:} \quad {}_{RY}\dot{Z}_{ab} = \frac{\dot{V}_a - \dot{V}_b}{\dot{I}_a - \dot{I}_b} \\ &\text{DZ-G: for phase a-g:} \quad {}_{RY}\dot{Z}_a = \frac{\dot{V}_a}{\dot{I}_a} \end{aligned}\right\} \tag{17.1}$$

(Relays for the other phases are described analogously.)

The composed quantities $(\dot{V}_a - \dot{V}_b)$, $(\dot{I}_a - \dot{I}_b)$ are called 'delta-voltages, delta-currents'. The composed quantities of Equation 17.1 have the dimensions of impedances $\dot{Z}(t) = \dot{V}(t)/\dot{I}(t)$, so that the

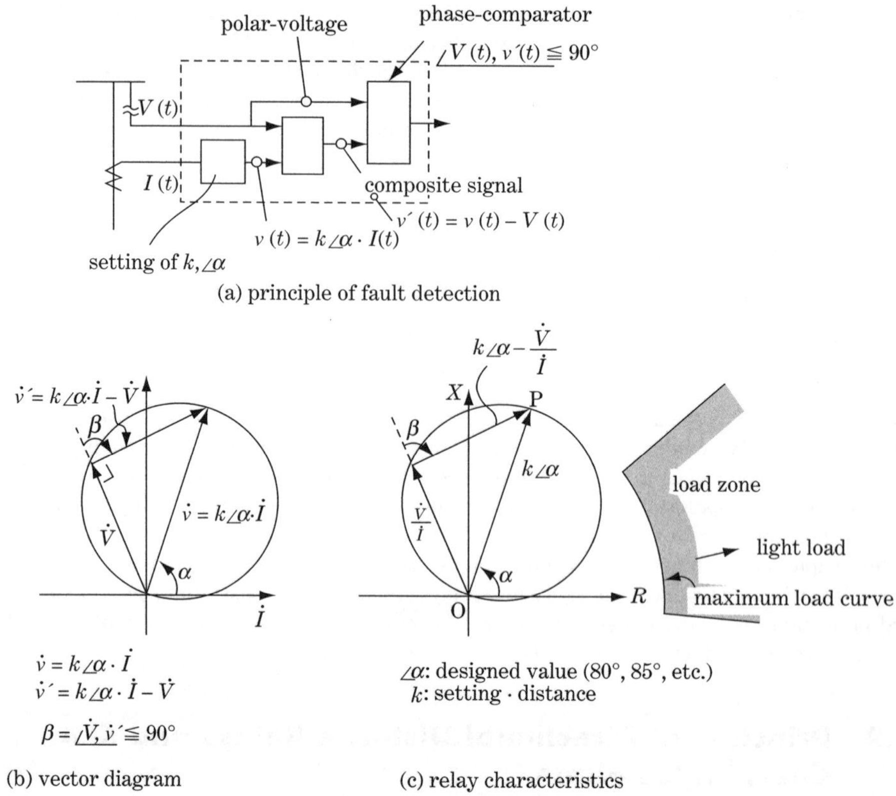

Figure 17.1 Mho-relay

characteristics of the relays are usually drawn as phenomena in *R–X* coordinates. Therefore we need first to investigate the definition of the *R–X* coordinates plane.

17.2.2 *R–X* coordinates and their relation to *P–Q* coordinates and *p–q* coordinates

An arbitrary impedance $\dot{V}/\dot{I} = \dot{Z} = R + jX$ can be plotted as a point (*R*, *X*) on an *R–X* coordinates plane, where $\dot{Z}$, *R*, *X* are mutually related to apparent power by the equations

$$\left.\begin{aligned} \dot{S} &= P + jQ = \dot{V} \cdot \dot{I}^* \\ \dot{Z} &= R + jX = \frac{\dot{V}}{\dot{I}} \end{aligned}\right\} \tag{17.2}$$

$$\left.\begin{aligned} P + jQ &= \dot{V} \cdot \frac{\dot{V}^*}{(R + jX)^*} = \frac{\dot{V}^2}{R - jX} \\ Z^* &= R - jX = \frac{V^2}{P + jQ} = \frac{1}{p + jq} \end{aligned}\right\} \tag{17.3}$$

where $p = P/V^2$, $q = Q/V^2$ (the same definitions as in Equation 12.18).

$R - jX$ is the inverse complex number of $P + jQ$ or $p + jq$. In other words, the R–X coordinates plane is defined as the inverse plane of the p–q coordinates plane, and a point (R, X) on the R–X coordinates plane is in one-to-one correspondence with a point (p, q) in the p–q coordinates plane. However, it must be noted that phase-unbalanced phenomena have to be treated for most cases in relay engineering, while p–q coordinates were treated as three-phase-balanced phenomena in Chapters 12–16.

17.2.3 Characteristics of DZ-Rys

DZ-Rys include various relays with different characteristics as shown in Figure 17.2, which have different names in practical engineering. We study DZ-Rys through the 'Mho-Ry' which is a relay possessing the base characteristics of the relay family.

17.2.3.1 Mho-Ry

Figure 17.1a is the fundamental structural design diagram of a Mho-Ry of electrostatic type. The signal voltage $V(t)$ and current $I(t)$ from the secondary terminals of the associated PT and CT are applied to the relay, and the new quantities $v(t)$, $v'(t)$ compose the detected part of the relay circuit as shown in the figure. Then, the phase angular difference between $v(t)$ and $v'(t)$ is continuously compared in the phase comparator part of the relay circuit. As seen in Figure 17.1b, if the condition $\beta \leqq 90°$ is satisfied, the vector relation between $V(t)$ and $I(t)$ must be within the circle. In other words, the relay which is designed not to operate at $\beta \geqq 90°$ but at $\beta \leqq 90°$ gives V–I characteristics of the circle as shown in figure. The relay would be operated whenever the vectors V, I satisfy the condition $\beta \leqq 90°$ in this figure.

Incidentally, all the equations of the V–I coordinates plane in Figure 17.1b can be divided by the current $\dot{I}(t)$, and the derived equations contain variables $\dot{V}/\dot{I}$, or $\dot{Z} = R + jX$. In other words, the V–I characteristics of the figure can be replaced with the R–X characteristics given in the R–X coordinates plane, and the resulting operating characteristics are also a circle with the dimensions of Ω, as shown in Figure 17.1c.

The relay installed at point O in the R–X coordinates plane will operate if the vector $\dot{V}/\dot{I}$ enters the inner zone of the circle. This means that the relay operates only when $\dot{V}/\dot{I}$ is in one direction and within a fixed length value (distance). The above explanation is the essence of DZ-Ry.

The design details of these relays, manufactured using advanced technology as mentioned above, are omitted in this book. Nevertheless, the relays have to detect faults exactly and within 15–40 ms delay, and incorrect operation cannot be allowed under serious transient voltage and current conditions probably with badly d.c. or harmonic superposed waveform distortion. The relays have to detect the

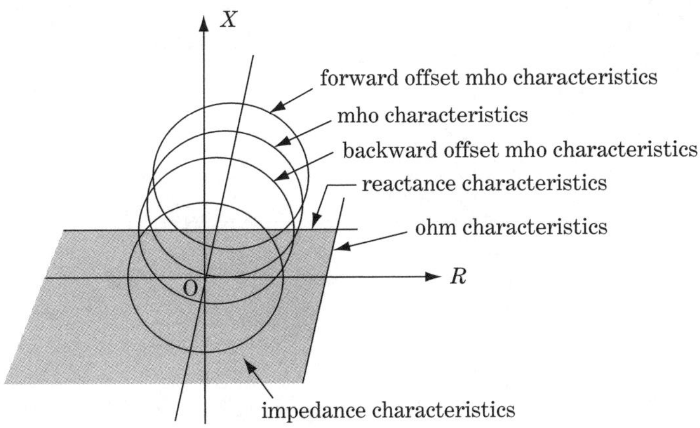

Figure 17.2 Directional distance relays, variation by operational characteristics

direction and distance of the fault even if the voltage V of V/I is almost zero during the fault at a close point, or even if some errors in the PT or CT exist, for example.

17.3 Impedance Locus in *R–X* Coordinates in Case of a Fault (under No-load Condition)

17.3.1 Operation of DZ(S)-Ry for phase b–c line-to-line fault (2ϕS)

Referring to Figure 17.3a, the behaviour of the relay installed for line 1 at the terminal m is examined, where a phase b–c line-to-line fault occurs at point f on line 1.

The equivalent circuit of this case is shown in Figure 17.3b. The current at point f is supplied from both sides of the terminal at m and n. The following equations are derived from the equivalent circuit,

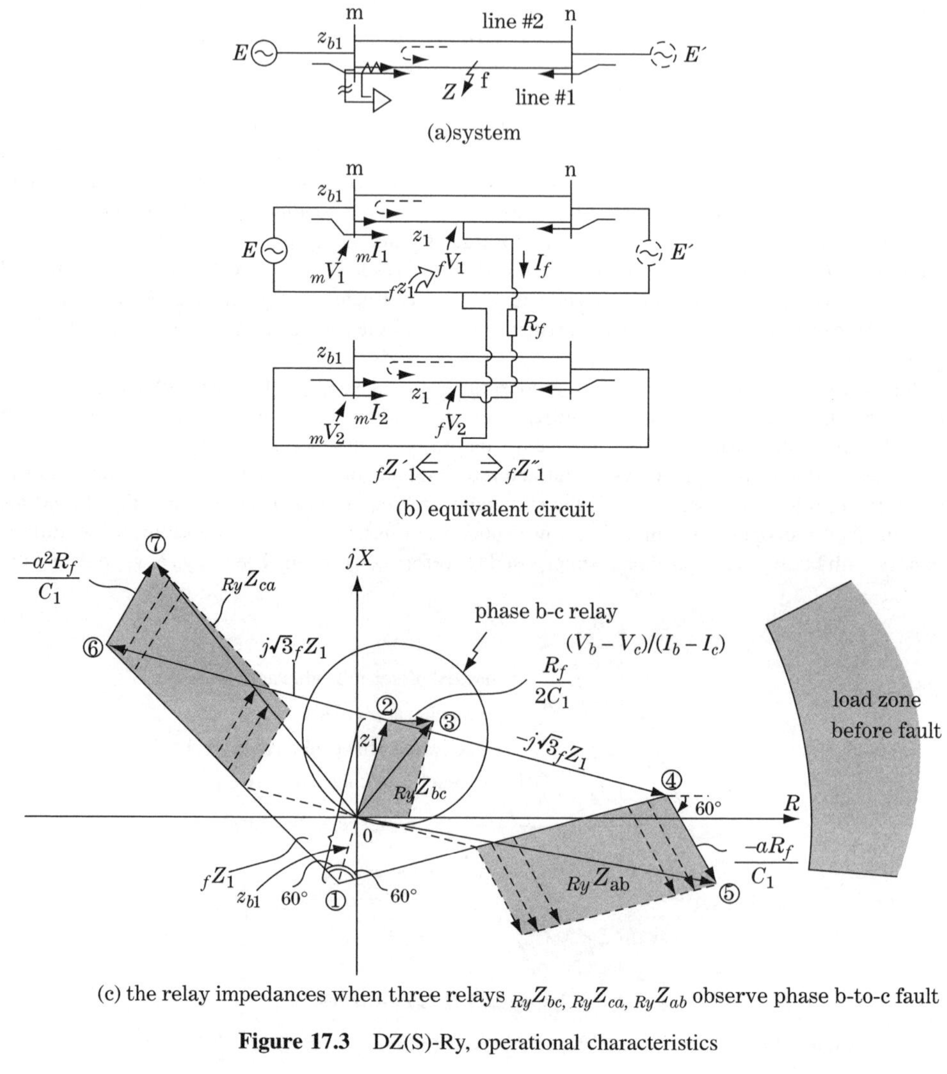

Figure 17.3 DZ(S)-Ry, operational characteristics

where we assume that the positive impedance and the negative impedance are the same (this assumption can be justified for most cases, because the difference between the positive- and negative-sequence reactances of the generator is diluted by the line reactances $X_{\text{line1}} = X_{\text{line2}}$):

$$\begin{aligned}
{}_fV_1 - {}_fV_2 &= R_f \cdot I_f \\
{}_fV_2 &= {}_fZ_1 \cdot I_f \\
{}_fV_1 &= {}_mV_1 - z_1 \cdot {}_mI_1 \\
{}_fV_2 &= {}_mV_2 - z_1 \cdot {}_mI_2 \\
{}_mI_1 &= C_1 I_f,\ {}_mI_2 = -C_2 I_f
\end{aligned} \tag{17.4}$$

The load current is zero. Then

$${}_mI_1 = -{}_mI_2 = C_1 \cdot I_f,\ C_1 = C_2$$

where

z_1: positive-sequence impedance from relay point m to fault point f
${}_fZ_1$: positive-(and negative-)sequence impedance looking into the circuit from point f
R_f: arc resistance
C_1: the ratio of ${}_mI_1$ to I_f in the positive-sequence circuit
C_2: the ratio of ${}_mI_2$ to I_f in the negative-sequence circuit
(C_1, C_2 are the vector coefficients of $(0-1)\angle\alpha$, where α is almost $0°$)

Since the positive- and negative-sequence reactances are equal, and the no-load condition is assumed, then $C_1 = C_2$ and ${}_mI_1 = -{}_mI_2$ in this case.

Now, the delta quantities $\Delta V = V_a - V_b$, $\Delta I = I_a - I_b$ are applied to the relay at point m, and the relays look for the fault by measuring $\Delta V/\Delta I = (V_a - V_b)/(I_a - I_b)$.

The measured impedances of each delta-phase relay ${}_{Ry}Z_{ab}, {}_{Ry}Z_{bc}, {}_{Ry}Z_{ca}$ are calculated as follows:

The impedance of the DZ-Ry for phase b-c line-to-line detection: ${}_{Ry}Z_{bc}$

$$\begin{aligned}
{}_{Ry}Z_{bc} &= \frac{{}_mV_b - {}_mV_c}{{}_mI_b - {}_mI_c} = \frac{{}_mV_1 - {}_mV_2}{{}_mI_1 - {}_mI_2} = \frac{({}_fV_1 + z_1 \cdot {}_mI_1) - ({}_fV_2 + z_1 \cdot {}_mI_2)}{{}_mI_1 - {}_mI_2} \\
&= z_1 + \frac{{}_fV_1 - {}_fV_2}{2 \cdot {}_mI_1} = z_1 + \frac{R_f I_f}{2C_1 I_f} = z_1 + \frac{R_f}{2C_1}
\end{aligned} \tag{17.5}$$

The impedance of the DZ-Ry for phase a-b line-to-line detection: ${}_{Ry}Z_{ab}$

$$\begin{aligned}
{}_{Ry}Z_{ab} &= \frac{{}_mV_a - {}_mV_b}{{}_mI_a - {}_mI_b} = \frac{{}_mV_1 - a \cdot {}_mV_2}{{}_mI_1 - a \cdot {}_mI_2} = \frac{({}_fV_1 + z_1 \cdot {}_mV_1) - a({}_fV_2 + z_1 \cdot {}_mI_2)}{{}_mI_1 - a \cdot {}_mI_2} \\
&= z_1 + \frac{{}_fV_1 - a \cdot {}_fV_2}{{}_mI_1 - a \cdot {}_mI_2} = z_1 + \frac{(1-a)\,{}_fV_2 + R_f I_f}{(1+a)\,{}_mI_1} \\
&= z_1 + \frac{(1-a)\,{}_fZ_1 I_f + R_f I_f}{-a^2 C_1 I_f} = z_1 + \frac{(a^2-a)\,{}_fZ_1 - a \cdot R_f}{C_1} \\
&= z_1 + \frac{-j\sqrt{3}\,{}_fZ_1}{C_1} - \frac{aR_f}{C_1}
\end{aligned} \tag{17.6}$$

The impedance of the DZ-Ry for phase c-a line-to-line detection: ${}_{Ry}Z_{ca}$

$$
\begin{aligned}
{}_{Ry}Z_{ca} = \frac{{}_mV_c - {}_mV_a}{{}_mI_c - {}_mI_a} &= z_1 + \frac{(a-a^2)\,{}_fZ_1 - a^2 \cdot R_f}{C_1} \\
&= z_1 + \frac{j\sqrt{3}\,{}_fZ_1}{C_1} - \frac{a^2 \cdot R_f}{C_1}
\end{aligned} \tag{17.7}
$$

Figure 17.3 shows the vector diagrams of ${}_{Ry}Z_{ab}, {}_{Ry}Z_{bc}, {}_{Ry}Z_{ca}$ that are derived in the R–X coordinates plane by the above equations and explained below.

Relay ${}_{Ry}Z_{bc}$: Equation 17.5. The relay ${}_{Ry}Z_{bc}$ at point O sees the impedance below as given by Equation 17.5 and Figure 17.3c.

$$\overline{0②}(z_1) + \overline{②③}(R_f/2C_1) = \overline{0③}$$

The length of $\overline{②③}$ is not long because R_f has a rather small value by nature, although C_1 varies as the value $C_1 = 0 - 1$.

Relay ${}_{Ry}Z_{ab}$: Equation 17.6. Referring to the equivalent circuit in Figure 17.3b, straight line $\overline{0②}$ (z_1: line impedance between f and m) and $\overline{②③}$ (the back impedance at m) can be written first in the R–X coordinates plane as shown in Figure 17.3. The straight line $\overline{①②}$ is of course $z_1 + z_{b1}$.

Now, we examine the special case of $C_1 = 1.0$, which means that all the current I_f at fault point f is supplied from point m through line 1 and is not supplied from point n. This means ${}_fZ_1'' = \infty$ in the positive-sequence equivalent circuit. Accordingly, the positive-sequence impedance looking into the circuit at point f (${}_fZ_1$) is equal to $z_1 + z_{b1}$. In other words, ${}_fZ_1 = z_1 + z_{b1}$ under the condition of $C_1 = 1.0$.

Next, the straight line $\overline{①②}$ can be multiplied by $-j\sqrt{3}$, and $-j\sqrt{3}\,{}_fZ_1$ is obtained as the new straight line $\overline{②④}$. This new line $\overline{②④}$ corresponds to the second term on the right in Equation 17.6 where $C_1 = 1$. The straight line $\overline{②③}$ corresponds to $R_f/2C_1$ under $C_1 = 1.0$, which can be multiplied by $-2a$ (turning 120° clockwise and doubling the length) to obtain the new line $\overline{④⑤}$. The straight line $\overline{④⑤}$ is $-aR_f/C_1$ and it corresponds to the third term on the right of Equation 17.6.

As a result, the derived straight line $\overline{0⑤}$ satisfies the relation below under the condition of $C_1 = 1$, whose right side is the same as that of Equation 17.6 under $C_1 = 1$:

$$\overline{0⑤}\,({}_{Ry}Z_{ab}) = \overline{①②}\,(z_1) + \overline{②④}\,(-j\sqrt{3}\,{}_fZ_1/C_1) + \overline{④⑤}\,(-aR_f/C_1)$$

In other words, the straight line 0⑤ in Figure 17.3c gives ${}_{Ry}Z_{ab}$ of Equation 17.6 under the condition of $C_1 = 1$.

The ratio C_1 can be varied between 0 and 1 depending on the fault location and the power source condition of both terminals m and n. The magnitudes of ${}_fZ_1/C_1$ and R_f/C_1 therefore have to be modified, so that the length of straight lines $\overline{②④}$ (then $\overline{①0}$) and $\overline{④⑤}$ may be expanded.

Relay ${}_{Ry}Z_{ca}$: Equation 17.7. In the same way, the relay impedance ${}_{Ry}Z_{ca}$ corresponding to Equation 17.7 can be drawn as shown in Figure 17.3c.

Now, if the phase b–c line-to-line fault occurs at point f on line 1 (where the line impedance between m and f is z_1), the DZ-Ry of each phase at the installed point m measures its own impedance zone within each parallelogram in Figure 17.3. Note that the relay ${}_{Ry}Z_{bc}$ measures the length of z_1 $\overline{0②}$ (the distance between m and f) exactly regardless of the magnitudes of R_f and C_1 (where the length of $\overline{②③}$ is small in comparison), and, accordingly, the relay will operate if the measured length z_1 is shorter than the diameter of the circle (the preset value). The relay does not detect faults beyond the previous set distance in the forward direction, nor faults in the backward direction.

Figure 17.3 also explains why the relay ${}_{Ry}Z_{ca}$ as well as ${}_{Ry}Z_{ab}$ do not detect the phase b–c line-to-line fault.

Incidentally, the magnitude of $P + jQ$ before the fault is within the capacity limit, while the relation between (P, Q) and (R, X) is defined by Equation 17.3. Then, the operating point (R, X) before the fault must exist on the furthest area from point 0 (0, 0). In other words, before the fault, the three relays are observing the load (R, X) in the furthest right area for the load flow in the point m to n direction (or farthest left area for the load flow in the opposite direction).

Whenever a phase b–c line-to-line fault occurs, the locus point (R, X) for each relay moves from the above-mentioned load area to the fault area of the parallelogram in Figure 17.3c, and at least the relay $_{Ry}Z_{bc}$ would detect the fault as a result, although $_{Ry}Z_{ca}$ and $_{Ry}Z_{ab}$ might not.

This theoretical visualization based on the *R–X* coordinates plane was first presented by **A. R. van C. Warrington** in his famous *AIEE Transactions* paper 'Performance of Distance Relays' in 1949. Following this seminal work, analytical methods of faulting phenomena and relay operation against various complicated phase-unbalanced faults were established and the technology of high-speed protection for large networks progressed remarkably. The faulting phenomena as well as the behaviour of the protective equipment based on directional distance relays cannot be appropriately described by any other method today, except for Warrington's analytical method.

17.3.2 Response of DZ(G)-Ry to phase a line-to-ground fault (1 ϕ G)

The equivalent circuit for a phase a line-to-ground fault (1 ϕ G) is given in Figure 17.4. In this case of DZ(G)-Ry detecting a phase-to-ground fault, the relay cannot accurately measure the line-to-ground fault by the simple introduction of V_a/I_a. Special countermeasures are required, because the zero-sequence circuit has very different impedance constants to that of the positive-sequence circuit, and, furthermore, a mutual reactance exists between parallel circuits of double circuit transmission lines.

From the equivalent circuit of Figure 17.4

$$\left.\begin{aligned}
&{}_mV_1 - {}_fV_1 = z_1 \cdot {}_mI_1 \\
&{}_mV_2 - {}_fV_2 = z_1 \cdot {}_mI_2 \\
&{}_mV_0 - {}_fV_0 = z_0 \cdot {}_mI_0 + Z_{0M} \cdot {}_mI_0' \\
&{}_fV_1 + {}_fV_2 + {}_fV_0 = 3R_f I_f \\
&{}_mI_1 = {}_mI_2 = C_1 I_f \ \text{(assuming load flow is zero)} \\
&{}_mI_0 = C_0 I_f \\
&{}_fV_2 = -{}_fZ_1 I_f, \qquad {}_fV_0 = -{}_fZ_0 \cdot I_f
\end{aligned}\right\} \tag{17.8}$$

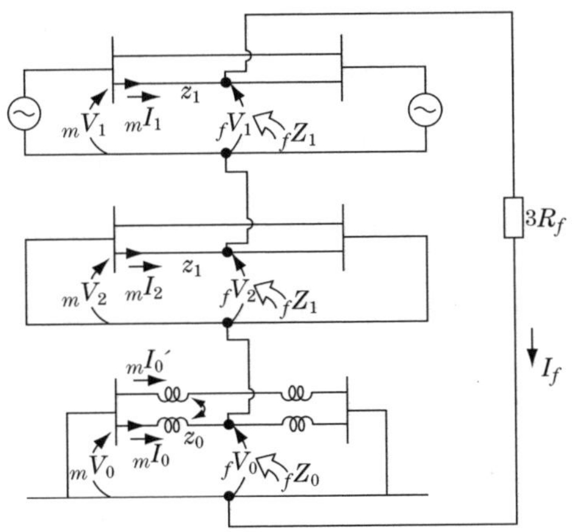

Figure 17.4 Equivalent circuit of phase a line-to-ground fault

where z_{0M}: zero-sequence mutual reactance between circuit 1 and 2 (the same as Z_{0M} in Figure 2.6 and Equation 2.20a)

$_f Z_1$, $_f Z_0$,: impedance at the fault point f looking into the circuit

Relay $_{Ry}Z_a$: Applying the above equations,

$$\left.\begin{aligned} {}_mV_a &= {}_mV_1 + {}_mV_2 + {}_mV_0 = z_1 \cdot {}_mI_1 + z_1 \cdot {}_mI_2 + z_0 \cdot {}_mI_0 + z_{0M} \cdot {}_mI'_0 + 3R_f I_f \\ &= z_1({}_mI_1 + {}_mI_2 + {}_mI_0) + (z_0 - z_1)_m I_0 + z_{0M} \cdot {}_mI'_0 + 3R_f I_f \\ &= z_1 \cdot {}_mI_a + (z_0 - z_1)_m I_0 + z_{0M} \cdot {}_mI'_0 + 3R_f I_f \end{aligned}\right\} \tag{17.9a}$$

This equation can be modified as follows:

$$\left.\begin{aligned} &\frac{{}_mV_a - (z_0 - z_1)_m I_0 - z_{0M} \cdot {}_mI'_0}{{}_mI_a} = z_1 + 3R_f \cdot \frac{I_f}{{}_mI_a} \\ &\text{where } {}_mI_a = (2C_1 + C_0)I_f \end{aligned}\right\} \tag{17.9b}$$

Equation 17.9b shows that the relay can measure the line-to-ground fault exactly by applying the composed voltages of the numerator on the left of Equation 17.9b to the relay instead of V_a. This is possible because the zero-sequence currents $_mI_0$, $_mI'_0$ of lines 1 and 2 are available in the same substation through the composition of the three-phase currents I_a, I_b, I_c and I'_a, I'_b, I'_c.

For DZ(G)-Ry, the measuring equation of the phase-to-ground relay is

$$_{Ry}Z_w = \frac{{}_mV_w - (z_0 - z_1)_m I_0 - z_{0M} \cdot {}_mI'_0}{{}_mI_w}$$

voltage quantity: $\{{}_mV_w - (z_0 - z_1)_m I_0 - z_{0M} \cdot {}_mI'_0\}$

current quantity: $\{{}_mI_w\}$

where $w = a, b, c$

$_mI_0 = (1/3)({}_mI_a + {}_mI_b + {}_mI_c)$, $\quad {}_mI'_0 = (1/3)({}_mI'_a + {}_mI'_b + {}_mI'_c)$

$-(z_0 - z_1)_m I_0$: zero-sequence compensation term of line 1

$-z_{0M} \cdot {}_mI'_0$: zero-sequence compensation term of the parallel circuit line 2

$_mI_0$, $_mI'_0$ can actually be derived from the residual circuit of three CT secondary circuits of composed values $1/3(I_a + I_b + I_c)$ and $1/3\,(I'_a + I'_b + I'_c)$

(17.10)

Relay $_{Ry}Z_a$: Utilizing the above equations,

$$_{Ry}Z_a = z_1 + 3R_f \cdot \frac{I_f}{{}_mI_a} = z_1 + \frac{3R_f}{2C_1 + C_0} \tag{17.11}$$

The relay $_{Ry}Z_a$ can measure (detect) phase a to grounding faults accurately by the countermeasure of zero-sequence current compensation.

Relay $_{Ry}Z_b$: The voltage and current quantities for this relay are calculated below under the conditions of Equation 17.8:

$$\left.\begin{aligned} &\text{voltage quantity: } \{{}_mV_b - (z_0 - z_1)_m I_0 - z_{0M} \cdot {}_mI'_0\} = ({}_fV_0 + z_0 \cdot {}_mI_0 + z_{0M} \cdot {}_mI'_0) \\ &\qquad + a^2({}_fV_1 + z_1 \cdot {}_mI_1) + a({}_fV_2 + z_1 \cdot {}_mI_2) - (z_0 - z_1)_m I_0 - z_{0M} \cdot {}_mI'_0 \\ &\qquad = z_1 \cdot {}_mI_b + {}_fV_b \\ &\text{current quantity: } {}_mI_b \end{aligned}\right\}$$

(17.12a)

Then, the impedance for the phase b relay observing the phase a line-to-ground fault is

$$ {}_{Ry}Z_b = z_1 + \frac{{}_fV_b}{{}_mI_b} \tag{17.12b}$$

Although this equation looks to be in good order, we need to examine the second term ${}_fV_b/{}_mI_b$ on the right:

$$\left.\begin{aligned} {}_fV_b &= {}_fV_0 + a^2\cdot {}_fV_1 + a\cdot {}_fV_2 = {}_fV_0 + a^2(3R_f I_f - {}_fV_2 - {}_fV_0) + a\cdot {}_fV_2 \\ &= (a-a^2)\,{}_fV_2 + (1-a^2)\,{}_fV_0 + 3a^2\cdot R_f I_f \\ &= \{(a^2-a)\,{}_fZ_1 + (a^2-1)\,{}_fZ_0 + 3a^2\cdot R_f\}I_f \\ {}_mI_b &= {}_mI_0 + a^2\cdot {}_mI_1 + a\cdot {}_mI_2 = {}_mI_0 - {}_mI_1 = -(C_1 - C_0)I_f \end{aligned}\right\} \tag{17.13a}$$

$$\left.\begin{aligned} \therefore\ {}_{Ry}Z_b &= z_1 + \frac{(a-a^2)\,{}_fZ_1 + (1-a^2)\,{}_fZ_0 - 3a^2\cdot R_f}{C_1 - C_0} \\ &= z_1 + \frac{j\sqrt{3}\,{}_fZ_1 - j\sqrt{3}\,a\cdot {}_fZ_0 - 3\,a^2\cdot R_f}{C_1 - C_0} \end{aligned}\right\} \tag{17.13b}$$

Relay ${}_{Ry}Z_c$

The equation below is derived in the same way:

$$\left.\begin{aligned} {}_{Ry}Z_c &= z_1 + \frac{(a^2-a)\,{}_fZ_1 + (1-a)\,{}_fZ_0 - 3a\cdot R_f}{C_1 - C_0} \\ &= z_1 + \frac{-j\sqrt{3}\,{}_fZ_1 + j\sqrt{3}a^2\cdot {}_fZ_0 - 3a\cdot R_f}{C_1 - C_0} \end{aligned}\right\} \tag{17.14}$$

The Equations 17.11, 17.13 and 17.14 specify the behaviour of relays ${}_{Ry}Z_a, {}_{Ry}Z_b, {}_{Ry}Z_c$ for the phase-a line-to-ground fault. Equation 17.11 tells us that the phase a relay ${}_{Ry}Z_a$ measures z_1 (namely, the direction and distance of the fault) appropriately. The operating zone of ${}_{Ry}Z_a$ can be drawn in the R–X coordinates plane by using C_1, C_0 and R_f as parameters. On the other hand, ${}_{Ry}Z_b$, ${}_{Ry}Z_c$ include ${}_fZ_0$ as well as C_1, C_2, R_f, so that they can be drawn as a point in the R–X coordinates plane for the case when ${}_fZ_0$ is given as the individual fault condition.

For the special case of $C_1 \doteqdot C_0$ the relays will encounter the following impedances:

$$\left.\begin{aligned} {}_{Ry}Z_a &= z_1 + \frac{R_f}{C_1} \\ {}_{Ry}Z_b &= z_1 + \infty \\ {}_{Ry}Z_c &= z_1 + \infty \end{aligned}\right\} \tag{17.15}$$

The phase-a line-to-ground fault can be accurately detected by ${}_{Ry}Z_a$, while ${}_{Ry}Z_b$, ${}_{Ry}Z_c$ do not operate in this case.

Table 17.1 shows the impedances that the six relays see for line-to-line faults and line-to-ground faults under various fault modes. The table is based on that given by A. R. van C. Warrington in the previously quoted AIEE paper. Each equation in the table can be derived, although this takes time.

17.3.2.1 DZ-Ry for high-impedance neutral grounded system

Finally in this section, an important fact should be pointed out: that is, the DZ-Ry for line-to-ground fault detection cannot be applied to a power system with non-effective (through resistive or reactive)

Table 17.1 Impedances observed by directional distance relays for various short circuits

Relays	3ϕS, 3ϕG	2ϕS (phase b to c)	1ϕG (phase a to ground)	2ϕG (phase b, c to ground)
${}_{Ry}Z_{ab}$	$z_1+\frac{R_f}{C_1}$	$z_1+\frac{(a^2-a)\,{}_fZ_1-aR_f}{C_1}$	$z_1+\frac{(a-a^2)\,{}_fZ_1+(1-a^2)({}_fZ_0+3R_f)}{3C_1}$	$z_1+\frac{r_f}{C_1}+\frac{3({}_fZ_1+r_f)({}_fZ_0+r_f+3R_f)}{C_1\{(1-a^2)({}_fZ+r_f)+(a-a^2)({}_fZ_0+r_f+3R_f)\}}$
${}_{Ry}Z_{bc}$	$z_1+\frac{R_f}{C_1}$	$z_1+\frac{R_f}{2C_1}$	∞	$z_1+\frac{r_f}{C_1}$
${}_{Ry}Z_{ca}$	$z_1+\frac{R_f}{C_1}$	$z_1+\frac{(a-a^2)\,{}_fZ_1-a^2R_f}{C_1}$	$z_1+\frac{(a^2-a)\,{}_fZ_1+(1-a)({}_fZ_0+3R_f)}{3C_1}$	$z_1+\frac{r_f}{C_1}+\frac{3({}_fZ_1+r_f)({}_fZ_0+r_f+3R_f)}{C_1\{(a-1)({}_fZ_1+r_f)+(a-a^2)({}_fZ_0+r_f+3R_f)\}}$
${}_{Ry}Z_a$	$z_1+\frac{R_f}{C}$	∞	$z_1+\frac{3R_f}{2C_1+C_0}$	$z_1+\frac{3\,{}_fZ_0+r_f+2R_f}{C_1-C_0}$
${}_{Ry}Z_b$	$z_1+\frac{R_f}{C_1}$	$z_1+\frac{{}_fZ_1-a^2R_f}{(a-a^2)C_1}$	$z_1+\frac{(a-a^2)\,{}_fZ_1+(1-a^2)\,{}_fZ_0-3a^2R_f}{C_1-C_0}$	$z_1+\frac{r_f\{a^2({}_fZ_1+r_f)+(a^2-a)({}_fZ_0+r_f+3R_f)\}-({}_fZ_1+r_f)(r_f+3R_f)}{C_1\{a^2({}_fZ_1+r_f)+(a^2-a)({}_fZ_0+r_f+3R_f)\}-C_0({}_fZ_1+r_f)}$
${}_{Ry}Z_c$	$z_1+\frac{R_f}{C_1}$	$z_1+\frac{{}_fZ_1-aR_f}{(a^2-a)C_1}$	$z_1+\frac{(a^2-a)\,{}_fZ_1+(1-a)\,{}_fZ_0-3aR_f}{C_1-C_0}$	$z_1+\frac{r_f\{a({}_fZ_1+r_f)+(a-a^2)({}_fZ_0+r_f+3R_f)\}-({}_fZ_1+r_f)(r_f+3R_f)}{C_1\{a({}_fZ_1+r_f)+(a-a^2)({}_fZ_0+r_f+3R_f)\}-C_0({}_fZ_1+r_f)}$

Note: z_1: positive-sequence line reactance from the relay to the fault point

${}_fZ_1$, ${}_fZ_0$: positive/zero-sequence impedances looking into the network at fault point

$C_1(C_0)$: the ratio of ${}_mI_1$ to I_f (complex number)

$C_1={}_mI_1/I_f$, $C_0={}_mI_0/I_f$

R_f, r_f : arc resistance (R_f, r in the case of 2ϕG corresponds to R, r in Table 3.1b 8)

neutral grounding. This is due to the following reasons. The ratio of $C_0 = {}_mI_0/I_f$ in Equation 17.8 would become very small and furthermore the angle might differ by almost 90° from that of C_1. Accordingly, in Equation 17.11, the second term on the right rather than the first term z_1 becomes dominant. Although this is a disadvantage of this relay in such fault detection, a 'grounding directional relay without the function of fault distance measurement' can be adopted (the explanation is omitted in this book).

17.4 Impedance Locus under Normal States and Step-out Condition

17.4.1 R–X locus under stable and unstable conditions

Figure 17.5a shows the power system we examine here, which operates under three-phase-balanced conditions. The induced voltages at points s and r are given by $\dot{e}_s$ and $\dot{e}_r$, and the electrical angular displacement is δ ($\dot{e}_s$ lags $\dot{e}_r$ by $\delta°$).

We define the ratio of voltage and current $\dot{v}/\dot{i}$ at each point s, m, n, r of the system expressed in symbols as ${}_s\dot{Z}, {}_m\dot{Z}, {}_n\dot{Z}, {}_r\dot{Z}$, namely

$$\left.{}_s\dot{Z} = \frac{\dot{e}_s}{\dot{i}},\ {}_m\dot{Z} = \frac{\dot{v}_m}{\dot{i}},\ {}_n\dot{Z} = \frac{\dot{v}_n}{\dot{i}},\ {}_r\dot{Z} = \frac{\dot{e}_r}{\dot{i}}\right\} \tag{17.16}$$

Between point m and point n,

$$\left.\begin{aligned} \dot{v}_m - \dot{v}_n &= \dot{z}_l \cdot \dot{i} \\ \therefore\ {}_m\dot{Z} - {}_n\dot{Z} &= \dot{z}_l \end{aligned}\right\} \begin{aligned} &\dot{z}_s, \dot{z}_l, \text{ etc.: circuit impedances} \\ &{}_m\dot{Z}, {}_n\dot{Z}, \text{ etc.: the impedances each relay observes} \end{aligned} \tag{17.17}$$

For the total system from point s to point r,

$$\left.\begin{aligned} &\left.\begin{aligned} \dot{e}_s &= \dot{e}_r + \dot{i}(\dot{z}_r + \dot{z}_l + \dot{z}_s) \\ \therefore\ {}_s\dot{Z} &= {}_r\dot{Z} + (\dot{z}_r + \dot{z}_l + \dot{z}_s) \end{aligned}\right\} \quad ① \\ &{}_m\dot{Z} = {}_s\dot{Z} - \dot{z}_s = {}_r\dot{Z} + (\dot{z}_r + \dot{z}_l) \quad ② \end{aligned}\right\} \tag{17.18}$$

Equations 17.17 and 17.18 are shown in Figure 17.5b as a vector diagram of impedances in the R–X coordinates plane. The equations and the diagram shows mutual relations of the impedances which the relay at each point sees. The original point (0, 0) of the coordinates can be selected arbitrarily and the point m is eventually selected in the figure. The relay ${}_m\dot{Z}$ at point m sees $\dot{z}_1 + \dot{z}_r$ in the forward direction and $\dot{z}_s$ in the backward direction. The relay ${}_n\dot{Z}$ at point n sees $\dot{z}_r$ in the forward direction and $\dot{z}_s + \dot{z}_l$ in the backward direction.

The induced voltages at point s and r are

$$\left.\begin{aligned} \dot{e}_s &= E_s e^{j\omega t} \\ \dot{e}_r &= E_r e^{j(\omega t + \delta)} \end{aligned}\right\} \tag{17.19}$$

The current through the line is

$$\dot{i} = \frac{E_s e^{j\omega t} - E_r e^{j(wt+\delta)}}{\dot{z}_r + \dot{z}_l + \dot{z}_s} = \frac{E_s}{\dot{z}_r + \dot{z}_l + \dot{z}_s}(1 - ke^{j\delta})e^{j\omega t},\ k = \frac{E_r}{E_s} \tag{17.20}$$

The impedances of generators s and r are contained in $\dot{z}_s$ and $\dot{z}_r$ respectively.

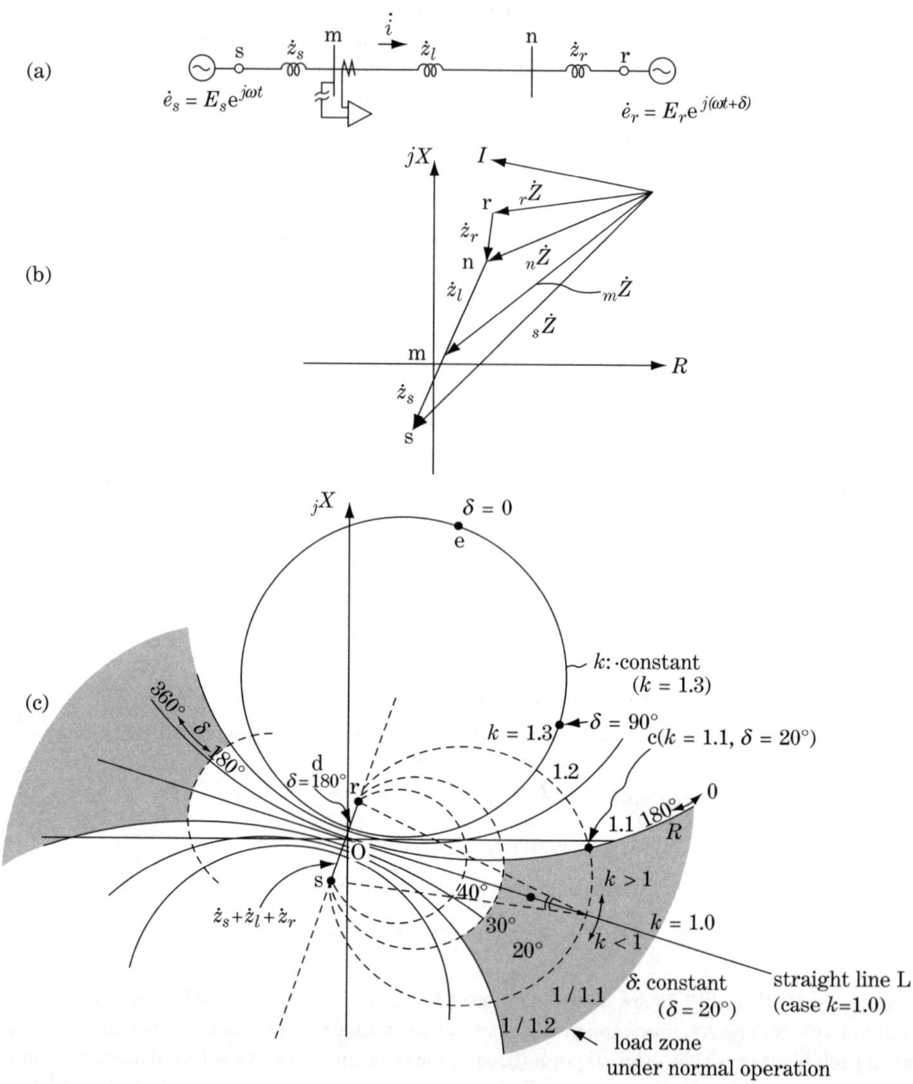

Figure 17.5 Impedance locus under load/step-out conditions (three-phase-balanced condition)

Accordingly, the impedances seen by the relays at each point under normal load conditions (without fault) are

$$\left.\begin{aligned}
&\text{point s: } {}_s\dot{Z} = \frac{\dot{e}_s}{\dot{i}} = (\dot{z}_r + \dot{z}_l + \dot{z}_s)\cdot\frac{1}{1-ke^{j\delta}} && ① \\
&\text{point m: } {}_m\dot{Z} = {}_s\dot{Z} - \dot{z}_s = (\dot{z}_r + \dot{z}_l + \dot{z}_s)\cdot\frac{1}{1-ke^{j\delta}} - \dot{z}_s && ② \\
&\text{where } k = \frac{E_r}{E_s}\ (k\text{: real number around } 1 \pm 0.1)
\end{aligned}\right\} \quad (17.21)$$

${}_s\dot{Z}$ of Equation ① is the impedance that the relay at s observes in the system under normal load conditions. A change in angle δ in the equation means a change in load flow or a power swing in the system. Equation ② means that the relay ${}_m\dot{Z}$ at point m observes the same phenomena at the different point m which is $\dot{z}_s$ distant from point s.

Now we try to draw Equation 17.21 as the impedance locus in the *R–X* coordinates plane by changing δ and k as parameters.

Equation 17.21① can be modified as follows:

$$\left.\begin{aligned} {}_s\dot{Z} &= \frac{\dot{A}}{1 - ke^{j\delta}} \\ \therefore\quad {}_s\dot{Z} - {}_s\dot{Z}\cdot ke^{j\delta} &= \dot{A} \\ \text{where}\, \dot{A} = \dot{z}_r + \dot{z}_l &+ \dot{z}_s\ (\dot{A}\ \text{is given as constant.}) \end{aligned}\right\} \tag{17.22}$$

From the equation the following conclusions are derived:

- ${}_s\dot{Z}$ is given as a circle locus by changing k under fixed δ (the k-circles).
- ${}_s\dot{Z}$ is also given as a circle locus by changing δ from 0° to 360°under fixed k (the δ-circles).

In order to prove that the ${}_s\dot{Z}$ become circles and to find the circle locus, we need to adopt a descriptive geometry approach, which is explained in detail in Supplement 1.

The impedance locus of ${}_s\dot{Z}$ by changing parameter k (the k-circles) and by changing parameter δ (the δ-circles) is shown in Figure 17.5c.

An arbitrary operating condition specified by k and δ can be written as the intersecting point of the circles ① and ②. As an example, point c ($k = 1.1, \delta = 20°$) in the figure is the intersecting point of circle ① ($k = 1.1$) and circle ② ($\delta = 20°$). The operating point c exists far from the original point (0, 0) in this case, because circle ① as well as circle ② are large at these values of k and δ.

It is clear that k in Equation 17.22 is a real number of around 0.9–1.1 and δ is about 0° to $\pm$ 40° under normal operating conditions.

17.4.1.1 The circle locus by changing k under fixed δ (the k-circles)

Three circles of this category for variable k under fixed $\delta = 20°, 30°, 40°$ are drawn in Figure 17.5c. To draw a circle, the circle is given as a circular arc with vertical angle $\delta°$, on a chord of straight line r–s ($\dot{z}_{rs}$ is the total impedance of the system).

Accordingly, this circular arc locus can be drawn rather easily in the *R–X* plane. The locus with $-\delta$ (i.e. inverse power flow) is given as the symmetry locus of $+\delta$ and exists on the left side of the plane. The locus exists far from the chord r–s under the condition of $\delta = 0°$ to $\pm$ 30°, but it rapidly approaches the chord r–s if δ is larger (i.e. heavy load operation or step-out).

17.4.1.2 The circle locus by changing k from 0° to 360° under fixed δ (the δ-circles)

The denominator of Equation 17.22 $(1 - e^{j\delta})$ gives a circular locus by changing δ from 0° to 360° under fixed k. The inverse term $1/(1 - e^{j\delta})$ also gives a circular locus by changing δ from 0° to 360° under fixed k. However, the proof and the drawing method are a little complicated and explained in detail in Supplement 1.

Summarizing the conclusion of Supplement 1 here, the circle is drawn as follows:

point O: the mid point of straight line r–s

line L: straight line meeting point O at a right angle

the centre point of the circle: the point which is $\{1/(k^2 - 1)\}\dot{z}_{rs}$ distant from point r on the extended straight line r–s the far diameter point: the point which is $\{1/(k-1)\}\dot{z}_{rs}$ distant from point r on the extended straight line r–s (the point e for $\delta = 0°$ in the figure)

the close diameter point: the point which is $\{1/(k+1)\}\dot{z}_{rs}$ distant from point r on the extended straight line r–s (the point d for $\delta = 180°$ in the figure).

The point O is written at point (0, 0) in the figure. The circle is quite large under the condition $k = 1.0 \pm 0.1$.

17.4.1.3 The existing area of the impedance locus under normal load conditions

The shading in Figure 17.5 shows the existing area of the impedance locus under the normal load condition of $k = 0.9 - 1.1$, $\delta = 0°$ to $\pm 30°$. This zone is of course far beyond the set range of the relays at points s, m, n and r.

17.4.1.4 The impedance locus under power-swing or step-out conditions

The system is now operating under normal conditions with fixed k (excitation E_s, E_r are constant) and δ around +20°. If the load flow were suddenly and largely increased, δ would be increased and the system would be caused to step out. As the generator r slips out, δ is increased as $\delta = 20° \rightarrow 60° \rightarrow 90° \rightarrow 180° \rightarrow 270° \rightarrow 360° \rightarrow 450° \rightarrow \cdots$. This phenomenon means that the impedance locus turns in a circle of category ②. Assuming constant k and a slipping speed of $360°/10$ sec, the locus is a circle around 10 sec.

With an increase of $\delta = 20° \rightarrow 60° \rightarrow 90°$, the locus quickly comes nearer to the point O and before long crosses the straight line r–s at $\delta = 180°$ around point O. Therefore, most (perhaps all) of the relays installed at points s, m, n and r would detect the locus within each operating zone and would be caused to operate. In other words, without reasonable countermeasures, the DZ-Rys installed at most of the stations in the network would operate and issue unnecessary tripping commands to a lot of breakers. Besides, it should be recalled that relays with extremely large setting circles may detect a heavy-load locus.

Most high-voltage substations today are equipped with a number of DZ-Rys per feeder and per phase for primary protection and/or back-up protection. Furthermore, most power systems have been continuing to grow into larger, complicated, meshed networks. Accordingly, coordination of the operating zone setting among lots of DZ-Rs in a total network is an important engineering matter for maintaining stable power system operation and requires practically minute work in each power system operation.

17.4.2 Step-out detection and trip-lock of DZ-Rys

DZ-Rys automatically detect large power-swing or step-out phenomena. Therefore we need to prepare particular countermeasures in protection equipment by which the DZ-Rys distinguish step-out (or large power-swing) phenomena from short-circuit faults and to avoid unnecessary breakers tripping by power swing or step-out. A typical example of the countermeasures is shown in Figure 17.6. Besides the mho-relay (M), the offset mho-relay (OM) is prepared as shown in Figures 17.6a and b. If the locus were to enter the operating zone by power swing, the relay (M) would be operated just before OM, and accordingly a tripping signal to the associated breaker would be locked in the sequence diagram (b). Whenever a fault occurs, M and OM will be forced to operate simultaneously so that the breaker tripping signal is dispatched immediately. Typical setting times are, for timer T_1, about 10 ms, and for T_2 about 1–2 sec.

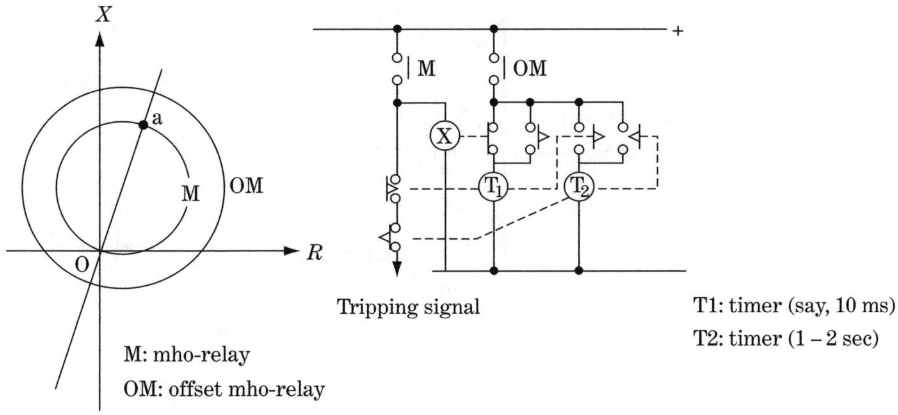

(a) (b)

Figure 17.6 Step-out detection and trip-lock sequence diagram

17.5 Impedance Locus under Faults with Load Flow Conditions

We have studied in the previous sections the impedance locus $\dot{Z}_{\text{load}}$ under normal load flow conditions as well as $\dot{Z}_{\text{fault}}$ under fault conditions with no-load flow. In this section, the impedance locus under fault conditions with load flow conditions is examined.

Large network systems today consist of many nodes and branches, so there is a great variety of power flow modes as well as fault current modes. It is not unusual, even at adjacent stations to the faulting point, that the detected fault current components are smaller than the load currents, and accordingly load flow cannot be neglected for the study of relay operation. This is the reason why we need to study the method to find the impedance locus under fault conditions with load flow.

The impedances we will look into at an arbitrary point are as follows:

$\dot{Z}_{\text{load}}$: impedance locus of load under normal conditions before the fault

$\dot{Z}_{\text{fault}}$: impedance locus of fault condition with no-load flow

$\dot{Z}_{\text{total}}$: superposed impedance locus under fault condition with load flow

The behaviour of a phase b distance relay is examined here, although the explanation can be applied to relays for any other phases.

For the phase b relay $\underline{(V_b/I_b)}$:

the impedance observed by the relay under ordinal load flow condition:

$$\dot{Z}_{\text{load}} = \frac{\dot{v}_b}{\dot{i}_{b\,\text{load}}} \qquad ①$$

the impedance observed by the relay for a fault under no-load flow condition:

$$\dot{Z}_{\text{fault}} = \frac{\dot{v}_b}{\dot{i}_{b\,\text{fault}}} \qquad ② \qquad (17.23)$$

(where $\dot{Z}_{\text{fault}}$ corresponds to $\dot{z}_r + \dot{z}_l + \dot{z}_s$ of the previous section)

the impedance observed by the relay for a fault under load flow condition:

$$\dot{Z}_{\text{total}} = \frac{\dot{v}_b}{\dot{i}_{b\,\text{total}}} = \frac{\dot{v}_b}{\dot{i}_{b\,\text{fault}} + \dot{i}_{b\,\text{load}}} \qquad ③$$

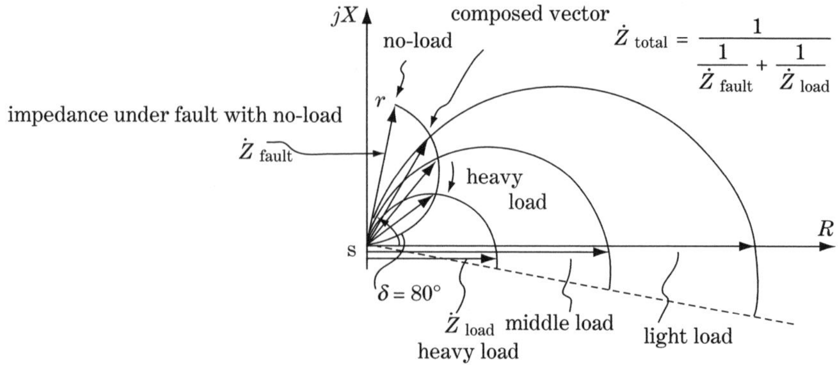

Figure 17.7 Impedance locus under fault with load flow

Substituting ①, ② into ③,

$$\dot{Z}_{total} = \frac{1}{\frac{1}{\dot{Z}_{fault}} + \frac{1}{\dot{Z}_{load}}} \tag{17.24}$$

We have already studied how $\dot{Z}_{load}$ and $\dot{Z}_{fault}$ are drawn in the *R–X* coordinates plane. Now we need to find a drawing method for $\dot{Z}_{total}$ from the already known $\dot{Z}_{load}$ and $\dot{Z}_{fault}$. The method to find complex-number vector $\dot{Z} = 1/\{1/\dot{A}) + (1/\dot{B})\}$ from given vectors *A*, *B* is detailed in Supplement 2.

Figure 17.7 shows the drawing method and the resulting impedance locus. The vectors $\dot{Z}_{fault}$ and $\dot{Z}_{load}$ (three cases from light load to heavy load are indicated) can be drawn in *R–X* coordinates at the beginning. Then, referring to Supplement 2, the vector $\dot{Z}_{total}$ can be composed as shown in Figure 17.7. The relay at point s (for example) sees the impedance of straight line s–r ($\dot{Z}_{fault}$) in the case of a fault with no load. However, if the load flow ($\dot{Z}_{load}$) is increased, the vector $\dot{Z}_{total}$ begins to fall down to the right in the figure and the magnitude diminishes rapidly, and accordingly the relays may fail to detect the fault under heavy load condition. All these things have to be taken into account in the application (selection and setting, for example) of DZ-Rys.

17.6 Loss of Excitation Detection by DZ-Rys

The severe behaviour and operational limit of generators in the weak (low-)excitation area were discussed in detail in Chapters 15 and 16. The utmost case, **loss of excitation**, would cause not only damage to the generator but also instability in the power system. Therefore if excitation of a generator is lost for some reason, the generator has to be tripped immediately. DZ-RYs fulfil the vital duty of protecting generators and preventing instability by detecting loss of excitation and commanding a generator tripping signal immediately.

17.6.1 Loss of excitation detection

The operating zone of the loss of excitation relay is set in the *R–X* plane as shown in Figure 17.8a as in common practice. That is, the operating zone in the *R–X* plane (the relay setting) is as follows:

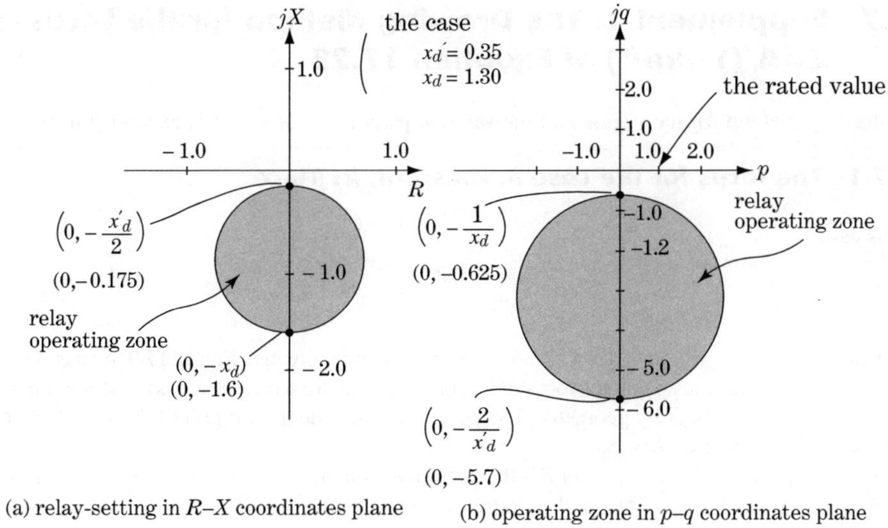

Figure 17.8 Loss of excitation relay

$$\left.\begin{array}{l}\text{diameter: straight line connecting } (0, -x_d) \text{ and } \left(0, -\dfrac{x'_d}{2}\right) \\[2ex] \text{centre of circle} \left(0, -\dfrac{x_d + \dfrac{x'_d}{2}}{2}\right), \text{ radius} \dfrac{x_d - \dfrac{x'_d}{2}}{2} \\[2ex] \text{the circle equation: } R^2 + \left\{X + \dfrac{x_d + \dfrac{x'_d}{2}}{2}\right\}^2 = \left\{\dfrac{x_d - \dfrac{x'_d}{2}}{2}\right\}^2 \end{array}\right\} \quad (17.25)$$

The above characteristics in the R–X plane can be inverse transformed to those in the p–q plane using Equation 17.3, and the resulting equation is also given as the circle below:

$$\left.\begin{array}{l}\text{diameter: straight line connecting} \left(0, -\dfrac{1}{x_d}\right) \text{and} \left(0, -\dfrac{2}{x'_d}\right) \\[2ex] \text{centre of circle} \left(0, -\dfrac{\frac{1}{x_d} + \frac{2}{x'_d}}{2}\right), \text{radius} \dfrac{\frac{2}{x'_d} - \frac{1}{x_d}}{2} \\[2ex] \text{the circle equation: } p^2 + \left\{q + \dfrac{\frac{1}{x_d} + \frac{2}{x'_d}}{2}\right\}^2 = \left\{\dfrac{\frac{2}{x'_d} - \frac{1}{x_d}}{2}\right\}^2 \end{array}\right\} \quad (17.26)$$

The loss of excitation point ($E_{fd} = 0$) is given by the point $(0, \ -1/x_d)$ as shown in Figure 16.3 in Chapter 16. Of course, the relay operating zone given in p–q coordinates by Equation 17.26 covers the loss of excitation zone described in Figure 16.3; therefore the relay setting zone from Equation 17.25 and Figure 17.8a in R–X coordinates is justified.

17.7 Supplement 1: The Drawing Method for the Locus $\dot{Z}=\dot{A}/(1-ke^{j\delta})$ of Equation 17.22

The drawing method for the vector $\dot{Z}$ in the variable parameters of k or δ is examined here.

17.7.1 The locus for the case δ: constant, k: 0 to ∞

In this case

$$\left.\begin{aligned} \dot{A} &= \dot{Z} - \dot{Z}\cdot ke^{j\delta} \\ \angle \mathrm{acb} &= \delta \end{aligned}\right\} \qquad (1)$$

Then a closed triangle with sides $\dot{A}$, $\dot{Z}$, $\dot{Z}e^{j\delta}$ can be drawn as shown in Figure 17.9. As δ is constant, then $\angle\mathrm{acb} = \delta$ is constant, so that the locus of $\dot{Z}$ is on a circular arc whose chord is the straight line a–b and the vertical angle is δ, by geometry. The point c is coincident with point b by $k = 0$, which is coincident with point a by $k = \infty$.

As a conclusion, the locus of vector $\dot{Z}$ is the circular arc on the chord A. The arc is a semi-circle for $\delta = 90°$, a short arc for $\delta = 180°$ and an extremely high arc for $\delta = 0°$

17.7.2 The locus for the case k: constant, δ: 0 to 360°

In this case, the locus of $\dot{Z}$ is also a circle, as shown in Figure 17.9. In this figure, the circle can be obtained as follows, although the proof by descriptive geometry is a little hard.

For the drawing method, the locus of $\dot{Z}$ is given by a circle whose diameter is the straight line $\overline{\mathrm{ed}}$, where points d and e are:

point d: the inner point of $\overline{\mathrm{ab}}$ = A divided in the ratio 1:k, namely $\overline{\mathrm{ad}}:\overline{\mathrm{db}}$ = 1:k

point e: the outer point of $\overline{\mathrm{ab}}$ = A divided in the ratio 1:k, namely $\overline{\mathrm{ea}}:\overline{\mathrm{eb}}$ = 1:k

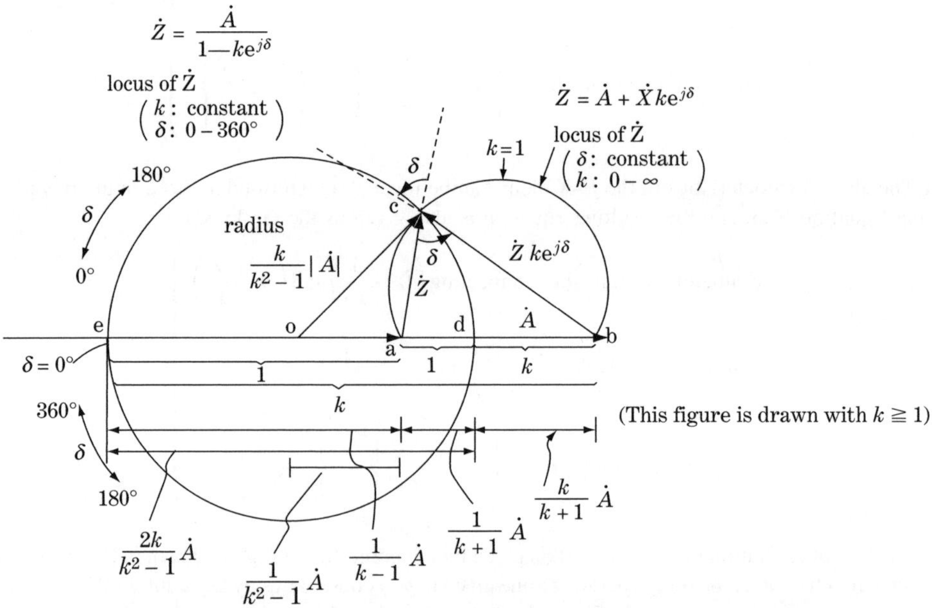

Figure 17.9 The drawing method of impedance locus $\dot{Z} = \dot{A}/(1-ke^{j\delta})$

Now, in proof of the above, the following relations exist among the points e, o, a, d, b on the same straight line:

$$\left.\begin{aligned}
&\overrightarrow{\mathrm{ad}} = \frac{1}{k+1}\dot{A},\ \overrightarrow{\mathrm{db}} = \frac{k}{k+1}\dot{A} \\
&\overrightarrow{\mathrm{ea}} : (\overrightarrow{\mathrm{ea}} + \dot{A}) = 1 : k \ \therefore\ \overrightarrow{\mathrm{ea}} = \frac{1}{k-1}\dot{A} \\
\text{diameter:}\quad &\overrightarrow{\mathrm{ed}} = \overrightarrow{\mathrm{ea}} + \overrightarrow{\mathrm{ad}} = \frac{2k}{k^2-1}\dot{A} \\
\text{radius:}\quad &\overrightarrow{\mathrm{od}} = \frac{k}{k^2-1}\dot{A} \\
&\overrightarrow{\mathrm{oa}} = \overrightarrow{\mathrm{od}} - \overrightarrow{\mathrm{ad}} = \frac{1}{k^2-1}\dot{A}
\end{aligned}\right\} \tag{2}$$

Looking at the triangle oac,

$$\overrightarrow{\mathrm{oc}} = \overrightarrow{\mathrm{oa}} + \dot{Z} = \frac{1}{k^2-1}\dot{A} + \frac{\dot{A}}{1-ke^{j\delta}} = \frac{k}{k^2-1}\cdot\frac{k-e^{j\delta}}{1-ke^{j\delta}}\dot{A} \tag{3}$$

As $\overline{\mathrm{oc}}$ includes δ in the equation, the length of $\overline{\mathrm{oc}}$ looks as if it might be affected by δ. However, the length of $\overline{\mathrm{oc}}$ is in fact not affected by δ because

$$\begin{aligned}
\left|\frac{k-e^{j\delta}}{1-ke^{j\delta}}\right| &= \left|\frac{(k-\cos\delta)-j\sin\delta}{(1-k\cos\delta)-jk\sin\delta}\right| = \sqrt{\frac{(k-\cos\delta)^2+\sin^2\delta}{(1-k\cos\delta)^2+(k\sin\delta)^2}} \\
&= \sqrt{\frac{k^2+1-2k\cos\delta}{k^2+1-2k\cos\delta}} = 1
\end{aligned} \tag{4}$$

Accordingly, for arbitrary δ,

$$|\overline{\mathrm{oc}}| = \frac{k}{k^2-1}\dot{A} \tag{5}$$

The equation shows that the length of $\overline{\mathrm{oc}}$ is not affected by δ under the condition of constant k, so the locus $\dot{Z}$ becomes a circle whose radius is $\overline{\mathrm{oc}}$.

For the drawing method of the circle as the locus of $\underline{\dot{Z}}$

<u>the center of the circle, point o</u>: (6)

the point on the extended straight line of $\overline{\mathrm{ab}}$ where the distance from a is

$$\frac{1}{k^2-1}\dot{A}$$

<u>far end e of the diameter</u>:

the point on the extended straight line of $\overline{\mathrm{ab}}$ where the distance from a is

$$\frac{1}{k-1}\dot{A}$$

<u>close end d of the diameter</u>:

the point on the extended straight line of $\overline{\mathrm{ab}}$ where the distance from a is

$$\frac{1}{k+1}\dot{A} \tag{6}$$

Figure 17.9 is eventually drawn for $k > 1.0$. In the case when $k \to 1.0$, the point o becomes more and more distant from a, and when $k = 1.0$ it at last converges to the straight line crossing the midpoint of a and b at right angles. In the case of $k < 1.0$, k can be replaced by $1/k$ in the above equation, and the locus moves to a symmetrical position on the straight line crossing the midpoint of a and b at right angles.

17.8 Supplement 2: The Drawing Method for Ż=1/(1/Ȧ+1/Ḃ) of Equation 17.24

The drawing method for the vector $\dot{Z}$ from the given vectors $\dot{A}$ and $\dot{B}$ is examined here in reference to Figure 17.10:

$$\dot{Z} = \frac{1}{\frac{1}{\dot{A}} + \frac{1}{\dot{B}}} \tag{1}$$

Accordingly,

$$\left.\begin{aligned} \dot{Z} &= \frac{\dot{A}}{1+\frac{\dot{A}}{\dot{B}}} = \frac{\dot{A}}{1+k\mathrm{e}^{-j\delta}} \\ \dot{Z} &+ \dot{Z}\cdot k\mathrm{e}^{-j\delta} = \dot{A} \\ \dot{A} &= \dot{B}\cdot k\mathrm{e}^{-j\delta} \end{aligned}\right\} \tag{2}$$

$$\left.\begin{aligned} \dot{Z} &= \frac{\dot{B}}{1+\frac{\dot{B}}{\dot{A}}} = \frac{\dot{B}}{1+\frac{1}{k}\mathrm{e}^{j\delta}} \\ \dot{Z} &+ \dot{Z}\cdot \frac{1}{k}\mathrm{e}^{j\delta} = \dot{B} \\ \dot{B} &= \dot{A}\cdot \frac{1}{k}\mathrm{e}^{j\delta} \end{aligned}\right\} \tag{3}$$

From Equation 2, vectors $\dot{A}$, $\dot{Z}$, $\dot{Z}\cdot k\mathrm{e}^{-j\delta}$ make a closed triangle △abc, where the vector $\dot{Z}\cdot k\mathrm{e}^{-j\delta}$ is obtained by making k times long of $\dot{Z}$ and by turning $\delta°$ clockwise. Accordingly, if δ is constant, the vertical angle $\angle \mathrm{aco} = \pi - \delta$ is constant, so point c is on the circular arc whose chord is $\dot{A}$ and the vertical angle is $(\pi - \delta)$.

In the same way, vectors $\dot{B}$, $\dot{Z}$, $\dot{Z}\cdot(1/k)\mathrm{e}^{j\delta}$ make a closed triangle △obc, where the vector $\dot{Z}\cdot(1/k)\mathrm{e}^{-j\delta}$ is obtained by making $\dot{Z}\cdot(1/k)$ times longer of $\dot{Z}$ and by turning $\delta°$ counterclockwise. Accordingly, if δ is constant, the vertical angle $\angle \mathrm{bco} = \pi - \delta$ is also constant, so point c is on the circular arc whose chord is $\dot{B}$ and the vertical angle is $(\pi - \delta)$.

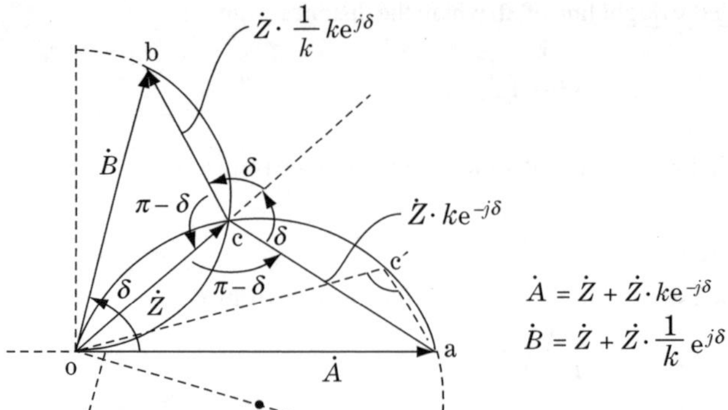

Figure 17.10 Drawing method of $\dot{Z} = 1/(1/\dot{A} + 1/\dot{B})$

As a result, it is proved that the vector $\dot{Z}$ (straight line $\overline{\text{oc}}$) is given by the crossing point c of two arcs whose chords are $\dot{A}$ and $\dot{B}$ and the vertical angles are $(\pi - \delta)$ for both arcs.

Incidentally, here

$$\overline{\text{ac}} : \overline{\text{co}} = \overline{\text{oc}} : \overline{\text{cb}} = k : 1 \qquad \text{and} \qquad \triangle\text{oac} \infty \triangle\text{boc} \tag{4}$$

In conclusion, $\dot{Z}_{\text{total}}$ in Equation 17.24 can be drawn from $\dot{Z}_{\text{load}}$ and $\dot{Z}_{\text{fault}}$ by putting $\dot{A} \rightarrow \dot{Z}_{\text{load}}$, $\dot{B} \rightarrow \dot{Z}_{\text{fault}}$, $\dot{Z} \rightarrow \dot{Z}_{\text{total}}$ in the above method.

Coffee break 9:Steinmetz, prominent benefactor of circuit theory and high-voltage technology

Another inventor of the **symbolic method by complex numbers** is **Charles Proteus Steinmetz** (1865–1923), but he should be remembered also as a great engineer who created the original form of today's high-voltage power system technology.

The first 'man-made lightning' created by Charles Proteus Steinmetz (1865–1923) in his lab in 1922

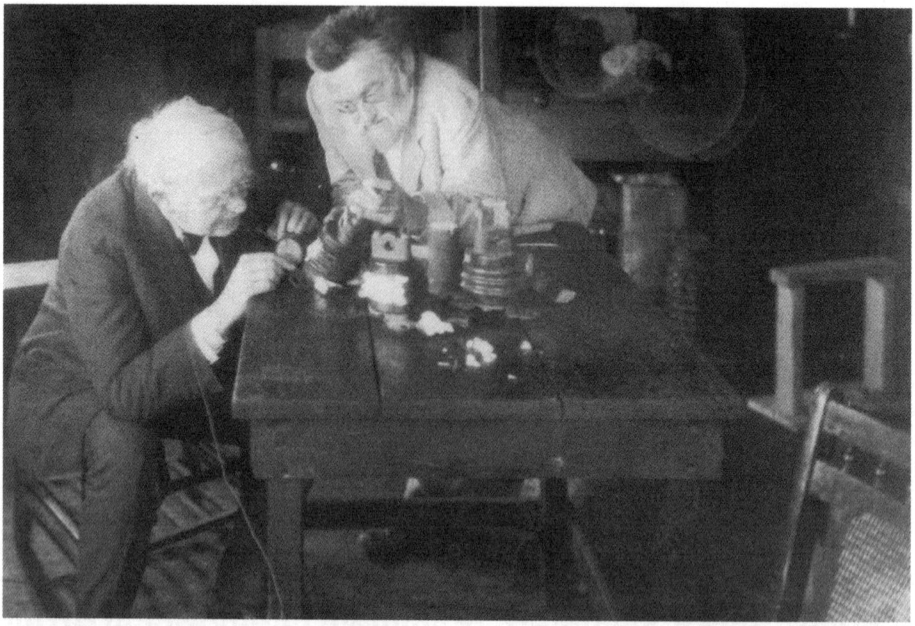

Edison and Steinmetz observing a lightning demonstration and examining a broken insulator and wooden pieces, 1922 (Courtesy of Schenectady Museum)

Steinmetz was born in Breslau, Germany, in 1865. When he was a student at the University of Breslau, he joined the Socialist Party and began to edit the *People's Voice* (this was the time of Bismarck). Steinmetz had to flee Germany for Switzerland and later came to the United States in 1889. Arriving in New York harbour, Steinmetz, a mathematical genius but crippled and penniless, was going to be sent back to Germany, but finally was allowed to go ashore due to the strenuous efforts of a friend. He obtained a job at the Osterheld and Eickenmeyer Company in Yonkers. Soon he was given the task to find a solution for the overheating of motors in trolley cars, before the idea of power loss was known. He solved the problem and his solution, 'the mechanism and the practical method for handling power-loss', were explained in the magazine *The Electrical Engineer* in 1892. The **law of hysteresis** or **Steinmetz's law** was the topic of a speech given by Steinmetz to the AIEE. Thomas Edison wanted to hire Steinmetz but then he began to work at the Edison General Electric plant in Lynn, Massachusetts. However, Edison General Electric and Thomson-Houston Co. were merged shortly afterwards and the newly formed General Electric Company was born in 1892; this was actually the time that Edison resigned from the business management side of the company.

In August 1893, Steinmetz presented the paper 'Complex quantities and their use in electrical engineering' at the International Congress in Chicago, independently of Kennelly. His paper included a generalized treatment of a.c. quantities by complex numbers and became better known than Kennelly's paper, although it appeared issued four months after Kennelly's 'Impedance'.

In 1894, Steinmetz was transferred to GE's main plant in Schenectady, NY. There, he expanded the opportunity for research and implementation of his ideas. He was appointed head of a new calculating department, his first job being to work on the company's proposal for building the new Niagara Falls hydro power station and power transmission of 26 miles (42 km) at 10 kV, 25 Hz. He immediately indoctrinated the engineers there into his mathematical design method.

During the next 20 years he prepared masterful papers, most of which were filled with equations. His prominent pioneer works by papers mostly of mathematical approach covers so widely circuit analysis of oscillating currents, ac-transient phenomena, wave and impulse current, magnetic losses, corona, arc discharge, theory of induction motor/generator, transmission and distribution, inductance of line conductors, ad-dc conversion by mercury rectifier, arc lamps and electric illuminants, vacuum discharge, surge phenomena, insulation and circuit protection and so on. He published several textbooks in order to educate the electrical engineering profession; in particular, *Theory and calculation of alternating current phenomena* (co-authored with Ernst J. Berg in 1897) and *Engineering mathematics*' (1911) became the standard texts at that time. It may be said that the original form of modern electrical engineering, in particular the approach method of **circuit analysis**, was settled by Steinmetz.

Steinmetz once said: 'I want to say that absolutely all the success I have had has been due to my through study of mathematics.' He was also a prolific inventor and originated nearly 200 patents for various electrical equipment.

Lightning phenomena, insulation of equipment, lightning protection – these were all areas of his particular interest.

One example of his achievements is the so-called **Steinmetz arrester** invented in 1909. He explained it as 'the arrester whose substantial circuit is composed with metal-oxide nonlinear-resistance and with series gap, which can absorb lightning-surge energy without interruption of power system operation.' Note that, today, we are still using Steinmetz's arrester of 100 years ago without any major change. In 1922, the year before his death, the world's first **man-made lightning** was created in his lab by a 120 kV **impulse generator** of his own design and construction which became the sensational turning point in advanced high-voltage and insulation technology by a quantitative approach. Again it is fair to say that the original form of today's high-voltage research practices was settled by Steinmetz.

Incidentally, Steinmetz made some predictions of the future, when housewives were using wood stoves, which included air-conditioning, thermostatically controlled heating and cooling, radio, television 'with the motion picture and the talking machine perfectly synchronized', electric cooking ranges and 'transmission line networks connecting many large central power stations'. All these predictions were later realized as the company's most important products. His greatest role in the company was as consultant to whom the company had great faith and he was referred to 'Supreme Court', to whom an engineer could go for professional advice.

He died in 1923, leaving behind a legacy and framework around which our electrical age has been built. Moreover, he turned out so many competent electricians as his students including Ernst Alexanderson (radio, ac-motor high frequency generator), William D. Coolidge (Tungsten ductile to better lamps and to X-rays tubes), Irving Langmuir (lamps, electron tubes), Edward M Hewlett (Suspension insulators). L.V. Bewley (surge phenomena) and R. H. Park. In the course of things, Schenectady became the Mecca of electricians after his death.

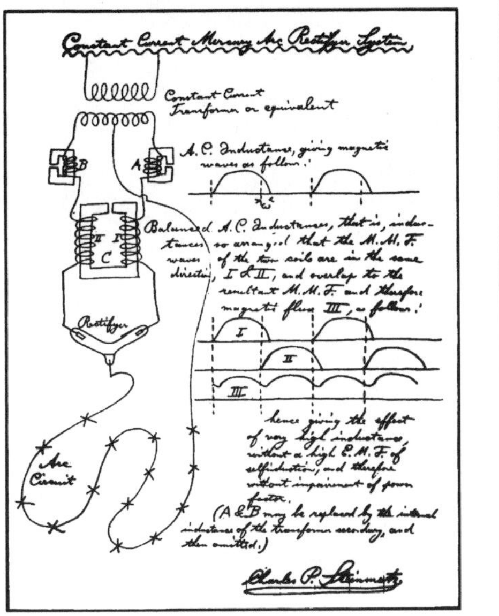

A sketch for mercury are rectifier circuit (1903)

18

Travelling-wave (Surge) Phenomena

We have focused in the previous chapters mainly on the behaviour of power frequency or transient phenomena in the lower frequency zone, but this is only a partial view of the entire image of power system behaviour. Our power system networks have profound dynamic characteristics containing behaviour in the very wide frequency zone from d.c. to surge phenomena, so it is vital to understand the fundamentals of power system networks from multi-dimensional viewpoints as much as possible. As a matter of fact, in most actual engineering work we have to wrestle with problems which cannot be reasonably solved from a one-sided viewpoint.

In this chapter the theory of electromagnetic waves (surge) on transmission lines is introduced as the first chapter to focus on higher frequency or surge phenomena.

18.1 Theory of Travelling-wave Phenomena along Transmission Lines (Distributed-constants Circuit)

Table 18.1 shows a frequency-diagram map of the phenomena of a power system classified by three categories of frequency for the sake of convenience:

① the lower order harmonic zone of zero to a few kilohertz

② the higher order harmonic zone of a few kilohertz to a hundred kilohertz

③ the surge order zone of megahertz and gigahertz.

The voltage and current characteristics of a single phase transmission line for an electromagnetic wave are expressed by the **distributed-constants circuit** shown in Figure 18.1, and the approximated circuit is the **concentrated-constants circuit** (*L*-circuit, *T*-circuit, π-circuit, etc.). Analytical treatment is far easier by concentrated-constants circuit although the error would become larger for higher frequency phenomena. Therefore evaluation of errors by such an approximation is always important, whenever we study higher harmonic phenomena by concentrated-constants circuits.

18.1.1 Waveform equation of a transmission line (overhead line and cable) and the image of a travelling wave

18.1.1.1 Introduction of travelling-wave equation (waveform equation)

Electromagnetic waves (or surges) travel along overhead transmission lines at a velocity of 300 000 km/s (equal to the velocity of light). The surge phenomena have to be examined by the distributed-constants circuit shown in Figure 18.1.

Handbook of Power System Engineering Yoshihide Hase
© 2007 John Wiley & Sons, Ltd

Table 18.1 Frequency map of power system phenomena (examples)

Frequency	Equivalent circuit of transmission lines	Application	Phenomena or analysis	Obstacle
d.c.		d.c. transmission electrochemical industry	d.c. component of fault current inrush current (transformer)	current-zero-missing (breaker) saturation (CT, relay)
0–50/60 Hz				very low harmonic waveform distortion (relay)
Power frequency 50/60 Hz	R–X circuit 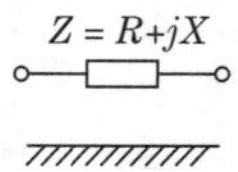	power utilization	<u>steady-state analysis</u>: P-Q-V-I analysis, fault analysis, <u>dynamic analyis</u>: P-Q-V stability, temporary overvoltage	overvoltage/overcurrent tripping voltage unstability/ power unstability
100 Hz – 500 Hz	L-circuit 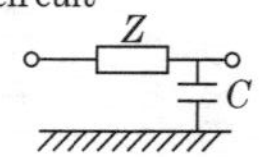		lower harmonics resonance waveform distortion AVR response analysis	voltage flickering (various loads, various controls) waveform distortion (relay) local heat vibration (generator/motor) capacitor overload
500 Hz – 1 kHz – 10 kHz	T-circuit, π-circuit 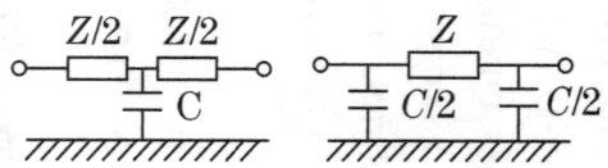	ripple-control (distribution line) refining industry	higher harmonic resonance	vibration, local heat (transformer, generator) noise protection
10 kHz – 100 kHz – 1 MHz	distribution circuit 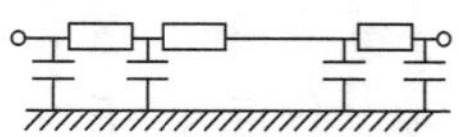	fault locator (pulse-radar type) power line carrier	corona noise transfer voltage (transformer) lightning surge switching surge (very fast) chopped wave	surge protection
10 MHz				
100 MHz				
1 GHz				

Note: Equivalent circuits should be appropriately selected considering length of transmission lines, frequency concerned and acceptable error. Concentrated circuits rather than distribution circuits are realistic approaches for most practical engineering applications, so error rate evaluation for concentrated equivalent circuits is quite important.

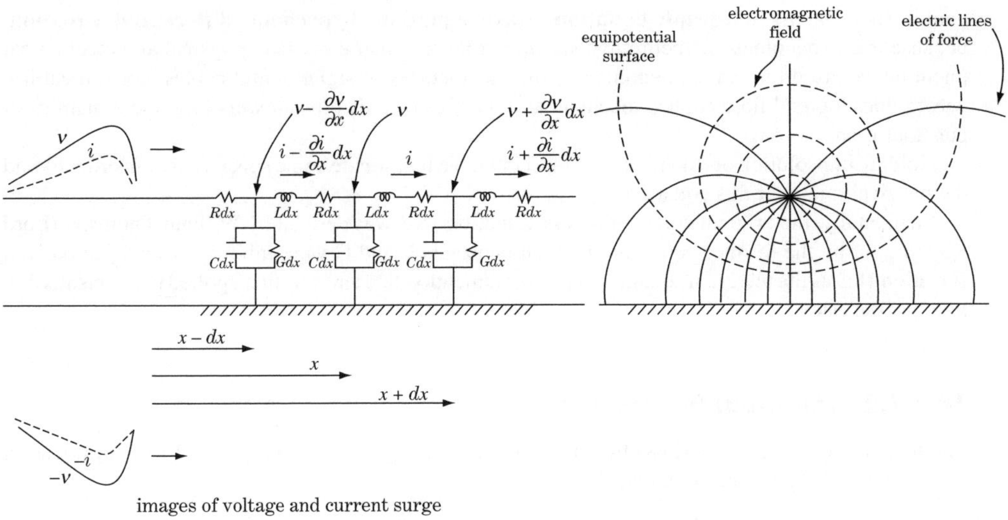

Figure 18.1 Distribution circuit of transmission line

In the figure, $v(x,t)$, $i(x,t)$ are the voltage and current at point x and time t, which is located at a distance of x from starting point s. The following equation can be derived for the minute section between point x and $x + \Delta x$:

$$\left.\begin{aligned} v(x,t) - \left((v(x,t) + \frac{\partial v(x,t)}{\partial x} dx) \right) &= (Ldx)\frac{\partial i(x,t)}{\partial t} + (Rdx)i(x,t) \quad ① \\ -i(x,t) + \left(i(x,t) - \frac{\partial i(x,t)}{\partial x} dx \right) &= (Cdx)\frac{\partial v(x,t)}{\partial t} + (Gdx)v(x,t) \quad ② \end{aligned}\right\} \tag{18.1}$$

$$\therefore \quad \left.\begin{aligned} -\frac{\partial v(x,t)}{\partial x} &= L\frac{\partial i(x,t)}{\partial t} + Ri(x,t) \quad ① \\ -\frac{\partial i(x,t)}{\partial x} &= C\frac{\partial v(x,t)}{\partial t} + Gv(x,t) \quad ② \end{aligned}\right\} \tag{18.2}$$

Partial differentiation of both sides of Equation 18.2 ① by $\partial/\partial x$ yields a new equation. Then, substituting Equation 18.2 ② into the new equation, the variable i can be eliminated and an equation in only variable v is derived. In the same way, another equation can be derived in only variable i. The resulting equations are

$$\left.\begin{aligned} \frac{\partial^2 v(x,t)}{\partial x^2} &= LC\frac{\partial^2 v(x,t)}{\partial t^2} + (LG + CR)\frac{\partial v(x,t)}{\partial t} + RG \cdot v(x,t) \quad ① \\ \frac{\partial^2 i(x,t)}{\partial x^2} &= LC\frac{\partial^2 i(x,t)}{\partial t^2} + (LG + CR)\frac{\partial i(x,t)}{\partial t} + RG \cdot i(x,t) \quad ② \end{aligned}\right\} \tag{18.3}$$

Each of these differential equations is called a **travelling-wave equation** and is essential as a theoretical starting point for surge phenomena of transmission lines. The equation also has other names such as the **telegraph equation**, **wave equation**, **hyperbolic differential equation**, because it also becomes the theoretical starting point for other electrical, physical and mechanical phenomena based on 'wave phenomena', such as theories of signal metal cables, coaxial cables, waveguides, optical fibre cables, mechanical dynamics of rods/pipes/towers (e.g. the dynamics of penstock), and so on.

Holding our solution of Equation 18.3 until later, we investigate some special cases in order to find the physical image of this equation.

Incidentally, the original equation before Equation 18.3 was derived by William Thomson (Lord Kelvin), but by the form of equation in which constants L and G were missing. Heaviside derived Equation 18.3 as the telegram equation of a communication line and was first applied to a transmission line by Blakesley.

18.1.1.2 The ideal (no-loss) line

The first case is the ideal (no-loss) line. If the losses are negligible, namely $R = 0$, $G = 0$, Equation 18.2 and 18.3 become the equations below:

$$\left.\begin{aligned} -\frac{\partial v(x,t)}{\partial x} &= L\frac{\partial i(x,t)}{\partial t} \\ -\frac{\partial i(x,t)}{\partial x} &= C\frac{\partial v(x,t)}{\partial t} \end{aligned}\right\} \tag{18.4}$$

$$\left.\begin{aligned} \frac{\partial^2 v(x,t)}{\partial x^2} &= LC\frac{\partial^2 v(x,t)}{\partial t^2} \\ \frac{\partial^2 i(x,t)}{\partial x^2} &= LC\frac{\partial^2 i(x,t)}{\partial t^2} \end{aligned}\right\} \tag{18.5}$$

The general solution of Equation 18.5 is

$$\left.\begin{aligned} &v(x,t) = v_1(x-ut) + v_2(x+ut) \\ &i(x,t) = \frac{1}{Z_0}[v_1(x-ut) - v_2(x+ut)] \\ &\text{where} \quad Z_0 = \sqrt{\frac{L}{C}} \quad \text{(surge impedance)}, \quad u = \frac{1}{\sqrt{LC}} \quad \text{(velocity)} \end{aligned}\right\} \tag{18.6}$$

That Equation 18.6 is the solution of Equation 18.5 can be easily proved by substituting Equation 18.6 into Equation 18.5 and by applying some theorems of differentiation, but to derive Equation 18.6 (the solution by D′ Alembelt) from Equation 18.5 is the work of mathematicians.

Now, although Equation 18.6 was found as the solution of the travelling-wave equation, we cannot say anything at this moment about v_1, v_2 for its characteristics or its relation with v. However, the dimension of $(x - ut)$ is distance and of t is time so that u should be velocity. Therefore, if we observe $v_1(x - ut)$ by moving along the transmission line with a velocity of u, we should observe the same $v_1(x - ut)$ unchanged.

In other words, $v_1(x - ut)$ runs along the line for a distance x at velocity u $v_2(x + ut)$ also runs along the line for a distance $(-x)$ at velocity u, that is in the opposite direction. Thus, $v_1(x - ut)$ is named the **forward wave** and $v_2(x + ut)$ the **backward wave** (or **reverse wave**). Equation 18.6 can be understood as the voltage at point x and time t, $v(x,t)$, expressed as the summation (or composite) of the forward wave and the backward wave. Equation 18.6 also tells us by analogy that the current $i(x,t)$ is expressed as the summation (or composite) of the forward wave $i_1(x-ut) = (1/Z_0)\cdot v_1(x-ut)$ and the backward wave $-i_2(x+ut) = (-1/Z_0)\cdot v_2(x+ut)$, while the polarity of the backward wave has a minus sign.

Figure 18.2 shows the physical image of the actual voltage v, forward voltage v_1 and backward voltage v_2, and the actual current i, forward current i_1 and backward current i_2 at point x and time t.

The phenomena at position x' and time t' are the same.

We need to recall some points about the physical meaning of the equations:

- We stated that voltage and current exist at point x and time t and named them as expressions of $v(x,t)$ and $i(x,t)$ (see Figure 18.1); however, we did not explain the reason why $v(x,t)$ and $i(x,t)$ exist. The equation was derived without any explanation of location and kind of power source (such as the switching-in of a generator or lightning surges). The equation is the specific characteristics of the transmission line itself which is not affected by outer circuit conditions or adding a power source.
- The actual existing voltage and current which we can measure on the transmission line are $v(x,t)$, $i(x,t)$, and the equation only says that the actual quantities $v(x,t)$ as well as $i(x,t)$ can be understood as the addition of the forward wave and the backward wave. In other words, it means that $v(x,t)$ can be decomposed into $v_1(x,t)$ and $v_2(x,t)$. The forward wave and the backward wave cannot be measured and as such they are the conceptual voltage and current. Accordingly, it is nonsense to discuss why and from where the forward wave or the backward wave would originate.
- In the case of the ideal (no-loss) line, the forward wave, the backward wave, as well as the original $v(x,t)$ and $i(x,t)$ will not be changed by waveform or velocity.

The forward wave $v_1(x-ut), i_1(x-ut)$ and the backward wave $v_2(x+ut), i_2(x+ut)$ can be specified by the condition of the power source (generator or lightning surge or switching surge) connected at the line's starting point s, the condition of the transmission line from point s to x, and the condition of the line after passing point x, as is explained later.

Incidentally, if a power source (a generator or lightning strike) is suddenly connected (or if any voltage or current is injected) at point x, the actual voltage and current surge begin to travel to both sides (the forward direction and the backward direction) of the line. All these quantities are measurable existing quantities and should not be confused with the above-mentioned conceptual v_1, i_1, v_2, i_2. This is discussed later in section 18.5.2.

18.1.1.3 The distortion-less line

The distortion-less line satisfies the following relation regarding the constants:

$$\alpha = \frac{R}{L} = \frac{G}{C} \tag{18.7}$$

In this case, Equations 18.2 and 18.3 are simplified as

$$\left.\begin{aligned} -\frac{\partial v(x,t)}{\partial x} &= L\left\{\frac{\partial i(x,t)}{\partial t} + \alpha i(x,t)\right\} \quad ① \\ -\frac{\partial i(x,t)}{\partial x} &= C\left\{\frac{\partial v(x,t)}{\partial t} + \alpha v(x,t)\right\} \quad ② \end{aligned}\right\} \tag{18.8}$$

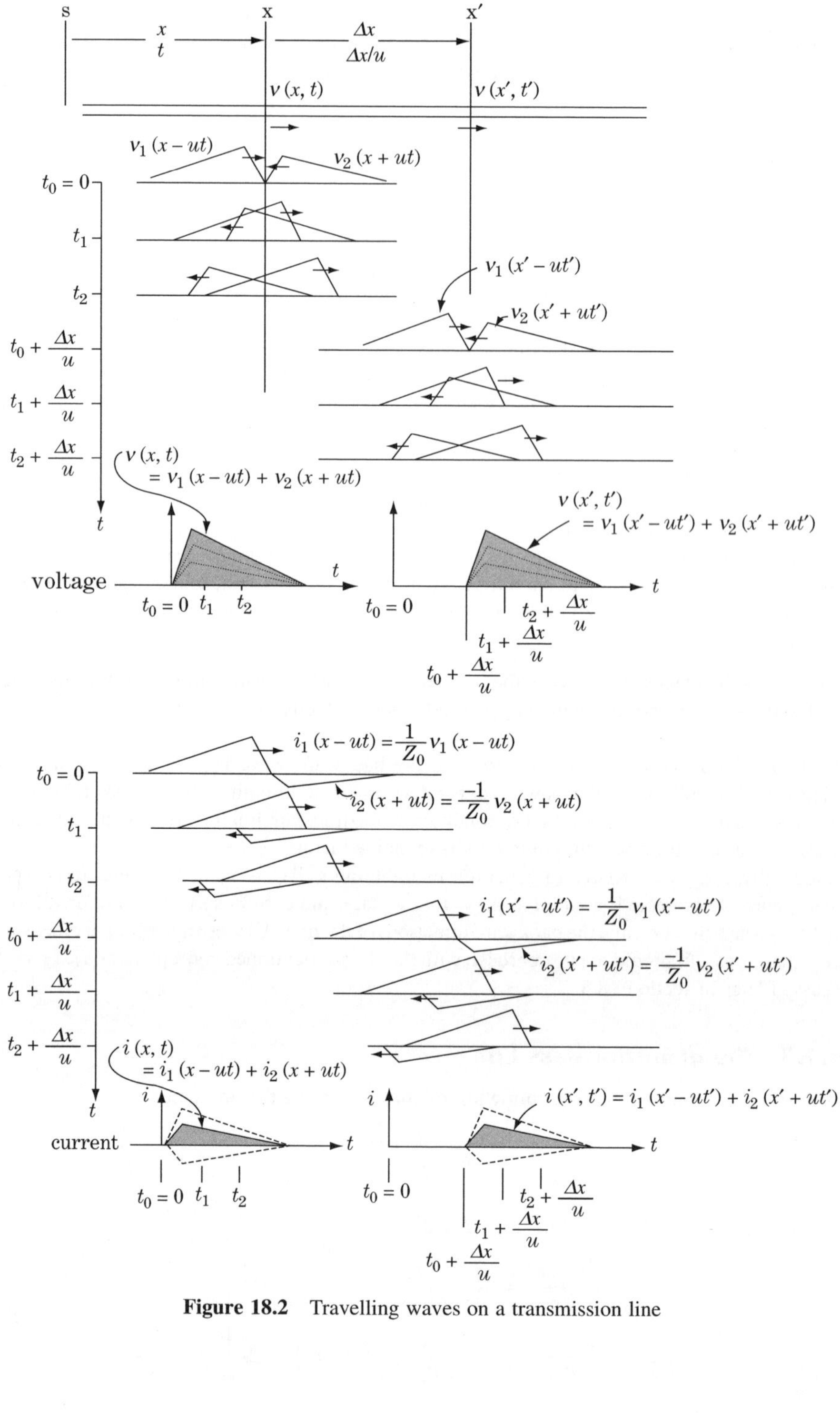

Figure 18.2 Travelling waves on a transmission line

$$\left.\begin{aligned}\frac{\partial^2 v(x,t)}{\partial x^2} &= LC\left\{\frac{\partial^2 v(x,t)}{\partial t^2} + 2\alpha\frac{\partial v(x,t)}{\partial t} + \alpha^2 v(x,t)\right\} \quad ① \\ \frac{\partial^2 i(x,t)}{\partial x^2} &= LC\left\{\frac{\partial^2 i(x,t)}{\partial t^2} + 2\alpha\frac{\partial i(x,t)}{\partial t} + \alpha^2 i(x,t)\right\} \quad ②\end{aligned}\right\} \tag{18.9}$$

The general solution of the equation is given below (see Supplement 1 for the derivation):

$$\left.\begin{aligned}v(x,t) &= e^{-\alpha t}\{v_1(x-ut) + v_2(x+ut)\} \quad ① \\ i(x,t) &= \frac{e^{-\alpha t}}{Z_0}\{v_1(x-ut) - v_2(x+ut)\} \quad ②\end{aligned}\right\} \tag{18.10}$$

This solution includes the attenuation term $e^{-\alpha t}$ in comparison with Equation 18.6. In this case, the forward wave as well as the backward wave of the voltage and current attenuate as a similar waveform along the line.

18.1.2 The general solution for voltage and current by Laplace transforms

We go back to Equation 18.2 and apply Laplace transforms. The equation written in the s domain $(d/dt \to s)$ is

$$\left.\begin{aligned}-\frac{dV(x,s)}{dx} &= (Ls+R)\cdot I(x,s) - L\cdot i(x,0) \\ -\frac{dI(x,s)}{dx} &= (Cs+G)\cdot V(x,s) - C\cdot v(x,0)\end{aligned}\right\} \tag{18.11}$$

where $v(x,0)$, $i(x,0)$ are the initial value of the voltage and current at point x and time $t=0$, respectively.

$V(x,s)$ or $I(x,s)$ can be eliminated from the above two equations:

$$\left.\begin{aligned}\frac{d^2V(x,s)}{dx^2} &= \gamma^2(s)\cdot V(x,s) + \varphi_v(x) \\ \frac{d^2I(x,s)}{dx^2} &= \gamma^2(s)\cdot I(x,s) + \varphi_i(x)\end{aligned}\right\} ① \tag{18.12}$$

where

$$\gamma(s) = \sqrt{(Ls+R)(Cs+G)} \quad \text{(the \textbf{propagation constant}) is} \quad ②$$

$$\left.\begin{aligned}\varphi_v(x) &= L\frac{di(x,0)}{dx} - C(Ls+R)\cdot v(x,0) \\ \varphi_i(x) &= C\frac{dv(x,0)}{dx} - L(Cs+G)\cdot i(x,0)\end{aligned}\right\} ③$$

$\gamma(s)$ can be modified as follows:

$$\left.\begin{aligned}\gamma(s) &= \sqrt{(Ls+R)(Cs+G)} = \sqrt{LC}\cdot\sqrt{\left(s+\frac{R}{L}\right)\left(s+\frac{G}{C}\right)}\\ &= \sqrt{LC}\cdot\sqrt{(s+\alpha+\beta)(s+\alpha-\beta)} = \sqrt{LC}\cdot\sqrt{(s+\alpha)^2-\beta^2}\end{aligned}\right\}$$

where

$$\left.\begin{aligned}\frac{R}{L} &= \alpha+\beta\\ \frac{G}{C} &= \alpha-\beta\end{aligned}\right\}\quad \left.\begin{aligned}&\textbf{attenuation constant}: &\alpha &= \frac{1}{2}\left(\frac{R}{L}+\frac{G}{C}\right)\\ &\textbf{wave length constant}: &\beta &= \frac{1}{2}\left(\frac{R}{L}-\frac{G}{C}\right)\\ &\textbf{velocity of propagation}: &u &= \frac{1}{\sqrt{LC}}\end{aligned}\right\}\quad ④$$

$\varphi_v(x)$, $\varphi_i(x)$ are the initial values at x at time $t = 0$ respectively, so that if the line is not charged before $t = 0$

$$\varphi_v(x) = 0,\ \varphi_i(x) = 0 \tag{18.13}$$

Then Equation 18.12 is

$$\left.\begin{aligned}\frac{d^2V(x,s)}{dx^2} &= \gamma^2(s)\cdot V(x,s) \quad ①\\ \frac{d^2I(x,s)}{dx^2} &= \gamma^2(s)\cdot I(x,s) \quad ②\end{aligned}\right\} \tag{18.14}$$

The general solution of the equation is

$$\left.\begin{aligned}V(x,s) &= A(s)e^{-\gamma(s)x} + B(s)e^{\gamma(s)x} \quad ①\\ I(x,s) &= \frac{1}{Z(s)}\{A(s)e^{-\gamma(s)x} - B(s)e^{\gamma(s)x}\} \quad ②\\ Z_0(s) &= \sqrt{\frac{Ls+R}{Cs+G}} = \frac{1}{Y_0(s)} \quad ③\end{aligned}\right\} \tag{18.15}$$

where

$Z_0(s)$: the **characteristic impedance** (the **surge impedance operator**)

$Y_0(s)$: the **characteristic admittance** (the **surge admittance operator**)

That Equation 18.15 ① is the solution of Equation 18.14 ① can be easily confirmed. That is, differentiating Equation 18.15 ① $V(x,s)$ twice by x gives the equation which satisfies Equation 18.14 ①.

Equation 18.15 is the general solution of voltage and current at arbitrary point x, and we need to add some initial conditions to find a concrete answer. The initial voltage $v(0,t)$ is injected at point $x = 0$ as a given value. Then

$$V(0,s) = \mathcal{L}[v(0,t)] \quad (\mathcal{L} \text{ is the Laplace transform}) \tag{18.16}$$

18.1.3 Four-terminal network equation between two arbitrary points

We now examine the voltage and current relations between two arbitrarily selected points s (distance $x = 0$) and r (distance x).

Putting $x = 0$ in Equation 18.15,

$$\left.\begin{array}{l}\left.\begin{array}{l}V(0,s) = A(s) + B(s)\\ I(0,s) = \dfrac{1}{Z_0(s)}\{A(s) - B(s)\}\end{array}\right\} ① \\ \left.\begin{array}{l}\therefore A(s) = \dfrac{1}{2}\{V(0,s) + Z_0(s)I(0,s)\}\\ B(s) = \dfrac{1}{2}\{V(0,s) - Z_0(s)I(0,s)\}\end{array}\right\} ②\end{array}\right\} \quad (18.17)$$

Substituting Equation 18.17 ② again into Equation 18.15,

$$\left.\begin{array}{ll}V(x,s) = \dfrac{e^{\gamma(s)x} + e^{-\gamma(s)x}}{2}\cdot V(0,s) - \dfrac{e^{\gamma(s)x} - e^{-\gamma(s)x}}{2}\cdot Z_0(s)I(0,s) & ①\\ I(x,s) = \dfrac{1}{Z_0(s)}\left\{-\dfrac{e^{\gamma(s)x} - e^{-\gamma(s)x}}{2}\cdot V(0,s) + \dfrac{e^{\gamma(s)x} + e^{-\gamma(s)x}}{2}\cdot Z_0(s)I(0,s)\right\} & ②\end{array}\right\} \quad (18.18a)$$

or as a matrix equation

$$\begin{bmatrix}V(x,s)\\ I(x,s)\end{bmatrix} = \begin{bmatrix}\cosh\gamma(s)x & -Z_0(s)\sinh\gamma(s)x\\ \dfrac{-1}{Z_0(s)}\sinh\gamma(s)x & \cosh\gamma(s)x\end{bmatrix}\cdot\begin{bmatrix}V(0,s)\\ I(0,s)\end{bmatrix} \quad (18.18b)$$

Also the inverse of Equation 18.18b is (see Supplement 2 for the procedure)

$$\begin{bmatrix}V(0,s)\\ I(0,s)\end{bmatrix} = \begin{bmatrix}\cosh\gamma(s)x & Z_0(s)\sinh\gamma(s)x\\ \dfrac{1}{Z_0(s)}\sinh\gamma(s)x & \cosh\gamma(s)x\end{bmatrix}\cdot\begin{bmatrix}V(x,s)\\ I(x,s)\end{bmatrix} \quad (18.19a)$$

Equations 18.18b and 18.19a are the equations of a four-terminal network for the distributed-constants circuit.

Equation 18.19a can be rewritten as follows by changing the symbols of the quantities with suffix s or r (meaning the sending point, the receiving point) and the distance between points s and r is l:

$$\begin{bmatrix}V_s(s)\\ I_s(s)\end{bmatrix} = \begin{bmatrix}\cosh\gamma(s)l & Z_0(s)\sinh\gamma(s)l\\ \dfrac{1}{Z_0(s)}\sinh\gamma(s)l & \cosh\gamma(s)l\end{bmatrix}\cdot\begin{bmatrix}V_r(s)\\ I_r(s)\end{bmatrix} \quad (18.19b)$$

where $V(x,s) = \mathcal{L}[v(x,t)]$, $I(x,s) = \mathcal{L}[i(x,t)]$. $\gamma(s)$ is given by Equation 18.12 ④, and $Z_0(s)$ is given by Equation 18.15 ③.

This is the four-terminal equation between points s and r in its popular form.

These equations show the exact relation of voltage and current quantities between the two arbitrary points, and so are important equations, because we can start most of the analysis from them instead of starting from the differential equations already introduced.

For the distortion-less line, this is a special case of Equations 18.12 and 18.19, namely

$$\left.\begin{array}{l}\left.\begin{array}{l}\alpha = \dfrac{R}{L} = \dfrac{G}{C} \quad \beta = 0, \quad \text{then}\\ Z_0 = \dfrac{1}{Y_0} = \sqrt{\dfrac{L}{C}}\\ \gamma(s) = \sqrt{LC}(s+\alpha) = \dfrac{s+\alpha}{u}\\ u = \dfrac{1}{\sqrt{LC}}\end{array}\right\}\end{array}\right\} \quad (18.20)$$

where

Z_0: surge impedance, Y_0: surge susceptance

$\gamma(s)$: propagation constant, u: velocity of propagation

For the ideal (no-loss) line,

$$\left.\begin{aligned} &\alpha = \beta = 0, \quad \text{then} \\ &Z_0 = \frac{1}{Y_0} = \sqrt{\frac{L}{C}} \\ &\gamma(s) = \sqrt{LC} \cdot s = \frac{s}{u} \\ &u = \frac{1}{\sqrt{LC}} \end{aligned}\right\} \tag{18.21}$$

We have now obtained the exact transmission line equations in the Laplace transformation domain, which can be applied to any transient analysis of every frequency zone (ignoring non-linear losses like corona loss, eddy-current loss).

By application of Laplace transforms, the symbol $s = d/dt$ is adopted, Ld/dt, Cd/dt are replaced by Ls, Cs, and furthermore algebraic manipulation can be applied.

If steady-state phenomena of angular velocity $\omega = 2\pi f$ (f does not necessarily mean commercial frequency) are to be treated, we can replace $s \to j\omega$ by that for the four-terminal network for steady-state phenomena.

Thus, for the distortion-less line

$$\gamma = \sqrt{LC}(j\omega + \alpha) = \frac{j\omega + \alpha}{u} \quad ① \tag{18.22}$$

and for the ideal (no-loss) line

$$\gamma = \sqrt{LC} \cdot j\omega = \frac{j\omega}{u} \quad ②$$

18.1.4 Examination of line constants

We examine below typical values of constants such as $\gamma(s)$, Z, u in regard to overhead transmission lines and power cables.

18.1.4.1 Overhead transmission lines

The working inductance L and the working capacitance C of a transmission line is given by Equations 1.9 and 1.35 and typical magnitudes are shown in Tables 2.1 and 2.2. These constants are determined only by the physical allocation of the line conductors (in other words, by the structure of the towers) and do not depend on induced voltages or frequency

Again ignoring corona loss and eddy-current loss and rewriting Equations 1.9 and 1.35 here, for the positive-sequence inductance

$$L = 0.4605 \log_{10} \frac{s_{ll}}{r} + 0.05\ [\text{mH/km}] \fallingdotseq 0.4605 \log_{10} \frac{s_{ll}}{r} \times 10^{-3} [\text{H/km}] \quad ①$$

and for the positive-sequence capacitance

$$C = \frac{0.02413}{\log_{10} \frac{s_{ll}}{r}} [\mu\text{F/km}] = \frac{0.02413}{\log_{10} \frac{s_{ll}}{r}} \times 10^{-6} [\text{F/km}] \quad ② \tag{18.23}$$

Applying this equation, $\gamma(s)$, Z, u of the ideal (no-loss) overhead transmission line ($R = G = 0$) are calculated below.

For the velocity of propagation

$$\left.\begin{aligned} u &= \frac{1}{\sqrt{LC}} = \frac{1}{\sqrt{0.4605 \times 10^{-3} \times 0.02413 \times 10^{-6}}} [\text{km/s}] \\ &= 300000 [\text{km/s}] = 300 [\text{m}/\mu\text{s}] \quad (\text{velocity of light } c \text{ in a vacuum}) \end{aligned}\right\} \tag{18.24}$$

for the surge impedance

$$\left.\begin{aligned}&Z=\sqrt{\frac{L}{C}}=\frac{1}{\sqrt{LC}}\cdot L=uL=\sqrt{\frac{0.4605\times10^{-3}}{0.02413\times10^{-6}}}\cdot\log_{10}\frac{S_{ll}}{r}=138\log_{10}\frac{s_{ll}}{r}[\Omega]\\&\left(\begin{array}{l}\text{if}\quad L=1\text{mH/km},\quad C=0.01\,\mu\text{F/km}\\\text{then}\quad Z=uL=300\,000\times(1\times10^{-3})=300\,\Omega\end{array}\right)\end{aligned}\right\}\tag{18.25}$$

and for the constant of propagation

$$\gamma(s)=\sqrt{LC}\cdot s=\frac{s}{u}\tag{18.26}$$

The logarithmic terms of L and C are cancelled by calculation of LC, so the velocity of propagation u becomes the velocity of light c unconditionally ($u=c$).

We learned in Section 2.5 and Tables 2.1 and 2.2 that the typical constants of overhead lines are $L=1$ mH/km and $C=0.01$ μF/km regardless of the rated voltage classes. As a result, typical surge impedance Z is around 300 Ω (the surge impedance of an overhead transmission line is practically 250–500 Ω).

18.1.4.2 Power cables

Referring to Table 2.3, we will calculate the constants for typical CV and OF cable lines.

For CV cable at 275 kV · 2000 square class (single core)

$$L=0.392\times10^{-3}\,\text{H/km},\quad C=0.25\times10^{-6}\,\text{F/km}$$

Accordingly,

$$u=101\,000\,\text{km/s}=101\,\text{m}/\mu\text{s},\quad Z=39.6\,\Omega\tag{18.27}$$

For CV cable at 154 kV · 800 square class (single core)

$$L=0.44\times10^{-3}\,\text{H/km},\quad C=0.19\times10^{-6}\,\text{F/km}$$

Accordingly,

$$u=109\,000\,\text{km/s}=109\,\text{m}/\mu\text{s},\quad Z=46.1\,\Omega\tag{18.28}$$

The surge impedance of the power cable is generally $15\sim50\,\Omega$, which means an order of magnitude less than that of the overhead transmission line, and the velocity of propagation u is approximately 100–140 m/μs, which is about half or less of that in an overhead transmission line (see Section 2.5).

18.2 Approximation of Distributed-constants Circuit and Accuracy of Concentrated-constants Circuit

We return to Equations 18.18 and 18.19 and continue our examination of the no-loss line for simplicity. Putting $R=0$, $G=0$ in Equations 18.12 and 18.19, the four-terminal network equation becomes

$$\left.\begin{aligned}&\begin{bmatrix}V(0,s)\\I(0,s)\end{bmatrix}=\begin{bmatrix}\cosh\gamma l & Z_0\sinh\gamma l\\\frac{1}{Z_0}\sinh\gamma l & \cosh\gamma l\end{bmatrix}\cdot\begin{bmatrix}V(l,s)\\I(l,s)\end{bmatrix}\\&\text{where}\quad\gamma=\sqrt{LC}\cdot s=\frac{s}{u},\quad Z_0=\sqrt{\frac{L}{C}},\quad R=0,\quad G=0\end{aligned}\right\}\tag{18.29}$$

The hyperbolic cosine and sine functions can be expanded below using Maclaurin's expansion theorem:

$$\left.\begin{aligned} \cosh \gamma l &= 1 + \frac{(\gamma l)^2}{2!} + \frac{(\gamma l)^4}{4!} + \frac{(\gamma l)^6}{6!} + \ldots \\ \sinh \gamma l &= \gamma l + \frac{(\gamma l)^3}{3!} + \frac{(\gamma l)^5}{5!} + \frac{(\gamma l)^7}{7!} + \ldots \end{aligned}\right\} \tag{18.30}$$

If we neglect the second term and other smaller terms on the right-hand side under the condition of smaller γl,
for $(\gamma l)^2/2 \ll 1$ or $\gamma l \ll \sqrt{2}$)

$$\left.\begin{aligned} &\begin{bmatrix} V(0,s) \\ I(0,s) \end{bmatrix} = \begin{bmatrix} 1 & Z_0\gamma l \\ \dfrac{\gamma l}{Z_0} & 1 \end{bmatrix} \cdot \begin{bmatrix} V(l,s) \\ I(l,s) \end{bmatrix} \fallingdotseq \begin{bmatrix} 1 & Lls \\ Cls & 1 \end{bmatrix} \cdot \begin{bmatrix} V(l,s) \\ I(l,s) \end{bmatrix} \\ &\cosh \gamma l \fallingdotseq 1, \quad \sinh \gamma l \fallingdotseq \gamma l = \frac{l}{u} \cdot s \end{aligned}\right\} \tag{18.31}$$

This is the equation of the **type-L concentrated-constants circuit** in Table 18.1.

Incidentally, when we study the single frequency characteristics of a line, we can let $s \to j\omega$ or $j2\pi f$. Then if we examine a transmission line with length l and single frequency f,

$$\gamma l = \frac{j\omega l}{u} = \frac{j2\pi fl}{u} \tag{18.32}$$

Accordingly, the accuracy-evaluation curves by line distance l and frequency f with a given smaller parameter of γl can be obtained.

Figure 18.3 is the graphic diagram of Equation 18.32 under the condition of $u = 300\,\mathrm{m}/\mu\mathrm{s}$ (overhead line) and $u = 150\,\mathrm{m}/\mu\mathrm{s}$ (power cable), and under the three conditions of $(\gamma l)^2/2 = 0.05, 0.1, 0.15$.

For example, an overhead line can be expressed as a type-L concentrated circuit within about 10% error ($(\gamma l)^2/2 \leqq 0.1$) for the phenomena within 200 Hz for 110 km or 400 Hz for 55 km, and for a power cable of 200 Hz for 55 km or 400 Hz for 27 km.

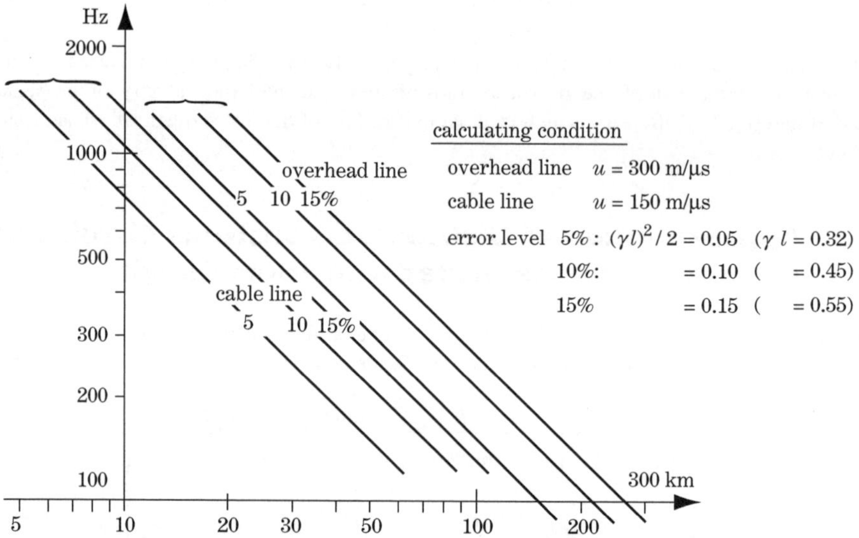

Figure 18.3 Accuracy evaluation of concentrated circuit by line length and frequency

18.3 Behaviour of Travelling Wave at a Transition Point

18.3.1 Incident, transmitted and reflected waves at a transition point

A point on a line at which there is an abrupt change of circuit constants (accordingly, an abrupt change of surge impedance) is called a **transition point**. Points such as an open- or short-circuited terminal, a junction with another line, a connecting terminal of a machine winding or capacitor as well as a short-circuit fault point are typical transition points. When a travelling wave on a line reaches a transition point, part of the wave is reflected back along the line, and part passes on to other sections of the circuit. The impinging wave is called the **incident wave** and the two waves arising at the transition point are called the **reflected wave** and **transmitted wave**, respectively.

We will now examine the behaviour of travelling-wave (surge) phenomena at the transition point as shown in Figure 18.4. The surge impedance Z_1 on the left-side line is abruptly changed to Z_2 at the transition point a. The behaviour when the incident wave from the left reaches the transition point a is explained by the following equations:

$$\left.\begin{array}{rl} \text{Incident wave} \quad e, i: & \dfrac{e}{i} = Z_1 \\ \text{Transmitted wave} \quad e_t, i_t: & \dfrac{e_t}{i_t} = Z_2 \\ \text{Reflected wave} \quad e_r, i_r: & \dfrac{e_r}{i_r} = Z_1 \\ \text{Continuity of voltages at both sides of point a}: & e + e_r = e_t \\ \text{Continuity of currents at both sides of point a}: & i - i_r = i_t \end{array}\right\} \tag{18.33}$$

The following equation is derived from Equation 18.33:
where

$$\left.\begin{array}{llll} e_r = \dfrac{Z_2 - Z_1}{Z_2 + Z_1} \cdot e & i_r = \dfrac{Z_2 - Z_1}{Z_2 + Z_1} \cdot i & \rho = \dfrac{Z_2 - Z_1}{Z_2 + Z_1}: & \textbf{the reflection operator} \\ e_t = \dfrac{2Z_2}{Z_1 + Z_2} \cdot e & i_t = \dfrac{2Z_1}{Z_1 + Z_2} \cdot i & \mu = \dfrac{2Z_2}{Z_1 + Z_2}: & \textbf{the refraction operator} \end{array}\right\} \tag{18.34}$$

The surge impedance is defined by Equation 18.15, where $\omega L \gg R$, $\omega C \gg G$ for no-loss line, then

$$Z_1 = \sqrt{\frac{L_1}{C_1}}, \quad Z_2 = \sqrt{\frac{L_2}{C_2}} \quad \text{for the ideal (no-loss) line.} \tag{18.35}$$

Simplification by ignoring losses of R, G is justified for most of the analysis of surge phenomena.

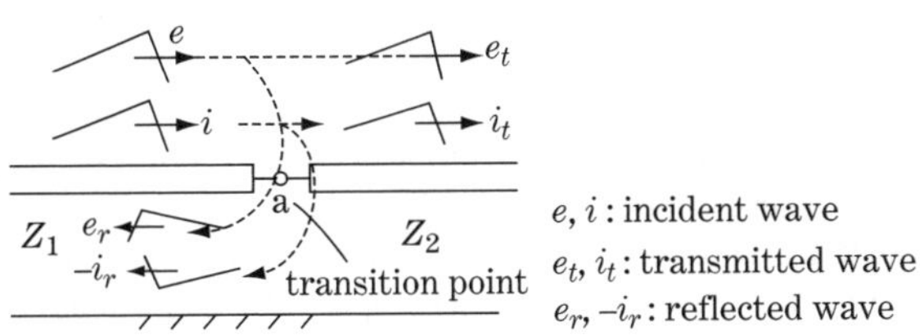

Figure 18.4 Behaviour of voltage and current travelling waves at a transition point

For the continuity of power at a transition point:
the power at the impinging side (left-hand side) is

$$P_A = (e + e_r)(i - i_r) = \left(\frac{2Z_2}{Z_1 + Z_2} \cdot e\right)\left(\frac{2Z_1}{Z_1 + Z_2} \cdot i\right) \tag{18.36}$$

and the power at the transmitted side (right-hand side) is

$$P_B = e_t i_t = \left(\frac{2Z_2}{Z_1 + Z_2} \cdot e\right)\left(\frac{2Z_1}{Z_1 + Z_2} \cdot i\right)$$

$$\therefore \quad P_A = P_B$$

Thus the continuity of power at the transition point is secured.

Note that there is an explanation where the current reflecting wave i_r is defined with opposite polarity sign, so that the equation is given by $e_r/i_r = -Z_1, \quad i + i_r = i_t$. The results are the same under either explanation.

18.3.2 Behaviour of voltage and current travelling waves at typical transition points

Figure 18.5 demonstrates the behaviour over time of impinging voltage and current travelling waves at four typical types of transition points:

- **Case 1 (the transition point is opened), $Z_2 = \infty$:** When the current travelling wave i reaches the transition point, the reflecting wave $i_r(= i)$ arrives at the transition point from the opposite direction (right-hand side) simultaneously so that the current at the point becomes $i - i_r = 0$. The reflecting voltage is $e_r = e$ (because e_r has the same polarity as i_r), so the resulting voltage becomes $e + e_r = 2e$.
- **Case 2 (the transition point is earth grounded), $Z_2 = 0$:** When the voltage travelling wave e reaches the transition point, the reflecting wave $e_r(= -e)$ arrives at the transition point from the opposite direction (right-hand side) simultaneously so that the voltage at the point becomes $e + e_r = 0$. The reflecting current is $i_r = -i$ (because i_r has the same polarity as e_r), so the resulting current becomes $i - i_r = 2i$.
- **Case 3 (the transition point is grounded through a capacitor), $Z_2 = 1/sC$:** The grounding through capacitance acts as if the terminal is directly grounded for the voltage incident wavefront (because $C \cdot de/dt$ is large), which acts as if the terminal is open for the voltage incident wave-tail (because $C \cdot de/dt \to 0$). Accordingly, the behaviour of this case can be drawn analogously to cases 1 and 2.
- **Case 4 (the transition point is grounded through a reactor), $Z_2 = sL$:** The grounding through inductance acts as if the terminal is open for the current incident wavefront (because $L \cdot di/dt$ is large), which acts as if the terminal is grounded for the current incident wave-tail (because $L \cdot di/dt \to 0$). Accordingly, the behaviour of this case can be drawn analogously to cases 1 and 2.

Figure 18.6 demonstrates the behaviour of travelling waves at a transition point, that is the junction of three lines with the same surge impedance (300 Ω).

The behaviour of travelling waves with a waveform of approximately a step function, and at typical transition points, can be drawn as was demonstrated in Figures 18.5 and 18.6. The above description is very helpful in the investigation of every kind of surge phenomena.

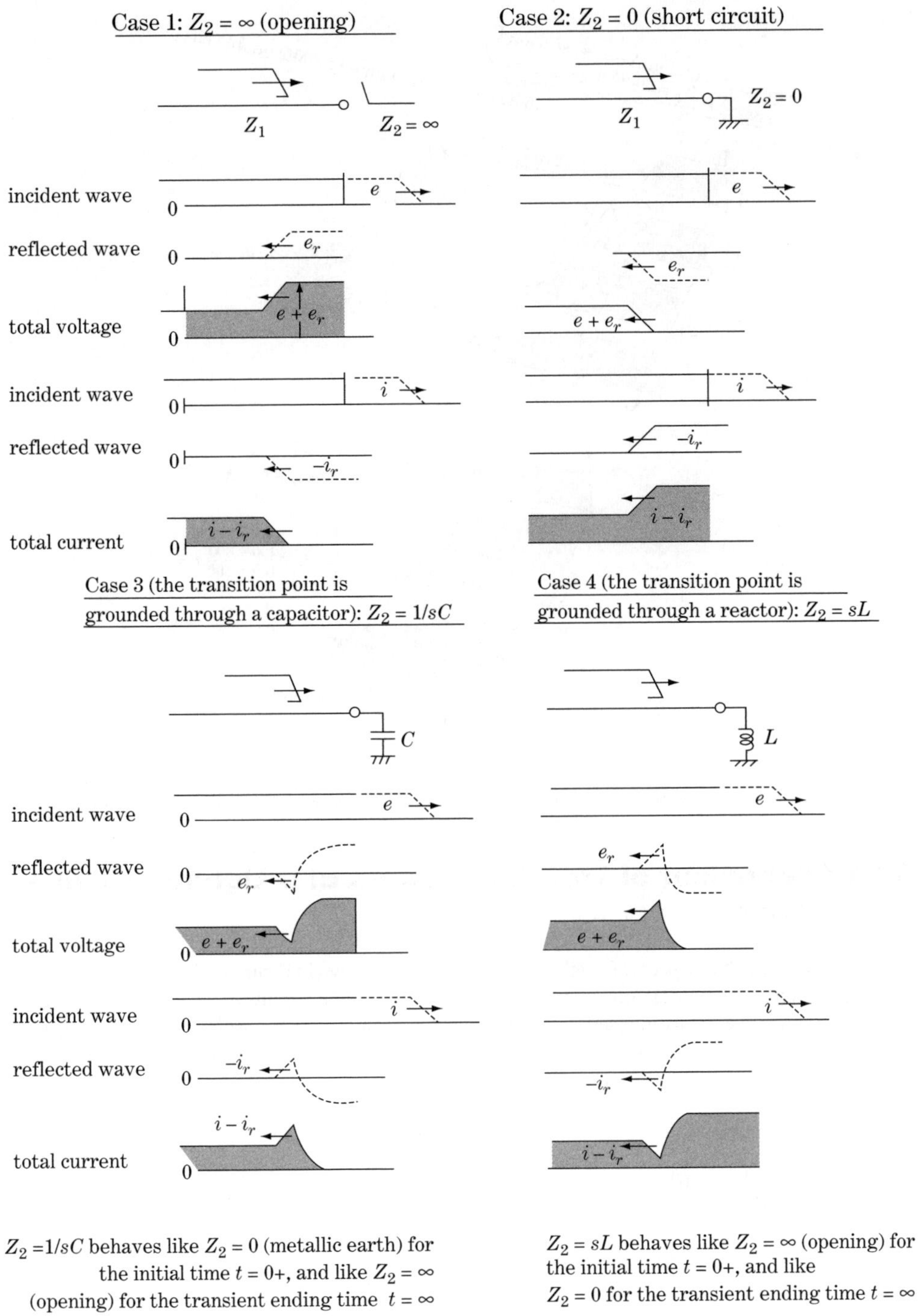

Figure 18.5 Behaviour of travelling waves at typical terminal conditions

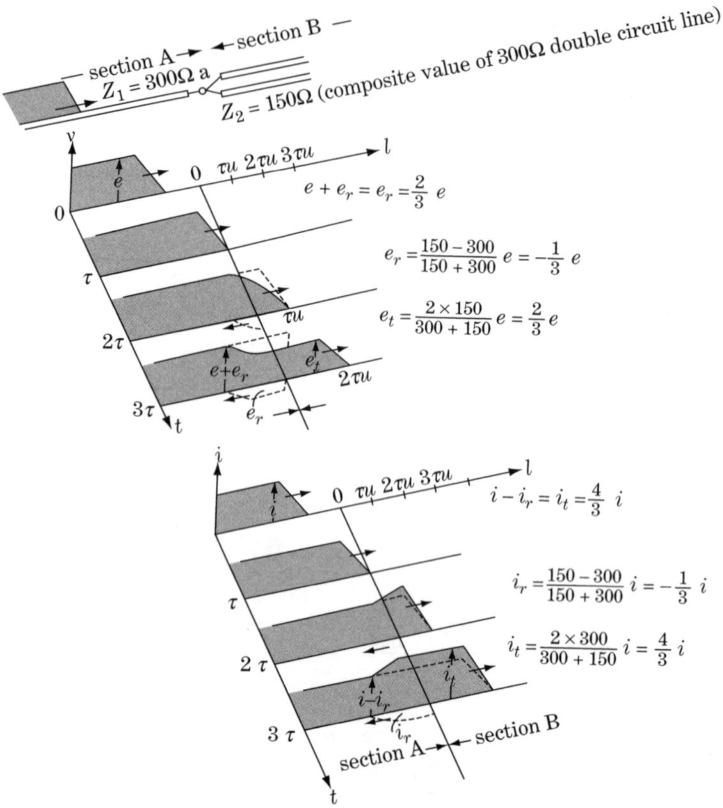

Figure 18.6 Behaviour of voltage and current travelling waves at a typical transition point

18.4 Behaviour of Travelling Waves at a Lightning-strike Point

Now, we investigate the aspects of a lightning-strike point as shown in Figure 18.7a. Point a in the figure is the transition point of surge impedance Z_1, Z_2, and we examine a transmitted surge voltage $E(t)$ or surge current $I(t)$ eventually injected at the transition point. This is to be considered as a travelling wave from a cloud (where Z_0 is the surge impedence of lightning pass) reaches the transition point a whose surge impedance is $Z_{total} = 1/(Z_1^{-1} + Z_2^{-1})$.

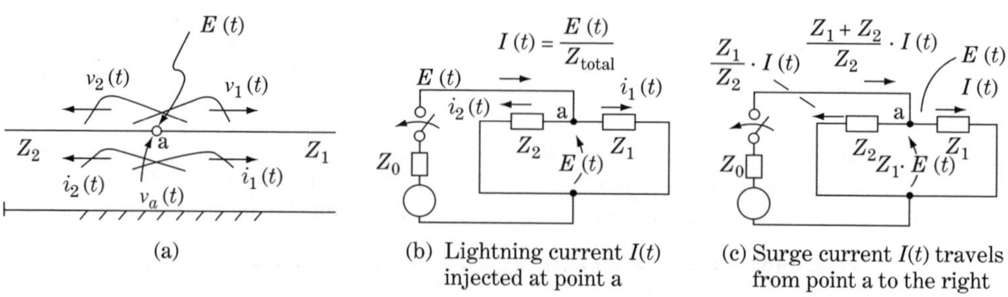

Figure 18.7 Behaviour of travelling waves at a lightning-strike point

We need to distinguish the following three expressions which are apt to be confused with each other:

- **Case 1: A lightning voltage $E(t)$ is transmitted at the transition point a:** This case corresponds to that in Figure 18.7b, in that source voltage $E(t)$ is suddenly injected at point a by closing a switch.

$$\left.\begin{array}{ll}
\underline{\text{Injected voltage at point a}} & \underline{\text{Injected current at point a}} \\
E(t) & I(t) = \dfrac{E(t)}{Z_{\text{total}}(t)} = \left(\dfrac{1}{Z_1} + \dfrac{1}{Z_2}\right) \cdot E(t) \\
\underline{\text{Travelling waves to the right}} & \underline{\text{Travelling waves to the left}} \\
\left.\begin{array}{l} v_1(t) = E(t) \\ i_1(t) = \dfrac{1}{Z_1} \cdot E(t) \end{array}\right\} & \left.\begin{array}{l} v_2(t) = E(t) \\ i_2(t) = \dfrac{1}{Z_2} \cdot E(t) \end{array}\right\}
\end{array}\right\} \qquad (18.37)$$

- **Case 2: A lightning current $I(t)$ is transmitted at the transition point a:** This case is the same as case 1, in that source voltage $E(t) = Z_{\text{total}} \cdot I(t)$ is injected at point a by closing a switch.

$$\left.\begin{array}{ll}
\underline{\text{Injected voltage at point a}} & \underline{\text{Injected current at point a}} \\
E(t) = Z_{\text{total}} \cdot I(t) = \dfrac{Z_1 Z_2}{Z_1 + Z_2} \cdot I(t) & I(t) \\
\underline{\text{Travelling waves to the right}} & \underline{\text{Travelling waves to the left}} \\
\left.\begin{array}{l} v_1(t) = E(t) = Z_{\text{total}} \cdot I(t) \\ i_1(t) = \dfrac{Z_2}{Z_1 + Z_2} \cdot I(t) \end{array}\right\} & \left.\begin{array}{l} v_2(t) = E(t) = Z_{\text{total}} \cdot I(t) \\ i_2(t) = \dfrac{Z_1}{Z_1 + Z_2} \cdot I(t) \end{array}\right\}
\end{array}\right\} \qquad (18.38)$$

- **Case 3: Travelling waves $E(t)$ and $I(t)$ caused at point a to the right direction:** This case corresponds to that in Figure 18.7c, in that source voltage $E(t) = Z_1 \cdot I(t)$ is injected at point a.

$$\left.\begin{array}{ll}
\underline{\text{Injected voltage at point a}} & \underline{\text{Injected current at point a}} \\
E(t) = v_1(t) = v_2(t) = Z_1 \cdot I(t) & i_1(t) + i_2(t) = \dfrac{Z_1 + Z_2}{Z_2} \cdot I(t) \\
\underline{\text{Travelling waves to the right}} & \underline{\text{Travelling waves to the left}} \\
\left.\begin{array}{l} v_1(t) = E(t) = Z_1 \cdot I(t) \\ i_1(t) = I(t) \end{array}\right\} & \left.\begin{array}{l} v_2(t) = E(t) = Z_1 \cdot I(t) \\ i_2(t) = \dfrac{1}{Z_2} \cdot v_2(t) = \dfrac{Z_1}{Z_2} \cdot I(t) \end{array}\right\}
\end{array}\right\} \qquad (18.39)$$

If lightning surge current $I(t)$ is injected at point a of the non-transition point of the transmission line, the above equations can be simplified by $Z = Z_1 = Z_2$. Accordingly the same surge currents of $\frac{1}{2}I(t)$ start travelling in both the right and left directions, and the induced voltage at point a, $v(t) = \{(1/2)I(t)\} \cdot Z = I(t) \cdot \{(1/2)Z\}$, also starts travelling in both directions.

Incidentally, the surge impedance $\sqrt{(L/C)}$ has a particular value belonging to each arbitrary minute section of the line, which is not the value proportional to line length. The line constants R, L, C also have particular values belonging to each arbitrary minute section of the line as a primary concept, which should be strictly distinguished from the concept of concentrated line constants R, L, C whose values are treated expediently as proportional to the line length l, although engineers who are familiar with handling circuits usually by the latter concept may confuse both concepts. Surge phenomena have always to be treated by the former concept, where R, L, C, $\sqrt{(L/C)}, 1/\sqrt{(LC)}$ are treated as distributed constants for each minute section. In other words, the above explanation in Figure 18.7 and by Equations 18.37–18.39 means that the surge voltage and current behaviour at the injection point a and the instantaneous timing $t = 0$ in Figure 18.7a can be explained by Figure 18.7b or c. It does not mean the subsequent behaviour after $t = 0$ (namely, $t > 0$) is explained by Figures 18.7b and c or by Equations 18.37–18.39, because the behaviour after $t = 0$ depends on the condition of all the line length.

Lastly, we studied Figure 18.7 by imagining a lightning strike as the power source. However, it should be stated that the explanation in this section is truly applied to switching phenomena, which are the main theme of Chapter 19.

18.5 Travelling-wave Phenomena of Three-phase Transmission Line

18.5.1 Surge impedance of three-phase line

We consider an arbitrary point p of a three-phase transmission line, where self- and mutual inductances (L_{aa}, L_{ab}, L_{ac}, etc.) as well as self- and mutual capacitances (C_{aa}, C_{ab}, C_{ac} etc.) exist for a minute section.

Now, if a current travelling wave $i_a(t)$ is running through the phase a conductor in a left to right direction, voltage travelling waves $v_a(t) = Z_{aa} \cdot i_a(t)$, $v_b(t) = Z_{ab} \cdot i_a(t)$, $v_c(t) = Z_{ac} \cdot i_a(t)$ ought to be running through the phase a, b, c conductors respectively in the same direction, where Z_{aa}, Z_{ab}, Z_{ac} are the surge impedances and are given by

$$Z_{aa} = \sqrt{L_{aa}/C_{aa}}, \quad Z_{ab} = \sqrt{L_{ab}/C_{ab}}, \quad Z_{ac} = \sqrt{\frac{L_{ac}}{C_{ac}}}.$$

This means that the matrix equation in regard to the accompanying voltage and current travelling waves and the surge impedance matrix can be derived.

Assuming a three-phase-balanced transmission line, the equation is

$$\underbrace{\begin{bmatrix} v_a(t) \\ v_b(t) \\ v_c(t) \end{bmatrix}}_{\boldsymbol{v}_{abc}(t)} = \underbrace{\begin{bmatrix} Z_s & Z_m & Z_m \\ Z_m & Z_s & Z_m \\ Z_m & Z_m & Z_s \end{bmatrix}}_{\boldsymbol{Z}_{abc}} \cdot \underbrace{\begin{bmatrix} i_a(t) \\ i_b(t) \\ i_c(t) \end{bmatrix}}_{\boldsymbol{i}_{abc}(t)} \quad \text{or} \quad \boldsymbol{v}_{abc}(t) = \boldsymbol{Z}_{abc} \cdot \boldsymbol{i}_{abc}(t) \quad ①$$

where $Z_s = \sqrt{\frac{L_s}{C_s}}$, $Z_m = \sqrt{\frac{L_m}{C_m}}$: self-surge imperdance, mutual surge impedance ②

$\boldsymbol{v}_{abc}(t)$: travelling-wave voltages

$\boldsymbol{i}_{abc}(t)$: travelling-wave currents

$\boldsymbol{Z}_{abc}(t)$: surge impedance matrix

(18.40)

The matrix equation can be of course transformed into symmetrical components:

$$\underbrace{\begin{bmatrix} v_0(t) \\ v_1(t) \\ v_2(t) \end{bmatrix}}_{\boldsymbol{v}_{012}(t)} = \begin{bmatrix} Z_s + 2Z_m & 0 & 0 \\ 0 & Z_s - Z_m & 0 \\ 0 & 0 & Z_s - Z_m \end{bmatrix} \cdot \begin{bmatrix} i_0(t) \\ i_1(t) \\ i_2(t) \end{bmatrix} = \underbrace{\begin{bmatrix} Z_0 & & \\ & Z_1 & \\ & & Z_2 \end{bmatrix}}_{\boldsymbol{Z}_{012}} \underbrace{\begin{bmatrix} i_0(t) \\ i_1(t) \\ i_2(t) \end{bmatrix}}_{\boldsymbol{i}_{012}(t)} \quad (18.41)$$

or $\boldsymbol{v}_{012}(t) = \boldsymbol{Z}_{012} \cdot \boldsymbol{i}_{012}(t)$

where

$$Z_1 = Z_2 = Z_s - Z_m = \sqrt{\frac{L_s}{C_s}} - \sqrt{\frac{L_m}{C_m}} \text{ : positive- (negative-)sequence surge impedance}$$

$$Z_0 = Z_s + 2Z_m = \sqrt{\frac{L_s}{C_s}} + 2\sqrt{\frac{L_m}{C_m}} \text{ : zero-sequence surge impedance}$$

$v_{012}(t)$: positive-, negative-, zero-sequence travelling-wave voltages
$i_{012}(t)$: positive-, negative-, zero-sequence travelling-wavecurrents

Now it is clear that the travelling-wave (surge) theory can also be treated by symmetrical components. Analogously, the $\alpha - \beta - 0$ method can be a useful analytical tool for surge phenomena.

18.5.2 Surge analysis by symmetrical coordinates (lightning strike on phase a conductor)

We examine the case where lightning from a thunder cloud directly strikes the phase a conductor of a transmission line at point p, and the strike current $I(t)$ of an approximate step function is transmitted into the phase a conductor, as shown in Figure 18.8a. Here ideal insulation of the line is assumed so that all the line structure withstands the induced surge voltages of each phase. The surge current of $I(t)/2$ begins to travel to the right as well as to the left. Considering the right-hand direction, travelling-wave voltages of $Z_s I(t)/2$, $Z_m \cdot I(t)/2$, $Z_m I(t)/2$ are induced on the phase a, b, c conductors, respectively, by the phase a current $I(t)/2$, and accordingly a voltage of $(Z_s - Z_m)I(t)/2$ is induced between the phase a and b conductors. In other words, these surge voltages begin to travel on each phase conductor. This aspect can be calculated by symmetrical coordinates as below.

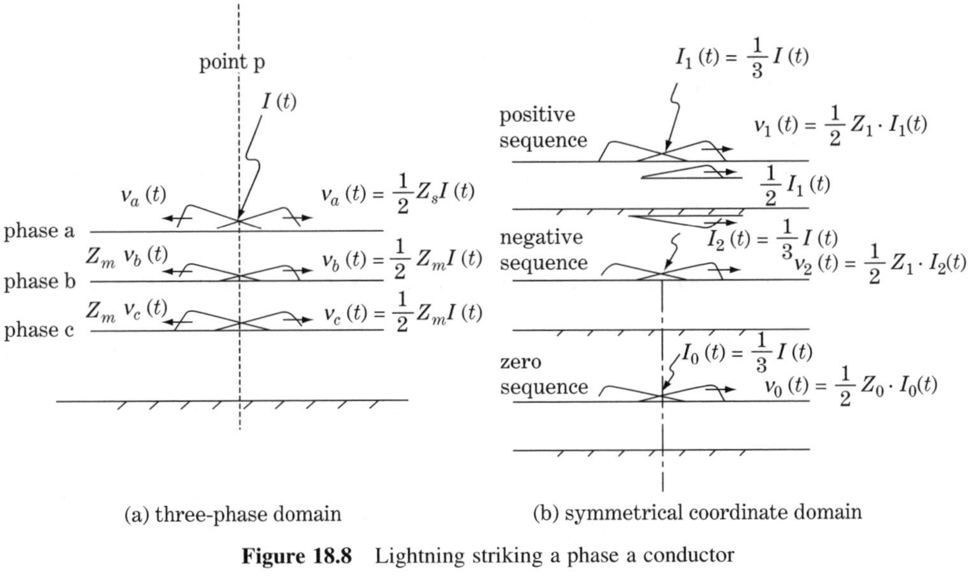

Figure 18.8 Lightning striking a phase a conductor

Surge current $I(t)$ is transmitted into the phase a conductor at point p. Then the surge currents are

$$\left.\begin{aligned} &I_a = I, \quad I_b = 0, \quad I_c = 0 \\ &\therefore \quad I_0 = I_1 = I_2 = \frac{1}{3} I(t) \end{aligned}\right\} \tag{18.42}$$

This means that the lightning current (1/3)I is equally injected into the point p of positive-, negative- and zero-sequence circuits as shown in Figure 18.8b.

Concerning the right direction, the travelling-wave current is one-half of $I(t)/3$ (i.e. $I(t)/6$) at each sequence circuit.

The travelling-wave voltages are

$$\left.\begin{aligned} v_1(t) &= Z_1\left(\frac{1}{2}I_1(t)\right) = \frac{1}{6}\cdot Z_1 I(t) \\ v_2(t) &= Z_1\left(\frac{1}{2}I_2(t)\right) = \frac{1}{6}Z_1 I(t) \\ v_0(t) &= Z_0\left(\frac{1}{2}I_0(t)\right) = \frac{1}{6}Z_0 I(t) \end{aligned}\right\} \tag{18.43}$$

Then the inverse transformed voltages are

$$\left.\begin{aligned} v_a(t) &= v_0(t) + v_1(t) + v_2(t) = \frac{1}{6}(2Z_1 + Z_0)I(t) \\ v_b(t) &= v_0(t) + a^2 v_1(t) + a v_2(t) = \frac{1}{6}(Z_0 - Z_1)I(t) \\ v_c(t) &= v_0(t) + a v_1(t) + a^2 v_2(t) = \frac{1}{6}(Z_0 - Z_1)I(t) \end{aligned}\right\} \tag{18.44}$$

or

$$\left.\begin{aligned} v_a(t) &= \frac{1}{2}Z_s I(t) \\ v_b(t) &= v_c(t) = \frac{1}{2}Z_m I(t) \end{aligned}\right\} \quad \text{where} \quad \left.\begin{aligned} Z_s &= \frac{1}{3}(2Z_1 + Z_0) \\ Z_m &= \frac{1}{3}(Z_0 - Z_1) \end{aligned}\right\} \tag{18.45}$$

Equations 18.44 and 18.45 gives the magnitudes of induced travelling-wave voltages at point p on each phase caused by the injection of transmitted current $I(t)$ to the phase a conductor (lightning hits the phase a conductor).

In actual practice, the insulation of the line could not probably withstand the induced surge voltages and faults would occur at some structural parts of the transmission line. However, this is a matter of insulation strength, which we examine in Chapters 20 and 21.

18.6 Line-to-ground and Line-to-line Travelling Waves

So called **line-to-ground travelling waves** and **line-to-line travelling waves** sometimes appear as useful concepts in practical engineering regarding surge phenomena. This is a kind of transformation of variables from a mathematical viewpoint. (Let us give them the interim names of '***l*–g and *l*–*l* transforms**'.)

Figure 18.9 illustrates the concept of such waves. That is, travelling waves v_a, v_b, v_c are transformed into newly defined travelling waves v_0 (line-to-ground travelling waves), v_{12} (phase a to phase b travelling waves) and v_{13} (phase a to phase c travelling waves). The current waves are of course defined in a similar way.

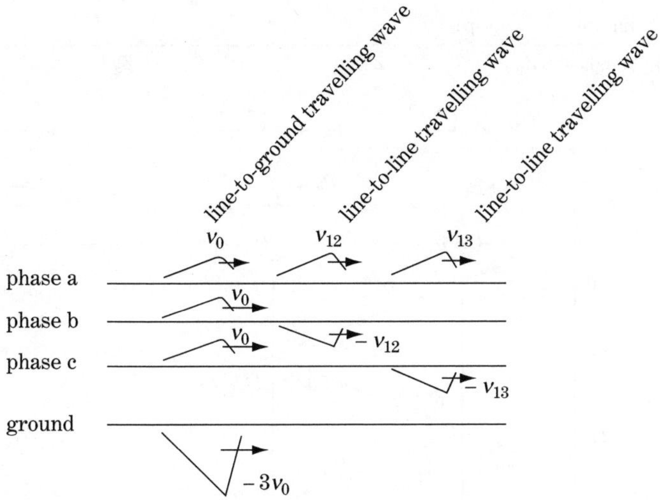

Figure 18.9 Line-to-ground and line-to-line travelling waves

The transformed equation is

$$\underbrace{\begin{bmatrix} v_a(t) \\ v_b(t) \\ v_c(t) \end{bmatrix}}_{\boldsymbol{v}_{abc}(t)} = \underbrace{\begin{bmatrix} 1 & 1 & 1 \\ 1 & -1 & 0 \\ 1 & 0 & -1 \end{bmatrix}}_{\boldsymbol{D}} \cdot \underbrace{\begin{bmatrix} v_0(t) \\ v_{12}(t) \\ v_{13}(t) \end{bmatrix}}_{\boldsymbol{v}_{l-g}(t)} \quad \text{or} \quad \boldsymbol{v}_{abc}(t) = \boldsymbol{D} \cdot \boldsymbol{v}_{l-g}(t) \quad \text{(the same form for currents)} \tag{18.46}$$

The inverse matrix equation is

$$\underbrace{\begin{bmatrix} v_0(t) \\ v_{12}(t) \\ v_{13}(t) \end{bmatrix}}_{\boldsymbol{v}_{l-g}(t)} = \underbrace{\frac{1}{3}\begin{bmatrix} 1 & 1 & 1 \\ 1 & -2 & 1 \\ 1 & 1 & -2 \end{bmatrix}}_{\boldsymbol{D}^{-1}} \cdot \underbrace{\begin{bmatrix} v_a(t) \\ v_b(t) \\ v_c(t) \end{bmatrix}}_{\boldsymbol{v}_{abc}(t)} \quad \text{or} \quad \boldsymbol{v}_{l-g}(t) = \boldsymbol{D}^{-1} \cdot \boldsymbol{i}_{abc}(t) \quad \text{(the same form for currents)} \tag{18.47}$$

Substituting Equation 18.40 ① and the current equation of Equation 18.46 into Equation 18.47, we have

$$\boldsymbol{v}_{l-g}(t) = \boldsymbol{D}^{-1} \cdot \boldsymbol{Z}_{abc} \cdot \boldsymbol{i}_{abc}(t) = (\boldsymbol{D}^{-1} \cdot \boldsymbol{Z}_{abc} \cdot \boldsymbol{D}) \cdot \boldsymbol{i}_{abc}(t) \qquad ①$$

$$\text{where} \quad \boldsymbol{D}^{-1} \cdot \boldsymbol{Z}_{abc} \cdot \boldsymbol{D} = \boldsymbol{Z}_{012} \qquad ②$$

$$\underbrace{\begin{bmatrix} v_0(t) \\ v_{12}(t) \\ v_{13}(t) \end{bmatrix}}_{\boldsymbol{v}_{l-g}(t)} = \underbrace{\begin{bmatrix} Z_s + 2Z_m & 0 & 0 \\ 0 & Z_s - Z_m & 0 \\ 0 & 0 & Z_s - Z_m \end{bmatrix}}_{\boldsymbol{D}^{-1} \cdot \boldsymbol{Z}_{abc} \cdot \boldsymbol{D} = \boldsymbol{Z}_{012}} \cdot \underbrace{\begin{bmatrix} i_0(t) \\ i_{12}(t) \\ i_{13}(t) \end{bmatrix}}_{\boldsymbol{i}_{l-g}(t)} \quad \text{or} \quad \boldsymbol{v}_{l-g}(t) = \boldsymbol{Z}_{012} \cdot \boldsymbol{i}_{l-g}(t) \qquad ③ \tag{18.48}$$

Equation ② is proved by a physical calculation of $\boldsymbol{D}^{-1} \cdot \boldsymbol{Z}_{abc} \cdot \boldsymbol{D}$. The resulting equation ③ means that the impedance Z_{012} in symmetrical components can be commonly adopted for that of '*l*–g and *l*–*l* transformation'.

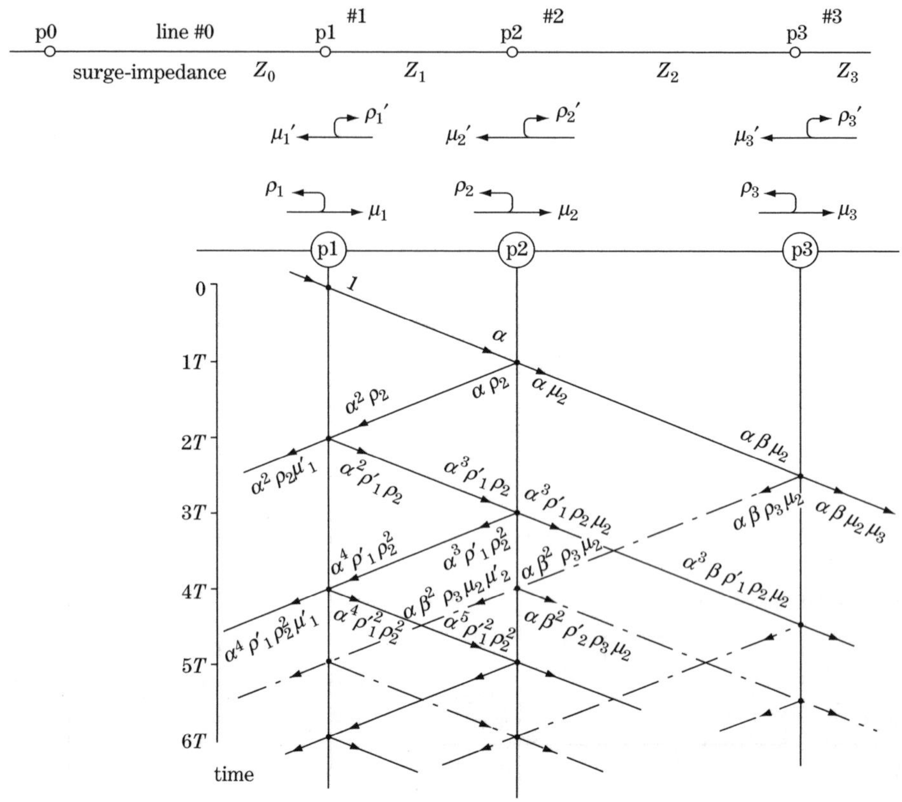

Figure 18.10 The reflection lattice

In the three-phase-balanced transmission line, the travelling waves v_a, v_b, v_c and i_a, i_b, i_c can be transformed into travelling waves v_0, v_{12}, v_{13} and i_0, i_{12}, i_{13}, where v_0, i_0 are called **phase-to-ground travelling waves**, and v_{12}, i_{12} as well as v_{13}, i_{13} are called **phase-to-phase (or line-to-line) travelling waves**. This is our so-called '*l*–g and *l*–*l* transformation' where the impedances of symmetrical components Z_{012} are commonly adopted.

For the phase-to-phase (or line-to-line) travelling waves v_{12}, i_{12} and v_{13}, i_{13}.

$$\left.\begin{array}{ll}\text{surge impedance: } Z_{l-l} = \sqrt{L_1/C_1} & = \sqrt{(\mathrm{L}_s - \mathrm{L}_m)/(\mathrm{C}_s + 3\mathrm{C}_m)} \\ \text{velocity of propagation: } u_{l-l} = 1/\sqrt{L_1 C_1} & = 1/\sqrt{(\mathrm{L}_s - \mathrm{L}_m)\cdot(\mathrm{C}_s + 3\mathrm{C}_m)}\end{array}\right\} ①$$

and for the line-to-ground travelling waves v_0, i_0

$$\left.\begin{array}{l}\text{surge impedance: } Z_{l-g} = \sqrt{L_0/C_0} = \sqrt{(\mathrm{L_s} + 2\mathrm{L_m})/\mathrm{C_s}} \\ \text{velocity of propagation: } u_{l-g} = 1/\sqrt{L_0 C_0} = 1/\sqrt{(\mathrm{L_s} + 2\mathrm{L_m})\cdot \mathrm{C_s}}\end{array}\right\} ② \qquad (18.49)$$

Incidentally,

$$Z_{l-l} < Z_{l-g}, \quad u_{l-l} > u_{l-g} \qquad (18.50)$$

These inequalities can be predicted from our knowledge of the characteristics of positive- and zero-sequence circuits. It is indeed true that the line-to-ground travelling waves v_0, i_0 are a little slower in velocity and attenuate a little faster in comparison with those in phase-to-phase (or line-to-line) travelling waves v_{12}, i_{12}.

Lastly, the '1–g and 1–1 transform' is obviously similar to the $\alpha - \beta - 0$ transformation, in that the line constants (L, C, R, G) as well as the surge impedances of symmetrical components are commonly applied. Analytical examination of travelling waves by the '1–g and 1–1 transformation' is very useful and the descriptions in Sections 23.5 and 23.6 are such examples.

18.7 The Reflection Lattice and Transient Behaviour Modes

18.7.1 The reflection lattice

We examine the surge phenomena shown in Figure 18.10, in which the lines 0, 1, 2, 3 (the surge impedances Z_0, Z_1, Z_2, Z_3, respectively) are connected in series.

Now the incident surge voltage is travelling along line 0 from left to right and arrives at the transition point p1 at time $t = 0$. Then the transmitted wave $e = 1$ (step waveform of value 1) begins to travel along line 1 from left to right. The surge arrives at point p2 with value α (α is the attenuation ratio by travelling one way along line 1, $\alpha = 0 - 1$) at time $t = l_1/u_1$ ($\equiv T$, the unit time of line 1), then the reflected wave (value $\alpha\rho_2$) and the transmitted wave (value $\alpha\mu_2$) appear and begin travelling. Therefore, the voltage at point p1 is zero for $t < T$, while it becomes $(\alpha + \alpha\rho_2)$ or $(\alpha\mu_2)$ for time $T \leqq t$. The travelling wave repeats such reflection/refraction at each transition point p1, p2, p3 until it disappears by attenuation so that **the reflection lattice** can be written as shown in Figure 18.9.

Assuming lines 0 and 2 are far longer than line 1, the surge voltages at the transition points p1, p2 can be derived as follows:

the point p1

$$\begin{aligned}
&\qquad (0-2)T \qquad\qquad (2-4)T \qquad\qquad\qquad (4-6)T \qquad\qquad\qquad (6-8)T \qquad\qquad (8-10)T\\
&v_1(t) = \{(1) + (\alpha^2\rho_2 + \alpha^2\rho_1'\rho_2) + (\alpha^4\rho_1'\rho_1^2 + \alpha^4\rho_1'^2\rho_2^2) + (\alpha^6\rho_1'^2\rho_2^3 + \alpha^6\rho_1'^3\rho_1^3) + (\alpha^8\rho_1'^3\rho_2^4 \\
&\qquad + \alpha^8\rho_1'^4\rho_2^4) + \cdots\}\\
&= [(1) + \alpha^2(1+\rho_1')\rho_2\{1 + (\alpha^2\rho_1'\rho 2) + (\alpha^2\rho_1'\rho_2)^2 + (\alpha^2\rho_1'\rho_2)^3 + (\alpha^2\rho_1'\rho_2)^4 + \cdots\}]\\
&\Rightarrow \text{converging to} \quad \frac{1+\alpha^2\rho_2}{1-\alpha^2\rho_1'\rho_2} \qquad \text{①}
\end{aligned}$$

the point p2

$$\begin{aligned}
&\qquad (0-1)T\,(1-3)T \qquad\quad (3-5)T \qquad\qquad\qquad (5-7)T \qquad\qquad\qquad (7-9)T\\
&v_2(t) = \{(0) + (\alpha + \alpha\rho_2) + (\alpha^3\rho_1'\rho_2 + \alpha^3\rho_1'\rho_2^2) + (\alpha^5\rho_1'^2\rho_2^2 + \alpha^5\rho_1'^2\rho_2^3) + (\alpha^7\rho_1'^3\rho_2^3 + \alpha^7\rho_1'^3\rho_1^4) + \cdots\}\\
&= \left[(0) + \alpha(1+\rho_2)\left\{1 + (\alpha^2\rho_1'\rho_2) + (\alpha^2\rho_1'\rho_2)^2 + (\alpha^2\rho_1'\rho_2)^3 + (\alpha^2\rho_1'\rho_2)^4 + \cdots\right\}\right]\\
&\Rightarrow \text{converging to} \quad \frac{\alpha(1+\rho_2)}{1-\alpha^2\rho_1'\rho_2} = \frac{\alpha\mu_2}{1-\alpha^2\rho_1'\rho_2} \qquad \text{②}
\end{aligned}$$

the point p2

The above v_2 obviously coincides with the equation below.

$$\begin{aligned}
&\qquad (0-1)T\,(1-3)T \quad (3-5)T \qquad (5-7)T \qquad\qquad (7-9)T\\
&v_2(t) = \{(0) + (\alpha\mu_2) + (\alpha\mu_2)(\alpha^2\rho_1'\rho_2) + (\alpha\mu_2)(\alpha^2\rho_1'\rho_2)^2 + (\alpha\mu_2)(\alpha^2\rho_1'\rho_2)^3 + \cdots\}\\
&\Rightarrow \text{converging to} \quad \frac{\alpha\mu_2}{1-\alpha^2\rho_1'\rho_2} \qquad \text{③}
\end{aligned}$$

where

Z_0, Z_1, Z_2, Z_3 : surge impedances of lines 0, 1, 2, 3 respectively.

Reflection operators ρ, ρ' and refraction operators μ, μ' are

$$\begin{aligned}
\rho_1 &= (Z_1 - Z_0)/(Z_1 + Z_0), \quad \mu_1 = 2Z_1/(Z_1 + Z_0)\\
\rho_1' &= (Z_0 - Z_1)/(Z_0 + Z_1), \quad \mu_1' = 2Z_0/(Z_0 + Z_1)\\
\rho_2 &= (Z_2 - Z_1)/(Z_2 + Z_1), \quad \mu_2 = 2Z_2/(Z_2 + Z_1) \qquad \text{etc.}
\end{aligned}$$

$T = l_1/u_1$: the unit time of line 1 (length : l_1, *velocity* : $u_1 = 1/\sqrt{L_1 \cdot C_1}$)

(18.51)

Note that $1 + x + x^2 + x^3 + x^4 + \ldots = 1/(1-x)$.

The voltages $v_1(t), v_2(t)$ for all times $t = (0 - n)T$ can be calculated by the above equations. The converged values of $v_1(t), v_2(t)$ mean the final values of transient voltages as $t = \rightarrow \infty$ (namely steady-state voltage); $v_2(t)$ can be derived either by equation ① or ②.

Note that the terms of the reflected waves from point p0 or p3 are not yet included in the equation. Accordingly, whenever the reflected wave from point p3 (or p0) arrive at point p2 (or p1), the related transmitted wave terms should be added to the above lattice equations according to the rules of lattice analysis.

Needless to say, if the length of the analytical line section Z_1 is relatively short in comparison with the length of the next sections Z_0, Z_2, the transient behaviour early on can be examined by Equation 18.51 without any additional reflected voltage terms.

18.7.2 Oscillatory and non-oscillatory convergence

Examining Equations 18.51 ①②, the equations for $v_1(t), v_2(t)$ are simple increasing arithmetic series under the condition $\rho'_1\rho_2 > 0$, and the oscillatory series under the condition $\rho'_1\rho_2 < 0$ over time. This can be summarized as follows.

For **case 1a** $Z_0, Z_2 > Z_1$, namely $\rho'_1, \rho_2 > 0$, or for **case 1b** $Z_0, Z_2 < Z_1$, namely $\rho'_1, \rho_2 < 0$. That is, $v_1(t)$, $v_2(t)$ are simple increasing non-oscillatory series, or in other words $dv_1(t)/dt > 0, dv_2(t)/dt > 0$. The voltage $v(t)$ at any arbitrary point within the section Z_1 is also non-oscillatory.

For **case 2a** $Z_0 > Z_1 > Z_2 \quad (\rho'_1 > 0, \quad \rho_2 < 0)$, or for **case 2b** $Z_0 < Z_1 < Z_2 \quad (\rho'_1 < 0, \quad \rho_2 > 0)$. That is, $v_1(t), v_2(t)$ are oscillatory series over time; in other words, the derivative $dv(t)/dt$ repeats plus and minus.

In Equation 18.51, if $\alpha = 1.0$ and $\rho'_1 = 1.0$, $\rho_2 = 1.0$ are assumed, the voltage finally diverges to an infinite value. Of course this is unrealistic, because the original transmitted traveling wave voltage is of limited tail length in stead of step wave from with infinitive tail length first, and furthermore the attenuation factor is actually the value of $0 < \alpha < 1.0$ so that $v(t)$ converges finally to a finite steady-state value. However, the converged value obviously becomes very large under the conditions of $\alpha \fallingdotseq 1.0$ and Z_0, $Z_2 \gg Z_1$.

This case may be compared with a landscape, where a narrow lowland Z_1 is surrounded by highland Z_0, Z_2 and water floods in from the highland Z_0 to the lowland Z_1. A violent water wave would thus be caused at Z_1. Of course, this violent behaviour would be reduced considerably if Z_2 were also lowland at a similar level or even lower.

18.8 Supplement 1: General Solution Equation 18.10 for Differential Equation 18.9

We introduce new variables $v'(x,t)$ for convenience as below:

$$v(x,t) = \mathrm{e}^{-\alpha t} \cdot v'(x,t) \tag{1}$$

Differentiating Equation 1 twice with respect to x,

$$\frac{\partial^2 v(x,t)}{\partial x^2} = \mathrm{e}^{-\alpha t} \cdot \frac{\partial^2 v'(x,t)}{\partial x^2} \tag{2}$$

Partial differentiation of Equation 1 twice with respect to t gives

$$\begin{aligned}\frac{\partial v(x,t)}{\partial t} &= -\alpha \mathrm{e}^{-\alpha t} \cdot v'(x,t) + \mathrm{e}^{-\alpha t} \cdot \frac{\partial v'(x,t)}{\partial t} \\ &= \mathrm{e}^{-\alpha t}\left\{-\alpha v'(x,t) + \frac{\partial v'(x,t)}{\partial t}\right\}\end{aligned} \tag{3}$$

$$\begin{aligned}\frac{\partial^2 v(x,t)}{\partial t^2} &= -\alpha e^{-\alpha t}\left\{-\alpha v'(x,t) + \frac{\partial v'(x,t)}{\partial t}\right\} + e^{-\alpha t}\left\{-\alpha \cdot \frac{\partial v'(x,t)}{\partial t} + \frac{\partial^2 v'(x,t)}{\partial t^2}\right\} \\ &= e^{-\alpha t}\left\{\alpha^2 v'(x,t) - 2\alpha \cdot \frac{\partial v'(x,t)}{\partial t} + \frac{\partial^2 v'(x,t)}{\partial t^2}\right\}\end{aligned} \tag{4}$$

Substituting Equations 1, 3, 4 into the right-hand side of Equation 18.9 ①,

$$\text{the right-hand side of Equation 18.9①} = LCe^{-\alpha t} \cdot \frac{\partial^2 v'(x,t)}{\partial t^2} \tag{5}$$

On the other hand, Equation 2 is just the same as the left-hand side of Equation 18.9 ①. Accordingly, Equation 18.9 ① for $v(x,t)$ is replaced by the equation below for $v'(x,t)$:

$$\frac{\partial^2 v'(x,t)}{\partial x^2} = LC\frac{\partial^2 v'(x,t)}{\partial t^2} \tag{6}$$

This equation is in the same form as Equation 18.5 so that the general solution is in the form of Equation 18.6, namely

$$v'(x,t) = v_1(x - ut) + v_2(x + ut) \tag{7}$$

Therefore the general solution of Equation 18.9 ① is

$$v(x,t) = e^{-\alpha t}\{v_1(x - ut) + v_2(x + ut)\} \tag{8}$$

The equation for the current can be derived analogously.

18.9 Supplement 2: Derivation of Equation 18.19 from Equation 18.18

Utilizing Equation 18.18 ①②, we derive ① ± ② · $Z_0(s)$. Then

$$V(x,s) + Z_0(s) \cdot I(x,s) = e^{-\gamma(s)x}\{V(0,s) + Z_0(s) \cdot I(0,s)\} \tag{1}$$

$$V(x,s) - Z_0(s) \cdot I(x,s) = e^{\gamma(s)x}\{V(0,s) - Z_0(s) \cdot I(0,s)\} \tag{2}$$

Transferring the exponential terms of both equations from the right to the left, and adding both equations, gives Equation 18.19.

19

Switching Surge Phenomena by Circuit-breakers and Line Switches

Power systems engineering cannot be discussed without mentioning the 'switching operation', which is closely related to overvoltages and high-frequency transient phenomena. Switching overvoltages caused by tripping/closing operations of circuit-breakers or line switches are inevitable phenomena which should be overcome by a combination of various specialized practical engineering countermeasures.

In this chapter, we study switching surges through a mathematical treatment first, because we believe this approach provides an essential base and is a shorter way to gain a better understanding of actual practical engineering.

19.1 Transient Calculation of a Single Phase Circuit by Breaker Opening

Calculation of transient phenomena is not easy, and a simple circuit even with just a few elements of *L*, *C*, *R* cannot generally be solved without proper approximations. However, engineers need to find transient solutions in practical engineering probably for more complicated circuits. Accordingly, 'to derive an accurate solution through a reasonable approximation' is an important part of practical engineering. Furthermore, engineers can generally discover the essentials of engineering practice through properly simplified model themes than through a more complicated large-scale theme. The calculation of transient behaviour by a breaker switching operation introduced in this chapter would be useful not only for a better understanding of switching phenomena but also as a good exercise in transient calculation techniques.

19.1.1 Calculation of fault current tripping (single phase circuit)

In the beginning, we carry out a transient calculation of the single phase circuit shown in Figure 19.1a in that a short-circuit fault occurred at point f and the breaker is going to be opened. Our objective is to find the transient voltage (switching surge voltage) across the two contacts, $v_{Br}(t)$ and that of the phase-to ground voltage $v_b(t)$, that are deeply related to the required breaker's tripping duties and the required insulation levels of the network.

© 2007 John Wiley & Sons, Ltd

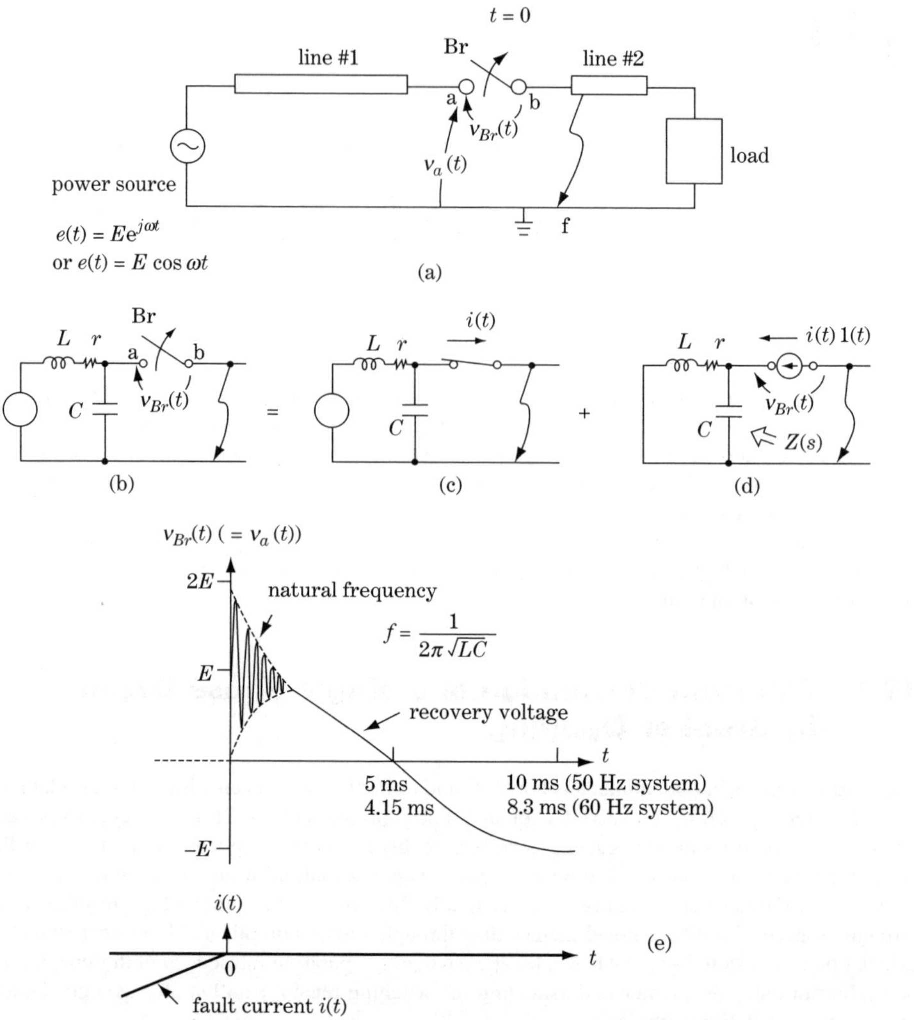

Figure 19.1 Transient calculation for single phase breaker tripping (the case of single source circuit)

The circuit of the calculation is given by Figure 19.1b where an L-type concentrated circuit is adopted. The transient calculation of this circuit can be solved as the superposition of the calculations in Figures 19.1c and d, by Thévenin's theorem.

19.1.1.1 Step 1: Calculation of steady-state term

Whenever a short-circuit fault occurs, a protective relay detects the fault within 1–3 cycles and dispatches a tripping signal to the associated circuit-breaker. The breaker begins to open the contacts immediately after receiving the signal and completes fault current tripping at 'the **time of current zero**' and within 2–3 cycles.

We assume that the transient term of the fault current is already attenuated at the timing of tripping completion, although actually the d.c. current component may not yet have disappeared (because the d.c. time constant could be of order 0.05–0.1 s).

The source voltage is

$$e(t) = \mathrm{Re}\{Ee^{j\omega t}\} = E\cos\omega t \tag{19.1}$$

The fault current $i(t)$ flowing through the breaker is almost not affected by C in the figure. Then

$$i(t) = \mathrm{Re}\left\{\frac{E}{j\omega L + r}e^{j\omega t}\right\} = \frac{E}{\omega L\left(1+\frac{r^2}{\omega^2 L^2}\right)}\left\{\sin\omega t + \frac{r}{\omega L}\cos\omega t\right\} \tag{19.2a}$$

Here we can utilize the condition $\omega L \gg r$ (the reason is explained later). Accordingly,

$$i(t) = \frac{E}{\omega L}\sin\omega t \tag{19.2b}$$

This is the initial current before breaker tripping.

19.1.1.2 Step 2: Calculation of transient term

The transient term can be calculated by applying Tevenin's law to the Figure 19.1d, that is, to insert the current source i(t) across the breaker contactors so that the flowing current just before switch opening are cancelled at t=0. Current breaking finishes at the time of '**current zero**' according to the nature of circuit-breakers (the reason is explained later) so that the initial current is given in the form of $i(t)\mathbf{1}(t) = \left(\frac{E}{\omega L}\right)\sin\omega t\,\mathbf{1}(t)$, which becomes zero at time $t = 0$. Also, the Laplace transform of $\sin\omega t$ is $\omega/\left(s^2+\omega^2\right)$. Then

$$\left.\begin{aligned}
&i(s) = \frac{E}{\omega L}\cdot\mathcal{L}[\sin\omega t] = \frac{E}{\omega L}\cdot\frac{\omega}{s^2+\omega^2} && \text{①}\\
&Z(s) = \frac{1}{\frac{1}{Ls+r}+Cs} = \frac{1}{C}\cdot\frac{s+\frac{r}{L}}{s^2+\frac{r}{L}s+\frac{1}{LC}} && \text{②}\\
&\therefore\ v_{Br}(s) = i(s)\cdot Z(s) = \frac{E}{LC}\cdot\frac{s+\frac{r}{L}}{(s^2+\omega^2)\left(s^2+\frac{r}{L}s+\frac{1}{LC}\right)} \equiv \frac{E}{LC}\cdot F(s) && \text{③}
\end{aligned}\right\} \tag{19.3}$$

Equation ② for $Z(s)$ is the impedance looking into the circuit across the contacts from the current source in the s domain as in Figure 19.1b. Accordingly, equation ③ is the voltage solution $v_{Br}(s)$ but in the s domain.

$F(s)$ defined in Equation ③ is

$$\left.\begin{aligned}
F(s) &= \frac{s+\frac{r}{L}}{(s^2+\omega^2)\left(s^2+\frac{r}{L}s+\frac{1}{LC}\right)} = \frac{s+2\alpha}{(s^2+\omega^2)(s^2+2\alpha s+u^2)}\\
&= \frac{s+2\alpha}{(s^2+\omega^2)\left\{(s+\alpha)^2+(u^2-\alpha^2)\right\}}\\
&\text{where}\quad u = \frac{1}{\sqrt{LC}},\qquad \alpha = \frac{r}{2L}
\end{aligned}\right\} \tag{19.4}$$

$F(s)$ has the denominator of fourth order in s, so the inverse transformation to the t domain is not necessarily easy.

Here we investigate the magnitudes of constants in preparation for our actually justifiable approximation.

Assuming $L = 1\,\text{mH/km}$, $C = 0.01\,\mu\text{F/km}$, $r = 0.01\,\Omega/\text{km}$, $f = 50\,\text{Hz}$, then

$$\left.\begin{aligned}
&\omega = 2\pi f = 314, \quad \omega^2 \fallingdotseq 10^5 \\
&\omega L = 314 \times 10^{-3}\Omega/\text{km} \fallingdotseq 0.3\,\Omega/\text{km} \\
&\frac{1}{\omega C} = \frac{1}{314 \times 10^{-8}}\Omega/\text{km} \fallingdotseq 3 \times 10^5\,\Omega/\text{km} \\
&\omega L \cdot \omega C \fallingdotseq 0.3 \times \frac{1}{3 \times 10^5} \fallingdotseq 10^{-6} \\
&u = \frac{1}{\sqrt{LC}} = \frac{1}{\sqrt{10^{-3} \times 10^{-8}}} \fallingdotseq 300\,000\,\text{km/s} \\
&2\alpha = \frac{r}{L} = \frac{0.01}{10^{-3}} = 10, \quad \alpha = 5
\end{aligned}\right\} ① \tag{19.5}$$

Accordingly, to reasonable accuracy,

$$\frac{1}{\omega C} \gg \omega L \gg r, \quad u \gg \alpha \quad ②$$

Then we can adopt the approximation $(u^2 - \alpha^2) \to u^2$ and the equation for $F(s)$ is

$$\begin{aligned}
F(s) &= \frac{s + 2\alpha}{(s+\omega)^2\{(s+\alpha)^2 + (u^2 - \underset{\nwarrow}{\alpha^2})\}} = \frac{s + 2\alpha}{(s + j\omega)(s - j\omega)(s + \alpha + ju)(s + \alpha - ju)} \\
&= \frac{k_1}{s + j\omega} + \frac{k_2}{s - j\omega} + \frac{k_3}{s + \alpha + ju} + \frac{k_4}{s + \alpha - ju}
\end{aligned} \tag{19.6}$$

The arrow $\nwarrow$ in the above equation indicates the part which is neglected as a reasonable basis for simplicity. (These arrows will be used as indicators of omission for reasonable simplifications.)

From the formula of the inverse Laplace transform

$$\mathcal{L}^{-1}\left(\frac{1}{s \pm a}\right) = e^{\mp at} \tag{19.7}$$

accordingly

$$\begin{aligned}
F(t) &= k_1 e^{-j\omega t} + k_2 e^{j\omega t} + k_3 e^{-(\alpha + ju)t} + k_4 e^{-(\alpha - ju)t} \\
&= \{(k_1 + k_2)\cos \omega t - j(k_1 - k_2)\sin \omega t\} \\
&\quad + e^{-\alpha t}\{(k_2 + k_4)\cos ut - j(k_3 - k_4)\sin ut\}
\end{aligned} \tag{19.8}$$

Now we still have work to do to find the coefficients k_1, k_2, k_3, k_4. The calculation procedure and the results are given in Supplement 1. Through the calculation, a total of 15 'arrows of omission' appear unexpectedly for only such the simple circuit. The general method to find these coefficients has already been explained in Supplement 1 of Chapter 10.

Consequently, the result obtained is

$$k_1 = k_2 = -k_3 = -k_4 = \frac{1}{2u^2} = \frac{LC}{2} \tag{19.9}$$

Then

$$\left.\begin{aligned}
&F(t) = LC(\cos\omega t - \mathrm{e}^{-\alpha t}\cos ut) && ①\\
&\therefore \quad v_{Br}(t) = v_a(t) = \frac{E}{LC}\cdot F(t) = E(\cos\omega t - \mathrm{e}^{-\alpha t}\cos ut)\\
&\qquad = E\left\{\underbrace{\cos\omega t}_{\text{steady-state term}} - \underbrace{\mathrm{e}^{-\frac{r}{2L}t}\cos\frac{1}{\sqrt{LC}}\cdot t}_{\text{transient term}}\right\} && ②\\
&\text{where} \qquad \text{natural oscillation frequency } f_0 = \frac{1}{2\pi\sqrt{LC}}\\
&\qquad \text{attenuation time constant } T = 2L/r
\end{aligned}\right\} \quad (19.10)$$

This is the solution of this circuit and the transient waveform of the equation is indicated in Figure 19.1e. The solution indicates the following conclusion:

- Current trip finishes at $t = 0$, at the time $i(t) = 0$ (**current-zero tripping**), and the source voltage at $t = 0$ is the peak value E (because the power factor of the fault current is almost 90° during the fault).
- The steady-state term of the voltage between contacts $v_{Br}(t)$ (**the recovery voltage** across the breaker contacts after breaker opening) is $E\cos\omega t$.
- The transient oscillation term appears just after tripping, which is of natural frequency f_0 and with attenuation time constant $T = 2L/r$, and it oscillates by $\pm E$ within the band 0–2E. As a result, the voltage before attenuation $v_{Br}(t)$ (**the transient recovery voltage** of the breaker just after breaker opening) reaches a magnitude of $2E$ in maximum.
- In this particular case, $v_{Br}(t) = v_a(t)$, because the voltage at point b is $v_b(t) = 0$. In other words, Equation 19.10 ② is also the phase overvoltage appearing on the left-hand side of the breaker terminal.

The resulting oscillation is the total of the repeated reflection and transmission of the travelling waves at the transition points.

19.1.2 Calculation of current tripping (double power source circuit)

The next circuit is shown in Figure 19.2a, in which double power sources exist at both terminal sides of the breaker. The calculation of this case is more complicated in comparison with the previous case, because the denominator of $F(s)$ here becomes an s function of sixth order. As a matter of fact, readers can count 48 'arrows of omission' throughout the calculation process of this problem. This is a typical example showing that proper approximation is a very important technique in practical engineering.

19.1.2.1 Step 1: Calculation of steady-state term

The generators are operating in synchronism with angular difference δ:

$$\left.\begin{aligned}
\dot{e}_1(t) &= E_1\,\mathrm{e}^{j\omega t}\\
\dot{e}'_1(t) &= E'_1\,\mathrm{e}^{j(\omega t-\delta)}
\end{aligned}\right\} \quad (19.11)$$

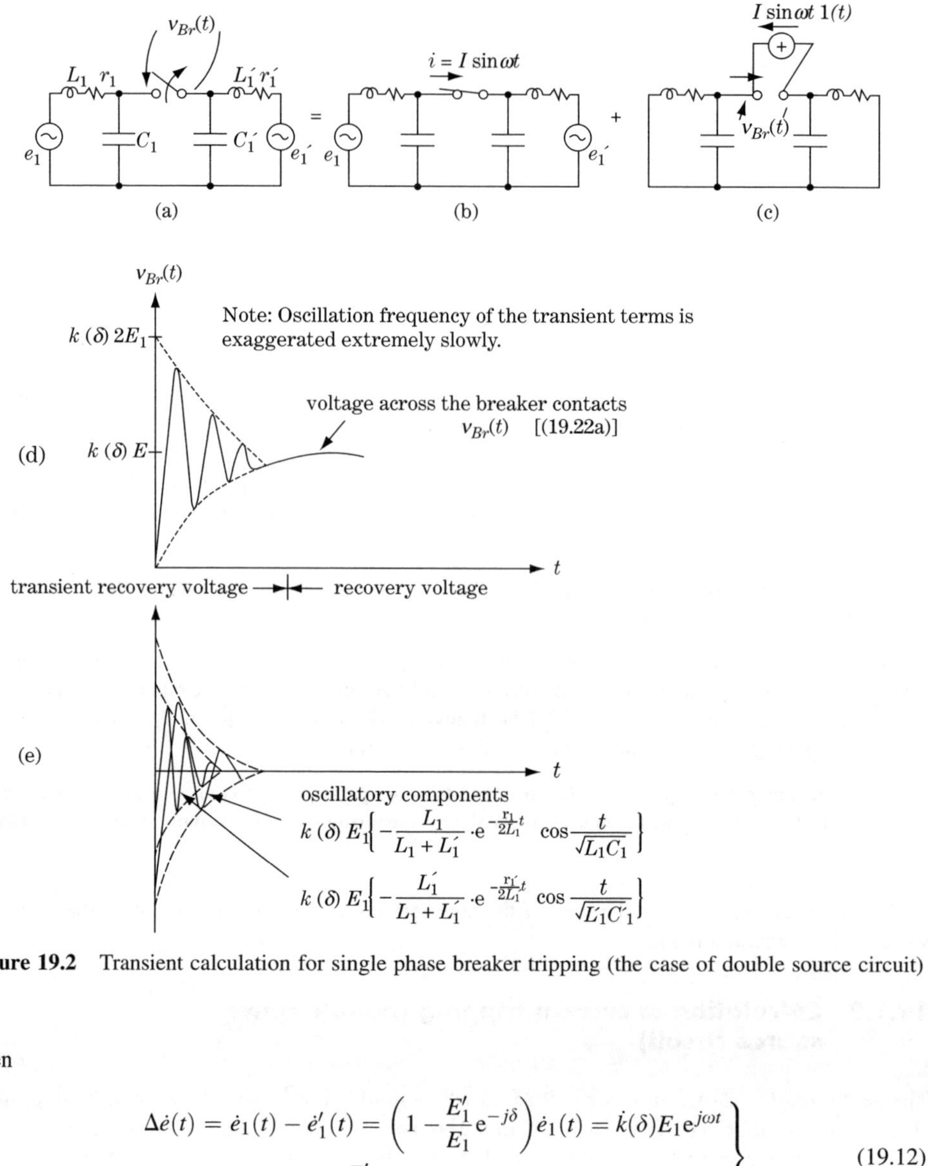

Figure 19.2 Transient calculation for single phase breaker tripping (the case of double source circuit)

Then

$$\left.\begin{aligned}\Delta\dot{e}(t) &= \dot{e}_1(t) - \dot{e}_1'(t) = \left(1 - \frac{E_1'}{E_1}e^{-j\delta}\right)\dot{e}_1(t) = \dot{k}(\delta)E_1 e^{j\omega t}\\ \dot{k}(\delta) &= 1 - \frac{E'}{E_1}e^{-j\delta}\end{aligned}\right\} \qquad (19.12)$$

The current $i(t)$ flowing through the breaker before tripping is not affected by the existence of C_1, and C_1', so the current equation is

$$\left.\begin{aligned}\dot{i}(t) &= \frac{\Delta\dot{e}(t)}{(j\omega L_1 + r_1) + (j\omega L_1' + r_1')} \fallingdotseq \frac{-j\Delta\dot{e}(t)}{\omega(L_1 + L_1')} = \dot{k}(\delta)\cdot\frac{-jE_1}{\omega(L_1 + L_1')}\cdot e^{j\omega t}\\ &= \dot{k}(\delta)\cdot\frac{E_1}{\omega(L_1 + L_1')}\cdot e^{j(\omega t - 90^\circ)}\end{aligned}\right\} \qquad (19.13)$$

Taking the real parts of all the above equations, for the power sources

$$\left.\begin{aligned} e_1(t) &= E_1 \cos \omega t \\ e_1'(t) &= E_1' \cos(\omega t - \delta) \end{aligned}\right\} \tag{19.14}$$

and for the fault current before tripping (steady-state current)

$$\left.\begin{aligned} i(t) &= \left|\dot{k}(\delta)\right| \cdot \frac{E_1}{\omega(L_1 + L_1')} \cos(\omega t - 90^\circ) = \left|\dot{k}(\delta)\right| \cdot \frac{E_1}{\omega(L_1 + L_1')} \sin \omega t \\ &\equiv I \sin \omega t \\ &\text{where } I = \left|\dot{k}(\delta)\right| \cdot \frac{E_1}{\omega(L_1 + L_1')}, \quad \dot{k}(\delta) = 1 - \frac{E_1'}{E_1} \mathrm{e}^{-j\delta} \\ &i(t) \text{ is 'current zero' at } t = 0 \end{aligned}\right\} \tag{19.15}$$

19.1.2.2 Step 2: Calculation of transient term

As shown in the Figure 19.2c the current $i(t)\mathbf{1}(t) = I \sin \omega t \cdot \mathbf{1}(t)$ is inserted at the position of the breaker terminals.

In the Laplace transform domain,

$$\left.\begin{aligned}
&\text{initial current} \\
&i(s) = I \cdot \frac{\omega}{s^2 + \omega^2} && ① \\
&\text{circuit impedances} \\
&\Sigma Z(s) = Z_1(s) + Z_1'(s) && ② \\
&Z_1(s) = \frac{1}{\dfrac{1}{L_1 s + r_1} + C_1 s} = \frac{1}{C_1} \cdot \frac{s + \dfrac{r_1}{L_1}}{s^2 + \dfrac{r_1}{L_1} s + \dfrac{1}{L_1 C_1}} = \frac{1}{C_1} \cdot \frac{s + 2\alpha_1}{(s + \alpha_1)^2 + u_1^2} && ③ \\
&Z_1'(s) = \frac{1}{\dfrac{1}{L_1' s + r_1'} + C_1' s} = \frac{1}{C_1'} \cdot \frac{s + \dfrac{r_1'}{L_1'}}{s^2 + \dfrac{r_1'}{L_1'} s + \dfrac{1}{L_1' C_1'}} = \frac{1}{C_1'} \cdot \frac{s + 2\alpha_2}{(s + \alpha_2)^2 + u_2^2} && ④ \\
&\text{transient recovery voltage across the contacts} \\
&v_{Br}(s) = i(s) \cdot \Sigma Z_1(s) = I\omega \left\{ \frac{1}{s^2 + \omega^2} (Z_1(s) + Z_1'(s)) \right\} \equiv I\omega \cdot F(s) && ⑤
\end{aligned}\right\} \tag{19.16}$$

$Z_1(s)$, $Z_1'(s)$ are of the same form as $Z(s)$ in Equation 19.3, so that $F(s)$ defined in Equation 19.16 ⑤ can be modified as

$$\begin{aligned}
F(s) &= \frac{1}{s^2 + \omega^2} \cdot \left\{ \frac{1}{C_1} \cdot \frac{s + 2\alpha_1}{(s + \alpha_1)^2 + u_1^2} + \frac{1}{C_1'} \cdot \frac{s + 2\alpha_2}{(s + \alpha_2)^2 + u_2^2} \right\} \\
&= \frac{s}{s^2 + \omega^2} \cdot \frac{C_1 \left\{ (s + \alpha_1)^2 + u_1^2 \right\} + C_1' \left\{ (s + \alpha_2) + u_2^2 \right\}}{C_1 C_1' \left\{ (s + \alpha_1)^2 + u_1^2 \right\} \cdot \left\{ (s + \alpha_2)^2 + u_2^2 \right\}} \\
&= \frac{s \{ C_1 (s^2 + u_1^2) + C_1' (s^2 + u_2^2) \}}{C_1 C_1' (s + j\omega)(s - j\omega)(s + \alpha_1 + ju_1)(s + \alpha_1 - ju_1)(s + \alpha_2 + ju_2)(s + \alpha_2 - ju_2)} \\
&= \frac{k_1}{s + j\omega} + \frac{k_2}{s - j\omega} + \frac{k_3}{s + \alpha_1 + ju_1} + \frac{k_4}{s + \alpha_1 - ju_1} + \frac{k_5}{s + \alpha_2 + ju_2} + \frac{k_6}{s + d_2 - ju_2}
\end{aligned} \tag{19.17}$$

$F(s)$ is the addition of the two terms whose denominators contain the fourth-order s function. Accordingly, $F(s)$ has a denominator of sixth order.

As shown in Supplement 2, coefficients $k_1 - k_6$ are derived after applying in total 40 'arrows of omission' and the results are

$$\left.\begin{aligned} k_1 = k_2 &= \frac{1}{2}(L_1 + L_1') \\ k_3 = k_4 &= -\frac{L_1}{2} \\ k_5 = k_6 &= -\frac{L_1'}{2} \end{aligned}\right\} \tag{19.18}$$

Then

$$\left.\begin{aligned} F(s) &= \frac{1}{2}(L_1 + L_1')\left\{\frac{1}{s+j\omega} + \frac{1}{s-j\omega}\right\} - \frac{L_1}{2}\left\{\frac{1}{s+\alpha_1+ju_1} + \frac{1}{s+\alpha_1-ju_1}\right\} \\ &\quad - \frac{L_1'}{2}\left\{\frac{1}{s+\alpha_2+ju_2} + \frac{1}{s+\alpha_2-ju_2}\right\} \\ &= (L_1 + L_1')\frac{1}{s^2+\omega^2} - L_1\frac{s+\cancel{\alpha_1}}{(s+\alpha_1)^2+u_1^2} - L_1'\frac{s+\cancel{\alpha_2}}{(s+\alpha_1)^2+u_2^2} \\ &= (L_1 + L_1')\frac{1}{s^2+\omega^2} - L_1\frac{s}{(s+\alpha_2)^2+u_1^2} - L_1'\frac{s}{(s+\alpha_2)^2+u_2^2} \end{aligned}\right\} \tag{19.19}$$

From the formula of the Laplace inverse transform

$$\left.\begin{aligned} \mathcal{L}^{-1}\left(\frac{s}{s^2+\omega^2}\right) &= \cos\omega t \\ \mathcal{L}^{-1}\left(\frac{s}{(s+\alpha_1)^2+u_1^2}\right) &= e^{-\alpha_1 t}\left(\cos u_1 t - \frac{\cancel{\alpha_1}}{u_1}\sin u_1 t\right) = e^{-\alpha_1 t}\cos u_1 t \\ \mathcal{L}^{-1}\left(\frac{s}{(s+\alpha_2)^2+u_2^2}\right) &= e^{-\alpha_2 t}\left(\cos u_2 t - \frac{\cancel{\alpha_2}}{u_2}\sin u_2 t\right) = e^{-\alpha_2 t}\cos u_2 t \end{aligned}\right\} \tag{19.20}$$

$$\therefore \quad \mathcal{L}^{-1}[F(s)] = (L_1 + L_1')\cos\omega t - L_1 e^{-\alpha_1 t}\cos u_1 t - L_1' e^{-\alpha_2 t}\cos u_2 t \tag{19.21}$$

Accordingly, the transient recovery voltage of the breaker is

$$\left.\begin{aligned} v_{Br}(t) &= I\omega\cdot\mathcal{L}^{-1}(F(s)) = |\dot{k}(\delta)|\cdot\frac{E}{L_1+L_1'}\cdot\mathcal{L}^{-1}[F(s)] \\ v_{Br}(t) &= |k(\delta)|\cdot E_1 \left\{\underbrace{\cos\omega t}_{\text{steady-state term}} \underbrace{- \frac{L_1}{L_1+L_1'}e^{-\frac{r_1}{2L_1}t}\cos\frac{1}{\sqrt{L_1C_1}}t}_{\text{transient term of the left circuit}} \underbrace{- \frac{L_1'}{L_1+L_1'}e^{-\frac{r_1'}{2L_1'}t}\cos\frac{1}{\sqrt{L_1'C_1'}}t}_{\text{transient term of the right circuit}}\right\} \\ &\text{where} \quad k(\delta) = 1 - \frac{E_1'}{E_1}e^{-j\delta} \end{aligned}\right\} \tag{19.22a}$$

The equation can also be written as

$$v_{Br}(t) = |k(\delta)| \cdot E_1 \left\{ \frac{L_1}{L_1 + L_1'} \left(\cos \omega t - e^{-\frac{r_1}{2L_1}t} \cos \frac{1}{\sqrt{L_1 C_1}} t \right) + \frac{L_1'}{L_1 + L_1'} \left(\cos \omega t - e^{-\frac{r_1'}{2L_1'}t} \cos \frac{1}{\sqrt{L_1' C_1'}} t \right) \right\} \tag{19.22b}$$

This derived equation should be considered to be very accurate because all the omissions in the processes were done after careful checks.

Figure 19.2d shows the transient voltages of Equations 19.22a and b, where the oscillation frequency of the transient terms exaggerated extremely slowly.

19.1.2.2.1 Evaluation of Equations 19.22a and 19.22b For the magnitude, the transient oscillation terms for the right- and left-side circuits appear just after tripping. The magnitude inside the braces ( ) in the equations reaches a maximum of 2 when the oscillation angles of the three terms coincide. Accordingly, $v_{Br}(t)$ would have a maximum of $|k(\delta)| \cdot 2E_1$.

Incidentally, assuming $E_1 = E_1' = 1$, $|k(\delta)| = \sqrt{2}$ for $\delta = 90°$ (tripping around the steady-state stability limit condition), the theoretical maximum value of the transient recovery voltage is $2\sqrt{2}E$. On the contrary, $|k(\delta)| = 0$ for the case of $\delta = 0°$ (no load tripping).

19.1.2.2.2 Rate of rise of voltage (RRRV – the Rate of Rise of Recovery Voltage) We have

	left-side circuit	right-side circuit
maximum peak value of transient voltages(line-to-ground voltages)	$2k(\delta)E_1 \cdot \dfrac{L_1}{L_1 + L_1'}$	$2k(\delta)E_1 \cdot \dfrac{L_1'}{L_1 + L_1'}$
natural oscillation frequency	$f_1 = \dfrac{1}{2\pi\sqrt{L_1 C_1}}$	$f_1' = \dfrac{1}{2\pi\sqrt{L_1' C_1'}}$

(19.23)

The total peak value of the transient terms of $v_{Br}(t)$ is a maximum of $2|k(\delta)| \cdot E_1$.

Assuming $L_1 = L_1' = 1\,\text{mH/km}$, $C_1 = C_1' = 0.01\,\mu\text{F/km}$ as typical values,

$$f_1 = f_1' = \frac{1}{2\pi\sqrt{10^{-3} \times 10^{-8}}} \fallingdotseq 50\,\text{kHz} \quad \therefore \text{ one-wavelength time} = 20\,\mu\text{s} \tag{19.24a}$$

Therefore the transient terms would rise from zero to the peak value $2k(\delta) \cdot E_1$ in about $5\,\mu\text{s}$ (quarter-wavelength time).

The approximate value of the RRRV is

$$2|k(\delta)| \cdot E_1 \times \frac{1}{5}\ [\text{kV}/\mu s] \tag{19.24b}$$

The exact value of RRRV can be calculated by differentiation of Equation 19.22a with respect to time $t = 0+$. Namely,

$$\frac{d}{dt}v_{Br}(t) = |k(\delta)| \cdot E_1 \left\{ -\omega \sin \omega t \right.$$
$$- \frac{L_1}{L_1 + L'_1}\left(-\frac{r_1}{2L_1} e^{-\frac{r_1}{2L_1}t} \cos \frac{1}{\sqrt{L_1 C_1}} t - e^{-\frac{r_1}{2L_1}t} \frac{1}{\sqrt{L_1 C_1}} \sin \frac{1}{\sqrt{L_1 C_1}} t \right) \tag{19.25a}$$
$$\left. - \frac{L'_1}{L_1 + L'_1}\left(-\frac{r'_1}{2L'_1} e^{-\frac{r'_1}{2L'_1}t} \cos \frac{1}{\sqrt{L'_1 C'_1}} t - e^{-\frac{r'_1}{2L'_1}t} \frac{1}{\sqrt{L'_1 C'_1}} \sin \frac{1}{\sqrt{L'_1 C'_1}} t \right) \right\}$$

In this equation, the first transient term in parentheses ( ) may be neglected, because r_1 and r'_1 are small.

Also, at time $t = 0 +$ (initial time just after tripping), the replacements below are possible:

$$\sin \omega t \to 0, \quad e^{-\frac{r_1}{2L_1}t} \to 1.0, \quad e^{-\frac{r'_1}{2L'_1}t} \to 1.0 \quad \text{at } t = 0+ \tag{19.25b}$$

Accordingly, $at + t = 0$

$$\frac{d}{dt}v_{Br}(t) = |k(\delta)| \cdot E_1 \left\{ \frac{1}{L_1 + L'_1}\sqrt{\frac{L_1}{C_1}} \sin \frac{1}{\sqrt{L_1 C_1}} t + \frac{1}{L_1 + L'_1}\sqrt{\frac{L'_1}{C'_1}} \sin \frac{1}{\sqrt{L'_1 C'_1}} t \right\} \tag{19.26}$$

The maximum value of RRRV is

$$\text{Max}\frac{d}{dt}v_{Br}(t) = |k(\delta)| \cdot \frac{E_1}{L_1 + L'_1} \underbrace{\left\{ \sqrt{\frac{L_1}{C_1}} + \sqrt{\frac{L'_1}{C'_1}} \right\}}_{\text{sum of surge impedances of both sides}} \tag{19.27}$$

RRRV (usually expressed in kV/μs) is a very important concept largely affecting the tripping duty of breakers. Equation 19.27 indicates that RRRV is directly proportional to the addition of surge impedances of both sides of lines.

19.2 Calculation of Transient Recovery Voltages Across a Breaker's Three Poles by 3ϕS Fault Tripping

The study of switching phenomena in three-phase circuits is vitally important not only for engineers who are involved with breakers, but also for engineers who are engaged with system overvoltages or large currents as well as insulation, regardless of each engineer's different interests or viewpoints. Considering the importance of such study, it may seem strange that a detailed description of the calculation of switching phenomena in three-phase circuits can seldom be found. In this section, we introduce the equations of transient voltages appearing at the first, second and third pole (phase) tripping. Readers could be acquainted with switching surge phenomena through the study of the mathematical treatments.

19.2.1 Recovery voltage appearing at the first phase (pole) tripping

Transient voltages and steady-state voltages appearing across the contacts (poles) just after breaker tripping are called the **transient recovering voltages** and **recovering voltages**, respectively. Before studying these voltages in detail, the recovering voltage by the first phase tripping can be observed intuitionally. The methods and the results are shown in Figure 19.3.

The figure indicates the aspects for fixing different fault cases for effective and non-effective neutral grounding systems, in that the steady-state voltage (the recovering voltage) appearing across the first tripping pole (the first tripping phase) is derived intuitionally.

19.2.1.1 Case 1: three-phase fault tripping

The voltages at the time just after the first phase a tripping are, for the solidly neutral grounding system,

$$v'_a = 0, \quad v_b = v'_b = 0, \quad v_c = v'_c = 0, \quad v_n = 0 \quad \therefore \quad v_a = E, \quad v'_{aa} = E$$

and for the neutral-opening system

$$v'_a = 0, \quad v_b = v'_b = 0, \quad v_c = v'_c = 0, \quad v_n = \text{indefinite}$$

Then the voltage at the midpoint of b and c is zero:

$$v_{ab} = v_a = 1.5E, \quad \therefore v_{aa} = 1.5E$$

19.2.1.2 Case 3: phase a′ and phase b grounding fault tripping

This is the case for grounding at both sides of the breaker.

For the solidly neutral grounding system

$$v'_a = 0, \quad v_b = v'_b = 0, \ v_n = 0 \quad \therefore v_a = 1E, \quad v'_{aa} = 1E$$

and for the neutral opening system

$$v'_a = 0, \quad v_b = v'_b = 0, \quad v_n = \text{indefinite} \quad \therefore v_{ab} = \sqrt{3}E, \quad v'_{aa} = \sqrt{3}E$$

19.2.1.3 Case 4: step-out tripping

The voltages v_a and v'_a may be in opposite directions just at the time of the first phase a tripping in the worst case.

For the solidly neutral grounding system

$$v_a = 1E, \quad v'_a = -1E, \quad v_n = 0 \quad \therefore v_a = 1E, \quad v'_a = -1E, \quad v'_{aa} = 2E$$

and for the neutral opening system

$$v_b = v'_b, \quad v_c = v'_c, \quad v_n = \text{indefinite}, \quad v'_n = \text{indefinite}, \quad v_{bc} = v_{b'c'} \quad \therefore v'_{aa} = 3E$$

The derived phase voltages and recovery voltages by this method would give a useful rough idea of the voltage behaviours. The magnitudes of these voltages would directly affect the tripping capabilities of the breakers as well as coordination of the insulation engineering in a total power system network.

Taking a general view, the phase overvoltages appearing as well as the breaker's recovery voltages are obviously smaller in solidly neutral grounding systems in comparison with non-effective neutral grounding systems. Also, breaker tripping under step-out conditions would be very heavy duty for the breakers.

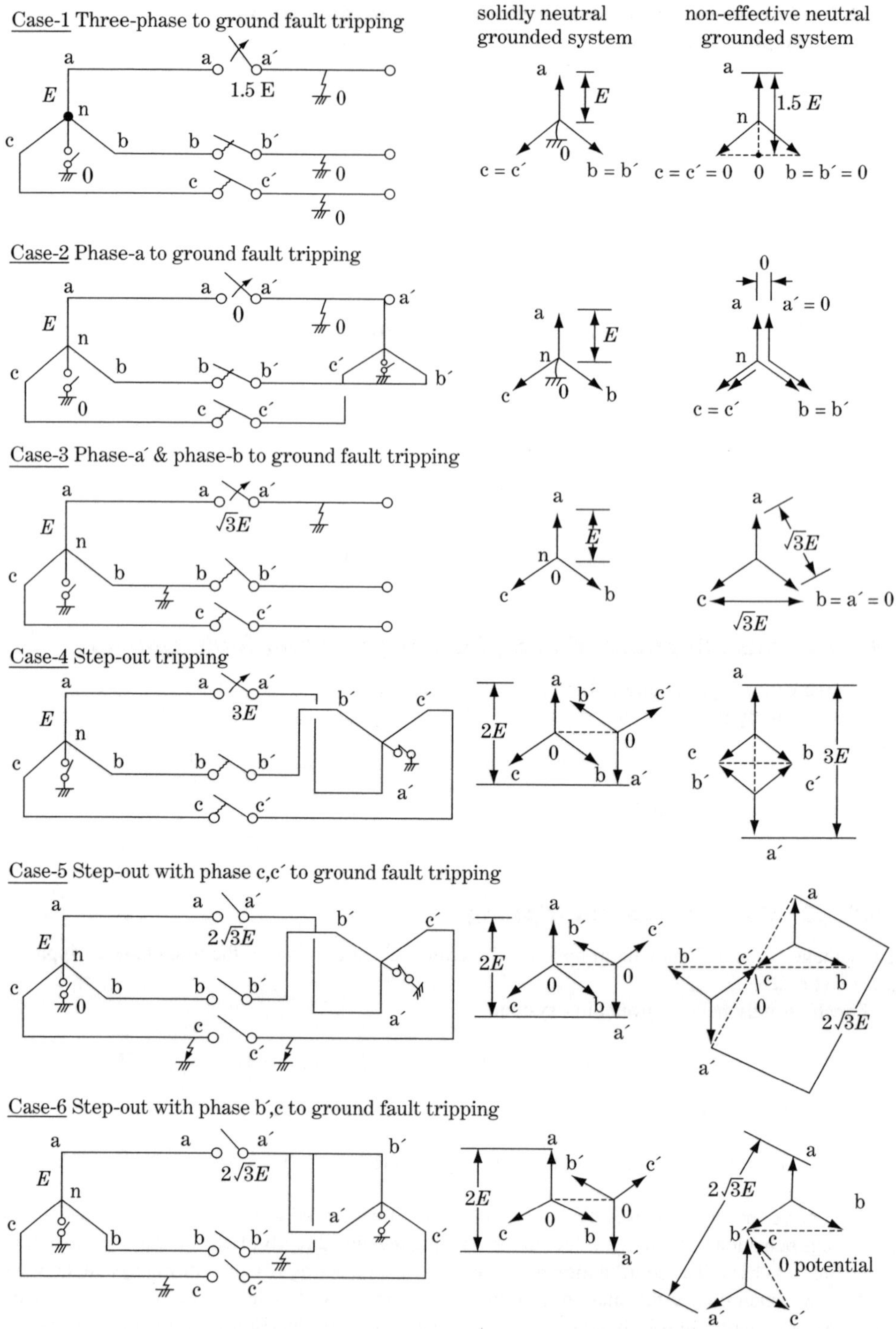

Figure 19.3 Recovery voltages appearing by the first pole (phase a) tripping, v'_{aa}

19.2.2 Transient recovery voltage across a breaker's three poles by 3ϕS fault tripping

We introduce the equations for the transient voltages appearing across the first, second and third poles when the breaker trips three-phase short-circuit fault currents as shown in Figure 19.4a. In the figure, a three-phase fault occurs at point f on line l' and the first phase tripping by the breaker phase a pole is going to be executed (of course, at the time of the phase a current zero). The transient calculation of Figure 19.4b can be solved as the superposition of the steady-state term in (c) and the transient term in (d).

19.2.2.1 The first pole tripping

19.2.2.1.1 Step 1: calculation of the steady-state current before phase a tripping We calculate the steady-states currents of the three-phase short-circuit fault before tripping. The current can be calculated only by the positive-sequence circuit. Accordingly, this case

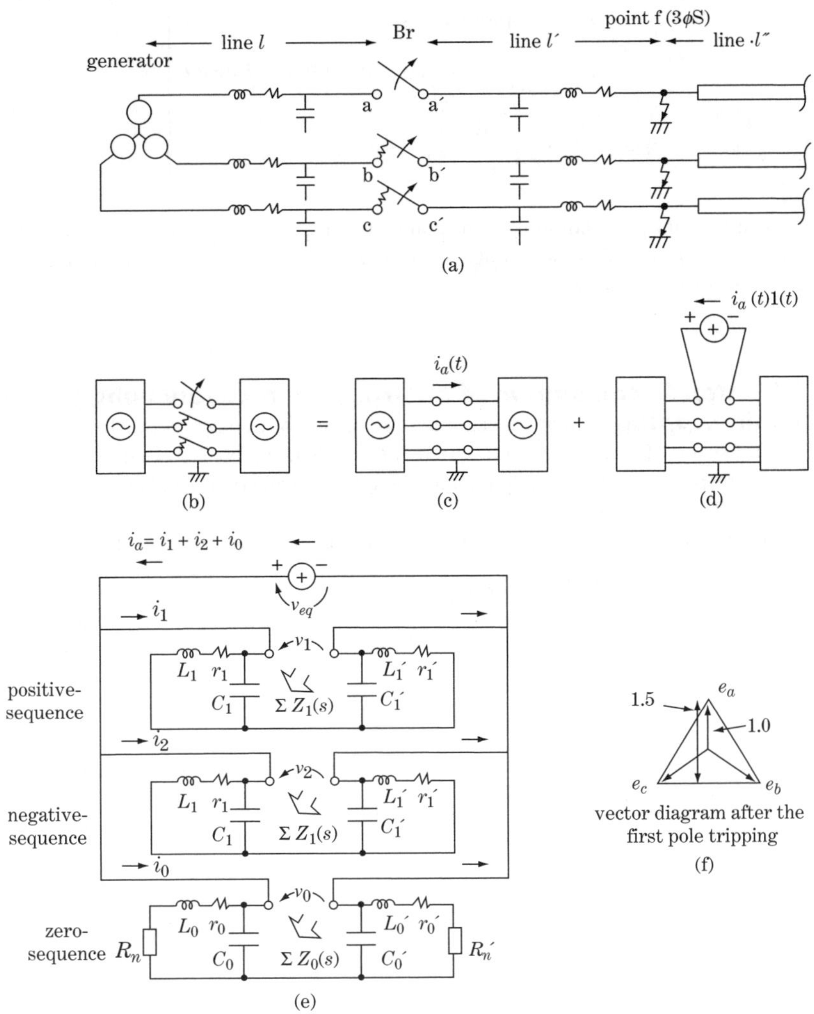

Figure 19.4 Calculation of transient recovery voltage (the first pole tripping)

is equivalent to the case of Section 19.1.2 and Figure 19.2 under the additional condition $e_1' = 0$. Therefore we can appropriate Equations 19.11–19.15 for use here. That is, for the phase a (positive-sequence) source voltage,

$$\dot{e}_a(t) = \dot{e}_1(t) = E\mathrm{e}^{j\omega t} \tag{19.28}$$

The current flowing through the breaker is given by Equation 19.13, where, putting $\Delta\dot{e}(t) = \dot{e}(t)$, $k(\delta) = 1$,

$$\dot{i}(t) = \frac{E\mathrm{e}^{j\omega t}}{(j\omega L_1 + r_1) + (j\omega L_1' + r_1')} \doteqdot \frac{E}{\omega(L_1 + L_1')}\mathrm{e}^{j(\omega t - 90^\circ)} \tag{19.29}$$

Taking the real part of the equation,

$$\left.\begin{aligned} e_1(t) &= E_1 \cos\omega t \\ i(t) &= \frac{E}{\omega(L_1 + L_1')}\cos(\omega t - 90^\circ) = I\sin\omega t \\ \text{where } I &= \frac{E}{\omega(L_1 + L_1')} \end{aligned}\right\} \tag{19.30}$$

The equations are of initial conditions corresponding to Figure 19.4c, where the current $i(t)$ is of 'current zero' at $t = 0$ and also the voltage is of peak value at $t = 0$ because the fault current is almost 90° lagging to the voltage.

19.2.2.1.2 Step 2: calculation of the transient recovery voltage just after the first pole tripping Now we begin the calculation of transient terms for Figure 19.4d, in that the current $I\sin\omega t \cdot \mathbf{1}(t)$ is suddenly inserted at $t = 0$. The equivalent circuit is given by Figure 19.4e, which corresponds to the equivalent circuit of symmetrical components in the case of phase a opening given in Table 3.2 [1B].

The related equations in the Laplace domain are derived as follows. The circuit equations are

$$\left.\begin{aligned} &\Sigma Z_1(s) = Z_1(s) + Z_1'(s) && ① \\ &\Sigma Z_0(s) = Z_0(s) + Z_0'(s) && ② \\ &\text{where} \\ &Z_1(s) = \frac{1}{\dfrac{1}{L_1 s + r_1} + C_1 s} = \frac{1}{C_1}\cdot\frac{s + \dfrac{r_1}{L_1}}{s^2 + \dfrac{r_1}{L_1}s + \dfrac{1}{L_1 C_1}} = \frac{1}{C_1}\cdot\frac{s + 2\alpha_1}{(s+\alpha_1)^2 + u_1^2} && ③ \\ &Z_1'(s) = \frac{1}{\dfrac{1}{L_1' s + r_1'} + C_1' s} = \frac{1}{C_1'}\cdot\frac{s + \dfrac{r_1'}{L_1'}}{s^2 + \dfrac{r_1'}{L_1'}s + \dfrac{1}{L_1' C_1'}} = \frac{1}{C_1'}\cdot\frac{s + 2\alpha_2}{(s+\alpha_2)^2 + u_2^2} && ④ \end{aligned}\right\} \tag{19.31}$$

where $Z_0(s)$, $Z_0'(s)$ are of the same form.

The voltage and current equations are

$$\left.\begin{aligned} &i(s) = \mathcal{L}[I\sin\omega t] = I\cdot\frac{\omega}{s^2+\omega^2} && ① \\ &v_{eq}(s) \equiv v_1(s) = v_2(s) = v_0(s) = i(s)\cdot\Sigma Z_{\text{total}} && ② \\ &\text{where}\quad \Sigma Z_{\text{total}} = \frac{1}{\dfrac{1}{\Sigma Z_1(s)}+\dfrac{1}{\Sigma Z_2(s)}+\dfrac{1}{\Sigma Z_0(s)}} && ③ \end{aligned}\right\} \tag{19.32}$$

$$\therefore v'_{aa}(s) = v_1(s)+v_2(s)+v_0(s) = 3\,i(s)\cdot\Sigma Z_{\text{total}} = 3I\frac{\omega}{s^2+\omega^2}\cdot\Sigma Z_{\text{total}} \tag{19.33a}$$

where ΣZ_{total} is

$$\left.\begin{aligned} &\text{for solidly neutral grounding system} \\ &\Sigma Z_1(s) \fallingdotseq \Sigma Z_2(s) \fallingdotseq \Sigma Z_0(s) && ① \\ \therefore\quad &\Sigma Z_{\text{total}} = \frac{1}{3}\cdot\Sigma Z_1(s) \\ &\text{for non-effective neutral grounding system} \\ &\Sigma Z_1(s) \fallingdotseq \Sigma Z_2(s) \ll \Sigma Z_0(s) && ② \\ \therefore\quad &\Sigma Z_{\text{total(s)}} = \frac{1}{2}\cdot\Sigma Z_1(s) \end{aligned}\right\} \tag{19.33b}$$

Accordingly,

$$\left.\begin{aligned} v'_{aa}(s) &= k\,i(s)\cdot\Sigma Z_1(s) \\ &= kI\frac{\omega}{s^2+\omega^2}\left\{\frac{1}{C_1}\cdot\frac{s+2\alpha_1}{(s+\alpha_1)^2+u_1^2}+\frac{1}{C'_1}\cdot\frac{s+2\alpha_2}{(s+\alpha_2)^2+u_2^2}\right\} \\ &\equiv kI\omega\cdot F(s) \\ \text{where}\quad & k \fallingdotseq 1 \text{ for solidly neutral grounding system} \\ & k \fallingdotseq 1.5 \text{ for non-effective neutral grounding system} \end{aligned}\right\} \tag{19.34}$$

This is the answer for the transient recovery voltage of the first tripped pole in the Laplace domain.

$F(s)$ defined in the equation is fortunately in the same form as the $F(s)$ in Equation 19.17, so we already know that the inverse transformed equation is given by Equation 19.21. Accordingly, for the transient recovery voltage of the first tripping pole,

$$\begin{aligned} v'_{aa}(t) &= \mathcal{L}^{-1}[v'_{aa}(s)] = kI\omega\cdot\mathcal{L}^{-1}[F(s)] = k\frac{E}{L_1+L'_1}\cdot\mathcal{L}^{-1}[F(s)] \\ &= kE\left\{\cos\omega t-\frac{L_1}{L_1+L'_1}e^{\frac{-r_1}{2L_1}t}\cos\frac{1}{\sqrt{L_1C_1}}t-\frac{L'_1}{L_1+L'_1}e^{-\frac{r'_1}{2L'_1}t}\cos\frac{1}{\sqrt{L'_1C'_1}}t\right\} \end{aligned} \tag{19.35a}$$

The term in the braces { } in the above equation has a maximum of 2, so the maximum value of the transient recovery voltage $v'_{aa}(t)$ (or the overvoltage $v_a(t)$ of terminal a, because $v'_a = 0$

in this case) is

$$\left.\begin{array}{ll}\text{Max.}\, v'_{aa}(t) = 2k \fallingdotseq 2E & \text{for solidly neutral grounding system} \quad ① \\ \qquad\qquad\qquad\quad = 3E & \text{for non-effective neutral grounding system} \quad ② \end{array}\right\} \tag{19.35b}$$

The equation of the steady-state voltage after the transient terms attenuate is $v'_{aa}(t) = kE\cos\omega t$, which coincides with case 1 in Figure 19.3.

19.2.2.1.3 RRRV (the maximum value) of $v'_{aa}(t)$

This is derived by differentiating Equation 19.35a with respect to t, and then following the same treatment as for Equations 19.25a–19.27. The result is

$$\text{Max}\frac{d}{dt}v_{Br}(t) = k \cdot \frac{E_1}{L_1 + L'_1} \underbrace{\left\{\sqrt{\frac{L_1}{C_1}} + \sqrt{\frac{L'_1}{C'_1}}\right\}}_{\text{sum of surge impedances of both sides}} \tag{19.36}$$

Note that this equation is in the same form as Equation 19.27 except for the coefficient k and $k(\delta)$, although Equation 19.36 is the case for fault current tripping while Equation 19.27 is the case for load flow tripping. The equation again indicates that the RRRV (the maximum value), which is an important indication of a breaker's fault current tripping duty, is closely related to the surge impedances of the transmission lines.

19.2.2.2 The second and third poles tripping

Our problem here is to solve the transient voltages across the second and third poles of the circuit shown in Figure 19.5a.

The voltages appearing on the second and third poles are more severe in a non-effective neutral grounding system rather than a solidly neutral grounding system. Thus we try to calculate the second/third pole fault tripping phenomena in the non-effective neutral grounding system.

19.2.2.2.1 Step 1: calculation of the steady-state current before phase b and c tripping (phase a is open)

We calculate the steady-state currents of the circuit in Figure 19.5b, referring to the equivalent circuit (for phase a conductor opening) for the symmetrical components given in Table 3.2 [1B], [1C], [1D], and under the condition $\Sigma Z_1 = \Sigma Z_2 \ll \Sigma Z_0$.

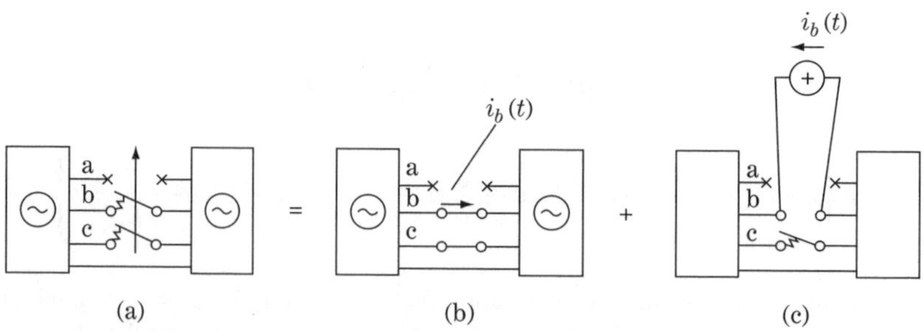

Figure 19.5 Calculation of transient recovery voltage (the second and third poles tripping)

The equations are

$$\left.\begin{aligned}
\dot{e}(t) &= Ee^{j\omega t} && ① \\
\dot{i}_1(t) &= \frac{\dot{e}(t)}{\Sigma Z_1 + \cfrac{1}{\cfrac{1}{\Sigma Z_2}+\cfrac{1}{\Sigma Z_0}}} = \frac{\dot{e}(t)}{\Sigma Z_1 + \cfrac{\Sigma Z_1}{\cfrac{\Sigma Z_1}{\Sigma Z_0}+1}} \fallingdotseq \frac{1}{2}\cdot\frac{\dot{e}(t)}{\Sigma Z_1} && ② \\
\dot{i}_2(t) &= \frac{-\Sigma Z_0}{\Sigma Z_2+\Sigma Z_0}\cdot \dot{i}_1(t) \fallingdotseq -\dot{i}_1(t) && ③ \\
\dot{i}_0(t) &= \frac{-\Sigma Z_2}{\Sigma Z_2+\Sigma Z_0}\cdot \dot{i}_1(t) \fallingdotseq 0 && ④ \\
&\text{where} \\
\Sigma Z_1 &= \Sigma Z_2 \ll \Sigma Z_0 && ⑤
\end{aligned}\right\} \qquad (19.37)$$

The steady-state solution is

$$\left.\begin{aligned}
\dot{i}_a(t) &= 0 \\
\dot{i}_b(t) &= -\dot{i}_c(t) = (a^2-a)\cdot \dot{i}_1(t) = -\sqrt{3}j \times \frac{1}{2}\cdot\frac{E}{(j\omega L_1 + r_1) + (j\omega L_1' + r_1')}e^{j\omega t} \\
&= -\frac{\sqrt{3}}{2}\cdot\frac{E}{\omega(L_1+L_1')}e^{j\omega t}
\end{aligned}\right\} \qquad (19.38)$$

Taking the real part,

$$\left.\begin{aligned}
e_a(t) &= E\cos\omega t \\
i_a(t) &= 0 \\
i_b(t) &= -i_c(t) = -\frac{\sqrt{3}}{2}\cdot\frac{E}{\omega(L_1+L_1')}\cos\omega t
\end{aligned}\right\} \qquad (19.39)$$

The current $i_b(=-i_c)$ is in inverse phase with the source voltage e_a, and accordingly leads the voltage v_{bc} by 90°. These are the initial currents of the second and third pole before tripping.

19.2.2.2.2 Step 2: calculation of the transient recovery voltage by the second and third poles (phase b and c) tripping.

The next calculation is to insert the above derived current i_b in the opposite direction (i.e. $-i_b$ in the forward direction) in Figure 19.5c. Then we change the polarity of the current i_b and i_c. Furthermore, we shift the timescale of Equation 19.39 by 90° (by replacing $\omega t \rightarrow \omega t - 90°$) in order to get the condition of 'current zero at new $t = 0$'.

Accordingly,

$$\left.\begin{aligned}
e_a(t) &= E\cos(\omega t - 90°) = E\sin\omega t \\
i_a(t) &= 0 \\
-i_b(t) &= +i_c(t) = \frac{\sqrt{3}}{2}\cdot\frac{E}{\omega(L_1+L_1')}\cos(\omega t - 90°) \\
&= \frac{\sqrt{3}}{2}\cdot\frac{E}{\omega(L_1+L_1')}\sin\omega t
\end{aligned}\right\} \qquad (19.40)$$

This is the initial value of the calculation after phase a is tripped. The currents $i_b = -i_c$ are of 'current zero' at $t = 0$ by the new timescale (which lags the former timescale by 90°).

Now the transient phenomena of the circuit in Figure 19.5c can be solved by the α–β–0 method, while solution by symmetrical components is actually impossible.

Referring to Table 6.2 #5, the equivalent circuit to be solved is obtained as shown in Figure 19.6, in which the $\alpha-0$-circuit and the β-circuit can be calculated separately. The initial value can be derived by transforming Equation 19.40 into the $\alpha-\beta$-0 domain:

$$\left.\begin{aligned} i_\alpha(t) &= \frac{-1}{3}(i_b(t) + t_c(t)) = 0 \\ i_\beta(t) &= \frac{\sqrt{3}}{3}(i_b(t) - i_c(t)) = \frac{E}{\omega(L_1 + L_1')}\sin\omega t \\ i_0(t) &= \frac{1}{3}(i_b(t) + i_c(t)) = 0 \end{aligned}\right\} \tag{19.41}$$

For the transient calculation of the α-0-circuit, the initial values of i_α and i_0 to be injected are zero because of the assumption $\Sigma Z_1 = \Sigma Z_2 \ll \Sigma Z_0$ so that $i_a(t) = i_0(t) = 0$. In other words, all the quantities are zero, that is $v_\alpha = v_0 = 0$ in the α–0-circuit.

For the transient calculation of the β-circuit, the current $i_\beta(t)$ is to be injected at time $t = 0$ and the equation in the Laplace domain is

$$\left.\begin{aligned} i_\beta(s) &= \frac{E}{\omega(L_1 + L_1')}\cdot\mathcal{L}[\sin\omega t] = \frac{E}{\omega(L_1 + L_1')}\cdot\frac{\omega}{s^2 + \omega^2} \\ v_\beta(s) &= i_\beta(s)\cdot\Sigma Z_1(s) = \frac{E}{L_1 + L_1'}\cdot F(s) \end{aligned}\right\} \tag{19.42}$$

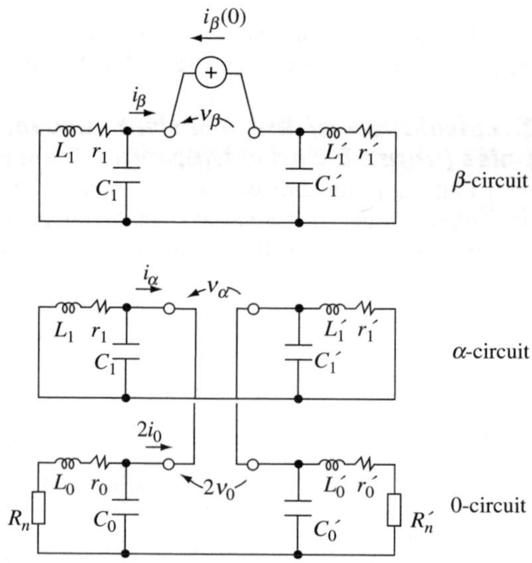

Figure 19.6 Calculation of the second and third poles tripping by the α–β–0 method

$\Sigma Z_1(s)$, $F(s)$ are again the same as in Equations 19.16 and 19.17 or 19.31. Then Equation 19.21 can be utilized for the inverse Laplace transform:

$$v_\beta(t) = \frac{E}{L_1 + L_1'} \cdot \mathcal{L}^{-1}[F(s)]$$
$$= \frac{E}{L_1 + L_1'} \cdot \left\{ (L_1 + L_1')\cos \omega t - L_1 e^{-\frac{r_1}{2L_1}t} \cos \frac{1}{\sqrt{L_1 C_1}} t - L_1' e^{-\frac{r_1'}{2L_1'}t} \cos \frac{1}{\sqrt{L_1' C_1'}} t \right\} \quad (19.43)$$
$$= E \left\{ \cos \omega t - \frac{L_1}{L_1 + L_1'} e^{-\frac{r_1}{2L_1}t} \cos \frac{1}{\sqrt{L_1 C_1}} t - \frac{L_1'}{L_1 + L_1'} e^{-\frac{r_1'}{2L_1'}t} \cos \frac{1}{\sqrt{L_1' C_1'}} t \right\}$$

For the inverse transform from the α–β–0 domain to the a–b–c domain, recalling that $v_\alpha(t) = v_0(t) = 0$,

$$v_{bb}'(t) = -v_{cc}'(t) = \frac{\sqrt{3}}{2} v_\beta(t)$$
$$= \frac{\sqrt{3}}{2} E \left\{ \cos \omega t - \frac{L_1}{L_1 + L_1'} e^{-\frac{r_1}{2L_1'}t} \cos \frac{1}{\sqrt{L_1 C_1}} t - \frac{L_1'}{L_1 + L_1'} e^{-\frac{r_1'}{2L_1'}t} \cos \frac{1}{\sqrt{L_1' C_1'}} t \right\} \quad (19.44)$$

In conclusion, the transient recovery voltage by the first pole tripping is given by Equation 19.35, and that by the second and third poles tripping is given by Equation 19.44. Figure 19.7 shows the waveform aspects, which are truly explained by the equations

the non-effective neutral grounding system	the transient recovery voltage	the recovery voltage
the first pole tripping	$3E$	$1.5E$
the second and the third pole tripping	$\sqrt{3}E$	$\frac{\sqrt{3}}{2}E$

The transient recovery voltage is larger from the first pole tripping rather than the second and third poles tripping and is a maximum of $3E$ in the case of the non-effective neutral grounding system.

For the solidly neutral grounding system, the transient recovery voltage and the recovery voltage by the first pole tripping is $2E$ and $1E$, respectively, and that of the second and the third poles is of smaller value although the calculation is omitted. (It can be calculated by the same method if ΣZ_1, ΣZ_0 are given.)

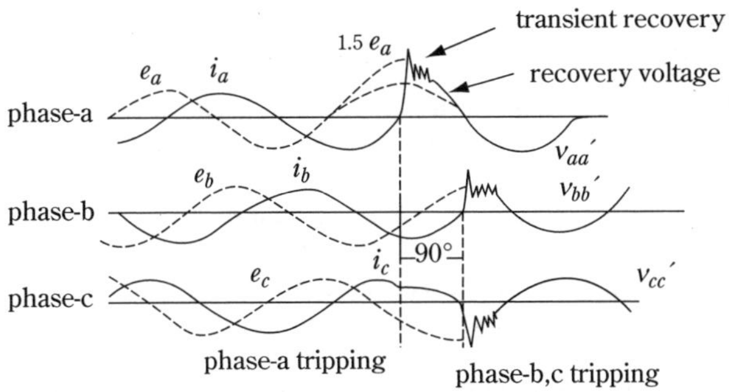

Figure 19.7 Transient recovery voltage caused by three-phase fault tripping

It should also be noted that the RRRV (the maximum value) in three phase circuits by the 3ϕS fault current tripping is given by Equation 19.36, which indicates that the surge impedances of the transmission line directly affect the breaker's duty.

19.3 Fundamental Concepts of High-voltage Circuit-breakers

The switching-off phenomena described in the previous sections are by so-called ideal circuit-breakers, which can trip fault current instantly at the time of 'current zero' without any accompanying arcs. The aspect of actual breaker tripping is a little different from the ideal case, because a current arc appears transiently across the leaving contacts.

19.3.1 Fundamental concept of breakers

A high-voltage circuit-breaker has a pair of contacts that are typically one fixed and one movable, although some breakers of the recently developed largest class may contain of double movable contacts. Just after receiving the trip signal (from the relay or manually), the movable contact begins to slide off and leave the fixed contact. In the case of typical SF_6-type breakers, the total stroke distance (wipe distance + departing distance) is 150–250 mm and the necessary moving time would be around 10 ms (less than one cycle).

At the instant when the movable contact leaves the fixed contact, a current arc (plasma of approximately 5000–10 000°C by thermal ionization) appears across the leaving contacts. The breaker is contrived to blow off the thermal ionized gas within a short time and complete tripping by around 20-60 ms (1–3 cycles).

The technology of breakers is designed to disperse and remove high-density ionized plasma gas within the small chamber space in a very short time – in contrast to the technology of nuclear fusion which is to contain such a high-density plasma gas within a small space. Furthermore, circuit-breakers have to be capable of tripping a fault current of 50 or 63 kA (rms) repeatedly without damaging (melting) the contacts. Although the technology of high-voltage circuit-breakers today is advanced, the long history of repeated failure and success is worthy of note among the other equipment in a power system network.

19.3.1.1 Principle of arc extinction

There are two typical interpretations in regard to the principle of **arc extinction**:

a) The **theory of insulation balance across the contacts**. That is, the keen struggle of the transient recovery voltage and RRRV appearing versus insulation recovery by sudden enlargement of the breaker's stroke distance and by forcing blow-off of the ionized arc.

b) The **theory of energy balance**. That is, the keen struggle of the supplied energy (the product of transient recovery voltage and leak current) versus the forced dispersing energy.

Regardless of the academic interpretations, successful current breaking means the capability of rapid insulation enlargement across the contacts to overcome the supplied arc energy from the power system in just a short time. Accordingly, it is understandable that the magnitude of transient recovery voltage and its initial rising speed (namely, RRRV) have become the two essential parameters for the breaker's tripping capability.

Breakers today can be classified as types of SF_6 **gas**, **air**, **oil** and **vacuum** by the different extinction media and each type has its own specific characteristics.

In regard to high-voltage, especially EHV and UHV, classes, SF_6-type breakers have become predominant.

19.3.1.2 SF_6 gas-type circuit breakers

Figure 19.8 shows some fundamentals of the SF_6 gas-type circuit-breaker for the UHV class as an example.

SF_6 gas is a thermodynamically stable gas with outstanding insulation and arc extinction characteristics. SF_6 gas at typically 4–10 atmospheres fills the extinction chamber whose structure is of a plunger-type puffer chamber (Figure 19.8b). The movable contact and the plunger rod form one body so that the gas is puffed out just after the contact begins to slide in order to extinguish ionized current arcs and to recover insulation across the opening contacts.

Figure 19.8c shows the typical gas-blowing characteristics of a puffer chamber. Breakers have to cut off the current at a single stroke within an effective puffer time. In other words, if a breaker cannot cut off the current within a specified time (say, 3 cycles), the breaker will be damaged by the thermal energy, which causes a tripping failure and furthermore the occurrence of a new phase fault at the broken breaker (of short-circuit or open-conductor modes).

The application limit of breakers described above is common for all types of breakers. In the case of oil-filled breakers, for example, oil gasification due to the arc temperature occurs in the chamber, so the effective tripping time is limited.

It should be noted that a **breaker fault** by tripping failure will have severe aspects on the power system for most cases, because several breakers are directly connected to the same bus and the associated breakers at adjacent substations have to be immediately tripped (by various back-up relays) to remove the original fault and the faulted breaker, possibly leading to other cascade faults or serious system disturbances such as line over-current, generator tripping by the I_2 limit or frequency limit, system instability, and so on.

19.3.2 Terminology of switching phenomena and breaker tripping capability

The duty of circuit-breakers is not limited to the role of tripping fault/load currents as switches of a network. Another important duty is to reduce the switching overvoltages (surge) within a certain limit, which is an essential part of insulation coordination. Bearing this in mind, we continue our study of the phenomena arising from breakers tripping.

The major terminology is as follows:

① **Transient recovery voltage:** The initial transient voltage appearing across the paired contacts just after the tripping action.

② **Ideal transient recovery voltage:** The theoretical transient recovery voltage by an ideal breaker without accompanying arcs.

③ **Recovery voltage:** The steady-state voltage of power frequency appearing across the paired contacts just after tripping.

④ **Peak value of transient recovery voltage:** The largest peak value among a few peak values of the oscillating transient recovery voltage.

⑤ **Initial peak value of transient recovery voltage (the first peak value):** The first peak value of the transient recovery voltage.

⑥ **Frequency of transient recovery voltage:** The oscillatory frequency (plural frequencies may be possible).

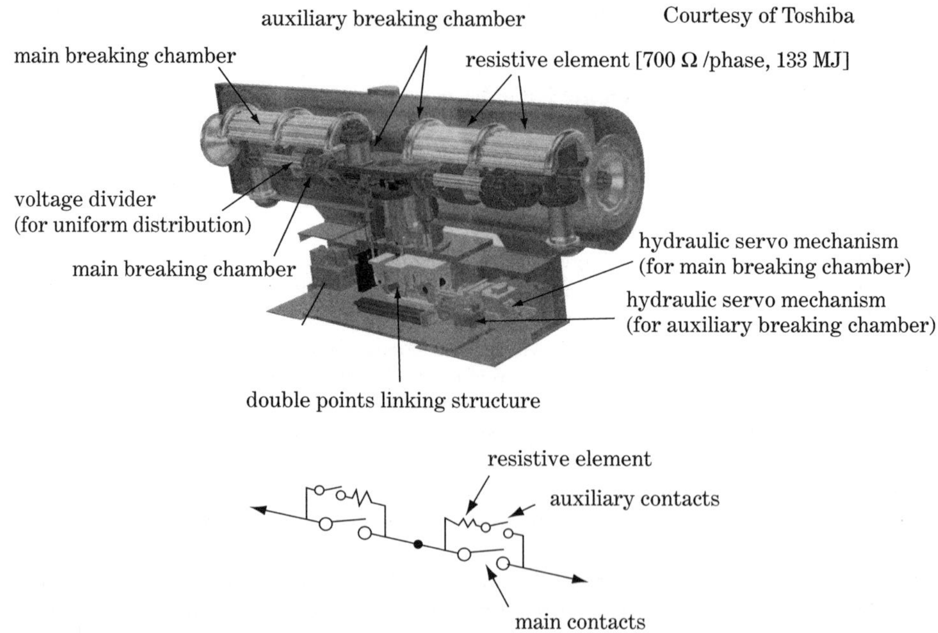

(a) Breaking chamber

One point breaking: 500 kV, 6000/8000 A, 63 kA, 2 cycle tripping

Double points breaking: 1000 kV, 6000/8000 A, 63 kA, 2 cycle tripping

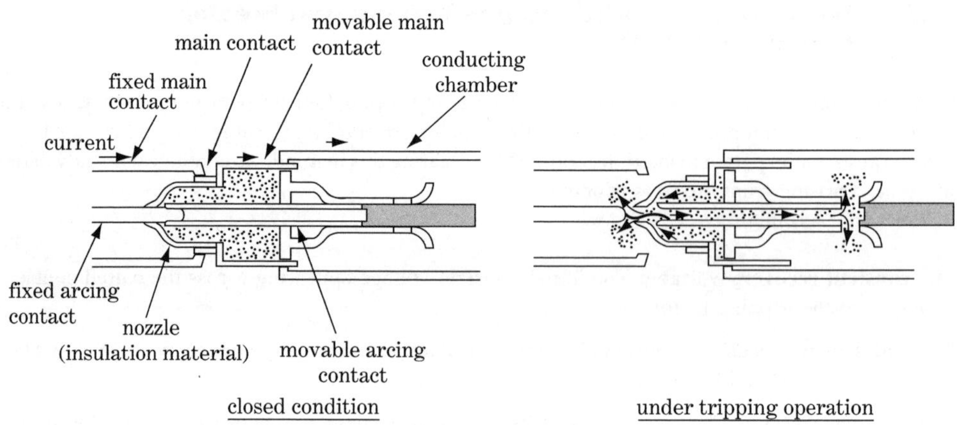

(b) principle of breaking chamber

Figure 19.8 SF_6 gas-type circuit-breaker (500–1000 kV class)

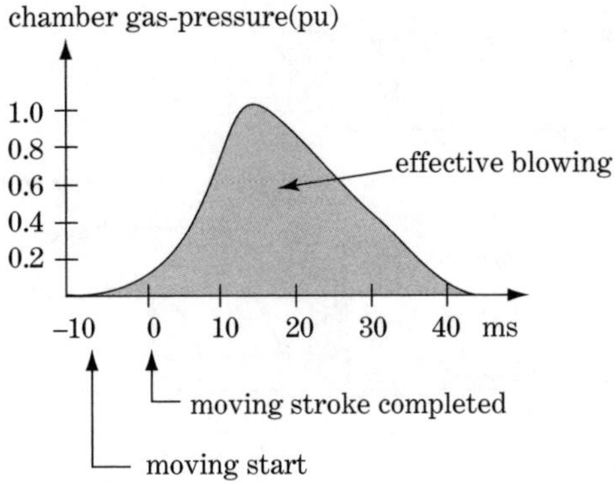

(c) gas pressure characteristics (effective blowing time)

Figure 19.8 (*Continued*)

⑦ **Amplitude ratio of transient recovery voltage:** The ratio of the transient recovery voltage to the peak value of the recovery voltage.

⑧ **RRRV:** The velocity of the voltage arising at the initial part of the transient recovery voltage (kV/μs).

⑨ **Reignition of arcs:** The reignition of arc currents within a quarter cycle (5 ms for 50 Hz, 4.1 ms for 60 Hz) of the initial current breaking.

⑩ **Restriking of arc:** The reigniting of arc currents after a quarter cycle (5 ms for 50 Hz, 4.1 ms for 60 Hz) of the initial current breaking. (The meaning of quarter cycle will be explained later.)

⑪ **Low-frequency extinction:** Arc current extinction at the time of low-frequency current zero.

⑫ **High-frequency extinction:** Arc current extinction at the time of high oscillatory frequency current zero.

All the above terms are important for a proper understanding of switching overvoltage phenomena. Readers should be able to grasp the meanings of ①–⑧ through a study of the calculation method of transient recovery voltages in the previous sections.

19.4 Actual Current Tripping Phenomena by Circuit-breakers

There are various voltage and current conditions for which the circuit-breakers are easy to break or poor at breaking, or the breakers may or may not cause severe switching overvoltages. We examine current tripping phenomena under typical different voltage and current conditions for an actual circuit-breaker.

19.4.1 Short-circuit current (lagging power-factor current) tripping

We studied the switching phenomena of the circuit shown in Figure 19.1 for the ideal breaker. However, actual fault current tripping by a breaker is a little different because a current arc appears

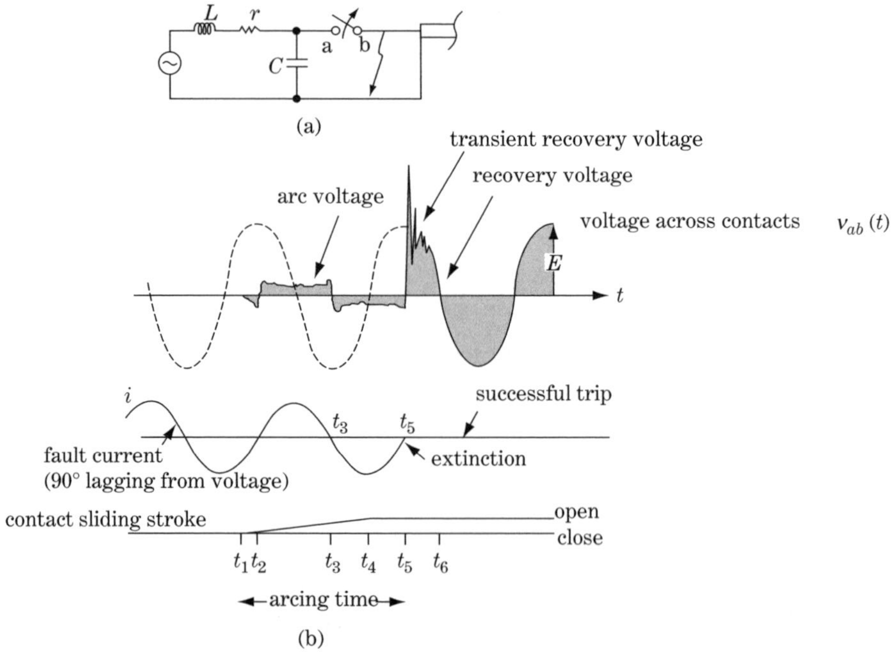

i At $t < t_1$ (before time t_1), fault current of 90° lagging power factor is flowing through the breaker, while the phase voltage is almost zero under the condition of the fault.

ii The movable contact begins to slide and leave the fixed contact at t_1, and simultaneously a current arc with smaller voltage drop $v_{ab}(t)$ appears across the contacts. The current arc will continue to flow until the timing of first or second 'current zero' (t_3 or t_5).

iii The current arc is not extinguished at t_3 (the first current zero) by chance, but is extinguished at t_5 (the second current zero). The polarity of the arc drop voltage is changed at the timing of every current zero.

iv The current arc is extinguished at t_5 (the second current zero), and simultaneously the transient recovery voltage $v_{ab}(t)$ as well as the surge overvoltage $v_a(t)$ at the terminal a appear.

v The breaker completes tripping at t_5 if reignition of arc does not occur.

Figure 19.9 Fault current tripping

transiently across the contacts. Figure 19.9a is the same circuit as in Figure 19.1, and where the actual voltage and current waveforms are also indicated. In this case, the breaker completes current tripping eventually at the time of the second 'current zero'. Detailed comments for the waveform are also written in Figure 19.9. The magnitude of the arc voltage drop would be, say, 5% or less of the peak value of the phase voltage.

Figure 19.10a shows a typical waveform of transient recovery voltage. Figure 19.10b is the explanation of transient recovery voltage versus the conceptual insulation withstanding voltage across the contacts, although the latter conceptual characteristics cannot be measured.

Incidentally, it is not easy to indicate the breaker's tripping capabilities by simple illustrated figures. Figure 19.10c is an attempt to illustrate the relation of the breaker's capability versus the expected duty required from the power system condition in regard to fault current tripping. The figure shows the coordinates of rated short-circuit capacity (MVA) of the breaker and the expected RRRV (kVμs) arising in the system. Two breakers of different capabilities are shown in the diagram. If the required system duty is given by the curve P_2, breaker 1 of the larger capability

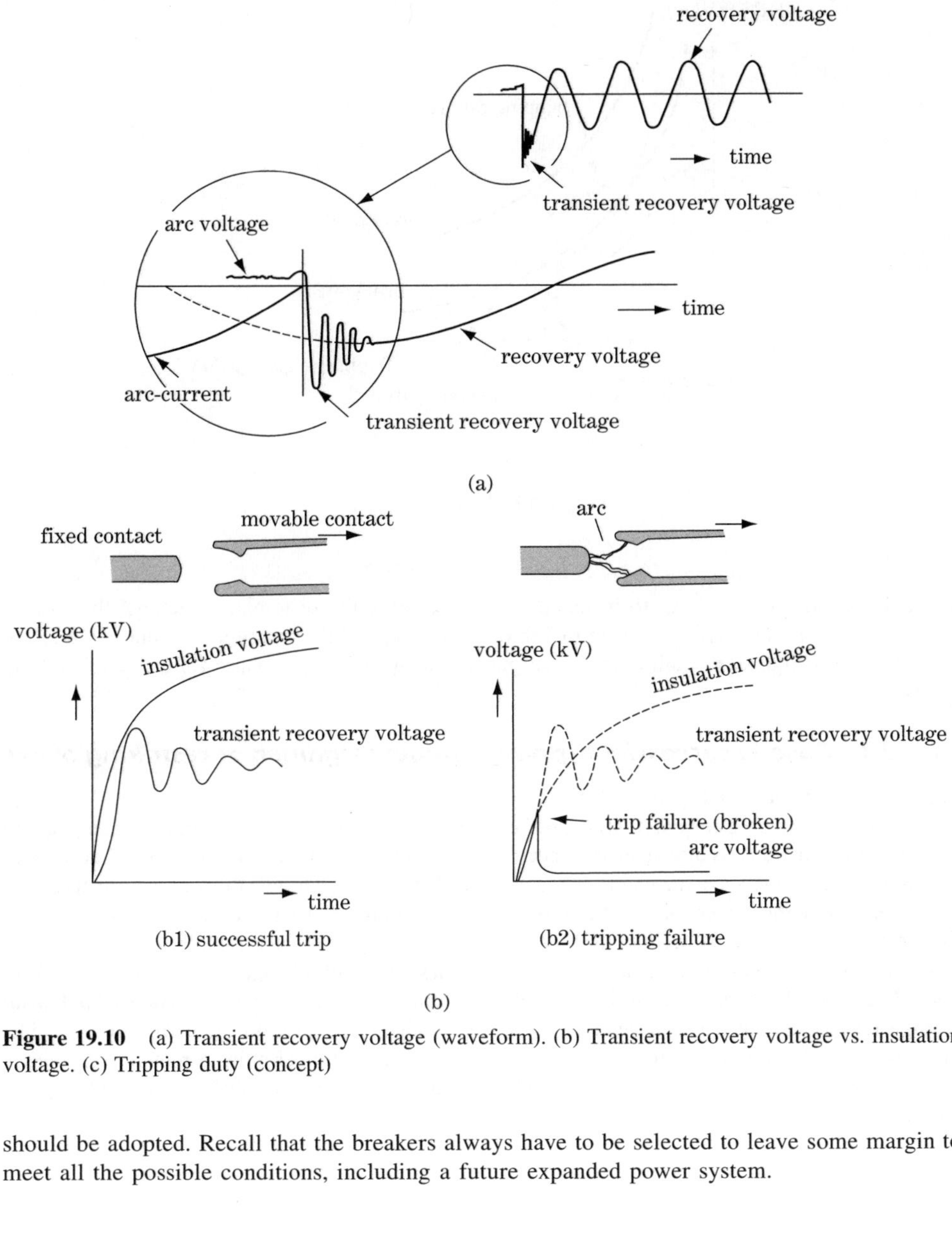

Figure 19.10 (a) Transient recovery voltage (waveform). (b) Transient recovery voltage vs. insulation voltage. (c) Tripping duty (concept)

should be adopted. Recall that the breakers always have to be selected to leave some margin to meet all the possible conditions, including a future expanded power system.

19.4.2 Leading power-factor small-current tripping

This is a very important case to understand unique switching phenomena and the delicate tripping characteristics of the breakers. Above all, the phenomena of the 'reignition of arcs' and 'restriking of arc' during the tripping procedure produce quite different results.

In Figure 19.11a, line section 2 is charged from the generator side while the opposite terminal end is open. Accordingly, a very small line charging current of leading power factor flows through the breaker. The breaker Br is going to be tripped to break 'the small leading power-factor current'

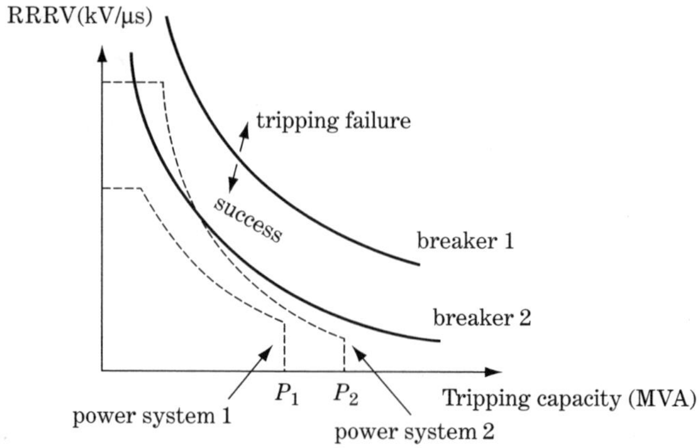

(c)

Figure 19.10 (*Continued*)

under this condition. We have to investigate two typical, quite different, switching phenomena by the breaker. Incidentally, referring to Table 2.2, the typical line charging current is 0.6 A/km for the 275 kV line, so a small current may mean 100 A or less per phase for a line of 100 km or less.

19.4.2.1 Case 1: successful tripping without reignition or restriking of arc

This case is explained by Figure 19.11b.

A small line charging current $i(t)$ flows before time t_1 (i.e. $t < t_1$) with 90° leading angular phase to the voltage E. Immediately after a breaker receives a tripping signal, the movable contact of the breaker begins to slide off and leave the fixed contact at t_1, while the small current still continues to appear as arc current. Next, the breaker breaks the current at the time of the first current zero t_2 (this is called the **low-frequency extinction**), because the breaker can quite easily break such a small current. After time $t_2 (t_2 < t)$, the system voltage $v_a(t)$ at the terminal point a sinusoidally changes ($v_a(t) = E \cos \omega t$) from E to $-E$. On the other hand, the charging d.c. voltage E remains on line section 2 as a **residual potential voltage**, because the current is 90° leading, so that the voltage value at the time of current zero t_2 is just the crest value E. In other words, the voltage of the breaker terminal point b remains unchanged as $v_b(t) = E$. Accordingly, the voltage across the contacts $v_{ab}(t)$ after t_2 would be increased from zero (at t_2) to $2E$ and soon decrease to zero.

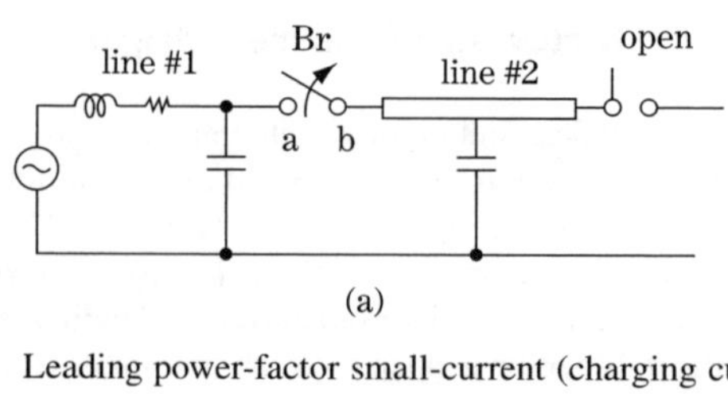

(a)

Figure 19.11 Leading power-factor small-current (charging current) tripping

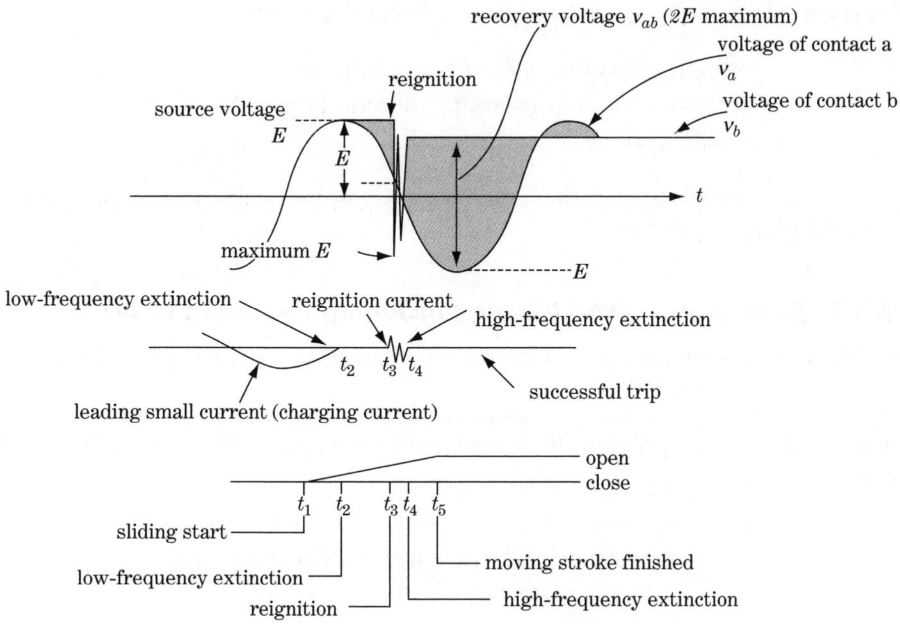

(b) successful tripping with reignition

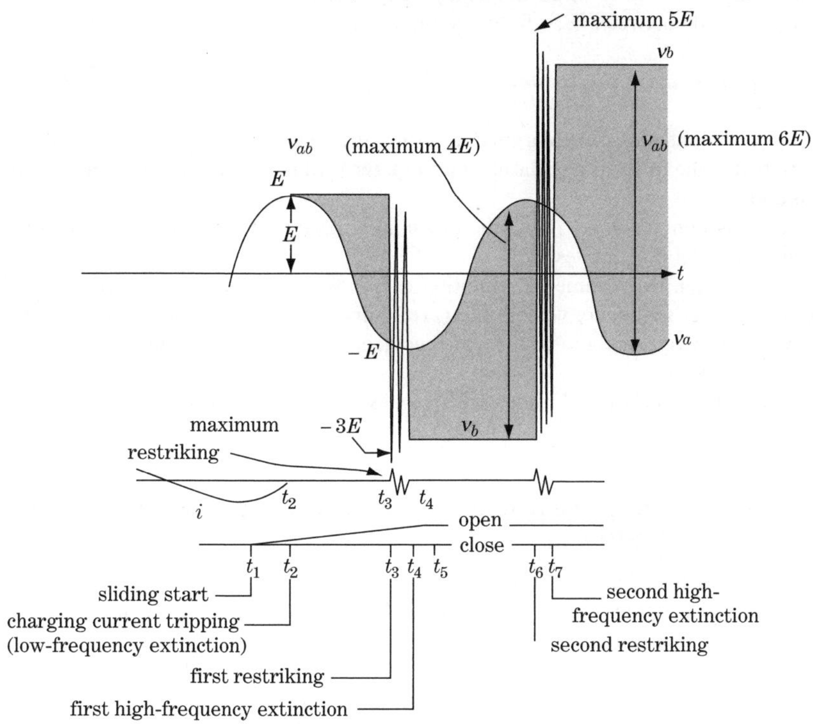

(c) tripping failure with restriking

Figure 19.11 (*Continued*)

The above behaviour after t_2 can be written symbolically as follows:

$$v_a(t) = E\cos\omega t = |E, -E| \quad \text{or} \quad v_a(t) \leqq E$$
$$v_b(t) = [E, -E] \quad (\text{where}[E, -E]\text{means between } E \text{ and } - E)$$
$$\therefore \ |v_{ab}(t)| \leqq 2E$$

The breaker tripping would be successfully completed because the breaker has enough capability to withstand the voltages within $2E$.

19.4.2.2 Case 2: successful tripping including reignition of arc

In the above process, let us assume the occurrence of **reignition** at t_3 which is a quarter cycle or less after t_2 as shown in Figure 19.11b.

When the reignition is caused at $t = t_3$, the potential voltage $v_b(t)$ at point b just before reignition is between E and 0, so that $v_{ab}(t)$ across the contacts is also between E and 0 for $t < t_3$. That is before reignition

$$\text{for } t < t_3 \quad v_a(t) = E\cos\omega t = [E, 0]$$
$$v_b(t) = [E, 0] \quad (\text{d} \cdot \text{c potential charging voltage})$$
$$\therefore \ |v_{ab}(t)| = E \text{ in maximum}$$

Then, the reignition voltage (high frequency free-oscillatory transient voltage) caused just after $t_3 \leq t$ is of magnitude between $[E, -E]$ by amplitude $\pm E$.

$$\text{for } t_3 < t < t_4 \quad v_a(t) = v_b(t) = [E, -E]$$

As the movable contact is still enlarging the stroke, the breaker easily breaks the small oscillatory current at t_4, that is the time of '**oscillatory current zero**' (this is called **high-frequency (current-zero) extinction**).

After arc extinction at $t = t_4 (t_4 \leqq t)$, the potential voltage $v_b(t)$ as well as source voltage $v_a(t)$ are kept within ordinal magnitudes E.

In conclusion, even if one time of reignition occurs within a quarter cycle after the initial arc extinction, the breaker's recovery voltage $v_{ab}(t)$ is $2E$ maximum. In other words, reignition does not cause severe transient voltages across the breaker contacts, so the breaker tripping will be successfully completed in this case.

'Reignition within one time through the type test' is usually permitted in the representative international standards for breakers.

19.4.2.3 Case 3: tripping phenomenon including restriking of arc (tripping failure)

This case is explained by Figure 19.11c.

If reignition (which we need to call '**restriking of arc**') is caused at t_3, which is at least a quarter cycle after t_2. before restriking

$$\text{for } t < t_3 \quad v_a(t) = E\cos\omega t = [0, -E]$$
$$v_b(t) = [E, 0] \quad (\text{d} \cdot \text{c charging voltages})$$
$$\therefore \ |v_{ab}(t)| = 2E \ \text{ in maximum}$$

Then, the restriking voltage (high frequency free-oscillatory voltage) after $t_3 \leq t$ is of magnitude between $[E, -3E]$ by amplitude $\pm 2E$.

$$\text{for } t_3 < t < t_4 \quad v_a(t) = v_b(t) = [E, -3E] \quad \text{(high frequency oscillatory voltage by amplitude } \pm 2E.)$$

Next, as the movable contact is still enlarging the stroke, the breaker breaks the oscillatory current at the time of 'oscillatory current zero' (at t_4; this is again **high-frequency extinction**.).

After arc extinction at $t = t_4(t_4 \leqq t)$,

$$\text{for } t_4 < t \quad v_a(t) = E\cos\omega t = [0, -E]$$

$$v_b(t) = [E, -3E] \quad \text{(d} \cdot \text{c potential charging voltage)}$$

$$\therefore \; E \leqq |v_{ab}(t)| \leqq 4E$$

In other words, one time of restriking may causes quite large restriking voltage of up to $4E$ across the contacts, and further $3E$ of phase voltages. Then, the second restriking would inevitably be caused at t_6.

$$\text{for } t_6 < t < t_7 \quad v_a(t) = v_b(t) = [5E, -3E]$$
$$\text{(transient restriking voltage of amplitude } \pm 4E \text{ in maximum)}$$

Further, if the arc is again extinguished at t_7, the phase voltage $v_b(t)$ at the terminal b may possibly become $5E$ *in maximum*, and the voltage $v_{ab}(t)$ across the contacts may become $6E$ in maximum.

If we assume the third restriking, $v_b(t)$ and $v_{ab}(t)$ could theoretically become so large magnitude of even $-7E$ and $8E$ in maximum, respectively by the similar analogy.

Of course, the breaker can not withstand such severe restriking voltages and the breaking chamber would be broken probably around the timing of the first or the second restriking. Besides, earth grounding faults would be newly induced on the breaker or on the other closely installed equipment.

Case 3 explains why the one-time occurrence of restriking may lead to quite a large transient recovery voltage $v_{ab}(t)$ as well as transient phase voltage (switching surge) $v_b(t)$ with high probability, by which tripping failure and/or a grounding fault may be caused and the breaker could be immediately broken, or by which successive restrikings are caused and further extreme voltages $v_{ab}(t), v_b(t)$ are induced from time to time, so that tripping failure and breakdown of the breaker would occur sooner or later. Furthermore, a new grounding failure by phase overvoltage on line section or on substation equipment would also be caused.

The results of restriking (case 3) and reignition (case 2) are significantly different. Obviously the occurrence of restriking may cause quite severe effects, so this actually means lack in the breaker's tripping capability, while reignition does not necessarily mean lack in capability.

It should be noted that restriking would be caused not only by tripping of small leading current, but also by tripping of ordinal fault current if the tripping capability is critical to the circuit condition,

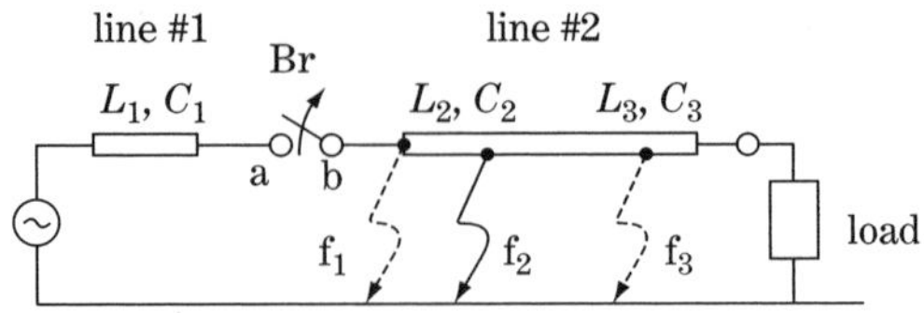

Case 1: The fault at point f_1 (very close to the breaker terminal b)

The voltage at the terminal b is actually zero just after the fault, namely $v_b(t) = 0$, accordingly $v_a(t) = v_{ab}(t)$ and the RRRV will not become relatively large in comparison with case 2 in spite of the fault current being larger than that of case 2.

Case 2: The fault at point f_2 (up to 10 km distant from the breaker terminal b)

The free oscillatory overvoltage $v_b(t)$ would appear at point b as the result of repeated travelling-wave voltage reflection between f_2 and b. In addition, the caused frequency would be higher and accordingly the RRRV appearing for the breaker's $v_{ab}(t)$ could also be larger, because the distance between f_2 and b is only a few to 10 km. (Assuming distance $\overline{bf_2} = 3$ km and the velocity is 300m/μs, one-way travelling time is 10 μs so that the caused natural frequency would be around 50 kHz.) Therefore, in total, this case could be more severe for the breaker tripping in comparison with case 1 in spite of the smaller fault current.

Case 3: The fault at point f_3 (rather distant from the breaker terminal b)

In comparison with case 2, the fault current is smaller. In addition, the similar oscillatory overvoltage would also appear in this case; however, the frequency would be lower because of the distance between f_3 and b. Therefore this case is easier for breaker tripping.

Figure 19.12 SLF (Short Line Fault) tripping

although the above explanations were made expediently for the processes of small leading current tripping.

In practical engineering, the necessary tripping duties of each type of breaker are confirmed by checking all the items in a type test against the specified authorized standards. The occurrence of restriking for any reason in such a type test is recognized as unsuccessful tripping (rejection by lack of the tripping capability), although one-time reignition may be permitted.

19.4.3 Short-distance line fault tripping (SLF)

We assume three fault points f_1, f_2, f_3 as shown in Figure 19.12.

In early 1950's, it was found that f_2 fault tripping (f_2 is a few kilometres distant from the breaker) could be more difficult than f_1 fault tripping, in spite of the smaller fault current through the breaker. This is because a higher RRRV would appear by the fault at f_2 as the result of repeated reflection of the travelling-wave voltages caused by the fault between point b and f_2. It was recognized that such aspects would often appear in cases of faults up to 10 km from the breaker point. Such a fault was named a **short line fault (SLF)**. With this recognition came the opportunity to make much of RRRV in addition to transient recovery voltage and recovery voltage in the history of circuit-breaker development.

19.4.4 Current chopping phenomena by tripping small current with lagging power factor

Breakers are specifically designed to extinguish large current arcs (of even 50 kA or 63 kA rms maximum), but then some smaller currents may be cut off before the time of current zero. This

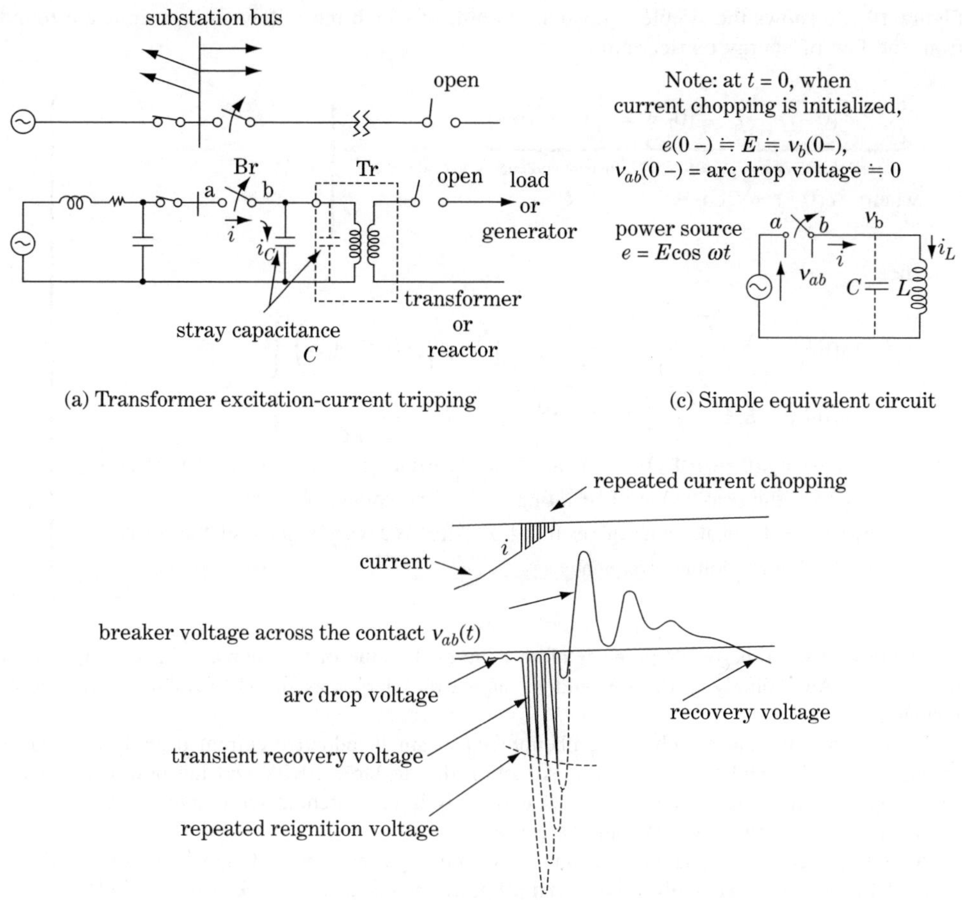

(a) Transformer excitation-current tripping

(c) Simple equivalent circuit

(b) Current chopping

Figure 19.13 Current chopping phenomena (small lagging current tripping)

phenomenon is called **current chopping**. This current chopping causes large transient recovery voltages physically because

$$v(t) = L \cdot \frac{di(t)}{dt} \to \infty$$

even if $i(t)$ is small.

A typical case is the tripping of transformer excitation current (90° lagging) as shown in Figure 19.13a, in which a small lagging current of the transformer excitation current is going to be tripped. Figure 19.13b shows the typical waveforms of current chopping, in that the small arc current is forcibly cut off just before its current zero (the voltage is almost the peak value in this timing) so that free oscillatory overvoltages appear. The resulting oscillatory frequency $f = 1/(2\pi\sqrt{LC})$ is also quite high because the leakage capacitance C (from the insulation structures of bus and feeders etc.) has a rather small value (say, 100–1000 pF).

Figure 19.13c shows the simple equivalent circuit, in which the following equations are found from the law of energy conservation:

$$\left.\begin{aligned}&\underbrace{\frac{1}{2}C\cdot e^2(0-)+\frac{1}{2}L\cdot i_L^2(0-)}_{\text{total stored energy at } t=0-}=\underbrace{\frac{1}{2}C\cdot v_b^2(0+)}_{\text{max stored energy by} C \text{ at } t=0+}\\&\text{where}\quad e(0-)=E\cos\omega t|_{t=0-}=E\end{aligned}\right\}\ ①$$

then

$$\left.\begin{aligned}&v_b(0+)=\pm\sqrt{\frac{L}{C}\cdot i_L^2(0-)+E^2}\qquad\left(>\sqrt{\frac{L}{C}}\cdot i_L(0-),E\right)\\&v_{ab}(0+)=E\pm\sqrt{\frac{L}{C}\cdot i_L^2(0-)+E^2},\qquad f=\frac{1}{2\pi\sqrt{LC}}\end{aligned}\right\}\ ②$$

where $v_b(0-), v_{ab}(0-), i(0-)$: the voltages and current at the time of current chopping

$v_b(0+)$: the peak value of resulting oscillatory voltage at point b

$v_{ab}(0+)$: the peak value of resulting transient recovery voltage of the breaker

f : oscillatory voltage frequency

(19.45)

The voltage at the time of chopping is almost the peak value of the source voltage $e(t)$, namely $v_b(0-)\fallingdotseq E$. Accordingly, $v_b(0+)$ would be at least as larger as the crest value of the source voltage E.

Furthermore, the current chopping phenomena of small inductive current tripping may cause repeated reignition and high-frequency extinction due to large RRRV and the powerful breaker capability for forced arc extinction, so that extremely large switching surge overvoltage $v_b(t)$ and transient recovery voltage $v_{ab}(t)$ could be caused.

An effective countermeasure to prevent such overvoltage and tripping failure by current chopping is the adoption of **breakers with resistive tripping**, which is general practice in today's higher voltage class breakers, in particular the EHV/UHV class. Breakers with resistive tripping/losing are discussed in Section 19.6.

19.4.5 Step-out tripping

As we have already seen in Figure 19.3, cases 4, 5 and 6, breaker tripping under the step-out procedure may cause large recovery voltages ($3E$ maximum) so it is one of the most severe tripping duties for the breaker. High-voltage breakers are usually assigned by the type test to have the capability to trip under step-out conditions, although most breakers are blocked from step-out tripping practically by protective relay equipment, as explained in Section 17.4.

19.4.6 Current-zero missing

Lastly, in this section, **current-zero missing phenomena** of breaker tripping should be recognized as an essential basis of breakers.

Our breaker is a so-called 'a.c. tripping breaker' capable of breaking the current flowing at the time of current zero within a few cycles. Therefore, breakers cannot trip d.c. offset current or, in other words, breakers can trip a.c. fault current only at the time of current zero within the duration of the effective tripping procedure (see Figure 10.8c). With this in mind, the following two cases should be considered.

19.4.6.1 Fault tripping with d.c. transient current with long time constant

We know experimentally that tripping failure of breakers by current-zero missing seldom occurs. This is partly because the time constant of the d.c. component i_{dc} in the fault current is rather small, fortunately for most of the network (say, $T = 2L/R = 0.01 - 0.05$ s), so that i_{dc} attenuates quickly at the beginning of the fault current.

However, the time constants of the power system network are apt to be larger for higher voltage systems. In particular, the time constant $T$of UHV systems may be around or even exceed 0.1 s, which may be near to the dangerous zone. Fortunately, most EHV/UHV breakers are of the resistive tripping type (see Section 19.6) which can be very effective in reducing time constants. This is another reason why tripping failure by current zero seldom occurs.

19.4.6.2 Magnetizing inrush current tripping of transformers

Magnetizing inrush currents including extremely distorted d.c. components might arise 1 or 10 s after initial excitation of transformers (see Figure 16.10). Therefore an attempt to trip inrush current forcibly may cause tripping failure of the breaker by current-zero missing, or severe overvoltages by current chopping ($= L \cdot di/dt$). Accordingly, a breaker should not be tripped for any reason until the inrush current actually disappears. Usually a trip-lock function is included in transformer protective equipment as typical practice, by which tripping is blocked for the initial few seconds just after the transformer is charged.

19.5 Overvoltages Caused by Breaker Closing (Close-switching Surge)

19.5.1 Principles of overvoltage caused by breaker closing

We now examine the overvoltages caused by breaker closing. In Figure 19.14 the breaker Br is going to be closed in order to charge the line. Just after the breaker receives the closing dispatch signal, the movable contact begins sliding to approach the fixed contact and completes mechanical closing within around 0.1 s. As the stroke approaches and just before the two contacts mechanically touch, preignition is started and the current arc begins to flow. The current arc will never be extinguished until the two contacts complete mechanical closing, because the stroke gap across the contacts is rapidly decreasing. Accordingly, breaker closing is electrically achieved at the time of ignition start. The timing of arc ignition may be accidental, unlike the timing of current zero for tripping. Also,

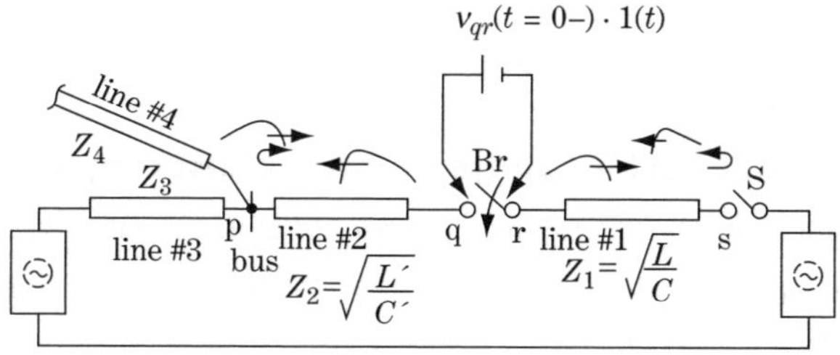

Figure 19.14 Transient overvoltages caused by breaker closing (breaker-closing surge)

so-called 're-extinction' is never caused during the closing process, though 'reignition or restriking' may be caused during the tripping process.

In Figure 19.14, $v_{qr}(0-)$ is the preignition voltage across the contacts q and r on initial stage of closing procedure. Now the transient phenomena caused by the breaker closing can be calculated by inserting the voltage with the opposite polarity $-v_{qr}(0-)$ across the points q and r as the sudden forced voltage source. At the instant of insertion (of arc ignition), the surge travelling waves $v_q(t)$, $i_q(t)$ and $v_r(t), i_r(t)$ begin to travel in the right and left directions, respectively. The surge equations can be written as follows:

$$\left.\begin{array}{ll} v_q(t) - v_r(t) = v_{qr}(0-), & i_q(t) = -i_r(t) \\ \dfrac{v_q(t)}{i_q(t)} = Z_2, & \dfrac{v_r(t)}{i_r(t)} = Z_1 \end{array}\right\} \quad (19.46a)$$

$\therefore$ surge in left direction; surge in right direction

$$\left.\begin{array}{l} \left.\begin{array}{l} v_q(t) = \dfrac{Z_2}{Z_1+Z_2} \cdot v_{qr}(0-) \\ i_q(t) = \dfrac{1}{Z_1+Z_2} \cdot v_{qr}(0-) \end{array}\right\} ① \quad \left.\begin{array}{l} v_r(t) = \dfrac{Z_1}{Z_1+Z_2} \cdot v_{qr}(0-) \\ i_r(t) = -\dfrac{1}{Z_1+Z_2} \cdot v_{qr}(0-) \end{array}\right\} ② \end{array}\right\} \quad (19.46b)$$

These equations give the initial travelling surge voltages and currents in both directions.

For the initial voltage across the breaker contacts just before preignition by closing, $v_{qr}(0-)$:

Case 1: points q and r are loop connected through another route or a parallel circuit :

At $t < 0$

$$v_{qr}(t) = E_1 e^{j\omega t} - E_2 e^{j(\omega t+\delta)} = E_1\left(1 - \frac{E_2}{E_1}e^{j\delta}\right)e^{j\omega t} \fallingdotseq E_1\left(1 - e^{j\delta}\right)e^{j\omega t} \leqq \sqrt{2}E_1$$

$\therefore v_{qr}(0-) = 1.4E$ maximum. ①

Case 2: the switch S at point s is opened :

The residual voltage of 0 to $\pm E$ may exist at line 1 unless the line is forcibly grounded. On the other hand, the phase angle of the generator source voltage is within 0° to 360°. Accordingly;

$v_{qr}(0-) = 2E$ maximum ②

Case 3: closing of inverse angular polarity (by operational mistake)

$v_{qr}(0-) = 2E$ maximum ③

Needless to say, breaker closing without synchronization has to be strictly avoided not only to prevent instability or generator shaft distortion, but also to prevent switching overvoltages and breaker failure.

(19.47)

19.5.1.1 Numerical check of surge voltage caused by breaker closing

In Figure 19.14, the following line conditions are assumed and attenuation is neglected:

- Overhead line 1: length 15 km, surge impedance $Z_1 = 300\,\Omega$, velocity $u_1 = 300\,\text{m}/\mu\text{s}$; switch S at the opposite terminal s is opened.
- Cable line 2: length 5 km, surge impedance $Z_2 = 30\,\Omega$, $u_2 = 150\,\text{m}/\mu\text{s}$.

- Overhead lines 3, 4: length infinite, surge impedance $Z_3 = Z_4 = 300\,\Omega$, $u_3 = u_4 = 300\,\text{m}/\mu\text{s}$.

For the initial surge in the left direction (from point q to p):

$$v_q(t) = \frac{Z_2}{(Z_1 + Z_2)} \cdot v_{qr}(0-) = \frac{30}{(300 + 30)} \cdot v_{qr}(0-) = 0.09 \cdot v_{qr}(0-)$$

$$i_q(t) = \frac{1}{(Z_1 + Z_2)} \cdot v_{qr}(0-) = \frac{1}{(300 + 30)} \cdot v_{qr}(0-) = 0.003 \cdot v_{qr}(0-)$$

- The coefficient of reflection at point p $= (150 - 30)/(150 + 30) = 0.67$.
- Total time going to and returning from points q, p $= 2 \times 5000/150\,\mu\text{s} = 66.7\,\mu\text{s} \therefore f = 15.0\,\text{kHz}$
- The voltage at point q at the time of arrival of the first reflected wave

$$(1 + 0.67) \cdot v_{qr}(0-) = 1.67 \cdot v_{qr}(0-)$$

For the initial surge in the right direction (between points r and s):

$$v_r(t) = \frac{Z_1}{Z_1 + Z_2} \cdot v_{qr}(0-) = 300/(300 + 30) \cdot v_{qr}(0-) = 0.91 \cdot v_{qr}(0-)$$

$$i_r(t) = \frac{1}{Z_1 + Z_2} \cdot v_{qr}(0-) = 1/(300 + 30) \cdot v_{qr}(0-) = 0.003 \cdot v_{qr}(t)$$

- The coefficient of reflection at open point s $= (\infty - 300)/(\infty + 300) = 1.0$
- Total time for going to and returning from points r, s $= 2 \times 15{,}000/300\,\mu\text{s} = 100\,\mu\text{s} \;\therefore f = 10\,\text{kHz}$
- The voltage at point r at the time when the first reflected wave arrives $(1 + 1) \cdot v_{qr}(0-) = 2.0 \cdot v_{qr}(0-)$

Higher frequency caused by short-distance reciprocating surge travel means higher steepness generally, while the attenuation caused by eddy-current losses and corona losses is rather quick. Recently developed gas-insulated switchgear (GIS) is generally of compact size in length, so that switching surges with extraordinarily higher frequency phenomena (say 1 GHz) could appear within a limited distance. We discuss this problem in Chapters 20 and 21.

19.6 Resistive Tripping and Resistive Closing by Circuit-breakers

19.6.1 Resistive tripping and closing

The structural practice of resistive tripping and resistive closing has been adopted for most EHV and UHV breakers. Diagrams of two typical practices are given in Figure 19.15. As shown in the figure, a breaker has one main contact frame s_1 and one auxiliary contact frame s_2 (resistive contacts) and the resistor element R (usually 300–500 Ω, short-duration time).

For the tripping sequence, in the original breaker closing condition, the main contact s_1 as well as the auxiliary contact s_2 are in the closed position so that all the current i flows through the main contact s_1. Just after the tripping signal is received, the main contact s_1 is opened first at t_1, then the current pass through the main contact is switched to the auxiliary pass through the resistive element and the

(example 1)

(example 2)

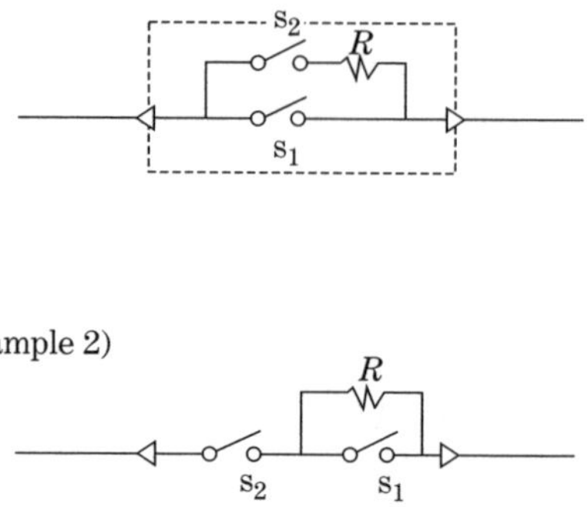

s_1: main contact frame

s_2: auxiliary (resistive) contact frame

tripping sequence

tripping signal → s_1 opening → s_2 opening

closing sequence

closing signal → s_2 closing → s_1 closing

Figure 19.15 Resistive tripping method and resistive closing method

auxiliary contact s_2, and the resistive restrained current i_{aux} flows through the auxiliary pass until s_2 is opened at t_2 (the time difference is, say, $\Delta T_1 = t_1 - t_2 = 10$ ms or less), thus completing the breaker tripping.

Although the duration ΔT_1 of the current i_{aux} flowing through the resistive element is only 10 ms, the resistive element has to withstand the temperature rise caused by the extremely large resistive loss $\int i_{aux}^2 \cdot R\,dt$.

For the closing sequence, in the original breaker opening condition, the main contact s_1 as well as the auxiliary contact s_2 are in the open position. Just after the closing signal is received, the auxiliary contact s_2 is closed first at t_1, then the resistive restrained current i_{aux} begins to flow through the auxiliary pass until s_1 is closed at t_2 (the time difference is, say, $\Delta T_2 = t_1 - t_2 = 10{-}20$ ms), completing the breaker closing.

Although the duration ΔT_2 of the current i_{aux} flowing through the resistive element is only 10 ms or less, the resistive element has to withstand the temperature rise caused by the extremely large resistive loss $\int i_{aux}^2 R\,dt$.

19.6.1.1 Reasons for adopting resistive tripping and closing

First, as a part of insulation coordination, the switching surge voltage by the breaker tripping/closing has to be restrained within a regulated upper limit level assigned to the duties of breakers by the associated standards or recommendations for coordinating system insulation.

Tables 21.2B, 2C are the IEC and IEEE standards for power systems over 200 kV, which indicate the permissible switching surge level. The permissible switching surge level based on these standards can be written usually as the multiple PU value (α) based on the crest value of the rated phase voltage.

For example, in the IEC standard (Table 21.2B) for a rated voltage 420 kV system, for the highest system voltage $V_{l\text{-}l} = 420$ kV (line-to-line, rms value), the crest value of the line-to-ground phase voltage is $(\sqrt{2}/\sqrt{3}) \cdot 420\,\text{kV} = 343\,\text{kV}$. The specified standard switching impulse-withstand voltage by the IEC for this operating voltage system is 850 kV or 950 kV or 1050 kV, corresponding to $\alpha = 2.48$, 2.77 or 3.06 times 343 kV, respectively.

Secondly, breakers of higher voltage and large current capacity (such as 50 kA, 63 kA), in particular of EHV/UHV, with a reasonable structure can be realized only by adopting the principles of resistive tripping/closing by which transient recovery voltages can be reduced.

Incidentally, as a matter of interest, we can say that resistive tripping/closing has always been an essential feature of the technology in the 100-year dramatic history of breaker development.

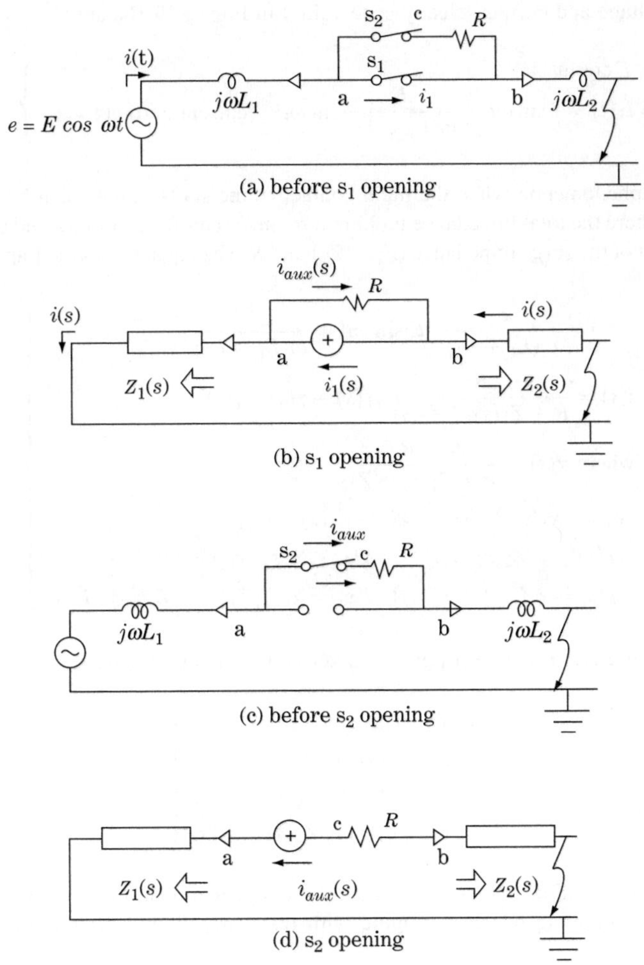

Figure 19.16 Calculation of breaker resistive tripping surge voltages

A resistive element of typically 300–500 Ω is adopted in order to satisfy the requirements of both the above reasons. Assuming $R = 500\ \Omega$ for a 500 kV breaker, the current flow i_{aux} of the auxiliary contact frame is $500/\sqrt{3} \div 500 \doteqdot 0.6\,\text{kA rms}$. Then the resistive element has to withstand thermal energy of $\int i_{aux}^2 \cdot Rdt \doteqdot (600)^2 \cdot 500\,\text{Watt} \cdot 10^{-2}\text{sec} = 1800\,\text{kJ}$, although the duration ΔT_1 is only about 10 ms. Ceramic resistors have been used recently because of their outstanding heat-resisting stable characteristics.

19.6.2 Overvoltage phenomena caused by tripping of breaker with resistive tripping mechanism

We examine the tripping phenomenon for the circuit in Figure 19.16a in which a grounding fault has occurred and the fault current (90° lagging) $i(t) = I \sin \omega t$ is flowing through the breaker. The main contact frame s_1 starts moving and breaks the through current at the time of current zero ($t = 0$). A little time ΔT later, the auxiliary contact frame s_2 breaks the current $i_{\text{aux}}(t)$ at the time of current zero.

The initial voltage and current (steady-state value) in Figure 19.16a are

$$\left.\begin{aligned} e(t) &= E\cos\omega t \\ i(t) &= i_1(t) = I\sin\omega t = \frac{E}{\omega(L_1+L_2)}\sin\omega t \quad \text{(current zero at } t=0) \end{aligned}\right\} \tag{19.48}$$

The transient phenomenon when the main contact frame starts moving can be calculated as in Figure 19.16b, where the total impedance looking into the circuit from points a and b is composed as the parallel circuit of the surge impedance $(Z_1 + Z_2)$ and R. The equation for the Laplace transform is

$$\left.\begin{aligned} i_1(s) &= \frac{E}{\omega(L_1+L_2)}\cdot\mathcal{L}[\sin\omega t] = \frac{E}{\omega(L_1+L_2)}\cdot\frac{\omega}{s^2+\omega^2} \\ i(s) &= \frac{R}{R+Z_1(s)+Z_2(s)}\cdot i_1(s) \equiv \alpha(s)\cdot i_1(s) \\ &\text{where } \alpha(s) = \frac{R}{R+Z_1(s)+Z_2(s)} \\ v_a(s) &= Z_1(s)\cdot i(s) = \alpha(s)\cdot Z_1(s)\cdot i_1(s) \\ v_b(s) &= -Z_2(s)\cdot i(s) = -\alpha(s)\cdot Z_2(s)\cdot i_1(s) \\ v_{ab}(s) &= \{Z_1(s)+Z_2(s)\}\cdot i(s) = \alpha(s)\{Z_1(s)+Z_2(s)\}\cdot i_1(s) \end{aligned}\right\} \tag{19.49}$$

The special case of a non-resistive-tripping breaker is obtained by letting $R \to \infty$. Then α (s)$\to$ 1.

$$\left.\begin{aligned} i(s) &= i_1(s) \\ v_a(s) &= Z_1(s)\cdot i_1(s) \\ v_b(s) &= -Z_2(s)\cdot i_1(s) \\ v_c(s) &= \{Z_1(s)+Z_2(s)\}\cdot i_1(s) \end{aligned}\right\} \tag{19.50}$$

In comparison with Equations 19.49 and 19.50, all the transient quantities $i(s)$, $v_a(s)$, $v_b(s)$, $v_{ab}(s)$ can be reduced by adopting resistive tripping with resistance R, and the rate of switching surge reduction is given by $\alpha(s)$:

$$\alpha(s) = \frac{R}{R+Z_1(s)+Z_2(s)} \tag{19.51a}$$

For the initial time interval before the reflected waves return from the transition point at the opposite terminal, simplification of $Z_1(s) \to \sqrt{L_1/C_1}$, $\quad Z_2(s) \to \sqrt{L_2/C_2}$ is possible and

$$\alpha = \frac{R}{R+\sqrt{\frac{L_1}{C_1}}+\sqrt{\frac{L_2}{C_2}}} \tag{19.51b}$$

Accordingly, the initial levels of the switching surges v_a, v_b, v_{ab}, i can be reduced by the rate of α in Equation 19.51b.

If the resistance R is selected to be smaller than $(Z_1 + Z_2)$, namely $R < (Z_1 + Z_2)$, the transient voltages can be reduced within a half. If smaller R is selected, the tripping duty of the main contact frame as well as the induced switching surge voltages can be reduced, though a relatively larger tripping duty would be required for the auxiliary contact frame.

Next, the auxiliary contact frame s_2 is opened with a time delay of about 10 ms from the opening of s_1, and the transient phenomenon is calculated as in Figures 19.16c and d. The transient term of i_{aux} at the opening of s_1 disappears quickly (because of the insertion of large R), and the initial i_{aux} at the time of s_2 opening is given by

$$\left.\begin{array}{l} i_{aux}(t) = \dfrac{E}{|R + j\omega(L_1 + L_2)|} \sin \omega t \fallingdotseq \dfrac{E}{R} \sin \omega t \quad \text{(current zero at } t = 0) \\ \text{where } R = 300 - 500\,\Omega \gg \omega(L_1 + L_2). \end{array}\right\} \tag{19.52}$$

The equation of the Laplace transforms is

$$\left.\begin{array}{l} i_{aux}(s) = \dfrac{E}{|R + s(L_1 + L_2)|} \cdot \dfrac{\omega}{s^2 + \omega^2} \fallingdotseq \dfrac{E}{R} \cdot \dfrac{\omega}{s^2 + \omega^2} \\ v_a(s) = Z_1(s) \cdot i_{aux}(s) \\ v_b(s) = -Z_2(s) \cdot i_{aux}(s) \\ v_c(s) = -\{R + Z_2(s)\} \cdot i_{aux}(s) \\ v_{ac}(s) = \{R + Z_1(s) + Z_2(s)\} \cdot i_{aux}(s) \end{array}\right\} \tag{19.53}$$

The above equations indicate the approximation

$$\left.\begin{array}{l} \text{the induced surge voltages} \\ v_a = \dfrac{Z_1}{R} E, \quad v_b = \dfrac{-Z_2}{R} E, \quad v_{ab} = \dfrac{Z_1 + Z_2}{R} E, \quad v_c = -\left(1 + \dfrac{Z_2}{R}\right) E \quad ① \\ \text{the current through the auxiliary contact frame} \\ i_{aux}(t) \fallingdotseq \dfrac{E}{R} \sin \omega t \quad ② \end{array}\right\} \tag{19.54}$$

Accordingly, tripping of the current i_{aux} by the auxiliary contact frame is relatively easy, because the magnitude of i_{aux} is small, and i_{aux} is almost in phase with the voltage so that the transient recovery voltage at the time of current zero of i_{aux} is also small. Then i_{aux} can be tripped by the relatively simple and small auxiliary contact frame.

Meanwhile, the heat energy which causes a temperature rise in the resistive element can be calculated by the equation

$$\int i_{aux}(t)^2 \cdot R dt \fallingdotseq i_{aux}^2 R \cdot \Delta T \tag{19.55}$$

$$\text{where } \Delta T \fallingdotseq 0.01 \text{ sec.}$$

The resistive element has to withstand the above heat energy and to maintain a relatively stable ohm value through the duration of ΔT.

19.6.3 Overvoltage phenomena caused by closing of breaker with resistive closing mechanism

In Figure 19.17, the main contact frame s_1 as well as the auxiliary contact frame s_2 are open when the breaker is initially opened. Breaker closing is started by s_2 first and is completed by s_1 closing with time delay ΔT.

The initial voltage $v_{ab}(0-)$ across the breaker terminals a–b is affected by the condition of the terminal d on the opposite side:

- When d is in the open condition,

$$v_d = -e \text{ to } 0 \text{ to } +e, \quad \text{then } v_{ab}(0-) = 2e \quad \text{maximum.}$$

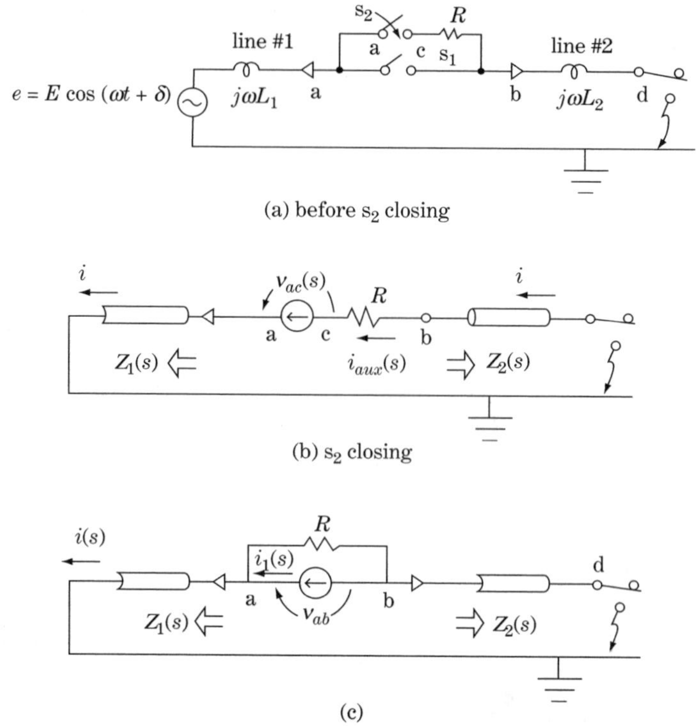

Figure 19.17 Calculation of breaker resistive closing surge voltages

- When d is under a grounding fault or manually grounded,

$$v_d = 0, \quad \text{then } v_{ab}(0-) = e \quad \text{maximum.}$$

- For the breaker closing under synchronization,

$$v_a, \quad v_b \quad \text{are in phase,} \quad \text{then} \quad v_{ab}(0-) \fallingdotseq 0.$$

Accordingly, the initial voltage can be written as $v_{ab} = k \cdot e$ (where $k = 0 - 2$) in our calculation, which is to be inserted between the points a and b in Figure 19.17b.

After the sliding action of s_2 starts, pre-ignition across s_2 would probably be caused before mechanical contact is complete, which is actually the time of 'electrical closing' because ignition continues without extinction until the mechanical contact of s_2 is finished. The 'timing of electrical closing $t = 0$' is a matter of chance.

Accordingly, the initial voltage across the breaker terminals at $t = 0$ is

$$\left.\begin{array}{c} v_{ac}(t) = k \cdot E\cos(\omega t + \delta) \\ \text{where } k = 0 - 2 \text{ (decided by the condition of line 2).} \end{array}\right\} \quad (19.56)$$

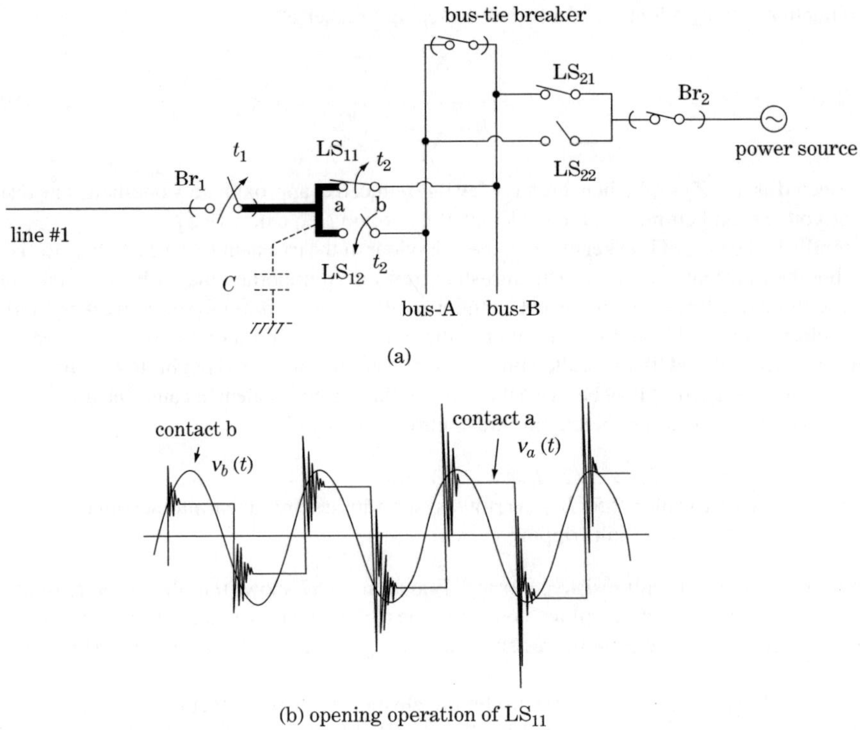

Figure 19.18 Switching surge caused by line switch (disconnecting switch) operation

The equation just after the 'electrical closing' of s_2 at $t = 0$ is

$$\left.\begin{aligned}
v_{ac}(s) &= k \cdot E \cdot \mathcal{L}[\cos(\omega t + \delta)] = kE \cdot \frac{s\cos\delta - \omega\sin\delta}{s^2 + \omega^2} & \text{①}\\
i_{\text{aux}}(s) &= \frac{1}{R + Z_1(s) + Z_2(s)} \cdot v_{ac}(s) & \text{②}\\
v_a(s) &= Z_1(s) \cdot i_{\text{aux}}(s) = \frac{Z_1(s)}{R + Z_1(s) + Z_2(s)} \cdot v_{ac}(s) & \text{③}\\
v_b(s) &= -Z_2(s) \cdot i_{\text{aux}}(s) = \frac{-Z_2(s)}{R + Z_1(s) + Z_2(s)} \cdot v_{ac}(s) & \text{④}\\
v_c(s) &= -\{R + Z_2(s)\} \cdot i_{\text{aux}}(s) = \frac{-\{R + Z_2(s)\}}{R + Z_1(s) + Z_2(s)} \cdot v_{ac}(s) & \text{⑤}
\end{aligned}\right\} \quad (19.57)$$

The special case of a non-resistive-closing breaker is obtained by letting $R \to 0$ in the above equations; then the denominator is changed as $\{R + Z_1(s) + Z_2(s)\} \to \{Z_1(s) + Z_2(s)\}$. Therefore, the resulting transient switching surge quantities $i(s)$, $v_a(s)$, $v_b(s)$, $v_c(s)$, $v_{ac}(s)$ can be reduced by adopting the resistive closing method, and the rate of switching surge reduction is given by $\beta(s)$ as

$$\beta(s) = \frac{Z_1(s) + Z_2(s)}{R + \{Z_1(s) + Z_2(s)\}} \quad (19.58a)$$

For the initial time interval before the reflected waves return from the transition points on the opposite sides, simplification of $Z_1(s) \to \sqrt{L_1/C_1}$, $Z_2(s) \to \sqrt{L_2/C_2}$ is possible. Then, the initial surge reduction rate by adoption of a resistive closing breaker is

$$\beta = \frac{\sqrt{\frac{L_1}{C_1}} + \sqrt{\frac{L_2}{C_2}}}{R + \sqrt{\frac{L_1}{C_1}} + \sqrt{\frac{L_2}{C_2}}} \tag{19.58b}$$

If R is selected as $R \doteq Z_1 + Z_2$, then the surge level is reduced to approximately one-half. The resulting transient voltages and currents can be calculated by solving Equation 19.57.

Generally in the case of breaker closing, the behaviour of the transient voltages v_a, v_b, v_{ab} is not so severe, but the transient current i_{aux} (the inrush current) is an important matter because the contacts have to withstand a large inrush current without melting, and furthermore the breakers under the resistive closing method have to withstand the thermal energy caused on the resistive element.

If the breaker is closed for the faulted line, the short-circuit current $i(t)$ begins to flow immediately, and the initial current $i(0+)$ may be twice the value of the a.c. component because the d.c. component may be superposed. Then the maximum inrush current is as follows:

$$\begin{aligned} &\text{theoretical largest value}: 2\sqrt{2} \cdot I_{\text{rate}} = 2.83 \cdot I_{\text{rate}} \\ &\text{practical largest value}: 2.5 \cdot I_{\text{rate}} \text{ (various system losses are taken into account)} \\ &\text{where } I_{\text{rate}} \text{ is the rated fault tripping current} \end{aligned} \tag{19.59}$$

Returning to the breaker with resistive closing, Equation 19.57 shows that the initial current i_{aux} is reduced by β, so the duty of the auxiliary contact frame to the inrush current is also reduced by β, while the accumulated thermal energy of the resistive element $i_{\text{aux}}^2 R \cdot \Delta T$ has to be treated as a breaker design factor.

Next, the voltage $v_{ab}(t) = R \cdot i_{\text{aux}}(t)$ just before closing the main contact frame is almost at steady state, because i_{aux} is already dominated by R. Accordingly, severe transient behaviour would not generally be caused, whose phenomena can be calculated as in Figure 19.17c, if necessary.

19.7 Switching Surge Caused by Line Switches (Disconnecting Switches)

A line switch (LS, also called a 'disconnecting switch') is a switching device which does not have the duty to break or to close load current or fault current. Its duty is to change the network connection by switching on/off under no-load circuit conditions. As shown in Figure 19.18 as an example, LSs are important members of the bus system and carry out their duties in combination with breakers. LSs can be operated only under the condition that all the series-connected breakers are open, and no current flows through the LS before and after switching is assured. (Therefore the **sequential interlocking facilities** in combination with breakers at the same substation are vitally important.) The LS is not required to operate quickly, so opening and closing are rather slow. The operating stroke time of a conventional open-air LS may be, say, 2–3 s, while that of an LS set in GIS would be, say, 0.5–1 s.

However, the LS has to trip '**leakage current**' which flows through a small leakage capacitance C in a neighbouring section, by which unique 'switching surge phenomena by LS' are caused.

Now we examine the mechanism of so-called **LS switching surge** as in Figure 19.18a. The substation here consists of a double bus structure (buses A and B), and feeder line 1 is connected to bus A (i.e. LS_{11} closed, LS_{12} open). Now the connection of feeder line 1 is going to be changed from bus A to bus B.

The operating process is follows:

- Time t_1: The breaker BR_1 is tripped (load current tripped).
- Time t_2: LS_{11} is opened. At this time, the line switch LS_{11} has to break the leakage current which is flowing across the leakage capacitance C of the small energized section connecting $Br_1/LS_{11}/LS_{12}$.

- Time t_3: LS_{12} is closed, so the bus connection changing processes of feeder line 1 from bus A to bus B are completed.

Concerning the behaviour of LS_{11} at time t_2 when LS_{11} is going to break the leakage current $i_c = j\omega C \cdot E$, the waveform of the voltage switching surge caused by LS_{11} opening is as shown in Figure 19.18b. The voltage at point b (one terminal of LS_{11}) is obviously a.c. system voltage $v_b = E\cos\omega t$ with a sinusoidal waveform, while the waveform for the voltage v_a at point a (another terminal of LS_{11}) is of repeated stepping form with quite high oscillatory transient terms.

19.7.1 LS switching surge: the mechanism appearing

Just after contact a leaves contact b, free oscillatory overvoltage v_a appears in the small section including point a, whose natural frequency is extraordinarily high because the travelling distance l is quite short (say, $l = 10$ m or less for a conventional substation, and a few meters or less for GIS). If the travelling length $l = 6$ m is assumed, the velocity $u \doteq 300$ m/μs and the natural frequency caused at the small section would be approximately 25 MHz. The oscillatory transient term of v_a may soon be attenuated, and v_a would remain on the small section as the d.c. charging voltage of the stray capacitance C. However, the voltage $v_b = E\cos\omega t$ still changes continuously, so the voltage v_{ab} across points a and b increases. On the other hand, the stroke speed of the LS is so slow that a second reignition is inevitably caused. The amplitude of the second oscillation may be larger than that of the first oscillation by chance. These reignitions would be repeated after some cycles, and the amplitude of the transient terms could become larger by repetition. However, the LS can finally break the charging current because the stroke distance of the LS is long enough to recover the insulation across the contacts.

The problem we need to consider is severe 'LS switching surge' attacks frequently on all the energized parts of the substation (buses A and B and all other connected equipment). The 'LS switching surge' could be even more severe than a lightning surge in practical engineering, especially for EHV and UHV substations, for the following reasons:

- **Physical surge level:** The magnitude of lightning surges arriving at a substation is repressed within a certain level by the arrester installed at each feeder terminal. In comparison with these repressed lightning surges, LS surges could be even larger in magnitude, higher in frequency and of a much longer duration, often in daily operation. This tendency is conspicuous for the EHV/UHV system, because it is generally designed with relatively a lower insulation level in total.
- **Protection hardness:** LS switching surges are caused inside the substation, so surge protection by arrestors or any other devices may not be easy. The LS switching surges appearing may affect nearby equipment directly without attenuation. Meanwhile, closely installed arresters (installed typically at the line feeder terminal point for lightning surge protection) have to withstand the thermal energy caused by long LS surges.
- **Chance:** LS switching surges appear quite often whenever any one of several LSs is operated.

We will discuss this matter again in Chapters 20 and 21.

Again, as in Figure 19.18a, capable LSs which cause fewer switching surges are of course important. Further, if a leakage resistance exists in parallel with the stray capacitance C, it could be an effective countermeasure to reduce the surge level. Adopting 'a gap-less arrester' at the small section may also be useful. The **gap-less arrester** could work as a kind of non-linear high-resistive device connected in parallel with the stray capacitance C, while the duty of the arrester might be severe.

Incidentally, the LS is usually equipped with the **earth grounding function** at its terminals, which should be grounded during maintenance work periods on the associated transmission line or any other associated part of the substation. The LS is usually equipped with a **manual earth grounding function** as typical practice.

19.8 Supplement 1: Calculation of the Coefficients k_1–k_4 of Equation 19.6

Coefficient k_1 can be calculated by putting $s = -j\omega$ in the equation $F(s) \cdot (s + j\omega)$ as explained in Section 10.5 of Chapter 10:

$$k_1 = F(s) \cdot (s + j\omega)|_{s=-j\omega} = \frac{-j\omega + 2\alpha}{-j2\omega \cdot (-j\omega + \alpha + ju)(-j\omega + \alpha - ju)}$$

$$= \frac{j}{2\omega} \cdot \frac{-j\omega + 2\alpha}{(u^2 - \omega^2 + \alpha^2) - j2\omega\alpha} \doteqdot \frac{-j}{2\omega} \cdot \frac{-j\omega}{u^2} = \frac{1}{2u^2}$$

In the same way

$$k_2 = F(s) \cdot (s - j\omega)|_{s=j\omega} \frac{-j}{2\omega} \cdot \frac{j\omega + 2\alpha}{(u^2 - \omega^2 + \alpha^2) + j2\omega\alpha} \doteqdot \frac{j}{2\omega} \cdot \frac{-j\omega}{u^2} = \frac{1}{2u^2} \qquad (1)$$

$$k_3 = F(s) \cdot (s + \alpha + ju)|_{s=-\alpha-ju} = \frac{j}{2u} \cdot \frac{-ju + \alpha}{(-u^2 + \omega^2 + \alpha^2) + j2u\alpha} \doteqdot \frac{j}{2u} \cdot \frac{-ju}{-u^2} \doteqdot -\frac{1}{2u^2}$$

$$k_4 = F(s) \cdot (s + \alpha - ju)|_{s=-\alpha+ju} = \frac{-j}{2u} \cdot \frac{ju + \alpha}{(-u + \omega^2 + \alpha^2) - j2u\alpha} \doteqdot \frac{-j}{2u} \cdot \frac{ju}{-u^2} = -\frac{1}{2u^2}$$

The arrows $\nwarrow$ are the 'arrows of omission' indicating neglected parts for reasons of simplicity.

19.9 Supplement 2: Calculation of the Coefficients k_1–k_6 of Equation 19.17

Here

$$k_1 = F(s) \cdot (s + j\omega)|_{s=-j\omega} = \frac{(-j\omega)\{C_1(-\omega^2 + u_1^2) + C_1'(-\omega^2 + u_2^2)\}}{C_1C_1'(-j2\omega)\{(-j\omega + \alpha_1)^2 + u_1^2\}\{(-j\omega + \alpha_2)^2 + u_2^2\}} \qquad (1)$$

$$= \frac{C_1u_1^2 + C_1'u_2^2}{2C_1C_1'u_1^2u_2^2} = \frac{1}{2C_1u_1^2} + \frac{1}{2C_1'u_2^2} = \frac{1}{2}(L_1 + L_1')$$

In the same way

$$k_2 = \frac{1}{2}(L_1 + L_1') = k_1 \qquad (2)$$

$$k_3 = F(s) \cdot (s + \alpha_1 + ju_1)|_{s=-(a_1+ju_1)}$$

$$= \frac{-(\alpha_1 + ju_1)[C_1\{(\alpha_1 + ju_1)^2 + u_1^2\} + C_1'\{(\alpha_1 + ju_1)^2 + u_2^2\}]}{C_1C_1'\{(\alpha_1 + ju_1)^2 + \omega^2\}(-j2u_1)\{(-\alpha_1 + ju_1 + \alpha_2)^2 + u_2^2\}} \qquad (3)$$

$$= -\frac{1}{2C_1u_1^2} = -\frac{L_1}{2}$$

$$k_4 = F(s) \cdot (s + \alpha_2 + ju_2)|_{s=-(\alpha_1-ju_1)} = -\frac{1}{2C_1u_1^2} = -\frac{L_1}{2} \qquad (4)$$

$$k_5 = k_6 = -\frac{L_1'}{2} \qquad (5)$$

These are the results given in Equation 19.18.

Coffee break 10: Fortescue's symmetrical components

Although the symbolic complex-number methods of Kennelly and Steinmetz made circuit analysis easier, the analysis of a three-phase circuit was still a problem because of the existence of mutual inductances and capacitances between the phases. Referring to the calculation in Table 7.2, for example, we cannot even write the circuit before solution without adopting symmetrical components. It is therefore worth learning a little about the history of symmetrical components.

The original (coordinate) method was developed by **Charles LeGeyt Fortescue** (1876–1936), a prominent engineer at Westinghouse. It was first disclosed in his paper 'Method of Symmetrical Coordinates Applied to the Solution of Polyphase Network', published in the *AIEE Transactions* in 1918, in which he treated the calculation of a small circuit including a rotating machine. In this paper he explained the method, using the symbols E, aE, a^2E as a decomposition of complex steady-state phasors of sinusoidal time functions, as the basis for the transformation of arbitrary three phase variables.

Charles LeGeyt Fortescue (1876–1936) (Courtesy of Westinghouse and IEEE History Center)

The application of symmetrical components first appeared by the following three papers. 'Synchronous Operation of Two Alternators through Unsymmetrical Impedances.' by S. Bekku (Electrical Institute of Japan, 1924)

'Calculation of short-circuit ground currents on three phase power networks, using the symmetrical coordinates' by S. Bekku, *Gen. Electr. Rev.*, 1925) (also in *Elektr. Z.*, 1925)

'Finding single-phase short-circuit currents on calculating boards' by R. D. Evans, *Electr. World*, 1925.

The second and the third papers were quoted by **Edith Clarke** in her famous book published in 1943 (Circuit Analysis of Power Systems, Volume I) as the first two papers in English on application of symmetrical components.

The following two books were probably the first published textbooks on symmetrical components:

Application of Symmetrical Components by Sadatoshi Bekku (1928, in Japanese)

Symmetrical Components as Applied to the Analysis of Unbalanced Electrical Circuit by C. F. Wagner and R. D. Evans (1933)

It is a little strange that papers or books on application of symmetrical coordinate method seldom appeared in years of 1918–1933 except the above quoted ones and a few other papers, despite its so real worth. It took 20 more years after 1918 for the true worth of symmetrical components to be noticed and the began to prevail among many electricians.

Fortescue seldom wrote about the applications of symmetrical component, probably because he had to concentrate on his research into lightning phenomena and high-voltage insulation, the most serious problems in 1920's. His style can be seen in his papers, for example:

'Lightning Discharges and Line Protective Measures' (*AIEE Trans.*, 1931)

'Counterpoises for Transmission lines' (*Electr. Eng.*, 1933)

'Counterpoise Test at Traffort' (*Electr. Eng.*, 1934)

L. L. Lewis, a student of Steinmetz and a great benefactor of lightning theory and insulation coordination, mentioned in the introduction to his famous book *The Protection of Transmission Systems against Lightning* (1949) the names of a few creditable contributors in the field of lightning technology before 1940, namely: **F. C. Peek** for theories of lightning, direct strike, ground wire; C. L. Fortescue for theories of direct strikes, ground wires and counterpoise wires; **L. V. Bewley** for theories on counterpoise wires and travelling waves.

Fortescue was born in York Factory, Manitoba. After his granduation he spent most of his professional career working for Westinghouse as a leading researcher competing with GE. The Charles LeGeyt Fortescue Scholarship was established in 1939 as a memorial in recognition of his valuable contributions to the field of electrical engineering.

- Voltage Imbalance - Magnitude Imb.
  - Phase Imb.

- Waveform Distortions

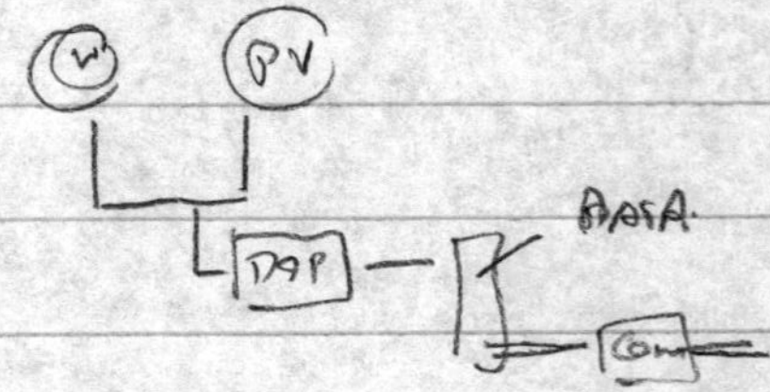

If I would like control [illegible] on DG what should I do?

Background.

What is passive network.
= advantage.
= disadvantage.

What is active network.
= advantage
= disadvantage.

What is Islanding. ?

What is power quality ⇒ What is it? Why it cause problem in DG?
= Transient — Impulsive
— Oscillatory
= Short duration - Sag
- Swell
- Interruption
= Long duration - Sustained Interruption
- Under Voltages

Table 2.1

$$F(t) = 1 - e^{-\frac{t}{T}}$$

t — 3 hr

T — Avg duration in hr

Eg.

$$F(t) \text{ of A Company} = 1 - e^{\left(\frac{-3}{2.38}\right)}$$

$$F(t) \text{ of A Company} = 1 - e^{(-1.2605)}$$

$$= 0.7164$$

Change to % = 100% × 0.7164

Chang to % = 71.64%

$\bar{\lambda} = 1.64$

$\bar{\lambda}$ = the expected number

Transient

Short

long

Poron Quality

- Variation

- Event.

20

Overvoltage Phenomena

We have examined the mechanisms of switching surges and other overvoltage phenomena in the previous chapters. In this chapter, we try to survey generally the various kinds of overvoltage behaviour including power frequency phenomena as well as switching surges and lightning surges. These determine the essential prerequisite conditions for coordinating the insulation of power system networks.

20.1 Classification of Overvoltage Phenomena

Table 20.1 lists our classification of overvoltages. We examine several different overvoltages along with this classification. The phenomena listed in item 2a, for example, may not usually be discussed because they seldom occur. However, engineers need to study the mechanism and to assure the absence of this sort of problem.

20.2 Fundamental (Power) Frequency Overvoltages (Non-resonant Phenomena)

Phenomena for four different power frequency overvoltages are listed below.

20.2.1 Ferranti effect

In the power system of Figure 20.1a, the vector diagrams for the cases of lagging and leading power factors are shown in Figures 20.1b and c. In the leading power-factor operation, the receiving terminal voltage v_r becomes larger than the sending terminal voltage v_s (i.e. $v_s < v_r$). The phenomenon in which v_r becomes larger than v_s is called the **Ferranti effect**. In fact, we have already discussed much of this in regard to leading power-factor load operation in Section 16.3.

20.2.1.1 Overvoltage by transmission line charging

The utmost case of pure capacitive load is transmission line charging from one terminal, whose vector diagram is shown in Figure 20.1d.

Handbook of Power System Engineering Yoshihide Hase
© 2007 John Wiley & Sons, Ltd

Table 20.1 Overvoltage phenomena

1. Fundamental (power) frequency overvoltages (temporary overvoltages)
1a. Ferranti effect
1b. Self-excitation of generator
1c. Overvoltages of unfaulted phases by one line-to-ground fault
1d. Sudden load tripping or load failure
2. Lower frequency harmonic resonant overvoltages
2a. Broad area resonant overvoltages (lower order frequency resonance)
2b. Local area resonant overvoltages
3. Switching surge
3a. Breaker closing overvoltages
3b. Breaker tripping overvoltages
3c. Switching surges by line switches
4. Overvoltage phenomena by lightning strikes
4a. Direct stroke to phase conductors (direct flashover)
4b. Direct stroke to overhead grounding wire or to tower structure (inverse flashover)
4c. Induced strokes (electrostatic induced strokes, electromagnetic induced strokes)
5. Overvoltages caused by abnormal conditions
5a. Interrupted ground fault of cable (in high-impedance neutral grounding system)
5b. Overvoltages induced on cable sheath (see Chapter 23)
5c. Touching of different kilovolt lines etc.

We calculate the overvoltage when the line is charged from sending terminal point s (the receiving terminal r is open). The four-terminal circuit equation between points s and r is given by Equation 18.19b. Now, putting $I_r = 0$ in the equation, we have

$$\left.\begin{aligned} &\frac{V_r(x)}{V_s(x)} = \frac{1}{\cosh \gamma(s) \cdot x} \quad \text{(ratio of overvoltages at the receiving terminal)} \\ &\text{where } \gamma(s) = \sqrt{(Ls+R)(Cs+G)} \end{aligned}\right\} \tag{20.1}$$

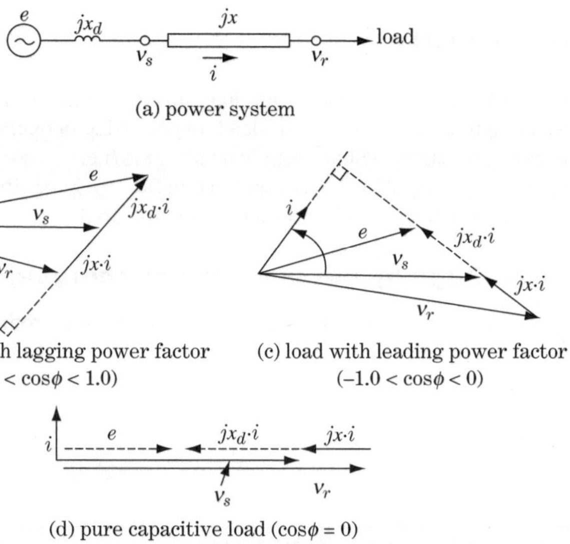

Figure 20.1 Vector diagrams under different power-factor loads

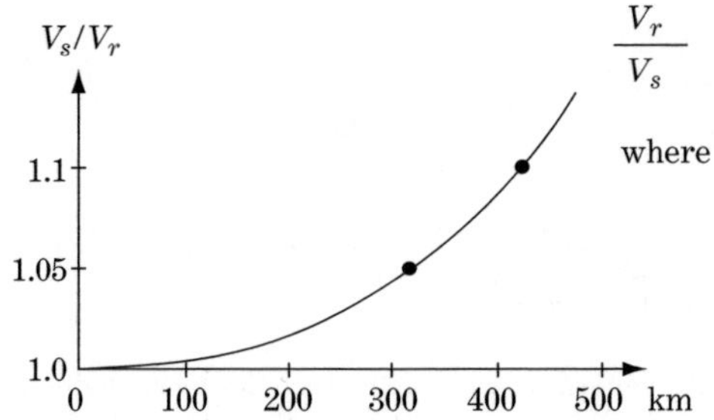

$$\frac{V_r}{V_s} = \frac{1}{\cos(2\pi f\sqrt{LC})}$$

where $L = 1$ mH/km, $C = 0.01$ μF/km, $f = 50$ Hz

Figure 20.2 Ferranti effect, ratio of voltage at receiving end

Neglecting R and G, and letting $s \to j\omega$ (because fundamental phenomena are investigated here),

$$\left.\begin{aligned} \cosh\gamma(s)\cdot x &= \cosh\left\{j\omega\sqrt{LC}\cdot x\right\} = \frac{1}{2}\left\{e^{j\omega\sqrt{LC}x} + e^{-j\omega\sqrt{LC}x}\right\} \\ &= \cos\left(\omega x\cdot\sqrt{LC}\right) \\ \therefore\ \frac{V_r}{V_s} &= \frac{1}{\cos\left(\omega x\cdot\sqrt{LC}\right)} \geqq 1.0 \end{aligned}\right\} \quad (20.2)$$

The equation is the rate of overvoltage caused at point r in comparison with the voltage at point s.

Figure 20.2 shows the calculated result of the equation under the condition of typical line constants $L = 1\,\text{mH/km}, C = 0.01\,\mu\text{F/km}$, 50 Hz. The rate of overvoltage at point r is 1.05 for line length $l = 312$ km, and 1.10 for $l = 432$ km. In the case of n parallel circuits, the constants are replaced as $L \to 1/n\cdot L,\ \ C \to n\cdot C$ so that $\sqrt{LC}$ and the rate V_r/V_s are not affected.

As we discussed in Chapter 18, the overvoltages by the Ferranti effect under relatively light load conditions might be rather severe, especially at a receiving substation in a big-city area where numbers of cable feeders are connected. An effective countermeasure to prevent excess overvoltage is to install reactor banks. ***V–Q* control** equipment based on the control of **reactor banks** and **on-load tap-changing transformers** is often adopted at key substations as another effective countermeasure to prevent excess overvoltages or to maintain stable voltages for 24 h.

20.2.2 Self-excitation of a generator

If a capacitive load is connected to a generator as shown in Figure 20.3a, overvoltage will be caused on the generator terminal even if E_{fd} is kept at zero. This phenomenon is called the **self-excitation phenomenon of a generator**.

We refer to Figure 16.8 in Chapter 16 in order to understand this phenomenon. Under leading power-factor operation, large terminal voltage e_G is apt to arise in spite of the small excitation jE_{fd}, as seen in Figure 16.8c.

Now we refer to Figure 20.3b, and examine the behaviour of a generator under the leading power-factor condition.

The generator is driven by the prime-mover under the open-terminal condition and is running at constant speed with excitation current zero ($jx_{ad}\cdot i_{fd} = E_f = 0$, in Equation 10.60 and Figure 10.5). However, small residual magnetism remains as a magnetic field of the rotor since termination of the last operation, even if the excitation current is zero. A small terminal voltage e_G also remains. If a very small capacitance C is connected to the generator under this condition, the same terminal voltage e_G will be maintained. Furthermore, if connected capacitance C is enlarged, e_G as well as leading current i in the armature coil will become larger; their characteristics can be written as curve a in Figure 20.3b.

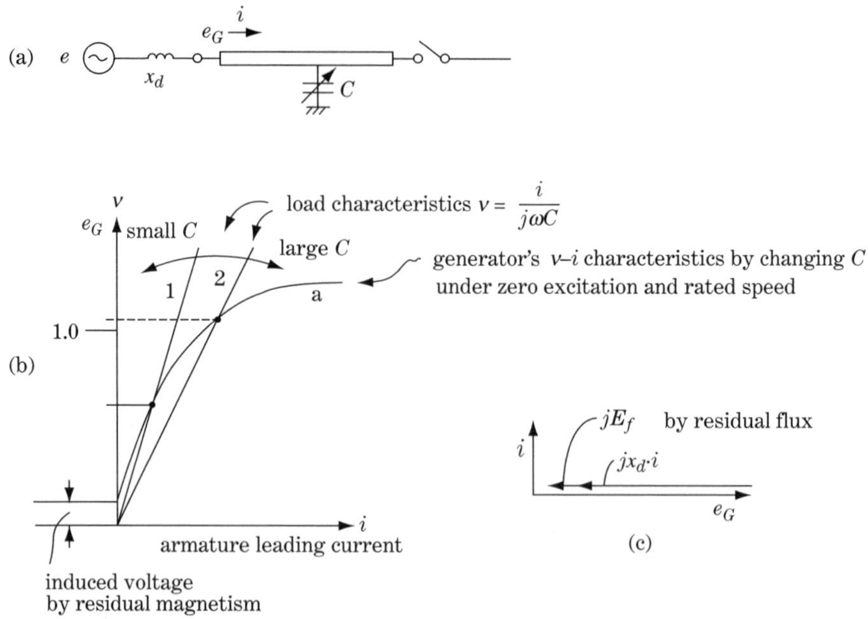

Figure 20.3 Self-excitation of a generator

This is a specific *v-i* characteristic of a generator under capacitive load with excitation current zero. As seen in the vector diagram of Figure 20.3c, current *i* leads the voltage by 90°. The characteristics begin to saturate naturally at around the rated voltage area. On the other hand, specific *v–i* characteristics of the capacitance *C* are given by $v = i/(\omega C)$ as shown in Figure 20.3b. The intersecting point of both characteristics is the actual operating point of the *C* load.

In conclusion, the self-excitation phenomenon of a generator is caused by connection of a pure capacitive load, in that extreme and sustained overvoltage (perhaps the ceiling voltages of the saturated zone) is caused. This condition also means critical overheating of the generator caused by the rapid increase of iron loss as a result of core saturation. Obviously, we need to avoid the self-excitation of generators.

20.2.3 Sudden load tripping or load failure

If some amount of load tripping or load failure occurs at a receiving substation, the voltage would be boosted by the following three reasons:

a) The generator terminal voltage at the sending station may be transiently boosted, so that the voltage at the receiving point also has to be boosted. Referring to Figure 10.5, or Equation 10.63, if the generator current i_{a1} suddenly decreases, the terminal voltage vector $e_{a1} = (E_{a1}e^{j\alpha_1})$ has to change unavoidably to a large vector closer to jE_f in magnitude and angle. In other words, the generator terminal voltage would be transiently boosted by the sudden decrease of load current, until the AVR decreased the excitation jE_f.

b) The impedance drop voltage of the transmission line would be decreased.

c) The total power factor of the load may be shifted a little in the leading power-factor direction so that the Ferranti effect only partially occurs.

20.2.4 Overvoltages of unfaulted phases by one line-to-ground fault

We have already studied the principles of this matter in Chapter 8, Section 8.2 and Table 8.2.

Referring to Figure 8.1 as well as Figure 21.1, overvoltage of unfaulted phases would become $0.8\,E$-$1.3\,E$ for the solidly neutral grounding system ($\delta = 0$-1) and $1.5\,E$-$1.9\,E$ for the non-effective neutral grounding system ($\delta \fallingdotseq 5$ to ∞). Figure 8.1 indicates the overvoltage ratio of the power frequency term, and the transient overvoltage ratio would become a little higher. (Transient overvoltage ratio can be derived in a similar curve format to that in Figure 8.1.)

Although the overvoltage ratio of unfaulted phases is quite large, every part of the network (including installed surge arresters) has to have enough insulation to withstand this overvoltage. Accordingly, this temporary overvoltage phenomenon is one of the important factors for the fundamental design of **insulation coordination**, which we discuss in detail in Chapter 21.

20.3 Lower Frequency Harmonic Resonant Overvoltages

If one inductance L and one capacitance C exist as elements of a circuit, series resonant frequency and parallel resonant frequency inevitably exist. Our concern is with series resonance phenomena of n-th order harmonic voltages and currents of the power system from the viewpoint of overvoltages and waveform distortion, whose condition is given by $Z = j2\pi n fL + 1/(j2\pi fC) \to 0$ (f: resonant frequency). We need to consider if such a series resonant condition could exist actually in our power network even at a low probability. Lower order harmonic phenomena especially (under, say, 1 kHz) may cause problems over a broad area and continue for a long period without attenuation. We need to assure the absence of these sorts of problems. Meanwhile, higher frequency voltages (over, say, 1 kHz) caused by resonant phenomena would not actually exist because such a component would be largely attenuated by corona loss and eddy-current loss of the transmission line even if it did exist.

20.3.1 Broad-area resonant phenomena (lower order frequency resonance)

Figure 20.4a is a simplified model connecting a generating plant and urban area with some cable lines. If we represent the overhead line by L and the cable lines by C, the circuit can be simply written as an LC series-connected circuit as shown in Figure 20.4b, although the load circuit is usually connected. We have already studied the Ferranti effect of these circuits as phenomena of power frequency. Nevertheless, we need to check the possibility of series resonance in other frequency zones under different conditions such as nominal load condition, or irregular conditions (phase fault, grounding fault, phase opening, reclosing time, etc.). At least we need to prove that such resonant phenomena would not occur during actual system operation.

20.3.1.1 Positive-sequence series resonance

We try to find a series resonance condition which may exist by chance in our power system. Let us assume the line constants below for the positive-sequence circuit of Figure 20.4b:

$$\left.\begin{array}{ll}\text{Overhead line : } 0.01\,\text{mH/km},\ l_1(\text{km}) & L = 10^{-5} \cdot l_1 \quad [\text{H}] \\ \text{Cable lines : } 0.33\,\mu F/\text{km},\ l_2(\text{km}),\ \text{numbers of circuits : } m, & C = 0.33 \times 10^{-6} \cdot l_2 \cdot m \quad [\text{F}]\end{array}\right\} \tag{20.3}$$

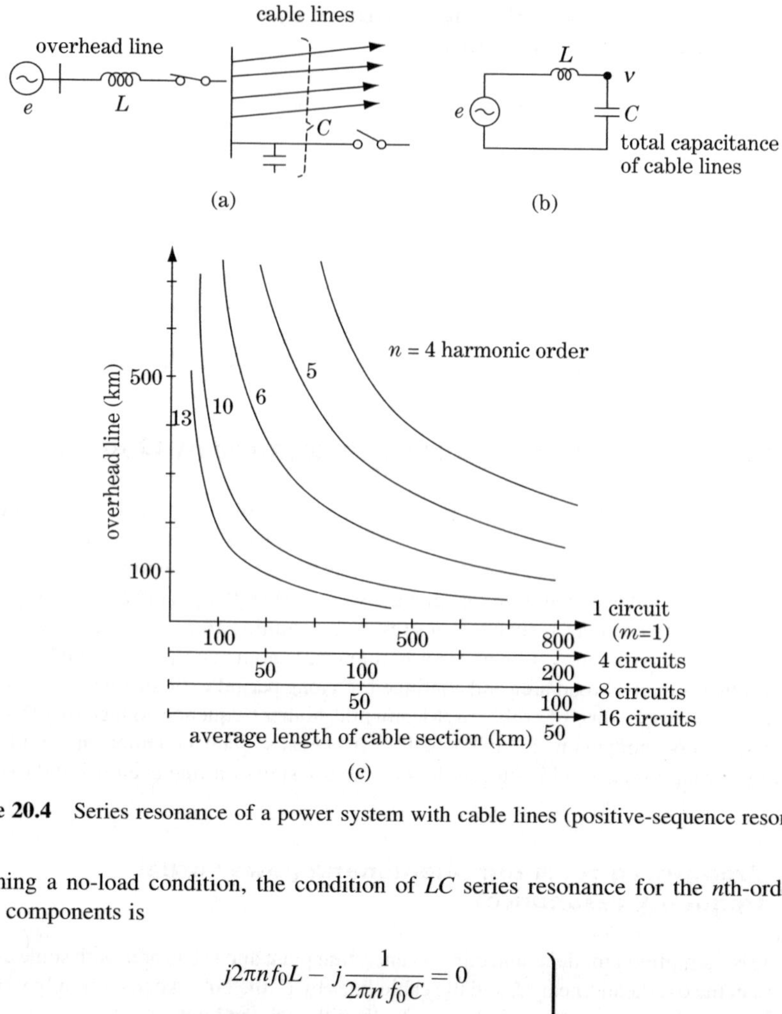

Figure 20.4 Series resonance of a power system with cable lines (positive-sequence resonance)

Assuming a no-load condition, the condition of *LC* series resonance for the *n*th-order higher harmonic components is

$$\left.\begin{aligned} & j2\pi n f_0 L - j\frac{1}{2\pi n f_0 C} = 0 \\ & \text{Then} \\ & (2\pi n f_0)^2 LC = (314n)^2 LC = 1.0 \\ & \therefore \quad 10^5 n^2 \cdot LC \fallingdotseq 1.0 \\ & \text{where } f_0 = 50\,\text{Hz} \end{aligned}\right\} \tag{20.4a}$$

Substituting 20.3 into 20.4 the resonant condition of the positive-sequence circuit is

$$n^2 \cdot l_1(l_2 m) \fallingdotseq 3 \times 10^6 \tag{20.4b}$$

Figure 20.4c is the series resonant condition derived from 20.4b. This figure gives us a rough idea of the existing zone of series resonance for the model system. Assuming that the length of the overhead line $l_1 = 200\,\text{km}$ and a total length of cable $l_2 = 400\,\text{km}$ or less, for example, resonant conditions do not exist fortunately for the lower frequency of $n = 2$–6.

Although the curves were derived under a somewhat unrealistic no-load condition, a similar condition may be caused by a sudden bus fault (all the connected breakers have to be tripped) in the key substation.

Recall that only the existing zone of resonant frequency in the system was checked in the above study, from which it may be concluded that natural resonance would seldom be caused in the system. However, we need to recognize that actual power system networks include not only generators of almost ideal sinusoidal source voltages, but also various loads and power conditioners which generate large rippled currents. If the frequency of the harmonic components of such a 'dirty forced current' matches the network natural frequency, abnormal overvoltages or waveform distortion could be significantly amplified in some local areas of the network. As a matter of fact, we have to recognize that harmonic currents of various frequency are apt to flow forcibly into the power system, so resonance phenomena have to be checked from the viewpoint of waveform distortion as well as overvoltage phenomena. (Waveform distortion is examined again in Chapter 22.)

20.3.1.2 *Series resonance under temporary conditions (faults, phase opening, reclosing time, etc.)*

We investigate the possibility of resonance under some irregular conditions including line-to-line fault modes, line-to-ground fault modes and phase opening modes (including dead-voltage time for faulting phase reclosing).

Referring to Tables 3.1 and 3.2 in Chapter 3 and Tables 6.1 and 6.2 in Chapter 6, we need to check if these conditions exist, that is the denominators of the equations in the tables would tend to values close to zero by chance.

For the modes of series resonance

$$\left.\begin{array}{lll} \text{normal condition :} & & Z_1 \to 0 \\ 2\phi G: & Z_1 + Z_0 \to 0, & Z_1 + 2Z_0 \to 0 \\ 1\phi G: & & 2Z_1 + Z_0 \to 0 \\ 1\phi \text{ opening :} & & Z_1 + Z_0 \to 0 \\ 2\phi \text{ opening :} & & Z_1 + 2Z_0 \to 0 \end{array}\right\} \qquad (20.5)$$

Perhaps we need not worry about the possibility of resonance caused by line-to-line and line-to-ground faults, because the resonance condition may be unlikely to occur by analogy to Figure 20.4c. Furthermore, these faults would be cleared within a few cycles, generally by the associated protective relays. However, phase-opening mode trouble may cause a little bother, because the protective relays may not be able to detect the phase open-mode faults exactly and would be able to continue without noticing.

In conclusion, it is worthwhile assuring the absence of these sorts of resonant conditions at several different points in the network.

20.3.2 Local area resonant phenomena

20.3.2.1 *Transformer winding resonant oscillation triggered by switching oscillatory surge*

In Figure 20.5, the transformer is going to be energized through the cable line. This circuit can be simplified to the parallel circuit of C by the cable and L by the excitation reactance of the transformer,

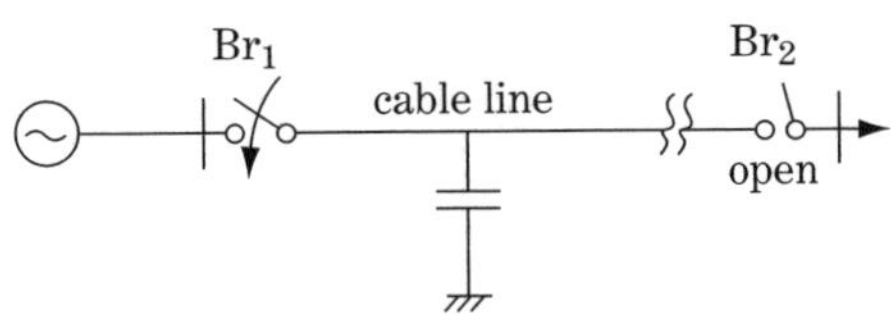

Figure 20.5 Resonance caused by transformer excitation

so that overvoltage would not be caused when the breaker Br_1 is closed. However, the resulting oscillatory inrush voltage and current could trigger a unique voltage oscillation in the transformer windings by chance.

Let us assume a cable line of 275 kV, 2000 sq (given in Table 2.3: 0.392 mH/km, $C = 0.25\mu F/km$, velocity 100 m/μs) and length 10 km.

The travelling time of the surge from the breaker Br to the transformer is 100 μs, so the oscillatory frequency of the switching surge is 5 kHz. The transient oscillation of this frequency zone (10th or lower order) will not be attenuated so quickly. Therefore a reasonable countermeasure is required at the transformer engineering side to avoid winding coil resonance by the frequency of the order above. This scheme is discussed again in Section 21.6.

Note that the **in-rush current of the transformer** is, regardless of the condition of the feeding line, an extremely waveform-distorted offset current which would continue for more than 10 seconds as shown in Figure 16.10. As an inrush current contains d.c. and lower order harmonics of slow attenuation, special consideration would generally be required in the practical engineering of transformers, breakers, generators, protective relays, etc.

20.3.2.2 Ferro-resonance caused by core saturation

Figure 20.6 explains the **ferro-resonance phenomenon**. As seen in diagram (c), if the operating voltage exceeds 1.1 pu of the rated voltage, it begins to saturate rapidly and the excitation current i greatly increases. In other words, the excitation impedance $Z_l (= jX_l)$ of the transformer has non-linear characteristics which rapidly decrease in value in the saturated zone. Figures 20.6a and b show the well-known 'series ferro-resonance circuit' and 'parallel ferro-resonance circuit' respectively. Both circuits have unique behaviour (although the explanation is omitted in this book) and we need to prevent its occurrence. The parallel ferro-resonance (b) may arise especially for a circuit with a large cable capacitance, while the series ferro-resonance (a) may arise especially for the irregular cases of unbalanced phase opening or lines with a series capacitor.

Ferro-resonance causes unstable, jumping overvoltage phenomena and, furthermore, extreme flux saturation leading to a severe temperature rise of the flux pass, all of which need to be avoided.

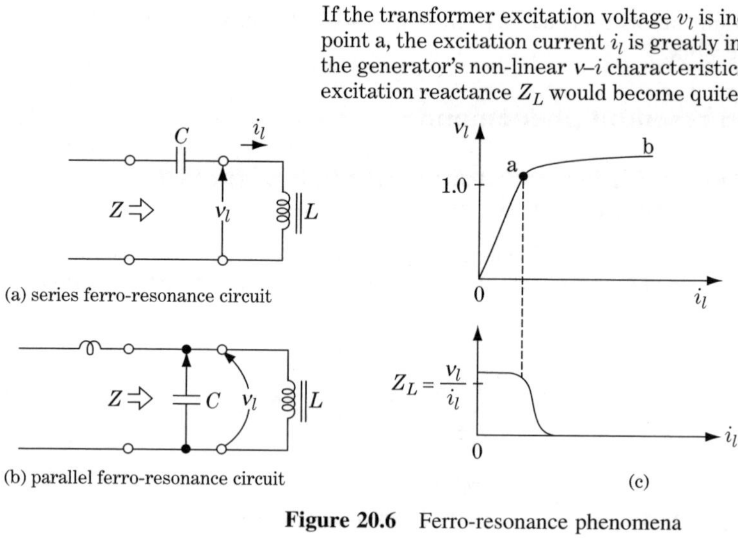

Figure 20.6 Ferro-resonance phenomena

20.3.3 Interrupted ground fault of cable line in a neutral ungrounded distribution system

This is a unique feature which would be limited to the distribution cable lines of a neutral ungrounded system (see Table 8.1). This system is a widely adopted one for distribution networks of 3–20 kV, whose advantage is that the grounding fault current for one phase to ground fault ($1\phi G$) can be greatly reduced to a value of only, say, 10 mA to 100 mA, that is 'zero amps'. With this system, noise interference on the communication lines can also be reduced because the zero-sequence circuit is actually open.

However, there is one disadvantage of this practice, namely the **interrupted ground fault of the cable lines**.

The solid insulation of a distribution cable may slowly deteriorate over a long period of operation. We assume that a minute pinhole crack appears in the cable insulation after this period of operation. A very small $1\phi G$ current flows through the pinhole across the insulation layer (called 'minute grounding' of perhaps 10 mA). However, this grounding current would be halted immediately and the insulation of the pinhole would soon be recovered, because the pinhole pass is not adversely affected by such a small current through the pinhole. The pinhole grounding would probably occur at a timing around the peak value of sinusoidal a.c. voltage, and would always be halted at the time of 'voltage zero'. Incidentally, the minute grounding fault current is almost of 90° lagging power factor, so halting the pinhole grounding current at the time of voltage zero gives rise to 'current chopping' phenomena, which cause large needle-shaped high-frequency transient overvoltages by the same mechanism explained in Section 19.4.

If such processes are caused once at an arbitrary point of the cable system, the same processes would be intermittently repeated at the same point over time so that deterioration of the pinhole would be slowly accelerated and finally lead to a permanent breakdown fault. Moreover, the above interrupted ground fault at one pinhole point intermittently generates needle-like high-frequency overvoltages, which would deteriorate the insulation of another part of the cable. Finally, all the parts of the cable line within a limited area would be frequently stressed by the high-frequency overvoltages caused at many pinhole points, and then cable deterioration would accelerate quickly and broadly.

Due to the excellent insulation technology of recent CV cables for distribution lines, interrupted ground faults may seldom occur today. However, such phenomena could occur on a neutral ungrounded system at any time and at any part, and furthermore may be initialized without any indication or notice.

Continuous observation or pre-detection of such overvoltages on a commercially operated system is perhaps almost impossible, so such phenomena, if they were to happen, would probably be noticed only after the cable lines have been damaged or seriously deteriorated.

20.4 Switching Surges

The operation of circuit-breaker and line switch opening (tripping) or closing inevitably causes severe switching surges. It is fair to say that such phenomena may be even more difficult to manage than lightning surges from some practical viewpoints. In order to understand the precise characteristics, we commence our discussion with a comparison of the characteristics with those of lightning surges.

Lightning surges are severe phenomena in nature. However, the following characteristics may mitigate the severity to some extent in comparison with switching surges:

Lightning strike

a) Rare occurrence.

b) Occurrence mostly on overhead transmission lines whose structures are made of metal (conductors, tower structures, arcing horns, etc.) and porcelain insulators, so that fault self-recovery characteristics are inherent (except a direct strike on a substation).

c) Very short surge duration of typical waveform $1.2 \times 50\,\mu s$.

d) Single action phenomena (mostly).

e) Attenuation through line travelling to substations expected in most cases.

f) Upper limit of incidental voltage level (can be controlled by arcing horns).

g) Relatively easy protection by advanced arresters installed at substations.

The characteristics of switching surges (SS for short) may be explained conversely. SS appear whenever any one of several breakers or line switches at the same substation are operated. SS occur in the vicinity of apparatus with no self-recovery insulation (generators, transformers, breakers, cables, etc.) in the same substation. SS are oscillatory transient surge of extremely high frequency with sharp waveforms and continuing for a long duration before attenuation or extinction. SS are often caused whenever switching operation is conducted. They may seldom attenuate before reaching weak insulation points at the substation because of the short travelling distance. Arresters may not be able to protect the weak insulation points at the substation because of the **relative distance of installed arresters** from the surge-generating breakers or line switches and the other equipment to be protected. Furthermore, SS would cause thermal heating of arrester–non-linear resistive elements and may even break the installed arresters by large switching energy.

20.4.1 Overvoltages caused by breaker closing (breaker closing surge)

We have already studied the mechanisms caused by breaker closing surges in Sections 19.5 and 19.6. In Figure 20.7, the transient phenomenon caused by the breaker closing means sudden insertion of the initial voltage $\{e_1(0)-e_2(0)\}\mathbf{1}(0)$ with opposite polarity across the breaker contacts. Accordingly the initially resulting voltage and current travelling waves are proportional to the initial voltage $\{e_1(0)-e_2(0)\}$ as shown by the equations in the Figure 20.7b. Successive phenomena will be affected by the circuit conditions, especially the included transition points.

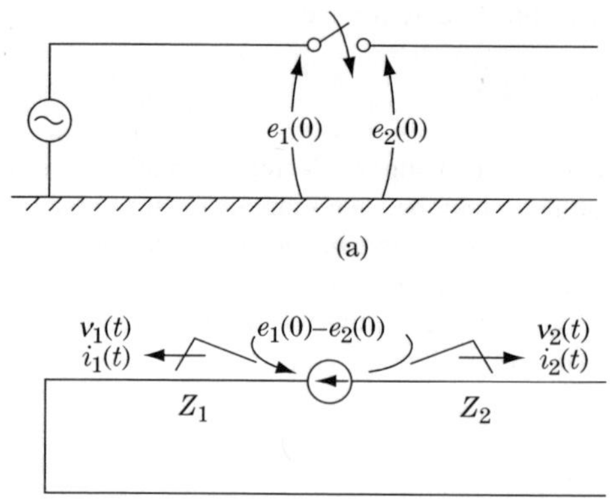

$$v_1(t) = Z_1 \cdot i_1(t) = \frac{Z_1}{Z_1 + Z_2} \cdot \{e_1(0)-e_2(0)\}$$

$$v_2(t) = Z_2 \cdot i_2(t) = \frac{-Z_2}{Z_1 + Z_2} \cdot \{e_1(0)-e_2(0)\}$$

$$i_1(t) = \frac{1}{Z_1 + Z_2} \cdot \{e_1(0) - e_2(0)\}$$

$$i_2(t) = \frac{-1}{Z_1 + Z_2} \cdot \{e_1(0) - e_2(0)\}$$

Figure 20.7 Breaker closing surge

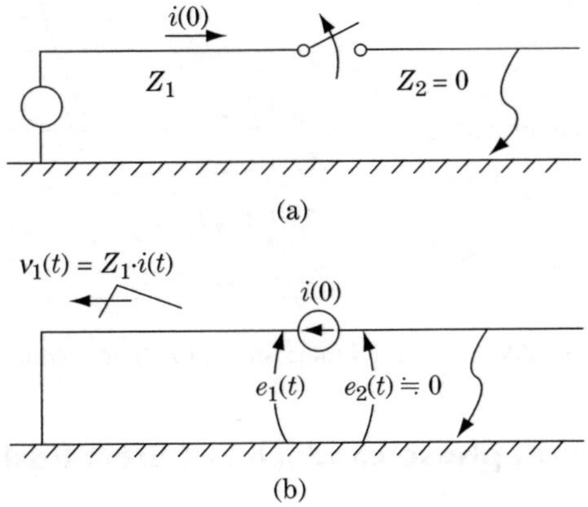

Figure 20.8 Breaker tripping surge

20.4.2 Overvoltages caused by breaker tripping (breaker tripping surge)

We have already studied the mechanisms of breaker tripping in Chapter 19 in detail and so only recall them here briefly. A typical case of fault current tripping is shown in Figure 20.8, where the initial surge voltage caused by breaker tripping can be calculated by the equations in diagram (b). The breaker's reignition surges also have to be taken into account. We already know that the breaker's restriking should be avoided not only to prevent breaker failure, but also to avoid the extremely serious overvoltages which may cause other insulation failures in the network.

20.4.3 Switching surge by line switches

LS switching surge voltages are unique phenomena, as was mentioned in Section 19.7 and shown in Figure 19.18, and insulation coordination in a substation is consequently very important. We discuss this matter in detail in Section 21.3.

20.5 Overvoltage Phenomena by Lightning Strikes

The mechanism of a lightning stroke is typically explained as follows.

When the electric field gradient at some point in a charged concentration of a cloud exceeds the breakdown value for moisture-laden and ionized air, an **electric streamer** darts out towards the ground but may soon halt after progressing perhaps a hundred metres. After a shot interval, the streamer again darts out and repeats the performance. This initial streamer progresses by a series of jumps called a **stepped leader**. Just after the point of the stepped leader reaches the ground, a heavy current (the **main stroke** of 1000–20 000 A) flows up the path blazed by the stepped leader with a velocity about one-tenth that of light.

Our concern in this book is with lightning which directly strikes electricity networks or causes surge voltages and current phenomena on them.

Lightning surge phenomena can be classified into three different stroke modes, which we discuss in turn.

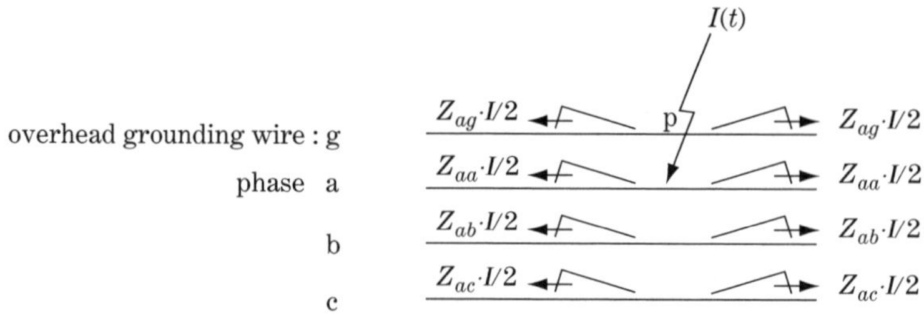

Figure 20.9 Direct lightning stroke to phase a conductor

20.5.1 Direct strike on phase conductors (direct flashover)

This is the case where a lightning stroke directly strikes one or more phase conductors. Figure 20.9 shows the case, in that the main stroke directly hits the phase a conductor and the current $I(t)$ is injected into it. The successive surge phenomena have already been discussed in detail in Sections 18.4 and 18.5.

The surge voltages $v_a = Z_{ag} \cdot (I/2)$, $v_g = Z_{gg} \cdot (I/2)$, etc., in the figure are induced because of the associated surge impedances. The induced surge voltage $v_{ag} = (v_a - v_g)$ would certainly exceed the insulation strength across the phase a conductor and point g (which may be the OGW, the top or arm of a tower, or arcing horn), flashover is caused and a phase a line-to-ground fault occurs. Furthermore, if the induced $v_{ab}(= Z_{ab} \cdot (I/2))$ exceeds the insulation strength across the phase a and b conductors, flashover occurs also between the two conductors causing a line-to-line fault. Of course, whenever a direct stroke occurs, line faults would be caused without exception.

20.5.2 Direct strike on OGW or tower structure (inverse flashover)

This is the case where a lightning stroke directly strikes the OGW or tower structure as shown in Figure 20.10a.

The induced surge voltage $v_{ga} = (v_g - v_a)$ would certainly exceed the insulation strength across point g and the phase a conductor. Flashover would be caused between point g (OGW, the top or arm of a tower, or arcing horn) and the phase a conductor.

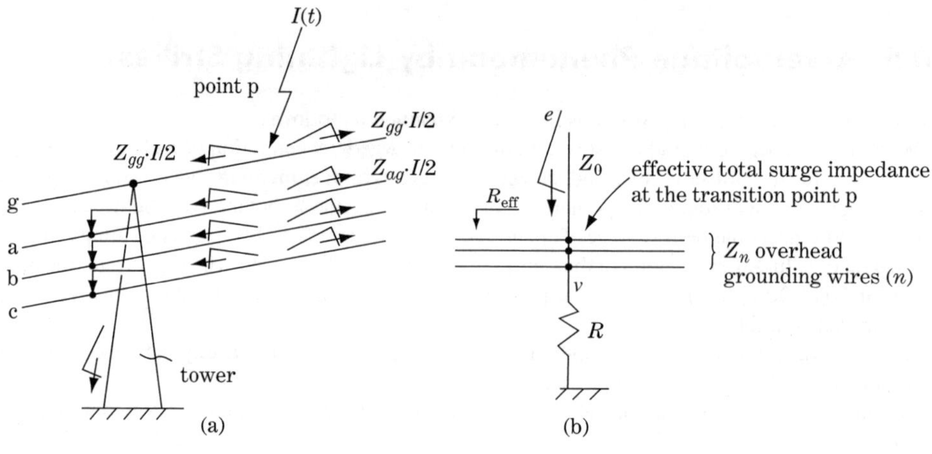

Figure 20.10 Direct stroke to overhead grounding wire or to tower structure

Assuming a lightning $e(t) = I(t) \cdot Z_0$ strikes at the transmission tower top directly as is shown Figure 20.10b, the stroke current $I(t)$ is divided into three passes (the tower and the both sides of the OGW), so that the induced voltage $v(t)$ is calculated as follows

$$v(t) = \frac{Z_g}{Z_0 + Z_g} \cdot e(t) = \frac{1}{\frac{1}{Z_o} + \frac{1}{Z_g}} \cdot \frac{e(t)}{Z_0} \equiv R_{\text{eff}} \cdot I(t) \tag{20.6}$$

where $$R_{\text{eff}} = \frac{1}{\frac{1}{Z_0} + \frac{1}{Z_g}} = \frac{1}{\frac{1}{Z_0} + \left(\frac{2}{Z_n} + \frac{1}{R}\right)}, \quad Z_g = \left(\frac{2}{Z_n} + \frac{1}{R}\right)^{-1}$$

R_{eff} : effective surge impedance

Z_0 : surge impedance of lightning pass through air from clouds

R : surge impedance of tower structure

Z_g : surge impedance of total overhead grounding wires(n)/tower circuit

$Z_n = Z_{gw}/n$, where Z_{gw}, Z_n are the surge impedances of OGW of one and n stripes respectively.

Trial calculation of the induced voltage at the top of the tower or on the OGW
Assuming $R = 300\,\Omega$, $Z_0 = 400\,\Omega$, $Z_n = Z_{gw} = 300\,\Omega$ (where $n = 1$),
Then the effective surge impedance $R_{\text{eff}} = 1/\{1/400 + 2/300 + 1/300\} = 80\,\Omega$.
If we assume $I(t) = 20\,\text{kA}$, the induced initial voltage v_g at the top is $v_g = 20\,\text{kA} \times 80\,\Omega = 1600\,\text{kV}$.

The power frequency voltage is of course superimposed on the above injected surge voltage. In the case when the rated line voltage is 400 kV, the power frequency voltages v_a, v_b, v_c have values of $\pm 400 \times \sqrt{2}/\sqrt{3} = \pm 327\,\text{kV}$ (peak value). Accordingly, the superposed voltage $v_{ga} = 1173 \sim 1927\,\text{kV}(3.6 - 5.89E)$, so flashover is inevitably caused.

If the number of OGWs is increased from one to n, the surge impedance Z_n is decreased $1/n$ times so that the total surge impedance R_{eff} can also be effectively decreased. Assuming $n = 3$ in the above calculation, then $Z_n = Z_g/3 = 100\,\Omega$, and accordingly $R_{\text{eff}} = 38\,\Omega$. That is, we can reduce the equivalent surge impedance by about half.

20.5.3 Induced strokes (electrostatic induced strokes, electromagnetic induced strokes)

These are called **induced lightning surges,** whose mechanism is shown in Figure 20.11.

Lightning may strike some point on the Earth at a short distance from a transmission line. This phenomenon can be considered as one virtual wire connecting a cloud and an earth point and a surge current $I(t)$ suddenly flowing along the virtual wire. The virtual wire has mutual capacitances and mutual inductances across the line conductors and the OGWs of the neighboured transmission line as a matter of course, so that capacitive induced voltage as well as inductive induced voltage appear on the phase conductors as well as on the OGWs.

20.5.3.1 *Capacitive induced lightning surges*

In Figure 20.11, a virtual conductor (named a) and one conductor (named b) exist. If the potential voltage of the cloud is $e(t)$, the capacitive induced voltage on conductor b is $v(t) = \{C/(C + C')\} \cdot e(t)$. The insulation may be broken down by the induced voltage and the resulting flashover is named the **capacitive induced lightning stroke**.

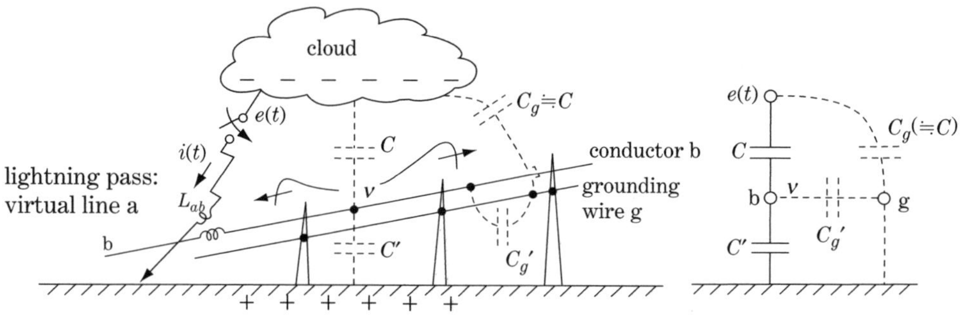

Figure 20.11 Induced lightning stroke

Now, we need to examine the function and the effect of the OGWs through a thought experiment.

In Figure 20.11, the virtual wire a (from the cloud to the Earth) and the line conductor b already exist. If one grounding wire g were additionally installed in parallel close to the conductor a, the equation of the capacitive induced voltage would be modified as follows:

capacitive induced voltage on conductor b

$$\left.\begin{array}{ll}\text{without grounding wire} : v = \dfrac{C}{C+C'} \cdot e(t) & ① \\ \text{with grounding wire g} : v \fallingdotseq \dfrac{C}{(C+C')+C_g'} \cdot e(t), \quad \text{where} C \fallingdotseq C_g \ll C' < C_g' & ② \\ \text{inductive induced voltage} : v = L\dfrac{d}{dt} i(t) & ③\end{array}\right\} \quad (20,7)$$

That is, the replaced new equation ② includes an additional C_g' in the denominator for the additional installation of the grounding wire. The conductor b and the grounding wire g are closely linked, so the relation of each capacitance size $C_g \fallingdotseq C < C' < C_g'$ is obviously justified. Accordingly, the capacitive induced voltage $v(t)$ can be reduced by the addition of the grounding wire g (the shielding effect). We discuss this matter again in Chapter 21.

20.5.3.2 Inductive induced lightning surges

As shown in Figure 20.11, mutual inductance L_{ab} exists between the virtual conductor a and the real conductor b, so that inductive surge voltage $v(t) = L_{ab} \cdot di(t)/dt$ is caused on conductor b by the lightning inrush current $i(t)$. The resulting flashover by the voltage is called the **inductive induced lightning surge**.

If we additionally install grounding wire g, L_{ab} can be reduced in the same way as we studied in Chapter 1. Therefore the grounding wire also can reduce the electromagnetic induced voltage.

Through the above study, we can imagine in regard to the grounding wire that the installation of OGWs can actually reduce L and increase C of the conductors so that the surge impedance $\sqrt{L/C}$ is also effectively reduced. This is discussed further in the next chapter.

21

Insulation Coordination

We have studied various kinds of overvoltage phenomena in the previous chapters from the viewpoint of their appearance. Now we need to study the total concept of insulation design of transmission lines, stations and all the installed equipment including the protection schemes for various overvoltages, namely the '**insulation coordination** of a power system' and the '**protection scheme for overvoltages**' as essential practical engineering of power systems.

We can imagine the problems our predecessors had when struggling with repeated failures caused by casual, irregular, over- or under-insulation design in the early twentieth century under the conditions of poor theoretical background, poor materials and application technology, poor experimental data, and with no commonly shared concept or guidelines to coordinate insulation strength as in today's standards or recommendations. Insulation coordination is a combined concept of 'physical phenomena', 'practical engineering and technology', as well as 'economy' based on much of the experience accumulated over a hundred years by our predecessors.

21.1 Overvoltages as Insulation Stresses

21.1.1 Conduction and insulation

Typical **conduction materials** are copper, aluminium and some other metals or alloys, and typical **insulation materials** are air, oil, SF_6 gas, porcelain, fibreglass, paper, plastic, cross-linked polyethylene, etc. Every small part of a power system is composed of a combination of conduction materials and insulation materials whose characteristics are very different. To obtain and control 'electricity' means to prepare an extremely large container called a power system, every part of which is made of a skilful **combination of different materials**.

In all materials, conduction is caused by the migration of charged particles. Conductors have large numbers of relatively free electrons, which will drift in an applied electric field. On the other hand, insulation materials (insulators) have very few free electrons. When electric stress in an insulation material is increased to a sufficiently high level, the resistivity along a path through the material will change from a high value to a value comparable with that of conductors. This change is the so-called **breakdown**.

In order to achieve our purpose, we need to have the following concepts from a practical engineering viewpoint:

- Magnitudes and characteristics of resulting overvoltages and over-currents.
- Required insulation strength of all the members comprising the power system.

© 2007 John Wiley & Sons, Ltd

- Countermeasures to reduce various overvoltages and to protect the insulation of lines and station equipment.

Insulation coordination is the total concept as well as the practical guideline for system insulation design, combining the above three concepts that are based on accumulated and firm theoretical, experimental, technical as well as economic data.

Each part of a power system has to have estimated possible impinging overvoltage levels and the required insulation withstanding voltage levels as set values in order to realize continuous long-life operation of individual equipment. Accordingly, we need to have the concept of **insulation strength** as the another feature of overvoltages, which is always a key theme of practical high-voltage engineering activities, although a detailed description of individual insulation strength is far beyond the purposes of this book.

The essential philosophy as well as concrete recommended practices based on practical data for insulation design have been issued as international and/or national standards (IEC, IEEE, ANSI, JEC, etc.), which include the characteristics of impinging overvoltages and methods to protect insulation as important guidelines of practical insulation design engineering. Today, insulation coordination is an established concept throughout the world, so the contents of standards issued by different organizations can be said to be substantially identical to each other.

21.1.2 Classification of overvoltages

We have studied overvoltages in the previous chapters mainly from the viewpoint of the mechanisms arising. Lightning is an overvoltage generated by external events, while the overvoltages generated by switching operations, fault occurrences, as well as ferro-resonance, load rejection, loss of ground, etc., are generated by internal events. Now we need to classify them again from the viewpoint of impinging overvoltages on the individual insulation structures.

Insulation strength of individual parts of the system or equipment is closely affected by the following factors of the impinging overvoltages:

- Characteristics of overvoltages (magnitude, shape, duration, polarity of the applied voltages)
- Insulation design of the electric field distribution in the insulation
- Type of insulation (gaseous, liquid, solid or a combination)
- Physical state of the insulation (temperature, pressure, mechanical stress, etc.)
- Operation and maintenance history of individual insulation materials.

The impinging overvoltages may exceed permissible insulation levels of individual parts of the system, so that these overvoltages should be reduced within permissible levels, or the insulated equipment should be safely protected against such overvoltages. This is essential to avoid insulation damage of equipment or to prevent possible undesirable system performance.

The **magnitudes**, the **wave shapes** (steepness of the voltages) and the **time duration** of overvoltages are important factors in regard to the stress on the insulation. Taking such important factors into account, the characteristics of overvoltages are generally classified into the following categories by authorities like the IEC and/or national standards bodies:

a) **Maximum continuous (power frequency) overvoltage**: U_s (MCOV). This can originate from the system under normal operating conditions.

b) **Temporary overvoltage** (TOV). This can originate from faults, switching operations such as load rejection, resonance conditions, non-linearity (ferro-resonance), or by a combination of these.

c) Slow-front overvoltages. These can originate from switching operations and direct lightning strikes to the conductors of overhead lines.

d) Fast-front overvoltages. These can originate from switching operations, lightning strikes or faults.

e) Very fast-front overvoltages. These can originate from faults or switching operations in gas-insulated switchgear (GIS).

The overvoltages of items (c)–(e) are the transient overvoltages. Now we examine these items below.

21.1.2.1 Maximum continuous (power frequency) overvoltages: U_s

Under normal operation, the power frequency voltages differ from one point of the system to another and vary in magnitude over time. For purposes of insulation design and coordination, continuous overvoltages should be considered the possible highest operating voltages under normal operation, which is usually defined as 'highest system voltage (the symbol U_s (kV)'.

In usual practice, U_s is presumed to have a value of 1.04–1.1 times system normal voltage. For example,

normal voltage phase-to-phase	highest system voltage U_s	representative TOV $(U_{rp} = k \cdot (1/\sqrt{3}) U_s)$
230 kV (phase-to-phase)	245 kV (phase-to-phase)	212 kV (phase to earth, $k = 1.5$)
500 kV	550/525 kV	476/455 kV ($k = 1.5$)
735 kV	765 kV	662 kV ($k = 1.5$)

21.1.2.2 TOVs and representative TOVs: U_{rp}

Typical situations that may give rise to TOVs include:

- Single line-to-ground faults
- Ferro-resonance
- Load rejection
- Loss of ground
- Long unloaded transmission lines (Ferranti rise)
- Coupled line resonance
- Transformer line inrush.

TOVs caused by all these situations are the result of some internal events caused by circuit connection changes, so the voltages tend to show unstable, slow transient behaviours to some extent and often include somewhat lower order harmonics or d.c. components, usually lasting a period of hundreds of milliseconds or longer, say a few minutes.

The highest TOVs for each of the above situations have to be presumed as the representative TOV U_{rp}, considering system configuration and operating practices at present and in the future. However, as a minimum, the overvoltages due to line-to-ground faults should be addressed, because they are typically the most significant and probably the highest TOVs for most cases. Single line-to-ground faults probably cause the largest power frequency overvoltages among various fault modes.

Courtesy of TEPCO

Double circuit lines, nominal voltage: 1000 kV,
maximum voltage: 1100 kV

Conductor wire ACSR 810 mm^2 (diameter 38.4 mm)
- 1258 A (continuous, in summer)
- eight-bundled conductors (w = 400 mm)
- maximum temperature 90°C

OPGW: double circuit 500 mm^2 × 2

Suspension insulator (320 mm type)
- double size 2 × 40 series connection
- horn gap: 6.3 m

Towers (average weight: 310 tons)
- average height: 111 m, height of lowest arm: 65 m
- phase-to-phase within same circuit: 20 m
- phase-to-phase between different circuits 33 m
- conductor height >32 m
- span average 632 m, maximum 1056 m

TOV (Temporary Overvoltage)
- 1 LG: 1.1 pu, LR (load rejection): 1.5 pu

LIWV (Lightning Impulse Withstand Voltage)
- GIS 2250 kV, Tr 1950 kV

SIWV (Switching Impulse Withstand Voltage)
- GIS 1550 kV, Tr 1425 kV

Figure 21.1 1000 kv design Double-Circuit Transmission Line (TEPCO, Kita-tochigi line, 427km)

Direct strike to the upper conductor (1998)

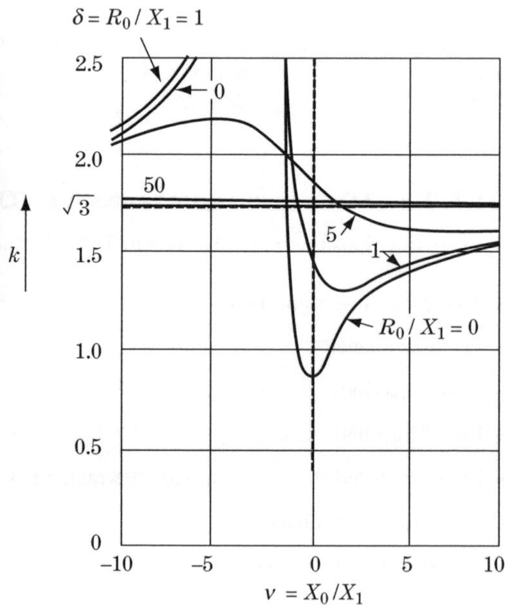

Figure 21.2 Overvoltage ratio of unfaulted phases during line-to-ground fault: k

21.1.2.2.1 Single line-to-ground faults Figure 21.2 is the enlarged curve of Figure 8.1 which is derived by fault analysis with power frequency. As in Chapter 8, the magnitudes of TOVs appearing on unfaulted phases during a line-to-ground fault are related closely to system grounding conditions (namely, $\delta = {}_fR_0/{}_fX_1$, $v = {}_fX_0/{}_fX_1$) and can be estimated by 'earth–fault factor k' (or 'coefficient of grounding', COG).

The factor k may be typically 1.5 for a solidly neutral grounding system, or $\sqrt{3}$ or higher for a resistance/reactance neutral grounding system.

21.1.2.2.2 *Load rejection* If a large load were suddenly separated (by the bus protective relay tripping in a large substation, for example), overvoltage by **load rejection** would be caused and continue, say, 10–200 s until voltage regulation by the AVR systems of power stations/substations is completed. Although the amplitudes of overvoltages depend on rejected load size and the network configuration, the amplitudes would be 1.2 pu or less in moderately extended systems due to the quick response characteristics of AVRs. However, overvoltages are apt to become larger in power systems of rather extended configurations. When full load rejection is caused at the receiving substation of a long transmission line end, the voltage may rise by a magnitude close to 1.5, because the Ferranti effect at the open receiving end would be added.

21.1.2.2.3 *Loss of ground* In high-resistive/reactive neutral grounded systems, special consideration is required. If one phase-to-ground fault occurs in the system and the neutral grounded transformer is tripped for some reason (by a back-up relay tripping, for example), this means **loss of ground** (assuming another neutral grounding transformer does not exist in the same system). As the potential of the system neutral point becomes free of the earth potential, unfaulted phase voltages become at least $\sqrt{3}$ pu during one phase to earth fault ($1\phi G$). Furthermore, even after one-phase-fault condition is removed, three-phase voltages as well as neutral point voltages fluctuate waywardly. In these conditions, numbers of installed arresters would probably be broken, which means extended new faults at arrester points, and/or earth faults would be caused at various locations.

Above all, loss of ground should be absolutely avoided. In high-resistive/reactive neutral grounding systems, neutral groundings by as many transformers as possible (at least by two transformers installed at different stations in one service area) are preferable.

The TOV caused by one phase to ground faults may be mostly presumed to be the representative largest power frequency overvoltage U_{rp} (which is usually expressed by rms kV value, phase-to-earth). Then

$$\text{representative TOV} = U_{rp} = k \cdot \left(1/\sqrt{3}\right) U_s$$

As a typical example,

$$\text{nominal voltage } 230\,\text{kV} \quad k = 1.5 \quad \text{TOV} = 1.5 \cdot \left(1/\sqrt{3}\right) \cdot 245 \doteqdot 212\,\text{kV}$$

$$735\,\text{kV} \quad k = 1.5 \quad \text{TOV} = 1.5 \cdot \left(1/\sqrt{3}\right) \cdot 735 = 662\,\text{kV}$$

21.1.2.3 *Slow-front overvoltages*

Slow-front overvoltages have durations of some tens to thousands of microseconds and tail durations of the same order of magnitude, and are oscillatory by nature. They arise generally in the occasions below:

- Line energization/re-energization (breaker closing/opening)
- Fault occurring/fault clearing
- Switching of capacitive/inductive current
- Load rejection
- Distant lightning strike (lightning strike wavefront flattened by travelling).

The voltage-arising mechanisms of all these have been studied in the previous chapters.

The representative voltage stress is characterized by the representative voltage amplitude and wave shape. Therefore, the **representative switching impulse voltage of 250/2500 μs** (time to peak 250 μs, and time to half-value on the tail 2500 μs, see Table 21.3 below) has been standardized as the common general concept throughout the world to represent standardized slow-front overvoltages.

Although the voltage-arising mechanisms of all the above have been studied in previous chapters, here are some additional comments.

For the switching of capacitive/inductive current, we have studied overvoltages caused by current chopping in Section 19.4, caused when the current power factor is almost zero. In particular, the following switching operations require special attention:

- Switching of unloaded cables or capacitor banks.
- Inductive current tripping (transformer magnetizing current tripping, for example).
- Arc-furnace load switching.
- Interruption of currents by high-voltage fuses.

The most useful and commonly used method of limiting the slow-front overvoltages is to adopt the resistive tripping/closing breakers we have studied already. Surge arrester protection against slow-front overvoltages will be discussed later.

21.1.2.4 Fast-front overvoltages

Typical fast-front overvoltages are of course lightning strikes, although they can originate also from switching operations. We have already learned that lightning overvoltages can be classified into direct strikes, back-flashovers and induced lightning strikes.

The induced lightning surges occur generally below 400 kV and so are of importance only for lower voltage systems of 100 kV. Back-flashover voltages are less probable on UHV systems of 500 kV or more, due to the high insulation withstanding values.

The representative wave shape of fast-front overvoltages is the well-known 1.2/50 μs wave (see Table 21.3 below).

21.1.2.5 Very fast-front overvoltages

Very fast-front overvoltages can originate from switching operations or from faults within GIS due to the fast breakdown of the gas gap and nearly undamped surge propagation within GIS, where the average distance between two adjacent transition points in the same GIS is very short. (If 7.5 m distance and $u = 300\,\mathrm{m}/\mu\mathrm{s}$ is assumed, the resulting travelling surges repeat 20 times every 1 μs, so the natural frequency is 20 MHz, and the first-front wavelength (quarter cycle) is 0.0125 μs.) However, the amplitudes of the surges would be rapidly dampened and flattened on leaving the GIS, so they are relieved to some extent at the external circuit of the GIS bushings.

The overvoltage shape is characterized by a very fast increase of the voltage to nearly its peak value, resulting in a front time below 0.1 μs. For switching operations this front is typically followed by an oscillation with frequencies of 1–20 MHz. The duration of very fast-front overvoltages would be less than 2–3 ms; however, 20 MHz and 3 ms means 60 000 times of beating stresses. Furthermore, they may occur several times. The magnitudes of overvoltage amplitude depend on the structure of the disconnector and on the adjacent structure of station equipment.

Very fast-front overvoltages can and have to be dampened/flattened to some extent, and the typical countermeasure is application of gap-less arresters, which means insertion of a non-linear high resistance in parallel across the phases and earth. The maximum amplitudes to 2.5 pu can be assumed to be achievable.

Due to faults within GIS, the connected equipment, in particular a transformer, is stressed by the overvoltages, which would contain frequencies up to 20 MHz, and the amplitude may exceed the breakdown voltages of the transformer without effective countermeasures.

21.2 Fundamental Concept of Insulation Coordination

21.2.1 Concept of insulation coordination

What is usually meant is the coordination or correlation of the transmission line insulation with that of the station apparatus, and perhaps the correlation of insulation of various pieces of apparatus and parts of the substation. The coordination of substation and equipment insulation is the protection of service and apparatus from overvoltages in excess of specified insulation withstanding values at optimum economy and reliabilities.

It is obvious that various kinds of reliable knowledge and accurate data based on a lot of experience and advanced technology had to be required in order for a reasonable concept of insulation coordination to be established, as follows:

- Investigation of mechanisms of various different overvoltages.
- Estimation of possible overvoltages on the line or in the substation.
- Countermeasures to reduce various overvoltages, including development of protective devices.
- Specification of insulation withstanding values for a transmission line.
- Specification of insulation withstanding values of the station apparatus, and so on.

Today, it is fair to say that the descriptions of all the standards relating to insulation coordination have converged on the same criteria. This great result has been achieved through a long process of evolution in the past hundred years. We study such this worldwide common concept in this chapter.

21.2.2 Specific principles of insulation strength and breakdown

The basic concept of the coordination or correlation of transmission line insulation and that of the station apparatus can be summarized as follows.

21.2.2.1 Insulation design criteria of the overhead transmission line

The basic criteria are:

- Flashovers caused by lightning strikes are allowed as fatal phenomena, while damage (damage to conductors, cracks in insulators, etc.) to the transmission line should be avoided.
- Technically as well as economically balanced insulation distance (clearance) is to be assured in the fundamental design, allowing some extent of failure rate caused by lightning strikes. Also, countermeasures should be adopted as much as possible to reduce the influence and frequency of effects on a substation.
- Flashover should not be caused by switching surges or by any sustained lower frequency overvoltages.

The above concept is based on the characteristics in atmospheric air of so-called insulation and cooling materials of infinite natural circulation type, so that, once broken, insulation would be restored (recovered) whenever the surge source disappears (**self-restoring insulation characteristics**).

The principal countermeasures are:

- To reduce the probability of lightning strike, and to limit the faulted circuits (in the case of multiple circuit transmission lines) and faulted phases as much as possible. (Adoption of overhead grounding wires and any other effective countermeasure.)
- To reduce the probability of back-flashover caused by lightning strikes on the overhead grounding wires or on the towers (surge impedance reduction of towers and overhead grounding wires).
- Countermeasures to relieve the travelling waves to some extent before reaching the substation terminal point.
- Insulation-level withstanding against switching surges from the substation.
- The idea to reduce the probability of simultaneous faults on plural circuits of the same route.
- Adoption of reclosing.

21.2.2.2 Insulation design criteria of the substation and substation apparatus

The basic criteria are that insulation of the station and the station apparatus should be protected to withstand lightning surges and switching surges, so that insulation failure of the apparatus and loss of station services over a long time should be avoided.

The principal countermeasures are:

- Countermeasures to reduce direct lightning strikes to the station as much as possible.
- To protect the substation and the station apparatus against transmitted lightning surge voltages from the overhead transmission lines without damage (arresters).
- To reduce switching surge levels and to protect the station apparatus against switching surges without damage. That is, flashover or insulation failure of equipment should not be caused by switching surges or by any sustained lower frequency overvoltages in the substation.

21.2.2.3 Insulation design criteria of the power cable line

The basic criteria are to protect the power cable against lightning surges, switching surges as well as fundamental frequency overvoltages. The power cable does not have the characteristics of self-restoring insulation, so it should be protected entirely from overvoltages in the same way as transformers or other substation apparatus.

All above criteria are just substantial images of the fundamentals of our power systems, and thus are the objectives of insulation coordination.

21.3 Countermeasures on Transmission Lines to Reduce Overvoltages and Flashover

We list in this section the major countermeasures usually adopted as concrete design for high-voltage transmission lines and substations in order to meet the design criteria described above. We have already studied in previous chapters the reasons why the adoption of each listed countermeasure is technically effective and can be justified.

Readers are requested to refer to the literature for those details exceeding the purposes of this book.

21.3.1 Countermeasures

21.3.1.1 Adoption of a possible large number of overhead grounding wires (OGWs, OPGWs)

The following effects are expected from the adoption of OGWs:

- By locating the **OGW** at the top of a tower, the probability that lightning directly strikes the phase conductors can be reduced (**shielding effect**, Section 20.5). A lightning strike directly on the OGW may occur, but most of the energy can be bypassed as strike current though the OGW and towers to earth.
- The surge impedance $\sqrt{L/C}$ of the line conductors can be reduced. (L is decreased largely and C is increased largely by installation of OGW and with increasing numbers of OGWs, Chapters 1, 3, 20). The absolute magnitude of induced surge voltages appearing on the phase conductors can be reduced so that the probability of phase faults by back-flashover or induced strikes is improved (Section 20.5). Also, alleviation of steep wave-front (surge-front flattening) of the lightning surge can be expected (Chapter 18).
- The time constants $T = 2L/r$ of the line can also be reduced so that the attenuation of traveling–wave or transient oscillation terms can be accelerated (Chapters 18, 19).
- Positive-, negative- and zero-sequence inductance $L_1 = L_2, L_0$ can be reduced, an important feature.
- Power frequency voltage drop by line reactance can be reduced (Chapters 1, 3).
- The stability limit as well as the power circle diagram can be improved (Chapters 12, 14).

Today, **OPGW** (grounding wire with optical fibre) has prevailed as OGW with the function of communication channel media.

21.3.1.2 Adoption of reasonable allocation and air clearances for conductors/grounding wires to assure insulation withstanding level and reduction of surge impedances

The phase conductors as well as the OGWs have to have sufficient clearance from each other so that the necessary insulation withstanding strength is maintained within some margin against the predicted largest short-duration overvoltages and switching surges coming from neighbouring substations. Of course, this is a trade-off between the probability of flashover failure and the construction cost of larger towers. The allocation of conductors and grounding wires is decided to result in a reasonably phase-balanced smaller L and smaller $\sqrt{L/C}$ within the necessary margin against **physical movement** (caused by **wind**, **heat expansion**, **galloping**, **sleet jumping**, etc.).

Details of **standard lightning impulse-withstanding voltages** are examined in Section 21.5.

21.3.1.3 Reduction of surge impedance of the towers

The **surge impedance of the towers** has to be reduced as much as possible, in that induced surge voltages on the transmission lines can be reduced, or the probability of striking can be reduced. In particular,

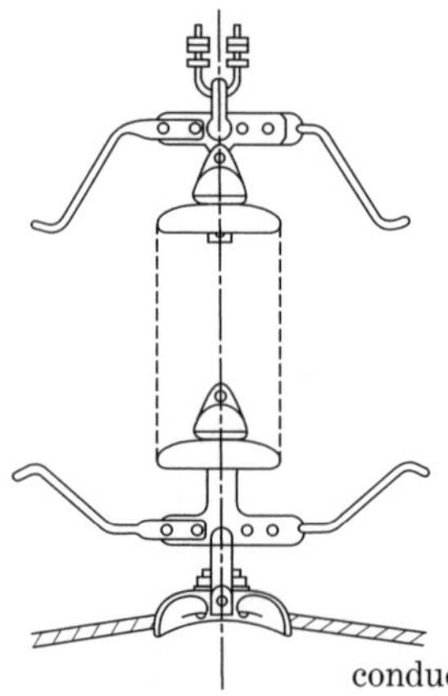

Figure 21.3 Arcing horn

back-flashover caused by direct strikes on the top of the towers or the grounding wires may be effectively reduced (Section 20.5). The magnitude of the tower surge impedance is related to the height of the tower and the resistivity of the ground, and may typically be 20–100 Ω for EHV class lines.

21.3.1.4 Adoption of arcing horns (arcing rings)

Arcing horns are a kind of air gap having self-restoring insulation characteristics. They are arranged in parallel with each insulator as a single body on every tower. Figure 21.3 is an example for a single conductor line. The duty and the purpose of arcing horns can be summarized as follows:

- Flashing overvoltages can be controlled by selection of the shape and air-gap length of the arcing horns so that the magnitudes of travelling surge voltages caused by the lightning can be limited by the flashover voltage of the arcing horn. The arcing horn may be said to be the intentionally arranged weak point of a conductor's insulation (**controllable limitation of surge voltage magnitude**).
- An arcing horn assembled together with an insulator can improve the potential gradients of the insulator in that the voltage distribution by series-connected individual porcelain pieces can be unified (**improving withstand voltage**).
- Flashover by arcing horn can avoid flashover along the surface of insulators, so that the insulators can be protected from damage against thermal shock (**protection of porcelain insulators**).

On the other hand, arcing horns have the limitation below:

- Arcing horns as well as any other part of the overhead transmission lines should not experience flashover by switching surges coming from adjacent substations.

The air-gap lengths of each arcing horn on the series-connected towers are generally arranged with equal lengths, while that on the first to third towers of outgoing feeder lines from the station may exceptionally be arranged with a little smaller gap length. Lightning overvoltage caused by striking the transmission line point within one or two spans from the station appears as impinging travelling surges to the station without attenuation or shape flattening. Accordingly, in order to alleviate such severe surges caused by a close-point direct strike, the arcing horns at the first few spans may be arranged with somewhat smaller gap lengths.

21.3.1.5 Adoption of unequal circuit insulation (double circuit line)

Line failure caused by lightning striking the transmission lines cannot be avoided; however, we need to reduce the situation first. Primarily, simultaneous double circuit faults should preferably be reduced. Furthermore, numbers of faulted phases of the same circuit should also be reduced as much as possible.

For this purpose, so-called **unequal circuit insulation** or **unequal phase insulation** has been partly applied, in that horn-gap lengths of a specified single circuit or phase are intentionally arranged within a short distance.

21.3.1.6 Adoption of high-speed reclosing

Automatic high-speed reclosing is an important practice to reduce the influence of lightning failure. Its essence is studied here, although it may not be directly related to insulation coordination.

In single phase reclosing for a phase a to ground fault (aϕG) (for example), immediately after the phase a conductor to earth fault occurs, the phase a pole contactors of both line terminal breakers are tripped within say 2–6 cycles (operating time of the relay + the breaker), that is single phase tripping. Then, although the phase a conductor is already separated from the station buses, the voltage v_a on the phase a conductor still remains, so the arc would also continue for some small duration (arcing time, say, 0.2–0.5 s), because in addition to the initial trapped potential charge (d.c.) the electrostatic (C-coupled) voltage is induced from unfaulted parallel phase voltages v_b, v_c. However, due to the outstanding self-restoring characteristics of natural air, the arc will soon be extinguished. Accordingly, if the phase a poles of the breakers of both terminals are reclosed after arc extinction, the faulted line again continues three-phase operation successfully.

The time from the one-pole tripping to reclosing is customarily called '**dead-voltage time**', which is a set value in the primary protective relay equipment for the line. The arcing time is apt to become longer for higher voltage systems, in particular UHV systems of 500 kV or more, so that dead-voltage time as a relay setting value has to be set longer. There is some presumption that in 1000 kV class power systems, self-arc extinction cannot be expected within a short time, so automatic reclosing may not be available without applying forced grounding switches.

For the classification of high-speed reclosing, let us assume double circuit line 1 (a,b,c) and line 2 (A,B,C):

- **Single phase reclosing**: This occurs on one line-to-ground fault (phase a reclosing against phase a fault).
- **Three-phase reclosing**: When a fault occurs on line 1, the three phase a–b–c poles are tripped and reclosed regardless of the fault modes on the line. This is obviously a practice which can be applied only for double circuit lines, or at least 'assured loop-connected lines'.
- **Multi-phase reclosing**: In the case of a fault of 'line 1 phase a and line 2 phase B', for example, reclosing is conducted for these two poles, because all phases A, b, C are still soundly connected even though this is s double circuit fault. This is quite an effective method in comparison with three-phase reclosing, because it can minimize line out of service, or relieve system instability. Also very accurate fault phase-detection by protective relays (preferably by phase differential protection) is required for this practices.

Incidentally, reclosing is allowed only when both terminal buses across the faulted line are operating in synchronization. Furthermore, possible sending power during the dead-voltage time has to be noted. Single phase reclosing on a single circuit line for example means one phase opening mode condition, as shown in Table 3.2 and Table 8.2 [8] [10]. Then reclosing should be allowed only when the presumed power flow during the dead-voltage time is within the stability limit under the condition of only the sound phases (refer to Section 14.4).

Also, we need to pay attention to the undesirable electrical and mechanical effects on the thermal generators caused by the appearance of negative- and zero-sequence currents during the reclosing dead-time.

21.4 Overvoltage Protection at Substations

Now we need to examine aspects of overvoltages at the substations and the various countermeasures to protect them against overvoltages or to reduce the stresses.

21.4.1 Surge protection by metal–oxide surge arresters

21.4.1.1 *The principle of surge protection by arresters*

Arresters are key devices to protect substations and station devices against lightning surges and whose surge protective capability actually decides the required insulation levels of the power system network. **Gap-less arresters** may also have the capability to reduce switching surges.

Typical high-voltage arresters are **metal–oxide surge arresters** having metal–oxide resistive disc elements with excellent non-linear *v–i* characteristics and thermal energy withstanding capability. The metal–oxide resistive elements are composed of a number of lapped disc elements, each of which is made from zinc oxide (ZnO) powdered material with some specially mixed inclusions, and, similar to pottery or porcelain is produced through a high-temperature baking procedure.

We examine first the fundamental principles of the arrester in Figure 21.4. In Figure 21.4a, the arrester is installed at point a, which is a transition point because the arrester impedance is at least

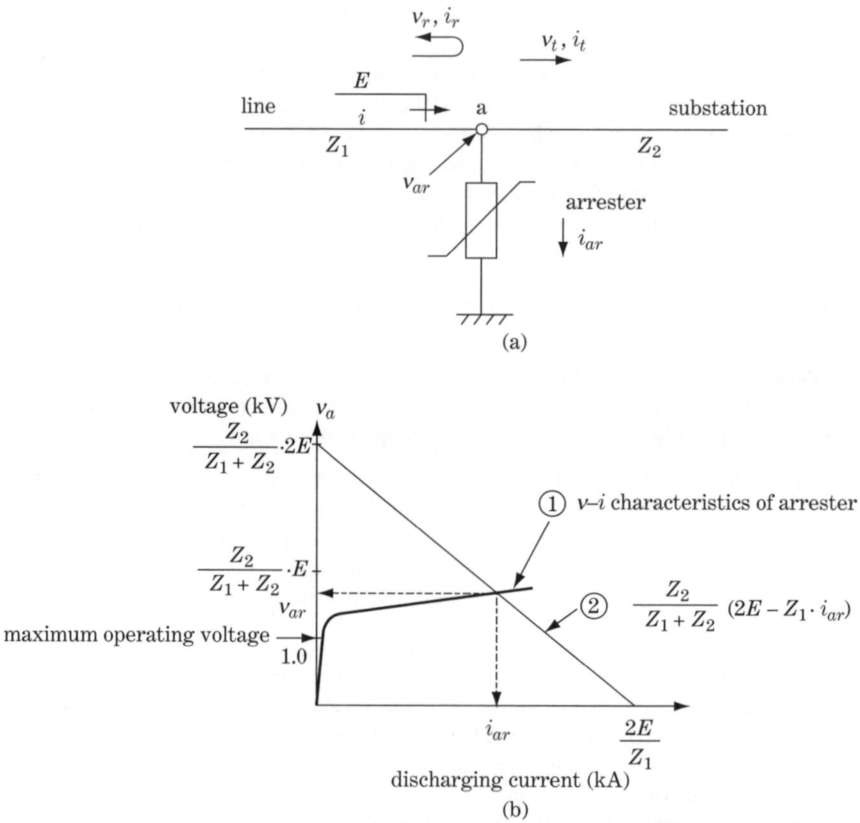

Figure 21.4 Principle of surge protection by arrester

connected. Now, incident overvoltage wave E comes from the left and passes through point a from the left-side line Z_1 to the right-side circuit Z_2 towards the station bus. The behaviour of the travelling waves at point a can be written by the following equations:

$$\left.\begin{array}{lrl} E+v_r=v_{ar}=v_t & \text{where} \quad E: & \text{incidental surge voltage from the left-side line} \\ i-i_r=i_t+i_{ar} & v_{ar}, i_{ar}: & \text{the terminal voltage and the current of the arrester at arrester point a} \\ E=Z_1 i & Z_{ar}: & \text{non-linear resistance of the arrester} \\ v_r=Z_1 i_r & v_t, i_t: & \text{the transmitted voltage and current at point a to the substation} \\ v_t=Z_2 i_t & Z_1: & \text{the surge impedance of the transmission line} \\ v_{ar}=Z_{ar}\cdot i_{ar} & Z_2: & \text{the surge impedance of the substation gateway at point a} \end{array}\right\} \tag{21.1a}$$

The relation between the arrester voltage v_{ar} and current i_{ar} is shown as the non-linear curve ① in Figure 21.3b, which represents the v–i characteristics of the arrester.

Eliminating v_r, i_r in the above equation, and by modification,

$$v_{ar}=v_t=\frac{Z_2}{Z_1+Z_2}(2E-Z_1 i_{ar}) \tag{21.1b}$$

This equation is written as the straight line ② in Figure 21.4b. The actual voltage v_{ar} at point a and the arrester current i_{ar} are given as those at the intersection of curve ① and the straight line ②. If the arrester does not exist, the voltage v_{ar} at point a would become $\{2Z_1/(Z_1+Z_2)\}\cdot 2E$ (then, E under the condition $Z_1=Z_2$), or $2E$ maximum under the condition $Z_2=\infty$ (i.e. the case when the feeding terminal is opened). However, if the arrester with appropriate non-linear v–i characteristics is installed at point a, surge voltage v_{ar} can be reduced to a smaller value than the original impinging surge value E and of course smaller than $2E$. On the other hand, in order to realize the above condition, the arrester is required to withstand the resulting extremely large thermal energy $\left(\int v_{ar}\cdot i_{ar}dt\right)$ without losing the original v–i characteristics of the resistive elements and without breaking. It is furthermore required to restore immediately the original electrical operating condition before the impinging surge.

Now, as a natural consequence of the above, the arrester has to have the following characteristics as its inherent duty requirement:

- Under power frequency operation with voltage U_s (MCOV) as well as of U_{rp} (representative TOV), the arrester has to have quite a high resistivity so that the arrester resistive elements can withstand the thermal stress caused by the small leakage current (**continuous current of arrester**, usually not more than a few milliamps, and probably 1 mA or less). For example, for a power system of nominal voltage 230 kV, $U_s=245\,\text{kV}$, $U_{rp}=k\cdot\left(1/\sqrt{3}\right)U_s=1.5\times\left(1/\sqrt{3}\right)\cdot 245=212\,\text{kV}$. Then, the arrester continuous current, assuming $i_{at}\leqq 1\,\text{mA}$ at $U_{rp}=212\,\text{kV}$, is $v_{ar}=212\,\text{kV}$, $i_{ar}\leqq 1\,\text{mA}$, $Z_{ar}\geqq 212\,\text{M}\Omega$ and the arrester thermal loss (stress) is $v_{ar}\cdot i_{ar}=212\,\text{kV}\cdot 1\,\text{mA}=212\,\text{W}$
- When the surge voltage and current arrive at the arrester point a, the arrester has to discharge surge current immediately without delay, and next, the voltage v_{ar} arising should be limited within a specified upper limit (**residual voltage**, or **discharge voltage**, U_{res}) during the passage of discharge. Then, as a result, the station equipment can be protected against the overvoltage.
- The arrester should withstand the thermal energy caused by the surge current through the arrester elements (**arrester discharge current**, say 10–100 kA), with thermally stable v–i characteristics.
- Immediately after the discharge current through the arrester disappears (probably within 50–100 μs), the following current i_{ar} by continuing power frequency voltage v_{ar} should become the original small leakage current within a few milliamps. Or, in other words, immediately after the surge voltage and current disappear, all the electrical characteristics have to be restored.

- The arrester has to have **switching surge discharging capability** within the specified levels. (The switching surge duty on metal–oxide arresters increases for higher system voltages.)

At the substation where the arrester is installed at the junction point with the transmission line, the appearing surge voltage (caused by the impinging lightning surge from the same line) can be limited to the **arrester discharge voltage** so that the station equipment having an insulation level exceeding the arrester discharge voltage (or residual voltage) can be protected by the arrester. This is the principle of station equipment protection by surge arrester.

21.4.1.2 Metal–oxide arresters

The fundamental configuration of these arresters can be classified as follows:

- **Gap-less arresters**
- **Series-gapped arresters**
- **Shunt-gapped arresters.**

Figure 21.5 shows a typical example of a porcelain-type metal-oxide gap-less arrester for high-voltage station use.

Before the appearance of gap-less arresters, probably around 1980, the configuration of highvoltage arresters was of the series-gap type without exception and arranged in series with non-linear resistive elements. It was mainly because the thermal duty of the resistive elements at the time was rather small that the elements could not withstand the thermal energy of continuous flowing leakage current caused outside the series gap by power frequency terminal voltages. Accordingly, the series-gap types used to be indispensable continuous devices to avoid thermal damage to resistive elements caused by continuous leakage current flowing through the elements.

However, it may be fair to say that nowadays most arresters for high-voltage systems are of the gapless type, in which the non-linear resistive element block is always directly charged by power frequency phase voltages so that the minute leakage currents flow through the arrester's resistive elements continuously. Of course, the resistive elements need to withstand the thermal stress caused by continuous thermal energy $\left(\int v_{ar} \cdot i_{ar} dt\right)$ from the leakage current and to maintain thermally stable v–i characteristics.

Incidentally, in the case of series-gapped arresters, the arcing ignition across the series gap is initiated by the surge voltage. Therefore arcing extinction across the gap immediately after the surge

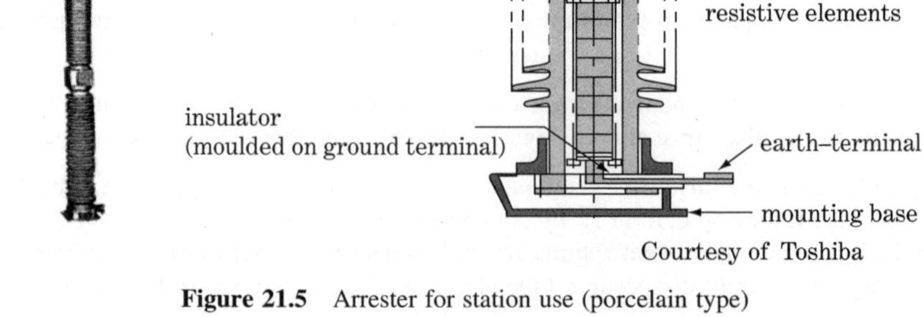

Courtesy of Toshiba

Figure 21.5 Arrester for station use (porcelain type)

current disappears under charging conditions by the continuing power frequency voltage is another important requirement for this arrester (**following current tripping duty**).

The technology to produce resistive discs with outstanding *v–i* characteristics and thermal energy withstanding duty has advanced remarkably in the last twenty years, and enabled the realization of gapless arresters with outstanding capabilities. It is worth remembering that, because this realization is based on advanced disc production skills, a drastic **decrease in the insulation levels** of transmission lines as well as of station equipment has been achieved.

Figure 21.6a shows the ***v-i* characteristic curve of arresters**. Under normal conditions, the *v-i* operating point goes back and forth between points a and b every power frequency cycle, where the largest current i_0 at a or b (the leakage current) has a very small value of 1–3 mA or even, say, 100–500 μA. If impinging surge voltage E arrives at the arrester terminal point at time t_1, the operating point (tracing along the curve) immediately goes past knee point c (c is the so-called '**initial discharging point**') at time t_2, then reaches the point of maximum voltage V_{max} at t_3 (the maximum voltage is called the '**residual voltage**'), and, next, arrives at the point of maximum current at t_4. As the surge current decreases, the voltage will decrease back towards the pre-surge level. In other words, as soon as the surge current disappears (possibly within 100 μs) the operating point moves to the points 0 and b. The return path is a little different to the outward path, because the characteristics of the resistive element are affected by the thermal effect.

Through all the above tracing process, a large discharging voltage of one polarity arises during time t_2–t_3–t_4–t_5 (the largest value of the discharge voltage at t_3 is the residual voltage).

Figure 21.6b is the arrester's '**residual voltage characteristics curve**' showing the protective level of the arrester. The vertical axis voltage is the residual voltage peak value which corresponds to the value of V_{max} at t_3 in Figure 21.6a. This curve actually indicates the protection capability of the arrester.

When lightning surge voltage and current arrive at the arrester terminal point, the surge voltage level is reduced by the arrester; however, discharge voltage (residual voltage) given by Figure 21.6b still arises. The value of the discharge voltage depends on the magnitude of discharge current (impulse current) flowing through the arrester i_{ar}.

The **surge current** through i_{ar} would mostly have values of 5–20–50 kA, while the presumed largest discharge current values could be 150 kA for the 400–500 kV, 100 kA for the 275–300 kV, 80 kA for the 160–230 kV, 60 kA for the 110–160 kV and 30 kA for the 60–90 kV class. Figure 21.6b is an example of quite advanced characteristics, by which the residual voltage would be distributed within a voltage range of 1.5 times the power frequency operating voltage for most cases, whereas larger discharge voltage of over two times, say 2–3.5 times, the operating voltage could also appear if quite a large impulse discharge current flows through the arrester.

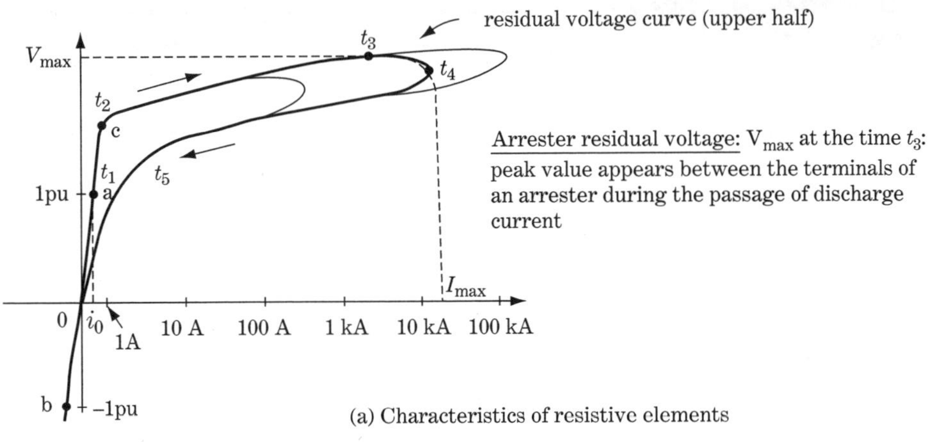

(a) Characteristics of resistive elements

Figure 21.6 Arrester characteristics

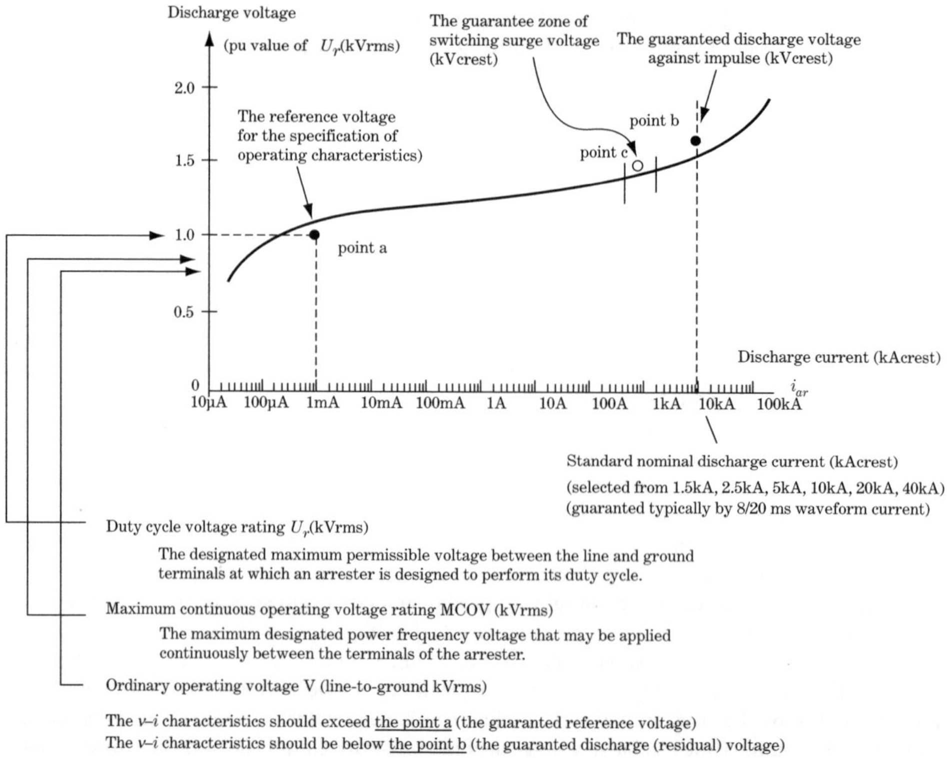

(b) Arrester v–i characteristic curve

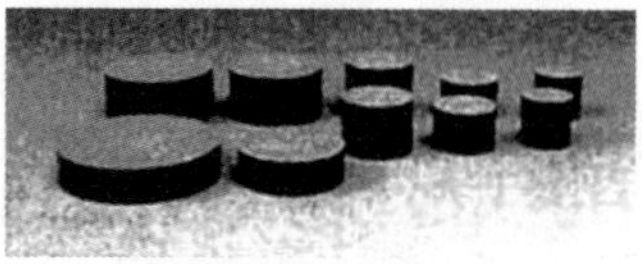

ZnO elements

Courtesy of Toshiba

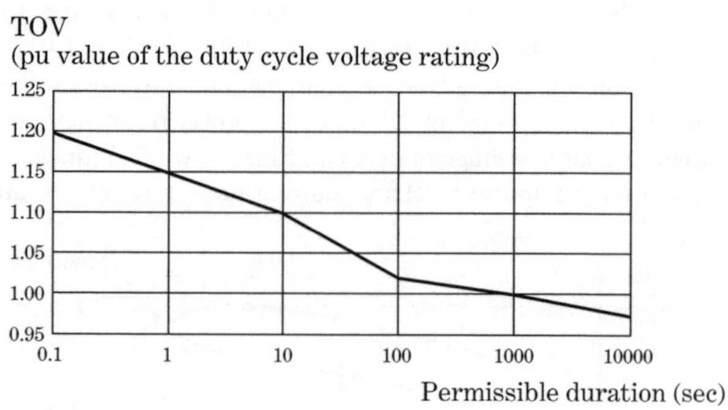

(c) TOV capability

Figure 21.6 (*Continued*)

Taking all the above into consideration, it may be concluded that, for insulation coordination of a substation and its equipment:

- Against impinging lightning surge to the station it is necessary:
  - to limit the arising station surge voltage within the **presumed largest residual voltages of the adopted arresters**
  - to realize the **withstanding insulation level of the station equipment** against the largest discharge voltage (residual voltage) for the presumed largest surge current.
- For power frequency voltages, the arrester should withstand the thermal stress of continuous leakage current caused by MCOV as well as TOV and of switching surge energy.

Figure 21.6c shows the typical TOV capability of arresters.

21.4.1.3 Classification and selection of arresters by ratings

We present here the general concept in regard to a guaranteed method for the characteristics of an arrester and the method of selection, although the detailed description of arrester standards may differ among national standards bodies.

Referring to Figure 21.6b, the arrester's essential v–i characteristics are guaranteed by the two points a and b.

Point a is the **reference voltage and current** (v_a (kVrms), i_a (mArms)) which are indicated by the supplier. It should be guaranteed that the v–i characteristics of an individual arrester at the reference current i_a (typically 1 mA) exceed the guaranteed reference voltage v_a, and the arrester has to withstand the thermal energy of the guaranteed power frequency continuous current i_a.

Point b is the **standard nominal discharge voltage and current** (v_b (kVcrest), i_b(kAcrest)). The standard nominal current i_b is specified as standard values like 1.5 kA, 2.5 kA, 5 kA, 10 kA, 20 kA, 40 kA for the purpose of arrester classification and for guaranteeing the discharge voltage characteristics. The discharge voltage of an individual arrester should be smaller than the guaranteed standard nominal discharge voltage v_b at the standard nominal discharging current i_b (tested typically with the 8/20 μs standard impulse wave current).

In practical engineering, individual arresters should be selected so that the guaranteed reference voltage (or the duty cycle voltage) exceeds the MCOV or the maximum TOV (U_{rp}).

High-voltage arresters of EHV/UHV classes are also assigned **switching surge durability** by the standards for the arresters, in that generally the **thermal energy absorbing capability** (kJ) at the specified discharge current is type tested. The guaranteed value can be additionally written as point c in Figure 21.6b.

In regard to switching surges, point c can be written as the guaranteed standard nominalbdischarge voltage and current (v_c (kVcrest), i_c (kA crest)) for the switching surges. The standard nominal currents for the switching surges are specified as standard values like 0.5 kA, 1 kA, 2 kAcrest. (A detailed description of the arrester's switching surge durability is omitted.)

21.4.2 Separation effects of station arresters

In regard to surge phenomena, the induced time-changing overvoltage of an arbitrary point is different from that of any other point in the same substation. Accordingly, the voltage at the transformer terminal (or at any other equipment) is different from that of the station arrester terminal. Generally we need to consider that the voltage at the protected insulation may possibly be higher than that at the arrester

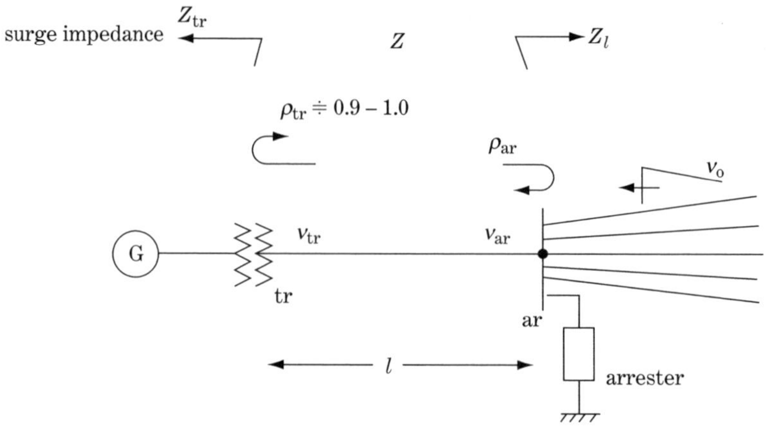

(a)

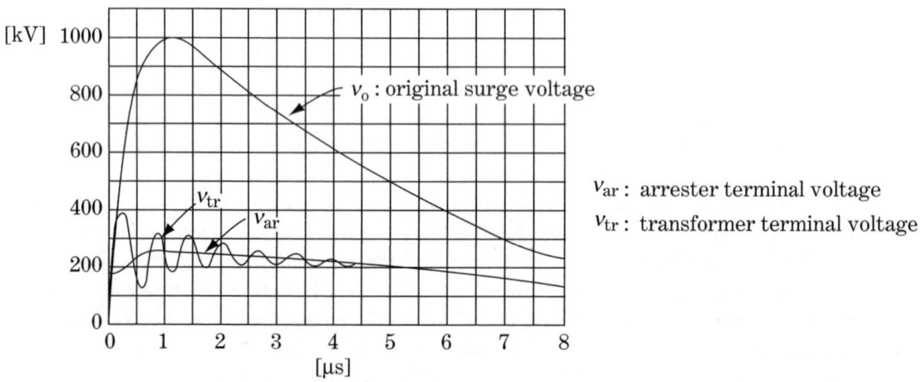

(b) surge voltages at a arrester point and a transformer terminals. (simulation)

(c) Typical values of surge impedances

overhead lines	: 300–500 Ω
cable lines	: 20–60 Ω
transformers	: 1000–10 000 Ω
rotating machinery	: 500–1500 Ω

Figure 21.7 Separation effects of arresters

terminals due to the travelling distance on connecting leads and the conductor circuit. This rise in voltage is called the **separation effect of an arrester**. This effect obviously lessens the surge protection performance of arresters. Referring to Figure 21.7a, the separation effects can be explained as the behaviour of travelling waves which is deeply linked with (1) the increasing rate of rise of incoming surge μ (kV/μs), (2) the distance l between the arrester and protective equipment (a transformer), and (3) the reflection factor ρ_{tr} of the equipment. The phenomena can be roughly calculated by the equation below, referring to Figures 21.7a, b and c.

We image that the transmitted lightning surge voltage at the arrester point with the time front $T_{front} \doteqdot 1.2\mu s$ and the initial steepness of α (kV/μs typically 200–500 kV/μs) appears on $t=0$, so the voltage during the initial small time interval of $0<t<T_{front}$ ($\doteqdot 1.2\mu s$) can be written $V_{ar}(t)=\alpha \cdot t$(kv). The surge voltage begins to travel to the transformer terminal (distance l) on $t=0$ and arrive on $t=l/u \equiv T$, so that the surge voltage at the transformer terminal (reflection factor ρ_{tr}) appears on $t=l/u$ (that is $t'=t-l/u=0$) as is the voltage form of $V_{tr}(t')=(1+\rho_{tr})\cdot\alpha\cdot t'$ (kV), where attenuation is neglected.

Assuming $l=60\,m$, $u=300\,m/\mu s$, namely $T=60/300=0.2\,\mu s$ for one way traveling. $Z_1=300\,\Omega$ (for station conductors), $Z_2=5000\,\Omega$ (for a transformer)

$$\rho_{tr}=(5000–300)/5000+300)=0.9 \text{ (reflection factor)}$$

$$V_{ar}(t)=\alpha\cdot t\,(kV) \quad \text{for } 0<t<2T=0.4\,\mu s$$

$$V_{ar}(t')=(1+\rho_{tr})\cdot\alpha\cdot t'\,(kV)=1.9\,\alpha\cdot t'\,(kV) \quad \text{for } 0<t'<2T=0.4\mu s$$

The equation shows that the transformer terminal voltage $V_{tr}(t')$ build up by $1+\rho_{tr}=1.9$ times of steepness from that of the arrester terminal voltage $V_{ar}(tr)$ during the initial time of up to $2T=0.4\mu s$.

Furthermore, V_{ar} continues to increase the magnitude until the interval of $t=T_{front}=1.2\,\mu s$ and reaches the maximum residual voltage, so that $V_{tr}(t')$ also continues to increase the magnitude for the interval of wave front $1.2\,\mu s$, while the waveform would become oscillatory mode after $t>2T=0.4\mu s$ because the negative reflection waves soon come back from the arrester point. As the result, the transformer terminal voltage $V_{tr}(t)$ could become totally larger magnitude than the arrester terminal voltage V_{ar} and be oscillatory mode by almost doubled steepness.

The above results indicate that the overvoltage stress to the transformer may be more severe than the arrester's protective level because of the separation effect by large destance l or by the large traveling time $T=1/u$. On the contrary, if the distance l is small, such a severe stress would not appear at the transformer terminal, because the negative reflected waves would soon come back from the arrester point and the V_{tr} and V_{ar} become almost the same.

In practical engineering, the arresters installed very close to the junction tower of a transmission line are quite meaningful as gateway barriers to protect entire substations against lightning surges from the transmission lines. However, these arresters may not be able to protect properly the transformers or any other facilities because of the separation effects.

This is the reason why important transformers or other equipment (including cables) at large stations are preferably protected by exclusively and closely installed **bespoke arresters**. Such arresters for individual transformers would also be very effective at reducing the overvoltage stresses caused by repeated switching surges or rare cases of direct lightning strikes on the station.

Figure 21.8 shows a typical example of GIS for a 500 kV out-door substation with double bus system, the breakers of which are of one-point-breaking type. The arresters are installed at each main transformer terminals as well as at transmission line feeding points.

21.4.3 Station protection by OGWs, and grounding resistance reduction

21.4.3.1 *Direct lightning strike on the substation*

Surge arresters are generally installed at the gateway point of the substation very close to the first tower of each feeding transmission line and well protect the substation against impinging lightning surges from the transmission lines. However, direct lightning strikes on the substation may be

Courtesy of TEPCO

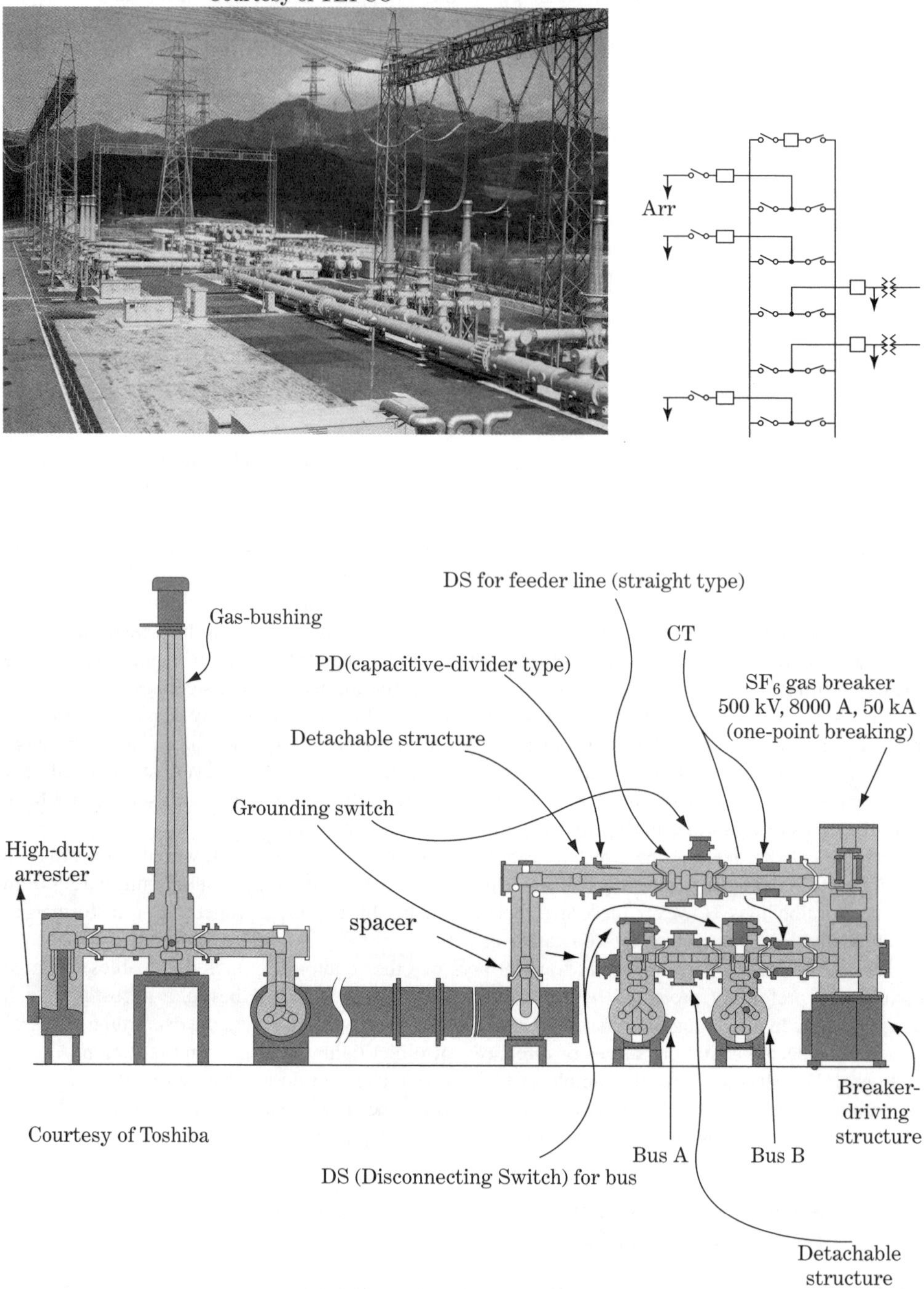

Figure 21.8 GIS substation (500 kV, 8000 A, 63 kA, double bus system)

possible. We cannot prevent such occasions, but can and need to minimize the probability of a strike on the station, or at least protect internal insulation failure (inner insulation failure) as much as possible.

Direct lightning strike on a substation would cause more severe aspects in comparison with a lightning strike on a transmission line, although the probability of occurrence may be smaller.

In case of lightning striking a transmission line, first, arcing horns would limit the magnitude of the surge voltage. Secondly, attenuation of the travelling waves before arrival at the substation would be expected for most cases. The arresters installed at the tower junction point and other points in the substation would appropriately protect the substation and the station apparatus.

On the contrary, the situation of a direct strike on the substation is quite different:

- It is impossible to protect perfectly external (air) insulation failure. Furthermore, the striking point and its aspects cannot be anticipated or limited because the physical configuration as well as electric circuit condition (distribution of surge impedance, for example) are quite complicated.
- Attenuation through long-line travelling cannot be expected.
- Various transition points and surge impedances exist within a narrow area, so unexpected voltage enlargement like the separation effect of arresters may occur.
- The surge energy may exceed the arrester's duty, which probably means cascade failure of the arrester or insulation parts of other equipment.

21.4.3.2 OGWs in station area

The major purpose of OGWs in the area of the substation is to reduce the probability of direct lightning strike on the substation as much as possible, although they cannot be a perfect countermeasure to protect a direct strike on the conductors. The OGWs can reduce the probability of a direct strike on the conductors by the shielding effect, first. Through and ample OGWs can reduce the probability of inverse flashover, second, because the surge impedance is also reduced or the stricken current would be bypassed.

21.4.3.3 Reduction of station grounding resistance and surge impedance

The reduction of the surge impedance at the substation is a vitally important countermeasure to protect or to reduce every kind of surge stress, in particular against direct lightning strike on the substation as well as switching surges and travelling surges from the lines. The reduction of **station surge impedance** is the most important step to reduce the surge impedance Z_2 in Figure 21.4 against incidental travelling surges, for example.

As a numerical check, assuming $Z_1 = 300\Omega$ as the line surge impedance, then the arrester terminal voltages are

$$\begin{aligned}\frac{2Z_2}{Z_1+Z_2}E &= 0.5\,E\,(\text{for } Z_2 = 100\Omega)\\ &= 0.28\,E\,(\text{for } Z_2 = 50\Omega)\\ &= 0.18\,E\,(\text{for } Z_2 = 30\Omega)\end{aligned} \tag{21.3a}$$

The above check clearly indicates the large effect of reducing surge impedance at the station.

In the case of a direct strike on the OGW in the substation, we can apply correspondingly the Equation 20.6 in Figure 20.10 and

$$R_{eff} = \frac{1}{\dfrac{1}{Z_0} + \dfrac{1}{Z_{OGW}} + \dfrac{1}{Z_2}} \tag{21.3b}$$

where Z_0: surge impedance of the lightning surge itself ($Z_0 \fallingdotseq 400\Omega$)
Z_{OGW}: total equivalent surge impedance of all the OGW wires
Z_2: surge impedance of the substation

Therefore, R_{eff} can obviously be reduced by a reduction in the station surge impedance Z_2, and accordingly back-flashover would also be reduced.

The resistive ohm value of the substation ground system should also be kept within a specified value for human safety. If ground resistance $R = 1\,\Omega$ and $I_g = 1000\,A$ are assumed, the induced voltage on the earth conductor could be $V_g = 1\,kV$, which is too large for the human body to withstand.

Typical practices for the ground system in substations involve **ground pilings or pipes/rods, ground conductors**, **ground meshes**, **ground mats** or a combination of these, and the resistive ohm values are designed to stay within specified values, say 0.5–1Ω by which the surge impedance of the station is also reduced.

Incidentally, some apparatus may be adopted in practical engineering where the external insulation (mostly the bushings) is designed to be a little weaker than the internal insulation in order to avoid internal faults.

21.5 Insulation Coordination Details

We have reached the stage where we can study the details of insulation coordination, taking all the above into account, and including material in previous chapters.

21.5.1 Definition and some principal matters of standards

21.5.1.1 The definition

First of all, the definitions of '**insulation coordination**' by the IEEE and IEC are as follows:

- The selection of insulation strength consistent with expected overvoltages to obtain an acceptable risk of failure (IEEE 1313.1-1996, Standard for insulation coordination).
- The selection of the dielectric strength of equipment in relation to the voltages which can appear on the system for which the equipment is intended and taking into account the service environment and the characteristics of the available protective devices (IEC 71-1, 1993, Insulation coordination).

The meanings of both definitions are the same, although the expressions are different. Needless to say, any other existing national standards (ANSI, JEC, etc.) would have definitions with the same meanings.

Table 21.1 summarizes the criteria in regard to the principal goals of insulation coordination which are probably recognized as common worldwide, and of course is a more concrete expression of the above definitions by the IEEE and IEC as well as by any other standards body with the same engineering expertise.

'Standards' may generally be a kind of engineering consensus or practical policy guidelines for industrial applications by their nature, but they are substantially based on expert detailed theories and technical facts obtained from a great deal of engineering experience in the field of practical application. This is the reason why all the standards in regard to power system insulation are almost the same for principal matters, although the terminology may differ. Readers should appreciate our intention to introduce 'the worldwide consensus and its technical background' instead of the quoted items from the IEEE or IEC standards, although the terminology and the figures are quoted mainly from these two representative standards.

21.5.2 Insulation configuration

Some important definitions for insulation configuration from the IEEE and IEC standards are as follows:

Insulation configuration:

- The complete geometric configuration of the insulation, including all elements (insulating and conducting) that influence its dielectric behaviour. Examples of insulation configurations are **phase-to-ground insulation**, **phase-to-phase insulation** and **longitudinal insulation** (IEEE 1313, similarly IEC 71-1).

Longitudinal insulation:

- An insulation configuration between terminals belonging to the same phase, but which are temporarily separated into two independently energized parts (open-switch device) (IEEE).
- An overvoltage that appears between the open contact of a switch (IEC).

External insulation:

- The air insulation and the exposed surfaces of solid insulation of equipment, which are both subject to dielectric stresses of atmospheric and other external conditions such as contamination, humidity, vermin, etc. (IEEE).
- The distances in atmospheric air, and the surfaces in contact with atmospheric air of solid insulation of the equipment which are subject to dielectric stresses and to the effects of atmospheric and other external conditions, such as pollution, humidity, vermin, etc. (IEC).

Internal insulation:

- Internal insulation comprises the internal solid, liquid, or gaseous elements of the insulation of equipment, which are protected from the effects of atmospheric and other external conditions such as contamination, humidity, and vermin (IEEE, IEC similarly).

Self-restoring insulation:

- Insulation that completely recovers its insulating properties after a disruptive discharge caused by the application of a test voltage; insulation of this kind is generally, but necessarily, **external insulation** (IEEE).

Table 21.1 The basic concept of insulation design criteria (general consensus)

	Overhead transmission lines			Substation and substation equipment		
	Lightning overvoltages	**Switching overvoltages**	**Temporary overvoltages**	**Lightning overvoltages**	**Switching overvoltages**	**Temporary overvoltages**
Matters of policy	*Insulation failures are unavoidable, but should be within reasonable failure rate *Flashover passes should be limited only through arcing horns so that damage to conductors or insulators is avoided	*Insulation failure should be entirely avoided without exception		*Insulation failures caused by lightning strikes to the overhead lines should be entirely avoided *Lightning strikes directly to the stations may be unavoidable, but countermeasures to minimize the probability of insulation failure, and in particular to protect internal insulation, are required as much as possible	*Insulation failures should be entirely avoided	
				*Power cables have to be entirely protected		
Major countermeasures	*OGW *Arcing horns *Tower impedance reduction	*Mainly dependent by the countermeasure at the substation. *The gap lengths of arcing horns should be set to avoid discharge by the switching surges or by any temporary overvoltages.		*OGW *Arresters *Station neutral grounding system (meshes, counterpoise)	*Arresters (gap-less) *Breakers (resistive tripping/ closing) *Station neutral grounding system (meshes, counterpoise)	*Shunt reactors and capacitors *Tap-changing transformers *AVR, AQR, V–Q control (local) * Load dispatching (total system) * V–Q control (total system)
Principles for withstanding voltages	*The discharge probability function of a self–restoring insulation based on a Gaussian cumulative frequency distribution is adopted. In typical practice, the voltages of nominalized deviation 3σ value based on the 50% discharge voltages are applied			*Station equipment has to be guaranteed by the withstanding insulation levels specified by the authorized standards such as in Tables 3.2A, B, C		

- Insulation which completely recovers its insulating properties after a disruptive discharge (IEC).

Non-self-restoring insulation:

- An insulation that loses its insulating properties or does not recover them completely after disruptive discharge caused by the application of a test voltage; insulation of this kind is generally, but not necessarily, **internal insulation** (IEEE).
- Insulation which loses its insulating properties, or does not recover them completely, after a disruptive discharge (IEC).

21.5.3 Insulation withstanding level and BIL, BSL

We have studied in the previous sections various kinds of overvoltages and the countermeasures to mitigate them or to protect insulation, in particular to protect internal insulation. Taking all this into consideration, concepts of **insulation strength** and **insulation-withstanding level**, and, furthermore, clear guidelines for the latter, that is **standard withstand voltages** or **standard insulation levels**, are introduced.

The guideline criteria have to be able to offer the important role of **the measures for insulation strength and withstanding level** at least for the following four important engineering procedures:

a) Measures **to determine or to select entire levels of insulation strength** of transmission lines and substations belonging to the same operating voltages of the individual power system.

b) Measures **to mitigate the actual possible overvoltage stresses** within the selected insulation levels.

c) Measures **to specify the required insulation strength** of individual equipment or facilities.

d) Measures **to prove the specified insulation-withstanding strength** by which required insulation levels of equipment or facilities can be tested and guaranteed.

The insulation strength or insulation withstanding voltage levels are expressed in terms of three representative categories of overvoltages, namely BIL, BSL and **the highest power frequency voltages,** as principal concepts of insulation coordination.

The IEEE definitions of BIL and BSL are as follows:

- **BIL** (Basic Lightning Impulse Insulation Level): The electrical strength of insulation expressed in terms of the crest value of a standard lightning impulse under standard atmospheric conditions. BIL may be expressed as either statistical or conventional.
- **BSL** (Basic Switching Impulse Insulation Level): The electrical strength of insulation expressed in terms of the crest value of a standard switching impulse. BSL may be expressed as either **statistical** or **conventional**.
- **Conventional BIL**: The crest value of a standard lightning impulse for which the insulation shall not exhibit disruptive discharge when subjected to a specific number of applications of this impulse under specified conditions, applicable specifically to non-self-restoring insulations.

- **Conventional BSL**: The crest value of a standard switching impulse for which the insulation does not exhibit disruptive discharge when subjected to a specific number of impulses under specified conditions, applicable to non-self-restoring insulations.

21.5.4 Standard insulation levels and their principles

Tables 21.2 A, B, C show the standard withstand voltages for power systems specified by IEEE 1313 and IEC 71-1.

The IEEE as well as IEC standards are divided into the following two parts:

a) IEEE, for the specified standard highest voltages:

- Class I: Medium (1–72.5 kV) and high (72.5–242 kV) voltages
- Class II: EHV and UHV: 242 kV

b) IEC, for specified highest voltages for equipment:

- Range I: Above 1 kV to 245 kV
- Range II: Above 245 kV

The contents as well as the background reasons of both standards can be said to be the same, although the terminologies are slightly different. Therefore, we show the above Class I by the IEEE and Range I by the IEC together in the same Table 21.2A. Tables 21.2B and C are their standards for voltages above 245 kV.

The standard withstanding BIL and BSL of equipment have to be proved also by overvoltage tests with the same specified wave shapes. The IEC and IEEE definitions for **the standard wave shapes for BIL and BSL tests** are the same, and are given in Table 21.2C:

- **Standard lightning impulse: 1.2/50 μs.**

  The wave shape having a time to peak of 1.2 μs and a time to half-value of 50 μs.

- **Standard switching impulse: 250/2500 μs.**

  The wave shape having a time to peak of 250 μs and a time to half-value of 2500 μs.

It is fair to say that the above outlines of insulation coordination by the IEC and IEEE standards including Tables 21.2A–D are almost the same as each other and it is because the above described outlines are based on theory and facts that they have been recognized as the worldwide consensus. Any other national standards or recommendations would actually also be equivalent.

21.5.5 Comparison of insulation levels for systems under and over 245 kV

In Table 21.2A for the systems under 245 kV, standard short-duration power frequency withstand voltages as well as standard lightning impulse withstand voltages are assigned. On the other hand, in Tables 21.2B and C for systems over 245 kV, standard switching impulse withstand voltages are assigned instead of short-duration power frequency withstand voltages. The reason for this may be explained as follows.

Table 21.2A IEC 71-1 and IEEE 1313: Standard withstand voltages for power systems of up to 245 kV

IEC 71-1, 1993-12 Insulation coordination Part 1: Definitions, principles and rules. Range 1: $1\,\text{kV} < U_m \leq 245\,\text{kV}$					
Highest voltage for equipment U_m		Standard short-duration power frequency withstand voltage		Standard lightning impulse withstand voltage	
IEEE 1313.1-1996 Insulation Coordination – Definitions, principles and rules, Class 1: $15\,\text{kV} < V_m \leq 242\,\text{kV}$					
Maximum system voltage (phase-to-phase) V_m		Low-frequency, short-duration withstand voltage (phase-to-ground)		Basic lightning impulse insulation level (phase-to-phase) BIL	
kV, rms value		kV, rms value		kV, crest value	
IEC	IEEE	IEC	IEEE	IEC	IEEE
3.6		10		20, 40	
7.2		20		40, 60	
12		28		60, 75, 95	
17.5	15	38	34	75, 95	95, 110
24	26.2	50	50	95, 125, 145	150
36	36.2	70	70	145, 170	200
52	48.3	95	95	250	250
72.5	72.5	140	95	325	250
			140		350
			140		350
123	121	(185)	185	450	450
		230	230	550	550
		(185)		(450)	
145		230	230	550	450
	145	275	275	650	550
			325		650
		(230)	230	(550)	550
170	169	275	275	650	650
		325	325	750	750
		(275)	275	(650)	650
		(325)	325	(750)	750
245	242	360	360	850	825
		395	395	950	900
		460	480	1050	950, 1050

21.5.5.1 *Systems under 245 kV (Table 21.2A)*

1a) There is a tendency for TOVs to be quite high. In particular, many lower voltage systems are non-effective neutral grounding systems, so the **ground–fault factor** (the ratio of the highest power frequency voltage on an unfaulted phase during a line-to-ground fault to the phase-to-ground power frequency voltage without the fault) is generally large.

Table 21.2B IEC 71-1: Standard withstand voltages for power system of over 245 kV, Range 2: $U_m > 245$ kV

	Standard switching impulse withstand voltage			
Highest Voltage for equipment U_m kV (rms value)	**Longitudinal insulation kV (peak value)**	**Phase-to-earth kV (peak value)**	**Phase-to-phase ratio to the phase-to-earth peak value**	**Standard Lightning impulse withstand voltage kV (peak value)**
300	750	750	1.5	850, 950
		850		950, 1050
362	850	850	1.5	950, 1050
		950		1050, 1175
420	850	850	1.6	1050, 1175
	950	950	1.5	1175, 1300
		1050	1.5	1300, 1425
525	950	950	1.7	1175, 1300
		1050	1.6	1300, 1425
		1175	1.5	1425, 1550
765	1175	1300	1.7	1675, 1800
		1425	1.7	1800, 1950
		1550	1.6	1950, 2100

Furthermore, especially in systems under 36 kV, various temporary and irregular electrical conditions (e.g. neutral terminal opening, conductor crossing from higher voltage systems) as well as mechanical damage to network facilities are apt to occur so that some reasonable margin for a.c. voltages may be required for social and security purposes. These are the reasons why the standard short-duration power frequency withstand levels are assigned relatively larger values.

1b) Switching impulse withstand levels (BSL) have no meaning in this class because they can be covered by BIL. Next, the ratio of $\mathrm{BIL}/\{(\sqrt{2}/\sqrt{3})U_m\}$ for this voltage class is larger than that for the higher voltage class of Tables 21.2B and C.

As a numerical check of $\mathrm{BIL}/\{(\sqrt{2}/\sqrt{3})U_m\}$:

$$\text{Table 21.2A}: 200/\{(\sqrt{2}/\sqrt{3})36\} = 6.80, \quad 325/\{(\sqrt{2}/\sqrt{3})72.5\} = 5.49,$$
$$850/\{(\sqrt{2}/\sqrt{3})245\} = 4.25$$
$$\text{Tables 21.2B, C}: 850/\{(\sqrt{2}/\sqrt{3})300\} = 3.47, \quad 1300/\{(\sqrt{2}/\sqrt{3})525\} = 3.03,$$
$$1800/\{(\sqrt{2}/\sqrt{3})800\} = 2.76$$

In other words, BILs of class 1 systems are relatively high, so BSLs must be relatively high. On the other hand, switching overvoltages are obviously proportional to the system operating voltages so that switching surges would be lower for the systems of this class. These are the reasons why BSL is omitted for the lower voltage systems of class 1.

Table 21.2C IEEE 1313: Standard withstand voltages for power systems of over 242 kV, Class 2: $V_m > 242$ kV.

Maximum system voltage (phase-to-phase) V_m kV, rms	Basic switching impulse insulation level (phase-to-ground) BSL kV, peak	Basic lightning impulse insulation level (phase-to-ground) BIL kV, peak
362	650	900
	750	975
	825	1050
	900	1175
	975	1300
	1050	
550	1300	1175
	1425	1300
	1550	1425
	1675	1550
	1800	
800	1300	1800
	1425	1925
	1550	2050
	1675	
	1800	

T_f: the time-to-crest value (virtual time)
T_t: the time-to-half value (virtual time)

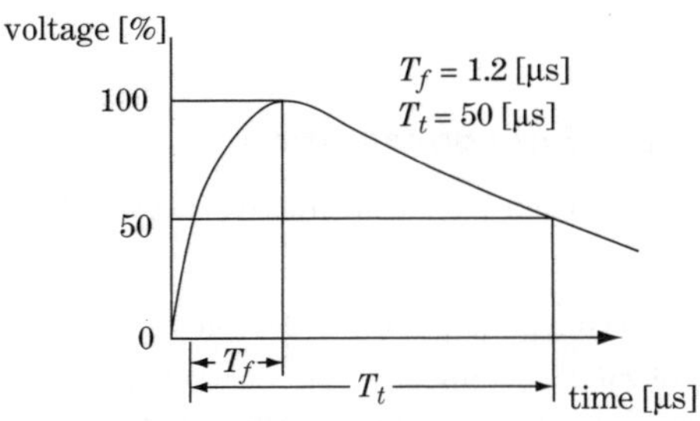

Standard lightning impulse test voltage 1.2/50 impulse

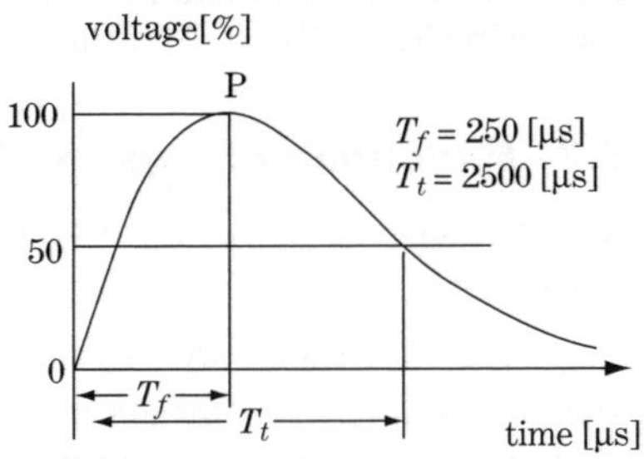

Standard switching impulse test voltage 250/2500 impulse

21.5.5.2 Systems over 245 kV (Tables 21.2B and C)

2a) Low-frequency short-duration withstand voltage is omitted, because, nowadays, all EHV and UHV class systems are treated under the solidly neutral grounding method (probably without exception) so that the ground fault factor is rather small.

2b) On the other hand, BSL is strictly assigned to this class. The reason can be explained by the inverse of item (1b) above. That is, first, BIL has been considerably reduced (due to the advance of arrester properties in particular) in EHV and UHV. Secondly, switching surges are proportional to the operating voltages and so are relatively large in comparison with the insulation levels. These are the reasons why BSL has to be assigned quite important items. In IEC 71-1, phase-to-phase and longitudinal BSL as well as phase-to-earth BSL are also assigned.

As a numerical check of BSL(phase-to-earth)/$\{(\sqrt{2}/\sqrt{3})U_m\}$:

	(i)BIL/$\{(\sqrt{2}/\sqrt{3})U_m\}$	(ii)BSL(phase-to-earth)/$\{(\sqrt{2}/\sqrt{3})U_m\}$
300 kV :	$(850-1050)/\{(\sqrt{2}/\sqrt{3})300\} = 3.47-4.29$,	$(750-850)/\{(\sqrt{2}/\sqrt{3})300\} = 3.06-3.48$
525 kV :	$(1175-1550)/\{(\sqrt{2}/\sqrt{3})525\} = 2.74-4.15$,	$(950-1175)/\{(\sqrt{2}/\sqrt{3})525\} = 2.22-2.74$
800 kV :	$(1800-2050)/\{(\sqrt{2}/\sqrt{3})800\} = 2.76-3.14$,	$(1300-1800)/\{(\sqrt{2}/\sqrt{3})800\} = 1.99-2.76$

This leads to the following:

- The BIL ratio to the operating voltage is remarkably low for EHV and UHV systems, which means the reduction of the insulation level has been realized; this owes much to advanced arrester technology to lower the protective levels.
- The BSL ratio is very close to the BIL ratio. Accordingly, the following points have to be carefully confirmed as essentially important for EHV and UHV systems:

– **EHV and UHV arresters** are required to limit the lightning impulse level to lower values, while withstanding the large switching surge (the capability to absorb large switching surge energy).

– **EHV and UHV breakers** are required to limit the switching surges to within the specified levels. For example, UHV breakers of 500 kV class are required to limit the BSL to within typically 2.5 by adopting resistive closing/tripping.

– Of course, all the other **EHV and UHV station equipment** has to guarantee the insulation levels specified in Tables 21.2B and C, including BSL.

21.5.5.3 Evaluation of degree of insulation coordination

The degree of coordination can be measured or evaluated by the **protective ratio (PR)** (IEEE PC62.22). That is,

$$\text{PR} = (\text{insulation withstanding level})/(\text{voltage at protected equipment})$$

or the ratio of the insulation strength of the protected equipment to the overvoltages appearing across the insulation. The **voltage at protected equipment** is equal to **arrester protective level**, if the **separation effect** is insignificant:

$$\text{PR}_l = \text{BIL}/\text{LPL} \qquad (\text{ratio for BIL : acceptable level } 1.15)$$

where LPL is the lightning impulse protective level

$$\text{PR}_s = \text{BSL}/\text{SPL} \qquad (\text{ratio for BIL : acceptable level } 1.2)$$

where SPL is the switching impulse protective level

The denominators of the above ratios are expected largest voltages possible. The insulation coordination of a power system can be evaluated by deriving the PR of individual lines and equipment by detailed overvoltage analysis.

Note that, besides BIL and BSL, a **chopped-wave overvoltage test** may be additionally introduced as an optional requirement for special conditions.

The standard chopped-wave impulse voltage shape is a standard lightning impulse that is intentionally interrupted at the tail by sparkover of a gap or other chopping equipment. Usually the time to chop is 2–3 μs. (Further details are omitted as the matter is beyond this book.)

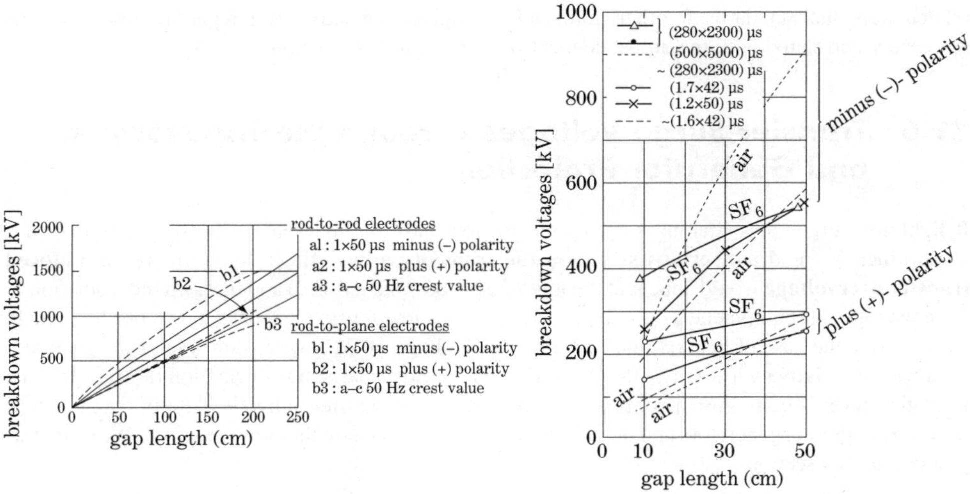

(a) Lightning impulse and a.c. flashover characteristics (P. Jacottet and W. Weicker, ETZ 1940)

(b) SF_6 gas flashover characteristics (T. Takuma, IEEE 1972)

Figure 21.9 Flashover characteristics (typical example) ((b) courtesy of Mr T. Takuma)

21.5.5.4 Breakdown voltage characteristics

All the transmission lines as well as station facilities and equipment have to be designed to satisfy the applied standard insulation levels. For this purpose, breakdown voltage characteristics under various conditions are required as **essential data of insulation characteristics**. The **breakdown characteristics** of any insulation materials (air, gas, oil, paper, porcelain, etc.) are affected by the **shapes of electrodes** and the **atmosphere** (pressure, humidity, temperature, etc.) so a huge amount of valuable experimental data has been accumulated from the past. Figure 21.9 is an example showing the breakdown characteristics of air and SF_6 gas.

21.5.5.5 Cable insulation

The power cables contain non-self-restoring insulation, so any breakdown of cable insulation would require extensive outage time for repairs at a high cost. Therefore insulation failure of power cables for network lines as well as for power station use should be avoided.

Cable circuits have a low surge impedance of typically 40–50 Ω so that surges coming from overhead lines will be reduced significantly at the line–cable junction. However, switching surges originating in the vicinity of cable lines at a substation may be largely reflected at the junction point to the transmission lines (see Section 18.7). Switching surge phenomena in the substation area are generally quite complicated. Metal–oxide gap-less arresters can provide excellent cable protection, while the arrester is required to absorb the relatively large thermal energy $\left(\frac{1}{2}CV^2\right)$ caused by the high-frequency oscillatory overvoltages.

Regardless, insulation coordination of the power system network including cable lines of large capacitance have to be carefully examined. The overvoltage behaviour of cable lines will be discussed again in Section 23.6.

Today, all the facilities of the primary circuit of a power system (transmission lines, power cable lines, power station/substation equipment and related configurations) are planned, specified, designed, manufactured, tested, installed and operated based on the insulation coordination standards and the related individual standards. In other words, all the engineering activities in regard to power systems are deeply correlated with the above concept of insulation coordination.

21.6 Transfer Surge Voltages Through the Transformer, and Generator Protection

If lightning surges or switching surges were to flow into the high-tension bushing terminal of a transformer, it could put **serious stress on the transformer windings**. Furthermore, an **induced transfer overvoltage** would appear at the low-tension bushing terminal and threaten the insulation of lower voltage side equipment. This is a serious problem in low-tension-side insulation coordination. In particular, in the case of generating plant, the rated voltage of the low-tension side of the generator terminals are relatively low (say, 10–30 kV) (in other words, a large transformation ratio) so that the insulation level is quite low in comparison with that of the high-tension side. Appropriate countermeasures to protect generators and other low-voltage equipment are therefore required. We study this problem in this section.

21.6.1 Electrostatic transfer surge voltage

The transformer we studied in Chapter 5 just had the characteristics of the power frequency based on inductance L of the coil windings, neglecting stray capacitances C. We need to consider that the behaviour of the transformer in the surge frequency zone is dominated by stray capacitance C instead of inductance L of the coil windings. Accordingly, we need to establish some equivalent circuit of the transformer based on capacitance C.

21.6.1.1 Equivalent circuit (single phase transformer)

Figure 21.10a shows a typical equivalent circuit for high-frequency phenomena. This is a distributed circuit with elemental capacitances from the high-tension (HT) coil to the low-tension (LT) coil as well as that from the LT coil to grounding earth, where the capacitances from the HT coil to grounding earth are not written because they can be treated as part of the connected HT outer circuit. As we are not applying powerful computers in this book, we need to simplify Figure 21.10a to 21.10b where the distributed capacitances are concentrated by C_{12} (total capacitance from the HT to LT coil) and C_2 (total capacitance from the LT coil to grounding earth). Then we examine two different cases in Figure 21.10b:

- Circuit 1 (switch S_1 closed, S_2 open): This is the case where terminals m and n are connected together, while the earth terminal is open. In the figure, the scale x shows the distributed coil position

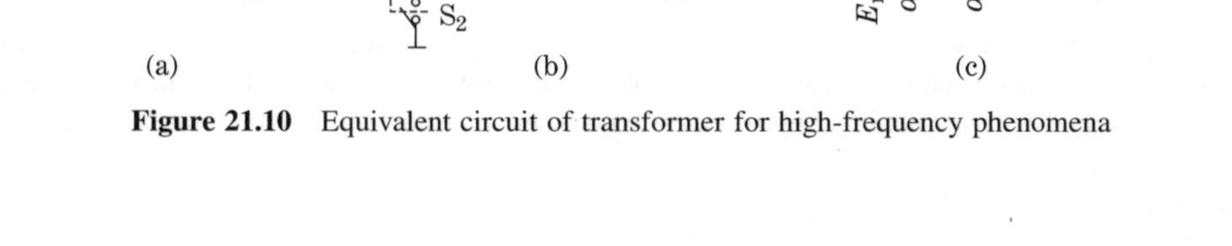

Figure 21.10 Equivalent circuit of transformer for high-frequency phenomena

being measured from point n to point m. When the surge voltage travelling from the HT line arrives at the connected points m and n as the magnitude of E, the surges then rush into the primary coil simultaneously from m and n and frontally meet point k so that the voltage at k obtains the largest value $E_{max}(=2E$ if attenuation is neglected). The initial voltage distribution in the HT coil is shown as curve a in Figure 21.9c. The averaged value of the distributed voltage can be written as $E=\alpha E$, where $\alpha=1.4-1.6>1.0$.

- Circuit 2 (switch S_1 open, S_2 closed): This is the case where terminal n is connected only to the earth and the surge E is injected at point m. The surge voltage E travels through the HT coil from point m to n so that the initial voltage distribution is as curve b in Figure 21.9c. The averaged value of the distributed voltage can be written as $E=\alpha E$, where $\alpha=0.5-0.7<1.0$.

As a result, we can presume that the averaged voltage αE is injected at point k of circuits 1 and 2 as follows:

$$\left.\begin{array}{l}\text{for circuit 1}:\alpha=1.4-1.6\\ \text{for circuit 2}:\alpha=0.5-0.7\end{array}\right\} \tag{21.4}$$

21.6.1.2 Calculation of electrostatic transfer surge voltage (single phase transformer)

Our problem is shown in Figure 21.11a, where surge voltages $E_m(t)$ and $E_n(t)$ arrive at terminals m and n simultaneously, and the equation for deriving the electrostatically induced voltage at the LT coil terminal is as follows:

$$\left.\begin{array}{l}{}_1v_m=E_m\\ {}_1v_n=E_n\end{array}\right\} \tag{21.5a}$$

Now we can transform the set voltages E_m and E_n into the line-to-ground travelling wave $E'(t)$ and line-to-line travelling wave $E''(t)$, and Figure 21.11a can be divided into the Figures 21.11b and c:

$$\left.\begin{array}{l}E_m=E'+E''\\ E_n=E'-E''\end{array}\right\}① \quad \left.\begin{array}{ll}\text{line-to-ground travelling wave} & E'=\frac{1}{2}(E_m+E_n)\\ \text{line-to-line travelling wave} & E''=\frac{1}{2}(E_m-E_n)\end{array}\right\}② \tag{21.5b}$$

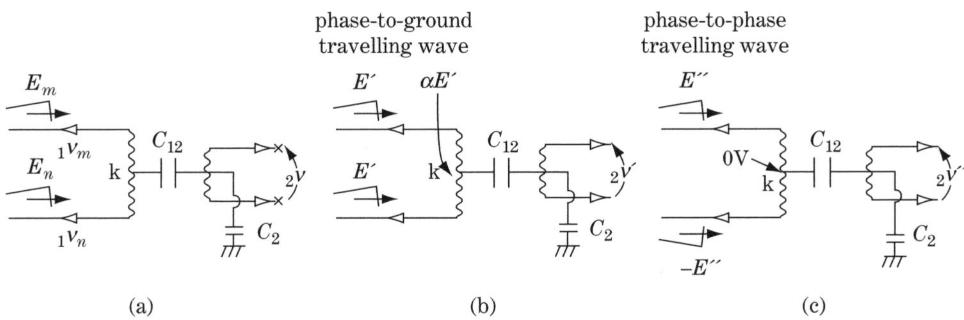

Figure 21.11 Transfer voltage from HT coil to LT coil

For Figure 21.11b, we can quote the result of circuit 1: that is, voltage $\alpha E'$ is charged at point k:

$$ {}_2v' = \frac{C_{12}}{C_{12}+C_2} \cdot \alpha E' = \frac{\alpha}{2} \cdot \frac{C_{12}}{C_{12}+C_2}(E_m + E_n) \tag{21.6a} $$

For Figure 21.11c, voltages E'' and $-E''$ are injected at points m and n respectively, so the frontal voltage at k becomes zero. In other words, this is a special case of $\alpha = 0$ on circuit 2. Namely,

$$ {}_2v'' = \frac{C_{12}}{C_{12}+C_2} \cdot 0 = 0 \tag{21.6b} $$

Thus the solution of Figure 21.11a is derived as the addition of the results by Figures 21.11b and c. Namely,

$$ {}_2v = {}_2v' + {}_2v'' = \frac{C_{12}}{C_{12}+C_2} \cdot \alpha E' = \frac{\alpha}{2} \cdot \frac{C_{12}}{C_{12}+C_2}(E_m + E_n) \tag{21.6c} $$

21.6.1.3 Calculation of electrostatic transfer surge voltage (three-phase transformer)

Figure 21.12 is a typical connection diagram of a main transformer for thermal or hydro-generating plants.

The surge voltages $E_a(t)$, $E_b(t)$, $E_c(t)$ arrive simultaneously at the HT terminal bushings. Our problem is to calculate the transfer surge voltages induced by the generator-side LT terminal bushings.

We can quote the result of Equation 21.6 for the initial transfer surge voltage induced at the LT side:

$$ \left.\begin{aligned} &\text{phase a} \quad {}_2v_a = \frac{\alpha}{2} \cdot \frac{C_{12}}{C_{12}+C_2}(E_a + {}_1v_n) \\ &\text{phase b} \quad {}_2v_b = \frac{\alpha}{2} \cdot \frac{C_{12}}{C_{12}+C_2}(E_b + {}_1v_n) \\ &\text{phase c} \quad {}_2v_c = \frac{\alpha}{2} \cdot \frac{C_{12}}{C_{12}+C_2}(E_c + {}_1v_n) \end{aligned}\right\} \tag{21.7} $$

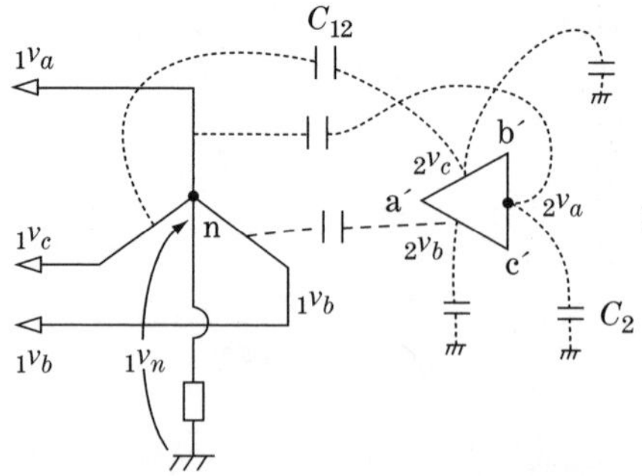

Figure 21.12 Stray capacitance between HT and LT windings

where ${}_1v_n$: surge voltage at the neutral point n
or by symmetrical components

$$\left.\begin{aligned}
{}_2v_0 &= \frac{1}{3}({}_2v_a + {}_2v_b + {}_2v_c) \\
&= \frac{\alpha}{2}\cdot\frac{C_{12}}{C_{12}+C_2}(E_0 + {}_1v_n) \\
{}_2v_1 &= \frac{\alpha}{2}\cdot\frac{C_{12}}{C_{12}+C_2}\cdot E_1 \\
{}_2v_2 &= \frac{\alpha}{2}\cdot\frac{C_{12}}{C_{12}+C_2}\cdot E_2
\end{aligned}\quad \text{where} \quad
\begin{bmatrix} E_0 \\ E_1 \\ E_2 \end{bmatrix} = \frac{1}{3}\begin{bmatrix} 1 & 1 & 1 \\ 1 & a & a^2 \\ 1 & a^2 & a \end{bmatrix}\cdot\begin{bmatrix} E_a \\ E_b \\ E_c \end{bmatrix}\right\} \tag{21.8}$$

Equations 21.7 and 21.8 are the resulting general equations giving the transfer surge voltages appearing at the bushings of the LT side, while the LT bushing terminals are in the open condition.

By applying the above derived equations, the equations for calculating the transfer voltages under different terminal conditions can be derived, which are summarized in Table 21.3.

Regarding the derivation processes shown in Table 21.3, the neutral point n should be carefully treated as a surge transition point. The transmittal wave coefficient at point n is obviously different for each case. For example, the coefficient of the voltage, v_n at the neutral point n is 2 for case 5, 2/3 for case 6, 4/3 for case 7, and so on, as explained in the table.

We can find the equation for calculating the transfer surge voltages appearing on the LT side under various different conditions on the HT voltage side using the above general equation and Table 21.3. The resulting equations for the seven cases can be summarized as the general equation below.

The transfer surge voltage from the HT to the LT side (the LT side is in open mode) is

$$ {}_2v = k\alpha\cdot\frac{C_{12}}{C_{12}+C_2}E \tag{21.9}$$

where α : given by Equation (21.4)
k : transfer coefficient given in Table 21.3

For example, for case 6 phase a, $k = 5/6$, and for phase b, $k = 2/6$, with

$$k\alpha\cdot\frac{C_{12}}{C_{12}+C_2} \ : \text{transfer voltage ratio}$$

Incidentally, the transfer voltages for each case with $C_{12} = 6000\,\text{pF}$, $C_2 = 3000\,\text{pF}$, $\alpha = 0.5$ (for the solidly neutral grounding system) and $\alpha = 1.5$ (for the neutral ungrounded system) are shown in the table as supplemental references. In the case of the high-resistive neutral grounding system, α is initially between 1.5 and 0.5, and soon (after the wavefront passes) becomes very close to 0.5.

Let us consider a trial calculation for a 275 kV class power station, with main transformer 275 kV/24 kV, y–Δ windings, solidly neutral grounding system. Then

$$C_{12} = 6000\,\text{pF},\quad C_2 = 3000\,\text{pF},\quad \alpha = 0.7,\quad k = \frac{1}{2}$$

From Equation 21.9

$$ {}_2v_a = \frac{\alpha}{2}\cdot\frac{6000}{6000+3000}E = \frac{1}{3}\alpha E = 0.23E \tag{21.10}$$

Table 21.3 Electrostatic transfer voltage (calculation formula)

	Surge voltage (conditions)	Transfer voltage from HT side to LT side (phase voltages, neutral voltage) $v = k\alpha \cdot \frac{C_{12}}{C_{12}+C_2}E$	Transfer voltage $\left(C_{12} = 6000[\text{pF}],\ C_2 = 3000[\text{pF}],\ k = \frac{C_{12}}{C_{12}+C_2} = \frac{2}{3}\right)$
Case 1 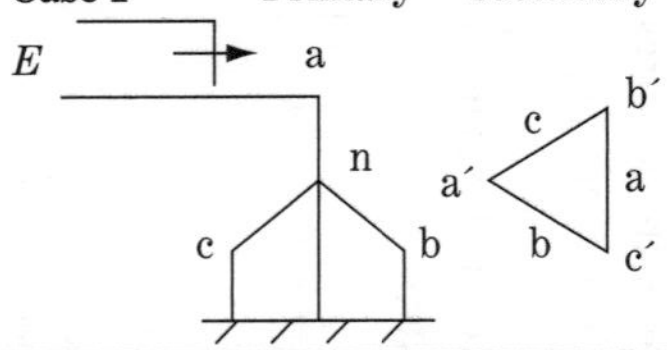	${}_1v_a = E,\ {}_1v_b = {}_1v_c = 0$ ${}_1v_n = 0$	${}_2v_a = \frac{\alpha}{2}\cdot\frac{C_{12}}{C_{12}+C_2}E,$ ${}_2v_b = {}_2v_c = 0$ $\therefore {}_2v_0 = \frac{\alpha}{6}\cdot\frac{C_{12}}{C_{12}+C_2}E$ $(k = 1/2)$	$\alpha = 0.5$ ${}_2v_a = \frac{0.5}{2}\cdot\frac{2}{3}E = \frac{1}{6}E$ ${}_2v_0 = \frac{0.5}{6}\cdot\frac{2}{3}E = \frac{1}{18}E$
Case 2 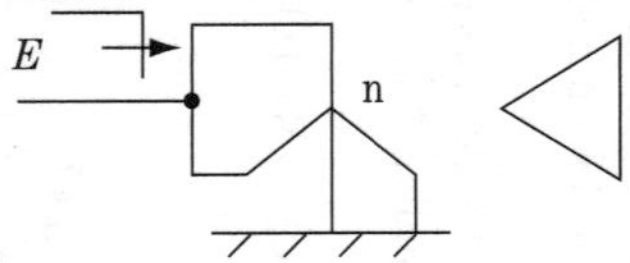	${}_1v_a = {}_1v_c = E,\ {}_1v_b = 0$ ${}_1v_n = 0$	${}_2v_a = {}_2v_c = \frac{\alpha}{2}\cdot\frac{C_{12}}{C_{12}+C_2}E,$ ${}_2v_b = 0$ $\therefore {}_2v_0 = \frac{\alpha}{3}\cdot\frac{C_{12}}{C_{12}+C_2}E$ $(k = 1/2, 1/3)$	$\alpha = 0.5$ ${}_2v_a = {}_2v_c = \frac{1}{6}E$ ${}_2v_0 = \frac{1}{9}E$
Case 3 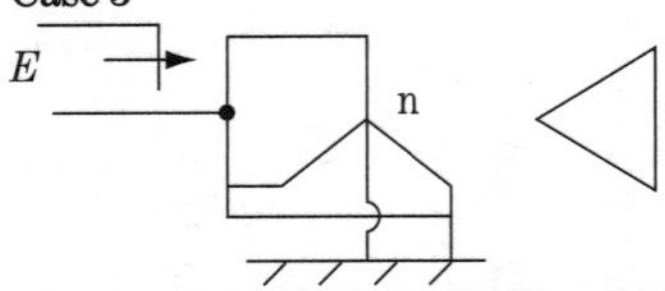	${}_1v_a = {}_1v_b = {}_1v_c = E$ ${}_1v_n = 0$	${}_2v_a = {}_2v_b = {}_2v_c = \frac{\alpha}{2}\cdot\frac{C_{12}}{C_{12}+C_2}E$ $\therefore {}_2v_0 = \frac{\alpha}{2}\cdot\frac{C_{12}}{C_{12}+C_2}E$ $(k = 1/2)$	$\alpha = 0.5$ ${}_2v_a = {}_2v_b = {}_2v_c = \frac{1}{6}E$ ${}_2v_0 = \frac{1}{6}E$
Case 4 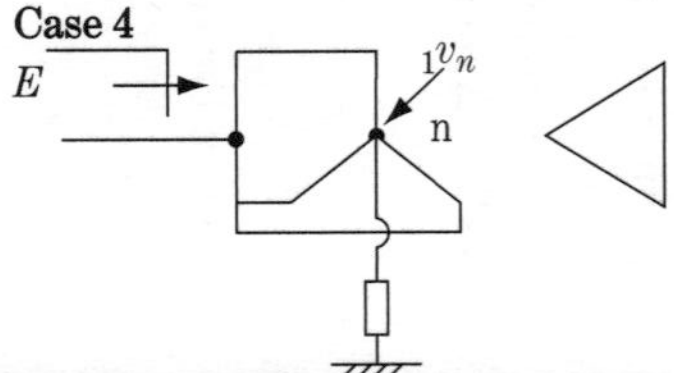	${}_1v_a = {}_1v_b = {}_1v_c = E$ ${}_1v_n$: voltage at point n $(E_n \neq 0)$	${}_2v_a; = {}_2v_b = {}_2v_c$ $= \frac{\alpha}{2}\cdot\frac{C_{12}}{C_{12}+C_2}(E + {}_1v_n)$ $\therefore {}_2v_0 = \frac{\alpha}{2}\cdot\frac{C_{12}}{C_{12}+C_2}(E + {}_1v_n)$ $(k = 1/2)$	early duration: the same as in case 5 later duration: the same as in case 2

Case 5 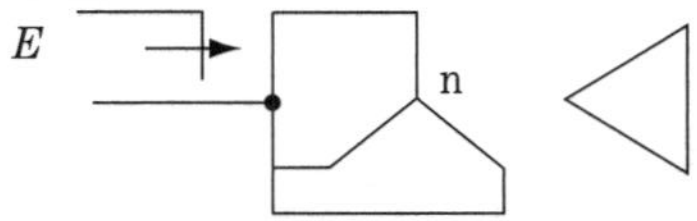	$_1v_a = {_1v_b} = {_1v_c} = E$ $_1v_n = 2E$ reflection factor at point n is 2	$_2v_a = {_2v_b} = {_2v_c} = \frac{3\alpha}{2} \cdot \frac{C_{12}}{C_{12}+C_2} E$ $\therefore {_2v_0} = \frac{3\alpha}{2} \cdot \frac{C_{12}}{C_{12}+C_2} E$ $(k = 3/2)$	$\alpha = 1.5$ $_2v_a = {_2v_b} = {_2v_c} = \frac{3 \times 1.5}{2} \cdot \frac{2}{3} E$ $= 1.5E$
Case 6 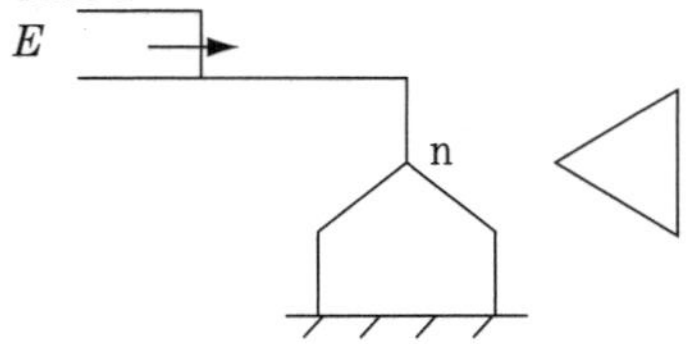	$_1v_a = E, {_1v_b} = {_1v_c} = 0$ $Z_1 : Z_2 = 2:1$ then $_1v_n = \frac{2Z_2}{Z_1 + Z_2} E = \frac{2}{3} E$ Z_1, Z_2: surge impedance before and after point n	$_2v_a = \frac{\alpha}{2} \cdot \frac{C_{12}}{C_{12}+C_2} \left(E + \frac{2}{3}E\right)$ $= \frac{5\alpha}{6} \cdot \frac{C_{12}}{C_{12}+C_2} E$ $_2v_b = {_2v_c}$ $= \frac{\alpha}{2} \cdot \frac{C_{12}}{C_{12}+C_2} \left(0 + \frac{2}{3}E\right)$ $= \frac{2\alpha}{6} \cdot \frac{C_{12}}{C_{12}+C_2} E$ $\therefore {_2v_0} = \frac{\alpha}{2} \cdot \frac{C_{12}}{C_{12}+C_2} E \ (k = 1/2, 5/6, 1/3)$	$_2v_0 = 1.5E$ $\alpha = 1.5$ $_2v_a = \frac{5}{6} E.$ $_2v_b = {_2v_c} = \frac{1}{3} E$ $_2v_0 = \frac{1}{3} E$
Case 7 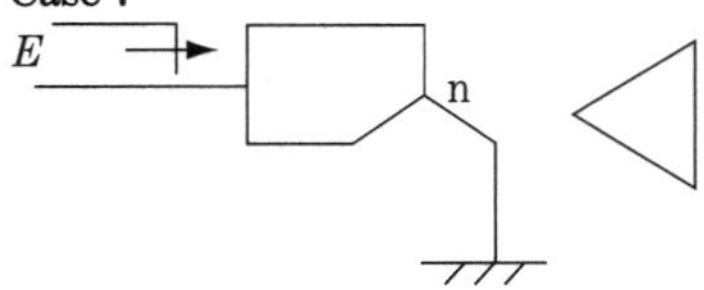	$_1v_a = {_1v_c} = E, {_1v_b} = 0$ $Z_1 : Z_2 = 1:2$ then $_1v_n = \frac{2Z_2}{Z_1 + Z_2} E = \frac{4}{3} E$	$_2v_a = {_2v_c}$ $= \frac{\alpha}{2} \cdot \frac{C_{12}}{C_{12}+C_2} \left(E + \frac{4}{3}E\right)$ $= \frac{7\alpha}{6} \cdot \frac{C_{12}}{C_{12}+C_2} E$ $_2v_b = \frac{\alpha}{2} \cdot \frac{C_{12}}{C_{12}+C_2} \left(0 + \frac{4}{3}E\right)$ $= \frac{4\alpha}{6} \cdot \frac{C_{12}}{C_{12}+C_2} E$ $_2v_0 = \alpha \frac{C_{12}}{C_{12}+C_2} E \quad (k = 1/2, 7/6, 4/6, 1)$	$\alpha = 1.5$ $_2v_a = {_2v_c} = \frac{1}{6} E$ $_2v_b = \frac{2}{3} E$ $_2v_0 = \frac{2}{3} E$

Note: k: coefficient of transfer voltage $k\alpha \cdot \frac{C_{12}}{C_{12}+C_2}$; ratio of transfer voltage α: refer to Equation 21.3.

The crest value of the normal phase voltage of the 275 kV side is $(\sqrt{2}/\sqrt{3}) \cdot 275 = 224.5\,\text{kV}$ and that of the normal phase voltage of the 24 kV side is $(\sqrt{2}/\sqrt{3}) \cdot 24 = 19.6\,\text{kV}$.

Now, assuming a surge of twice the value of 224.5 kV, then ${}_2v_a = 2.0 \times 224.5 \times 0.23 = 103\,\text{kV}$. The derived magnitude is indeed 5.3 times 19.6 kV and certainly threatens the insulation of the generator side without effective countermeasures, although the equation is derived under the LT bushings open condition.

We also need to recognize that each case in the table is often caused by real surge modes. In the case of lightning striking the connected transmission line, the incidental surges E_a, E_b, E_c come together at the station regardless of the fault phases modes, so they include the equal components E which correspond to case 3 or 4 or 5, for example. Another example is switching surges caused by the first pole closing of a breaker, which corresponds to case 1. All the cases shown in Table 21.3 are realistic phenomena.

21.6.1.4 Transfer voltage arriving at the generator terminal

Next, we need to calculate the transfer voltages arriving at the generator terminals when a generator (the surge impedance Z_g) is connected to the transformer. The circuit is shown in Figure 21.13a, where the incidental voltage is $k\alpha \cdot e(t)$ from Equation 21.9.

The circuit equations in the Laplace domain are

$$\begin{aligned} &\{k\alpha \cdot e(s) - {}_2v(s)\} \cdot sC_{12} = i(s) = i_2(s) + i_g(s) \\ &{}_2v(s) = \frac{i_2(s)}{sC_2} = i_g(s) \cdot Z_g \quad (Z_g\ : \text{surge impedance on the generator}) \end{aligned} \tag{21.11}$$

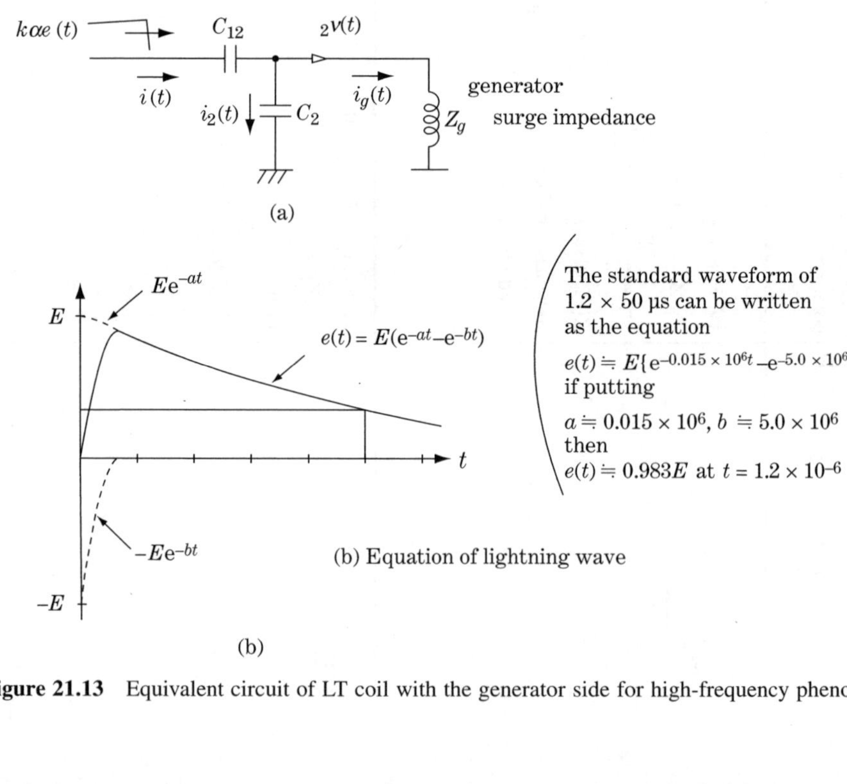

Figure 21.13 Equivalent circuit of LT coil with the generator side for high-frequency phenomena

$$\left.\begin{aligned} \therefore\ {}_2v(s) &= \frac{sC_{12}\cdot k\alpha\cdot e(s)}{s(C_{12}+C_2)+\dfrac{1}{Z_g}} = \frac{C_{12}}{C_{12}+C_2}\cdot\frac{s}{s+\delta}\cdot k\alpha\cdot e(s) \\ \text{where}\quad \delta &= \frac{1}{(C_{12}+C_2)Z_g} \end{aligned}\right\} \tag{21.12}$$

Next, the incidental surge voltage coming from the HT side $e(t)$ may have a waveform similar to the virtual standard waveform, which is expressed by $e(t) = E(\mathrm{e}^{-at} - \mathrm{e}^{-bt})$ as shown in Figure 21.13b.

The incident surge is

$$\left.\begin{aligned} e(t) &= E(\mathrm{e}^{-at} - \mathrm{e}^{-bt}) \\ e(s) &= E\left(\frac{1}{s+a} - \frac{1}{s+b}\right) \end{aligned}\right\} \tag{21.13}$$

We calculate the LT side voltage ${}_2v(s)$, ${}_2v(t)$:

$$\begin{aligned} {}_2v(s) &= k\alpha E\frac{C_{12}}{C_{12}+C_2}\cdot\left\{\frac{1}{s+a}\cdot\frac{s}{s+\delta} - \frac{1}{s+b}\cdot\frac{s}{s+\delta}\right\} \\ &= k\alpha E\frac{C_{12}}{C_{12}+C_2}\cdot\left\{\frac{1}{\delta-a}\left(\frac{\delta}{s+\delta} - \frac{a}{s+a}\right) - \frac{1}{\delta-b}\left(\frac{\delta}{s+\delta} - \frac{b}{s+b}\right)\right\} \\ &= k\alpha E\frac{C_{12}}{C_{12}+C_2}\left\{\frac{(a-b)\delta}{(\delta-a)(\delta-b)}\cdot\frac{1}{s+\delta} - \frac{a}{\delta-a}\cdot\frac{1}{s+a} + \frac{b}{\delta-b}\cdot\frac{1}{s+b}\right\} \end{aligned} \tag{21.14}$$

The transfer surge voltage at the generator bushing terminals is

$$\begin{aligned} {}_2v(t) &= k\alpha E\frac{C_{12}}{C_{12}+C_2}\left\{\frac{(a-b)\delta}{(\delta-a)(\delta-b)}\mathrm{e}^{-\delta t} - \frac{a}{\delta-a}\mathrm{e}^{-at} + \frac{b}{\delta-b}\mathrm{e}^{-bt}\right\} \\ &= k\alpha E\frac{C_{12}}{C_{12}+C_2}\left\{\frac{-\delta}{a-\delta}\mathrm{e}^{-\delta t} + \frac{a}{a-\delta}\mathrm{e}^{-at} - \mathrm{e}^{-bt}\right\}, \quad \text{where } a, \delta \ll b \end{aligned} \tag{21.15}$$

where $a \fallingdotseq 0.015\times 10^6$, $b \fallingdotseq 5.0\times 10^6$ for the virtual standard waveform of $1.2\times 50\,\mu\mathrm{s}$ as the incidental voltage (see Figure 21.11b).

In addition, as is shown in the trial calculation below, δ is smaller than a. Accordingly, $\delta < a \ll b$.

Therefore, at the time around the crest value, $t = 1$–$1.2\,\mu\mathrm{s}$, we can put $\mathrm{e}^{-at}\to 1$, $\mathrm{e}^{-\delta t}\to 0$, $\mathrm{e}^{-bt}\to 0$ and $\delta/(a-\delta)\to 0$.

The transfer surge voltages from the HT side at the generator terminals are

$$\left.\begin{aligned} {}_2v(t) &\fallingdotseq k\alpha E\frac{C_{12}}{C_{12}+C_2}\left\{\frac{a}{a-\delta}\cdot\mathrm{e}^{-at} - \frac{\delta}{a-\delta}\mathrm{e}^{-\delta t}\right\} \\ \text{where}\quad a \gg \delta &= \frac{1}{(C_{12}+C_2)Z_g} \end{aligned}\right\} \tag{21.16a}$$

Therefore

$$\left.\begin{aligned} {}_2v(t) &= k\alpha E\frac{C_{12}}{C_{12}+C_2}\cdot \mathrm{e}^{-at} && \text{①}\\ &\fallingdotseq k\alpha E\frac{C_{12}}{C_{12}+C_2} \quad \text{for} \quad t=0+ && \text{②}\\ \text{where}&\\ \alpha &= 0.5 \text{ for solidly neutral grounded system}\\ &= 1.5 \text{ for neutral ungrounded system}\\ &= \text{between } 0.5 \text{ and } 1.5 \text{ for high-resistive neutral grounded system}\\ k &= 1/3, 1/2, 2/3, 5/6, 1, 7/6 \text{ depending on the transformer neutral}\\ &\quad \text{connection see Table 21.3)}\\ E\,(\text{kVcrest}) &: \text{surge transmitted voltage from HV bushing terminal} \end{aligned}\right\} \quad (21.16\text{b})$$

As a trial calculation for a surge impedance of generators typically $Z_g = 30 - 300\,\Omega$ (Z_g is smaller for generators of larger MVA capacity), putting $C_{12} = 6000\,\text{pF}$, $C_2 = 3000\,\text{pF}(C_{12} + C_2 = 9000 \times 10^{-9}\,\text{F})$ and $Z_g = 30\text{–}300\,\Omega$, then $\delta \fallingdotseq 3000 - 300\,(< a \ll b)$.

Accordingly,

$${}_2v(t) = 0.67k\,\alpha\,E\frac{6000}{6000+3000} = 0.447k\,\alpha E$$

The resulting Equation 21.16b means that the transfer voltages from the HT side cannot be reduced by connection of the generator surge impedance. Appropriate countermeasures to protect the LT-side insulation against the transfer voltage are essential.

21.6.2 Generator protection against transfer surge voltages through transformer

An appropriate countermeasure is required to protect generators against the transfer surge voltage coming from the HT side, because the rated voltages of the generators are perhaps 10–35 kV, so the insulation level of the LT side against surge is relatively lower.

The typical countermeasure is to install a so-called **'surge absorber' at the LT bushing terminal** of the transformer for each phase, which is a parallel circuit of a capacitor and arrester, as shown in Figure 21.14.

By adding capacitance C as the surge absorber, Equation 21.16 is modified as $C_2 \rightarrow C_2 + C$ so that

$$\frac{C_{12}}{C_{12}+C_2} \rightarrow \frac{C_{12}}{C_{12}+(C_2+C)} \tag{21.17}$$

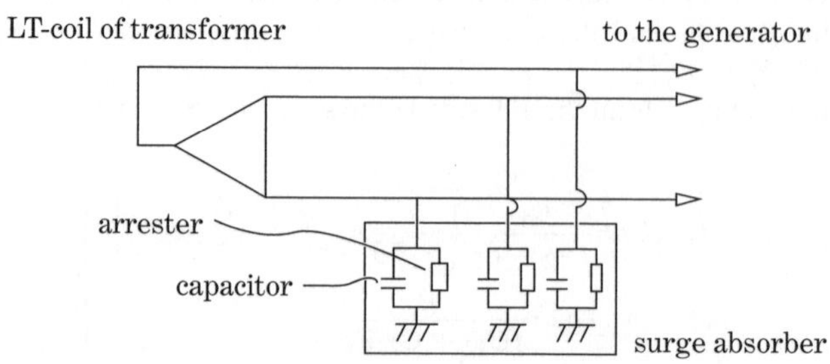

Figure 21.14 Surge absorber installed at LT bushing terminal

Accordingly, the transfer voltages can be considerably reduced by adding C of value larger than that of C_{12}, C_2 (namely, $C \gg C_{12}, C_2$). Typically, $C = 0.1 - 0.5\,\mu\text{F}$ may be a reasonable range. The arrester in parallel is to relieve the steep wavefront.

21.6.3 Electromagnetic transfer voltage

A transfer voltage to the LT side by electromagnetic coupling also arises. Voltages caused by magnetic coupling can be written simply as Mdi/dt, so the transfer voltage would be roughly proportional to the transformation ratio, T-ratio $= n_{\text{HT}}/n_{\text{LT}}$, the turns ratio of the HT and LT coils of the transformers.

Considering a virtual standard waveform of $1.2 \times 50\,\mu\text{s}$ as the incident voltage E at the HT side and a current surge with a similar waveform, the transfer voltage to the LT side in an initial time duration of $t = 0$–$1.2\,\mu\text{s}$ would be roughly (T-ratio)·E.

In the case of a 275 kV/24 kV transformer (T-ratio 11.5) as an example, the incidental surge voltage E on the HT-side line would be transferred to the LT side by $(1/11.5)E = 0.087E$. A less serious threat arises from the absolute value as well as the relatively gradual wave shape in comparison with the electrostatic transfer voltages. Electrostatic, rather than electromagnetic, coupling plays the lead role in transfer voltage phenomena.

21.7 Internal High-frequency Voltage Oscillation of Transformers Caused by Incident Surge

A transformer is a very compact piece of apparatus in that HT, medium-tension (MT) and LT coils are concentrically and tightly arranged surrounding the laminated silicon steel core and each coil is composed of a number of winding sections. Therefore, when we examine transient voltage and current phenomena on the internal coils, we need to establish the equivalent circuit as a distributed circuit with a number of L and C elements. Internal transient oscillations would obviously be caused by the impact of an external surge voltage wave, so effective suppression of such transient oscillatory voltages as well as effective insulation design for each sectional part of the coils are vital. This is a unique problem that transformer engineers have to overcome in the total design engineering of coil allocation and coil insulation strength.

21.7.1 Equivalent circuit of transformer in EHF domain

Figure 21.15 shows sketches of a typical large-capacity transformer. As seen in Figures 21.15a and b, the LT, MT and HT coils are concentrically arranged in this order as cylinder coils. Each coil is assembled from a numbers of series-connected sectional windings. These figures also show typical coil winding structures named 'multi-layer cylindrical windings' and 'disc windings'.

Now we examine the voltage behaviour of a transformer's internal coil when an incident surge voltage E arrive at the HT bushing terminal. Figure 21.16a is the equivalent circuit of the HT coil against surge or high oscillatory frequency phenomena, although resistance as well as susceptance are ignored and the mutual couplings with MT and LT coils are not written. In the case of a disc-winding coil, each section of the ladder circuit means each disc winding is connected in series.

21.7.2 Transient oscillatory voltages caused by incident surge

The initial conditions of the circuit just after the surge arrives can be very accurately expressed by the Figure 21.16b, which is equivalent to Figure 21.16a in regard to the C distribution, although

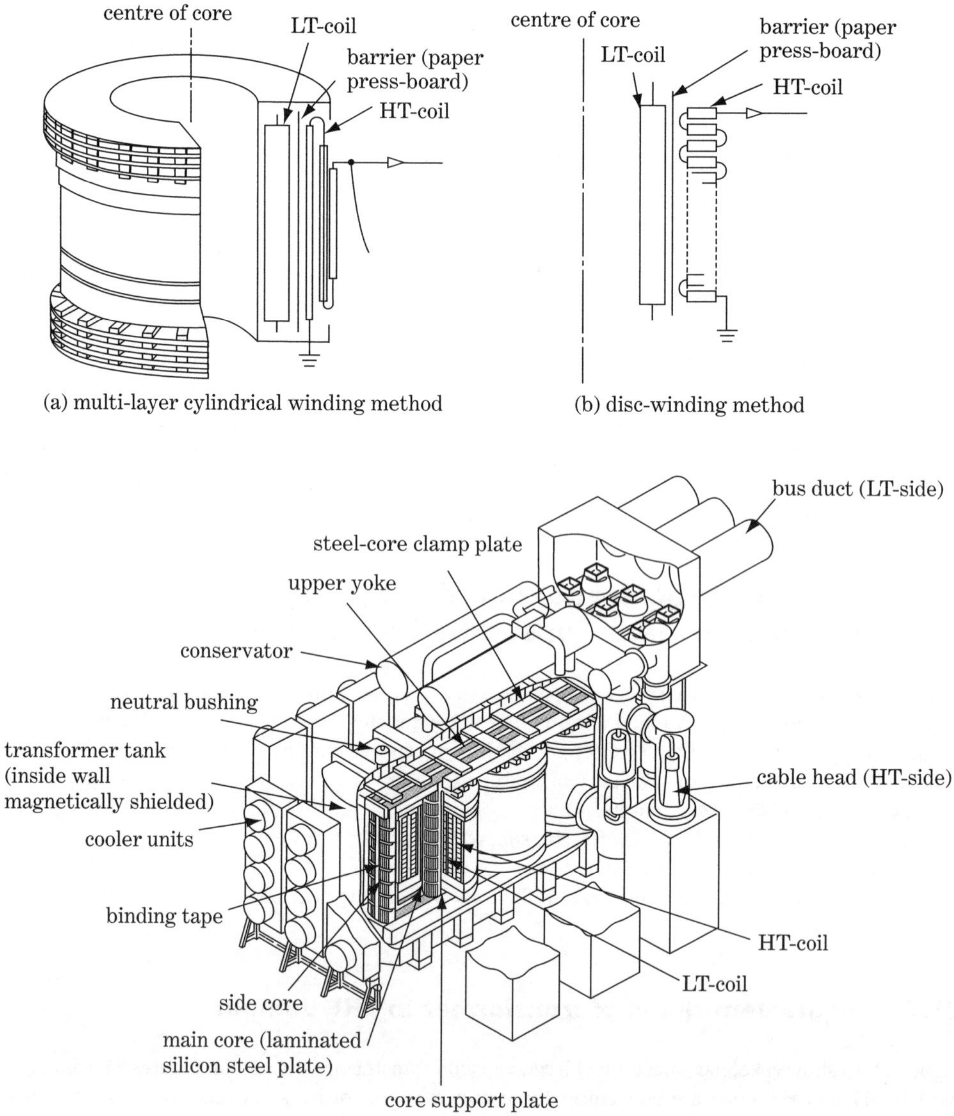

(a) multi-layer cylindrical winding method (b) disc-winding method

(c) structure

Figure 21.15 Large-capacity transformer for power station use (y–Δ connection)

all the inductances are ignored. On the other hand, the final conditions after the transient terms disappear would be expressed as the circuit (a) but ignoring all the C elements, namely by Figure 21.16d.

Figure 21.16b is a very accurate equivalent circuit of the HT coil for EHF (extra high frequency) phenomena. Now we will calculate the surge voltage distribution of this circuit where

C: the total capacitance from the HT coil to earth-ground
Cdx: the capacitance of one disc(between x and $x+dx$) to earth-ground
K: the total series capacitance of HT coil from the HT terminal bushing to neutral terminal
K/dx: the series capacitance across disc(between x and $x+dx$)
i_c: the current flowing through Cdx
i_k: the current flowing through K/dx

The equation at the winding section x to $x+dx$ is

$$\left.\begin{aligned} i_k &= \frac{K}{dx}\cdot\frac{\partial}{\partial t}\Big\{\underbrace{v}_{\text{voltage at } x} - \underbrace{\left(v+\frac{\partial v}{\partial x}\right)}_{\text{voltage at } x+dx}\Big\} & \therefore\quad i_k &= -K\frac{\partial^2 v}{\partial t\partial y} \\ \underbrace{-\frac{\partial i_k}{\partial x}\cdot dx}_{\text{reduced current of } i_k \text{ through } dx} &= \underbrace{Cdx\cdot\frac{dv}{dt}}_{\text{leakage current through } Cdx} & \therefore\quad -\frac{\partial i_k}{\partial x} &= C\frac{\partial v}{\partial t} \end{aligned}\right\} \tag{21.18}$$

Eliminating i_k from both equations,

$$K\frac{\partial}{\partial t}\cdot\frac{\partial^2 v}{\partial x^2} = C\frac{\partial v}{\partial t} \tag{21.19}$$

then

$$\left.\begin{aligned} K\frac{\partial^2 v}{\partial x^2} &= Cv \\ \frac{\partial^2 v}{\partial x^2} &= \alpha^2 v \quad \text{where} \quad \alpha = \sqrt{\frac{C}{K}} \end{aligned}\right\} \tag{21.20}$$

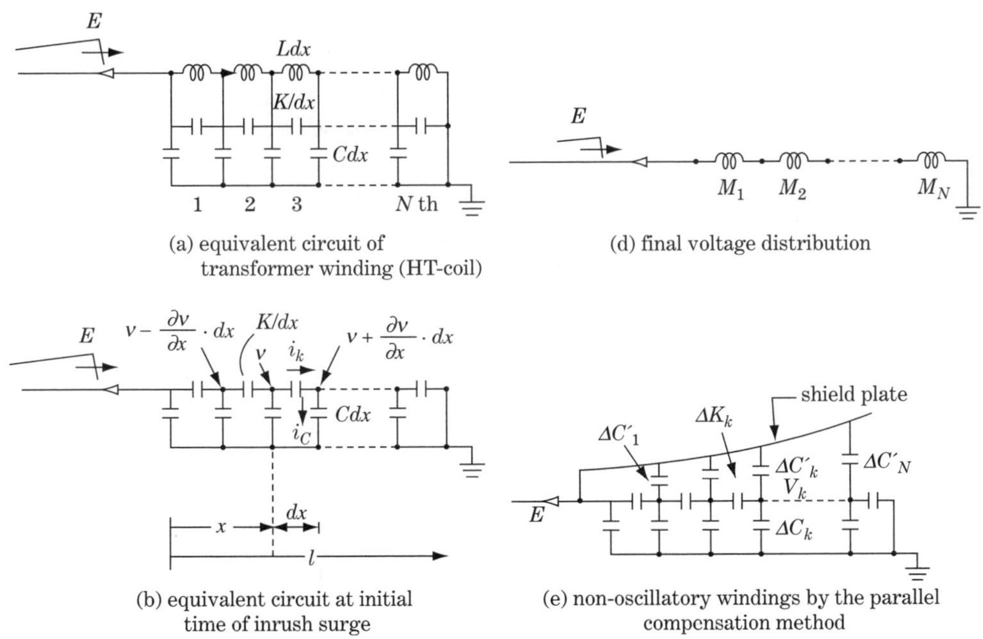

Figure 21.16 Internal surge behaviour of transformer

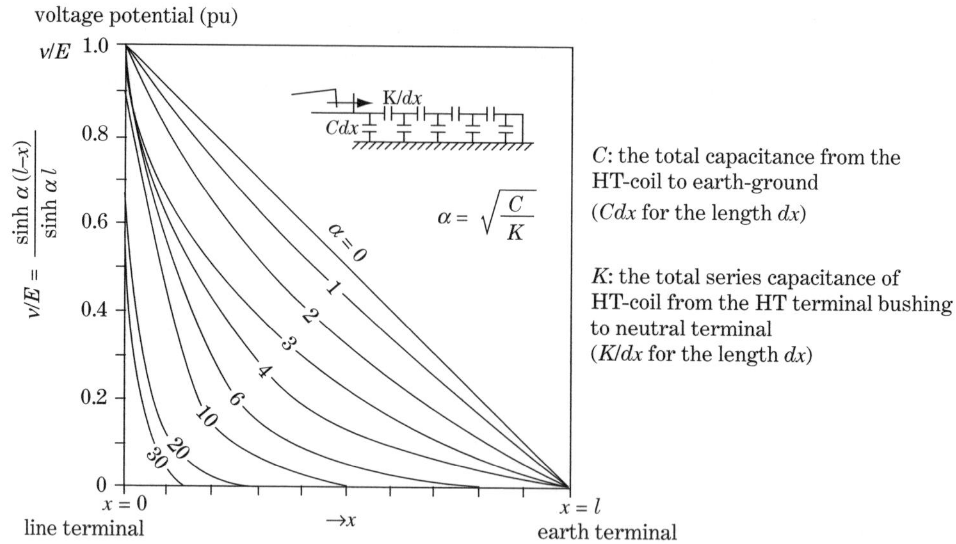

(c1) initial surge voltage distribution

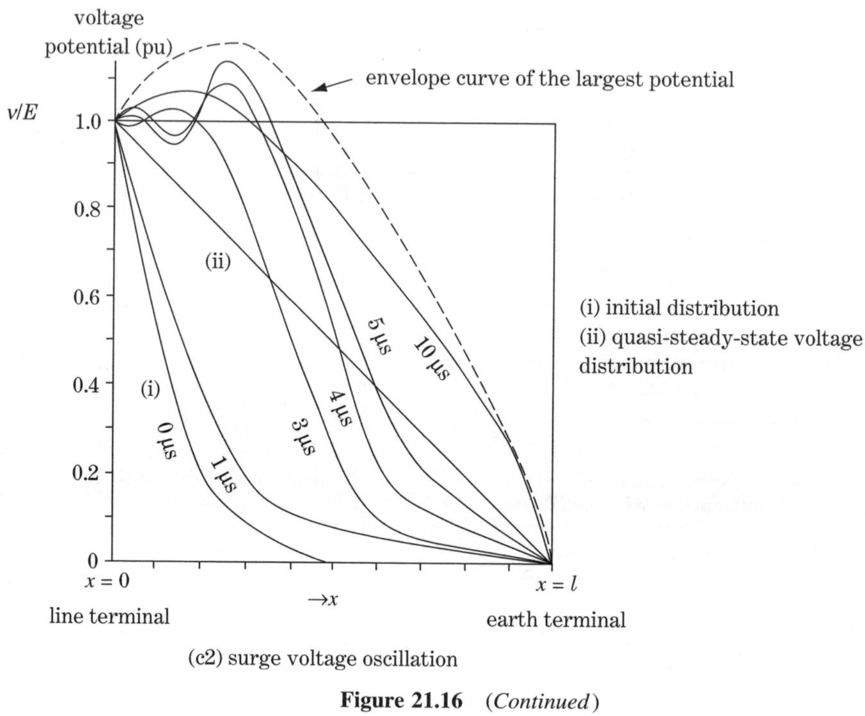

(c2) surge voltage oscillation

Figure 21.16 (*Continued*)

The general solution of the above partial differential equation is

$$v = A\cosh\alpha x + B\sinh\alpha x \tag{21.21}$$

where A and B are determined by the terminal conditions (refer to the supplement for the proof).

The terminal conditions are

$$\left.\begin{aligned} v &= 0 \quad \text{at} \quad x = l \qquad \therefore\ 0 = A\cosh\alpha l + B\sinh\alpha l \\ &\qquad\qquad\qquad\qquad\quad = \frac{A+B}{2}\mathrm{e}^{\alpha l} + \frac{A-B}{2}\mathrm{e}^{-\alpha l} \\ v &= E \quad \text{at} \quad x = 0 \qquad \therefore\ E = A\cosh 0 + B\sinh 0 = A \end{aligned}\right\} \tag{21.22}$$

$$\begin{aligned} \therefore A &= E \\ B &= -E\frac{\mathrm{e}^{\alpha l} + \mathrm{e}^{-\alpha l}}{\mathrm{e}^{\alpha l} - \mathrm{e}^{-\alpha l}} = -E\frac{\cosh\alpha l}{\cosh\alpha l} \end{aligned} \tag{21.23}$$

Substituting Equation 21.23 into 21.21 and modifying,

$$\left.\begin{aligned} v(x,0) &= E\frac{\mathrm{e}^{\alpha(l-x)} - \mathrm{e}^{-\alpha(l-x)}}{\mathrm{e}^{\alpha l} - \mathrm{e}^{-\alpha l}} = E\frac{\sinh\alpha(l-x)}{\sinh\alpha l} \\ \alpha &= \sqrt{\frac{C}{K}} \end{aligned}\right\} \tag{21.24}$$

This equation gives the initial voltage distribution of the HT coil at $t = 0$, where $v(x, t)$ means the transient voltage on coil position x and at time t. Figure 21.16c1 shows the curve of the initial voltage distribution $v(x, 0)$ at $t = 0$ along with the coil position x and with the parameter α, which was derived from Equation 21.24.

The figure indicates that at initial time $t = 0$, surge voltage E cannot be uniformly charged to each coil section except only in the exceptional case under the condition $\alpha = 0$ (unrealistic case of $C = 0$). If α is larger (larger C and smaller K), most of the surge voltage is charged unequally on the coil sections close to the HT bushing. If $\alpha = 10$, for example, quite a large voltage stress of approximately $0.8E$ is placed on the first 20% of coil sections close to the line terminal bushing.

The potential gradient is

$$\frac{\partial v}{\partial x} = -E\frac{\alpha\cosh\alpha(l-x)}{\sinh\alpha l} \tag{21.25}$$

and

$$\left.\frac{\partial v}{\partial x}\right|_{x=0} = -E\frac{\alpha\cosh\alpha l}{\sinh\alpha l} \tag{21.26}$$

The equation shows the potential gradient of the initial distribution is largest at the high-voltage terminal.

Let us investigate the behaviour a little more. Just after the incident surge of probably triangular waveform with a short wavefront and long wave-tail (say, 1.0×50 to $200\ \mu\mathrm{s}$, for example) is charged on the HT coil, the voltage at each position $v(x, t)$ begins to oscillate from the initial distribution $v(x, 0)$. Figure 21.16c2 shows the oscillatory behaviour of the voltage distribution just after the initial condition. The surge voltage at each coil section would repeat some over-swing oscillation across the straight line (ii). Soon, the transient voltage oscillation would be decreased over time and disappear, so the voltages converge to the final distribution of the straight line (ii), which is called the **quasi-steady-state voltage distribution**.

The incident voltage E can be treated as the d.c. component in the time duration of the long wave-tail so that $dv/dt \to 0$ at this time, although v is still large. Accordingly, at the end of the long voltage wave-tail, $Cdv/dt \to 0$ and $K/(dv/dt) \to \infty$, or in other words the capacitive elements act as like open-circuit elements. Therefore all the C and K branched circuits in Figure 21.16a can be ignored during this time and the final voltage distribution can be derived from Figure 21.16d. The voltage distribution during this time converges to a uniform distribution, that is the distribution of (ii) in Figure 21.16c2.

The final distribution by in Figure 21.16d is expressed by the equation below, for disc coil number 1–N,

$$E = \sum_{k=1}^{N} \Delta v_k \qquad ①$$
$$\Delta v_k = \left(\sum_{j=1}^{N} L_{kj}\right)\frac{di}{dt} \equiv M_k \frac{di}{dt} \qquad ② \qquad (21.27)$$

where

$k = 1$–N: the number of disc windings
i: the current flowing through each winding (the same current)
Δv_k: shared voltage by kth disc winding
L_{kj}: mutual inductance between kth and jth disc windings
$M_k = \sum_{j=1}^{N} L_{kj}$: total summation of self- and mutual inductances of kth disc winding (the specific value of each disc winding)

At the time of the final distribution condition (quasi-steady-state voltage distribution), the voltage oscillation is terminated so that the shared voltage of each disc Δv_k has a constant value, and accordingly di/dt is constant from Equation ②. In other words, at the time of the quasi-steady-state voltage distribution, the surge voltage oscillation is terminated, whereas **the surge current i is still increasing** at constant speed. Although this surge current continues to increase a little, sooner or later it will stop and soon disappear because it is resistively attenuated.

It can be concluded from Figure 21.16c2 that the surge voltage distribution $v(x,t)$ of the coil sections initiates internal oscillation from the initial distribution of Figure 21.16c1 and would repeat oscillatory over-swing across the quasi-steady-state voltage distribution line (ii) and soon terminate voltage oscillation, while the surge current still continues to increase and then disappears. The envelope curve of the voltage oscillatory distribution is also shown in Figure 21.16c2. In all events, transformer design with excess non-uniform initial voltage distribution should be avoided in order to avoid excess concentrated stress on a few coil sections around the HT bushing side and to reduce oscillatory surge voltage.

21.7.3 Reduction of internal oscillatory voltages

We need to make the initial distribution curve as much as possible coincident with the final distribution curve (the straight line of $\alpha = 0$) by reducing α; in other words, by decreasing C or by increasing K.

However, to decrease C is actually impossible because enlarging the distance from the winding to the core/tank/other windings cannot be done realistically. K also cannot be increased because the winding discs are already very closely arranged.

The widely applied effective countermeasure to reduce oscillation is known as the **parallel compensation method** of stray capacitances.

21.7.3.1 Non-oscillatory windings by the parallel compensation method

In Figure 21.16e, the shield ring plate with the voltage potential of the HT bushing terminal (often called the **rib shield**) is additionally arranged. The distance between each winding disc and the rib shield is closer near the HT terminal and far apart around the neutral terminal (i.e. $\Delta C_1' > \cdots > \Delta C_k' > \cdots > \Delta C_N'$) (parallel capacitance compensation method).

At the initial time when surge E is charged, the current through ΔC_k is supplied directly through $\Delta C_k'$, so the current through direct capacitance ΔK_k for each disc becomes uniform, which means the initial shared voltage by each disc winding is kept almost equal. We can write this symbolically as follows.

At the initial timing, for the parallel charging currents for each winding disc,

$$\Delta C_k' \frac{\partial (E - v_k)}{\partial t} = \Delta C_k \frac{\partial v_k}{\partial t} \qquad \text{(where} \quad k = 1.2, \cdots \mathrm{N})$$

$$\therefore \quad \frac{\partial v_k}{\partial t} = \frac{1}{\dfrac{\Delta C_k}{\Delta C_{k'}} + 1} \cdot \frac{\partial E}{\partial t} \equiv \delta_k \cdot \frac{\partial E}{\partial t} \tag{21.28}$$

Then, if the δ_k at each winding disc are designed almost equal for each other ($\delta_1 \fallingdotseq \delta_2 \fallingdotseq \cdots \fallingdotseq \delta_k \fallingdotseq \cdots$), the initial voltage rising velocity at each winding disc ($\delta v_k / \delta t$) can be kept in almost equal, so that the initial surge voltage is uniformly distributed.

The oscillatory voltage behaviour on the transformer winding is closely affected by the required insulation of the transformer. However, the phenomenon is not usually included in the concept of the term 'insulation coordination' because it is not necessarily related to the insulation coordination of the power system.

In other words, this is considered as an insulation matter which the transformer engineers or the suppliers have to solve for individual transformer products.

21.8 Oil-filled Transformers Versus Gas-filled Transformers

Power transformers of large capacity (say, over 60 MVA) are usually oil-filled transformers because they have been utilized as the standard type in many years, although small-capacity transformers (say, typically 0.5–40 MVA) have been utilized as of dry type (air or SF_6 gas insulation/coolant) as well as oil-filled type. This is because oil is available as the one material having the characteristics of 'superior electrical insulation' and 'superior thermal loss discharging (coolant)'. Accordingly, it was believed for a long time that oil could not be technically replaced by SF_6 gas for large-transformer applications, because SF_6 gas has quite poor characteristics for thermal capacity (or thermal conductivity), in spite of its outstanding insulation characteristics. This fact is in stark contrast to the engineering history of circuit-breakers, because SF_6-gas-type breakers have been widely used in place of oil-filled type/ air-blast-type breakers over the last half century.

However, SF_6-gas-filled transformers with a large capacity of 300 MVA · 275 kV were first utilized in the mid-1990s in the underground EHV substations of high buildings in Tokyo, whose fundamental structures were of the familiar disc windings but filled with gas instead of oil. These achievements led to a breakthrough by changing the conservative concept of large-capacity/EHV transformers only as of oil-filled type.

Needless to say, oil-filled transformers have one major weak point: that is, the severe damage which would inevitably be caused to the transformer if a breakdown fault of the internal coil were to occur, and furthermore possibly influence the installed surroundings.

Whenever a short circuit occurs in the coil insulation of oil-insulation-type transformers, liquid oil around the fault is immediately gasified by the arcing temperature, so the internal gauge pressure of the tank would be rapidly increased. The oil–gas pressure increase would continue until fault tripping by the related breakers is completed, so the accumulated pressure may reach a very high level, even exceeding the mechanical withstanding strength of the tank within just a few cycles (50–100 ms), in particular the critical strength of the tank-cover cramping. As a result, a hot-oil-blasting overflow or even fire from the oil burning may be caused in the worst cases. Of course, the shape of the coils would be deeply distorted.

On the contrary, in the case of a short-circuit fault in an SF_6-gas-insulated transformer, the physical damage caused by the internal short-circuit fault would probably be limited to a narrow spot on the coil where a breakdown arcing pass would be produced, so the concentrically arranged HT/MT/LT coils might not be badly deformed, although carbonization of insulation tapes/press-board barriers and the copper conductor melting in a limited area would be caused. In other words, SF_6 gas transformers do not produce serious blasting or fire even in the case of an internal short-circuit fault. This is obviously a big advantage from the safety point of view, in particular for receiving substations located in city areas, whether the substations are outdoors or in-house, or under high-rise buildings.

The reasons why the above breakthrough was realized can be summarized by the following three points:

a) Extensive study of thermal discharge (conductivity) characteristics as well as insulation/breakdown characteristics of SF_6 gas under various shapes of coils, gas-flowing passes and gas-flow speed based on a detailed mathematical simulation approach and experimental model tests, and finally well-investigated smart coil structure and gas-flow pass design.

b) Adoption of 0.4 MPa (4 atm) SF_6 gas pressure.

c) Application of class F insulation materials (maximum temperature 130°C) instead of class A (105°C) for coil taping based on polyethylene teraphthalate (PET) films.

Oil flow depends on its liquid viscosity characteristics, so oil may not flow easily through very narrow passes, while SF_6 gas can flow easily through even narrow passes because its gas-flow distribution does not depend on viscosity. This characteristic can be said to be an advantage of SF_6 gas in comparison with oil. However, SF_6 gas may flow through various passes in an unbalanced way, in contrast to a joule-loss distribution which should be cooled at a continuous withstanding temperature (130°C). This is obviously a disadvantage of SF_6 gas in comparison with oil as a coil coolant material.

Furthermore, the thermal capacity of SF_6 gas under 1 atm pressure is only 1/200 = 0.5%, while that under 4 atm pressure is still only 2.4/200 = 1.2% in comparison with that of oil. (see Table 21.4). These are the reasons why accurate analysis and careful gas-flow pass design are required. Adoption of 4 atm pressure gas is also quite a valuable improving factor in order to realize effective cooling by SF_6 gas, although its thermal capacity is still much smaller than that of oil.

Figure 21.17 shows 400 MVA/330 kV/132 kV SF_6 gas transformers installed in Australia. Large-capacity SF_6 gas insulation transformers can be realized only by fully involving the above three countrermeasures. As can be seen in Figure 21.17, cylindrical tanks, instead of the traditional box type, are generally advantageous for gas transformers because the tank needs to withstand the high pressures. Gas insulation transformers also have a smaller volume and can save on installation space in comparison with oil insulation transformers of the same capacity, in particular their height, because the oil conservator, oil-absorbable saucer, fireproof barrier, etc., can be removed.

Large-capacity SF_6 gas transformers should rapidly prevail in the near future, in particular as large transformers installed in urban areas, regardless of the location of the transformers. SF_6-gas-filled

Table 21.4 Comparison of thermal capacity

	Ratio of thermal capacity
Oil	200
SF_6 gas: 0.125 MPa-g	1
SF_6 gas: 0.40 MPa-g)	24

Note: 0.1 MPa-g (megapascal to gravity) = 1 atm = 1000 mb approximately.

Courtesy of Transgrid (Australia)/Toshiba

Haymarket substation (Australia)

primary: 330 kV ± 10% (21 taps) 400 MVA
secondary: 138.6 kV 400 MVA
tertiary: 11 kV 20 MVA
rated gas pressure: 0.43 MPa-g (20°)
heat exchanges: gas-to-water cooling

Figure 21.17 A 400 MVA SF_6-gas-insulated transformer (with On-load-tap-changer)

shunt reactors of 150 MVA have also been utilized, and should prevail for the same reasons for gas transformers.

21.9 Supplement: Proof that Equation 21.21 is the solution of Equation 21.20

Differentiating v of Equation 21.21 twice,

$$v = A\cosh\alpha x + B\sinh\alpha x = \frac{A}{2}(e^{\alpha x} + e^{-\alpha x}) + \frac{B}{2}(e^{\alpha x} - e^{-\alpha x})$$

$$\therefore \quad \frac{\partial v}{\partial x} = \frac{A+B}{2}\alpha e^{\alpha x} - \frac{A-B}{2}\alpha e^{-\alpha x}$$

$$\therefore \quad \frac{\partial^2 v}{\partial x^2} = \frac{A+B}{2}\alpha^2 e^{\alpha x} + \frac{A-B}{2}\alpha^2 e^{-\alpha x} = \alpha^2 v$$

Therefore Equation 21.21 satisfies Equation 21.20.

Coffee break 11: Edith Clarke, the prominent woman electrician

The $\alpha - \beta - 0$ method originally appeared in the following papers written by **Edith Clarke** (1883–1959):

'Determination of Voltages and Currents during Unbalanced Faults', *GE Rev.*, 1937

'Overvoltages Caused by Unbalanced Short Circuits', *AIEE Trans.*, 1938 (co-authored by **C. N. Weygandt and C. Concordia**)

She was born in Howard County, Maryland, and aged 18 she attended Vassal College and concentrated on mathematics and astronomy, during a time when it was almost unheard of for a woman to acquire a college degree. After graduation she enrolled at the University of Wisconsin, and worked during the summer vacations for AT&T as a 'computer' (as the position was called). She then entered MIT, where she earned her masters degree in electrical engineering, becoming the first woman to do so there.

She then worked at GE Schenectady as the first professionally employed female engineer from 1919 to 1945. Her paper 'Three phase multiple conductor circuits' (1932) was judged the best paper of the year in the AIEE Northern District. In 1937 and 1938 she presented the above two papers in which she disclosed her $\alpha - \beta - 0$ method. Another paper, 'Stability Limitations of Long–Distance Alternating Current Power Systems', was awarded 'the best paper of the year in the AIEE'. She wrote 19 papers in all and published the well-known books *Circuit Analysis of Power Systems* (Volume 1 in 1943, Volume 2 in 1950) in which she wrote 'Components of current answering to the description of α, β and 0, although not so named, were used in a method developed by W.W. Lewis and published in 1917, to determine system currents and voltages during line–to ground faults'. That is the method shown in Figure 6.7. However it is true that she originated α-β-0 method by her mathematically clear definition as is seen in the chapter 6 Equation 6.1 and 6.2.

Her another contribution was introduction of various kinds of 'quick estimation tables and charts' in her books as the result of profound mathematical treatments, which were highly valued for the practical engineering of power system network in the era of before computer.

After retiring from GE, she became professor of engineering at the University of Texas for 1945–1956. In 1959 she died in Olney, Maryland, at the age of 76. She was also the first woman to be elected a Fellow of the AIEE.

Edith Clarke (1883–1959)

22

Waveform Distortion and Lower Order Harmonic Resonance

Continuous as well as temporary waveform distortion may have been slowly but steadily increasing for most power systems. It may be a problem in the same category as air pollution or water contamination because an effective simple solution does not exist. Electrical engineers should understand the true nature of waveform distortion if only for this reason.

22.1 Causes and Influences of Waveform Distortion

22.1.1 Classification of waveform distortion

Table 22.1 classifies waveform distortion by various causes, namely active causes and passive causes.

In addition to conventional loads like electric furnaces and rectifiers, most of the broadly prevailing recent loads depending on power semiconductor applications have theoretically the characteristics of 'harmonic generators'. The power network facilities as applications of power electronics have similar characteristics.

We need to remind ourselves that power electronic (semiconductor) applications still prevail, and will probably tend to be further accelerated in future. Typical examples of **power electronic application loads** may be the advanced technology of 'speed-controlling equipment' and 'position-controlling equipment'. High-speed elevators in tall buildings and most of the precision instruments adopted by industrial manufacturers would be loads in this category, in that the rotating speed and timing of large or small high-speed motors are quite accurately and smoothly switching controlled by power electronic devices, where the voltage and/or current is forced to be repeatedly switched on and off, so the harmonic components would be more or less generated.

UPS (Uninterrupted Power Supply equipment) is the name for power sourcing systems of electronic type, essentially required for every infrastructure system (broadcasting equipment, telecommunication equipment, computer networking systems, survers for internet communication, etc.), that are capable of supplying stable power of so-called 'constant voltages and constant frequency continuously' even if unexpected voltage drops of short duration (within 100 ms) are caused by short-circuit faults on the power system.

Besides the conventional hydro/thermal/nuclear generating systems, **distributed-type smaller generating systems** have prevailed in recent years. Miniature hydro-generation, solar generation, wind generation, fuel cell generation, small gas-turbine generation, secondary battery generation, etc., are often called 'smaller, new-energy-type generating systems'. All of these small generating systems can be connected to the large power system only through electronic **power conditioners**, which are kinds of power electronic frequency changers.

Handbook of Power System Engineering Yoshihide Hase
© 2007 John Wiley & Sons, Ltd

Table 22.1 The background of waveform distortion (occurrence of low-order harmonics)

(a) Active causes (active generation of harmonic voltages and currents)

(a1) Causes by loads (continuous or periodic)

- Intermittent rushing-type loads (trains, rolling machines)
- High-frequency-type loads (induction furnaces)
- Discharging-type loads (welder machines, electric furnaces)
- Rectification loads (rectifiers, various power conditioners)
- Powered semiconductor-type application loads (speed-control-type loads; elevators)
- UPS (Uninterrupted Power Supply equipment; power sources for infrastructures of computers/data processing servers/telecommunications/broadcasting, etc.)
- Battery chargers

(a2) Causes by power network (continuous)

- Converters for d.c. transmission lines, frequency changers
- Power conditioners for distributed-type small generating systems (for solar, wind, fuel cell, small gas turbine, secondary battery, etc.)

(b) Passive causes (resonant phenomena under special conditions)

(b1) Negative-/zero-sequence or d.c. currents flowing through rotating machines

Causes by loads (continuous)

- Single phase distribution loads
- Unbalanced three-phase loads (electric furnaces, trains)

Causes by power network

- Negative-/zero-sequence voltages and currents caused by network imbalance (continuous)
- Inrush current of transformers (temporary)
- Dead-voltage time of single phase reclosing (temporary)
- Breaker tripping failure (temporary)

(b2) Saturation

- Saturable non-linear-type loads (continuous)
- Ferro-resonance phenomena (temporary)

(b3) *LC* resonance under special conditions of power network

- Local resonance under normal operation (continuous)
- Local resonance under fault conditions (temporary)

The d.c. transmission system is another example, in that countermeasures to reduce harmonic components such as multi-phase (12- or 24-phase) power converting methods, power filters, etc., are generally adopted.

Table 22.2 is a summary of various effects which may be caused by waveform distortion. Figure 22.1 is a typical example of the continuous or temporary waveform distortion of load voltage and current.

It is fair to say that all this **power electronic application equipment** shares the characteristics of 'harmonic generators' on one hand, but ironically shares the characteristics of 'apt to go wrong' more or less by waveform distortion, on the other hand. Generally, excessive waveform distortion adversely affects most control equipment as well as measuring equipment. Capacitor banks for improving the power factor may be thermally overheated, because $I = j2\pi\, fC \cdot v$ would be increased by harmonic components of larger f. Electrical appliances driven by single phase motors with supplementary capacitors may be forced into pulsatory operation or thermally overheat the capacitors. Power network facilities are also badly affected by harmonic components or waveform distortion.

If the ratio of harmonic content in the voltage or current exceeds 5 or 10%, various serious effects could result. Waveform distortion has perhaps been slowly but steadily increasing for most of the area on one hand, and broadly effective means of reducing it do not exist from practical viewpoints, on the other hand.

Table 22.2 Various influences of waveform distortion

(a) Influences on loads
- **Shunt capacitors**: rise of thermal losses or damage by thermal overheating
- **Industrial motors**: disorders of pulsating torque
- **Single phase a.c. motors using capacitors for rotating field production**: disorder of rotating field mechanism using capacitors
- **Control equipment**: increase of error
- **Illumination**: flickering

(b) Influences on power networks
- **Generators**: thermal overheating of rotor surface etc., pulsating torque
- **Capacitors for power-factor control**: thermal overheating
- **Transformers**: local heat, mechanical vibration of cores and windings
- **PT, CT**: degradation of accuracy
- **Protective relays**: malfunction, inaccurate operation
- **Control equipments (AVR etc.)**: irregular operation caused by voltage signal detection error
- **Energy (watt-hour) meters**: measuring error

(c) Interference to communication network

(d) Interference to power line carrier system (of lower frequency carrier type)
- **Load management system**

22.1.2 Causes of waveform distortion

We present below various complements of the effects of waveform distortion on power network facilities, listed in Table 22.2b.

22.1.2.1 Generators and large industrial motors

As discussed in Chapters 10 and 16, our generators are so-called 'fundamental frequency positive-sequence voltage and current generators', so they have weak points to negative- or zero-sequence voltages/currents or to d.c. or harmonic components:

- If a negative-sequence current flows into a generator, the 3rd, 5th, 7th, etc., odd-number harmonics are caused.
- If a d.c. (offset) component current flows into a generator, the 2nd, 4th, 6th, etc., even-number harmonics are caused.

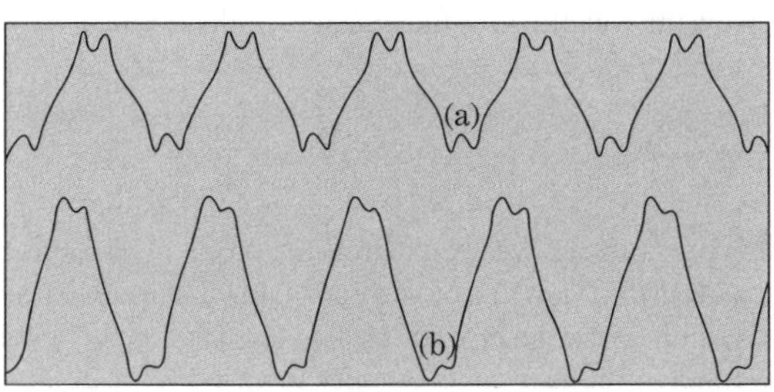

Figure 22.1 Waveform distortion of voltage and current (oscillogram recorded experimentally on a local system)

Special attention has to be paid to reduce continuous negative-/zero-sequence currents. The items below concern the most essential matters:

- To keep the **phase imbalance rate** of three-phase overhead transmission lines within a certain reasonable limit.
- To keep a total three-phase balance of many single phase loads mainly in the distribution network.
- Special countermeasures for the large imbalanced industrial loads or d.c. transmission lines etc. (three-phase load balancing over time as much as possible, **local power filter, multi (12-, 24-)-phase conversion,** etc.).

22.1.2.2 Special phenomena caused by saturation and resonance

Waveform distortion cannot be discussed separately from the phenomena of saturation as well as resonance, which we have already examined in Chapter 20. Saturation and resonance may cause not only overvoltages but also unexpected abnormal phenomena, so we have to remove the causes in advance. However, effective countermeasures would not exist were it not for responsible engineers.

22.2 Fault Current Waveform Distortion Caused on Cable Lines

Unique phenomena of fault current waveform distortion have been recognized on urban network systems where some cable lines exist. This is a kind of fault current distortion caused by free-energy oscillation between large capacitances C of cable lines and inductances L of overhead transmission lines, by which serious technical problems with high-speed protective relays (in particular, current differential relays and directional distance relays) may occur. These unique phenomena should be especially recognized because cable lines have been increasing remarkably in our modern power system networks.

The phenomena are investigated below.

22.2.1 Introduction of transient current equation

Figure 22.2a shows a power system in which overhead transmission line 1 and cable line 2 are connected at the power receiving substation q, and a three-phase fault occurs at point f which is length x distant from the point q. Now the waveform of the transient fault current at points p, q, r during the fault is investigated here.

22.2.1.1 Step 1

As this is study of transient phenomena, we need to treat the problem by Laplace transforms. Referring to Figure 18.4, an overhead transmission line of within 100 km and a cable line of within 50 km can be treated as a concentrated circuit within an error of 10% or less for the phenomenon of 500 Hz or lower frequency. Therefore we adopt the concentrated circuit shown in Figure 22.2b as the equivalent circuit. Incidentally, we assume for simplicity that terminal r is open, while similar phenomena can be observed even if point r is connected to the load.

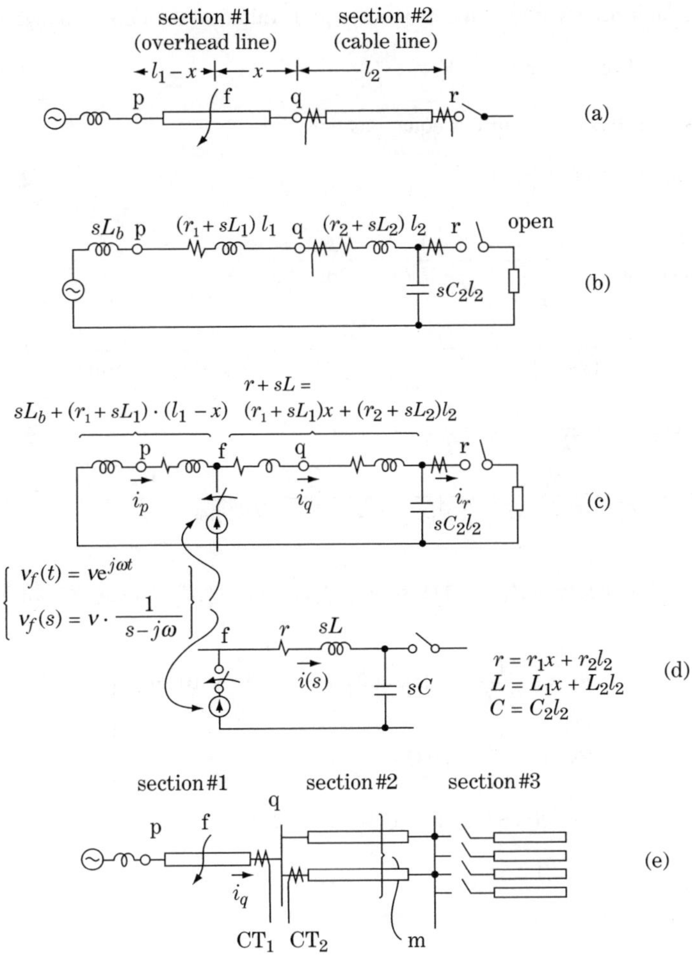

Figure 22.2 Transient fault calculation

22.2.1.2 *Step 2*

Now, the line-to-ground fault occurred at time $t = 0$ at point f between p and q. The transient phenomenon of the fault can be calculated by Figure 22.2c, where initial source voltage $v_f(t)$ is switched in at point f by Thévenin's theorem. The transient current at the point q can be calculated by Figure 22.2d, which is the right half of Figure 22.2c. That is,

$$\left.\begin{aligned} L &= L_1 x + L_2 l_2 \\ C &= C_2 l_2 \\ r &= r_1 x + r_2 l_2 \end{aligned}\right\} \tag{22.1}$$

We assume the fault occurs at the time of voltage peak value, so the initial voltage is

$$\dot{v}_f(t) = V_f e^{j\omega t}, \quad v_f(s) = V_f \frac{1}{s - j\omega} \qquad ①$$

and the circuit impedance and current equations are

$$Z(s) = r + sL + \frac{1}{sC} \qquad ②$$

$$\therefore \frac{1}{Z(s)} = \frac{1}{L} \cdot \frac{s}{s^2 + \frac{r}{L}s + \frac{1}{LC}} = \frac{1}{L} \cdot \frac{s}{s^2 + 2\alpha s + u^2}$$

$$= \frac{1}{L} \cdot \frac{s}{(s+\alpha)^2 + (u^2 - \alpha^2)} \fallingdotseq \frac{1}{L} \cdot \frac{s}{(s + \alpha + ju)(s + \alpha - ju)} \qquad ③$$

$$\text{where} \quad \alpha = \frac{r}{2L}, \quad u = \frac{1}{\sqrt{LC}} \qquad ④$$

$$i(s) = \frac{v_f(s)}{Z(s)} = \frac{V_f}{L} \cdot \frac{s}{(s - j\omega)(s + \alpha + ju)(s + \alpha - ju)} \equiv \frac{V_f}{L} \cdot F(s) \qquad ⑤ \qquad (22.2)$$

Assuming the surge impedance $Z_2 = 20\,\Omega$, surge velocity $u = (1/2)c = 150\,000$ km/s for the power cable, then

$$\left. \begin{aligned} u_2 &= \frac{1}{\sqrt{L_2 C_2}} = 150\,000\,\text{km/s} = 150 \times 10^6\text{m/s} \\ Z_2 &= \sqrt{\frac{L_2}{C_2}} = 20\,\Omega \\ r_2 &= 0.03\,\Omega/\text{km} = 0.3 \times 10^{-4}\Omega/\text{m} \end{aligned} \right\} \qquad (22.3)$$

Accordingly,

$$\left. \begin{aligned} L_2 &= \frac{Z_2}{u_2} = \frac{20}{150 \times 10^6} = 0.133 \times 10^{-6}\,\text{H/m} = 0.133\,\text{mH/km} \\ C_2 &= \frac{1}{Z_2 u_2} = \frac{1}{20 \times 150 \times 10^6} = 0.33 \times 10^{-9}\,\text{F/m} = 0.33\,\mu\text{F/km} \\ \alpha_2 &= \frac{r_2}{2L_2} = \frac{0.3 \times 10^{-4}}{2 \times 0.133 \times 10^{-6}} = 1.13 \times 10^2 \end{aligned} \right\} \qquad (22.4)$$

Also $\omega = 2\pi \times 50 = 314$.
Then

$$\left. \begin{aligned} u_2 &\gg \alpha_2, \omega \\ u &\gg \alpha, \omega \end{aligned} \right\} \qquad (22.5)$$

This is the reason why we can ignore α^2 in the denominator of Equation 22.2b③.
For the calculation of $F(s)$ defined in Equation 22.2b⑤

$$F(s) = \frac{s}{(s - j\omega)(s + \alpha + ju)(s + \alpha - ju)} = \frac{k_1}{s - j\omega} + \left\{ \frac{k_2}{s + \alpha + ju} + \frac{k_3}{s + \alpha - ju} \right\} \qquad (22.6)$$

The coefficients k_1, k_2, k_3 can be calculated by the same method with Equations 19.6 and 19.18 and the associated supplement. The result is

$$\left.\begin{aligned}
k_1 &= F(s)\cdot(s-j\omega)|_{s=j\omega} = \frac{j\omega}{(j\omega+\alpha)^2+u^2} = \frac{j\omega}{u^2} = j\omega LC \\
k_2 &= F(s)\cdot(s+\alpha+ju)|_{s=-(\alpha+ju)} = \frac{-(\alpha+ju)}{2ju\{\alpha+j(u+\omega)\}} = \frac{j}{2u} = j\frac{\sqrt{LC}}{2} \\
k_3 &= F(s)\cdot(s+\alpha-ju)|_{s=-(\alpha-ju)} = \frac{\alpha-ju}{2ju\{\alpha-j(u-\omega)\}} = \frac{-j}{2u} = -j\frac{\sqrt{LC}}{2}
\end{aligned}\right\} \tag{22.7}$$

The arrowbn ↖ shows the negligible part based on Equation 22.5. Thus

$$\frac{j}{2u}\left\{\frac{1}{s+\alpha+ju} - \frac{1}{s+\alpha-ju}\right\} = \frac{1}{(s+\alpha)^2+u^2} \tag{22.8a}$$

$$\therefore F(s) = \frac{j\omega}{u^2}\cdot\frac{1}{s-j\omega} + \frac{1}{u}\cdot\frac{u}{(s+\alpha)^2+u^2} \tag{22.8b}$$

$$\left.\begin{aligned}
\mathcal{L}^{-1}\left(\frac{1}{s-j\omega}\right) &= e^{j\omega t} \\
\mathcal{L}^{-1}\left(\frac{u}{(s+\alpha)^2+u^2}\right) &= e^{-\alpha t}\sin ut \\
u &= \frac{1}{\sqrt{LC}}
\end{aligned}\right\} \tag{22.9}$$

Therefore, Equation 22.2b⑤ is

$$\begin{aligned}
\dot{i}(t) &= \frac{V_f}{L}\left\{\frac{j\omega}{u^2}e^{j\omega t} + \frac{1}{u}e^{-\alpha t}\sin ut\right\} = V_f\cdot\omega C e^{j(\omega t+90^\circ)} + \frac{1}{\sqrt{L/C}}e^{-\alpha t}\sin ut \\
\therefore \quad i(t) &= \mathrm{Re}[\dot{i}(t)] = V_f\Bigg\{\underbrace{\omega C\sin\omega t}_{\text{steady-state term}} + \underbrace{\frac{1}{\sqrt{L/C}}e^{-\frac{r}{2L}t}\sin\frac{t}{\sqrt{LC}}}_{\text{transient term}}\Bigg\}
\end{aligned} \tag{22.10}$$

The second term on the right-hand side is the transient term, which is of amplitude $V_f/\sqrt{L/C}$, oscillatory frequency $f = 1/(2\pi\sqrt{LC})$ and attenuation time constant $T = 2L/r$.

22.2.2 Evaluation of the transient fault current

The derived equation $i(t)$ is the current i_q in Figure 22.2c whose line constants are given by Equation 22.1. Now let us examine the above result from the viewpoint of practical engineering.

22.2.2.1 *Case 1: The single cable circuit line (fundamental case)*

Referring to Figure 22.2e, this is the fundamental case where the cable section is of a single circuit line. The waveform-distorted transient current (the transient term) is

$$\left.\begin{aligned}
&\text{magnitude}: i_q = \frac{V_f}{\sqrt{L/C}} = \frac{V_f}{Z_{\text{surge}}} \\
&\text{oscillatory frequency}: f = \frac{1}{2\pi\sqrt{LC}} \\
&\text{attenuation time constant}: T = 2L/r
\end{aligned}\right\} \tag{22.11}$$

where

$$\left.\begin{aligned} L &= L_1 x + L_2 l_2 \\ C &= C_2 l_2 \\ r &= r_1 x + r_2 l_2 \end{aligned}\right\}$$

The magnitude of the current distortion can be grasped as a function of l_1 (the length of the overhead line), l_2 (the length of the cable section), and x (the distance from point q to the fault point f on the overhead line).

22.2.2.2 Case 2: The plural (n) parallel cable circuits line

In this case, modifications of the cable sections are $L_2 \to (1/n)\,L_2, C_2 \to nC_2$, accordingly. The surge impedance Z_{surge} increases $1/n$ times while the natural frequency of the specific cable section is not changed.

Now, in regard to the total system, the constants should be replaced as follows in Equation 22.11:

$$\begin{aligned} L = L_1 x + L_2 l_2 &\to L = L_1 x + \frac{L_2 l_2}{n} \\ C = C_2 l_2 &\to C = n\,C_2 l_2 \end{aligned} \qquad (22.12)$$

In the case of the fault at point f very close to the point q $(x = 0)$, the cable section

$$\left.\begin{aligned} &\underline{\text{one circuit}} && \underline{n\text{ circuits}} \\ &Z_{\text{surge}} = \sqrt{\frac{L_2}{C_2}} && Z_{\text{surge}} = \frac{1}{n}\sqrt{\frac{L_2}{C_2}} \\ &i_q = \frac{V_f}{\sqrt{L_2/C_2}} && i_q = n\cdot\frac{V_f}{\sqrt{L_2/C_2}} \\ &f = \frac{1}{2\pi l_2\sqrt{L_2 C_2}} && f = \frac{1}{2\pi l_2\sqrt{L_2 C_2}} \\ &T = \frac{2L_2}{r_2} && T = \frac{2L_2}{r_2} \end{aligned}\right\} \qquad (22.13)$$

That is, under the condition of a fault at $x = 0$, the transient fault current i_q would become n times larger proportional to the number of parallel cable circuits, while the oscillatory frequency as well as the time constant would not change.

22.2.2.3 Case 3: Additional cables with a third section

This is the case where cables of the third section are additionally connected as in Figure 22.2e.

In this case the total capacitance of the third section C_3 would be added to C_2 so that the transient current i_q becomes larger and of lower frequency. The situation would not change much even if transformers existed between the second and third sections.

22.2.2.4 Numerical check

Replacing Equation 22.12 in 22.11, the magnitude and the frequency of the distorted transient fault current i_q can be calculated as functions of n and x. Figure 22.3 shows the calculated result under typical lengths and constants of the lines.

The result indicates that a distorted transient fault current of quite large magnitude and lower frequency order is caused in the urban networks where many cable routes and lines exist. The voltage at point q would also be badly distorted, although the calculation is omitted.

Figure 22.4 is an oscillograph of a line-to-line (phase a to b) fault which was conducted as an artificial fault test on a real power network under conditions similar to that of the above calculation.

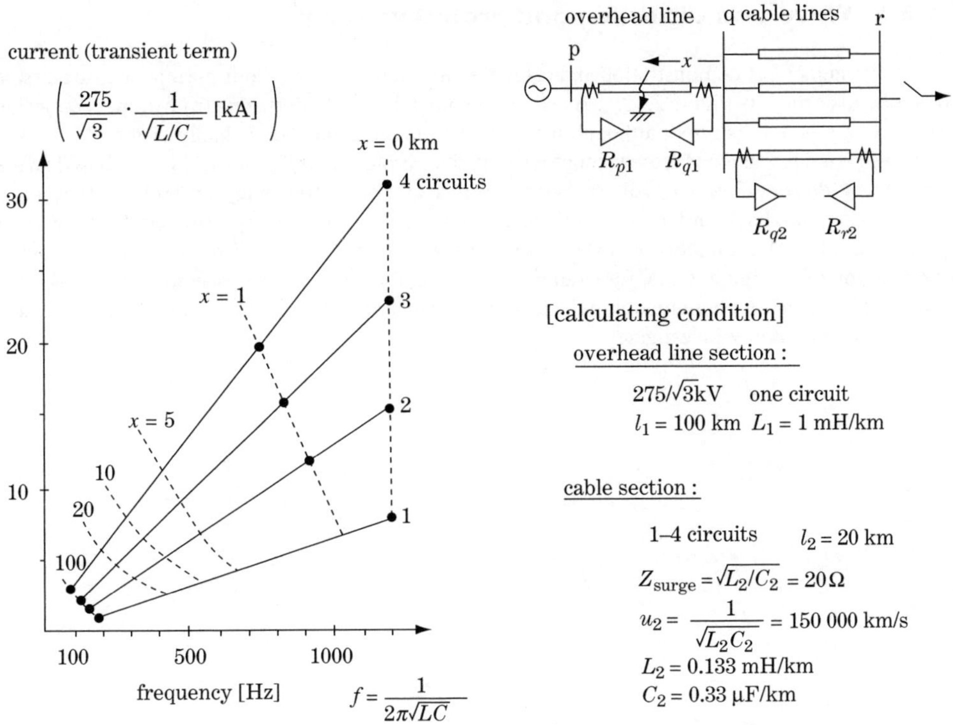

Figure 22.3 Fault calculation of power system with cables and overhead lines

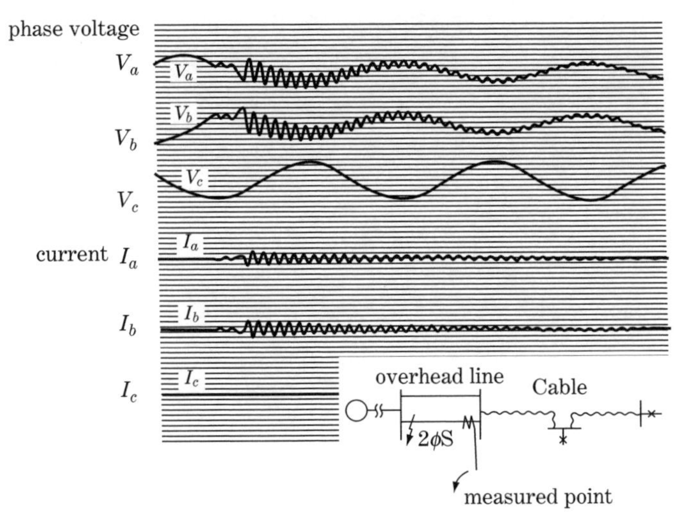

Figure 22.4 Waveform distortion of fault current (simulation)

22.2.3 Waveform distortion and protective relays

The above numerical demonstration indicates the appearance of the fault current with excessive waveform distortion. In Figure 22.2, we need to remember that the fault current from the generator side measured at p is not distorted, although the current from the cable side is badly distorted.

As we hinted in Chapter 17, the technology of various protective relays is more or less based on the principle of detection of the voltages and currents at power frequency. Therefore, the above phenomena could badly affect the various high-speed protective relays (in particular, current differential relays and directional distance relays) for sections 1 and 2 of the network. Assuming a fault at point f in Figure 22.2e, for example, the relays for section 2 'can see only badly distorted current'. This is an important matter in the practical engineering of protective relays which engineers would have to carefully investigate.

23

Power Cables

Power cables for 10–60 kV began to prevail around the year of 1910 in Europe and North America. Today, power cable networks are widely used as essential social infrastructures of high-voltage meshed networks and distribution networks in urban areas all over the world.

It is fair to say that today's power systems cannot be discussed without a substantial knowledge of power cables. In this chapter, the fundamentals of power cables are studied, which are considered to be essential knowledge for most practical engineers regardless of their specialties or responsibilities.

23.1 Power Cables and Their General Features

23.1.1 Classification

High-voltage power cables of 30 kV or more can be classified into two types by the difference of insulation materials, and are called simply CV cables and OF cables.

The CV cable (Cross-linked polyethylene insulated with Vinyl sheathed cable) has another name, XLPE (Cross-Linked Poly-Ethylene). The main insulation is composed of the solid organic material XLPE.

In the continuous production processes, colloidal high-viscosity polyethylene in a temperature range of 120–130°C is extruded to cover the moving core conductor and then chemical molecular cross-linking is reacted in a hardening process by cooling with water.

For the OF (Oil-Filled) cable, the main insulation is oil-impregnated insulation papers. The conductor is wrapped in several layers of the paper in a continuous process and is dried in a vacuum chamber to remove humidity, and then is impregnated with insulating oils. Under the operating conditions, the cable is filled with pressurized oil which is supplied from a pressure tank installed at the end or jointing terminals.

Typical structures and roles of each laminated layer of CV and OF cables are shown in Figures 23.1a–c. Also, representative standards of the required testing voltages for the main insulation and outer covering are given in Tables 23.1a and b.

The practical application of OF cables for high voltages of 30 kV or more began to spread in the era around 1910. Conversely, the first applications of CV cables were realized around 1970.

CV cable is a modern development of the outstanding organic material XLPE realized by advanced chemical materials production technology, which has various outstanding properties such as excellent electrical insulation characteristics, mechanical strength, bending flexibility, chemical stability, heat-withstanding capability, etc., and outstanding production technology including continuous extrusion and cross-linking reaction processes.

© 2007 John Wiley & Sons, Ltd

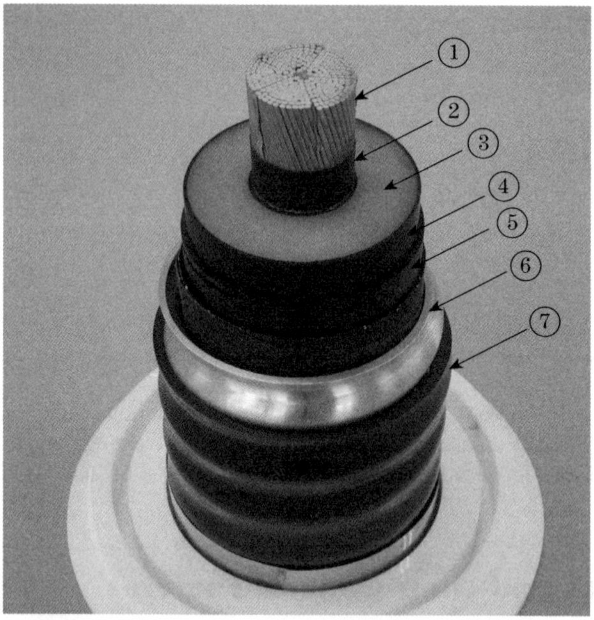

Courtesy of
Exsym corporation

Figure 23.1a CV cable

① **Conductor:** This consists of stranded, compacted copper or aluminium wires. For large sizes of conductor (typically 1000–2500 mm^2), Millikan construction or a compact segmental stranded conductor is used to reduce skin and proximity effects and a.c. conductor resistance.

② **Conductor screen:** This is employed to reduce excessive electrostatic stress or to smooth electric fields on voids between conductor and insulation. Polyethylene or other polymers with carbon black are extruded directly under the insulation.

③ **Cross-linked polyethylene insulation:** XLPE is the insulation material of CV cable, whose molecular structure is cross-linking polyethylene. XLPE is a chemically stable material and has outstanding properties of electrical insulation strength, mechanical strength (tension, compression, twisting, bending–flexibility, etc.) as well as stable thermal characteristics. Polyethylene has characteristics modified at 105–110°C and becomes fluid with high viscosity at over 120°C. Therefore, in the manufacturing process, the conductor can be continuously covered by XLPE by extrusion at a temperature around 120–125°C. The three layers of conductor screen (②), insulation (③) and insulation screen (④) are generally extruded simultaneously in one process so that irregularities (gap, crevices, voids, particles, etc.) can be avoided between the layers. Chemical cross-linking is treated in the cooling process after extrusion. Conversely, the allowable continuous operating temperature of CV cable is 90°C maximum.

④ **Insulation screen:** This is employed to reduce excessive stresses on voids between the insulation and metallic screen. Polyethylene or other polymers with carbon black are extruded directly on the insulation.

⑤ **Cushion layer:** For cables with a metallic sheath, cushion layers are applied between the insulation screen and metallic sheath for absorbing the expansion and contraction of the core by heat cycles. Semiconducting tape and/or fabric tape woven with copper wire are commonly used.

⑥ **Metallic screen:** This is used to confine the electric field within the cable. Copper tape is used for low- or medium-voltage cables (33 kV and under) and copper wires or metallic sheath are used for high- or ultrahigh-voltage cables (66 kV and over).

⑦ **Outer covering:** This is employed to protect the cable from mechanical damage or chemical deteriorating factors. PVC or PE is used for its good weather-resistant, abrasion-resistant and chemical-resistant properties. When a fire-retardant property is required, PVC is employed. In the case of a cable without a metallic sheath, plastic tape laminated with metal foil is often employed underneath the outer covering to prevent ingress of water.

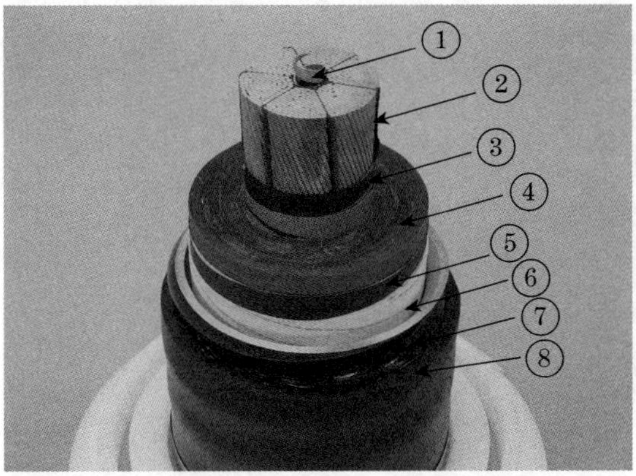

Courtesy of
Exsym corporation

Figure 23.1b OF cable

① **Oil duct:** Insulation oil fills the oil duct which is fed from the oil pressure tank installed at the cable end terminal (or jointing point for long lines). The oil pressure is always kept at a constant value regardless of the expansion or contraction of the insulating oil caused by cable temperature fluctuations. A steel strip open spiral is typically applied as the oil duct.

② **Conductor:** This consists of stranded, compacted copper or aluminium wires. For large sizes of conductor (typically 1000–2500 mm^2), Millikan construction or a compact segmental stranded conductor is used to reduce skin and proximity effects and a.c. conductor resistance.

③ **Conductor screen:** This is employed to reduce excessive electrostatic stress or to smooth electric fields on voids between the conductor and insulation. Carbon paper or metallized carbon paper is typically adopted.

④ **Oil-impregnating insulation paper:** Insulation of oil-filled cable is by a complex insulator composed of paper and oil. Several layers of insulation paper are wrapped around the conductor, dried in a vacuum chamber to remove humidity and impregnated with mineral or synthetic insulating oil. For ultrahigh-voltage cable, a synthetic blend of alkyl benzenes may be used for its superior electrical properties.

⑤ **Metallic screen:** Non-magnetic metal tape or metallized carbon paper is generally adopted as the metallic screen which is employed mainly for the purpose of electrical shielding.

⑥ **Binding tape:** This is applied to protect the metallic screen from external damage in the manufacturing process, and furthermore has the function to electrically conduct charging current to the metallic sheath.

⑦ **Metallic sheath:** This is made of corrugated aluminium or of lead, and has the function to maintain oil pressure, to protect the cable structure from external damage and, furthermore, to act as a conductor for the electrical earth–fault return pass.

⑧ **Outer covering:** This is employed to protect the cable from mechanical damage or chemical deteriorating factors. PVC or PE is used for its excellent weather-resistant, abrasion-resistant and chemical-resistant properties. When a fire-retardant property is required, PVC is employed.

Today, CV cables have been mostly adopted for lower voltage classes of 60/70 kV and distribution networks, while CV as well as OF are adopted together for higher voltages of 100–400 or 500 kV according to individual users' requirements.

OF cable is based on the traditional insulation method of 'paper and oil', so it is generally considered a very reliable product, while on the other hand it has some disadvantages from a practical application viewpoint that mainly concern maintenance work around the oil pressure tank and fear of fire. CV cable is quite an advanced 'large-size product' of even 2000–2500 m in length, which is made entirely by 'solid insulation'. Because it is based on solid insulation without oil, maintenance work is quite simple, but advanced and careful production technology/management is required in order to achieve chemically/mechanically quite stable and homogeneous insulation, or

Courtesy of Exsym Corporation

Figure 23.1c Various CV and OF cables

to avoid small particles, voids, irregular shapes, and so on. Advanced technology is also required for cable jointing.

In the past, before 1980–1985, CV cable was considered less reliable in comparison with OF cable, especially for higher voltage classes. However, today, the idea has taken root that CV cables are already as reliable as OF cables even for EHV classes. Today, due to advances in CV production technology and accumulated successful applications, CV cables have become the principal cables even for 200–400 kV classes. UHV lines of a.c. 500 kV CV cable lines as well as ±500 kV d.c. OF submarine cable lines have been put into practice in Japan as the highest voltage applications.

23.1.2 Unique features and requirements of power cables

Power cable has unique features among members comprising today's power systems in comparison with other station equipment or overhead line conductors. These features may be explained by the following three viewpoints.

Table 23.1a Test voltages of CV cable

	Rated voltage	Highest voltage for equipment	Value of U_0 for determination of test voltages	Voltage test (routine test)		Impulse voltage test (type test)	Switching impulse voltage test (type test)
	U kV	U_m kV	U_0 kV	$2.5U_0$ kV	Duration [min]	kV	kV
IEC 60840	45 to 47	52	26	65	30	250	—
	60 to 89	72.5	36	90	30	325	—
	110 to 115	123	64	160	30	550	—
	132 to 138	145	76	190	30	650	—
	150 to 162	170	87	218	30	750	—
IEC 62067	220 to 230	245	127	318	30	1050	—
	275 to 287	300	160	400	30	1050	850
	330 to 345	362	190	420	60	1175	950
	380 to 400	420	220	440	60	1425	1050
	500	550	290	580	60	1550	1175

U_0: the rated power frequency voltage between conductor and core screen for which the cable and its accessories are designed (rms).
U_m: the highest line-to-line voltage (rms) of the system that can be sustained under normal operating conditions (rms).
Besides, partial discharge test at 1.5 U_0 to confirm non-detectable discharge shall be carried out.
Notes:

1. U_0 is the same as $U/\sqrt{3}$, or the close value exceeding $U/\sqrt{3}$.
2. Impulse and switching voltage tests are performed as items of type test using a sample cable at least 10 m in length.
3. The applied waveform. Lightning impulse test: time to peak T_f: 1–5 μs, time to half value T_t: 40–60 μs. Switching impulse test: time to peak $T_f = 250\,\mu s \pm 20\%$, time to half-value $T_t = 2500\,\mu s \pm 60\%$.
4. The test voltages for OF cables can be considered as the same as in the table, although they are slightly different due to historical reasons.

23.1.2.1 *Variety of installation environment*

The largest feature of the power cable can be explained by the keywords 'extremely long and non-self-insulation-restoring products', and 'variety of installation environment.' Rolling and rerolling by large drums have to be repeated in the production, transportation and installation procedures because they

Table 23.1b Withstanding voltages of outer-covering layers

Rated voltages [kV]	Impulse withstanding voltages [kV]
~33	–
66–74	50
110–187	65
220–275	75

Note: The table is quoted from JEC 3402 (1990) in order to show the sensitive values of the withstanding voltages of the outer covering. The standard IEC 229 (1982) 'Test on oversheath' indicates the routine tests of 'Ad hoc voltage of 8 kV/mm of specified nominal thickness of the extruded oversheath for 1 min'.

are quite long. In particular, the installation environment is very varied, as illustrated by the following keywords picked randomly from this field:

- ground surface/soil-buried/duct/trench/tunnel/overhead/sea-bottom/river-bottom/sea-bottom-varied/hanging, etc.
- flat/slope/vertical
- hard/soft/muddy basis
- hot/cold/chemical-dirty, contamination
- tensile-tension/press/bending/mechanical beat, expansion/contraction
- soil vibration/water current/sea tide/wind power, etc.
- unequal-sink/earthquake/flood/fire, and so on.

Cable installation is practically based on large civil construction projects covering various and large environmental areas. Furthermore, fault pre-detection and repairing or retrying installations are not generally easy because of the specific nature of cables and of the installed environment.

23.1.2.2 Non-existence of 'rated capacity' or 'rated current'

The operating temperature of power cable is determined as the result of heat balance between heat generation (caused by the resistive current losses of conductors, the metallic sheath and dielectric losses of insulators) and heat discharge from the outer-covering surface to the installed environment. Heat generation is determined only by cable structure and electric condition (load current, sheath connection), whereas heat discharge is widely affected by the environmental conditions of cable installation. Accordingly, the calculation of heat balance and the temperature rise has to be conducted section by section as part of important installation design in order to maintain cable operating temperature within an allowable upper limit (typically 90°C for continuous current, 105°C for temporally over-current and 250°C for fault current within a few seconds).

This is the reason why the specific rated capacity or rated current does not exist for power cable. The allowable current limit has to be determined by cable conductor size and the individual conditions of cable installation.

23.1.2.3 Complexity of metallic sheath circuit and outer-covering insulation

Metallic sheath has an important role as the mechanical container to protect the delicate insulation and the electrical earth–fault current return pass by which induced current/voltage/noise interference with the outer environment can be reduced. The outer covering also has an important role to protect the cable from various mechanical, chemical, thermal, electrical stresses from the environment over a long time, even 30 years, and has allowable electrical and thermal upper limits. Figure 23.2b shows the typical insulation withstanding level of the outer covering.

Accordingly, we have to investigate not only the voltage/current behaviour on the conductors and main insulation, but also the induced voltage/current on the metallic sheath and the outer covering.

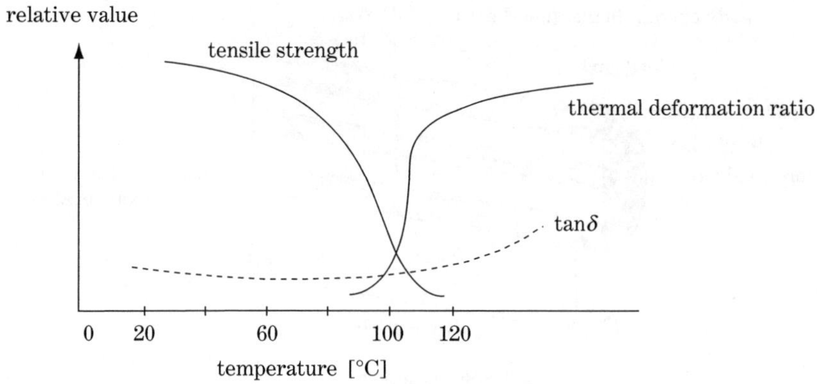

Figure 23.2 Physical characteristics of cross-linked polyethylene (XLPE)

We learned in Chapter 21 that cable lines have to be perfectly protected against various overvoltages from the viewpoint of insulation coordination. This is taken to be understood with the knowledge of the unique features for power cables described above.

23.2 Circuit Constants of Power Cables

Now, let us examine the constants of power cables referring to Table 2.3 in Chapter 2.

23.2.1 Inductances of cables

Power cable electrically consists of concentric cylinders with three layers of metal conductor, insulation material and metallic sheath as Figure 23.3a. When current flows through the conductor, sheath current is induced by the mutual inductance between the conductor and the sheath circuit. The induced flux by this conductor current appears partly within the cable sheath as ϕ_{in} and partly beyond the cable sheath as ϕ_{out}. Furthermore, ϕ_{out} would partly interlink the cable conductors/metallic sheaths of two other parallel-installed phases so that mutual inductances exist between the conductors/metallic sheaths of different phase cables. In other words, the cable line consists of three cables of **single phase type** as well as cable of **three-phase type** and can be expressed as a six-parallel-wire inductive circuit consisting of conductors a, b, c and metallic sheaths a′, b′, c′ and installed in parallel under ground–earth along with the earth surface as shown in Figure 23.3b.

Incidentally, cables of three-phase type or **triplex type** (the three twisted cables are bound) are well-balanced three-phase circuits by nature. In the case of a three single-phase-type cable circuit, the installed layout would obviously affect it somewhat as a phase-unbalancing factor. However, a three single-phase-type cable circuit with an **equilateral triangular installed layout** can be considered to be a well-balanced three-phase circuit. Also, long line cables with **cross-bond connection** (see Section 23.4) can be considered almost three-phase balanced. We assume here a three-phase-balanced circuit for simplification.

Referring to the equations of inductances, Equation 1.3–1.6 of Chapter 1, the related equations in regard to the three-phase-balanced cable inductances can be written as follows, where the earth effect is already taken into account:

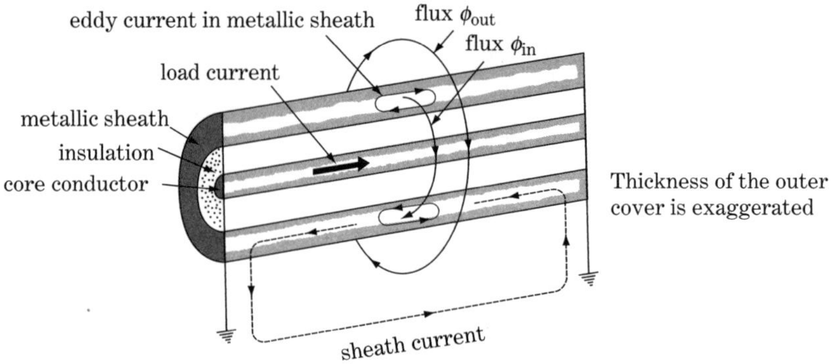

(a) Cable structure

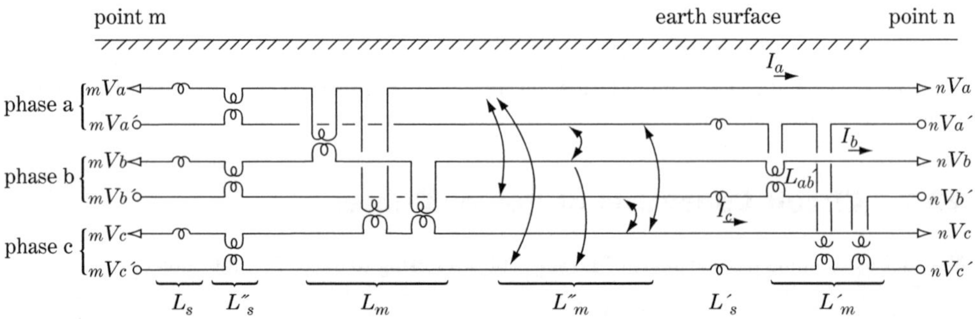

(b) Inductances

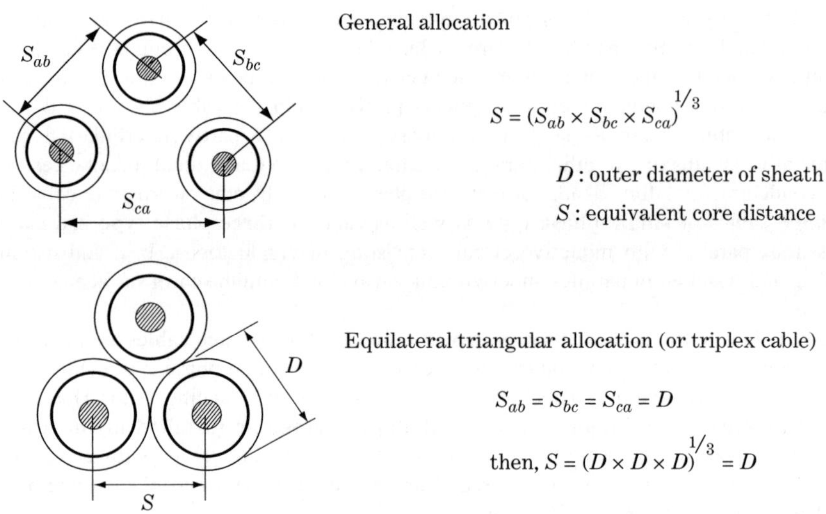

(c) Three-phase allocation

Figure 23.3 Inductance of cable

conductor voltages

$$\underset{{}_m\boldsymbol{V}_{abc}}{\begin{bmatrix} {}_m v_a \\ {}_m v_b \\ {}_m v_c \end{bmatrix}} - \underset{{}_n\boldsymbol{V}_{abc}}{\begin{bmatrix} {}_n v_a \\ {}_n v_b \\ {}_n v_c \end{bmatrix}} = \underset{\boldsymbol{Z}_{abc}}{\begin{bmatrix} Z_s & Z_m & Z_m \\ Z_m & Z_s & Z_m \\ Z_m & Z_m & Z_s \end{bmatrix}} \underset{\boldsymbol{I}_{abc}}{\begin{bmatrix} I_a \\ I_b \\ I_c \end{bmatrix}} + \underset{\boldsymbol{Z}''_{abc}}{\begin{bmatrix} Z''_s & Z''_m & Z''_m \\ Z''_m & Z''_s & Z''_m \\ Z''_m & Z''_m & Z''_s \end{bmatrix}} \underset{\boldsymbol{I}'_{abc}}{\begin{bmatrix} I'_a \\ I'_b \\ I'_c \end{bmatrix}} \quad ①$$

metallic sheath voltages

$$\underset{{}_m\boldsymbol{V}'_{abc}}{\begin{bmatrix} {}_m v'_a \\ {}_m v'_b \\ {}_m v'_c \end{bmatrix}} - \underset{{}_n\boldsymbol{V}'_{abc}}{\begin{bmatrix} {}_n v'_a \\ {}_n v'_b \\ {}_n v'_c \end{bmatrix}} = \underset{\boldsymbol{Z}''_{abc}}{\begin{bmatrix} Z''_s & Z''_m & Z''_m \\ Z''_m & Z''_s & Z''_m \\ Z''_m & Z''_m & Z''_s \end{bmatrix}} \begin{bmatrix} I_a \\ I_b \\ I_c \end{bmatrix} + \underset{\boldsymbol{Z}'_{abc}}{\begin{bmatrix} Z'_s & Z'_m & Z'_m \\ Z'_m & Z'_s & Z'_m \\ Z'_m & Z'_m & Z'_s \end{bmatrix}} \underset{\boldsymbol{I}'_{abc}}{\begin{bmatrix} I'_a \\ I'_b \\ I'_c \end{bmatrix}} \quad ②$$

${}_m\boldsymbol{V}_{abc}, {}_n\boldsymbol{V}_{abc}, \boldsymbol{I}_{abc}$: voltages and currents of metal conductors

${}_m\boldsymbol{V}'_{abc}, {}_n\boldsymbol{V}'_{abc}, \boldsymbol{I}'_{abc}$: voltages and currents of metal sheaths(earth connected at n)

where

$\boldsymbol{Z}_{abc} = j\omega\boldsymbol{L}_{abc}$ impedance matrix of three-phase conductors

$\boldsymbol{Z}'_{abc} = j\omega\boldsymbol{L}'_{abc}$ impedance matrix of three-phase sheaths

$\boldsymbol{Z}''_{abc} = j\omega\boldsymbol{L}''_{abc}$ mutual impedance matrix between conductors and sheaths

(23.1)

The equations can be transformed into the symmetrical components equations

conductor voltages

$$\underset{{}_m\boldsymbol{V}_{012}}{\begin{bmatrix} {}_m V_0 \\ {}_m V_1 \\ {}_m V_2 \end{bmatrix}} - \underset{{}_n\boldsymbol{V}_{012}}{\begin{bmatrix} {}_n V_0 \\ {}_n V_1 \\ {}_n V_2 \end{bmatrix}} = j\omega \underset{\boldsymbol{L}_{012}}{\begin{bmatrix} L_s+2L_m & & \\ & L_s-L_m & \\ & & L_s-L_m \end{bmatrix}} \cdot \underset{\boldsymbol{I}_{012}}{\begin{bmatrix} I_0 \\ I_1 \\ I_2 \end{bmatrix}} + j\omega \underset{\boldsymbol{L}''_{012}}{\begin{bmatrix} L''_s+2L''_m & & \\ & L''_s-L''_m & \\ & & L''_s-L''_m \end{bmatrix}} \cdot \underset{\boldsymbol{I}'_{012}}{\begin{bmatrix} I'_0 \\ I'_1 \\ I'_2 \end{bmatrix}} \quad ①$$

metallic sheath voltages

$$\underset{{}_m\boldsymbol{V}'_{012}}{\begin{bmatrix} {}_m V'_0 \\ {}_m V'_1 \\ {}_m V'_2 \end{bmatrix}} - \underset{{}_n\boldsymbol{V}'_{012}}{\begin{bmatrix} {}_n V'_0 \\ {}_n V'_1 \\ {}_n V'_2 \end{bmatrix}} = j\omega \underset{\boldsymbol{L}''_{012}}{\begin{bmatrix} L''_s+2L''_m & & \\ & L''_s-L''_m & \\ & & L''_s-L''_m \end{bmatrix}} \cdot \underset{\boldsymbol{I}_{012}}{\begin{bmatrix} I_0 \\ I_1 \\ I_2 \end{bmatrix}} + j\omega \underset{\boldsymbol{L}'_{012}}{\begin{bmatrix} L'_s+2L'_m & & \\ & L'_s-L'_m & \\ & & L'_s-L'_m \end{bmatrix}} \cdot \underset{\boldsymbol{I}'_{012}}{\begin{bmatrix} I'_0 \\ I'_1 \\ I'_2 \end{bmatrix}} \quad ②$$

${}_m\boldsymbol{V}_{012}, {}_n\boldsymbol{V}_{012}, \boldsymbol{I}_{012}$: voltages and currents of metal conductors

${}_m\boldsymbol{V}'_{012}, {}_n\boldsymbol{V}'_{012}, \boldsymbol{I}'_{012}$: voltages and currents of metal sheaths

L_s, L_m: self- and mutual inductances among three conductors

L'_s, L'_m: self- and mutual inductances among three sheaths

L''_s: mutual inductance between the conductor and the sheath of the same phase

L''_m: mutual inductance between one conductor and one sheath of different phases

(23.2)

Equation 23.2 ① shows that the positive-sequence inductance is given by $(L_s - L_m)$ under the condition $I'_1 = 0$ (sheath current zero), whereas some correction is necessary under the condition $I'_1 \neq 0$. The situation is the same for the zero-sequence inductance $(L_s + 2L_m)$.

Also in Equation 23.1 ①, we can imagine a special case in which the current I goes out through the phase a conductor and comes back through the phase b conductor so that no other currents exist on the phase c conductor, or on sheaths of three-phase cables, as well as on the earth circuit.

The equations are

$$I_a = -I_b = I, \quad I_c = 0, \quad I'_a = I'_b = I'_c = 0$$
$$\therefore \ {}_mV_a - {}_nV_a = j\omega(L_s - L_m)I$$

This is the same case as in Figure 1.2 of Chapter 1 and the inductance $(L_s - L_m)$ is of course the **working inductance** which is given by Equation 1.9. Accordingly, the working inductance of cable phase conductors is

$$L_s - L_m = 0.4605\log_{10}\frac{S}{r} + 0.05 \quad [\text{mH/km}] \tag{23.3}$$

where S is the average phase-to-phase distance of three parallel conductors (centre to centre)

$$S = (S_{ab}\cdot S_{bc}\cdot S_{ca})^{1/3} \ \text{(refer to Figure 23.3c)}$$
$$r: \text{radius of each conductor } (= d/2, \text{ where } d \text{ is the diameter})$$

In addition, $(L_s + 2L_m)$ is the inductance which is the same as that measured by the method of Figure 1.2b.

The sheath terminals are generally earth grounded at both end terminals as well as at the middle jointing terminals for long-distance cable lines, whereas exceptionally only one end terminal may be earth grounded for a short-distance line (typically for factory or in-house cable lines). Thus we need to check both cases.

Case-1: for the metallic sheath earth grounded at one terminal point n (for short-distance lines), this is the case where ${}_n\boldsymbol{V}'_{abc} = 0$, $\boldsymbol{I}'_{abc} = 0$ so that ${}_n\boldsymbol{V}'_{012} = 0$ and $\boldsymbol{I}'_{012} = 0$ in Equations 23.1 and 23.2. Accordingly,

$$\left.\begin{array}{ll}
\text{equation of main circuit} & \\
\left.\begin{array}{l}
{}_mV_1 - {}_nV_1 = j\omega L_1 I_1 \\
{}_mV_2 - {}_nV_2 = j\omega L_1 I_2 \\
{}_mV_0 - {}_nV_0 = j\omega L_0 I_0
\end{array}\right\} & ① \\
\left.\begin{array}{ll}
\text{positive-/negative-sequence inductance:} & L_1 = L_2 = L_s - L_m \\
\text{zero-sequence inductance:} & L_0 = L_s + 2L_m
\end{array}\right\} & ② \\
\text{induced sheath voltages at open sheath end point m} & \\
\left.\begin{array}{l}
{}_mV'_1 = j\omega(L''_s - L''_m)I_1 \\
{}_mV'_2 = j\omega(L''_s - L''_m)I_2 \\
{}_mV'_0 = j\omega(L''_s + 2L''_m)I_0
\end{array}\right\} & ③
\end{array}\right\} \tag{23.4}$$

Case-2: For the metallic sheath earth grounded at both terminal points m and n (general case), This is the case where ${}_mV'_{abc} = 0, {}_nV'_{abc} = 0$; then ${}_mV'_{012} = 0, {}_nV'_{012} = 0$ in Equations 23.1 and 23.2. Accordingly, currents I'_1, I'_2, I'_0 can be eliminated from both equations, then

$$\left.\begin{array}{l}
\text{equation of main circuit} \\
\left.\begin{array}{l} {}_mV_1 - {}_nV_1 = j\omega L_1 I_1 \\ {}_mV_2 - {}_nV_2 = j\omega L_2 I_2 \\ {}_mV_0 - {}_nV_0 = j\omega L_0 I_0 \end{array}\right\} \quad ① \\
\left.\begin{array}{l}
\text{positive-/negative-sequence inductance:} \\
L_1 = L_2 = (L_s - L_m) - \delta_1 \quad \text{where} \quad \delta_1 = -\dfrac{(L''_s - L''_m)^2}{L'_s - L'_m} \\
\text{zero-sequence inductance:} \\
L_0 = (L_s + 2L_m) - \delta_0 \quad \text{where} \quad \delta_0 = -\dfrac{(L''_s + 2L''_m)^2}{L'_s + 2L'_m} (\delta_1, \delta_0 \text{ are correction factors})
\end{array}\right\} \quad ② \\
\text{induced sheath currents through the metallic sheath between points m and n} \\
\left.\begin{array}{l}
I'_1 = -\dfrac{L''_s - L''_m}{L'_s - L'_m} \cdot I_1 \\
I'_2 = -\dfrac{L''_s - L''_m}{L'_s - L'_m} \cdot I_2 \\
I'_0 = -\dfrac{L''_s + 2L''_m}{L'_s + 2L'_m} \cdot I_0
\end{array}\right\} \quad ③
\end{array}\right\} \qquad (23.5)$$

To summarize cable inductance:

- The inductance of a cable line is affected by the dimensions of cables, the mutual allocation of three-phase conductors and the terminal condition of the metallic sheath.
- Working inductance $(L_s - L_m)$ is derived from Equation 23.3, whose variable parameters are the average distance S between the conductors and the radius r of the conductors.
- If the metallic sheath is open at one terminal, positive-sequence impedance is given by working inductance, namely $L_1 = L_s - L_m$. However, L_1 is a little smaller than the working inductance under the metallic sheath current-flowing condition because of correction factors δ_1, δ_0 by Equation 23.5 ②.

As a numerical check for working inductance $L_s - L_m$ by Equation 23.3:

- Case 1: $r = 25$ mm, $S = 75$ mm, $L_s - L_m = 0.4605 \log_{10}(75/25) + 0.05 = 0.269\,\text{mH/km}$
- Case 2: $r = 25$ mm, $S = 100$ mm, $L_s - L_m = 0.327\,\text{mH/km}$
- Case 3: $r = 25$ mm, $S = 150$ mm, $L_s - L_m = 0.408\,\text{mH/km}$

The working inductances of cables are typically 0.15–0.4 mH/km. Positive-sequence inductance L_1 would become much smaller than working inductance because most cable lines are metallic sheath earth grounded at plural terminal points. Accordingly, the typical value is $L_1 = 0.1 - 0.3\,\text{mH/km}$, which is 1/5–1/10 times smaller than that of overhead lines. (Refer to Section 6.2 and Table 2.2.)

23.2.2 Capacitance and surge impedance of cables

The capacitance of a three-phase cable line can be expressed as in Figure 23.4a, so it can be written as in the three-phase π-circuits of Figure 23.4b. Assuming the circuit constants are phase balanced for

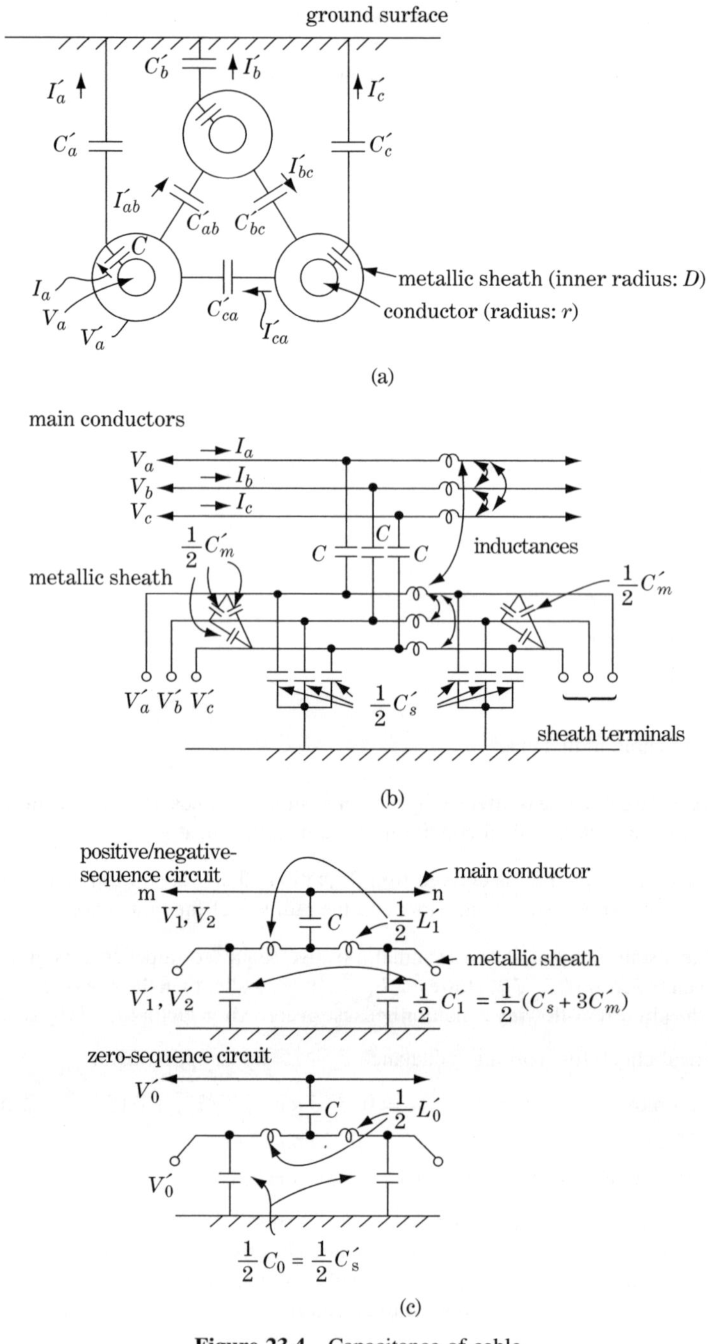

Figure 23.4 Capacitance of cable

simplicity, the symmetrical equivalent circuit of Figure 23.4c can be derived from Figure 23.4b in the same way as that for Figure 2.7.

The cable conductor is contained by the metallic sheath of an electrically concentric cylinder and the gap is filled with insulation material of dielectric constant ε. If electric charge $+q$ is given to the conductor and $-q$ is given to the metallic sheath, all the electric lines of force start from the conductor and reach the metallic sheath through radial pass, so that are not affected by the electrical conditions on the outer side of the sheath. Therefore, the capacitance C across the conductor and the metallic sheath is the typical case of a concentric cylinder which is quoted in most textbooks on electromagnetism and given by the equation

$$C = \frac{2\pi\varepsilon_0}{\log_e \frac{D}{r}} = \frac{\varepsilon_0}{2\times 9\times 10^9 \log_e \frac{D}{r}} [\mathrm{F/m}] = \frac{0.02413\varepsilon_0}{\log_{10}\frac{D}{r}} \quad [\mu\mathrm{F/km}] \tag{23.6}$$

where
r: radius of the conductor
D: radius of the sheath (inner radius)
ε_0: dielectric constant of insulation material

The equation corresponds with Equation 1.27 whose matrix size is 1 × 1, namely $\boldsymbol{v}_\mathrm{a} = \boldsymbol{p}_\mathrm{aa} \times \boldsymbol{q}_\mathrm{a}$. The circuit equation of Figure 23.4a–c can be written as follows. For the a–b–c domain

$$\left.\begin{array}{ll}
\text{leakage current} & \\
\left.\begin{array}{l} I_a = j\omega C(V_a - V_a') \\ I_b = j\omega C(V_b - V_b') \\ I_c = j\omega C(V_c - V_c') \\ \text{from conductor to sheath} \end{array}\right. & \text{①} \\
\begin{array}{ll} I_a' = j\omega C_a' \cdot V_a' & I_{ab}' = j\omega C_{ab}' \cdot (V_a' - V_b') \\ I_b' = j\omega C_b' \cdot V_b' & I_{bc}' = j\omega C_{bc}' \cdot (V_b' - V_c') \\ I_c' = j\omega C_c' \cdot V_c' & I_{ca}' = j\omega C_{ca}' \cdot (V_c' - V_a') \\ \text{from sheath to earth} & \text{from sheath to sheath} \end{array} & \text{②}
\end{array}\right\} \tag{23.7}$$

and for the 0–1–2 domain

$$\left.\begin{array}{ll}
\text{leakage current} & \\
\begin{array}{l} I_1 = j\omega C(V_1 - V_1') = j\omega C_1' V_1' \\ I_2 = j\omega C(V_2 - V_2') = j\omega C_1' V_2' \\ I_0 = \underbrace{j\omega C(V_0 - V_0')}_{\text{from conductor to sheath}} = \underbrace{j\omega C_0' V_0'}_{\text{from sheath to earth}} \end{array} & \text{①} \\
\text{induced sheath voltage} & \\
\left.\begin{array}{l} V_1' = \dfrac{C}{C + C_1'} \cdot V_1 \\ V_2' = \dfrac{C}{C + C_1'} \cdot V_2 \\ V_0' = \dfrac{C}{C + C_0'} \cdot V_0 \\ \text{where} \quad C_1' = C_s' + 3C_m', \quad C_0' = C_s' \end{array}\right\} & \text{②}
\end{array}\right\} \tag{23.8}$$

Note that the metallic sheath end terminals are earth grounded generally at both ends or at least at one end (for short-distance lines), so $V'_a = V'_b = V'_c = 0$ and $V'_0 = V'_1 = V'_2 = 0$. Consequently, Figure 23.7① and Figure 23.8① are simplified as follows.

Then, in the a–b–c domain,

$$\begin{bmatrix} I_a \\ I_b \\ I_c \end{bmatrix} = j\omega C \begin{bmatrix} V_a \\ V_b \\ V_c \end{bmatrix} \tag{23.9}$$

and in the 0–1–2 domain

$$\begin{bmatrix} I_0 \\ I_1 \\ I_2 \end{bmatrix} = j\omega C \begin{bmatrix} V_0 \\ V_1 \\ V_2 \end{bmatrix} \tag{23.10}$$

That is, the capacitance of a cable line is $C_1 = C_2 = C_0 = C$ where C is given by Equation 23.6 in the symmetrical components domain.

The stray capacitance C is determined only by the specific physical size of D and r of the cable and the dielectric constant ε of the insulation material layer, so it is not affected by the cable installation layout; inductance L is, however, affected by the layout, so the value of C is given in the specification supplied by the cable manufacturer.

As a result, the surge impedance and the travelling-wave velocity of the cable are

	surge impedance	surge velocity
positive-sequence circuit	$Z_{1\text{surge}} = \sqrt{\dfrac{L_1}{C}}$	$u_1 = \dfrac{1}{\sqrt{L_1 C}}$
zero-sequence circuit	$Z_{0\text{surge}} = \sqrt{\dfrac{L_0}{C}}$	$u_0 = \dfrac{1}{\sqrt{L_0 C}}$

(23.11)

Numerical examples are shown in Table 2.2 and Section 2.5.

23.3 Metallic Sheath and Outer Covering

23.3.1 Role of metallic sheath and outer covering

High-voltage power cable is produced and transported as coiled cable on a drum and the length of cable per drum is typically 500–2000 m. These cable production units are connected in series through cable joints for the route whose span is longer than the production unit length. Sheath terminals are prepared at each end of this production unit.

There are various types of power cables due to the variety of electrical requirements as well as mechanical and installed surroundings requirements (including submarine cable). Accordingly, the metallic sheath as well as the outer covering also have various structures. The metallic sheath of high-voltage CV cables or of OF cables over 60 kV consists generally of a corrugated aluminium sheath or a lead sheath (and with steel or copper wires as reinforced members if necessary) so that the cable can be electrically considered as a metal cylinder containing concentric allocated conductor and insulation materials.

The major roles of the **metallic sheath** are:

- Containing and protection of insulation materials (cross-linked polyethylene or oil-immersed paper layers)
- Isolation from the outer environment (air, water, contaminants)
- Sharing mechanical strength
- Return pass of fault currents, continuous unbalanced sheath currents
- Electrical shield (electrostatic shield, electromagnetic shield).

The major roles of the **outer covering** are as follows:

- To protect the cable from mechanical damage or chemical deterioration and to isolate the metallic sheath in order to secure against electric shock by touching.
- The outer covering is a kind of insulation material made typically of polyvinyl chloride (PVC) or polyethylene (PE), so it should be protected electrically against induced sheath voltages. Furthermore, the outer covering has to be protected against temperature rise caused as a result of the heat imbalance of joule-loss generation by the main conductor current plus metallic sheath current and heat radiation to earth.
- Possesses outstanding physical characteristics with weather-resistant, abrasion-resistant and chemical-resistant properties and a fire-retardant property.

23.3.2 Metallic sheath earthing methods

A power cable is equipped with earth-connecting metallic sheath terminals at each end of the production unit. Engineering practices and their principles in regard to metallic sheath terminal earthing is investigated here.

23.3.2.1 Double sheath end terminals earthing method (solid-sheath-bonding method)

The features are as follows:

- Stationary sheath voltage is kept by the earth potential (0 V) practically at any point along the cable length. Accordingly, the outer covering is released from sheath voltage stresses. (The situation is different for surge phenomena.)
- Sheath currents induced by unbalanced continuous load or by temporary induced current flow cause thermal losses on the metallic sheath. Accordingly, the outer covering as well as the insulation layer are thermal heated by the main conductor current plus the sheath current. The thermal heat of the metallic sheath by the sheath current could be a disturbing factor of thermal heat diffusion from the main conductor to the earth.

The **solid-sheath-bonding method** is an typical appellation of practices in which plural sheath terminals are earth connected. (refer the Section 23.4)

23.3.2.1.1 Sheath current caused by electromagnetic induction

The sheath currents I'_1, I'_2, I'_0 are calculated from Equation 23.5 ③. As a closed circuit is composed through the metallic sheath and the earth ground, sheath current I'_1 is always induced by I_1. I'_2, I'_0 would also be induced if I_2 or I_0 existed.

The temperature rise in the metallic sheaths caused by sheath current as well as by conductor current have to be considered from the viewpoint of heat balance of the cable, in that main insulation materials as well as outer-covering materials have to be kept within allowable temperatures.

23.3.2.1.2 *Electrostatic induction*

The sheath potential is kept at earth voltages at both end terminals so that serious electrostatic voltage would not be caused in this case.

23.3.2.2 *Single sheath end terminal earthing method (single-sheath-bonding method)*

The features are as follows:

- Sheath voltage on the open-sheath terminal or in the closer zone appears by electrostatic induction, which would be calculated by Equation 23.4 ③. The outer covering has to be protected against the induced voltage in regard to electrical insulation as well as corrosion. Therefore this practice can be adopted only for short-length cable line, typically in-house cable lines.
- Sheath currents do not flow through the sheath, so thermal heat generation on the metallic sheath is not caused.

Referring to Equation 23.4 ③, the induced power frequency voltage on the sheath can be calculated by the equation below for the sheath voltage under normal conditions (power frequency)

$$E = j\omega(l'_s - l'_m) \cdot I \quad [\mathrm{V/km}] \tag{23.12}$$

where

$l'_s - l'_m$: mutual inductance per km between the conductor and the sheath [Ω/km]

I: conductor current [A]

As a numerical check, the nominal induced sheath voltage by the single-sheath-bonding method is, for the conditions $f = 50$ Hz, $(l'_s - l'_m) = 2\,\mathrm{mH/km}$,

$$E = j2\pi \cdot 50 \cdot 2 \times 10^{-3} \cdot I = 0.628 \cdot I \quad [\mathrm{V/km}]$$
$$\therefore\ E = 62.8[\mathrm{V/km}] \quad \text{for}\, I = 100[\mathrm{A}],\ \text{or}\, E = 314[\mathrm{V/km}] \quad \text{for}\, 500[\mathrm{A}]$$

This is quite a large voltage from the viewpoint of human security as well as the capability of the outer covering insulation, so the single-sheath-bonding method can be adopted only for short-distance lines of within a few hundred meter.

23.4 Cross-bonding Metallic-shielding Method

23.4.1 Cross-bonding method

In regard to long-distance high-voltage cable lines with single-conductor-type power cables and with a number of cable jointing points, the constants of both the main conductors L_{abc}, C_{abc} and the sheath circuits L'_{abc}, C'_{abc} would be forced to become quite unbalanced. In particular, the imbalance in L'_{abc}, C'_{abc} causes **unbalanced continuous sheath currents**, leading to undesirable temperature rises in the cables even if the main conductor's three-phase currents are balanced. Moreover, the surge voltages transmitted to the main conductors for any reason would induce quite serious unbalanced **surge voltages on the sheath metals** of each section, by which the insulation of the outer-covering layer could be damaged or broken. The widely adopted practical countermeasure to overcome this problem is the **cross-bonding metal-shielding method**. The simply called **cross-bonding connection** is a kind of transposition of three-phase sheath circuits as shown in Figure 23.5a, by which cable sheath ends are earth connected at every three sections and transposition of the sheath connection is conducted within the three sections.

Figure 23.5b shows details around the jointing boxes of the coupled three sections, where four jointing points I, II, III, IV are shown. The sheath terminal ends are earth grounded at jointing boxes I and IV, while at jointing boxes II and III the sheath terminal ends are cross-connected without earth grounding. In other words, the cable sheath terminals are earth grounded at the jointing boxes every three spans, and twice cross-connected at the cross-bonding jointing boxes in the three spans. Under this practice, the inductance L'_{abc} and the capacitance C'_{abc} of the sheath circuits can be approximately three-phase balanced over the total length of the three spans, although they may not necessarily be balanced at each longitudinal section. The major expected effects are obviously (1) reducing induced sheath voltages; (2) restraining the temperature rise of insulators; and (3) restricting interference with the environment. One disadvantage of this practice is that the sheath terminals are earth grounded longitudinally at every three spans instead of at every span. Therefore, surge voltage and current protection of the sheath circuit is vitally important, because the interval of earth connection is three times longer.

Now we examine surge phenomena arising on the cable conductors and sheath circuits.

23.4.2 Surge voltage analysis on the cable sheath circuit and jointing boxes

Surge phenomena arising on the sheath circuit should be carefully investigated, because electromagnetic and electrostatic coupling densities between the cable conductor and the metal sheath are quite high for all the frequency zone.

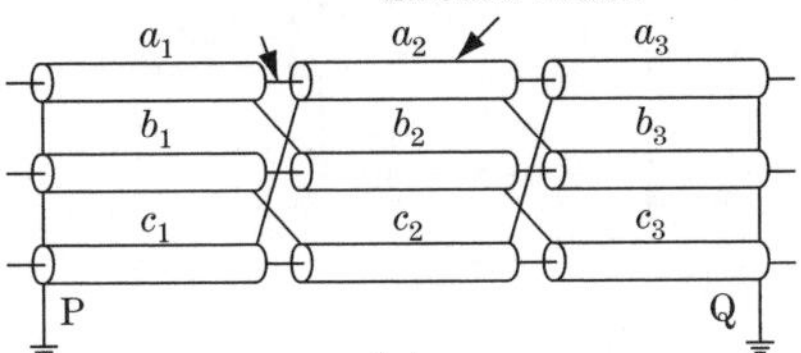

(a)

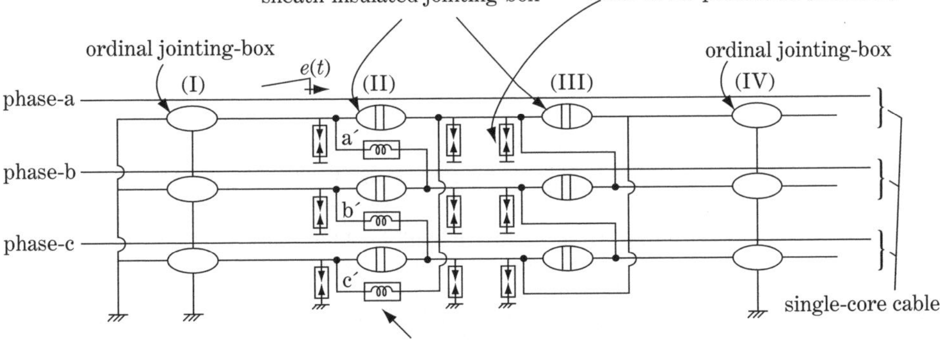

(b) Cross-bonding metallic-shielding method

Figure 23.5 Cross-bonding metallic shielding method

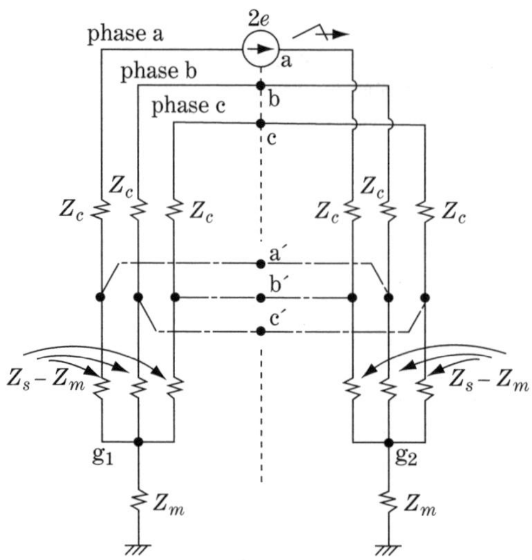

a, b, c: conductor jointing points

a´, b´, c´: metallic sheath jointing points

Z_c: surge impedance across the conductor and metallic sheath

Z_s: self-surge impedance across the metallic sheath and earth

Z_m: mutual surge impedance across the sheaths of different phases

(c) the equivalent circuit-1 of insulation box II

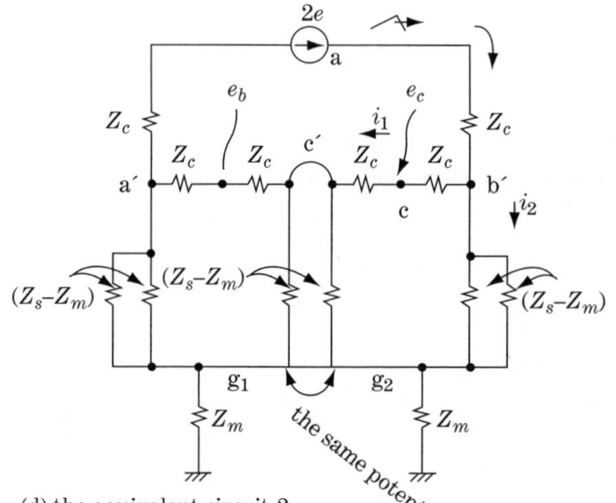

(d) the equivalent circuit-2

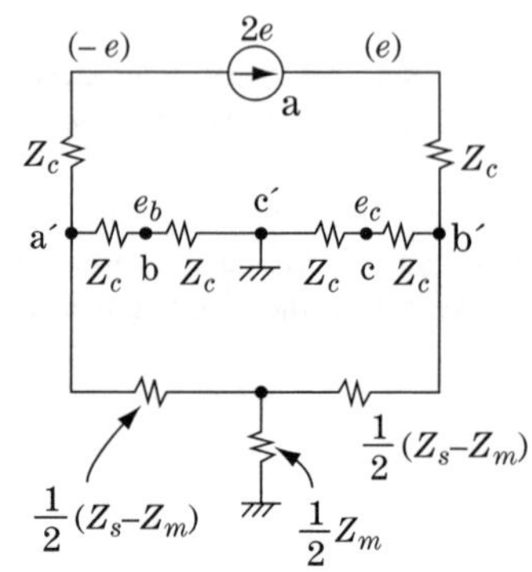

(e) the equivalent circuit-3

Figure 23.5 (*Continued*)

We undertake below a surge voltage analysis of the sheath ends at the intermediate jointing box II with a cross-bond connection which would be caused by the incident surge voltage $e(t)$ from the main conductors.

In Figure 23.5b, incident surge voltage e travelling through the phase a conductor arrives at the insulated jointing box II. The surge impedance circuit of the jointing box can be written as in Figure 23.5c where surge source voltage $2e$ is inserted at point a.

Incidentally, the reason why the source voltage $2e$, instead of e, is to be inserted in this case are understandable by the explanation of Figure 23.6 and Equation 23.16 in the next section.

Figure 23.5c is vertically symmetrical, so it can be modified to Figure 23.5d. Next, the latter figure is also symmetrical, so points g_1 and g_2 should have the same potential. Accordingly this figure is simplified as in Figure 23.5e.

The impedance looking down into circuit from points a′ and b′ is

$$Z_{b'a'} \equiv \{(4Z_c)//(Z_s - Z_m)\} = \frac{4Z_c(Z_s - Z_m)}{4Z_c + (Z_s - Z_m)} \tag{23.13}$$

The voltage between points a′ and b′ is

$$\left.\begin{aligned} e'_b - e'_a &= \frac{Z_{b'a'}}{2Z_c + Z_{b'a'}} \cdot 2e = \frac{(Z_s - Z_m)}{4Z_c + 3(Z_s - Z_m)} \cdot 4e \equiv m \cdot 4e \\ \text{where} \quad m &= \frac{Z_s - Z_m}{4Z_c + 3(Z_s - Z_m)} \end{aligned}\right\} \tag{23.14}$$

This equation means that the initial surge voltage of magnitude $4me$ appears between the points a′ and b′. This voltage is shared by the four Z_c in Figure 23.5b. Accordingly, the surge potentials at each point are $2me$ for point b′, me for point c, zero for point c′ and $-me$ for points b. Totally, we have the solution as in the equations below.

The initial induced surge arising on the conductors and the sheath of each phase at the jointing box II is, for transmitted wave voltages,

$$\begin{array}{ll} \text{the conductor voltage} & \text{the sheath voltage} \\ \left.\begin{aligned} e_a &= e \\ e_b &= -me \\ e_c &= me \end{aligned}\right\} ① & \left.\begin{aligned} e'_a &= -2me \\ e'_b &= 2me \\ e'_c &= 0 \end{aligned}\right\} ② \end{array} \tag{23.15a}$$

Furthermore, we need to consider the longitudinal induced voltages across the right and left terminals of the insulated jointing box II in Figure 23.5b, as follows:

$$\begin{array}{llll} \text{sheath box phase a} & e'_a = -2me & e'_b = 2me & e'_a - a'_b = -4me \\ \text{sheath box phase b} & e'_b = 2me & e'_c = 0 & e'_b - e'_c = +2me \\ \text{sheath box phase c} & e'_c = 0 & e'_a = -2me & e'_c - e'_a = +2me \\ & \text{left-side terminals} & \text{right-side terminals} & \text{across voltage} \end{array} \tag{23.15b}$$

$$m = \frac{Z_s - Z_m}{4Z_s + 3(Z_s - Z_m)}$$

In the figure, the voltage $4me$ would appear across the left and right terminals of the phase a insulating box. Arresters as well as current restraining devices (inductances) have to be adopted as shown in this figure.

As a numerical check, although the surge impedances of cables are affected by the design structure of the cable and the installed conditions, the individual differences are rather small. Let us assume $Z_c = 15\,\Omega$, $Z_s = 25\,\Omega$, $Z_m = 13\,\Omega$. In this case, m can be derived from Equation 23.14 as $m = 0.125$ and, accordingly:

- the surge voltage of the conductors (transmitted wave) is $e_a = e$, $e_b = -0.125e$, $e_c = +0.125e$
- the surge voltages of the sheaths are $e'_a = -0.25e$, $e'_b = +0.25e$, $e'_c = 0$
- the surge voltage across the insulated joint box is $e'_a - e'_b = -0.5\,e$, $e'_b - e'_c = +0.25\,e$, $e'_c - e'_a = +0.25\,e$.

The above values of sheath voltage of $0.25e$ as well as the voltage across the jointing box of $0.5e$ are relatively large values, so without appropriate protection by the arresters, the insulation of the cable

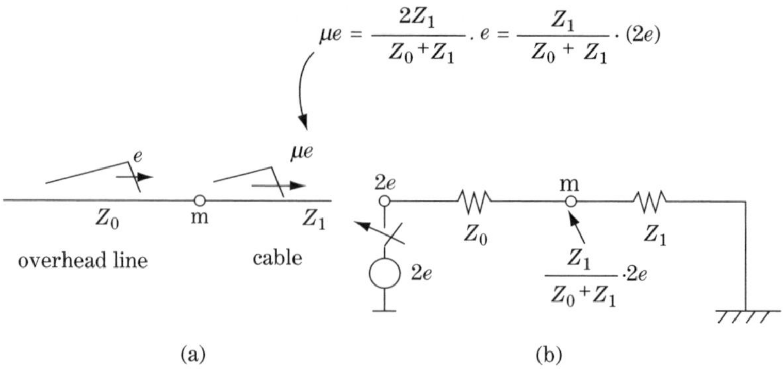

Figure 23.6 The transition point of overhead line and cable

outer-covering layer or that across the cable joint sheath would be damaged by conductor surge e. Figure 23.5b also shows the practice of surge protection by arresters.

23.5 Surge Voltages Arising on Phase Conductors and Sheath Circuits

In Figure 23.6a, lines Z_0 and Z_1 are connected at the transition point m, and surge voltage e is travelling on line Z_0 (surge impedance Z_0 may be an overhead line) from left to right and arrives at m. Immediately the transmitted wave voltage μe begins to travel on line Z_1 (may be a cable line), where $\mu = 2Z_1/(Z_0 + Z_1)$ is the transmitted wave operator from Z_0 to Z_1. Thus

$$\mu \cdot e = \frac{2Z_1}{Z_0 + Z_1} \cdot e = \frac{Z_1}{Z_0 + Z_1} \cdot (2e) \tag{23.16}$$

This equation shows that the transmitted wave voltage μe in Figure 23.6a is given by the same equation for the voltage at point m in Figure 23.6b in which source voltage $2e$ is going to be switched to the circuit. In other words, the transmitted wave μe in Figure 23.6a can be calculated as the voltage of the source voltage $2e$ divided by Z_0 and Z_1.

Bearing the above explanation in mind, we examine the surge phenomena arising in the cable conductors and the metallic sheaths as shown in Figure 23.7a.

23.5.1 Surge voltages at the cable connecting point m

Figure 23.7a is the diagram we are going to examine, where incident surge e is coming from the overhead transmission line Z_0 to the cable line. For this condition, we need to calculate surge voltages e_1 (conductor voltage across the cable insulation layer) and e_2 (sheath voltage across the outer layer) at the jointed cable terminal point m. The voltages can be calculated analogously to Equation 23.16 as follows.

The situation of surge impedances around point m and the equivalent circuit are as in diagrams b1 and b2.

The voltage appearing across the cable insulation layer (**coaxial mode wave**) is

$$e_1 = \frac{2Z_{12}}{Z_0 + Z_{12} + \dfrac{Z_2 Z_{s1}}{Z_2 + Z_{s1}}} \cdot e \quad \text{where} \quad \frac{Z_2 Z_{s1}}{Z_2 + Z_{s1}} = (Z_2 // Z_{s1})$$

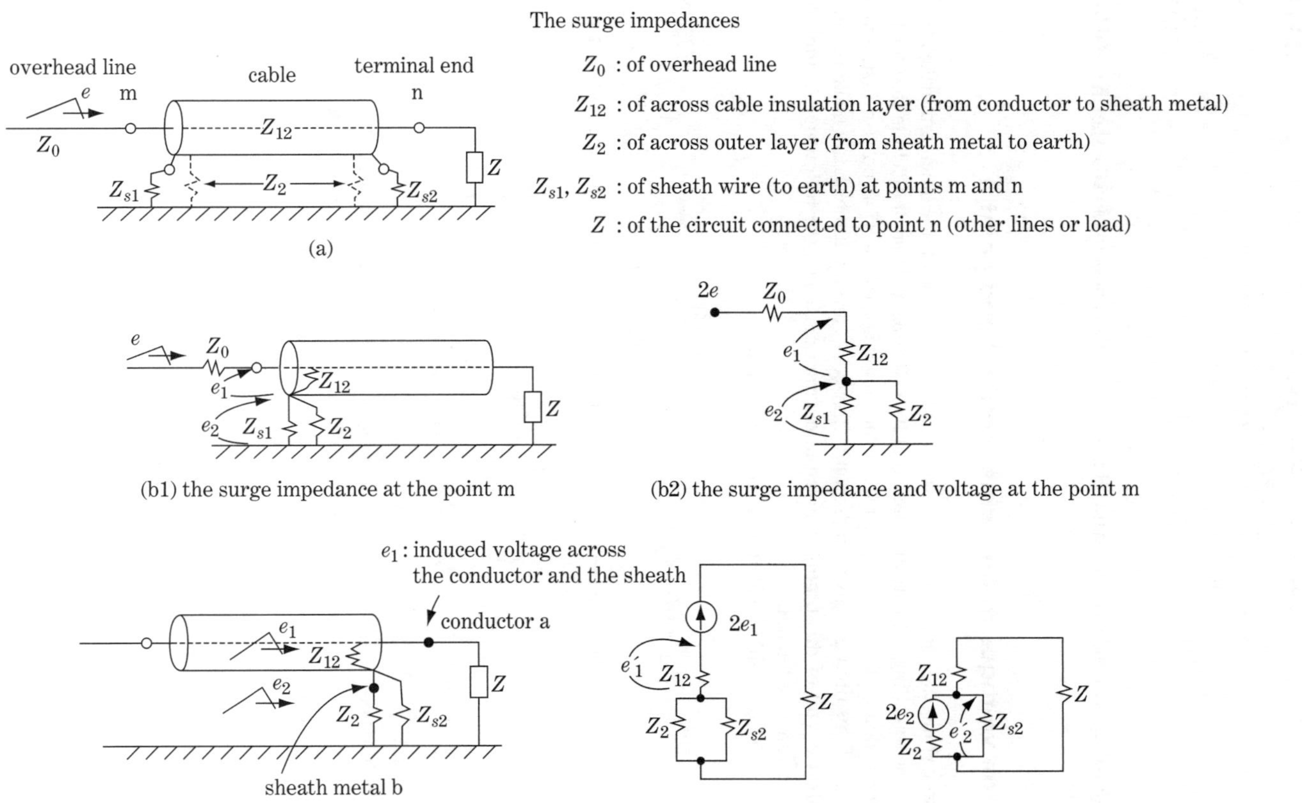

(b1) the surge impedance at the point m

(b2) the surge impedance and voltage at the point m

(c1) the surge impedance around the point n (the cable end)

(c2) the surge impedance and voltage distribution across the insulation layer at the point n (coaxial mode)

(c3) the surge impedance and voltage distribution across the outer layer at the point n (line-to-ground mode)

Figure 23.7 Surge voltages arising on the cable conductors and the metallic sheath

The voltage across the outer covering (**sheath to ground mode wave**) is

$$e_2 = \frac{2 \cdot \dfrac{Z_2 Z_{s1}}{Z_2 + Z_{s1}}}{Z_0 + Z_{12} + \dfrac{Z_2 Z_{s1}}{Z_2 + Z_{s1}}} \cdot e \tag{23.17b}$$

or

$$e_2 = \frac{Z_2 // Z_{s1}}{Z_{12}} \cdot e_1 \tag{23.17c}$$

The above surge voltages e_1 and e_2 begin to travel along the cable conductor and the sheath metal from left to right.

23.5.2 Surge voltages at the cable terminal end point n

The induced surge voltages e_1 and e_2 at point m travel on the cable towards the other terminal end point n (diagram c1). The travelling velocity of e_1 may be 130–150 m/μs and that of e_2 a little slower (due to the line-to-ground mode). The surge e_1 arrives at the conductor (point a) at point n first, and the equivalent circuit of this timing is given in diagram c2. Next, e_2 arrives at the sheath metal (point b), and the equivalent circuit of this timing is given in diagram c3. Then, the following equations are derived where we ignore attenuation.

The voltage across the cable insulation layer at point n is

$$e_1' = \frac{2Z_{12}}{Z_{12} + (Z_2 // Z_{s2}) + Z} \cdot e_1 = \frac{2Z_{12}}{Z_{12} + \dfrac{Z_2 Z_{s2}}{Z_2 + Z_{s2}} + Z} \cdot e_1 \tag{23.18a}$$

The voltage across the cable outer-covering layer (between the sheath and the earth) at point n is

$$\left.\begin{aligned} e_2' &= \frac{2Z_r}{Z_2 + Z_r} \cdot e_2 \\ \text{where} \quad Z_r &= \{Z_{s2} // (Z_{12} + Z)\} = \frac{Z_{s2}(Z_{12} + Z)}{Z_{s2} + (Z_{12} + Z)} \end{aligned}\right\} \tag{23.18b}$$

The cable insulation layer has to withstand the surges e_1, e_1' and their following reflected waves. Of course, the above equations give only the initial transferred wave voltages before reflection, so the actual voltage at arbitrary time t after arrival should be calculated by the lattice method.

If our concern is the stress on the insulation layer, we can assume that the sheath is 'ideal earth grounded'; in other words, the surge impedance of the sheath is zero ($Z_{s1} \to 0$, $Z_{s2} \to 0$), neglecting the earth mode travelling surge e_2, e_2' on the outer-covering layer. In this case, only the travelling surges of the coaxial mode e_1, e_1' exist, and e_1, e_1' can be calculated as pessimistic (large) values:

$$\left.\begin{aligned} e_1 &= \frac{2Z_{12}}{Z_0 + Z_{12}} \cdot e \\ e_1' &= \frac{2Z_{12}}{Z_{12} + Z} \cdot e \\ &\text{where} \quad Z_{s1} \to 0, \quad Z_{s2} \to 0 \\ &Z : \text{surge impedance of outer circuit} \end{aligned}\right\} \tag{23.19}$$

23.6 Surge Voltages on Overhead Line and Cable Combined Networks

23.6.1 Overvoltage behaviour on cable line caused by lightning surge from overhead line

Now we examine the overvoltage behaviour of the cable system when a lightning surge (incidental surge) is injected from the connected overhead line, as shown in Figure 23.8.

Referring to Figure 18.10 and Equation 18.51, the overvoltages appearing at the junction points m and n can be calculated as follows:

$$e_t = \mu \cdot e = \frac{2Z_1}{(Z_0 + Z_1)} \cdot e \quad \text{(the transmitted wave voltage at point m)} \qquad ①$$

the point m

$$V\,(0-2)T \qquad (2-4)T \quad (4-6)T \quad (6-8)T \quad (8-10)T$$

$$v_m = e_t[(1) + \alpha^2(1+\rho_1')\rho_2\{1 + (\alpha^2\rho_1'\rho_2) + (\alpha^2\rho_1'\rho_2)^2 + (\alpha^2\rho_1'\rho_2)^3 \cdots\}] \Rightarrow \frac{1+\alpha^2\rho_2}{1-\alpha^2\rho_1'\rho_2} \cdot e_t \qquad ②$$

the point n

$$(0-1)T \qquad (1-3)T \quad (3-5)T \quad (5-7)T \quad (7-9)T$$

$$v_n = e_t[(0) + \alpha(1+\rho_2)\{1 + (\alpha^2\rho_1'\rho_2) + (\alpha^2\rho_1'\rho_2)^2 + (\alpha^2\rho_1'\rho_2)^3 \cdots\}] \Rightarrow \frac{\alpha(1+\rho_2)}{1-\alpha^2\rho_1'\rho_2} \cdot e_t \qquad ③$$

(23.20)

It is obvious from the above equations that v_m, v_n would become larger if ρ_1', ρ_2 were positive large values (namely, Z_0, $Z_2 > Z_1$) and would become the largest under the conditions $Z_0 \gg Z_1$, $Z_2 \gg Z_1$.

We need then to investigate Figure 23.8 as a typical severe case with cable $Z_1 = 30\,\Omega$, overhead line $Z_0 = 300\,\Omega$ (sufficiently long), $Z_2 = \infty$ (open end).

Accordingly,

$$\text{the point m}: \rho_1' = (Z_0 - Z_1)/(Z_0 + Z_1) = 1(300-30)/(300+30) = +0.82$$

$$\text{the point n}: Z_2 = \infty \text{ (open end)}: \ \rho_2 = (Z_2 - Z_1)/(Z_2 + Z_1) = 1$$

$$e_t = 2 \times 30/(300+30) \cdot e = 0.182e \qquad ①$$

$$(0-2)T \quad (2-4)T \quad (4-6)T \quad (6-8)T \quad (8-10)T$$

$$v_m = e_t[(1) + 1.82\alpha^2\{1 + (0.82\alpha^2) + (0.82\alpha^2)^2 + (0.82\alpha^2)^3 \cdots\}] \Rightarrow (1+\alpha^2)/(1-0.82\alpha^2) \cdot e_t \qquad ②$$

the point n

$$(0-1)T \quad (1-3)T \quad (3-5)T \quad (5-7)T \quad (7-9)T$$

$$v_n = e_t[(0) + 2\alpha\{1 + (0.82\alpha^2) + (0.82\alpha^2)^2 + (0.82\alpha^2)^3 \cdots\}] \Rightarrow 2\alpha/(1-0.82\alpha^2) \cdot e_t \qquad ③$$

(23.21)

The equation means that v_m and v_n repeat reflection at both terminals and the values are simply increased over time towards the final converged value. Assuming $\alpha = 1.0$ (attenuation zero), the converged value of v_m and v_n is $11.1e_t$, which is a large and severe value.

Fortunately, however, the above severe values are unrealistic for the following reasons:

a) This is a calculation in which the incidental surge voltage e of step-wave form continues infinitely. The lightning surges are not with infinitive tail length, so that they would disappear within

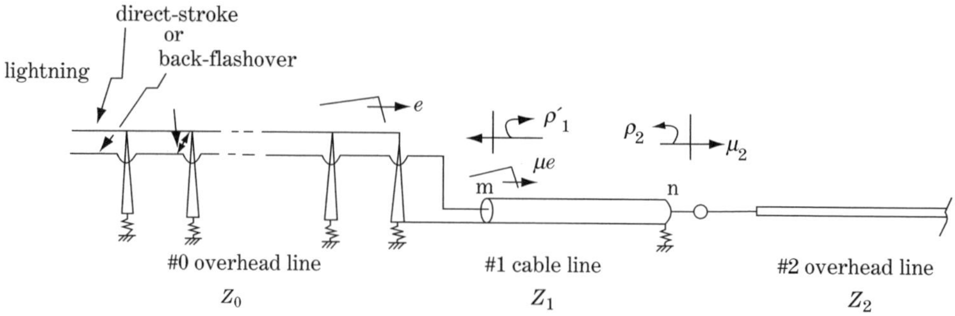

Figure 23.8 Lightning surge on the cable line

100–200 μs. Therefore the step voltage in the above calculation should be replaced by the realistic impinging waveform of a typically standard impulse wave of $1.2 \times 50\,\mu s$ (see Figure 21.10b).

b) $Z_0 > Z_1$. Accordingly the transmitted operator μ at point m from the overhead line (Z_0) to the cable (Z_1) is less than 1.0 ($\mu = 0.182$ in the above calculation). In other words, the original surge voltage e diminishes to the first transmitted wave voltage $\mu \cdot e$.

c) The diminishing effect by attenuation operator $\alpha\,(0 < \alpha < 1)$ may be significant. The converged values calculated by Equation 23.20 are 3.45, 1.70 and 1.33 for $\alpha = 0.8$, 0.6 and 0.4, respectively. The resistive constants R and G of the cable act as attenuating factors of the cable.

d) In normal operation of the network, smaller Z_2, Z_0 are expected, which would significantly reduce the voltages. If the cable terminal point n is connected, for example, to five parallel circuits of overhead lines, Z_2 would be around $300/n = 60\,\Omega$; accordingly $\rho_2 = 0.33$, so v_m, v_n are significantly reduced. If the double circuit cable line is connected to point n, then $Z_2 \doteqdot 30/2 = 15\,\Omega$ and ρ_2 has a further negative value between 0 and -1. If ρ'_1 or ρ'_2 is negative (i.e. if $Z_0 < Z_1$ or $Z_2 < Z_1$) the v_m, v_n would diminish very quickly under oscillatory modes.

Regardless, the above investigation indicates that transient surge voltage phenomena on a cable line caused by an incidental lightning surge become more severe by value and duration when the surge impedances of the connected adjacent lines (Z_0, Z_2) are larger than that of the cable line Z_1 (namely, $Z_0, Z_2 \gg Z_1$). As cable insulation does not have self-restoring characteristics, detailed surge analysis and the necessary protection (typically by arresters) against incidental surges are essential in practical engineering.

23.6.2 Switching surges arising on cable line

Now we examine switching surge phenomena arising on a cable line as shown in Figure 23.9a, where cable 1 may be installed within the power station yard or may be a part of the outside network.

When the breaker is closed, a transient switching surge appears on the adjacent cable line and the transmitted voltage appearing at point m is $\{2Z_1/(Z_0 + Z_1)\} \cdot e_0$ (e_0 is the voltage (crest value) across the breaker contacts at $t = 0-$). As the station bus probably has a number of parallel circuits, Z_0 (the resulting total surge impedance of the station side) would be small; accordingly, the above voltage v_m at point m would become almost $2e_0$. Furthermore, large and long duration of the transient surge is anticipated for larger Z_2 (surge impedance of another connected line).

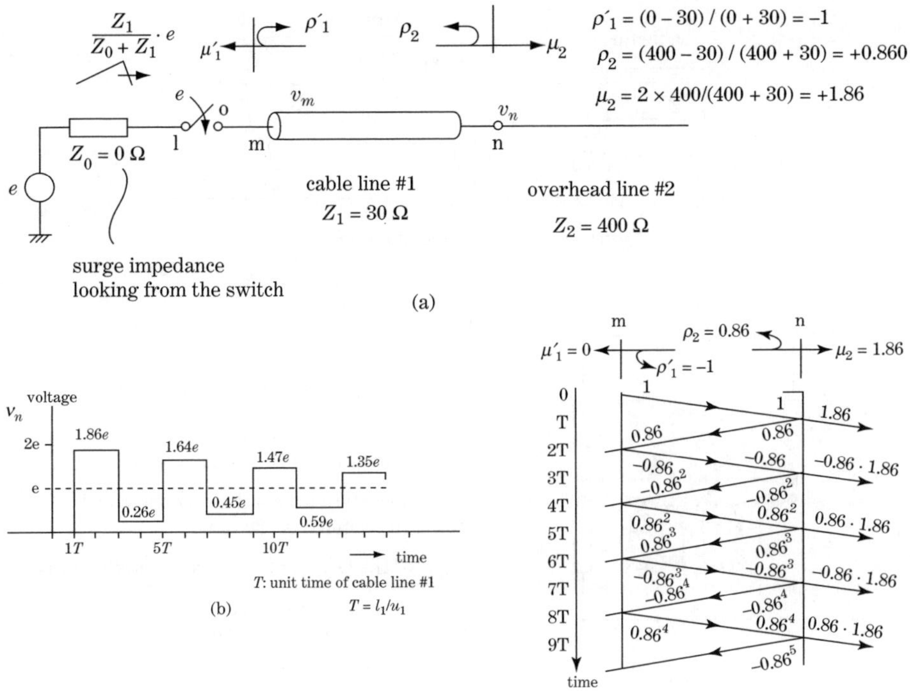

Typical waveform of switching surge (10 kHz to 1 MHz) calculation by Eq. 18 · 51:

$$v_n = (1+\rho_2)\cdot\{1+(\rho_1'\rho_2)+(\rho_1'\rho_2)^2+(\rho_1'\rho_2)^3+(\rho_1'\rho_2)^4+\cdots\}\cdot e$$

Figure 23.9 Switching surge on the cable and overhead line

Figure 23.9b shows the calculated result for v_n at point n under the conditions $Z_0 = 0$, $Z_1 = 30\,\Omega$, $Z_2 = 400\,\Omega$. The calculation indicates that an oscillatory switching surge of maximum 1.86 times the nominal voltage (crest value) is induced on the cable line. The voltage level would become more severe if the terminal n were opened. The transmitted wave voltage at the terminal point m is $v_m = 2e$ under the condition of $Z_0 = 0$.

Actual switching surges appearing on the cable may be more complicated and probably include multi-frequency components because the surge impedance circuits of the station/outer lines are generally also complicated, though the mechanism illustrated by the above investigation is unchanged. Of course, we need to confirm in practical engineering that the insulation of the cables (insulation layer and outer covering) is certainly protected against switching surges. The cable should have an insulation level to withstand the maximum switching surge level specified by authorized standards or recommendations as shown in Table 23.1a as one example.

In particular, in the case of a long cable line with a cross-bonding jointing connection, the insulation of the outer covering should be carefully investigated, because the metallic sheath terminals are earth grounded only every three jointed spans.

23.7 Surge Voltages at Cable End Terminal Connected to GIS

A GIS (Gas-Insulated Substation) contains all the functional equipment (bus conductors/breakers/line switches/arrester/PT/CT/bushings, etc.) in a hermetically sealed metal container filled with SF_6 gas

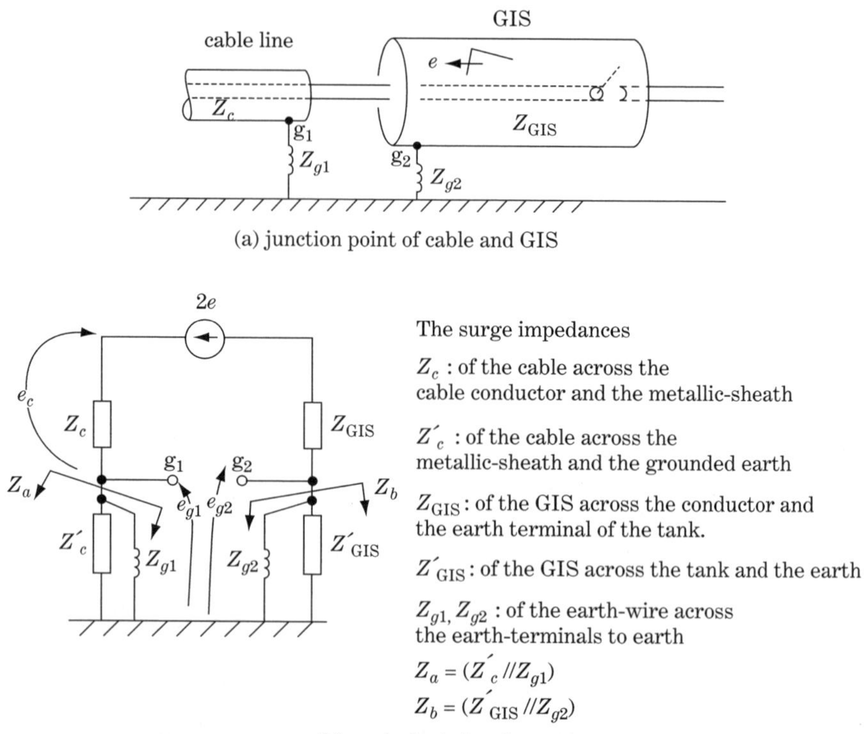

Figure 23.10 Surge voltages arising on the cable sheath by the switching operation of GIS

probably at a few atmospheres pressure (typically 4 atm) as an insulating material (see Figure 21.8). GISs are of extremely compact design in comparison with conventional open-air substations and, furthermore, a number of (phase-unbalanced) feeders are connected through the bus conductors. In other words, the GIS is quite a complicated surge impedance circuit containing various transition points in a narrow three-dimensional space. Therefore, the behaviour of switching surges or transferred lightning surges in the GIS containers includes high-frequency components (even at 10 MHz) caused by short travel distances between the transition points. This is the reason why '**very fast-front voltages**' have been designated in the standards for insulation coordination (see Section 21.1).

Switching surge voltages with very fast fronts caused by a breaker in the GIS (time of wavefront, $T_f = 0.1 - 1\ \mu s$) may cause particular problems to the GIS-connected cable lines.

A typical phenomenon is indicated in the system of Figure 23.10a. Referring to Figure 23.10b, the incidental switching surge voltage e from the GIS is divided by the various surge impedances at the cable jointing point, where abnormal surge overvoltages e_{g1} and e_{g2} appear. The voltage e_{g1} may damage the outer covering, or the voltage $(e_{g1} - e_{g2})$ may damage the longitudinal insulation cylinders in the **jointing box**. These voltages (the first wave before reflection) can be calculated as follows.

The surge voltages are: across the cable conductor to the sheath

$$e_c = \frac{Z_c}{Z_c + Z_a + Z_{GIS} + Z_b} \cdot 2e$$

across the cable sheath

$$e_{g1} = \frac{Z_a}{Z_c + Z_a + Z_{\text{GIS}} + Z_b} \cdot 2e$$

across the GIS conductor to the grounding terminal

$$e_{\text{GIS}} = \frac{Z_{\text{GIS}}}{Z_c + Z_a + Z_{\text{GIS}} + Z_b} \cdot 2e$$

across the GIS grounding terminal to earth

$$e_{g2} = \frac{-Z_b}{Z_c + Z_a + Z_{\text{GIS}} + Z_b} \cdot 2e \qquad (23.22)$$

and across the insulated cylinder of the cable terminal box

$$e_{g1-g2} = \frac{Z_a + Z_b}{Z_c + Z_a + Z_{\text{GIS}} + Z_b} \cdot 2e$$

where

$$Z_a = \frac{Z'_c Z_{g1}}{Z'_c + Z_{g1}} \quad \text{(surge impedance across the sheath terminal to earth)}$$

$$Z_b = \frac{Z_{\text{GIS}} Z_{g2}}{Z_{\text{GIS}} + Z_{g2}} \quad \text{(surge impedance across the GIS grounding terminal to earth)}$$

Detailed computer analysis of such phenomena is possible by the lattice method if necessary. However, manual calculation of 'the duration of the first wave' by the above equations gives us a general concept of the behaviour, because the attenuation of reflected waves is quite fast for such high-frequency phenomena, so the first wave would contain dominant components at least for the absolute values.

As the last sentence of Chapters 18–23, we would like to emphasize that high-voltage transient phenomena on actual large and complicated power systems can be recognized only as understandable phenomena for small and simple power systems, and, furthermore, any available detail computer analysis, for example of large power systems, has been realized by the accumulation of simple theories as discussed in these chapters. The capability of so-called hand calculation may be a proof of engineering basis.

Coffee break 12: Park's equations, the birth of the d–q–0 method

In the d–q–0 transformation method, synchronous generators and motors can be written as accurate equations, and furthermore generators can be connected to other 'solid-state or non-rotating facilities' as the circuit equational forms of large power network systems.

In the short history of the theory of rotating machine, the first man to be recognized may be **A. Blondel**, who presented a paper 'Synchronous motors and converters' (*AIEE Trans*. 1913) in which he originally treated salient-pole machines by resolving the fundamental space component of the armature mmf into two fundamental space components. Next came **R. E. Doherty** and **C. A. Nickle** who together presented their paper 'Synchronous machines I and II' (*AIEE Trans*. 1926) in which they began their analysis by resolving the armature current into direct-axis and quadrature-axis components of current. Then, **R. H. Park** published his first paper in 1928.

Park was a leading staff member of the GE Schenectady Lab. He wrote five important papers:

a) 'The Reactances of Synchronous Machine', *AIEE Trans*., 1928

b) 'Definitions of an Ideal Synchronous Machine and Formula for the Armature Flux Linkages', 1928

c) 'Two-reaction Theory of Synchronous Machines: Generalized Method of Analysis. Part I', *AIEE Trans*., 1929

d) 'System Stability as a Design Problem', AIEE, 1929

e) 'Two-reaction Theory of Synchronous Machines. Part II', *AIEE Trans*., 1933.

Park wrote in his paper a) that the concept of 'transient' should be divided into ' transient' and 'sub-transient', and he proposed the name and symbol for sub-transient reactance x''.

Park's equation appeared in paper c) in 1929. His famous equation was written in this paper as follows:

$$e_d = p\varphi_d - ri_d - \varphi_q p\theta$$
$$e_q = p\varphi_q - ri_q - \varphi_d p\theta$$
$$e_0 = -(r_0 + x_0 p)i_0$$

These equations are the same as Equation 10.29 although Park wrote them with Heaviside's operationals.

Incidentally, in Chapter 10 of this book, after the Park's equation, Equation 10.37, was derived, the additional complicated per unitization procedure of Equations 10.38–10.46 was required to obtain elegant per unitized equations and the smart equivalent circuit in Figure 10.4. This is the process where the equivalent mutual reactance between armature and rotor circuits for the d-axis as well as q-axis becomes a reciprocal. This procedure by smart base selection was first presented by **M. W. Schulz** in his 1948 paper 'A Simplified Method of Determining Instantaneous Fault Currents and Recovery Voltages in Synchronous Machines'.

Papers with titles concerning stability began to appear around 1925. The mid-1920s might be considered the dawn of the 'modern power system with long transmission and interconnection'.

Today, Park's equations are widely applied in combination with symmetrical components as important initial equations of power system analysis. However, we would like to be reassured that all engineers who are familiar with using the equations already understand their physical meaning.

24

Approaches for Special Circuits

First, some network conditions with unique transformers are investigated in this chapter. Then, calculations of mis-connected abnormal circuits are demonstrated as the final section of this book, because they may be good exercises for readers who may often have to find their own solutions in their engineering activities. The key factor always required by engineers in their practical engineering activities is to go forward logically, step by step.

24.1 On-load Tap-changing Transformer (LTC Transformer)

The LTC transformer has an on-load tap-changer which is connected in series to the neutral side terminal of primary (HT) windings as shown in Figure 24.1a, in which the number of turns of the primary windings (accordingly, tap ratio) can be changed (typically 0 to $\pm$ 10%) under on-load operation.

LTC transformers are typically installed at larger power receiving substations in urban zones and are operated as key transformers to regulate voltages as well as reactive power around the station area, which would probably be automatically controlled by AVR or AQR equipment in combination with shunt reactor banks and/or capacitor banks.

As the turn ratio of the transformer is changed under tap-changing operation, the per unit method explained in Chapter 5 should be modified. We study in this section how we can treat the network, including the LTC transformer, by the PU method. All the symbols in this chapter are the same as in Chapter 5.

The three-winding transformer of turn numbers ${}_pN' : {}_sN : {}_tN$ shown in Figure 5.2a is described by Equation 5.12a using the base quantities of Equation 5.7. Now, we imagine a new condition where the primary turn is modified as ${}_pN \rightarrow {}_pN'$. This condition can be expressed by the equation below.

For turn numbers ${}_pN' : {}_sN : {}_tN$ and base quantities

$$\left.\begin{aligned} &\frac{{}_pV_{\text{base}}}{{}_pN} = \frac{{}_sV_{\text{base}}}{{}_sN} = \frac{{}_tV_{\text{base}}}{{}_tN} = \frac{{}_pV'_{\text{base}}}{{}_pN'} \\ &{}_pI_{\text{base}} \cdot {}_pN = {}_sI_{\text{base}} \cdot {}_sN = {}_tI_{\text{base}} \cdot {}_tN = {}_pI'_{\text{base}} \cdot {}_pN' \\ &k \equiv \frac{{}_pN'}{{}_pN} = \frac{{}_pV'_{\text{base}}}{{}_pV_{\text{base}}} = \frac{{}_pI_{\text{base}}}{{}_pI'_{\text{base}}} = 0.9 \text{ to } 1.1 \end{aligned}\right\} ①$$

the primary voltage and current ${}_pV, {}_pI$ are, for the PU value by the base quantities ${}_pV'_{\text{base}}, {}_pI'_{\text{base}}$ (${}_pN'$ turns base),

$$\left.{}_p\overline{V}' = \frac{{}_pV}{{}_pV'_{\text{base}}} \quad {}_p\overline{I}' = \frac{{}_pI}{{}_pI'_{\text{base}}}\right\} ②$$

Handbook of Power System Engineering Yoshihide Hase
© 2007 John Wiley & Sons, Ltd

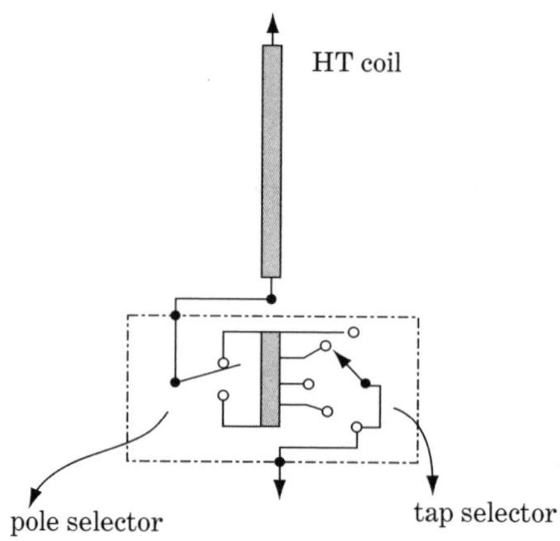

(a) on-load tap changer

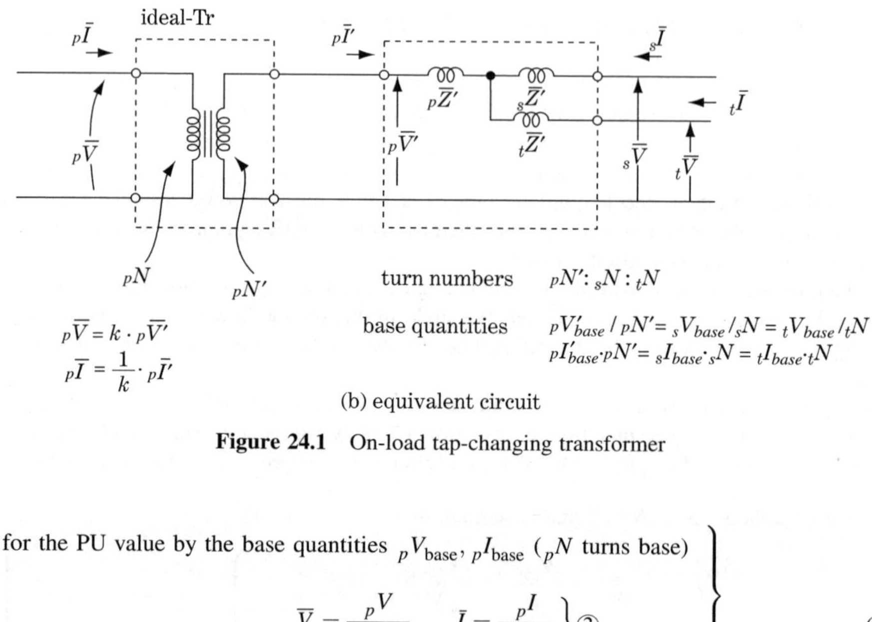

(b) equivalent circuit

Figure 24.1 On-load tap-changing transformer

and for the PU value by the base quantities $_pV_{\text{base}}, {}_pI_{\text{base}}$ ($_pN$ turns base)

$$_p\bar{V} = \frac{_pV}{_pV_{\text{base}}} \quad {}_p\bar{I} = \frac{_pI}{_pI_{\text{base}}} \Bigg\} ③ \tag{24.1}$$

The equivalent circuit of this condition is written on the right-hand side of Figure 24.1b. The equivalent impedances $_pZ', {}_sZ', {}_tZ'$ may be slightly changed from the original $_pZ, {}_sZ, {}_tZ$.

The relation between both base quantities is

$$\frac{_pV_{\text{base}}}{_pN} = \frac{_pV'_{\text{base}}}{_pN'}, \quad {}_pI_{\text{base}} \cdot {}_pN = {}_pI'_{\text{base}} \cdot {}_pN'$$

Then

$$k = \frac{_pN'}{_pN} = \frac{_pV'_{\text{base}}}{_pV_{\text{base}}} = \frac{_pI}{_pI'_{\text{base}}} = 0.9 - 1.1 \tag{24.2}$$

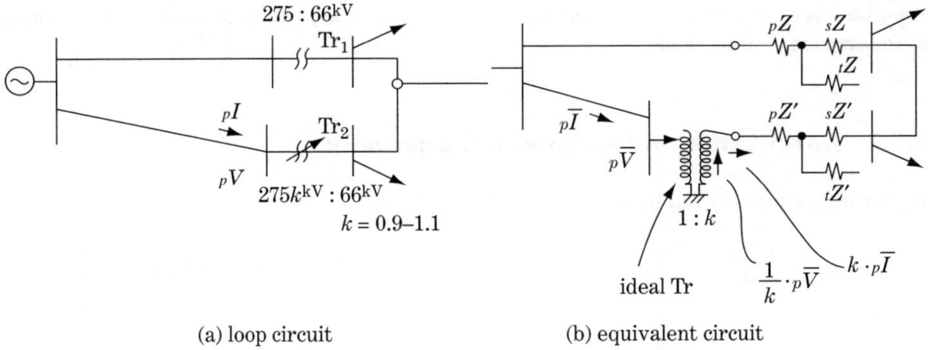

Figure 24.2 Loop circuit

Accordingly,

$$\left.\begin{aligned} {}_p\overline{V} &= \frac{{}_pV}{{}_pV_{\text{base}}} = \frac{{}_pV'_{\text{base}}}{{}_pV_{\text{base}}} \cdot \frac{{}_pV}{{}_pV'_{\text{base}}} = k \cdot {}_p\overline{V}' \\ {}_p\overline{I} &= \frac{{}_pI}{{}_pI_{\text{base}}} = \frac{{}_pI'_{\text{base}}}{{}_pI_{\text{base}}} \cdot \frac{{}_pI}{{}_pI'_{\text{base}}} = \frac{1}{k} \cdot {}_p\overline{I}' \end{aligned}\right\} \tag{24.3}$$

Equation 24.3 is just the equation of the ideal transformer with turn numbers ${}_pN : {}_pN'$, which can be additionally written as on the left-hand side of Figure 24.1b. In other words, this figure is just the equivalent circuit of the LTC transformer under the condition of the primary turn ${}_pN'$ because the figure satisfies all the equations in Equations 24.1–24.3. The tap ratio is generally specified to within 10%, so $k = 0.9 - 1.1$. The reactances ${}_{p-s}X$, ${}_{p-t}X$, ${}_{s-t}X$ for each tap number are indicated generally on the name-plate of the individual transformer.

In conclusion, Figure 24.1 or Equations 24.1–24.3 with the ideal transformer have to be adopted for circuit analysis of power frequency phenomena such as power flow analysis or *P–Q–V* stability analysis.

Figure 24.2a is a typical example of a system including a loop circuit, in which transformer Tr_2 with the tap-changer may be operated by a different winding ratio to that of Tr_1. The power frequency phenomena of this system can be calculated by the circuit in Figure 24.2b. All the necessary equations can be derived from this equivalent circuit.

24.2 Phase-shifting Transformer

In the case of networks with loop circuits, the current flow for each line is distributed proportionally to the inverse ratio of the line impedances, which may not necessarily be an appropriate distribution in regard to the rated capacity limit of each line, *P–Q–V* regulation/stability, load flow economy, and so on. The phase-shifting transformer (or simply phase-shifter) can change the current flow distribution actively for the above loop system. Needless to say, the phase-shifter can control the power flow of the meshed circuit network system as well as of the parallel loop circuit system if the installed location and the MVA bank capacity are planned appropriately.

Figure 24.3a is a typical connection diagram of the phase-shifter which is installed at the location where the primary- and secondary-side rated voltages are the same. As is shown in the figure, the phase-shifting transformer bank consists of one **three-phase regulating transformer** and one

three-phase series-connected transformer, whose windings are connected in series directly through the outer bushings or oil ducts.

24.2.1 Introduction of fundamental equations

The following equations can be derived from Figure 24.3a:

$$\left.\begin{aligned}
&\frac{{}_1\boldsymbol{V}_{abc}}{N}=\frac{\boldsymbol{v}_{abc}}{n} \qquad N\{{}_1\boldsymbol{I}_{abc}-{}_2\boldsymbol{I}_{abc}\}+n\boldsymbol{i}_{abc}+n_g\boldsymbol{i}_g=0\\
&\frac{\boldsymbol{V}'_{abc}}{N'}=\frac{\boldsymbol{v}'_{abc}}{n'} \qquad \text{where } \boldsymbol{i}_g=\begin{bmatrix} i_g\\ i_g\\ i_g\end{bmatrix}\\
&\boldsymbol{V}'_{abc}={}_1\boldsymbol{V}_{abc}-{}_2\boldsymbol{V}_{abc} \qquad N'\cdot{}_2\boldsymbol{I}_{abc}+n'\cdot\boldsymbol{i}'_{abc}=0\\
&\underbrace{\begin{bmatrix} & 1 & -1\\ -1 & & 1\\ 1 & -1 & \end{bmatrix}}_{\boldsymbol{\phi}}\cdot\underbrace{\begin{bmatrix} \boldsymbol{v}_a\\ \boldsymbol{v}_b\\ \boldsymbol{v}_c\end{bmatrix}}_{\boldsymbol{v}_{abc}}=\underbrace{\begin{bmatrix} \boldsymbol{v}'_a\\ \boldsymbol{v}'_b\\ \boldsymbol{v}'_c\end{bmatrix}}_{\boldsymbol{v}'_{abc}},\quad \underbrace{\begin{bmatrix} \boldsymbol{i}_a\\ \boldsymbol{i}_b\\ \boldsymbol{i}_c\end{bmatrix}}_{\boldsymbol{i}_{abc}}=\underbrace{\begin{bmatrix} & 1 & -1\\ -1 & & 1\\ 1 & -1 & \end{bmatrix}}_{\boldsymbol{\phi}}\cdot\underbrace{\begin{bmatrix} \boldsymbol{i}'_a\\ \boldsymbol{i}'_b\\ \boldsymbol{i}'_c\end{bmatrix}}_{\boldsymbol{i}'_{abc}}
\end{aligned}\right\} \tag{24.4}$$

Accordingly,

$$\left.\begin{aligned}
&\{1-k\boldsymbol{\phi}\}\cdot{}_1\boldsymbol{V}_{abc}={}_2\boldsymbol{V}_{abc} \qquad k=\frac{n}{N}\cdot\frac{N'}{n'}\\
&{}_1\boldsymbol{I}_{abc}+k_g\cdot\boldsymbol{i}_g=\{1+k\boldsymbol{\phi}\}\cdot{}_2\boldsymbol{I}_{abc} \qquad k_g=\frac{n_g}{N}
\end{aligned}\right\} \tag{24.5a}$$

or

$$\begin{aligned}
&{}_2V_a={}_1V_a-k({}_1V_b-{}_1V_c) \qquad {}_1I_a+k_g\cdot i_g={}_2I_a+k({}_2I_b-{}_2I_c)\\
&{}_2V_b={}_1V_b-k({}_1V_c-{}_1V_a) \qquad {}_1I_b+k_g\cdot i_g={}_2I_b+k({}_2I_c-{}_2I_a)\\
&{}_2V_c={}_1V_c-k({}_1V_a-{}_1V_b) \qquad {}_1I_c+k_g\cdot i_g={}_2I_c+k({}_2I_a-{}_2I_b)
\end{aligned} \tag{24.5b}$$

The regulating transformer generally has taps and polarity switches (typically of on-load changing) so that k can be specified as a plus or minus value and typically may be $k=+0.1$ to 0 to -0.1. The vector diagrams of the phase-shifter under the three-phase-balanced condition is shown in Figure 24.3b.

Now we transform Equation 24.5 into symmetrical components:

$$\left.\begin{aligned}
&\{1-k\cdot(\boldsymbol{a}\times\boldsymbol{\phi}\times\boldsymbol{a}^{-1})\}\cdot{}_1\boldsymbol{V}_{012}={}_2\boldsymbol{V}_{012}\\
&{}_1\boldsymbol{I}_{012}+k_g\begin{bmatrix} i_g\\ 0\\ 0\end{bmatrix}=\{1+k\cdot(\boldsymbol{a}\times\boldsymbol{\phi}\times\boldsymbol{a}^{-1})\}\cdot{}_2\boldsymbol{I}_{012}\\
&(\boldsymbol{a}\times\boldsymbol{\phi}\times\boldsymbol{a}^{-1})=\begin{bmatrix} & & \\ & a^2-a & \\ & & a-a^2\end{bmatrix}=\begin{bmatrix} & & \\ & -j\sqrt{3} & \\ & & j\sqrt{3}\end{bmatrix}
\end{aligned}\right\} \tag{24.6}$$

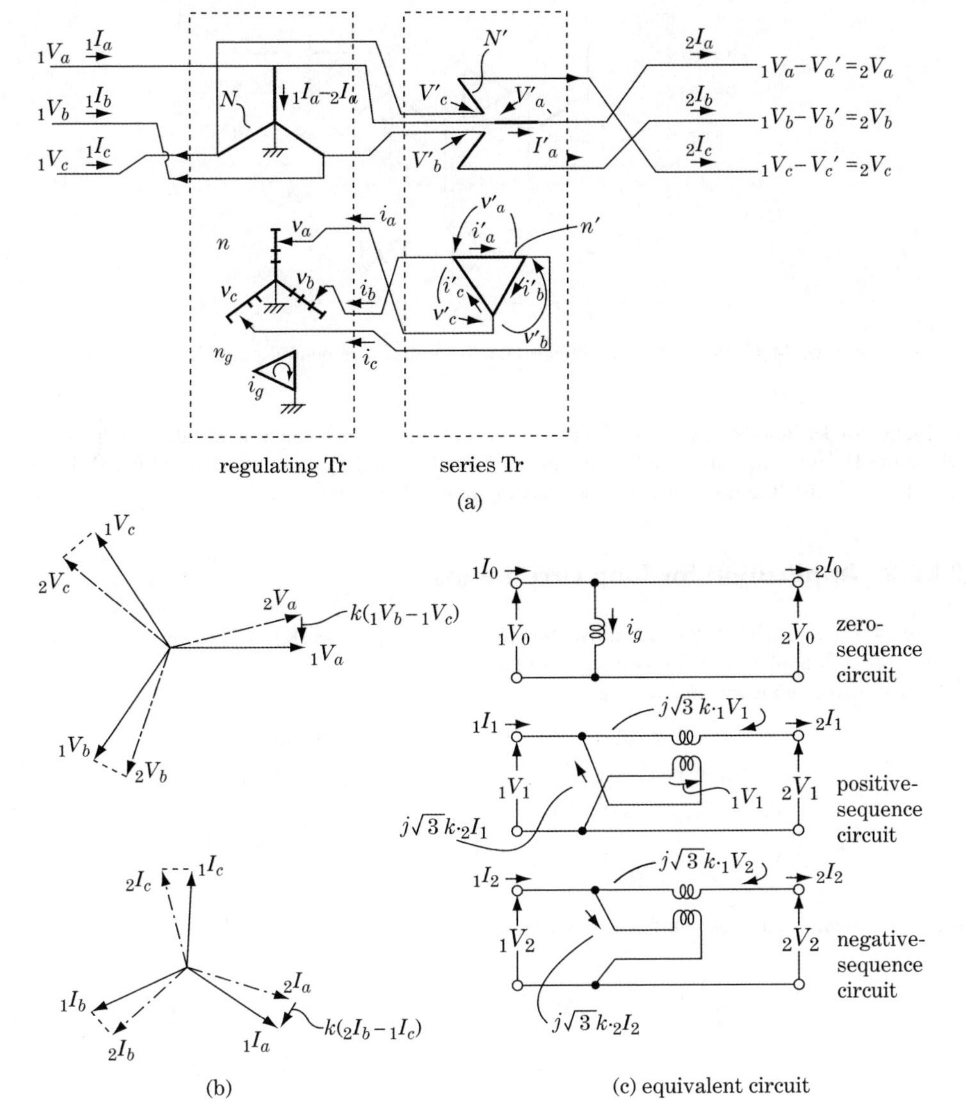

Figure 24.3 Phase-shifting transformer (for solidly grounded system)

Then

$$\left.\begin{array}{r} {}_1V_0 = {}_2V_0 \\ \left(1 + j\sqrt{3}k\right) \cdot {}_1V_1 = {}_2V_1 \\ (1 - j\sqrt{3}k) \cdot {}_1V_2 = {}_2V_2 \end{array}\right\} ① \quad \left.\begin{array}{r} {}_1I_0 + k_g i_g = {}_2I_0 \\ {}_1I_1 = \left(1 - j\sqrt{3}k\right) \cdot {}_2I_1 \\ {}_1I_2 = \left(1 + j\sqrt{3}k\right) \cdot {}_2I_2 \end{array}\right\} ② \Bigg\} \tag{24.7}$$

From this equation the symmetrical equivalent circuit in Figure 24.3c is derived, which is to be adopted in the analysis of power frequency phenomena.

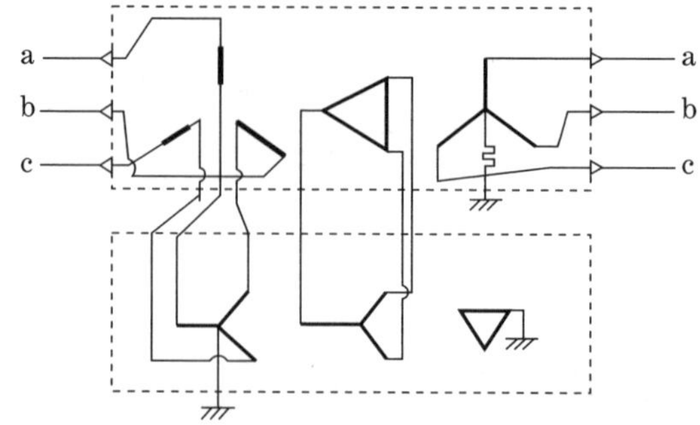

Figure 24.4 Phase-shifting transformer (for impedance neutral grounded system)

Figure 24.4 is another example of a phase-shifting transformer which is installed at the location where the HT windings are solidly grounded and the LT windings are resistive grounded. In other words, this is just 'the transformer which has the power flow control function'.

24.2.2 Application for loop circuit lines

Figure 24.5 is a positive-sequence equivalent circuit of a loop network in which the above described phase-shifter is applied at the receiving terminal r.

The positive-sequence equations are

$$\left.\begin{array}{c}\text{Route 1}\\ \dot{V}_{s1}-\dot{V}_{rr1}=\dot{I}_1\cdot\dot{Z}_1\\ \dot{V}_{rr1}-\dot{V}_{r1}=-j\sqrt{3}k\cdot\dot{V}_{rr1}\\ \therefore\ \dot{V}_{s1}-\dfrac{1}{1+j\sqrt{3}k}\cdot\dot{V}_{r1}=\dot{I}_1\cdot\dot{Z}_1\end{array}\right\}① \quad \left.\begin{array}{c}\text{Route 2}\\ \dot{V}_{s1}-\dot{V}_{r1}=\dot{I}'_1\cdot Z'_1\ ②\\ \left.\begin{array}{l}\dot{Z}_1=jx_1\\ \dot{Z}'_1=jx'_1\end{array}\right\}\ ③\end{array}\right\} \tag{24.8}$$

For the case where a phase-shifter exists in route 1

$$\left.\begin{array}{l}\dot{I}_1=\dfrac{1}{\dot{Z}_1}\left\{\dot{V}_{s1}-\dfrac{1}{1+j\sqrt{3}k}\cdot\dot{V}_{r1}\right\}\\ \dot{S}_{s1}=P_{s1}+jQ_{s1}=\dot{V}_{s1}\cdot\dot{I}_1^*=\dfrac{1}{\dot{Z}_1^*}\left\{\dot{V}_{s1}^2-\dfrac{1}{1-j\sqrt{3}k}\cdot\dot{V}_{s1}\cdot\dot{V}_{r1}^*\right\}\end{array}\right\} \tag{24.9}$$

For the case where a phase-shifter does not exist in route 1 (equivalent to the case of $k=0$)

$$\dot{S}_{s1,k=0}=P_{s1}+jQ_{s1}=\frac{1}{\dot{Z}_1^*}\left\{\dot{V}_{s1}^2-\dot{V}_{s1}\cdot\dot{V}_{r1}^*\right\} \tag{24.10}$$

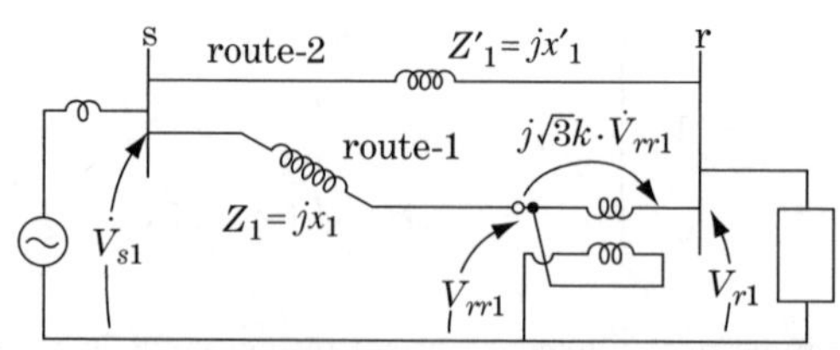

Figure 24.5 Application for loop circuit (positive-sequence circuit)

The difference in the apparent power by the above two cases is

$$\left.\begin{aligned}
\Delta\dot{S}_{s1} &= \dot{S}_{s1} - \dot{S}_{s1,k=0} = \frac{-j\sqrt{3}k}{1-j\sqrt{3}k}\cdot\frac{\dot{V}_{s1}\cdot\dot{V}_{r1}^*}{\dot{Z}_1^*} = \frac{3k^2 - j\sqrt{3}k}{3k^2+1}\cdot\frac{\dot{V}_{s1}\cdot V_{r1}^*}{Z_1^*}\\
&\doteqdot (3k^2 - j\sqrt{3}k)\cdot\frac{\dot{V}_{s1}\cdot\dot{V}_{r1}^*}{-jx_1}\\
&= \frac{\sqrt{3}k + j3k^2}{x_1}\cdot\dot{V}_{s1}\cdot\dot{V}_{r1}^* \doteqdot \frac{\sqrt{3}k}{x_1}\cdot\dot{V}_{s1}\cdot\dot{V}_{r1}^* \qquad ①\\
&\text{Putting}\\
&\dot{V}_s = |V_s|\angle\delta^\circ,\ \dot{V}_r = |V_r|\angle 0^\circ\\
&\text{then } \Delta\dot{P}_{s1} + \Delta j\dot{Q}_{s1} \doteqdot \frac{\sqrt{3}k}{x_1}\cdot|\dot{V}_{s1}\cdot\dot{V}_{r1}^*|\cdot(\cos\delta + j\sin\delta) \qquad ②
\end{aligned}\right\} \tag{24.11}$$

The approximation in the process is based on $k = +0.1$ to -0.1. The power flow of Equation 24.11 on route 1 can be controlled by the existence of the phase-shifter.

The network can be assumed in that line 1 is longer than line 2 so that the reactance of line 1 is larger than that of line 2 ($jx_1 > jx_2$) whereas the line capacity of line 1 is far larger than that of line 2. The power flow of this circuit may not be appropriate because the power flow distribution of line 2 becomes larger despite the far smaller current capacity.

The phase-shifter can solve the inadequate current distribution by changing the current flow distribution with time. Furthermore, adequate current distribution control can reduce the angular difference across points s and r so that the phase-shifter can improve system stability to some extent.

24.3 Woodbridge Transformer and Scott Transformer

Electric train loads are typically fed by a single phase circuit, so these are fatally unbalanced loads from the viewpoint of a three-phase network. The Woodbridge-type as well as the Scott-type transformer have special windings in order to mitigate the load imbalance.

24.3.1 Woodbridge winding transformer

The connection diagram of the Woodbridge transformer is shown in Figure 24.6a. This is a three-phase three-windings transformer whose S (secondary) and T (tertiary) windings are of delta connection and of the same design except that their winding terminal connection is specially arranged. The voltage-transformed power is fed out from the A terminal as well as from the B terminal. The voltage boost-up transformer Tr_2 (turn ratio $1 : \sqrt{3}$) is also connected to the B terminal. The A and B terminals would be typically fed out to feeding sections 1 and 2 (the up- and down-train lines, for example), both of which have approximately a similar load by capacity as well as load pattern over time.

The equations are

$$\left.\begin{aligned}
{}_pn\cdot{}_pI_a + {}_sn({}_sI_a + {}_tI_a) &= 0\\
{}_pn\cdot{}_pI_b + {}_sn({}_sI_b + {}_tI_b) &= 0\\
{}_pn\cdot{}_pI_c + {}_sn({}_sI_c + {}_tI_c) &= 0
\end{aligned}\right\} \qquad \left.\begin{aligned}
\therefore k\cdot{}_pI_a + {}_sI_a + {}_tI_a &= 0\\
k\cdot{}_pI_b + {}_sI_b + {}_tI_b &= 0\\
k\cdot{}_pI_c + {}_sI_c + {}_tI_c &= 0
\end{aligned}\right\}$$

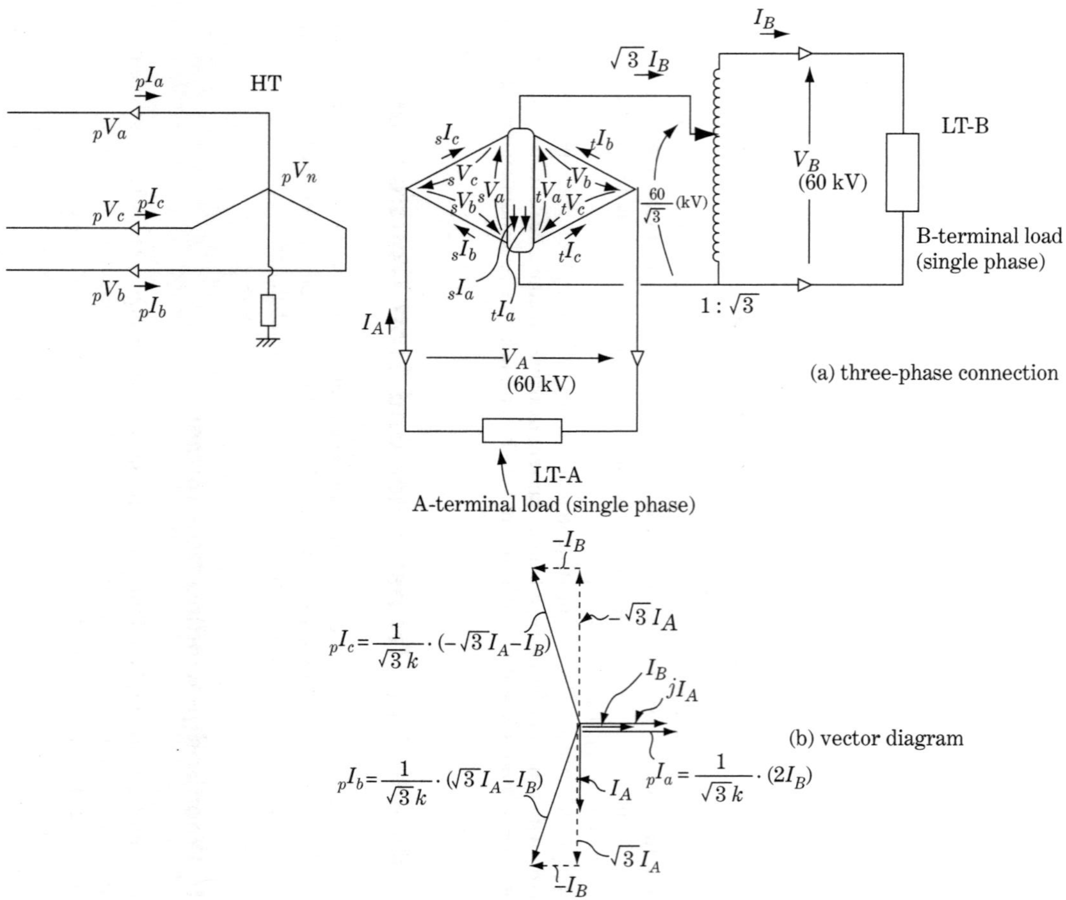

Figure 24.6 Woodbridge transformer

where

$${}_pn\text{: turns of the primary winding}$$

$${}_sn\text{: turns of the secondary and tertiary windings,} \quad k = \frac{{}_pn}{{}_sn} \tag{24.12}$$

$$\left.\begin{array}{ll} \text{A-terminal current} & \\ I_A = {}_sI_c - {}_sI_b = {}_tI_c - {}_tI_b & ① \\ \text{B-terminal current} & \\ \sqrt{3}I_B = ({}_sI_c - {}_sI_a) + ({}_tI_b - {}_tI_a) & ② \end{array}\right\} \tag{24.13}$$

Eliminating ${}_tI_b, {}_tI_c$ from Equations 24.12 and 24.13,

$${}_sI_b - {}_sI_c = \frac{-k}{2}\left({}_pI_b - {}_pI_c\right) \tag{24.14}$$

Using the above equations, the current I_A, I_B can be written as a function of the primary current as follows:

$$\begin{aligned} I_A &= \frac{k}{2}\left({}_pI_b - {}_pI_c\right) \\ \sqrt{3}I_B &= ({}_sI_c - {}_sI_a) + ({}_tI_b - {}_tI_a) \\ &= \frac{k}{2}\left(2\,{}_pI_a - {}_pI_b - {}_pI_c\right) \end{aligned}$$

namely

$$\left.\begin{array}{l} \begin{bmatrix} I_A \\ \sqrt{3}I_B \end{bmatrix} = \dfrac{k}{2}\begin{bmatrix} 0 & 1 & -1 \\ 2 & -1 & -1 \end{bmatrix} \cdot \begin{bmatrix} {}_pI_a \\ {}_pI_b \\ {}_pI_c \end{bmatrix} \\ \begin{bmatrix} I_A \\ \sqrt{3}I_B \end{bmatrix} = \dfrac{k}{2}\begin{bmatrix} 0 & 1 & -1 \\ 2 & -1 & -1 \end{bmatrix} \cdot \begin{bmatrix} 1 & 1 & 1 \\ 1 & a^2 & a \\ 1 & a & a^2 \end{bmatrix} \cdot \begin{bmatrix} {}_pI_0 \\ {}_pI_1 \\ {}_pI_2 \end{bmatrix} = \dfrac{k}{2}\begin{bmatrix} 0 & -j\sqrt{3} & j\sqrt{3} \\ 0 & 3 & 3 \end{bmatrix} \cdot \begin{bmatrix} {}_pI_0 \\ {}_pI_1 \\ {}_pI_2 \end{bmatrix} \end{array}\right\} \tag{24.15}$$

Accordingly,

$$\begin{bmatrix} I_A \\ I_B \end{bmatrix} = \frac{\sqrt{3}}{2}k\begin{bmatrix} 0 & -j & j \\ 0 & 1 & 1 \end{bmatrix} \cdot \begin{bmatrix} {}_pI_0 \\ {}_pI_1 \\ {}_pI_2 \end{bmatrix} \tag{24.16}$$

Then

$$\left.\begin{array}{l} \left.\begin{array}{l} I_A = -j\dfrac{\sqrt{3}}{2}k\left({}_pI_1 - {}_pI_2\right) \\ I_B = \dfrac{\sqrt{3}}{2}k\left({}_pI_1 + {}_pI_2\right) \end{array}\right\} ① \\ \text{inverse equations} \\ \left.\begin{array}{l} {}_pI_1 = \dfrac{1}{\sqrt{3}k}(I_B + jI_A) \\ {}_pI_2 = \dfrac{1}{\sqrt{3}k}(I_B - jI_A) \\ {}_pI_0 = 0 \end{array}\right\} ② \\ \left.\begin{array}{l} {}_pI_a = \dfrac{1}{\sqrt{3}k}(2I_B) \\ {}_pI_b = \dfrac{1}{\sqrt{3}k}\left(\sqrt{3}I_A - I_B\right) \\ {}_pI_c = \dfrac{1}{\sqrt{3}k}\left(-\sqrt{3}I_A - I_B\right) \end{array}\right\} ③ \end{array}\right\} \tag{24.17}$$

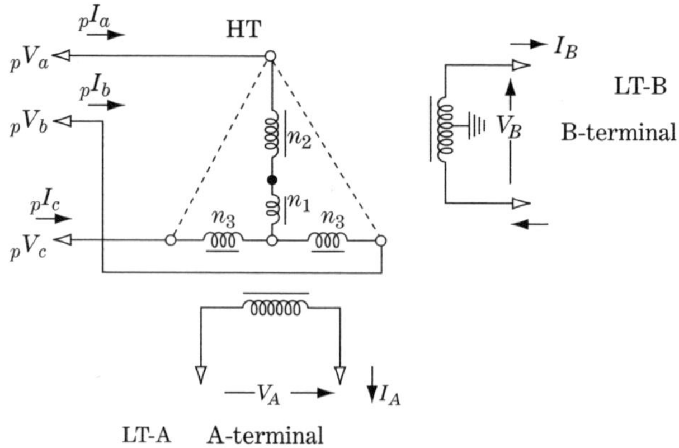

Figure 24.7 Scott transformer

The equations indicate the following:

- $_pI_0 = 0$: zero-sequence current is always zero regardless of the load conditions.
- $_pI_2 = 0$: under the condition of $jI_A = I_B$. That is, the negative-sequence currents can be reduced to zero if the magnitudes of the A-terminal load and B-terminal load are equal, or it can be reduced if the both load values are similar over time, whereas $_pI_2$ would become a large value of $_pI_1 = {_pI_2}$ under the condition $I_A = 0$ or $I_B = 0$.

24.3.2 Scott winding transformer

The connection diagram for the Scott transformer is shown in Figure 24.7, where the turn ratios of the primary windings are $n_1 : n_2 : n_3 = 1 : 2 : \sqrt{3}$.

Anticipating the conclusion, this transformer gives exactly the same equation as Equation 24.17, although the derivation is omitted in this book.

Obviously the Woodbridge transformer is better than the Scott transformer from a design and manufacturing viewpoint, because the former is just a three-phase-balanced $y - \Delta - \Delta$ winding design, except for the terminal connection treatment, while the latter is of non-symmetric design, although both transformers are completely equivalent from the application viewpoint. In these situations, Scott transformers might actually be succeeded by Woodbridge transformers yet seemed to be historical monuments.

24.4 Neutral Grounding Transformer

Larger industrial factories in various application fields such as chemicals, metal furnaces, steel, electrical appliances, etc., often adopt power-receiving transformers with delta windings for lower voltage sides, in order to isolate the factories' zero-sequence circuits from that of the power system networks. However, the factories' in-house networks have to be neutral grounded to obtain in-house insulation coordination. Neutral grounding transformers are adopted for such cases as special transformers to earth–ground the zero-sequence circuit of the in-house networks.

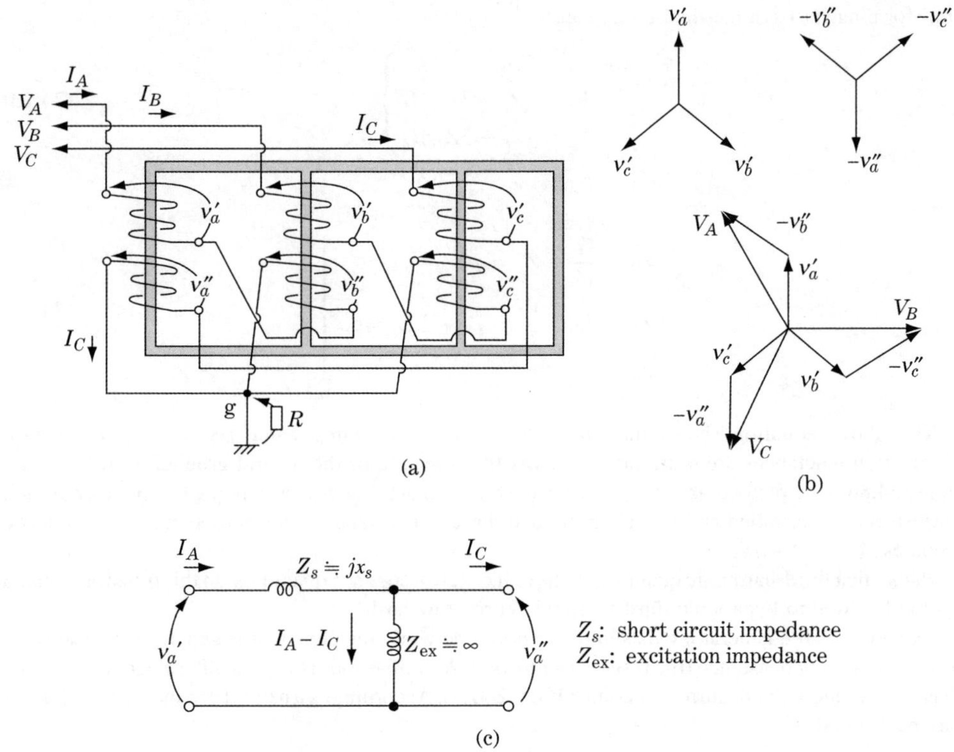

Figure 24.8 Neutral grounding transformer

Figure 24.8a shows the winding connection diagram. Figure 24.8c is the equivalent circuit marking the first core pole, where all equations are unitized.

The related equations are

$$\left.\begin{aligned} v_a' - v_a'' &= Z_s I_A \\ v_b' - v_b'' &= Z_s I_B \\ v_c' - v_c'' &= Z_s I_C \end{aligned}\right\} ① \qquad \left.\begin{aligned} v_a'' &= Z_{ex}(I_A - I_C) \\ v_b'' &= Z_{ex}(I_B - I_A) \\ v_c'' &= Z_{ex}(I_C - I_B) \end{aligned}\right\} ②$$
$$\left.\begin{aligned} V_A &= v_a' - v_b'' \\ V_B &= v_b' - v_c'' \\ V_C &= v_c' - v_a'' \end{aligned}\right\} ③ \qquad (24.18)$$

Eliminating v_a', v_a'', etc.,

$$\left.\begin{aligned} V_A &= Z_S I_A + Z_{ex}(2I_A - I_B - I_C) \\ &= Z_S I_A + 3Z_{ex}(I_A - I_0) \\ &= (Z_s + 3Z_{ex})I_A - 3Z_{ex}I_0 \end{aligned}\right\} \qquad (24.19a)$$

$$\therefore \quad \left.\begin{aligned} V_A &= (Z_s + 3Z_{ex})I_A - 3Z_{ex}I_0 \\ V_B &= (Z_s + 3Z_{ex})I_B - 3Z_{ex}I_0 \\ V_C &= (Z_s + 3Z_{ex})I_c - 3Z_{ex}I_0 \end{aligned}\right\} \qquad (24.19b)$$

Transforming into symmetrical components,

$$\left.\begin{aligned} V_0 &= Z_s I_0 \\ V_1 &= (Z_s + 3Z_{ex})I_1 \\ V_2 &= (Z_s + 3Z_{ex})I_2 \end{aligned}\right\} \tag{24.20}$$

and

$$\left.\begin{aligned} Z_0 &= \frac{V_0}{I_0} = Z_s = jx_s \\ Z_1 &= Z_2 = \frac{V_1}{I_1} = Z_s + 3Z_{ex} \doteqdot \infty \\ Z_{ex} &\doteqdot \infty \end{aligned}\right\} \tag{24.21}$$

The above equation shows that $jx_0 = jx_s$, but on the other hand $jx_1 = jx_2 \doteqdot \infty$ because the extation reactances are quite large. This is the principle of the neutral grounding transformer. Figure 24.8b is the vector diagram explaining the function of the neutral grounding transformer. If such a transformer is installed and is earth grounded through resistance R, the zero-sequence impedance becomes $Z_0 = 3R + jX_s$.

Recall that third-harmonic quantities behave like zero-sequence quantities, so this transformer also has the function to bypass the third-harmonic current to earth.

Incidentally, if a generator exists in the in-house network, the generator is generally neutral earth grounded through resistance (the resistive value of $3\ R = 50-200\ \Omega$ is typically selected so that the phase-to-grounding fault current becomes $100-200$ A). Accordingly, a neutral grounding transformer may be omitted.

24.5 Mis-connection of Three-phase Orders

As the final section of this book, calculation of the circuit with phase mis-connection is tried.

24.5.1 Cases

24.5.1.1 Case 1: phase a–b–c to a–c–b mis-connection (figure 24.9)

Our problem is shown in Figure 24.9a. If the two balanced circuits are connected normally, the voltages and currents at the connecting point are given by the equations

$$\left.\begin{aligned} i_1 &= \frac{E_1 - E_1'}{Z_1 + Z_1'} \\ i_2 &= i_0 = 0 \end{aligned}\right\} ① \qquad \left.\begin{aligned} i_a &= \frac{E_1 - E_1'}{Z_1 + Z_1'} \\ i_b &= a^2 i_a, i_c = a i_a \end{aligned}\right\} ② \tag{24.22}$$

Now we examine the case of a wrong connection. The related equations are

$$\underset{\mathbf{v}_{abc}}{\begin{bmatrix} v_a' \\ v_b' \\ v_c' \end{bmatrix}} = \underset{\mathbf{v}'_{acb}}{\begin{bmatrix} v_a' \\ v_c' \\ v_b' \end{bmatrix}}, \quad \underset{\mathbf{i}_{abc}}{\begin{bmatrix} i_a \\ i_b \\ i_c \end{bmatrix}} = \underset{\mathbf{i}'_{acb}}{\begin{bmatrix} i_a' \\ i_c' \\ i_b' \end{bmatrix}} \tag{24.23}$$

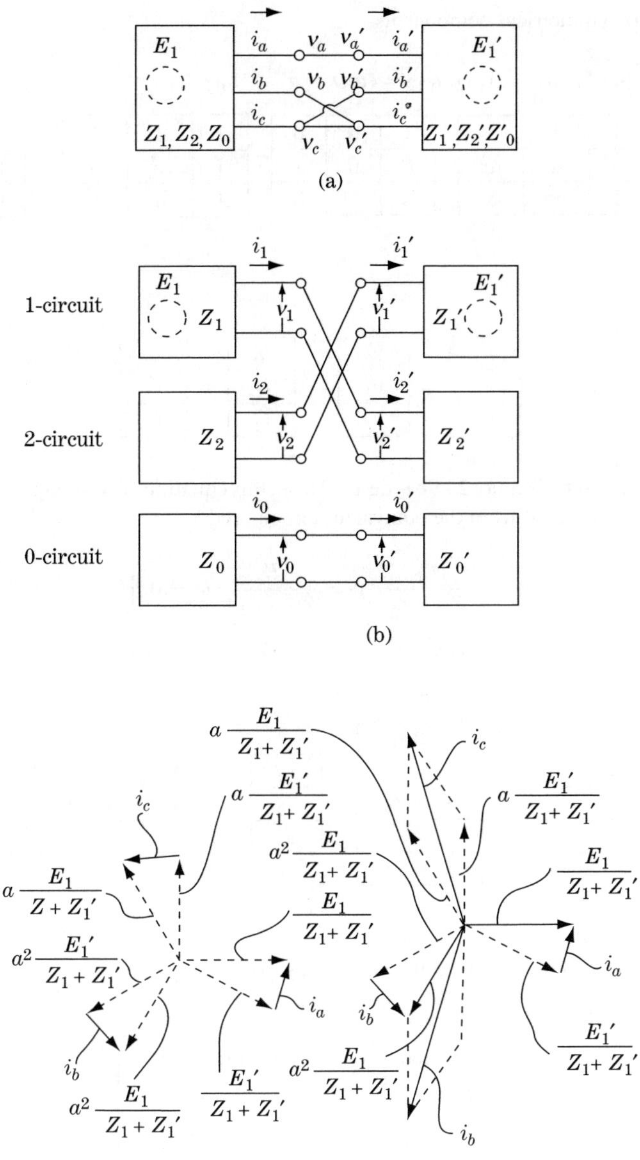

(c1) normal connection (c2) mis-connection

Figure 24.9 Phase a–b–c to a–c–b mis-connection

Then

$$\left.\begin{array}{l}\underset{\boldsymbol{v}'_{acb}}{\begin{bmatrix} v'_a \\ v'_c \\ v'_b \end{bmatrix}} = \underset{\boldsymbol{k}_1}{\begin{bmatrix} 1 & 0 & 0 \\ 0 & 0 & 1 \\ 0 & 1 & 0 \end{bmatrix}} \cdot \underset{\boldsymbol{v}'_{abc}}{\begin{bmatrix} v'_a \\ v'_b \\ v'_c \end{bmatrix}}, \quad \underset{\boldsymbol{i}'_{acb}}{\begin{bmatrix} i'_a \\ i'_c \\ i'_b \end{bmatrix}} = \underset{\boldsymbol{k}_1}{\begin{bmatrix} 1 & 0 & 0 \\ 0 & 0 & 1 \\ 0 & 1 & 0 \end{bmatrix}} \cdot \underset{\boldsymbol{i}'_{abc}}{\begin{bmatrix} i'_a \\ i'_b \\ i'_c \end{bmatrix}} \\ \therefore \quad \boldsymbol{v}'_{acb} = \boldsymbol{k}_1 \cdot \boldsymbol{v}'_{abc} \qquad \boldsymbol{i}'_{acb} = \boldsymbol{k}_1 \cdot \boldsymbol{i}'_{abc} \end{array}\right\} \quad (24.24)$$

Transforming into symmetrical components,

$$\left.\begin{aligned}&\boldsymbol{v}_{012} = \boldsymbol{a}\cdot\boldsymbol{v}_{abc} = (\boldsymbol{a}\cdot\boldsymbol{k}_1\cdot\boldsymbol{a}^{-1})\cdot\boldsymbol{v}_{012},\ \boldsymbol{i}_{012} = (\boldsymbol{a}\cdot\boldsymbol{k}_1\cdot\boldsymbol{a}^{-1})\cdot\boldsymbol{i}_{012}\\ &\boldsymbol{a}\cdot\boldsymbol{k}_1\cdot\boldsymbol{a}^{-1} = \frac{1}{3}\begin{bmatrix}1&1&1\\1&a&a^2\\1&a^2&a\end{bmatrix}\cdot\begin{bmatrix}1&0&0\\0&0&1\\0&1&0\end{bmatrix}\cdot\begin{bmatrix}1&1&1\\1&a^2&a\\1&a&a^2\end{bmatrix} = \begin{bmatrix}1&0&0\\0&0&1\\0&1&0\end{bmatrix}\end{aligned}\right\} \tag{24.25}$$

namely

$$\begin{bmatrix}v_0\\v_1\\v_2\end{bmatrix} = \begin{bmatrix}v'_0\\v'_2\\v'_1\end{bmatrix},\quad \begin{bmatrix}i_0\\i_1\\i_2\end{bmatrix} = \begin{bmatrix}i'_0\\i'_2\\i'_1\end{bmatrix} \tag{24.26}$$

The equivalent circuit of Figure 24.9b is derived from this equation. The voltages and currents at the connecting point are derived from the equivalent circuit, and

$$\left.\begin{aligned}&i_1 = \frac{E_1}{Z_1 + Z'_2} \quad i_2 = \frac{-E'_1}{Z'_1 + Z_2} \quad i_0 = 0\\ &v_1 = i_1 Z'_2 \qquad v_2 = -i_2 Z_2 \quad v_0 = 0\end{aligned}\right\} \tag{24.27}$$

$$\begin{bmatrix}i_a\\i_b\\i_c\end{bmatrix} = \begin{bmatrix}1&1&1\\1&a^2&a\\1&a&a^2\end{bmatrix}\cdot\begin{bmatrix}0\\ \dfrac{E_1}{Z_1+Z'_2}\\ \dfrac{-E'_1}{Z'_1+Z_2}\end{bmatrix},\quad \begin{bmatrix}v_a\\v_b\\v_c\end{bmatrix} = \begin{bmatrix}1&1&1\\1&a^2&a\\1&a&a^2\end{bmatrix}\cdot\begin{bmatrix}0\\ \dfrac{E_1 Z'_2}{Z_1+Z'_2}\\ \dfrac{E'_1 Z_2}{Z'_1+Z_2}\end{bmatrix} \tag{24.28}$$

These are the voltages and currents at the connecting point.
Accordingly,

$$\left.\begin{aligned}&\text{normal connection}\\ &i_a = \frac{E_1 - E'_1}{Z_1 + Z'_1}\\ &i_b = \frac{a^2(E_1 - E'_1)}{Z_1 + Z'_1}\\ &i_c = \frac{a(E_1 - E'_1)}{Z_1 + Z'_1}\end{aligned}\right\} ① \quad \left.\begin{aligned}&\text{mis-connection}\\ &i_a = \frac{E_1 - E'_1}{Z_1 + Z'_1}\\ &i_b = \frac{a^2 E_1 - aE'_1}{Z_1 + Z'_1}\\ &i_c = \frac{aE_1 - a^2E'_1}{Z_1 + Z'_1}\end{aligned}\right\} ② \tag{24.29}$$

where $Z_1 = Z_2, Z'_1 = Z'_2$.

As can be seen in the vector diagram of Figure 24.9c2, large negative-sequence current appears, and accordingly extremely large unbalanced phase currents i_b, i_c arise. Of course, such situations would seriously affect generators, most of the other station equipment and every kind of load.

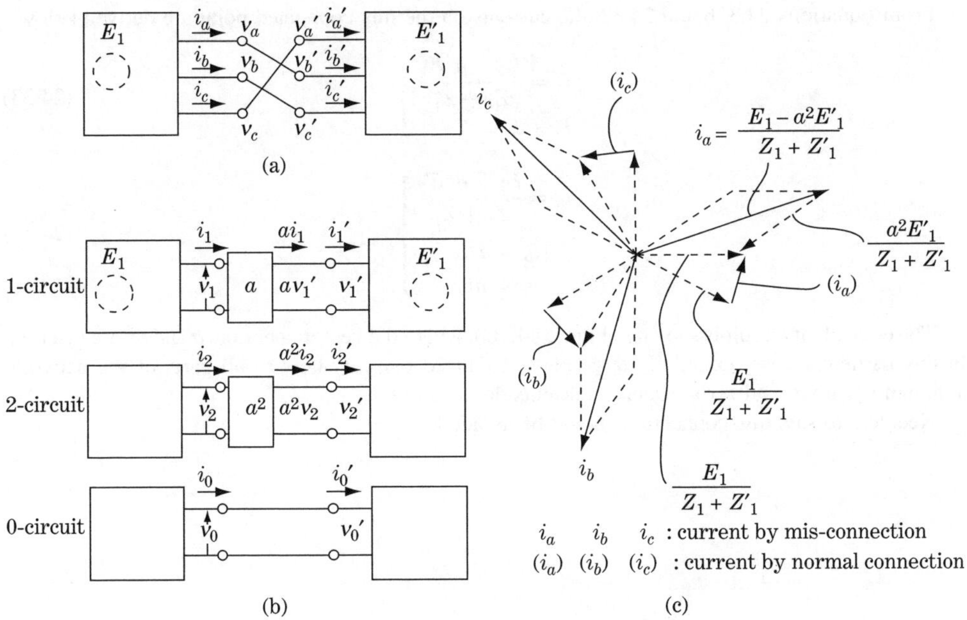

Figure 24.10 Phase a–b–c to b–c–a mis-connection

24.5.1.2 Case 2: phase a–b–c to b–c–a mis-connection (figure 24.10)

The next problem is shown in Figure 24.10a. The related equations are

$$\left.\begin{aligned}
\underset{\boldsymbol{v}_{abc}}{\begin{bmatrix} v_a \\ v_b \\ v_c \end{bmatrix}} = \underset{\boldsymbol{k}_2}{\begin{bmatrix} & 1 & \\ & & 1 \\ 1 & & \end{bmatrix}} \cdot \underset{\boldsymbol{v}'_{abc}}{\begin{bmatrix} v'_a \\ v'_b \\ v'_c \end{bmatrix}}, \quad \underset{\boldsymbol{i}_{abc}}{\begin{bmatrix} i_a \\ i_b \\ i_c \end{bmatrix}} = \underset{\boldsymbol{k}_2}{\begin{bmatrix} & 1 & \\ & & 1 \\ 1 & & \end{bmatrix}} \cdot \underset{\boldsymbol{i}'_{abc}}{\begin{bmatrix} i'_a \\ i'_b \\ i'_c \end{bmatrix}}
\end{aligned}\right\} \quad (24.30)$$

Transforming into symmetrical components,

$$\left.\begin{aligned}
&\boldsymbol{v}_{012} = \left(\boldsymbol{a}\cdot\boldsymbol{k}_2\cdot\boldsymbol{a}^{-1}\right)\cdot\boldsymbol{v}_{012}, \quad \boldsymbol{i}_{012} = \left(\boldsymbol{a}\cdot\boldsymbol{k}_2\cdot\boldsymbol{a}^{-1}\right)\cdot\boldsymbol{i}_{012} \\
&\text{where } \boldsymbol{a}\cdot\boldsymbol{k}_2\cdot\boldsymbol{a}^{-1} = \begin{bmatrix} 1 & 0 & 0 \\ 0 & a^2 & 0 \\ 0 & 0 & a \end{bmatrix}
\end{aligned}\right\} \quad (24.31a)$$

namely

$$\left.\begin{aligned}
v_0 &= v'_0 & i_0 &= i'_0 \\
v_1 &= a^2v'_1 & i_1 &= a^2i'_1 \\
v_2 &= av'_2 & i_2 &= ai'_2
\end{aligned}\right\} \quad (24.31b)$$

The equations of the right- and left-side circuits are

$$\left.\begin{aligned}
v_0 &= -Z_0i_0 & v'_0 &= Z'_0i'_0 \\
v_1 &= E_1 - Z_1i_1 & v'_1 &= E'_1 + Z'_1i'_1 \\
v_2 &= -Z_2i_2 & v'_2 &= Z'_2i'_2
\end{aligned}\right\} \quad (24.32)$$

From Equations 24.31b and 24.32, the currents on the mis-connected point are derived below:

$$\left.\begin{aligned} i_1 &= \frac{E_1 - a^2 E_1'}{Z_1 + Z_1'} \\ i_2 &= i_0 = 0 \end{aligned}\right\} \tag{24.33}$$

$$\left.\begin{aligned} i_a &= \frac{E_1 - a^2 E_1'}{Z_1 + Z_1'} \\ i_b &= a^2 i_a \\ i_c &= a i_a \end{aligned}\right\} \tag{24.34}$$

The equivalent circuit is shown in Figure 24.10b, where the vector operators a and a^2 are inserted. In this particular case, extremely large phase-balanced currents flow to all parts of the network, although negative- and zero-sequence currents do not exist.

Needless to say, mis-connection should be avoided.

Coffee break 13: Power system engineering and insulation coordination

The First World War was fought in 1914–1918 mainly on the European mainland, its end marking the time when the United States came to the fore. It may be considered likely that this period also became a turning point in the history of electricity. The most prominent achievements in regard to electricity before this time had been made in European countries, in particular Britain and Germany, while in the 1920s the stronghold of technology for electricity and power seemed to move overseas to the United States, in particular to Schenectady, New York.

The 1920s–1930s were the time of recovery for European countries, but the time of amazing growth in the United States for power system networks and their technology. Many of Steinmetz's students, like R. H. Park and E. Clarke, and some Westinghouse employees like C. L. Fortescue, always stayed in the same place.

A typical example can be seen in the research area of lightning and **insulation coordination**. **W. W. Lewis**, one of the most famous students of Steinmetz, wrote in his well-known book *The protection of transmission systems against lightning* (1949), recollecting the situation of the 1920s in the United States, as follows:

> In 1924 I had occasion to investigate the failure of a transformer on a power system in Pennsylvania, and found that there were no lightning arresters, that the transmission line had no overhead-ground-wires, and furthermore that it had very high insulation compared with that of the transformer. Up to this time there had been ***no thought of coordinating the line and apparatus insulation***, and practically ***nothing was known about the impulse strength*** of either.

He also wrote, 'In 1926 the Wallenpaupack–Siegfried 220 kV line in Pennsylvania was put into operation without OGW and without arresters.'

However, at length, in 1928, two AIEE papers on insulation coordination were published. In 1928 a cathode-ray oscilloscope was adapted to lines above 220 kV and the first oscillograph of the voltage due to direct lightning strike was obtained in 1930.

The theories of overvoltages (lightning, switching surges, voltage oscillation in the transformers, etc.) and the countermeasures to protect against surge (OGW, tower earthing, counterpoise, arrester, horn gaps, reclosing, etc.) or to strengthen the insulation of apparatus, and a detailed discussion to create appropriate guidelines for insulation coordination, were quite extensively undertaken in the 1930s mainly in the United States. F. W. Peek, W. W. Lewis, C. L. Fortescue, L. V. Bewley and J. H. Hagenguth were the leading characters. The mathematical theory of surge phenomena written by L. V. Bewley in his book *Travelling waves on transmission systems* (1940, final revised edition 1950) is completely fresh, even today.

We power system engineers would do well to recollect the great history of our predecessors and their contribution to our own successes today.

W.W. Lewis **(Courtesy of Schenectady museum)**

Appendix A
Mathematical Formulae

1. Formulae of trigonometric functions and hyperbolic functions

$$\sin^2\alpha + \cos^2\alpha = 1$$
$$\sin(\alpha \pm \beta) = \sin\alpha\cos\beta \pm \cos\alpha\sin\beta$$
$$\cos(\alpha \pm \beta) = \cos\alpha\cos\beta \mp \sin\alpha\sin\beta$$
$$\cos(\alpha + \beta) + \cos(\alpha - \beta) = 2\cos\alpha\cos\beta$$
$$\cos(\alpha + \beta) - \cos(\alpha - \beta) = -2\sin\alpha\sin\beta$$
$$\sin 2\alpha = 2\sin\alpha\cos\alpha$$
$$\cos 2\alpha = \cos^2\alpha - \sin^2\alpha = 1 - 2\sin^2\alpha = 2\cos^2\alpha - 1$$
$$\cos^2\theta + \cos^2\left(\theta - \frac{2\pi}{3}\right) + \cos^2\left(\theta + \frac{2\pi}{3}\right) = \frac{3}{2}$$
$$\sin^2\theta + \sin^2\left(\theta - \frac{2\pi}{3}\right) + \sin^2\left(\theta + \frac{2\pi}{3}\right) = \frac{3}{2}$$

Euler's formula:

$$e^{jy} = \cos y + j\sin y \qquad e^{-jy} = \cos y - j\sin y$$
$$e^z = e^{x+jy} = e^x(\cos y + j\sin y)$$
$$\sin x = \frac{e^{jx} - e^{-jx}}{2j} \qquad \cos x = \frac{e^{jx} + e^{-jx}}{2} \qquad \tan x = \frac{e^{jx} - e^{-jx}}{e^{jx} + e^{-jx}}$$
$$\sinh x = \frac{e^x - e^{-x}}{2} \qquad \cosh x = \frac{e^x + e^{-x}}{2} \qquad \tanh x = \frac{e^x - e^{-x}}{e^x + e^{-x}}$$

2. Formulae of series expansion

$$e^x = 1 + \frac{x}{1!} + \frac{x^2}{2!} + \frac{x^3}{3!} + \cdots$$
$$(1+x)^r = 1 + \frac{rx}{1!} + \frac{r(r-1)x^2}{2!} + \frac{r(r-1)(r-2)x^2}{3!} + \cdots$$
$$\sin x = x - \frac{x^3}{3!} + \frac{x^5}{5!} - \frac{x^7}{7!} + \cdots$$
$$\cos x = 1 - \frac{x^2}{2!} + \frac{x^4}{4!} - \frac{x^6}{6!} - \cdots$$
$$\sinh x = x + \frac{x^3}{3!} + \frac{x^5}{5!} + \frac{x^7}{7!} + \cdots$$
$$\cosh x = 1 + \frac{x^2}{2!} + \frac{x^4}{4!} + \frac{x^6}{6!} + \cdots$$

Handbook of Power System Engineering Yoshihide Hase
© 2007 John Wiley & Sons, Ltd

3. Formula of differential equations

a. If $y = f(ax)$ then

$$\frac{dy}{dx} = \frac{df(ax)}{d(ax)} \cdot \frac{d(ax)}{dx} = a \cdot \frac{df(ax)}{d(ax)}$$

b. If f and g are functions of x, then

$$\frac{d}{dx}\{f(x) \cdot g(x)\} = \left\{\frac{d}{dx} f(x)\right\} \cdot g(x) + f(x) \cdot \left\{\frac{d}{dx} g(x)\right\}$$

$$\frac{d}{dx}\left\{\frac{f(x)}{g(x)}\right\} = \left\{\left(\frac{d}{dx} f(x)\right) \cdot g(x) - f(x) \cdot \left(\frac{d}{dx} g(x)\right)\right\} / g^2(x)$$

c. If $z = f(u)$, $u = g(x)$, then

$$\frac{dz}{dx} = \frac{dz}{du} \cdot \frac{du}{dx}$$

d. If $z = f(u, v)$, $u = g(x)$, $v = k(x)$, then

$$\frac{dz}{dx} = \frac{\partial z}{\partial u} \cdot \frac{du}{dx} + \frac{\partial z}{\partial v} \cdot \frac{dv}{dx}$$

e. If $z = f(u, v), u = g(x, y), v = h(x, y)$

then $\dfrac{\partial z}{\partial x} = \dfrac{\partial z}{\partial u} \cdot \dfrac{\partial u}{\partial x} + \dfrac{\partial z}{\partial v} \cdot \dfrac{\partial v}{\partial x}$

$y = f(x)$	$\frac{dy}{dx}$
x^n	nx^{n-1}
a^x	$a^x \log_e a$
$\log x$	$1/x$
$\sin x$	$\cos x$
$\cos x$	$-\sin x$
e^x	e^x
e^{ax}	$a \cdot e^{ax}$

4. Formulae of Laplace transformation

Equation of definitive transformation:

$$F(s) = \mathcal{L}[f(t)] = \int_0^\infty f(t) e^{-st} dt$$

$$f(t) = \mathcal{L}^{-1}[F(s)] = \frac{1}{2\pi j} \int_{c-j\infty}^{c+j\infty} e^{st} F(s) ds$$

$$\mathcal{L}^{-1}\left[\frac{1}{s}\right] = 1(t), \mathcal{L}^{-1}\left[\frac{1}{s \pm a}\right] = e^{\mp \alpha t} \cdot 1(t)$$

$$\mathcal{L}^{-1}\left[\frac{s + \alpha}{(s + \alpha)^2 + \omega^2}\right] = e^{-\alpha t} \cos \omega t \cdot 1(t)$$

$$\mathcal{L}^{-1}\left[\frac{\omega}{(s \pm \alpha)^2 + \omega^2}\right] = e^{-\alpha t} \sin \omega t \cdot 1(t)$$

$$(s + \alpha - j\omega)(s + \alpha + j\omega) = (s + \alpha)^2 + \omega^2$$

$$\therefore \mathcal{L}^{-1}\left[\frac{A\angle\theta}{s + \alpha - j\omega} + \frac{A\angle - \theta}{s + \alpha + j\omega}\right]$$

$$= 2Ae^{-\alpha t} \cos(\omega t + \theta) \cdot 1(t)$$

$f(t)$	$F(s) = \mathcal{L}[f(t)]$
$1(t > 0)$	$\frac{1}{s}$
$\frac{t^{n-1}}{(n-1)!}$	$\frac{1}{s^n}$
$e^{\mp \alpha t}$	$\frac{1}{s \pm \alpha}$
$\sin \omega t$	$\frac{\omega}{s^2 + \omega^2}$
$\cos \omega t$	$\frac{s}{s^2 + \omega^2}$
$e^{-\alpha t} \sin \omega t$	$\frac{\omega}{(s + \alpha)^2 + \omega^2}$
$\sinh \omega t$	$\frac{\omega}{s^2 - \omega^2}$
$\cosh \omega t$	$\frac{s}{s^2 - \omega^2}$
$e^{-\alpha t} \cos(\omega t \pm \theta)$	$\frac{(s + \alpha) \cos \theta \mp \omega \sin \theta}{(s + \alpha)^2 + \omega^2}$
$e^{-\alpha t} \sin(\omega t \pm \theta)$	$\frac{\pm(s + \alpha) \sin \theta + \omega \cos \theta}{(s + \alpha)^2 + \omega^2}$
$2Ae^{-\alpha t} \cos(\omega t + \theta)$	$\frac{A\angle\theta}{s + \alpha - j\omega} + \frac{A\angle-\theta}{s + \alpha + j\omega}$

Appendix B
Matrix Equation Formulae

Matrix equation analysis is an essential approach for many kinds of analysis including multiple variables of three or more, not only as computational tools but also as logical steps. Explained below are some essential points in regard to **matrix analysis** from a practical viewpoint; for further mathematical details readers should refer to specialized works of mathematics.

(a) Matrix $l \times m$ with row l and column m

$$\boldsymbol{A} = \underset{\text{row } l}{\downarrow} \overset{\text{column} \; \rightarrow m}{\begin{array}{|c|c|c|c|} \hline A_{11} & A_{12} & \cdots & A_{1m} \\ \hline A_{21} & A_{22} & \cdots & A_{2m} \\ \hline \vdots & \vdots & \cdots & \vdots \\ \hline A_{l1} & A_{l2} & \cdots & A_{lm} \\ \hline \end{array}} \tag{1}$$

(b) Multiplication of two matrices $\boldsymbol{A}$ (of row l and column m) and $\boldsymbol{B}$ (of row m and column n). For example, multiplication of $\boldsymbol{A}$ (3×2) and $\boldsymbol{B}$ (3×3):

$$\underbrace{\begin{array}{|c|c|} \hline A_{11} & A_{12} \\ \hline A_{21} & A_{22} \\ \hline A_{31} & A_{32} \\ \hline \end{array}}_{\boldsymbol{A}} \cdot \underbrace{\begin{array}{|c|c|c|} \hline B_{11} & B_{12} & B_{13} \\ \hline B_{21} & B_{22} & B_{23} \\ \hline \end{array}}_{\boldsymbol{B}} = \underbrace{\begin{array}{|c|c|c|} \hline A_{11}B_{11} + A_{12}B_{21} & A_{11}B_{12} + A_{12}B_{22} & A_{11}B_{13} + A_{12}B_{23} \\ \hline A_{21}B_{11} + A_{22}B_{21} & A_{21}B_{12} + A_{22}B_{22} & A_{21}B_{13} + A_{22}B_{23} \\ \hline A_{31}B_{11} + A_{32}B_{21} & A_{31}B_{12} + A_{32}B_{22} & A_{31}B_{13} + A_{32}B_{23} \\ \hline \end{array}}_{\boldsymbol{A} \cdot \boldsymbol{B}} \tag{2}$$

(c) If and only if the number of columns of matrix $\boldsymbol{A}$ and the number of rows of matrix $\boldsymbol{B}$ are equal, the multiplied new matrix $\boldsymbol{A} \cdot \boldsymbol{B}$ of row number l and column number $n(l \times n)$ can exist as follows:

$$\begin{array}{ll} \text{column} & \quad m \qquad n \qquad\qquad n \\ \text{row } l & \downarrow \overset{\rightarrow}{\boldsymbol{A}} \cdot m \downarrow \overset{\rightarrow}{\boldsymbol{B}} = l \downarrow (\boldsymbol{A} \overset{\rightarrow}{\cdot} \boldsymbol{B}) \end{array} \tag{3}$$

(d) A matrix whose numbers of rows and columns are the same is called a **square matrix**.

© 2007 John Wiley & Sons, Ltd

(e) A square matrix whose diagonal elements are one and all other elements are zero is called a **unit matrix**. The unit matrix is usually written by the symbol **1**.

For example, the unit matrix of 3×3:

$$\mathbf{1} = \begin{bmatrix} 1 & 0 & 0 \\ 0 & 1 & 0 \\ 0 & 0 & 1 \end{bmatrix} \quad (4)$$

Multiplication of a square matrix and a unit matrix is possible and the resulting matrix is unchanged. For example,

$$\boldsymbol{C} = \boldsymbol{C} \cdot \mathbf{1} = \mathbf{1} \cdot \boldsymbol{C} \quad (5)$$

(f) If a multiplied matrix $\boldsymbol{C} \cdot \boldsymbol{D}$ of two square matrices $\boldsymbol{C}$ and $\boldsymbol{D}$ become a unit matrix **1**, the square matrices $\boldsymbol{C}$ and $\boldsymbol{D}$ are called **inverse matrices** of each other:

$$\boldsymbol{C} \cdot \boldsymbol{D} = \mathbf{1} \quad (6a)$$

$\boldsymbol{D}$ may be written as $\boldsymbol{C}^{-1}$, and $\boldsymbol{C}$ may be written as $\boldsymbol{D}^{-1}$:

$$\boldsymbol{C} \cdot \boldsymbol{C}^{-1} = \mathbf{1}, \quad \boldsymbol{D} \cdot \boldsymbol{D}^{-1} = \mathbf{1} \quad (6b)$$

Typical examples of inverse matrices are the operational matrices $\boldsymbol{a}, \boldsymbol{a}^{-1}$ of symmetrical components $\boldsymbol{\alpha}, \boldsymbol{\alpha}^{-1}$ of the $\alpha - \beta - 0$ method, and $D(t), D^{-1}(t)$ of the d–q–0 method.

Another example:

$$\underbrace{\begin{bmatrix} 1 & 0 & 3 \\ 2 & 4 & 1 \\ 1 & 3 & 0 \end{bmatrix}}_{\boldsymbol{C}} \times \underbrace{\begin{bmatrix} -1 & 3 & -4 \\ \frac{1}{3} & -1 & \frac{5}{3} \\ \frac{2}{3} & -1 & \frac{4}{3} \end{bmatrix}}_{\boldsymbol{C}^{-1}} = \underbrace{\begin{bmatrix} 1 & 0 & 0 \\ 0 & 1 & 0 \\ 0 & 0 & 1 \end{bmatrix}}_{\mathbf{1}}$$

(g) Two square matrices $\boldsymbol{C}$ and $\boldsymbol{E}$ are of the same size as each other in number of rows and columns. Generally, the commutative law cannot be satisfied between $\boldsymbol{C}$ and $\boldsymbol{E}$:

$$\boldsymbol{C} \cdot \boldsymbol{E} \neq \boldsymbol{E} \cdot \boldsymbol{C} \quad (7)$$

If either $\boldsymbol{C}$ or $\boldsymbol{E}$ is a unit matrix, or if $\boldsymbol{C}$ and $\boldsymbol{E}$ are inverse matrices of each other, the commutative law is exceptionally satisfied:

$$\boldsymbol{C} \cdot \mathbf{1} = \mathbf{1} \cdot \boldsymbol{C} = \boldsymbol{C} \quad (8)$$

$$\boldsymbol{C} \cdot \boldsymbol{C}^{-1} = \boldsymbol{C}^{-1} \cdot \boldsymbol{C} = \mathbf{1} \quad (9)$$

(h) The following equation is always satisfied when the matrices $\boldsymbol{A},\boldsymbol{B},\boldsymbol{C}$ satisfy the condition of (c):

$$\boldsymbol{A} \cdot \boldsymbol{B} \cdot \boldsymbol{C} = (\boldsymbol{A} \cdot \boldsymbol{B}) \cdot \boldsymbol{C} = \boldsymbol{A} \cdot (\boldsymbol{B} \cdot \boldsymbol{C}) \quad (10)$$

(i)

$$\boldsymbol{A} \cdot \boldsymbol{B} + \boldsymbol{A} \cdot \boldsymbol{C} = \boldsymbol{A} \cdot (\boldsymbol{B} + \boldsymbol{C}) \quad (11)$$

$$\boldsymbol{A} \cdot \boldsymbol{B} + \boldsymbol{D} \cdot \boldsymbol{B} = (\boldsymbol{A} + \boldsymbol{D}) \cdot \boldsymbol{B} \quad (12)$$

(j)

$$\left.\begin{aligned} Z_1 &= A_{11}y_1 + A_{12}y_2 \\ Z_2 &= A_{21}y_1 + A_{22}y_2 \\ Z_3 &= A_{31}y_1 + A_{23}y_2 \end{aligned}\right\} \tag{13}$$

The above equation can be written symbolically as follows:

$$\underbrace{\begin{array}{|c|}\hline Z_1 \\ \hline Z_2 \\ \hline Z_3 \\ \hline \end{array}}_{\boldsymbol{Z}} = \underbrace{\begin{array}{|c|}\hline A_{11}y_1 + A_{12}y_2 \\ \hline A_{21}y_1 + A_{22}y_2 \\ \hline A_{31}y_1 + A_{32}y_2 \\ \hline \end{array}}_{\boldsymbol{A}\cdot\boldsymbol{y}} = \underbrace{\begin{array}{|c|c|}\hline A_{11} & A_{12} \\ \hline A_{21} & A_{22} \\ \hline A_{31} & A_{32} \\ \hline \end{array}}_{\boldsymbol{A}} \cdot \underbrace{\begin{array}{|c|}\hline y_1 \\ \hline y_2 \\ \hline \end{array}}_{\boldsymbol{y}} \text{ or symbolically } \boldsymbol{Z} = \boldsymbol{A}\cdot\boldsymbol{y} \tag{14}$$

where

$$\begin{array}{ccccc} 1 & & 2 & & 1 \\ \rightarrow & = & \rightarrow & & \rightarrow \\ 3\downarrow \boldsymbol{Z} & & 3\downarrow \boldsymbol{A} & \times & 2\downarrow \boldsymbol{y} \end{array}$$

(**k**) Besides Equation 14, there is another equation

$$\underbrace{\begin{array}{|c|}\hline y_1 \\ \hline y_2 \\ \hline \end{array}}_{\boldsymbol{y}} = \underbrace{\begin{array}{|c|}\hline B_{11}x_1 + B_{12}x_2 + B_{13}x_3 \\ \hline B_{21}x_1 + B_{22}x_2 + B_{23}x_3 \\ \hline \end{array}}_{\boldsymbol{B}\cdot\boldsymbol{x}} = \underbrace{\begin{array}{|c|c|c|}\hline B_{11} & B_{12} & B_{13} \\ \hline B_{21} & B_{22} & B_{23} \\ \hline \end{array}}_{\boldsymbol{B}} \cdot \underbrace{\begin{array}{|c|}\hline x_1 \\ \hline x_2 \\ \hline x_3 \\ \hline \end{array}}_{\boldsymbol{x}} \text{ or } \boldsymbol{y} = \boldsymbol{B}\cdot\boldsymbol{x} \tag{15}$$

The variables y_1, y_2 can be deleted from Equations 14 and 15 by the matrix procedure

$$\left.\begin{aligned} &\boldsymbol{Z} = \boldsymbol{A}\cdot\boldsymbol{y} = \boldsymbol{A}\cdot(\boldsymbol{B}\cdot\boldsymbol{x}) = (\boldsymbol{A}\cdot\boldsymbol{B})\cdot\boldsymbol{x} \equiv \boldsymbol{C}\cdot\boldsymbol{x} \\ &\text{where} \qquad \boldsymbol{C} = \boldsymbol{A}\cdot\boldsymbol{B} \end{aligned}\right\} \tag{16a}$$

or

$$\begin{array}{|c|}\hline Z_1 \\ \hline Z_2 \\ \hline Z_3 \\ \hline \end{array} = \begin{array}{|c|c|c|}\hline C_{11} & C_{12} & C_{13} \\ \hline C_{21} & C_{22} & C_{23} \\ \hline C_{31} & C_{32} & C_{33} \\ \hline \end{array} \cdot \begin{array}{|c|}\hline x_1 \\ \hline x_2 \\ \hline x_3 \\ \hline \end{array} \qquad \boldsymbol{Z} = \boldsymbol{C}\cdot\boldsymbol{x} \tag{16b}$$

where $\boldsymbol{C}$ is

$$\begin{array}{ccccc} 3 & & 2 & & 3 \\ \rightarrow & = & \rightarrow & & \rightarrow \\ 3\downarrow \boldsymbol{C} & & 3\downarrow \boldsymbol{A} & \cdot & 2\downarrow \boldsymbol{B} \end{array}$$

and

$$\begin{aligned} \boldsymbol{C} \equiv \begin{array}{|c|c|c|}\hline C_{11} & C_{12} & C_{13} \\ \hline C_{21} & C_{22} & C_{23} \\ \hline C_{31} & C_{32} & C_{33} \\ \hline \end{array} &= \boldsymbol{A}\cdot\boldsymbol{B} = \begin{array}{|c|c|}\hline A_{11} & A_{12} \\ \hline A_{21} & A_{22} \\ \hline A_{31} & A_{32} \\ \hline \end{array} \cdot \begin{array}{|c|c|c|}\hline B_{11} & B_{12} & B_{13} \\ \hline B_{21} & B_{22} & B_{23} \\ \hline \end{array} \\ &= \begin{array}{|c|c|c|}\hline A_{11}B_{11} + A_{12}B_{21} & A_{11}B_{12} + A_{12}B_{22} & A_{11}B_{13} + A_{12}B_{23} \\ \hline A_{21}B_{11} + A_{22}B_{21} & A_{21}B_{12} + A_{22}B_{22} & A_{21}B_{13} + A_{22}B_{23} \\ \hline A_{31}B_{11} + A_{32}B_{21} & A_{31}B_{13} + A_{32}B_{22} & A_{31}B_{13} + A_{32}B_{23} \\ \hline \end{array} \end{aligned} \tag{17}$$

where $C_{11} = A_{11}B_{11} + A_{12}B_{12}$, $\quad C_{12} = A_{11}B_{12} + A_{12}B_{22}$, etc.

(**l**) Given matrix equation

$$\boldsymbol{C} = \boldsymbol{A} \cdot \boldsymbol{B} \tag{18}$$

to derive the matrix $\boldsymbol{B}$ from the above equation we left-multiply by $\boldsymbol{A}^{-1}$

$$\begin{aligned} &\boldsymbol{A}^{-1} \cdot \boldsymbol{C} = \boldsymbol{A}^{-1} \cdot \boldsymbol{A} \cdot \boldsymbol{B} = (\boldsymbol{A}^{-1} \cdot \boldsymbol{A}) \cdot \boldsymbol{B} = \boldsymbol{B} \\ &\therefore\ \boldsymbol{B} = \boldsymbol{A}^{-1} \cdot \boldsymbol{C} \end{aligned} \tag{19a}$$

To derive matrix $\boldsymbol{A}$ from the above equation we right-multiply by $\boldsymbol{B}^{-1}$

$$\begin{aligned} &\boldsymbol{C} \cdot \boldsymbol{B}^{-1} = \boldsymbol{A} \cdot \boldsymbol{B} \cdot \boldsymbol{B}^{-1} = \boldsymbol{A} \cdot (\boldsymbol{B} \cdot \boldsymbol{B}^{-1}) = \boldsymbol{A} \\ &\therefore\ \boldsymbol{A} = \boldsymbol{C} \cdot \boldsymbol{B}^{-1} \end{aligned} \tag{19b}$$

Generally

$$\boldsymbol{A} \neq \boldsymbol{B}^{-1} \cdot \boldsymbol{C}, \quad \boldsymbol{B} \neq \boldsymbol{C} \cdot \boldsymbol{A}^{-1}$$

(**m**) Given the matrix equation

$$\boldsymbol{P} = \boldsymbol{L} \cdot \boldsymbol{M} \cdot \boldsymbol{N} \tag{20}$$

where ***P, L, M, N*** are square matrices, then the following equations can be derived by applying the process of (l) twice:

$$\left.\begin{aligned} \boldsymbol{L} &= \boldsymbol{P} \cdot \boldsymbol{N}^{-1} \cdot \boldsymbol{M}^{-1} \\ \boldsymbol{M} &= \boldsymbol{L}^{-1} \cdot \boldsymbol{P} \cdot \boldsymbol{N}^{-1} \\ \boldsymbol{N} &= \boldsymbol{M}^{-1} \cdot \boldsymbol{L}^{-1} \cdot \boldsymbol{P} \end{aligned}\right\} \tag{21}$$

(**n**)

$$\boldsymbol{F} + \boldsymbol{G} = \boldsymbol{G} + \boldsymbol{F} \tag{22}$$

(**o**) If

$$(\boldsymbol{F} + \boldsymbol{G}) \cdot \boldsymbol{Q} = \boldsymbol{K} \tag{23}$$

then

$$\boldsymbol{F} + \boldsymbol{G} = \boldsymbol{K} \cdot \boldsymbol{Q}^{-1} \quad \therefore \boldsymbol{F} = \boldsymbol{K} \cdot \boldsymbol{Q}^{-1} - \boldsymbol{G}, \quad \boldsymbol{G} = \boldsymbol{K} \cdot \boldsymbol{Q}^{-1} - \boldsymbol{F} \tag{24}$$

(**p**)

$$\left.\begin{aligned} m_1 \downarrow \overset{\overset{l}{\rightarrow}}{\boldsymbol{W}_1} &= m_1 \downarrow \overset{\overset{n_1}{\rightarrow}}{\boldsymbol{R}_{11}} \times n_1 \downarrow \overset{\overset{l}{\rightarrow}}{\boldsymbol{V}_1} + m_1 \downarrow \overset{\overset{n_2}{\rightarrow}}{\boldsymbol{R}_{12}} \times n_2 \downarrow \overset{\overset{l}{\rightarrow}}{\boldsymbol{V}_2} \\ m_2 \downarrow \overset{\overset{l}{\rightarrow}}{\boldsymbol{W}_2} &= m_2 \downarrow \overset{\overset{n_1}{\rightarrow}}{\boldsymbol{R}_{21}} \times n_1 \downarrow \overset{\overset{l}{\rightarrow}}{\boldsymbol{V}_1} + m_2 \downarrow \overset{\overset{n_2}{\rightarrow}}{\boldsymbol{R}_{22}} \times n_2 \downarrow \overset{\overset{l}{\rightarrow}}{\boldsymbol{V}_2} \end{aligned}\right\} \tag{25}$$

The above two matrix equations can be combined into one matrix equation:

$$\begin{bmatrix} m_1 \downarrow \overset{\overset{l}{\rightarrow}}{\boldsymbol{W}_1} \\ m_2 \downarrow \overset{\overset{l}{\rightarrow}}{\boldsymbol{W}_2} \end{bmatrix} = \begin{bmatrix} m_1 \downarrow \overset{\overset{n_1}{\rightarrow}}{\boldsymbol{R}_{11}} & m_1 \downarrow \overset{\overset{n_2}{\rightarrow}}{\boldsymbol{R}_{12}} \\ m_2 \downarrow \overset{\overset{n_1}{\rightarrow}}{\boldsymbol{R}_{21}} & m_2 \downarrow \overset{\overset{n_2}{\rightarrow}}{\boldsymbol{R}_{22}} \end{bmatrix} \cdot \begin{bmatrix} n_1 \downarrow \overset{\overset{l}{\rightarrow}}{\boldsymbol{V}_1} \\ n_2 \downarrow \overset{\overset{l}{\rightarrow}}{\boldsymbol{V}_2} \end{bmatrix} \tag{26}$$

Needless to say, Equation 26 can be divided into the two equations of Equation 25.

(q) The new matrix in which the row and column elements of the original matrix $\boldsymbol{A}$ are interchanged is called the transposed matrix of $\boldsymbol{A}$. It is written by the symbol ${}^tA$.
If

$$\boldsymbol{C} = \boldsymbol{A} \cdot \boldsymbol{B} \tag{27}$$

then

$${}^t\boldsymbol{C} = {}^t\boldsymbol{B} \cdot {}^t\boldsymbol{A} \tag{28}$$

In regard to matrix $\boldsymbol{C}$ in Equation 17, the transposed matrix is

$${}^t\boldsymbol{C} \equiv \begin{bmatrix} C_{11} & C_{21} & C_{31} \\ C_{12} & C_{22} & C_{32} \\ C_{13} & C_{23} & C_{33} \end{bmatrix} = \underbrace{\begin{bmatrix} B_{11} & B_{21} \\ B_{12} & B_{22} \\ B_{13} & B_{23} \end{bmatrix}}_{{}^t\boldsymbol{B}} \cdot \underbrace{\begin{bmatrix} A_{11} & A_{21} & A_{31} \\ A_{12} & A_{22} & A_{32} \end{bmatrix}}_{{}^t\boldsymbol{A}} = {}^tB \cdot {}^tA \tag{29}$$

Analytical Methods Index

© 2007 John Wiley & Sons, Ltd

Components Index

Handbook of Power System Engineering Yoshihide Hase
© 2007 John Wiley & Sons, Ltd

Subject Index

Handbook of Power System Engineering Yoshihide Hase
© 2007 John Wiley & Sons, Ltd